Primates in History, Myth, Art, and Science

Non-human primates (hereafter just primates) play a special role in human societies, especially in regions where modern humans and primates co-exist. Primates feature in myths and legends and in traditional indigenous knowledge. Explorers observed them in the wild and brought them, at great cost, to Europe. There they were valued as pets and for display, their images featured in art and architecture, and where they were literally teased apart by scientists. The international team of contributors to this book draws these different perspectives together to show how primates helped humans better understand their own place in nature. The book will be of interest to undergraduate and graduate students as well as scholars in disciplines ranging from anthropology to art history.

Key features:

- Includes contributions from an international team of historians and natural scientists
- Integrates various perspectives and perceptions of primates across time and place
- Summarizes the place of primates in science, art, and culture
- Includes rare early illustrations

Primates in History, Myth, Art, and Science

Edited by
Cecilia Veracini and Bernard Wood

CRC Press
Taylor & Francis Group
Boca Raton London New York

CRC Press is an imprint of the
Taylor & Francis Group, an **informa** business

Designed cover image: The Alice and Nasli Heeramaneck Collection, Gift of Alice Heeramaneck, 1981.

First edition published 2024
by CRC Press
2385 NW Executive Center Drive, Suite 320, Boca Raton FL 33431

and by CRC Press
4 Park Square, Milton Park, Abingdon, Oxon, OX14 4RN

CRC Press is an imprint of Taylor & Francis Group, LLC

Library of Congress Cataloging-in-Publication Data
Names: Veracini, Cecilia, editor. | Wood, Bernard A., editor.
Title: Primates in history, myth, art, and science / [edited by] Cecilia Veracini and Bernard Wood.
Description: First edition. | Boca Raton : CRC Press, 2024. | Includes bibliographical references and index.
Identifiers: LCCN 2023042622 (print) | LCCN 2023042623 (ebook) | ISBN 9781032710877 (hardback) | ISBN 9781138198395 (paperback) | ISBN 9781315270371 (ebook)
Subjects: LCSH: Primates—History. | Primates—Mythology. | Human-animal relationships—History.
Classification: LCC QL737.P9 P658 2024 (print) | LCC QL737.P9 (ebook) | DDC 599.8—dc23/eng/20240125
LC record available at https://lccn.loc.gov/2023042622
LC ebook record available at https://lccn.loc.gov/2023042623

ISBN: 9781032710877 (hbk)
ISBN: 9781138198395 (pbk)
ISBN: 9781315270371 (ebk)

DOI: 10.1201/b21963

Typeset in Times
by codeMantra

Contents

List of Box Details

Editors

Cecilia Veracini is an associated researcher at the Centre for Public Administration and Public Policies, Institute of Social and Political Sciences, Universidade de Lisboa, Portugal, and she also serves as an adjunct professor at the Faculty of Anthropology within the same institute. She graduated with a degree in Biological Science from the University of Pisa (Italy) and received a Ph.D. degree in Anthropological Sciences and a Ph.D. degree in History of Science from the Florence and Pisa Universities. She served some years as Assistant Professor at the Florence and Pisa Universities and worked as collaborator at various institutions in different countries including Brazil, the US, and Spain. Her publications include papers in national and international peer reviewed journals and book chapters. She is co-editor of the books *History of Primatology: yesterday and today. The Mediterranean Tradition* (2019), and *Peoples, nature and environments: learning to live together* (2020).

Bernard Wood is the University Professor of Human Origins at George Washington University. Previously, he was the Courtuald Professor of Anatomy in the University of London and the Derby Professor of Anatomy in the University of Liverpool. In 1968, he joined Richard Leakey's first expedition to the Turkana Basin, Kenya, and he subsequently joined the group of researchers working on the hominins recovered from Koobi Fora in Northern Kenya. In addition to his paleo-anthropological research, he has a long-standing interest in primate comparative anatomy, with a focus on the ability of hard- and soft-tissue anatomy to recover the recent evolutionary history of the extant apes, and on the history of primate comparative anatomy. He is the co-author of research articles on many aspects of comparative anatomy, and the author, or co-author, of 20 books, including *Food Acquisition and Processing in Primates* (1984), *Major Topics in Primate and Human Evolution* (1986), and photographic atlases of the musculoskeletal anatomy of gorillas (2010), gibbons and siamangs (2012), chimpanzees (2013), orangutans (2013), and bonobos (2017).

Contributors

Malak A. Alghamdi
King Saud bin Abdulaziz University for Health Sciences
Jeddah, Saudi Arabia

Alessio Anania
Unione Bolognese Naturalisti
Bologna, Italy

Giulio Barsanti
Università di Firenze
Florence, Italy

Ana Lucia Camphora
Center for Contemporary Equine Studies - CCES-
California, US

Giuseppe Donati
Oxford Brookes University
Oxford, United Kingdom

Cybelle Greenlaw
Trinity College
Dublin, Ireland

Philip Lutgendorf
University of Iowa
Iowa City, Iowa

Marco Masseti
Istituto Zooprofilattico della Sicilia
Sicily, Italy

Ryan McRae
George Washington University
Washington, DC

Rowan M. Sherwood
University of Michigan
Ann Arbor, Michigan

Cecilia Veracini
Universidade de Lisboa
Lisbon, Portugal

Bernard Wood
George Washington University
Washington, DC

Box Details Contributors

Adrian A. Barnett
Department of Natural Sciences
Middlesex University
Hendon, London, UK

Carla Soraia Soares de Castro
Universidade Federal da Paraíba, Centro de Ciências Aplicadas e Educação Departamento de Engenharia e Meio Ambiente, Paraíba, Brasil

Susana Costa
CIAS (Centro de Investigação em Antropologia e Saúde)
Universidade de Coimbra
Portugal

Dante Martins Teixeira
Museu Nacional - Universidade Federal do Rio de Janeiro
Rio de Janeiro, Brazil

Marco Vespa
The Hebrew University of Jerusalem
Jerusalem, Israel

Acknowledgements

The editors express their appreciation of the important role our colleague Rui Diogo played in the early stages of this project. His vision and initiative were important elements in shaping the contents of the book and setting it on the road to completion.

We also emphasize the debt we owe to the many scholars (authors and reviewers) whose work is included in this book. Without their enterprise and hard work, there would be no story of non-human primates to tell.

Introduction

> The commercial and scientific traffic in monkeys and apes is a traffic in meaning as well as in animal lives. (Haraway, 1989: 1)

This book deals with non-human primates (hereafter just primates), the animals that are our closest living relatives. It sets out how and when we encountered them, and our interactions with them across times, places, and cultures. Combining historical, scientific, and cultural perspectives, the book covers subjects as diverse as how primates feature in the myths and beliefs of societies across the world, how they were and are perceived, valued, and used by those societies; how they feature in ancient history; and how they were affected by the modern European expansion. It combines all these sources to demonstrate the history of primates is inescapably associated with our own, giving rise to the interdependent and inseparable natural-cultural spaces we see today.

Primates play a fundamental role as seed and pollen dispersers in the diversity and dynamics of tropical and subtropical forest ecosystems (Chapman, 1995; Sussman et al., 2013). They are and were important actors in human cultures, both in countries where they co-exist with human communities and in locations far from their places of origin. Recent interpretations see animals as historical agents capable of influencing and determining the course of human events (Hribal, 2007; Rees, 2017), and show how human and non-human interactions have played fundamental roles in contemporary and past human societies (e.g., Fudge, 2006; Brantz (ed.), (2010); Cockram & Wells (eds.) 2017). Studies undertaken in the last two decades addressing broad, cross-disciplinary questions have questioned human exceptionalism on both ethical, historical, and philosophical grounds (e.g., Corbey & Lanjouw (eds.), 2013; Gillespie & Collard, 2015; Poirier et al. (eds.), 2022) and emphasized the complex interactions between people, ecosystems, and animals (e.g., Fuentes & Baynes-Rock, 2017; Oma & Goldhahn, 2020). Primates evolved in close proximity to prehistoric human societies, often sharing the same ecological niche and often cohabiting in integrated socio-ecological spaces (Fuentes & Wolfe (eds.), 2002; Sengupta & Radhakrishna, 2018). This has resulted in mutually and complex entanglements between humankind and the other primates, with reciprocal adaptations across time and space. Primates may have both positive (e.g., as seed dispersers) and negative (e.g., by spreading zoonotic diseases, and crop-raiding) influences on humans, which reciprocally influence other primates' lives, even shaping their evolution (Rodrigues et al., 2023).

Today, when almost half of all primate taxa are threatened with extinction (Estrada et al., 2017, 2020), it is especially important to study how they and modern humans have co-existed in the past. Through recorded time, people have used primates as sacred presences, as well as sources of food and companionship (Urbani et al., 2022 for review), but the European expansion of the modern period began a period of unprecedented human pressure on many populations of non-human primates, as well as on the flora and fauna on which they depend (e.g., Miller, 2000; Teixeira & Papavero, 2010; Pádua, 2010; Veracini, 2017). This European expansion and the consequent "1500 Columbian exchange" (Crosby, 2003) are transformative events in the Anthropocene (Gibbard et al., 2022). Studying how these events impacted primates may help us find solutions to the imminent threat of extinction. The past and present ethno-primatology studies featured in this book that emphasize the roles played by primates in local cultures, combined with the extensive knowledge of the natural history of primates developed over the millennia by traditional communities, are important resources for helping to prevent the collapse of wild populations of primates (cf. Estrada et al., 2022).

This book shows how primates permeate human life, from metaphysics to politics and economy, from science to art; these animals and the meaning attributed to them are fundamental to what it means to be "human." Because they are our closest relatives, we share more of our physiognomy, physiology, and behavior with them than with any other group of animals. The resemblance causes us to reflect on our characteristics, our life and nature, and our spirituality (Corbey & Theunissen (eds.), 1995; Corbey, 2005). Throughout history, they have played a role in religion, art, music, and literature, and many other cultural manifestations of mankind (e.g., McDermott, 1938; Janson, 1952; Morris, 2013; Cohen, 2017; Urbani et al., 2022), including using them for food, traditional medicine, labor, and even as sex slaves (Alves et al., 2013; Veracini, 2018).

The first and second sections of this book, which focus on how we and the other primates interacted in different continents and in different historical periods, show how primates have featured in various cultures and places throughout history. Special attention is paid to the role of primates in Western society and in the cultures of the Near East, from the Greco-Roman period to the Renaissance.

The later sections of the book deal mainly with the natural history of primates and how that understanding evolved in the modern period up to the present day. By critically reviewing interpretations provided by scholars since the Middle Ages, they document how the scientific representation of primates has changed. This period covers the time when Western scientists were "discovering" primates, including the great apes, whose proximity to modern humans had philosophical, religious and socio-political implications that we are still grappling with. In the West, the study and interpretation of primates has never been neutral

(Barsanti, 1989; Corbey, 2005; Sebastiani, 2015) but has been shaped by the socio-cultural needs and motivations of the historical moment (Haraway, 1989). As Poskett (2022) and other scholars have suggested, this history, like that of the natural sciences in general, is not just a Western history, or a product of European culture, but it involved contributions from many other cultures, especially the knowledge of local people who had co-existed with non-human primates for millennia. This input, which resulted from the European expansion, was inevitably intertwined with many lamentable aspects of the behavior of colonial powers, such as slavery and imperialism.

Exotic animals arriving in Europe beginning in the 16th century were mediators of contact between Europe and the other continents. Primates, which were especially influential because of their anthropomorphic bodies and human-like behavior, promoted changes in the notion that they were contemptible caricatures of people. The process of recognition of the true nature of primates, and in particular the nature of the anthropoid apes, occupies a special place in the history of Western science (Malone, 2022). The notions of human exceptionalism and the binary division between nature and culture have been carefully deconstructed in recent decades (Ingold, 2004; Descola, 2010). As stated by Corbey (in Malone, 2022), "Redefining humans as animals with a direct kinship to and ecological entanglement with other primates and animals … is a recent development, not just in the history of primate studies but also in western consciousness at large."

This volume is dominated by contributions from historians of science with a special interest in primates and how they are seen through art and ethnography, but it also includes contributions from comparative anatomists, zoologists, anthropologists, and primatologists. Traditionally, these disciplines rarely exchanged ideas or perspectives, but it is our hope that this book will inspire others to explore the rich contribution that primates have made to their close human relatives.

We hope we have helped shed light on the fascinating but also problematic and conflicting relationship between us, *Homo sapiens*, and our closest living relatives.

REFERENCES

Alves, R. R. N., Souto, W. M. S., Barboza, R. R. D., & Bezerra, D. M. M. (2013). Primates in traditional folk medicine: world overview. In *Animals in Traditional Folk Medicine*. edited by R. Alves and I. Rosa. Berlin/Heidelberg: Springer, pp. 135–70.

Barsanti, G. (1989). L'orang-outan déclassé. Histoire du premier singe à hauteur d'homme (1780-1801) et ébauche d'une théorie de la circularité des sources. *Bulletins et Mémoires de la Société d'Anthropologie de Paris*, 1(3), 67–104.

Brantz, D. (ed.) (2010). *Beastly Natures: Animals, Humans, and the Study of History*. Charlottesville: University of Virginia Press.

Chapman, C. A. (1995). Primate seed dispersal: Coevolution and conservation implications. *Evolutionary Anthropology*, 4, 74–82.

Cockram, S. & Wells, A. (2017). *Interspecies Interactions. Animals and Humans between the Middle Ages and Modernity*. Taylor & Francis.

Cohen, S. (2017). Ars simia naturae: The Animal as Mediator and Alter Ego of the Artist in the Renaissance. *Explorations in Renaissance culture*, 43(2), 202–231.

Corbey, R. (2005). *The Metaphysics of Apes: Negotiating the Animal-Human Boundary*. Cambridge University Press.

Corbey, R. & Lanjouw, A. (orgs) (2013). *The Politics of Species: Reshaping Our Relationships with Other Animals*. Cambridge: Cambridge University Press.

Corbey, R. & Theunissen, B. (eds.) (1995). *Ape, Man, Apeman: Changing Views since 1600*. Leiden: Department of Prehistory of Leiden University.

Crosby, A. W. (2003). *The Columbian Exchange: Biological and Cultural Consequences of 1492* (Vol. 2). Greenwood Publishing Group.

Descola, P. (2010). *Diversité de Natures Diversité de Culture*. Paris: Bayard Éditions.

Estrada, A., Garber, P. A., & Chaudhary, A. (2020). Current and future trends in socio-economic, demographic and governance factors affecting global primate conservation. *PeerJ*, 8, e9816.

Estrada, A., Garber, P. A., Gouveia, S., et al. (2022). Global importance of Indigenous Peoples, their lands, and knowledge systems for saving the world's primates from extinction. *Science Advances*, 8(31), eabn2927.

Estrada, A., Garber, P. A., Rylands, A. B., et al. (2017). Impending extinction crisis of the world's primates: Why primates matter. *Science Advances*, 3(1), e1600946.

Fudge, E. (2006). *Brutal Reasoning: Animals, Rationality, and Humanity in Early Modern England*. Ithaca and London: Cornell University Press.

Fuentes, A. & Baynes-Rock, M. (2017). Anthropogenic landscapes, human action and the process of co-construction with other species: Making anthromes in the Anthropocene. *Land*, 6(1), 15.

Fuentes, A. & Wolfe, L. D. (eds.) (2002). *Primates Face to Face: The Conservation Implications of Human-Nonhuman Primate Interconnections*. Cambridge: Cambridge University Press.

Gibbard, P., Walker, M., Bauer, A., et al. (2022). The anthropocene as an event, not an epoch. *Journal of Quaternary Science*, 37(3), 395–399.

Gillespie, K. & Collard, R. C. (eds.). (2015). *Critical Animal Geographies: Politics, Intersections and Hierarchies in a Multispecies World*. Routledge.

Haraway, D. J. (1989). *Primate Visions: Gender, Race, and Nature in the World of Modern Science*. New York, NY: Routledge.

Hribal, J. C. (2007). Animals, agency, and class: Writing the history of animals from below. *Human Ecology Review*, 101–112.

Ingold, T. (2004). Beyond biology and culture. The meaning of evolution in a relational world. *Social Anthropology*, 12(2), 209–221.

Janson H. W. (1952). *Ape and Ape Lore in the Middle Ages and the Renaissance*. The Warburg Institute, University of London.

Malone, N. (2022). *The Dialectical Primatologist: Past, Present and Future Life in the Hominoid Niche*. Routledge.

McDermott, W. C. (1938). *The Ape in Antiquity*. The Johns Hopkins University studies in Archaeology.

Miller, S. W. (2000). *Fruitless Trees: Portuguese Conservation and Brazil's Colonial Timber*. Stanford University Press.
Morris, D. (2013). *Monkeys*. London: Reaktionbook
Oma, K. A. & Goldhahn, J. (2020). Human-animal relationships from a long-term perspective. *Current Swedish Archaeology*, 28(1), 11–22.
Pádua, J. A. (2010). European colonialism and tropical forest destruction in Brazil: The environmental dimension beyond economic history. In *Environmental History: As If Nature Existed.* edited by J. McNeill, J. A. Pádua and M. Rangarajan. Delhi: Oxford University Press.
Poirier, N., Nocella, A. J., & Bernatchez, A. (2022). *Emerging New Voices in Critical Animal Studies: Vegan Studies for Total Liberation*. Peter Lang.
Poskett, J. (2022). *Horizons: A Global History of Science*. UK: Penguin.
Rees, A. (2017). Animal agents? Historiography, theory and the history of science in the Anthropocene. *BJHS Themes*, 2, 1–10.
Rodrigues, M. A., McKinney, T., & Waters, S. (eds.). (2023). *Primates in Anthropogenic Landscapes: Exploring Primate Behavioural Flexibility across Human Contexts*. Springer International Publishing.
Sebastiani, S. (2015). Challenging boundaries. Racism Analysis, 105.
Sengupta, A. & Radhakrishna, S. (2018). The hand that feeds the monkey: Mutual influence of humans and rhesus macaques (Macaca mulatta) in the context of provisioning. *International Journal of Primatology*, 39, 817–830.
Sussman, R. W., Tab Rasmussen, D., & Raven, P. H. (2013). Rethinking primate origins again. *American Journal of Primatology*, 75(2), 95–106.
Teixeira, D. M. & Papavero, N. (2010). O Trafico de Primatas brasileiros nos séculos XVI e XVII. In *Mamiferos de Restingas e Manguezais do Brasil.* edited by M. L. Pessoa, Tavares W. Correa and S. Siliciano. Rio de Janeiro: SBMz, Sociedade Brasileira de Mastozoologia. Museu Nacional, pp. 253–82.
Urbani, B., Youlatos, D., & Antczak, A. T. (2022). *World Archaeoprimatology*. Cambridge University Press.
Veracini, C. (2017). Non-human primate trade in the age of discoveries: European importation and its consequences. In *Environmental History in the Making*. edited by C. Joanaz, E. de Melo, E. Vaz and L. M. Costa Pinto. Berlin: Springer.
Veracini, C. (2018). Interactions between humans and the hamadryas baboon in Pre-Islamic and Medieval Arabia. *Abstracts 6th Iberian Primatological Congress*, 89, 214. doi: 10.1159/000488668.

Section I

Lore and Mythology of Non-human Primates since Antiquity

1 South, Southeast, and East Asia

Philip Lutgendorf
University of Iowa

How can there be a relationship between men and monkeys?

Valmiki,
Ramayana 5.33.2 (Goldman & Sutherland Goldman, 1996: 201)

The question above is posed at a crucial moment by the heroine of one of the most beloved and influential narratives of South and Southeast Asia, the *Ramayana* ("journey of Rama", **Fig. 1.1**). The earliest known version of this story, attributed to the sage and poet Valmiki, is a Sanskrit epic in some 24,000 couplets that is now dated by most academic scholars to roughly the last centuries BC. A text that highlights, through dramatic and memorable narrative situations, some of the strictures and dilemmas of the key cultural concept of *dharma* (often translated by such terms as "duty," "virtue," "morality," "religion," and "social order"), it tells of the unjust banishment of Prince Rama, rightful heir to the throne of Ayodhya, and his adventures while in forest exile, accompanied by his beloved wife, Sita, and his younger brother Lakshmana. Following the banishment itself, the central crisis of the narrative is the kidnaping of Sita by Ravana, the monstrous sorcerer-king of the island of Lanka, where he hides the princess in his citadel. Despondent over the loss of his wife, Rama sets out with Lakshmana to find her, and soon meets and befriends a race of fabulous, talking monkeys who become his allies in the quest. While searching for Sita in the far south of the subcontinent, one of these monkeys—Hanuman, the able counselor to their king—heroically leaps across the vast sea to Lanka and, after a long search and many adventures, finds the princess imprisoned in a garden. Hidden in a tree in diminutive form (for he can alter his size at will), Hanuman witnesses Sita's steadfast resistance to Ravana's demands to become his wife and her subsequent desperate resolve, after Ravana returns to his palace, to take her own life rather than endure his blandishments and the torment of her separation from Rama. It is at this desperate moment—the absolute nadir of the protagonists' fortunes and the approximate midpoint of the Sanskrit epic—that the monkey reveals himself to the princess, identifying himself as her husband's messenger. This elicits her puzzled question to him, which is soon answered by an eloquent recital by the monkey (who is also a skilled storyteller) of the prince's adventures since her kidnaping. His speech instills the hope of rescue in Sita's heart and so initiates the gradual upward arc of the narrative toward her husband's climactic battle with Ravana, reunion with her, and triumphant return to his kingdom following his exile's term.

FIGURE 1.1 Rama and Hanuman; popular poster, artist unknown, late 20th century. (Author's collection.)

But Sita's question is also an appropriate epigram for this chapter, which will consider the "relationship"—in the original Sanskrit, literally "coming together" (*samagama*), a word that can also connote erotic intimacy—of human beings and monkeys in several Asian cultures, from ancient times to the present. The princess's incredulity at the alliance between a noble Aryan ("high-born," "cultured") human and a race of furry, long-tailed forest creatures points to one aspect of that relationship: a tendency common to many human societies (and well attested by other contributors to this volume) to literally "look down on" their closest primate relations, to disdain them as "fallen" or "debased" caricatures of *Homo sapiens*, and to feel discomfort or even fear due to their perceived liminal and boundary-challenging nature. Such attitudes are not altogether absent in the

DOI: 10.1201/b21963-2

responses to anthropoid primates found in Asian societies, and a number of examples of them will be noted in the pages that follow. But they coexist, in these cultural zones that were the natural habitat of numerous non-human primate species, with other responses that are more ambivalent, or indeed that are admiring, respectful, and even worshipful. For it is in the cultures of South, Southeast, and East Asia that we encounter the two most celebrated and protean simian hero-figures (and, in some contexts, deities) of world narrative: the monkey general Hanuman and the monkey king Sun Wu-k'ung—and, incidentally, the much-debated question of a possible "relationship" between these two must also, in due course, be considered.

I will begin with some general observations on attitudes toward non-human primates in Asian cultures, particularly those that have been deeply influenced by the worldviews of Hinduism, Buddhism, and Taoism. I will then undertake a journey (recalling the epic one of Rama and his simian helper) through Asian monkey lore that will proceed geographically from the Indian subcontinent (the "western paradise" to which Sun Wu-k'ung journeyed with his human companion, the Buddhist monk Tripitaka), to its far eastern extent in China and Japan, making a stopover in the island archipelago and adjacent coastal regions of Southeast Asia (which, as an intermediate contact-zone, was once tellingly labeled "Indo-China" by geographers, though it has its own distinct cultural features and monkey lore). This west-to-east journey will also be roughly chronological in charting the development and dissemination of the most elaborate, enduring, and culturally significant lore that features simian characters. But I must, first of all, admit my limitations: I am a cultural historian of South Asia, somewhat knowledgeable in Rama traditions, particularly through their later Hindi-language retellings and author of a substantial study on Hanuman (Lutgendorf, 2007, 2000). For Southeast and East Asian materials, I must rely on secondary sources and the assistance of other scholars, as indeed I must for the Sanskrit and Prakrit texts of the subcontinent that precede the period of my specialization, as well as the many relevant Indian vernacular works of later times in languages other than Hindi. These deficiencies may lead me to errors of fact or interpretation; hopefully, the scholarly sources I cite can remedy some of these for interested readers. My shortcomings may also lead to inconsistencies of spelling, since I am dealing with transliterations from numerous languages, for several of which there is more than one convention of Romanization. For this, I apologize in advance to those who know better.

A pertinent general observation regarding the regions under consideration is that their dominant indigenous worldviews—with their central concepts of *dharma* and *tao* (cosmic order, balance, and harmony) and their frequent idealization of *ahimsa* (non-injury to living beings) do not subscribe to the Abrahamic traditions' notion of a sharp divide between human beings and animals, nor to the idea that only the former possess souls or that humankind alone was made "in the image of God"—a doctrine that often resulted in non-human primates being regarded with particular disdain and aversion. Instead, Asian cultures have generally understood that animals (and plants) have souls, and they have also posited fairly permeable barriers between species of beings—for example, through beliefs in transmigration, metamorphosis, possession, or potential intermarriage. Additionally, with a few exceptions, Asian societies have not regarded non-human primates as suitable for hunting or consuming as food and have, indeed, largely avoided deliberately killing them. Even when the animals' pilfering or destruction of human possessions, occasional aggression, and unconcealed sexual behavior caused them to be treated with wariness, embarrassment, or ridicule, this did not develop into the attitude of generic condemnation found, for example, in the Christian notion of Satan as *simia Dei* ("God's monkey") or of monkeys themselves as the "likeness of the devil" (Janson, 1952: 20; cf. Veracini, Ch. 7 this volume). More commonly, Asian attitudes toward anthropoid primates have included a broad spectrum of responses that are in some cases quite species-specific (for example, in the Chinese notion of "good gibbons and evil macaques"; Zhang, 2015) and that can as readily include amusement at their human-like social behavior or trained performances in captivity, as awe and admiration at their beauty, agility, or harmony with their environment. It is when such awe shades into the realm of the numinous—as in the Chinese judgment of certain gibbons as *shen* ("holy," "sacred") or the Hindu reverence for langurs as relatives of Hanuman—that Anglo-European discourse, especially of the post-Reformation and "Enlightenment" periods, has displayed particular discomfort and indeed disgust, as it has with any worldview that does not restrict sacrality to a single, monolithic, and transcendent divine entity. Where human-animal interactions were concerned, this prejudice gave rise to such 18th- and 19th-century academic neologisms as "zoolatry" (the worship of animals), "theriomorphism" (cults of deities in animal form), and indeed the phrase "monkey-worship," which became a trope for the primitive and debased. Later, Anglophone Indian academics would themselves adopt this term as a convenient explanation for Hanuman's tremendous popularity, projecting it (despite a lack of substantial evidence) onto "tribal" communities within their own society or even onto all human cultures in the distant past, though they could never rid it of its disparaging overtones (Nagar, 1995: 217–221; Sen, 1987: 43–46; cf. Lutgendorf, 2007: 12–24, 335–344). It is as important to keep the ideological foundations of such terminology in mind when considering the complexity of Asian responses to non-human primates, as it is to understand the particular contexts in which the numinous qualities of these beings were foregrounded.

1.1 SOUTH ASIA

The Indian subcontinent may well claim to be the fountainhead of heroic simian lore in Asia, and the earliest specimen of this (apart from a small number of clay monkey figurines—possibly more whimsical than votive—from the third millennium BCE, unearthed in mature Harappan or Indus Civilization sites in modern Pakistan and northwestern India) occurs in a single, enigmatic poem. It is found in the tenth *mandala* ("collection, cycle") of the *Rigveda,* the earliest surviving text of Indo-European literature, which most scholars now date to the mid-to-late second millennium BCE. Since most of the thousand-plus poems in this anthology (usually called "hymns" in English-language scholarship) were evidently liturgical and intended to be chanted during rituals of fire sacrifice, hymn 10.86, with its bawdy, even obscene banter and imagery, has long puzzled scholars. It is set as a dialogue between three or possibly four speakers: the god Indra (the warrior-chieftain who headed the Vedic pantheon); his wife, who is here called "Indrani" (the feminine form of his name); and a strange being, found in no other Vedic hymn, called "bull monkey" or "virile monkey" (*vrishakapi*)—at one point, another voice, thought by some to be of the monkey's own wife, joins in. Long regarded (in the words of a famous Indologist) as "the strangest poem in the *Rigveda*" (O'Flaherty, 1981: 257), the dialogue seemingly presents a divine domestic quarrel in which Indrani complains about the indiscretions of Indra's "friend," Vrishakapi, including the defiling of sacrificial *soma* offerings (a mysterious, intoxicating substance equated with the nectar of immortality) that should rightfully go to Indra, and, in one especially enigmatic passage, of having sexual designs on Mrs. Indra herself. The monkey (and possibly his wife) then joins the conversation, with explicit sexual banter that caused Victorian translators to resort to euphemisms or outright deletions. The hymn ends with some sort of husband-wife reconciliation accompanied by a reaffirmation of Indra's friendship with the animal.

Wildly variant interpretations of the hymn have ranged from taking it as an allegorical record of climatic and astronomical phenomena to reading it as a "fertility charm" intended to restore potency, via the "virile monkey" to a depleted Indra (and his human votaries), to seeing it as a parody of the horse sacrifice (*ashvamedha*), one of the most solemn and elaborate rituals of the late-Vedic period, that was likewise intended to restore potency to a compromised king and in the context of which the seemingly aberrant hymn might have been chanted (Joshi, 1978: 115–124; O'Flaherty, 1981: 261; Jamison, 1996: 65–88). The central role of (what J. R. Joshi called) the "bold, lascivious monkey" seems crucial to the ludic and even transgressive spirit of the hymn, especially when placed in the context of later Indian animal lore that generally classed such creatures as lowly and impure in the hierarchical bestiary of earth and cosmos (Jamison, 1996: 77). However, two further observations about this obscure text are pertinent here. The first is that the monkey's "defilement" of offerings may reflect the concerns of sacrificial priests in a setting (the Punjab region of northwestern India, where the Vedic culture seems to have been geographically centered) in which rhesus macaques (*Macaca mulatta,* cf. **Box Details 1.1**) probably abounded—as they do today—so that their penchant for investigating, sampling, and stealing human foodstuffs would have been warily noted. The second is the hymn's frisson of potential inter-species miscegenation, expressed in the wife's apparent warning to her divine husband that the "bull-monkey" has erotic designs on her. This points to a wider human pattern of gendered thinking, in patriarchal societies, about men and monkeys—in which human women may be sexually approached and sometimes sexually satisfied or even impregnated by oversexed male monkeys, who thus threaten the dominance of human males and the purity of their lineages. Such speculation and anxiety about this especially uncomfortable sort of "coming-together" of different species—found in texts as widely separated in space, time, and media as Chinese Han dynasty stories (circa 200 BCE–220 CE) of "a demonic ape abducting women" (Walker, 1998: 87) and the Hollywood movie *King Kong* (1933, and later remakes)—may make its first appearance here. It will continue to resonate in later, more popular, and less enigmatic Asian tales of interspecies relations and will also influence, in one way or another, the personalities of the two great cultural heroes Hanuman and Sun Wu-k'ung.

FIGURE 1.2 *Macaca mulatta*, Katmandu, Nepal. (Photo Gali Di Giulio.)

BOX DETAILS 1.1: RHESUS MACAQUE, *MACACA MULATTA* ZIMMERMANN, 1780 (FIG. 1.2)

Cecilia Veracini
Universidade de Lisboa

The Rhesus macaque is one of the best-known species of Old World monkeys or Catarrhini. This species has the widest geographic range of any non-human primate, occupying a great diversity of altitudes and a great variety of habitats, from grasslands to arid and forested areas, but is also found close to human settlements. It is native to South, Central and Southeast Asia: Afghanistan; Bangladesh; Bhutan; Southern China; Hainan; India; Laos; Myanmar; Nepal; Pakistan; Thailand; Vietnam (cf. Mittermeier et al., 2013: 645). These macaques live in large complex multi-male-multifemale groups of about 20–200 individuals, with resident lineages of female kin. Females tend not to leave the social group and have highly stable matrilineal hierarchies in which a female's rank is dependent on the rank of her mother. Juvenile male macaques at four to five years of age are driven out of their natal groups by the dominant male. This species has been used extensively in medical and biological research (the US imported hundreds of thousands of macaques from India between the 1950s and late '70s). In 1974, China began captive breeding and was exporting them by 1984 (Mittermeier et al., 2013: 646). Colonies were also established at the Caribbean Primate Research Center (University of Puerto Rico) on the island of Cayo Santiago. Confiscation for laboratory testing is a mostly localized threat, but it is considerable in certain areas. Moreover, trapping for the pet trade and entertainment still occurs in Nepal, North India, and Afghanistan. In Lao PDR and Vietnam, the major threat to the species is hunting, although the loss of forest in river valleys has also likely impacted the species (Timmins et al., 2008).

Scholars seeking a pre-*Ramayana* source for the Hanuman character have pursued several leads apart from the bewildering Vedic hymn; for example, seeking to identify him as originally a *yaksha* or earth-sprite, a class of supernatural beings associated, in late-Vedic times, with geographical features such as trees, ponds, and boulders. *Yakshas* were occasionally depicted in animal or half-animal forms and were propitiated, often with blood sacrifices, for mundane favors (Bulcke, 1960: 402; Narula, 1991: 20–24). But although there are resonances of such worship in Hanuman's later cult, there is no historical evidence of the veneration of a *yaksha* in simian form. Buddhist sources believed to be roughly contemporary with the composition of the Sanskrit *Ramayana* have likewise been combed in search of possible precursors to Valmiki's monkey hero. The voluminous *Jataka* literature (didactic and entertaining "birth stories" of previous incarnations of the Buddha) indeed contains a lengthy version of the Rama story, but this narrative focuses on the theme of Prince Rama's Buddha-like renunciation of his kingdom and features no kidnaping of his wife—and hence no demons or monkeys. The latter are not uncommon in other *Jatakas*, however, appearing in more than 20 tales, and in one (story number 407), a "great monkey" (*mahakapi*) uses his body to form a bridge so that his companions may reach a mountain to escape danger; in the process, he is badly injured and later dies. This exceptionally altruistic monkey is identified as a bodhisattva or future Buddha. More commonly, however, monkeys in Indian Buddhist tales "represent all that is undesirable in human nature" (Sattar, 1990: 29). Judgment of their behavior as foolish, duplicitous, and capricious would lead to the well-known metaphor, later found throughout the East Asian Buddhist world, of the "mind-monkey" (or "monkey mind"), which flits restlessly from one thought or desire to another and must be tamed through meditation—a theme that will eventually figure in the narrative of the Chinese monkey hero.

The genesis of Prince Rama's simian allies, and the necessity of his "coming-together" with them, has its traditional explanation within the *Ramayana* itself: they are *devas* or celestial gods who take birth on earth as monkeys in order to assist the god Vishnu, incarnate as the human Rama, in his appointed task of slaying Ravana. This demonic titan was oppressing the universe after having wrested a boon from the creator god Brahma that he could not be slain by any of the many kinds of divine and semi-divine beings that he thought might pose a threat to him. But Ravana, in asking his boon, had neglected to mention what he perceived to be the "weak, insignificant" categories of human beings and monkeys, thus opening a loophole that would lead to his eventual demise. Accordingly, the gods were dispatched to the forests of southern India to sire powerful offspring on female monkeys, and the prominent heroes in the story are indeed regularly identified as the sons of various deities; thus Sugriva is the offspring of Surya (the sun), and Valin of Indra (*Ramayana* 1.16). Hanuman's own conception is from the seed of Vayu, mighty god of the wind and vital breath, bestowed on a somewhat reluctant celestial courtesan named Anjana who has been cursed to take birth as a monkey for a previous indiscretion. The accounts of his birth and childhood (to be discussed shortly) receive special treatment in two narrations within the poem (*Ramayana* 4.65.8–18; 7.35–36).

The poem's furry warriors are identified by a sizeable glossary of terms that encompass all of the standard Sanskrit words for "monkey"—the most common is *vanara* ("forest being"), followed by *kapi* and *hari* (both connoting a "tawny-colored" creature), as well as rarer epithets like *shakhamriga* ("branch animal"), *plavaga* ("leaper"), and *pingalaksha* ("yellow-eyed"). Their simian physiognomy (which would become problematic in some later discourse) is never in doubt, and the poem abounds in descriptors which are at times suggestive of the two most common species found in the subcontinent, the "tawny" rhesus macaque and the gray langur (*Semnopithecus* spp.)—often dubbed the "Hanuman langur"—as well as the rarer lion-tailed macaque (*Macaca silenus*, **Fig. 1.3**) and possibly the western Hoolock gibbons (*Hoolock hoolock*), the only ape species native to the subcontinent and now restricted to upper Assam and Bangladesh. Unlike apes, however, the *Ramayana's* monkeys definitely have tails, which they use effectively as weapons in the climactic battle, along with their sharp teeth and nails, as well as hefted boulders and tree limbs. Being demi-divine, they have shape-shifting abilities similar to those of the poem's demonic antagonists, yet they deploy these only rarely (as, for example, when Hanuman assumes the form of a Brahman mendicant for his first meeting with Rama and Lakshmana; *Ramayana* 4.3). Another magical ability—of hyperbolically expanding and contracting the monkey body and its tail—is periodically used to great effect by Hanuman. Yet superhuman traits do not efface the essential "monkeyness" (*kapitva*) of Rama's allies, and the poet often reminds readers of their friskiness, emotional instability, and uncontrolled sexuality; though again, the poem's most important monkey will rise above all of these shortcomings.

From his first appearance in the epic's fourth sub-book, Hanuman assumes a prominent role in the narrative, guiding Rama and his brother to their meeting and friendship-pact with Sugriva, the deposed monkey monarch whom Rama will reinstate in exchange for his assistance in locating Sita—for although Hanuman is sometimes mistakenly labeled a "monkey king" in apparent confusion with the later Chinese hero, he is in fact simply an adviser and courtier in the Indian narrative, albeit one who quickly eclipses his royal master in narrative significance. Subsequently, Hanuman joins in the search for the kidnaped princess, and in the beloved fifth sub-book, *Sundarakanda* ("the beautiful book"), he triumphantly leaps the sea to find her, challenges and battles the forces of her captor, and torches Ravana's capital city with his tail, which the demons have set afire. This sub-book, in which Rama is almost entirely absent from the narrative, has elicited special enthusiasm from epic audiences. Its dramatic events, focused on Hanuman and Sita, not only constitute the traditional "heart" of the poem, but also represent "a kind of epiphany of Hanuman…to the status of a divinity" (Goldman & Sutherland Goldman, 1996: 40). Once Sita has been located, Hanuman joins in the building of the great causeway that permits Rama's army to cross the sea and lay siege to Lanka, and during the ensuing battle he slays several prominent enemy warriors in single combat. More crucially, he undertakes a journey to the distant Himalayas (using his flight-like power of leaping, part of his patrimony from Vayu) to bring back miraculous healing herbs that save the lives of wounded warriors in Rama's army, including the prince's beloved younger brother Lakshmana. This episode highlights both Hanuman's swiftness and his strength, for since he is unable to identify the precise herb prescribed by the battlefield physician, he uproots an entire mountain and carries it back with him (*Ramayana* 6.89)—the image of Hanuman flying through the air with a mountain peak balanced effortlessly on his upturned palm has become one of his most ubiquitous Indian icons **(Fig. 1.4 and Fig. 1.5).**

FIGURE 1.3 The lion-tailed macaque, *Macaca silenus*. (Photo by Cecilia Veracini.)

FIGURE 1.4 (*Ramayana* 6.89)—the image of Hanuman flying through the air with a mountain peak balanced effortlessly on his upturned palm has become one of his most ubiquitous Indian icons.

FIGURE 1.5 Sandstone sculptures of Hanuman bearing the mountain, outside a temple sculptor's studio in Varanasi, India, 1990. (Author's photo.)

After the defeat of Ravana, Hanuman continues to serve as Rama's special envoy (to Sita, informing her of her husband's victory and her imminent release, and to Rama's brother Bharata, who is awaiting Rama's return from exile) and reverently participates in Rama's consecration as restored ruler of Ayodhya. Blessed by both Rama and Sita—as well as, earlier in the story, by all the celestial gods—Hanuman achieves virtual physical immortality, being destined to remain on earth for as long as Rama's story endures (*Ramayana* 7.40.16–18; 7.108.33–35).

South Asian storytellers and audiences do not seem to have been content with these exploits, however, and the past two millennia have witnessed not only a proliferation of Rama narratives in virtually all languages of the subcontinent—the conventional formula "three hundred *Ramayanas*" being possibly an understatement (Ramanujan, 1991)—but a comparable growth in tales focusing on Hanuman, who has emerged as (borrowing a Hollywood category) the saga's "best supporting actor." These narratives, which seem to have proliferated after about 1000 CE and especially in the 18th through 20th centuries, commonly enhance Hanuman's status by making him not merely the "son of the wind" (one of his most common epithets) and servant and devotee of Vishnu-Rama, but also an *avatara* or incarnation of Shiva, another of the most important gods of Hinduism. They also supply him with a lengthy "biography" that both precedes and postdates the traditional Rama narrative, generating an epic-like and still growing cycle of tales that I have dubbed (following one modern Hindi author) the "*Hanumayana*" (Lutgendorf, 2007: 113–235).

One popular tale is frequently interpolated into the standard *Ramayana* storyline, during the final phase of the war in Lanka. Ravana, knowing he is facing defeat, summons from the depths of the netherworld a super-demonic brother or double, endowed with more awesome powers than his own. This *Ahiravana* ("snake Ravana," sometimes called *Mahiravana*, "earth Ravana") uses sorcery to kidnap Rama and Lakshmana while they are sleeping under heavy guard in the monkey camp and spirits them away to his nether kingdom, where he prepares to offer them as human sacrifices to his tutelary tantric goddess. When Rama's troops realize what has happened, they despair, but Hanuman rises (or rather, sinks) to the occasion, plunging into the earth through the stalk of a lotus and making his way to Ahiravana's realm. Many adventures ensue, perhaps the most striking of which is an encounter, at the gate of the nether citadel, with a powerful monkey guard who is Hanuman's own double, and who indeed identifies himself as "the son of Hanuman," named *Makaradhvaja* ("fish-banner"). Since the South Asian Hanuman is (in most narratives) a confirmed bachelor and celibate ascetic, this revelation shocks him, until he learns that this "son" was born from a fish-mother who had swallowed a drop of Hanuman's sweat or phlegm during his flight over the sea. Loyal to his own master Ahiravana, Makaradhvaja wrestles with his father, but loses; later, after Hanuman's slaying of Ahiravana, he is consecrated by his dad as the new ruler of the subterranean realm. The signal feature of this popular narrative—which is at least a thousand years old and exists in forms ranging from book-length poems to comics and a television serial—is its unembarrassed "doubling" of the Rama narrative, with the kidnaping of Sita here replaced by that of the two princely brothers, and Rama's simian "helper" elevated to the starring role. The human hero becomes a passive victim (in some versions, lying in a box with his brother, in a magically induced coma, as they await sacrificial death) and the spotlight shifts to Hanuman, whose wily, trickster-like ways are a better match for the "snaky" sorcerer of the underworld (Lutgendorf, 2004). As I will show, this shift in focus has striking echoes in the lore of Southeast and East Asia.

Since, according to traditional Hindu time, Rama is supposed to have lived in the second age of the current cosmic time-cycle—roughly 900,000 years ago—Hanuman's life on earth has been long indeed, and the popular stories in the *Hanumayana* give him both an adventurous childhood and an extended (and still continuing) old age. The former includes the story (already twice-told in Valmiki's epic) of his attempt, as a hungry infant, to devour the rising sun, which he mistakes for a ripe fruit. In punishment for his audacity, he is mortally wounded by the thunderbolt of Indra, which mars his countenance and gives him his most common Sanskrit name, *Hanu-man* ("one having a prominent or disfigured jaw"—an apparent allusion to simian prognathism). The sulky revenge of Hanuman's father, Vayu, who withholds his vital breath and begins to suffocate the cosmos, then causes the celestial gods to revive the monkey-child and gift him with their own powers. Thus endowed, young Hanuman goes on a rampage of monkey-like destruction of the forest hermitages of human sages, who in turn curse him to forget his supernatural powers until, as an adult monkey in Rama's service, he will be reminded of them.

Another popular childhood story of non-Valmikian and indeed fairly recent origin has young Hanuman disguise himself as the trained monkey of a traditional street performer (who is likewise the disguised god Shiva, and hence Hanuman's own alter-ego); after performing a "monkey dance" before the toddler Rama and his brothers, he is adopted as a royal pet and so participates in Rama's entire childhood, until dispatched (by his omniscient young master) to the southern forests to await the prince's wanderings in exile. Whereas the older childhood stories emphasize the rampant and potentially dangerous energy (*shakti* is the key Indic term) of the "over-reaching" simian, who boldly leaps into the heavens to challenge the gods and then wreaks havoc on earth, the "dancing monkey" tale highlights simian cuteness as well as Hanuman's emotional dedication and exemplary service (encoded in the word *bhakti*). These complementary or even paradoxical themes of unleashed "power" and submissive "devotion"—which, in their usual Indic forms, conveniently rhyme—resonate through much of South Asian Hanuman lore and are often cited by contemporary worshipers in explaining his immense popularity.

The post-Rama phase of the *Hanumayana* narrative includes numerous tales that bring Hanuman into the fourth and current age of the cosmic cycle—the dismal and discordant Kaliyuga, when *dharma* is at its lowest strength. They feature his encounters both with major heroes and religious figures (including his mighty half-brother Bhima from the epic *Mahabharata*, who is likewise sired by the wind god), as well as with humble devotees who, even today, believe that he is present wherever the Rama story is narrated, and who regularly report his miraculous and timely interventions in their lives (Lutgendorf, 2007: 279–297).

Although it is possible that modest shrines to Hanuman—for example, as a "boundary guardian" (*dvarapala*) of villages and fields, a role he still regularly performs—may have existed from ancient times, evidence of larger sculpted images intended for worship and of buildings to house them appears largely in the second millennium CE. A standard explanation for this in modern scholarship is that veneration of the "great hero" (another of Hanuman's epithets) was a defensive Hindu response to the militant incursions and conquests of Islamic rulers that date to roughly the same period. In my larger study of Hanuman, I question this argument, now tinged with the xenophobia of Hindu nationalism, and point instead to the social upheaval that likewise occurred during this period and that brought greater prosperity and prestige to lower-class and low-caste communities, for whom Hanuman—the monkey who rose to immortality and godhood through his own efforts—was a patron deity and emblem of "upward mobility" (Lutgendorf, 2007: 61–88). In any case, the real explosion of temple-building to Hanuman is datable only to the colonial period (mid-18th century onward) and included both small and large shrines erected by diverse patrons, ranging from merchants and matrons to sadhus, politicians, and wrestlers—for Hanuman is also the patron god of athletes. In the late-20th century, the monkey god's sculptural icons grew like his own legendary body, as new, mainly middle-class patrons seemingly vied with one another to have (as one man enthusiastically told me) "the biggest Hanuman in India," resulting in several colossi of 30 m or more (**Fig. 1.6**; Lutgendorf, 1994). To such sanctuaries, contemporary worshipers flock—especially on Tuesdays and Saturdays, when Hanuman is believed to offer special protection against malefic planetary influences—and their walls echo with the chanting of "forty verses to Hanuman" (*Hanuman chalisa*), a praise poem attributed to the 16th-century poet-saint Tulsidas that has become one of the most widely recited ritual texts of Hinduism (Lutgendorf, 2007: 92–102).

FIGURE 1.6 Colossal statue of Hanuman in Delhi, India, 2006. (Photo by Cynthia Humes, used with permission.)

BOX DETAILS 1.2: HANUMAN, GRAY OR SACRED LANGURS, *SEMNOPITHECUS* SPP. (FIG. 1.7)

Cecilia Veracini
Universidade de Lisboa

The langurs (Cercopithecidae, Colobinae) of the genus *Semnopithecus* (meaning in Ancient Greek,

FIGURE 1.7 *Semnopithecus entellus* with a statue of Hanuman; Chitrakut, Madhya Pradesh, India. (Photo by the author.)

"revered, holy monkey"), are the most widespread langurs of the Indian Subcontinent stretching from the Himalayas in the north to Sri Lanka in the south, and from Bangladesh in the east to Pakistan in the west. According to the recent classification, this genus is composed of eight species that have different coloration (Mittermeier et al., 2013: 733–739). They live in large multimale-multifemale, unimale-multifemale, and all-male groups. Social hierarchies exist for all group types and females within a group are matrilineally related. These langurs who feed on leaves, shoots, fruits, flowers, tree exudates, barks etc., can adapt to a variety of habitats and altitudes (up to 4,000 m) and also to human settlements, such as villages, towns, and areas with housing or agriculture. Some of them (e.g,. *Semnopithecus entellus*) are considered sacred and are revered and fed in the vicinity of temples. All the species of this genus have conflicts with local people for their crop raids. Other threats include intensive agriculture and habitat loss. The species with a very restricted geographical range are threatened with extinction (e.g., *Semnopithecus ajax, Semnopithecus vetulus*) or are vulnerable (*Semnopithecus hypoleucos, Semnopithecus johnii*) (cf. the entry *Semnopithecus* IUCN Red List, 2018).

Before leaving the subcontinent and leaping eastward over the sea, I must cite one additional set of texts: the "alternative" Rama-stories produced by Jain teachers beginning in the early centuries AD. Proponents of a numerically small but influential religion that emphasized absolute *ahimsa* or non-injury to living beings and hence rejected the Vedic sacrificial tradition, offered their lay followers, who included both kings and merchants, distinctive Jain versions of the popular epic that was increasingly being claimed by their Brahman rivals. Jain narrators could thus entertain their sectarian listeners with a beloved and popular tale while assuring them that the majority of its tellers had gotten it all wrong. Such revisionist retellings—some 45 of which exist, from the 2nd to the 17th century CE (Kulkarni, 1990: 12–14)—usually make two distinctive interventions where Hanuman is concerned. The first is to deny his ***kapitva*** or "monkeyhood"—because Jain standards of cleanliness deemed such beings repugnant and relegated them to the "low" animal-plant realm in the hierarchy of living beings—and to make him instead a human-looking demi-divine being known as a *vidyadhara* ("bearer of wisdom"). Such magical creatures were a staple of ancient Indian entertainment lore, akin to the fairies of Elizabethan romance; they were long-lived, capable of flight, and could change their shape at will. Hanuman, the Jain storytellers said, belonged to a specific *vidyadhara* lineage that had adopted a monkey as its heraldic sign and fought battles with magical "nooses" that resembled monkeys' tails. But monkeyness was also entertaining, and so, in some Jain versions, *vidyadhara*-Hanuman would metamorphose into simian form at key moments, and so could still use his flaming tail to burn Lanka.

The Jain storytellers' other innovation was likewise a slap-in-the-face to the coalescing Hindu narrative, which stressed Hanuman's chastity and rigid self-control in his devoted service to his master. For their Hanuman was a *bon vivant* and playboy, with numerous wives and several extra-marital liaisons to his credit, even after he entered, somewhat reluctantly, into an alliance with Prince Rama. Indeed, Jain narrators explained, he was irresistible to women because he was 1 of the 24 *kamadevas* (embodiments of the god of love and lust) who take birth in every cosmic time cycle. Thus, when he was challenged, while in search of Sita, by a female demon guarding the gates of Lanka (here, Valmiki's monkey knocked the lady unconscious) the handsome *vidyadhara* promptly seduced and bedded her. The Jain laity could enjoy such juicy monkey business in their tales, however, because they knew that, according to Jain teachings, the *kamadeva* would ultimately wise up, renounce the world, and become a salvation-seeking mendicant—as Hanuman, after the victory over Ravana and lots more fun, eventually does in Jain tales (Cort, 1993: 197). As I have noted, many Jains were merchants, and their religion flourished in the southern kingdoms of the subcontinent during precisely the period (5th–12th centuries CE) when these regions were most actively engaged in maritime commerce with the islands and coastal regions of Southeast

Asia. One durable if intangible Indian export was the Rama story in its many versions, and the next section will make clear that the Jains' concupiscent Hanuman, if ultimately rejected by most South Asian narrators, would find appreciative audiences—and more sexual conquests—further to the east.

The tendency for the principal human protagonist in the *Ramayana* saga to gradually assume a "supporting" role in relation to his simian helper may already be detected in South Asian literary tales and folklore, and it is certainly apparent today in the sheer number and ubiquity of Hanuman temples and shrines, which far outnumber those of his master—despite the pious claim, by many Hindus, that Hanuman merely epitomizes the perfect servant and model devotee of the "ideal man" of *dharma*, Rama (**Fig. 1.7**). In comparable narratives that became popular in Southeast Asia and China, however, the balance of power and prominence would unequivocally shift in favor of the simian partner in the relationship, and it is to these regions and stories that I now turn.

1.2 SOUTHEAST ASIA

The islands and coastal regions of Southeast Asia are home to nearly 100 extant species of non-human primates, including numerous varieties of tailed, "Old World" monkeys, —ground-dwelling macaques (*Macaca* spp.), arboreal langurs (genera *Presbytis, Trachypithecus, Pygathrix, Simias, Rhinopithecus, Nasalis*), and prosimians of many taxa (Tarsiers and Lorises)—as well as more than a score of species of apes, comprised primarily of the many species of gibbons (Family Hylobatidae), but also including two great apes: the tree-dwelling orangutans of Borneo and Sumatra, *Pongo* spp. Recorded folklore and practice with respect to these animals suggest patterns that replicate some of the ambivalence shown by human populations elsewhere in Asia. On the one hand, there is lore that conflates a "fall" into simian status with ancestral human misbehavior (Janson, 1952: 106; Barnard, 1995: 77; Knight, 1999: 626); on the other, there are reported traditions of respect and deference being accorded to simian populations as embodiments of the spirits of ancestors (Riley & Priston, 2010: 5). The crop-raiding activities of several species, especially the sometimes aggressive macaques, were long a matter of concern to subsistence agriculturists, yet peasant farmers generally were reluctant to kill the animals, and though a few hunter-gatherer groups would eat certain monkeys, Buddhists in mainland Southeast Asia and Hindus (on Bali) generally abhorred this practice, and Muslims, the dominant religious group in Indonesia and Malaysia, considered their flesh to be *haram*, or forbidden by Islamic law (cf. Veracini & Aghamadi, this volume).

The most widely-documented monkey lore of Southeast Asia—attested in literature, visual art, and performance traditions—reflects the spread and indigenization of the Rama narrative in the region, a process that probably began early in the Common Era, concurrent with a robust maritime trade that linked mercantile cities on the southern coast of South Asia, via numerous ports of call in Southeast Asia, with places as distant as Quanzhou in southern China. Because shipping in this vast area relied on the annual cycle of monsoon winds, it necessitated periods of residence of as long as half a year by sailors and merchants in destination ports, while they awaited the reversing of prevailing winds that would permit their return journey (Walker, 1998: 17–21). Such temporary communities sometimes put down roots, intermarrying with local women, building temples, and engaging the services of Hindu, Buddhist, and Jain religious experts brought from the subcontinent, including exponents of sacred narratives. The rise of "Indianized" states and kingdoms such as Funan, on the Mekong Delta (1st–6th centuries AD), Srivijaya, on the island of Sumatra (7th–11th centuries), and the Khmer Empire (which flourished in what is now Cambodia, Laos, and Thailand between the 11th and 15th centuries) was accompanied by the development of a hybrid culture within which the Rama story—probably in multiple versions that reflected Hindu, Buddhist, and Jain retellings—played a notable role. The story developed a range of local variants, some of which eventually crystalized into major literary works in regional languages, such as the *Ramayana kakawin* (the latter term referring to the genre of long poems) in Old Javanese, attributed to the 10th century poet Yogishvara and showing the influence of Shiva-worshiping traditions of South India; the *Hikayat seri Ram* ("Chronicle of Lord Rama") in Malay (circa 13th century); and the much later Thai *Ramakien* ("Glory of Rama"), written under the patronage of (and allegedly partly by) King Rama I in about 1800, but reputedly based on older texts that were lost in the sack of the Thai capital, Ayutthaya (the Thai form of the Sanskrit name Ayodhya), in 1767 by a Burmese army. These and other written adaptations are believed to reflect centuries of oral storytelling and performance, particularly in dance dramas and shadow-puppet plays (known in Javanese as *wayang*) that became popular throughout much of the region, many of which featured stories inspired by the ancient Indian epics.

The Rama story was also a subject for local sculptors and other visual artists, reflected in major sculptural programs at such temple complexes as Prambanam in Java (9th century CE) and Angkor Vat (12th century). The growing influence of Theravada Buddhism in the mainland kingdoms of Southeast Asia and later of Islam on the Malay Peninsula and the islands of Indonesia did not eliminate the popularity of the story; instead, Rama was again idealized as a bodhisattva by Buddhist storytellers and as a brave and noble warrior-king (though not, of course, as a divine incarnation) by Muslims. The attraction of his simian allies seems to have further contributed to his tale's entertainment value. The wrestling match between the monkey king Valin and his brother Sugriva (which Rama, pursuant to his friendship pact with the latter, would end by slaying the former) was a favorite subject in Khmer sculpture. This was perhaps a warning against familial discord since this deadly duel climaxes the epic's most dramatic example of fraternal rivalry, contra the code of *dharma* so memorably adhered to by the human characters in the saga, but projected onto its monkeys and demons.

Another favorite subject for sculptural representation among the Khmers was the story, originating in the Sanskrit epics and *puranas* (anthologies of "ancient lore"), of the competitive churning of the primordial milk ocean by the gods and their demonic first cousins in order to extract and consume the "cream" or nectar of immortality. In several famous bas-reliefs that depict this tug-of-war, involving the cosmic serpent as rope and the world-mountain, Meru, as the churning rod, a heroic monkey figure appears at the (literal) tail-end of the divine team, seemingly playing a prominent role in its victory. Since no such figure appears in most Indian versions of the story, scholars have long speculated on the monkey's identity, with most researchers favoring Hanuman (who enjoys, as will be seen, unusual prominence in Southeast Asian Rama lore), but one recent study offers evidence that the intended triumphant simian may have been Valin. For this monkey monarch's death at Rama's hands did not tarnish his reputation for extraordinary strength and heroic deeds, including the slaying of several of the gods' demonic enemies and some South Indian narratives, like the Southeast Asian reliefs, give him a role in the ocean-churning story (Narayanan, 2013).

A consistently celebratory take on simian behavior can be found in the representation of Hanuman in Southeast Asian literature, art, and performance, for in several of the Rama traditions of this region his role seems to eclipse that of his human master. At times, Hanuman's relationship with Rama transcends friendship and devotion to take on, so to say, a genetic dimension. In a variant on his birth story found in Javanese, Laotian, Malay, and Thai sources, Rama, while searching for Sita in the forest, consumes a magic fruit that transforms him into a monkey for three years, in the course of which he himself mates with Anjana to sire Hanuman. In another Malay variant, Rama and Sita (prior to her kidnaping) bathe in an enchanted forest pond that turns them both into monkeys. They mate, and Sita conceives while in simian form. Lakshman finds them, however, and plunges them into a second pool that undoes the spell. Rama then extracts the unborn fetus from Sita's womb and it is transported by the wind to the monkey Anjana's open mouth. In still another Malay variant, Rama, alone in the forest, ejaculates on seeing the beautiful Anjana, and the wind carries his semen into her mouth (Govindchandra, 1976: 239–249; 258). These tales (which would shock most contemporary Indian audiences) appear to be variants on Shaiva stories popular in southern India, for this sometimes transgressive deity may scorn normal societal standards and in a number of tales lusts after various females, or assumes an animal form, together with his loving wife, Parvati, to engage in frisky amorous play. The stories' assertion of Hanuman's "blood" relationship to Rama also echoes a more sanitized Indian variant, datable to at least the 15th century, in which Anjana's impregnation occurs as a result of swallowing a bit of sweet-rice porridge, dropped into her mouth by a vulture who snatched it from the divine vessel presented to Rama's father by Agni, god of fire, at the climax of King Dasharatha's "son-producing" sacrifice. Since the rest of the porridge, given to the king's three chief queens, results in the births of Rama and his brothers, Hanuman likewise enjoys a fraternal relationship with the princes while retaining his mother's simian form (Lutgendorf, 2007: 130–131; 183).

The transformation, by Jain storytellers apparently uncomfortable with Rama having "bestial" cohorts, of Valmiki's sternly-devoted monkey into an anthropomorphic playboy seems to have held particular appeal for Southeast Asian storytellers and audiences, who had no problem, however, with his being in monkey form. The result is a veritable simian Casanova who manages to admirably serve his human master while simultaneously chalking up a series of amorous conquests that delighted regional audiences. The guardian goddess of Lanka (sometimes transformed into the daughter of its gatekeeper) was one such dalliance, and Cambodian and Thai narrators added another tale of Punnakay (also called Benyakai), a niece of Ravana, who was sent on a secret mission to weaken Rama's resolve to fight and instead ended up in the arms of the seductive simian (Lutgendorf, 2016: 150; 192).

Similarly, the (usually anonymous) female fish who conceives Hanuman's unlikely "son" in subcontinental tellings of the Ahiravana episode—and then soon dies at the hands of nether fishermen who are surprised to discover a robust young monkey in her innards—takes on, in the maritime and deltaic cultures of Southeast Asia, more character and characteristics. Often, she is called "golden fish" (***suvannamaccha***) and is a mermaid-like beauty who is queen of all marine creatures and sometimes a daughter of Ravana **(Fig. 1.8).** Most commonly her encounter with Hanuman begins when she is sent by Ravana along with her piscine army to disrupt the construction of the causeway to Lanka. In a Malay variant, Hanuman uses his tail to agitate the water around the causeway, stirring up silt and causing crabs and fish to surface, whereupon his cohorts greedily seize and devour them. But in Cambodian and Thai narratives, Hanuman's own "appetite" is satisfied only through his seduction of the golden fish-maiden, with whom he falls in love (Govindchandra, 1976: 257). She eventually succumbs to his charms, and the son who results from this liaison is a marvelous hybrid—a sort of "mer-monkey," with a simian upper body and the tail of a fish—a striking visual motif, popular in certain Thai art and handicrafts, that literally "bridges" the fauna of the forest and the sea (Bulcke, 1999: 440). In Southeast Asian variants of the Ahiravana story, it is of course this hybrid son whom Hanuman will encounter in the netherworld at the bottom of the ocean.

In Southeast Asian representations of Rama stories, the depiction of simians is rarely naturalistic or species-specific, and their stylized, fanged faces in fact resemble those

FIGURE 1.8 Hanuman romances Suvannamaccha; from a mural at Wat Phra Kaew, Bangkok; Wikimedia Commons]: https://en.wikipedia.org/wiki/Suvannamaccha#/media/File:072_Ramakien_Murals_(9150815520).jpg.

of the saga's demonic antagonists, although audiences clearly discerned a difference—the feral aggression of the monkey soldiers was deployed on the side of good. Almost inevitably, they were given tails—signaling monkeys and not apes. But in modern Thai and Cambodian art and performance, Hanuman is frequently depicted and referred to as a "white monkey," and this association may reflect the influence of ancient Chinese lore about mystical white- and golden-colored gibbons, which will be discussed in the next section.

1.3 EAST ASIA

The relationship of human beings to non-human primates in Japan has been the subject of considerable research in modern times, both by outsiders and by Japanese anthropologists and primatologists. Studies by the latter, beginning in the mid-20th century, focused on the physiology and social behavior of the Japanese macaque (*Macaca fuscata*) indigenous to three of the four main islands of Japan and commonly known as the "snow monkey," because its habitation includes areas that receive considerable snowfall in winter, making it the most northern-dwelling and cold-adapted of wild primate species (cf. Mittermeier et al., 2013: 647). Intensively studied, in part because of its "special" status in Japanese culture—it is regularly referred to as "our national monkey" and is the only animal referred to in speech with the honorific suffix ***-san***, normally reserved for humans (Knight, 2003: 85–86)—this greyish-tan, red-faced macaque lives in matrilineal groups and engages in such human-like activities as collective bathing in hot springs during cold weather and washing of food to remove grit (e.g. cf. Itani, 1958; Kawamura, 1959) Japanese primatologists famously developed an "empathetic and intuitive" attachment to their research subjects, and their labs sometimes conducted "monkey funerals" for deceased macaques (Haraway, 1989: 244, 251–252; cf. also De Waal, 2001).

According to anthropologists and cultural historians, although the ***saru*** (the usual Japanese term for "monkey") has not occupied a starring role in the kinds of epic narratives found in the other Asian cultures I am surveying, it has nevertheless generated a good deal of lore that reveals the characteristic ambivalence I have noted elsewhere. In the indigenous religious tradition of Shinto, a "monkey deity" (*saru gami*) named Saruta Biko appears in texts datable to at least the 8th century and is associated with liminal spaces that separate villages and their cultivated fields from surrounding forests and mountains. He is identified as an ***otsukai***, "a sacred messenger…acting as a mediator between the *kami* spirits and human beings" (Knight, 1999: 626) and is said to especially play this role in relation to the powerful Mountain Deity. In addition, he has a strong connection with horses and is invoked in rites intended to protect them from disease—a role that may reflect the influence of older Chinese traditions (Asquith, 1995: 313; Ohnuki-Tierney, 1987: 48–50; 1995: 301). Yet, according to anthropologists Emiko Ohnuki-Tierney and John Knight, collectors of primate lore in Japan have been guilty of oversimplifying a more complex range of responses and also of depicting them as static and unchanging. According to Ohnuki-Tierney (1987: 6–7), the fact that, to the Japanese, the *saru* is:

> the animal considered to be most similar and therefore closest to humans…has made it in turn a revered mediator during early periods in history and a threat to the human-animal boundary in later periods, fostering ambivalence that is expressed by mocking the animal.

She also notes the macaque's association, during feudal times, with the despised category of "special status people"—Japan's own class of outcastes—who, like it (and like so-called "untouchables" in South Asia) may paradoxically mediate between human beings and supernatural powers and yet also be stigmatized as bearers of pollution (Ohnuki-Tierney, 1987: 75–100). The many instances in Japanese folklore of cross-species transformation (a feature of Chinese lore as well) rarely involve monkeys, yet those that do nearly always involve the one-way

FIGURE 1.9 *Luohan*, hanging scroll, with gibbons (*Hylobates lar*), Liu Songnian, 1207. (National Palace Museum, Image of Public Domain (https://en.wikipedia.org/wiki/Monkeys_in_Chinese_culture#/media/File:Liu_Songnian-Luohan.jpg).)

metamorphosis of a human being who has incurred divine punishment (Knight, 2003: 85–86; Ohnuki-Tierney, 1995: 300). Knight's ethnographic research, although questioning Ohnuki-Tierney's neat periodization of evolving responses to non-human primates on the islands, identifies a similar ambivalence, particularly in late-20th-century reactions of rural Japanese to the troops of macaques that increasingly raided agricultural fields and orchards (Knight, 1999, 2003: 84–122). I will return to this agonistic aspect of human-non-human primate relations in my concluding section.

The most comprehensive study of references to animals in the early literature of China portrays a culture in which "the demarcation of the human and animal realms was not perceived to be permanent or constant, and the fixity of species was not self-evident or desirable"—indeed, one in which the natural world was "a negotiated reality" (Sterckx, 2002: 5, 9). In early Taoist thought, both humans and animals were said to possess "blood and *qi*"—the latter term (sometimes transliterated *ch'i*) referring to "breath" or "life force"—though the *qi* of animals was said to be "coarse" and that of humans "refined" (Sterckx, 2002: 73–75). Yet according to Roel Sterckx's survey, specific references to non-human primates were (with one exception, to be discussed shortly) comparatively rare—e.g., the inclusion of a monkey as the 9th of the 12 zodiacal animals of the Chinese calendar cycle (Sterckx, 2002: 66–67). The evolving pictographic characters for non-human primates were generally not species-specific, but reflected consciousness of these creatures' resemblance to humans, albeit with tails (Sterckx, 2002: 37; van Gulik, 1967: 18–22; Zhang, 2015: 217). Although, in time, words came into use that may be translated as "ape" (*yuan*, traditionally referring to tailless gibbons, but see Niu, 2023) and "monkey" (*hou*, usually referring to macaques), they were not always used consistently and, in general, ancient Chinese authors showed little interest in the anatomical specificities of non-human primate species. Instead, they tended to emphasize the presence or absence of "morality" and "ethical" behavior among them (Zhang, 2015: 217, 223–224).

Viewed from this perspective, gibbons and macaques acquired stereotypic and enduring characteristics in early Chinese thought. The gibbons—arboreal, mainly plant-eating apes who live in small family groups in remote and forested mountains—came to be regarded as the well-behaved "gentlemen" of the animal world, given to filial piety, deferential behavior, protective care of offspring, and slow, discriminating eating. Moreover, the suppleness of their long limbs, by means of which they swooped gracefully through the treetops was thought to indicate proficiency at absorbing *qi* from the atmosphere, making them models for Taoist recluses in search of immortality, Their distinctive "singing"—of melodic sequences of pentatonic notes especially voiced in the early morning—became a trope, in the celebrated poetry of the Tang Dynasty (7th–10th centuries CE), for contemplative melancholy, evoking the loneliness experienced by travelers on boats moving down the gorge of the Yangtze River, where gibbons could be heard (but rarely seen) calling out on the heights (van Gulik, 1967: 40–45) (**Fig. 1.9**). Rhesus macaques, on the other hand (closely related to South Asian varieties)—ground-dwelling, usually omnivorous, tailed, and generally living in larger bands—were characterized as "garrulous, unruly, and not family-oriented." Living in closer proximity to people than the aloof gibbons, they were also seen as destructive pests: "they gnaw at unripe fruit and throw it away after the first bite. They steal people's food and fill it in their cheek pouches. They trample down or uproot young plants and tree shoots in their habitat mountains" (Zhang, 2015: 220–222). Relatively easy to capture and capable of being trained as street performers—in which role they were commonly seen—macaques were viewed as "clever" rather than "wise," and as "comical and silly" rather than (like gibbons) solemn and mysterious (Zhang, 2015: 220–222).

Although these stereotypes would prove enduring—as a recent computer-aided study of some 10,000 Chinese texts spanning two millennia has confirmed (Zhang, 2015: 216)—they also admitted notable exceptions, suggesting once again the ambivalence of human responses to non-human primates. Thus the "sacred" gibbon could also be stigmatized, in narrative poems of the 1st–4th centuries CE from the Sichuan region, as an aggressive abductor of human women, on whom it sired offspring who "are all like men in appearance" (Walker, 1998: 56–57). This lustful, "demonic ape" was sometimes identified as white in color, and hence resembles the "white monkey" of Southeast Asian Rama stories, and may even have influenced their developing portrayal of Hanuman as an erotic hero. He would reappear, again in Sichuan, in the period of the Song Dynasty (10th–13th centuries) as the nemesis of a popular hero god named Erlang, a skilled archer who was eventually able to defeat him. Coincidentally, the Song period also saw the growth of the maritime trading city of Quanzhou, in Fujian Province on the southern coast, where archaeological evidence reveals that a colony of Tamil traders from South India flourished by the 12th century and were sufficiently involved in cultural patronage to have left more than 200 pieces of Hindu sculpture (Walker, 1998: 48–50). Roughly a century later, a Chinese-built temple in the same city would feature a monkey-headed deity who, though Hanuman-like in physiognomy (and endowed with a tail), was evidently not the Indian god, but an indigenous folk-deity, elevated from among the lowly, much-maligned macaques, to earn, through his own extraordinary achievements, the status of both a Taoist immortal and a devout disciple of the Buddha—in short, a prototype of Sun Wu-k'ung, also known as the "Monkey King" (Walker, 1998: 80).

This endearing character—a monkey who is at once mischievous and wise, comical and heroic—is now best known through the great "hundred-chapter" Ming Dynasty novel *Hsi-yu Chi* (or *Xiyouji*, "Record of the Westward Journey"; in Anthony Yu's celebrated English rendering, *The Journey to the West*), one of the most beloved works of premodern Chinese fiction. Published anonymously in 1592, this sprawling masterpiece (sometimes attributed to one Wu Cheng'en) was inspired by the historical account of the sixteen-year pilgrimage to India of the Buddhist monk Hsuan-tsang (or Xuanzang, 596–664 CE), which it expanded with a blend of Taoist, Buddhist, and Confucian traditions—and, quite possibly, elements inspired by Rama tales. To the Tang-period narrative, repeatedly retold since Hsuan-tsang's own memoir, supernatural elements gradually accrued, including four traveling companions: a monkey, a pig, a dragon-turned-horse, and a water spirit transformed into a mace-bearing bodyguard. But in the earliest versions of the fictionalized tale, now dated to roughly the 13th century, the monkey is the pilgrim's only named companion (notably, he assumes the disguise of a human scholar for his initial meeting with the monk, a motif that echoes Rama and Hanuman's first encounter; Walker, 1998: 72); by the time of the *Hsi-yu Chi,* he had come to dominate the tale so thoroughly that Arthur Waley's greatly abridged 1943 translation was titled simply *Monkey* (Yu, 1977: ix, 5–9).

The novel opens with a brief cosmogony that culminates in a vision of a primordial stone egg atop a mountain, from which emerges a plucky and opportunistic monkey—significantly, the wind or breath of life (father of Hanuman) plays a role in shaping this form and releasing it from its stone matrix (Walker, 1998: 7). Rays of light flash from Monkey's diamond eyes and penetrate the highest heaven, where they disturb the supreme celestial bureaucrat, the Jade Emperor. Worse transgressions soon follow: after establishing himself as king of his species, Monkey goes on a quest to acquire immortality, becoming an apprentice to the Taoist patriarch Subodhi. The latter scoffs at Monkey's prodigious leaps, which the simian calls "cloud-soaring," noting that they are nothing like the true flying-on-clouds practiced by human immortals. He declines to teach this secret to Monkey but instead offers ("in accordance with your form") what he calls the "cloud-somersault," a serviceable-enough technique that enables Monkey to travel 108,000 miles in a single jump (Walker, 1998: 9–10). The Monkey soon unleashes mayhem, stealing a king's arsenal and compelling the dragon lords of the four oceans to give him a magic weapon, the "compliant iron rod." Faced with numerous complaints, the Jade Emperor sends his generals in an unsuccessful attempt to defeat the upstart and then tries to co-opt him with a minor appointment in the heavenly stables (for in China, as in Japan, the monkey deity had a special link with horses). The Monkey soon rejects his lowly position and devours the heavenly peaches of immortality, becoming Sun Wu-k'ung, "the monkey awake to emptiness" (alluding to the Mahayana Buddhist doctrine of the void). When he reveals his intention of overthrowing the Jade Emperor and assuming his throne, Buddha himself is summoned. Challenged to display his special leaping-cum-flying skill, Monkey travels in an instant to the farthest reaches of the universe, where he scrawls graffiti on one boundary column and urinates at the base of another; when he returns, he finds both marks of defilement on the Buddha's fingers, proving that he has never left the Perfect One's palm; its fingers then become strong bonds that imprison Monkey beneath a mountain. Penitent at last, he is offered liberation after five centuries provided he becomes a monk and accompanies a scripture-seeking pilgrim to the western paradise (**Fig. 1.10**).

The bulk of the novel concerns this journey and features a long series of supernatural challenges, nearly all of which are overcome through Monkey's fantastic strength, courage, and wit, as well as his persistent (and usually well-founded) disobedience to the novel's human "hero," the monk Tripitaka (the fictionalized Hsuan-tsang), whom translator Yu characterized as "joyless and humorless . . . dull of mind and peevish in spirit, his muddle-headedness matched only by his moral pusillanimity" (Yu, 1977: 44). Indeed, throughout the novel, the portrayal of the monk seems intended "to stress his absolute need for . . . the protective guidance of Sun Wu-k'ung" (Yu, 1977: 44). Monkey thus emerges as the novel's true hero, and though periodically allegorized in Buddhist terms as the restless and impetuous "monkey of the mind," whose instability must be "tamed" by a higher power (as in the Indian *Jataka* tales), Sun Wu-k'ung displays traits more suggestive of a culture hero who embodies the pragmatism, pluck, and resourcefulness so admired by the Chinese. Introducing himself to Tripitaka (in the first of many self-eulogies), he proudly catalogs some of the same supernatural powers (in Sanskrit, *siddhis*) for which Hanuman is known:

FIGURE 1.10 Monkey King Sun Wu-k'ung in an illustration from a Chinese edition of *Hsi-yu Chi*. (Wikimedia Commons.)

> I, old Monkey, possess the ability to subdue dragons and tame tigers, and the power to overturn rivers and stir up oceans. . . . If I want to be big, I can fill the universe; if I want to be small, I can be smaller than a piece of hair. In sum, I have boundless ways of transformation and incalculable means of becoming visible or invisible. (Yu, 1977: 302)

Not surprisingly, post-revolutionary scholarship in the People's Republic tended to downplay the Buddhist framework of the story and to view Monkey (whom Chairman Mao is said to have particularly admired) as a "magnificent heroic character . . . whose defiance and love of independence we have come to cherish" and whose "subjugation" by a Buddhist monk constituted a "basic flaw" in the narrative (Yu, 1977: 54).

The *Hsi-yu chi*'s apparent parallels to the Rama story have occasioned much debate since 1920 when a Japanese scholar first proposed that the Indian tale had influenced the Ming novel (Mair, 1989: 704; Yu, 1977: 10–11). As Victor Mair has noted, the resemblances are too numerous to be merely coincidental, particularly if one assumes the *Ramayana* to refer not simply to the Valmiki poem (as those who argue against Indian influence have tended to do) but to a broader oral and vernacular storytelling tradition (Mair, 1989: 662–665). A Chinese translation of a Sanskrit collection of Buddhist tales, produced in Jianye (modern Nanking) in 251 CE, includes a version of the Indian epic that in fact follows Valmiki closely but gives an even more prominent role to Hanuman, whom the narrating Buddha, at tale's end, reveals to have been an incarnation of Indra and also the future Buddha Maitreya (Ibid. 675–679). Ramayana influence probably entered China not only via the silk road from Central Asia and Tibet, but by the southern sea route through the Malay archipelago, and it is notable that the earliest Monkey stories, from the Tang dynasty, feature a "white monkey" hero who conquers both demons and damsels, and thus resembles the playful and promiscuous Hanuman of Southeast Asian Rama tales (Ibid. 670–671, 697). Of course, like monsoon winds, stories can move in more than one direction, and it is also possible that features of comparatively late Southeast Asian Rama tales were influenced by Taoist-inspired motifs from China. Thus, in the 18th century Thai *Ramakien,* the "white monkey" Hanuman possesses a magic trident that he can reduce to a "diamond hair" and hide near his chest, seemingly echoing the magical iron rod that Monkey shrinks to the size of a pin and conceals in his ear (Sattar, 1990: 80, n. 63). Yet the trait of carnal appetite, for both food and sex, is absent from the Ming novel's hero; instead, it is transferred to another of Tripitaka's traveling companions, a muckrake-toting pig, leaving Sun Wu-k'ung, like (the South Asian) Hanuman, chastely devoted to his master's mission.

In light of strong circumstantial evidence for some degree of Ramayana influence on the *Hsi-yu chi*, it is worth considering what was at stake for the Chinese scholars who so fervently argued against it, in one case castigating a colleague who advocated it for his alleged "slave mentality . . . worshiping everything that comes from abroad and denigrating his own native country" (Mair, 1989: 708). In China, Sun Wu-k'ung came to be considered a national treasure and the thought that he might have been "derivative" or "imported" from elsewhere was deeply offensive. A comprehensive 1998 study, drawing on both archeological and textual evidence, offered a reasonable compromise, concluding that that there indeed existed in China "an indigenous monkey legend dating back to the Chu kingdom, which became the fundamental basis for the development of Sun Wukong," and that it was the presence of this legend that allowed Chinese storytellers to accept "foreign" narrative strands, "including fragmentary elements of the Indian Rama saga" (Walker, 1998: 77–78).

Certainly, the Chineseness of Monkey has never been at issue for audiences, either of the novel or of the popular performance genres that draw on its story. Chinese operas regularly include the crowd-pleasing episode of Sun Wu-k'ung's theft of the peaches of immortality from the Jade Emperor's garden, and this and other Taoist motifs that pervade the story may indeed reflect an old and indigenous tradition of an overreaching monkey who leaps between earth and heaven, which became inflected, in time, with motifs from a renowned and multiform South and Southeast Asian narrative. Despite the centuries-long effort of the classical literati to suppress non-elite and vernacular literature and lore (Mair, 1989: 730–731), and the post-revolutionary crackdown on religious expression on the mainland, a Taoist folk cult of Monkey survived into recent times on Taiwan and in Singapore, where he was revered as a trickster, esoteric preceptor, healer (especially of children and horses), and as an exorcist (Elliot, 1950; Schipper, 1995). Such beliefs and practices, at least, appear to have been more inspired by Chinese responses to the region's native species of macaques and gibbons than by the influence of foreign tales.

1.4 CONCLUDING REFLECTIONS

During my research on Hanuman, I assembled more than 20 substantial modern sources for "*Hanumayana*" lore, mainly in Hindi, ranging from paperback bazaar digests of prayers, rituals, and stories, to a newly-composed epic poem, a forty-episode TV serial, and two "autobiographies"—texts in which the divine monkey purportedly narrated his own long and colorful life story. One tome, in decidedly Indian English, ran to more than 800 pages. Besides recounting many of the narratives found in other sources, its author (an octogenarian retired from military service) included accounts of several personal experiences as well as some provocative musings. In one of these, he highlighted some of the environmental crises now facing our planet—air and water pollution—and then proposed, as one solution to them, the recognition of Hanuman, son of the wind and best of monkeys, as "Lord of Environment and Ecology" (Sundd, 1998: 433–438).

I was intrigued, even moved, by this proposition, which was unique in the body of material I was studying. Indeed, in years of reading and listening to religious discourses at Hindu shrines, I had never encountered such an idea, or even such acknowledgement of the environmental decay that, in almost half of century of travel to India and periods of residence there, I had seen increase dramatically. Following the "liberalization" of the Indian economy in the early 1990s, the country's middle class had grown (by various estimates) to between two and three hundred million people, all aspiring toward the kind of consumer lifestyle enjoyed by Euro-Americans. At the same time, problems of air and water pollution and of solid waste disposal grew, cities expanded exponentially, and the country's very limited forest cover steadily shrank. Such problems are shared by most nations in the densely inhabited regions that have been the focus of this chapter, and one of their consequences is increased levels of endangerment for non-human primate species that depend on forest cover (such as gibbons, orangutans, and langurs); for those with a greater acquired tendency to share the settled landscape with humans, there is often an increased level of competition for basic resources. The resultant inter-species conflict has added a new dimension to the ambivalence with which non-human primates have typically been viewed, though without entirely effacing characteristically Asian religio-cultural perspectives.

This is especially the case with macaques. The second most widespread (in many habitats and different climate regions) primate genus on earth (after humans), comprising 22 species found from Japan to southern Europe (cf. Mittermeier et al., 2013: 629–648), macaques have long had close but also at times competitive interactions with humankind. Noting their "extremely advanced" social structure and impressive geographical range (which, in the Pleistocene era, extended to the British Isles, from which they were apparently ousted by *Homo sapiens*), one researcher wryly observed, "I am convinced that had not man arrived on the primate evolutionary scene, these ubiquitous creatures would have ruled the world" (Napier, 1976: 21, 96). In India, although the monkey often linked to Hanuman in visual art and folklore is the black-faced and grey-furred langur, the species most often encountered in settled areas are the smaller, reddish-faced, tawny-furred *M. mulatta* and (in the south) the bonnet macaque, *Macaca radiata*. They too partake in some measure of the general respect that awareness of the Rama story bestows on all simians, and it is not uncommon to hear Indians jokingly refer to a troop of them as "Rama's monkey army." But popular Western accounts of the sacredness accorded to such creatures have often been exaggerated—for instance, the claim that "no true Hindu would dare to lift a finger against one in defense of his property" (Morris & Morris, 1966: 23). Hindus, "true" or otherwise, have in fact been lifting fingers—not to mention rocks, heavy bamboo staves, and sometimes more deadly weapons—for millennia in defense of their crops and other property against simian marauders. Citing the joint research of Indian and American primatologists, Donna Haraway noted "a deep ambivalence among local populations toward the monkeys. Villagers were usually unwilling to kill the animals…but they were often pleased to see them trapped and removed" (Haraway, 1989: 262). For decades, such trapping was the first step in a lucrative global trade in Indian primates that supplied laboratories in Europe and America. During the 1950s and '60s, the kidneys of macaques were used to produce polio vaccine, resulting in an enormous "harvest" of these animals, and large numbers of them also died painfully in radiation experiments conducted by the U.S. Department of Defense; the revelation of these monkey-fates in Indian media caused a public outcry that finally lead to a ban on simian exports in 1978 (Haraway, 1989: 259–60).

In the early 21st century, as Delhi mushroomed into a megalopolis of more than 20 million people, officials had to contend with growing numbers of complaints regarding monkey attacks, and indeed with troops of macaques (from an urban forested reserve) invading the stately blocks of the government's Central Secretariat and even Parliament House. After unsuccessful attempts to chase them away by bringing in langurs as "monkey police," the administration resorted to wholesale trapping and transport into hoped-for "forest exile." Soon, villagers in the Himalayan foothills were complaining about trucks from the capital dumping, in the dead of night, hundreds of monkeys in their midst. To make matters worse, these "migrant" macaques were perceived to be very different from those the villagers had tolerated for generations; these were urban, "street-smart" macaques with aggressive attitudes and awful manners. They intimidated and drove away their own country cousins and wreaked havoc on local fields, markets, and homes. On closer examination, anthropologist Radhika Govindrajan found that the "migrant monkey" furor mirrored the anxiety of hill people over a simultaneous exodus of mountain youth seeking more lucrative jobs and an influx of ethnically different human migrants from the plains who were beginning to dominate the local economy (Govindrajan, 2015, 2018: 90–118).

In rural Japan, decades of commercially oriented forestry resulted in the clearing of mixed woods (containing wild fruit trees that provided food to macaques) and their replacement with conifers, "useful" to humans but not to monkeys, who then began venturing in larger numbers into surrounding fields, orchards, and villages. In the 1980s and early 1990s, Japanese media presented sensational reports of "monkey damage" (*engai*), labeling the perpetrators "criminals" (*bannin*) and, in articles with headlines like "Monkey or Man?," used the vocabulary of human warfare to describe a supposed life-or-death struggle. As a result, government-sponsored "culling" resulted in the killing of approximately 5,000 macaques annually between 1983 and 1992 (Knight, 1999: 628–630). Yet anthropologist John Knight found that although some rural Japanese complained about the new level of simian depredations, many were deeply disturbed or angered by the killing of monkeys, pointed out that monkey damage was exaggerated by the media, or even "reserved" special fruit trees for monkeys or put out extra food for them on the edges of their property, especially in winter (Ibid. 633–634). In addition, some villagers showed understanding, as the mass media did not, of the underlying ecological causes of the "monkey war"—loss of habitat and food supplies, followed by social disintegration within monkey bands due to the selective "culling" of young, aggressive macaques who were seen to lead "raiding parties." Again, these dysfunctional processes oddly echoed human patterns in industrialized Japan: the devaluation of rural life and the out-migration of the young and able-bodied to the cities. As Knight observed, "*Engai* monkeys seem to reflect the social disorder of the villages they raid" (Ibid. 635; on similar responses to macaque raiding in Sulawesi, Indonesia, see Riley & Priston, 2010).

Will the ancient and richly nuanced "relationship between men and monkeys" in South, Southeast, and East Asia be characterized, in the future, by increased conflict and the extinction of many species, or by new patterns of coexistence and conservation? The answer will surely depend in part on programs of land and resource management as well as on improving the lives of both rural and urban people—but to be successful, all such efforts will have to be informed by local knowledge, needs, and traditions. In the most admiring assessments of their Asian observers, non-human primates have regularly been credited with a special capacity to mediate between humankind and (a variously conceived) "natural" realm. In Japan, Saru-gami was associated with the god of the forested mountains, for whom he served as a messenger to human beings. In China, Sun Wu-k'ung seemed most at home on his own "Flower-Fruit Mountain," with its waterfall and forests, where he "played" happily with his simian subjects. In one episode from the *Hsi-yu Chi*, he "restored" the mountain after an ecological catastrophe caused by the excesses of human hunters (Yu, 1978: 33–39). The Monkey King's response of grief and rage might likewise be interpreted (at least from our present perspective) as dramatizing the need for restraint in human exploitation of the earth's resources. His acts of environmental restoration (he obtained "sweet, divine water . . . to wash his mountain and make it green again" and planted a variety of trees on the ravaged slopes) offer a hopeful vision of renewal, as well as a welcome break from the tendentious piety of his human master (Yu, 1978: 39). In India, Hanuman has long been associated with liminal spaces that mark the boundary between human habitation and unsettled "jungle," as well as with the Himalaya and other mountain ranges and the (vanishing) forests, where his (sometimes) honored cohorts, the langur and macaque, still dwell. Moreover, he has, through his father Vayu, a special identification with the very air—the vital breath of all life—that is so often compromised in Asian megacities. Associating this super-natural primate—a matchless mediator between the animal, human, divine, and demonic realms—with "ecology," a relatively new concept even in English, may be just another leap for one who already plays so many roles. But it could also be a small step forward, for his human fellow travelers, in their quest for solutions to the self-generated woes that now menace us all.

REFERENCES

Asquith, P. J. (1995). "Of Monkeys and Men: Cultural Views in Japan and the West." In R. Corbey, ed. *Ape, Man, Apeman*, 309–325. Leiden: Leiden University Department of Prehistory.

Barnard, A. (1995). "Monbodo's Orang Outang and the Definition of Man." In R. Corbey, ed. *Ape, Man, Apeman*, 71–85. Leiden: Leiden University Department of Prehistory.

Bulcke, C. (1960). The Characterization of Hanuman. *Journal of the Oriental Institute of Baroda* 9.4: 393–402.

Bulcke, C. (1999). *Rāmkathā, utpatti aur vikās* (in Hindi). Revised edition, first published in 1950. Prayag (Allahabad): Hindi Parishad.

Cort, J. (1993). "An Overview of the Jaina Purāṇas." In W. Doniger, ed. *Purāṇa Perennis*, 185–206. Delhi: Sri Satguru Publications.

De Waal, F. (2001). *The Ape and the Sushi Master, Cultural Reflections by a Primatologist.* Basic Books (Perseus Books Group).

Elliot, A. J. A. (1950). *Chinese Spirit-Medium Cults in Singapore*. London: London School of Economics and Political Science.

Goldman, R. and Sutherland Goldman, S. (1996). *The Rāmāyaṇa of Vālmīki, Vol. V: Sundarakāṇḍa*. Princeton, N.J.: Princeton University Press.

Govindchandra, R. (1976). *Hanumān ke devatva tathā mūrti kā vikās* (in Hindi). Prayag (Allahabad): Hindi Sahitya Sammelan.

Govindrajan, R. (2015). Monkey Business: Macaque Translocation and the Politics of Belonging in India's Central Himalayas. *Comparative Studies of South Asia, Africa and the Middle East* 35.2: 246–262.

Govindrajan, R. (2018). *Animal Intimacies*. Chicago: University of Chicago Press.

Haraway, D. (1989). *Primate Visions*. New York: Routledge.

Itani, J, (1958). On the Acquisition and Propagation of a New Food Habit in the Troop of Japanese Monkeys at Takasakiyama. *Primates* 1: 84–98.

Jamison, S. W. (1996). *Sacrificed Wife, Sacrificer's Wife*. New York: Oxford University Press.

Janson, H. W. (1952). *Apes and Ape Lore in the Middle Ages and the Renaissance*. London: Warburg Institute, University of London.

Joshi, J. R. (1978). *Minor Vedic Deities*. Poona: University of Poona.

Kawamura, S. (1959). The Process of Sub-human Culture Propagation among Japanese Macaques. *Primates* 2: 43–60.

Knight, J. (1999). Monkeys on the Move: The Natural Symbolism of People-Macaque Conflict in Japan. *Journal of Asian Studies* 58.3: 622–647.

Knight, J. (2003). *Waiting for Wolves in Japan*. Oxford: Oxford University Press.

Kulkarni, V. M. (1990). *The Story of Rāma in Jain Literature*. Ahmedabad: Saraswati Pustak Bhandar.

Lutgendorf, P. (1994). My Hanuman Is Bigger Than Yours. *History of Religions* 33.3: 211–245.

Lutgendorf, P. (2000). "City, Forest, and Cosmos: Ecological Perspectives from the Sanskrit Epics." In C. K. Chapple and M. E. Tucker, eds. *Hinduism and Ecology*, 269–289. Cambridge, MA: Harvard University Press.

Lutgendorf, P. (2004). "Hanuman's Adventures Underground: The Narrative Logic of a *Ramayana* 'Interpolation'." In M. Bose, ed. *The Ramayana Revisited*, 149–163. New York: Oxford University Press.

Lutgendorf, P. (2007). *Hanuman's Tale, the Messages of a Divine Monkey*. New York: Oxford University Press.

Lutgendorf, P. (2016). "Ally, Devotee, and Friend." In F. McGill, ed. *The Rama Epic, Hero, Heroine, Ally, Foe,* 144–181. San Francisco: Asian Art Museum.

Mair, V. H. 1989. "Suen wu-kung=Hanumat? The Progress of a Scholarly Debate." In *Proceedings of the Second International Conference on Sinology*, 659–752. Taipei: Academia Sinica.

Mittermeier, R. A., Rylands, A. B. and Wilson D.E. (2013). *Handbook of the Mammals of the World: Volume 3,* Primates. Barcelona: Lynx Edicions.

Morris, R. and Morris, D. (1966). *Men and Apes*. New York: McGraw Hill.

Nagar, S. L. (1995). *Hanumān in Art, Culture, Thought, and Literature*. New Delhi: Intellectual Publishing House.

Napier, J. (1976). *Monkeys without Tails*. New York: Taplinger Publishing.

Narayanan, V. (2013). Who Is the Strong-Armed Monkey Who Churns the Ocean of Milk? *Udaya, Journal of Khmer Studies* 11: 3–28.

Narula, J. (1991). *Hanuman: God and Epic Hero*. Delhi: Manohar.

Niu, K. (2023). Reexamining the Evidence for the Original Identity of Yuan: A Reply to Fan et al. (2022). *International Journal of Primatology* 44.1: 175–177.

O'Flaherty, W. D. (1981). *The Rig Veda*. Harmondsworth: Penguin Books.

Ohnuki-Tierney, E. (1987). *The Monkey as Mirror.* Princeton: Princeton University Press.

Ohnuki-Tierney, E. (1995). "Representations of the Monkey (Saru) in Japanese Culture." In R. Corbey, ed. *Ape, Man, Apeman*, 297–308. Leiden: Leiden University Department of Prehistory.

Ramanujan, A. K. (1991). "Three hundred *Rāmāyaṇas*." In P. Richman, ed. *Many Rāmāyaṇas*, 22–49. Berkeley and Los Angeles: University of California Press.

Riley, E. and Priston, N. (2010). Macaques in Farms and Folklore: Exploring the Human-Nonhuman Primate Interface in Sulawesi, Indonesia. *American Journal of Primatology* 71: 1–7.

Sattar, A. (1990). *Hanumān in the Rāmāyaṇa of Vālmīki*. Ph.D. dissertation, University of Chicago.

Schipper, K. M. (1995). *The Monkey King Epic in Chinese Popular Religion*. Lecture, Center for the Intercultural Study of Literature and Society, University of Leiden.

Sen, D. (1987). *The Bengali Ramayanas*. Delhi: Amar Prakashan. Reprint; first published in 1920.

Sterckx, R. (2002). *The Animal and the Daemon in Early China*. Albany, NY: State University of New York Press.

Sundd, M. (1998). *Goswami Tulsidas-ji's Devised Sri Sankat Mochan Hanuman Charit Manas*. New Delhi: Aravali Books International.

Timmins, R. J., Richardson, M., Chhangani, A. and Yongcheng, L. (2008). *Macaca mulatta*. The IUCN Red List of Threatened Species 2008:e.T12554A3356486#.https://dx.doi.org/10.2305/IUCN.UK.2008.RLTS.T12554A3356486.en.

van Gulik, R. H. (1967). *The Gibbon in China, an Essay in Chinese Animal Lore*. Leiden: E. J. Brill.

Walker, H. S. (1998). *Indigenous or Foreign? A Look at the Origins of the Monkey Hero Sun Wukong*. Sino-Platonic papers no. 81. Philadelphia, PA: Department of East Asian Languages and Civilizations; University of Pennsylvania.

Yu, A., ed. & trans. (1977–1983). *The Journey to the West*, 4 vols. Chicago: University of Chicago Press.

Zhang, P. (2015). Good Gibbons and Evil Macaques: A Historical Review on Cognitive Features of Non-human Primates in Chinese Traditional Culture. *Primates* 56: 215–225.

2 Continental Africa

2.1 MONKEYS IN EGYPT AND THEIR INFLUENCE IN THE MEDITERRANEAN WORLD

Cybelle Greenlaw
Trinity College

Non-human primates featured prominently in Egyptian religion and culture from the Old Kingdom to the Ptolemaic Period, and even in the Predynastic Period (c. 5000–3000 B.C.E.) baboons had great religious significance. Later, smaller monkeys, such as grivets (*Chlorocebus aethiops*) and patas monkeys (*Erytrocebus* spp.)[1] were kept as pets and appeared in a variety of media. During this time span of three millennia, representations of primates and their functions in daily life changed and evolved. Their popularity as symbols of wealth in tomb chapel scenes grew, or diminished, according to the economic and political landscape of the period, and new developments in the production of faience led to innovations in the portrayal of the animals. The growing popularity of the myth of Osiris may have promoted the frequent depiction of the baboon as a manifestation of Thoth. However, the use of monkeys in the Old Kingdom should be viewed as a continuation of the late predynastic and early dynastic periods.

Numerous faience and limestone figurines found at Hierakonpolis, Abydos, and Elephantine attest to the existence of baboon cults even in early dynastic times (**Fig. 2.1**). The early figurines are believed to represent, Hedj-Wer, or the Great White One, the earliest identifiable baboon deity (Friedman, 1998:16). Its name was discovered on an ivory label of the first dynasty ruler, Semerkhet, dating from c. 2900 BCE. (Spencer, 1993: 67, fig. 46; Wilkenson, 1999: 285). Although Hedj-Wer's function is uncertain, some scholars have suggested that early kings may have been portrayed as baboons, or that baboons represented deified ancestors (Friedman, 1995: 25–26). The species depicted in the figurines is difficult to determine. Hedj-Wer merged with the lunar god, Thoth, whose cult center was Hermopolis, sometime early in the Old Kingdom (Lurker, 1980: 121). During this early period, Thoth was most often depicted in his anthropomorphic, ibis-headed form, but baboons were used as visual metaphors to represent qualities associated with him, such as wisdom, knowledge, judgment, writing, and excellence (Wilkinson, 1999: 157). In later periods, Thoth was strongly associated with the hamadryas (*Papio hamadryas*), but skulls found at the Royal Cemetery at Hierakonpolis (HK-6), which date from the Naqada III period, have been identified as olive baboons (*Papio anubis*) (Linseele, 2003).

Other baboon deities are mentioned in the Pyramid Texts (cf. Faulkner, Trans. 1998), dating from the Old Kingdom. The *bnti* (baboon) is named as the son of Re in

FIGURE 2.1 Baboon. Early Dynastic, ca. 3100–2649 B.C.E. From Egypt, Northern Upper Egypt, Abydos, Osiris temple, Chamber M69, Egypt Exploration Fund excavations, 1902–03. Faience. Dimensions: H. 7.9 × W. 3.7 × D. 5.2 cm (3 1/8 × 1 7/16 × 2 1/16 in.). Credit Line: Gift of Egypt Exploration Fund, 1903. https://collectionapi.metmuseum.org/api/collection/v1/iiif/547487/1215072/main-image

[1] Ancient artifacts are often fragmentary and the quality of artistic representations varies significantly. It is not always possible to identify the exact species depicted. Monkeys with a black face surrounded by white fur are believed to be grivet monkey (*C. aethiops*) because they would most likely have been imported from the modern-day region of Sudan, Eritrea, and Ethiopia. There are several recognized members of the genus *Chlorocebus*, and, although another subspecies could be depicted, *Chlorocebus aethiops* is the most likely. For that reason, monkeys that appear to belong to the genus will be referred to as grivets. Other authors may refer to the same representations as green monkeys (*C. sabaeus*) or vervets (*C. pygerythrus*), but these subspecies are less likely. Patas monkeys can also be divided between the more common *E. patas*, and the Blue Nile patas monkey (*Erythrocebus poliphaeus*). For the purpose of this section, the inclusive term 'patas' will be used.

DOI: 10.1201/b21963-3

Utterance 363 and two deified baboons, *bnti.wi*, again sons of Re, are mentioned in Utterance 569, in an exhortation to the god to allow the coming of the deceased king. These *bnti* may be the same baboons believed to sit in the eastern sky to adore the rising sun. Although baboons are generally viewed in a positive manner, their negative characteristics were also acknowledged. The baboon in hieroglyphic script could be used as a determinative for aggression (Jannsen & Jannsen, 1989: 21). Furthermore, the modern word 'baboon' may have come from an aggressive deity called Baba (Wilkinson, 2000: 73), alternatively known as Babwi or Babi, whose original manifestation in the Pyramid Texts was that of a baboon. Utterance 549 of the Pyramid Texts is a reproach of his thievery (Faulkner, Trans. 1998):

> Get back, Babi, red of ear and purple of hindquarters! You have taken the thigh joint of your goddess to your mouth!

The god's red ears may serve to identify Babi's baboon manifestation as a hamadryas, whose ears are red or pink. The color may also emphasize the sexual potency and aggression of the male baboon. In Egyptian art, red could represent fire, blood, life, and regeneration, as well as harsh desert areas, anger, and destruction. Red ink was used to write the hieroglyph for 'evil,' and other hostile deities, such as Seth and Apophis, were associated with the color.

Smaller monkeys and maneless baboons (either female hamadryas baboons or Anubis baboons of either sex) do not seem to have had any religious significance during the Old Kingdom. However, they were represented in the form of glazed composition and steatite amulets. Squatting monkeys holding babies were likely designed for female wearers, while amulets of walking monkeys, dating from the First Intermediate Period, may have been worn for their sexual connotations (Andrews, 1994: 67–70). Monkeys were also associated with women through their portrayal on, or as, containers. Several Old Kingdom pithemorphic vessels depict a female grivet clasping a baby to her breast (**Fig. 2.2**). Most vessels are made of alabaster, and some bear the names of Sixth Dynasty monarchs (Fischer, 1993: 1–3). It is noteworthy that fragments of an alabaster mother grivet vessel were found as far away as Mycenae, in the Peloponnese (Karetsou et al., 2000: 253–4). Mother monkeys are frequently depicted wearing the bracelets, anklets, and collars of domestic pets. The vessels often served as perfume containers, which, as Fischer (1993: 9) notes, seems appropriate because both myrrh and monkeys were imported from Punt, most likely located in modern-day Eritrea.

Some of the most informative evidence of the roles monkeys played in Egyptian daily life, however, comes from tomb chapel scenes, beginning in the Fourth Dynasty. Hamadryas and olive baboons, grivets, and patas monkeys all appear in these scenes, and all four were clearly kept as pets. Common themes include monkeys and their keepers, monkeys under their owner's chair, and monkeys engaged in a variety of human activities (Vandier d'Abbadie, 1964: 147–177). The first category involves servants holding monkeys by a leash, or simply accompanying the animals. The guardians may be ordinary men or dwarves. Some scenes, such as the painted relief from the Fifth Dynasty tomb of Ti, portray the monkeys as twice the size of the dwarf handler, even though the morphology of the animal clearly indicates a smaller species. Indeed, dwarves are not portrayed with baboons at all (Dasen, 1993: 116). The large size of the monkey most likely indicates its perceived value or importance. Another common depiction of smaller monkeys was under the owner's chair. Numerous scenes show grivets sporting collars, bracelets, and anklets similar to those on the alabaster vessels. The monkeys are often portrayed munching on fruit placed before them in baskets while the owner sits above. The third category comprises monkeys engaged in diverse human activities for the sake of entertainment. For the most part, these are scenes of music and dance.

FIGURE 2.2 Monkey with baby. Old Kingdom, Dynasty 6, Reign of Pepi I, ca. 2289–2255 B.C.E. Medium: Travertine (Egyptian alabaster), paint. Dimensions: H. 13.7 cm (5 3/8 in.) Credit Line: Purchase, Joseph Pulitzer Bequest, Fletcher Fund, and Lila Acheson Wallace, Russell and Judy Carson, William Kelly Simpson, and Vaughn Foundation Gifts, in honor of Henry George Fischer, 1992. https://collectionapi.metmuseum.org/api/collection/v1/iiif/543908/1151879/main-image

By the First Intermediate Period or Early Minoan II–III in the Aegean, Egyptian-influenced monkeys began to appear in Crete in the form of ivory stamp seals. One of the earliest comes from Trapeza-Lasithi and depicts a monkey sitting in a human pose, using its tail as a stool (see Masseti & Veracini, this volume). While Minoans were beginning to welcome this exotic creature, the image would not become popular until much later.

After the collapse of the Old Kingdom in Egypt, monkeys did not figure in tomb chapel scenes again until the 12th Dynasty and 2nd Intermediate Period, when scenes were more prevalent in the tombs of nomarchs at Beni-Hasan and public officials at el-Kab. This decline was probably the result of troubles at the end of the Old Kingdom and social changes brought about by the 12th Dynasty kings. The rise of a middle class may have caused monkeys to lose some of their appeals as symbols of luxury used to ornament wealthy households (Vandier d'Abbadie, 1965: 12). One scene, however, that does deserve attention comes from the tomb of Knumhotep at Beni-Hasan. It depicts three baboons, probably anubis because of their brown fur and lack of mane, eating figs in a tree, while two men on either side of the tree collect fruit in a basket. This is the first example of a theme used extensively in the New Kingdom, the exact nature of which is unclear. Many scholars (e.g., see Deputte & Anderson, 2009) argue that the baboons are helping to pick the fruit and were trained specifically for this purpose. Houlihan, though, refutes the suggestion, claiming monkeys have never been trained to pick soft fruits. He also maintains that because the monkeys are eating the figs, they are probably helping themselves, not the men below. While he acknowledges that pig-tail macaques, (*Macaca nemestrina*) have long been employed as assistants in coconut harvests in parts of Southeast Asia, he claims they do not assist with the harvest of soft fruit (Houlihan, 1997: 32–37). However, there have been experiments in Thailand to train pig-tail macaques to assist with the harvest of tamarinds and mangoes. The monkeys are instructed to climb trees the height of a three-storey house and shake the branches. While in the trees, the monkeys are allowed to eat as much fruit as they like (about monkeys picking fruits see Masseti & Veracini, this volume).

Because of the democratization taking place in society, new stone-working techniques became available to artists outside the palace. Semi-precious stones imported from the eastern and western deserts inspired the creation of elegant jewelry and glyptic art. Grivet and maneless baboon amulets that had been produced in glazed composition and steatite could now be carved from amethyst and carnelian (Andrews, 1994: 66). Faience, however, remained an important medium, and many apotropaic animal figures, including baboons, continued to be produced in this material. These faience figures often depict the same protective demons that appear on magic knives or wands (**Fig. 2.3**). These objects, only produced during the Middle Kingdom, were made from hippopotamus ivory and resemble throw sticks used in bird hunting. Some have abrasions on the tip, indicating they may have been used to scratch lines or protective circles in the earth. Clay models of the wands were sometimes interred with the deceased (Pinch, 1994: 78). The primary function, though, was to protect pregnant women, new mothers, and children (Bourriou, 1988: 114; Weingarten, 1991: 4). Another material that became available during the Middle Kingdom was anhydrite, also known as blue marble. The stone was used almost exclusively for toilet jars, most of which are unadorned. A few, though, incorporate the forms of ducks or monkeys. Some are pithemorphic in the style of Old Kingdom alabaster jars (**Fig. 2.4**), but most are decorated with smaller monkeys or baboons stretched across the outside of the container (Terrace, 1966: 57–63) (**Fig. 2.5**). The monkey's association with vessels, cosmetics, and women is also apparent in several tomb chapel scenes, in which grivets hold cosmetic jars for their mistresses (Vandier d'Abbadie, 1965: 181–84).

FIGURE 2.3 Wand. Middle Kingdom. Dynasty 12–13, ca. 1981–1640 B.C.E. Egypt. Medium: Hippopotamus ivory. Dimensions: L. 33.1 × W. 5.3 × H. 12.4 × Th. 1 cm (13 1/16 × 2 1/16 × 4 7/8 × 3/8 in.). Credit Line: Theodore M. Davis Collection, Bequest of Theodore M. Davis, 1915. Accession Number: 30.8.218. https://collectionapi.metmuseum.org/api/collection/v1/iiif/545740/1220970/main-image

In funerary practices, the myth of Osiris was growing in importance. Horus and Thoth, who figured prominently in the tradition, began to be represented on the east and west sides of coffins (Wilkinson, 1999: 71). An aspect of the conflict between Osiris and Seth that captured the Egyptians' imagination was the trial and role of Thoth as advocate. Scenes of judgment began to be depicted regularly on coffins, a practice that would continue into the New Kingdom (Boylen, 1922: 25).

By the 18th Dynasty, monkeys began to appear in a wider variety of non-religious art, which was largely the result of

FIGURE 2.4 Baboon ointment jar. Middle Kingdom. Dynasty 13-17, ca. 1800–1550 B.C.E. Said to be from Northern Upper Egypt, Girga (Thinis). Medium: Anhydrite. Dimensions: H. 13.3 cm (5 1/4 in.); W. 5.9 cm (2 5/16 in.); D. 8.2 cm (3 1/4 in.). Credit Line: Rogers Fund, 1910. Accession Number: 10.176.54a, b. https://collectionapi.metmuseum.org/api/collection/v1/iiif/544035/1227399/main-image

FIGURE 2.5 Ointment bowl with monkeys in relief. Middle Kingdom–early New Kingdom. Dynasty 13–17, ca. 1800–1550 B.C.E. Egypt. Medium: Anhydrite. Dimensions: H. 4.3 cm (1 11/16 in.); W. 9.8 cm (3 7/8 in.); Dia. of inner rim 8.5 cm (3 3/8 in.). Credit Line: Theodore M. Davis Collection, Bequest of Theodore M. Davis, 1915. Accession Number: 30.8.139. https://collectionapi.metmuseum.org/api/collection/v1/iiif/544037/1166211/main-image

the active colonization that took place after the departure of the Hyksos rulers. Numerous scenes of tribute bearers bringing hamadryas and anubis baboons, grivets and other exotic animals to the Pharaoh boast of Egypt's control of lands to the south. Elements of humor can be found in some of the scenes, such as the Nubian procession from the tomb of Rekh-mi-Re, in which a small grivet climbs the neck of a giraffe (**Fig. 2.6**).

In addition to their value as exotics, monkeys began to be enjoyed for their entertainment value once again. Many New Kingdom ostraca depict monkeys engaged in human activities, particularly in association with music and dance. Political satire may even be found in limestone toys from Amarna. A statuette of a monkey driving a chariot pulled by another monkey is thought to be an unfavorable representation of the heretic king, Akhenaten, who regularly rode through the city with his family. Another unfinished statuette that depicts monkeys engaged in grooming activities may have been created to satirize scenes of the king kissing his daughters (Frankfort & Pendlebury, 1933: 99).

Hamadryas baboons had also developed strong connections to the moon through their association with Thoth and another lunar god, Khonsu, whose usual form was that of a mummified man or child, wearing the sidelock of youth and a crescent or full moon on his head. Thoth became a nocturnal double of the solar god, Re, allowing sunlight to remain in darkness. Khonsu's function related to the phases of the moon and the rhythm of the tides (Lachaud, 1995: 92). His form became conflated with that of Thoth, and blue faience figurines of baboons wearing the lunar disk may represent either god. Blue is often associated with both the heavens and primal flood, and both lunar and solar objects are often blue (Wilkinson, 1999: 107–108). In his lunar aspect, Thoth is often represented in the form of a blue faience baboon, wearing the lunar disc and carrying the wedjat eye.

The image of the baboons raising their arms to greet the sun also became far more prevalent in the New Kingdom. Sun hymns, such as this, describe their functions:

FIGURE 2.6 Facsimile from the tomb of Rekhmir. Artist: Nina de Garis Davies (1881–1965). New Kingdom. Dynasty 18, ca. 1504–1425 B.C.E. Original from Egypt, Upper Egypt, Thebes, Sheikh Abd el-Qurna, Tomb of Rekhmire (TT 100). Medium: Tempera on paper. Dimensions: Facsimile: H. 47.5 × W. 58.5 cm (18 11/16 × 23 1/16 in.); Framed: H. 49.1 × W. 60 cm (19 5/16 × 23 5/8 in.); Scale. 1:1. Credit Line: Rogers Fund, 1931. Accession Number: 31.6.40. https://collectionapi.metmuseum.org/api/collection/v1/iiif/544613/1222392/main-image

> The baboons that announce Re when this great god is to be born again about the sixth hour in the netherworld. They appear for him after they have come into existence. They are at both sides of this god until he rises in the eastern horizon of the sky. They dance for him, they jump gaily for him, they sing form him, they sing praises for him, they shout for him. When this great god appears before the eyes of [all humankind] then these hear the speech of jubilation of the Wetenet-country [=baboons?]. They are those who announce Re on heaven and earth. (Houlihan, 1996: 96)

Te Velde remarks that the jumping and singing of the baboons may denote ecstatic behavior, as does the portrayal of the baboons as ithyphallic (1988: 130–1). Baboons also gained importance in funerary iconography during the New Kingdom. The human head of Hapi, one of the four sons of Horus represented on canopic jars, was replaced with that of a hamadryas from the 19th Dynasty. Vignettes of spell 125 from the Book of the Dead depict Thoth in his anthropomorphic, ibis-headed form as an advocate of the dead, recording the results as the heart of the deceased is weighed in the balance, but often his presence is reinforced in the form of a baboon sitting on top of the balance.

After the weighing of the heart, Thoth functioned as an interpreter for the deceased. Te Velde cites this tradition in evidence:

> I am Thoth and I speak to you the language of Re as herald. They spoke to you before my words were understood... I give god's offerings to the gods and invocation offerings to the blessed dead. I am Thoth who ascribes truth to the Ennead and all that comes from my mouth comes into being like (all that comes from the mouth) of Re...(1988: 136).

He was, therefore, an intermediary between the human and divine realms, who aided Egyptians in their most important transitional phase, the rite of passage from this world to the next.

Because of his association with language and writing, Thoth was also considered the special patron and protector of scribes. Many small monuments from the New Kingdom portray scribes sitting with Thoth in his baboon manifestation to demonstrate their closeness and importance to the god (**Fig. 2.7**). In some cases, the baboon is actually clinging to the scribe's head, which may signify "scribe of Thoth" or "excellent scribe" (Wilkinson, 1999: 157).

Babi was still an active and fearsome deity in the New Kingdom. He was often conflated with the chaos god, Seth, by this time (Derchain, 1952: 34–35). In the story of Horus and Seth, Babi delivers a terrible insult to Re-Harakhti and is chased off by the Ennead. An address from the Book of the Dead, spell 125, contains a plea from the deceased to be rescued from Babi, who feeds on the entrails of nobles. Indeed, the god's pronounced aggression may define him as an irrational, violent counterpart of Thoth.

By the first quarter of the 18th Dynasty, the period roughly corresponding to Late Minoan IA on Crete, the monkey had become more prevalent in the art of Aegean civilizations as well. It should be noted, though, that no direct physical evidence exists that monkeys were ever imported to Crete or Thera, so iconographical evidence is our only means of understanding the Minoan attitude toward these exotic creatures. Monkeys appear on a small number of seal stones and six wall paintings from Crete and Thera, most of which date to the Late Minoan period. Most scholars accept that the image was adopted from Egypt and that some of the monkeys in wall paintings are a cercopithecid species, in particular of the genus *Chlorocebus* (cf. Masseti, 1980; Groves, 2008; Masseti & Veracini, this volume), but the identification is not so straightforward. Some of the monkeys do have these characteristics, and Egypt was certainly the most powerful influence on the Minoan adoption of simian imagery. However, in many cases, the artists produced hybrids. In at least one painting, an Asian species, the rhesus macaque (*Macaca mulatta*) may be depicted, and it could have been introduced to the Aegean through Mesopotamia (the species identification of Minoan frescos is further discussed in Masseti & Veracini, this volume).

It is worth remarking that Aegean monkeys are always blue. The use of blue may have been adapted from Egyptian figurines. As mentioned above, both lunar and solar baboons were commonly depicted in blue. Additionally, Cretan monkeys have realistic bodies, while those of Theran monkeys are highly anthropomorphic. The gender

FIGURE 2.7 Thoth as baboon on the head of the scribe. New Kingdom, Ramesside. Dynasty 19–2, ca. 1295–1070 B.C.E. From Egypt, Upper Egypt, Thebes, Sheikh Abd el-Qurna. Medium: Granodiorite. Dimensions: H. 22 cm (8 11/16 in). Credit Line: Rogers Fund, 1929. Accession Number: 29.2.16. https://collectionapi.metmuseum.org/api/collection/v1/iiif/544748/1213742/main-image

is never identifiable—a clear departure from Egyptian prototypes.

The earliest painting, dating from Middle Minoan IIIA, is the Saffron Gatherer from the Early Keep at Knossos. The species is difficult to identify because of the fragmentary nature of the painting, but it does have the slender limbs of a small monkey. While it may be tempting to immediately identify the monkey as a grivet, it is worth noting that another fragment from the scene has been discovered, and it is a pink rounded muzzle, similar to that of the rhesus macaque (*M. mulatta*). The fragment is oriented in the opposite direction of the Saffron Gatherer, which indicates that at least two monkeys were present in the scene. It is also worth noting that a lapis lazuli necklace, HM Y 146, with seated monkey figurines was found at Knossos, which may suggest some knowledge of the rhesus macaque since lapis lazuli would have been imported from the rhesus macaque's Westernmost habitat, roughly corresponding to modern-day Afghanistan. Of course, lapis lazuli was also imported to Egypt, where it had significant religious connotations. The stone could well have been imported to Crete via Egypt, but the identification of the monkey as a rhesus macaque remains speculative. Because the so-called saffron gatherer appears to be picking a crocus, it can be compared with a scene from Xeste 3, room 3a from Thera.

This painting, like all those remaining to be discussed, dates from Late Minoan IA and depicts a highly anthropomorphic monkey handing saffron to a seated female figure. The fact that a griffin stands behind her suggests that she is a goddess. The monkey has some unusual characteristics that aid in identifying the species. The only extant fragment of the face depicts a long-pink muzzle, similar to that of a hamadryas baboon. Furthermore, its tail is short in relation to its body, which ends mid-calf with a tufted tip, again reminiscent of the hamadryas. In the nearby rocky landscape, girls of various ages are picking crocuses. The scene is set in a room that was probably designed for religious functions, and, as Marinatos notes, each girl is depicted in a different stage of the crocus gathering process (1987: 123). On the north wall, a girl is seated on a rock, holding her bleeding right foot, which may symbolize blood lost during menstruation, childbirth, or more likely, the loss of virginity. This notion is reinforced by the fact that saffron has been used to ease pain associated with these events (p.132). The girl's hairpin has also fallen off, releasing her backlock, which Davis suggests may have been a symbol of virginity (1986: 402). The scene appears to be a rite of passage, and Marinatos is probably correct in asserting that the monkey is making the offering of saffron on behalf of the girls (1987: 124). A similar function, as we have seen, was also associated with Thoth in his role of advocate of the deceased. In each case, humans are shown in a liminal state, and both cultures, therefore, seem to have required an intermediary figure to intercede for them at uncertain periods of their lives. In both cases, it appears to be a baboon that fulfills this role.

In the House of Frescoes at Knossos, monkeys are shown in a naturalistic, but not realistic, landscape, hunting for and eating eggs (cf. Masseti, 1980). These monkeys are roaming free, unrestrained by belts or collars. Some have well-preserved faces, revealing a reddish area around the eyes, running down toward the nose, underlined by a dark stripe above fluffy, white cheeks. The long tails indicate a smaller species, possibly an artistic hybrid. The presence of papyri in the landscape reflects an Egyptian influence. This scene finds a counterpart in the Beta 6 painting from Thera, in which long-tailed monkeys jump and swing through a rocky landscape. Despite their anthropomorphic bodies, their activities are entirely simian, and no human is present in either painting. Unlike the figure in Xeste 3, room 3a, these monkeys have long, curling tails and black facial skin, highly reminiscent of grivet monkeys. Although landscape scenes exist in Egyptian paintings, monkeys are not present in them. This combination seems to be an entirely Aegean creation.

The last two Theran paintings are very fragmentary, but certain aspects are worthy of comment. In the fragments from Xeste 3, room 4, a monkey is playing a harp or lyre and at least one other is engaged in a sword fight or dance. Marinatos suggests that because this room is connected to Xeste 3, room 3a by pier and door partition, the monkeys may be holding a contest in honor of the goddess (1984: 113). It seems more likely, however, that the scene is borrowed from the Egyptian theme of monkeys as entertainers. The fragment from the North Magazines depicts a monkey with a grivet-like face, raising its arms in the manner of the hamadryas in its role of Sun Greeter, as it faces an architectural structure, probably an altar. In this variation of the Egyptian theme, the monkey retains its function as a greeter but loses its association with the sun. Again, these monkeys are genderless, so clearly the Minoans had no interest in adopting the ithyphallic aspect of the animals.

After the destruction of Thera, monkeys disappeared entirely from Aegean iconography and did not reappear until the late Geometric period. The Mycenaeans apparently had little interest in adopting the image (cf. Masseti & Veracini, this volume). Nevertheless, monkeys continued to be represented in Egyptian art, and several Middle and Neo Assyrian kings mentioned receiving tribute in the form of exotic animals from the West, including monkeys (Dunham, 1985: 236–40). Around 750–700 BCE, monkey-like, seated figurines were beginning to appear at various sites throughout the Mediterranean, including Elis, Laconia, Arcadia, Euboea, Rhodes and Central Italy (Langdon, 1990: 407). It has been suggested that this was a motif intended for votive offerings, since all those which come from a recorded context were found in sanctuaries (1990: 413).

Further Eastern influences can be found in the Phoenician silver bowls from Kourion and from Praeneste, both dating to the late 8th century and depicting the same story of an 'ape' hunt. McDermot identified the primate in question as a gorilla (1938: 235). Of course, this is highly improbable (see also Masseti & Veracini, this book). Nine separate episodes are present in the scene:

1. The prince leaves his town in a chariot
2. He stops to shoot a stag
3. He pursues the wounded animal
4. The charioteer attends the horse while the prince flays the stag
5. The prince makes an offering to a winged deity
6. The prince is attacked by an ape-like creature, but a winged goddess saves him
7. On land, the prince pursues the creature
8. The prince kills the creature
9. The prince returns home in triumph

Some scholars believe this scene reflects a lost Phoenician fable or epic (Markoe, 1985: 67–8). This seems plausible enough, considering the long tradition of fables in Egypt and the East, and it is around this time that references to the better-known Aesopic fables turn up in the writings of Greek authors, such as Hesiod and Archilochus. In fact, the later, writing in the early to mid-7th cent. BCE is the first to make reference to a monkey fable. This is an extremely fragmentary epode, relating to a fable in which a monkey is duped by a fox (Lasserre, 1959, frag. 224).

Within Egypt, primates continued to play vital roles in many aspects of daily life. After the New Kingdom, monkeys became even more closely associated with women's fertility. Numerous small faience plaques depict monkeys sitting on or being held by Bes or a voluptuous nude woman with an exotic hairstyle (**Figs. 2.8** and **2.9**). Other amulets represent monkeys without any accompanying figures. These objects have never been securely dated, and estimates range from the Third Intermediate Period to the Ptolemaic Period. Bulté, however, has recognized certain similarities to faience from the Libyan Period, and most of the figures were found in the eastern Delta, where Bubastis and Tanis flourished (Yoyotte in Bulté, 1991: 8). The monkey depicted is always one of the smaller species, usually a vervet or grivet. Bulté identified three main thematic groupings that included figurine plaques of Bes,

FIGURE 2.8 Bes rattle. Third Intermediate Period. Dynasty 22–25, ca. 945–664 B.C.E. Medium: Faience. Dimensions: H. 21.6 cm (8 1/2 in.); W. 8.6 cm (3 3/8 in.); Th. 1.7 cm (11/16 in.). Credit Line: Purchase, 2007 Benefit Fund and Lila Acheson Wallace and Diane Carol Brandt Gifts, 2015. Accession Number: 2015.11. https://collectionapi.metmuseum.org/api/collection/v1/iiif/671414/1535215/main-image

FIGURE 2.9 Nude female holding monkeys. Third Intermediate Period–Late Period. Dynasty 22–26, ca. 945–525 B.C.E. Egypt. Medium: Pale blue faience. Dimensions: H. 20.7 × W. 8 × D. 3.3 cm (8 1/8 × 3 1/8 × 1 5/16 in.). Credit Line: Purchase, Fletcher Fund and The Guide Foundation Inc. Gift, 1966. Accession Number: 66.99.71. https://collectionapi.metmuseum.org/api/collection/v1/iiif/549698/1457646/main-image

vervet figurines that may have had connections to Thoth, and female figurines linked to Bastet. The groupings are related to iconographical themes in scenes depicting the nurturing of newborn infants, music and animals, toilette, and plants (1991: 11).

Another amulet that remained popular from the Late Period through the Ptolemaic is the hypocephalus. This was a circular, wooden amulet, inscribed with spell 162 from the Book of the Dead, that was placed under the head of the deceased. It often contained a variety of solar images, including baboons. The four sons of Horus were also commonly represented, along with Thoth, Amun-Re, Isis, Nephthys, and Horus (Ikram & Dodson, 1998: 144).

Animal cults also grew in popularity by the Late Period, and most mummified monkeys and baboons date from this period or later. The main sources of these mummies are the cemeteries near the temple complexes of Saqqara and Tuna el-Gebel, which encompasses the western cemetery of Hermopolis Magna. Hundreds of thousands of ibises were mummified at these sites and left at the temples as votive offerings to Thoth. A much smaller number of baboons, only a few hundred, were mummified as votive offerings (1998: 136).

Emery, who began excavating at Saqqara in 1964, found that the ibis galleries had been intentionally situated in the area previously occupied by Third Dynasty officials. He hypothesized that the location may have been selected because of its proximity to the tomb associated with Imhotep, the master builder of Djoser's pyramid, who had been deified in the Late Period (Emery, 1965: 8; Smith, 1974: 27). By the Ptolemaic Period, Imhotep became conflated with Aslepius, son of Hermes, who, in turn, became conflated with Thoth. Ostraca found at the entrance of the ibis gallery reference Thoth, Thrice Great, and other areas of the site contained references to Imhotep (Smith, 1974: 28). The baboon catacomb at Saqqara comprises upper and lower galleries, dating from c. 400–200 BCE and c. 200–30 BCE respectively. Inside the galleries are hundreds of niches, once sealed with limestone slabs containing the name of the animal, its place of origin, the date of burial, and a prayer for its survival in the afterlife. The panels indicated that mature baboons kept in the Temple of Ptah-under-his-Moringa-tree as living incarnations of the god. When the baboon died, he became "Osiris the Baboon" One inscription named a baboon the "Hearing Ear," a title assigned to deities with oracular functions (Smith, 1974: 41–2).

The northern entrance to the tomb of Petosiris, the temple administrator at Hermopolis Magna, dating from the reign of Ptolemy I, depicts scenes of Osiris the Baboon and Osiris the Ibis. To the north, a temple built in the name of Alexander IV is dedicated to Osiris the Baboon. An annex, containing a cage for a sacred baboon in connection with the god's oracular function, was added to the temple's southern side after a violent earthquake. Excavations in earlier parts of the galleries even uncovered an intact hamadryas baboon mummy, dating from the Persion Period (Kessler & Nur el-Din, 2002: 36–38). The discovery of the hamadryas is noteworthy. All baboon sculptures from North Saqqara depict this species, but Goudsmit and Brandon-Jones discovered that during the Ptolemaic Period, other species were used as substitutes, perhaps because the hamadryas baboons were less plentiful. They recovered the remains of 169 monkeys in the catacomb and found that 146 were Anubis baboons, 21 were Barbary macaques, and 2 were vervet/grivet monkeys. Surprisingly, there is no evidence of the importation of Barbary macaques to Egypt before this time (2000: 112).

A few monkey mummies post-dating the New Kingdom were just pets. One of the best-known was a mummy interred with Maetkare, the God's Wife of Amun, that was once believed to be her illegitimate child. In the 1970s, X-ray images revealed the mummy to be a pet monkey. Although it was first identified as a baboon, it is now believed to be a grivet or patas monkey (Ikram & Dodson, 1998: 329).

2.2 NORTH AFRICA AND SUB-SAHARAN AFRICA: INSIGHTS INTO HUMAN AND NON-HUMAN PRIMATE INTERACTIONS

Cecilia Veracini
Universidade de Lisboa

> ...when they see it [chimpanzee] they have great compassion: they never harm it or strike it because
> they consider it the soul of their forefathers
>
> (Manuel Alvares, 17th century)

2.2.1 Introduction

After the rapid political and social changes which took place in the last century across the African continent, traditional cultures were moulded by contacts with the outside world, and now African peoples have to adjust their mode of subsistence to an ever-changing world, including political, economic, and now climate, changes. Despite the impact of the slave trade, colonialism and racism on the life, dignity and rights of Africans, African writers, poets and artists show us that myths and traditional religions endure, which:

> unless Christianity and Islam fully occupy the whole person as much as, if not more than, traditional religious do, most converts to these faiths will continue to revert to traditional practices [...] thus traditional beliefs continue to be found in any African society. They are not formulated into a systemic set of dogmas, [...] each generation takes them up with modifications suitable to its own historical situation or needs. (Mbiti 1990: 3)

The African continent hosts many cultures and religious beliefs, and its peoples speak thousands of languages, yet African mythologies all share a core belief that each individual is made of spirit and flesh (Willis, 2005: 264), the former being distinctive and eternal (Thomas & Alanamu, 2018). This vision sees the world as both a material reality and a concept. The world is divided into two spheres: the subterranean one, accessible to traditional healers, seers, and divine beings, which is invisible, and is where geniuses and ancestors live. The other, visible, sphere is the earth and the sky where supreme beings (creators) live. The earth is occupied by humans, animals and plants, and African cosmologies operate on rules that govern every area of an individual's life.

The Western concept of a nature/culture dichotomy is foreign in many African cultures. Nature is neither dominated by, nor socially separated from, humans, with the latter always involved in negotiations with "the other beings surrounding them and the forces which fill them with life" (Geslin, 2002: 115). For many African people, animals and plants constitute human food, but many wild animals are sacred in religious concepts, as either the living-dead or human spirits or messengers who brought news from God (Mbiti, 1990: 51; Olupona, 1993). These species, which are 'emblems' or 'totems,'[2] cannot be killed except for cultural rites, or in self defence. They helped to form the world and the human form (Willis, 1994: 277), and killing them risks misfortune for the killer or their family (Sifuna, 2012).

Non-human primates, for many traditional African peoples, were/are also regarded as sacred, or totemic animals, according to Western anthropologists' definition of the term. A totem is an animal from which a community traces its ancestry, a guardian spirit and the terrestrial incarnation of a divine being. Totems play key roles in the identities of individuals, clans, and ethnic groups. Spirits do not have concrete shapes and features, but it is believed that when they appear to people, they may look like human beings, animals, or take other forms, and they can change suddenly (Mbiti, 2015: 79). In traditional African societies, the human form is not always important, and all beings are forever changing their shapes:

> The passage from human to animal form was considered one of the many possible transitions such as death, the passage from boy to man, from girl to woman. You became what you eat, you became what you are eaten by. [...] Among many tribal peoples, the ability to take the form of an animal was the sign of a great shaman. (Sax, 2001: xi–xii)

Totemism, and some taboos, are seen by many scholars as ecological adaptations or social regulations based on magico-religious motivations. Traditional societies, which have extensive biodiversity and ecological knowledge, maintain a profound connection to nature through their social and religious values (e.g., Descola, 2005; Gadgil et al., 1993; Eyong, 2007; Angwafo & Chuo, 2016;). Although parts of Africa are reserved to protect wildlife from humans, traditional knowledge about wildlife, and the spiritual knowledge associated with it, is maintained through practice, culture, and imagination (Bortolamiol et al., 2018). It is crucial that regional conservation efforts incorporate indigenous knowledge and respect the rights of traditional communities (Ogar et al., 2020).

These themes appear in many of the ethnic groups we discuss in this chapter, which cannot possibly be an exhaustive survey. For each geographical region, we examine cases that are more emblematic than others. We are aware

[2] We use this concept in its generic and simple form (see https://www.britannica.com/topic/totemism-religion accessed 22th April, 2018), but the meaning of totemism has changed many times since its first formulation in the 19th century (for an overview see: Willis (org.), 2005). African religions interpreted within a western reductionist vision, have been reviewed within new paradigms (*e.g.* Mages, 1997; Mbiti, 1990, 2015).

of the risk of extinction most non-human African primates have been facing in recent years, but it is not the purpose of this review to discuss complex issues related to the potential disappearance of many African primate species.

2.2.2 North Africa[3] and Sahara Desert

The conditions in North Africa and the Sahara Desert are presently not conducive to the survival of many non-human primates (primates hereafter), although in historical times this part of Africa was wetter, and the vegetation richer than today (see below). Three taxa are presently found in this part of Africa: (a) the Barbary Macaque, *Macaca sylvanus* (L. 1758), (**Fig. 2.10**), the only surviving primate in Africa north of the Sahara Desert. Its present distribution is the remnant of a more widespread population which, up to the last glacial episode, inhabited large areas in Europe and Africa (cf. **Box Details 2.1**); (b) Baboons are represented today by a few populations of the olive baboon, *Papio anubis,* Lesson 1827 (**Fig. 2.11**), which inhabit some mountainous massifs of the Sahara in Niger and Chad, from the central Sahara southwards (cf. Masseti & Bruner, 2009: 14; Mittermeier et al., 2013: 603), but recent studies provide evidence to a northern extension of their current geographical range. In the Holocene, baboons also inhabited the mountain ranges of the Tassili n'Ajjer (Algeria) and the Tadrart Acacus region in South-Western Fezzan (Libya); (c) *Erythrocebus patas* (Schreber, 1775), one of the three species currently recognized in the genus *Erythrocebus* (Trouessart, 1897). The so-called red or patas monkeys are ground-dwelling primates distributed over semi-arid areas of West Africa and into Eastern Africa. They are adapted to life in open savanna and semi-desert habitats, and isolated populations of *E. patas* survive in the Aïr massif (Nord Niger) and Ennedi Massif (Chad) (Mittermeier et al., 2013: 671). The range of the genus *Erythrocebus* extended to the southern Sahel, and in the past patas monkeys occupied northern areas and perhaps Egypt, in what is now desert (Osborn & Osbornová, 1998).

BOX DETAILS 2.1: THE BARBARY MACAQUES OF GIBRALTAR

Cecilia Veracini

FIGURE 2.10 Group of Barbary Macaques on the Gibraltar Rock. (Photo Cecilia Veracini.)

3 North Africa refers to the territory consisted of Algeria, Egypt, Libya, Morocco, Sudan, Tunisia, and Western Sahara, but sometimes it is used as synonymous of Northern Africa. This last term is generally accepted referring to the northernmost region of the African continent stretching from the Atlantic shores of Morocco in the west to the Suez and the Red Sea in the east, including the Sahara Desert. In this section it will not be considered the Egypt that is treated separately.

The Barbary Macaque (also called Barbary ape) has been the best-known Old World monkey in the Western World since Antiquity. The only macaque found in Africa, it is characterized by the absence of a tail (it has a vestigial tail) as in true apes. It lives in great mixed groups of several females and males, with social groups of up to 100 individuals. Troops are matriarchal, with their hierarchy determined by lineage to the lead female. Unlike other macaques, the males participate in rearing the young. The Barbary Macaque is the only surviving primate in Africa north of the Sahara Desert, and the present distribution is the remnant of a much more widespread population which, up to the last glacial episode, inhabited vast areas in Europe and Africa (Fa, 1999; Camperio Ciani, 1986). In the past, until historic times, it was widespread throughout North Africa from Morocco to western Tunisia, but its current distribution is limited to small relict patches of cedar forest and scrub in mountainous areas of Algeria and Morocco (Mittermeier et al., 2013: 629); it is extinct in Tunisia (Butynski et Al., 2008). Overall, the population of this species is estimated to have declined at a rate exceeding 50% over the last three generations (twenty-four years). *M. sylvanus* is listed as endangered by the IUCN Red List 2020 (Wallis et al., 2020).

The Barbary Macaque population in Gibraltar, a British possession in the Iberian Peninsula since 1704 (Zeuner, 1952), is the oldest established colony of free-range monkeys in Europe. These animals also live in *La Montaigne des Singes* (Kintzheim, France), a nature reserve of 60 acres of forest where 200 macaques roam freely; it is also possible to see Barbary Macaques in a semi-free environment in Germany, too (Affenberg, Salem). In the Upper Rock area of the Gibraltar Nature Reserve, 300 animals are distributed in five social troops. They are currently managed by the Gibraltar Ornithological and Natural History Society (GONHS), but until 1991 the macaque population was under the care of the Royal Artillery and later (1957) of the Gibraltar Regiment, who carefully controlled the population (Garcia, 1979: 10). Veterinary care is provided by the Gibraltar Veterinary Clinic: the animals are caught on a regular basis to check their health status; they receive a daily supply of fresh water and vegetables, fruit and seeds as a supplement to food resources they find in the natural environment (Mediterranean forest) of the Rock. In 1923, the secretary of State for the Colonies, under pressure from the inhabitants of Gibraltar, agreed to the elimination of the animals because of the frequent raids into the town and the depredation of fruit and vegetables, and only five monkeys were left (Garcia, 1979: 5; Ryan, 1999: 25). In 1942 (during World War II), there were just seven. A popular belief holds that as long as the Barbary Macaques exist on Gibraltar, the territory will remain under British rule. Winston Churchill ordered their numbers be replenished from forest fragments in both Morocco and Algeria, probably because of this traditional belief. A telegram was sent to Gibraltar:

> The prime minister has expressed some concern as to the welfare of Barbary apes on Gibraltar about which he has heard disquieting rumours he is most anxious that they should not be allowed to die out. I have received directions from the Prime Minister that the establishment of apes should be no less than 24 and that every effort should be made to achieve this number as soon as possible and maintain it as a minimum thereafter. (Garcia, 1973: 8)

By the end of 1944, numbers had increased to 16. The monkeys were generally named after VIPs: in 1950 one male was named Winston, in honor of the Prime Minister, later others Charles and Anna, after Prince Charles and Princess Anne. Although most populations in Africa are facing a decline, the population of monkeys in Gibraltar is growing, and for these reasons, a small number of them are sometimes sent to other places (e.g., to a safari park in Scotland, personal comm.). From 1957 to 1974 forty monkeys were exported to British, US, and Canadian zoos. The Barbary Macaques have become the top tourist attraction in Gibraltar, but keepers strictly monitor their contact with large numbers of tourists, to help preserve the integrity of their social groups. The origin of the macaques of Gibraltar is still controversial. Macaques may have been introduced by the Moors under the Saracen Tariq, who captured and fortified Gibraltar in 710–711 CE. However, since no mention of monkeys was made during the period of Spanish occupancy between 1492 and 1704, it is possible that the Moorish importation died out, subsequently being replaced by new stock (cf. Masseti & Bruner, 2009: 41). An interesting Moorish tradition says that the animals came over from Africa by an underground passageway (cf. Garcia, 1979: 24; Ryan, 1999: 25).

FIGURE 2.11 Adult male *Papio anubis*. (Photo Marco Alesci.)

2.2.2.1 Barbary Macaques

There is evidence that from the Neolithic period onwards, Barbary Macaques interacted in some way with human populations. Neolithic remains of Barbary Macaques are known from the coastal Algerian site of Boulevard Bru and the upper part of Djebel Thava (**Fig. 2.12**), in the Saharan Atlas (Masseti & Bruner, 2009: 37), but they are not represented in the Rock Art of North Africa. They seem to have a different profile from that of other animals (Camps, 1990). Animals that were heavily hunted such as rhinoceros, elephants, and buffalos are rarely found in Neolithic sites, yet they are commonly illustrated in Rock Art. Other hunted animals such as the wild boar (*Sus scrofa*) are also rarely represented. A cultural or religious filter probably operated, because Rock Art mostly ignores species such as hyenas or macaques, which must have been frequently seen by the artists (Camps, 1990: 46). Primates are also poorly represented in the art of the Sahara dating from the Mesolithic to historical times, which suggests they could have been regarded as symbolic animals (see later).

Historical sources attest that since antiquity the Berbers (Amazigh), the ethnic group indigenous to North Africa, considered the Barbary Macaque as either an edible, or a sacred and totemic animal, depending on local customs (Veracini, 2020). As reported by McDermott (1938: 57), the first reference to the Barbary Macaque is by Herodotus (c. 484–425 BCE), who mentioned people of ancient Numidia called Zayeces and Gyzantes and their custom of eating monkeys: "they painted themselves red and fed on the meat of monkeys that the nearby mountains provided in abundance" (*History*, 1920, vol. IV: 193–94). These populations, called also Byzantes in other sources (Mannert, 1842: 661), were probably nomad Berber tribes who lived in the territories south of Carthage (from the mountains of Tébessa to the border with Algeria), in what is now Tunisia (cf. Joleaud, 1931: 134; Wheeler, 1854: 554–56). The habit of hunting and eating monkeys in this area and in the surrounding regions (present-day North-East Algeria and North Tunisia) seems to have remained as a form of sustenance and traditional feeding until the 19th century, although these animals had already become rare (Joleaud, 1931: 134). Subsequent sources (cf. Veracini, 2020) indicate that monkeys were routinely consumed in the 19th century on the Algerian coast (in the today town of Skikda) where "Arabs very much appreciated their meat" (Gervais, 1848, 1854: 98), although this is considered *haram* (impure) by the Koran.

A different attitude to Barbary Macaques was seen in other Algerian territories. According to the Roman writer Diodorus Siculus (c.90 BCE–27 BCE), the Barbary Macaque was sacred in some Algerian territories, where they were neither hunted nor eaten. In his description of the expedition of Agathocles (tyrant of Syracuse, Sicily) against the Carthaginians in 310 BCE, Siculus told of the capture of one of the three so-called monkey cities by his lieutenant, Eumachus. These cities were probably located west of Carthage (though their exact location is unknown) and monkeys were common. For this reason, the Greeks named them *Pithecusae*, that is "the cities of monkeys" (cf. Gilliies, 1820: 211; McDermott, 1938: 58) and

> the monkeys lived in the same houses as the men, being regarded among them as gods, just as dogs are among the Egyptians. [...] For any who killed this animal, as if he had committed the greatest sacrilege, death was established as the penalty. (Diodorus Siculus, 1954, Book XX, s58)

Moreover, the Greek Timotheus of Gaza (active during the reign of the Eastern Roman Emperor Anastasius, 491–518 CE) in his work *De animalibus* (cf. Bodenheimer & Rabinowitz, 1949), claimed that in Numidia (which roughly corresponded to the north part of present-day Algeria) and Mauretania (in modern times Morocco and Algeria), monkeys were honored as highly as in the "cities of monkeys".

The worship of macaques did not prevent people in North Africa from capturing and trading them, in both ancient and modern times (Veracini, 2020 for review). Islamic influence may have further contributed to these monkeys being regarded in a negative way in territories where macaques had previously been respected. In the mountainous areas of the Atlas, where Berber traditional populations lived in close contact with macaques, the animals featured in their cultural beliefs. Local oral traditions reported by Joleaud in the 1930s show how Barbary Macaques were considered to be human beings who had fallen into disgrace, and thus transformed into monkeys. The cosmogony of the Kabyle people (a Berber ethnic group in the north of Algeria) which probably precedes that of the monotheistic religions (Dahbia & Chaker, 2004), tells that the First Mother of the World—the deity who gave origin to all things—used different methods to create animals. For example, she shaped some out of wheat flour, while others derived from the metamorphosis dictated by the rage of the First Mother of the World, as was the case when a young boy was transformed into a monkey. The bond between people and the Barbary Macaque survives in contemporary Kabyle folk tradition: monkeys are associated with mountains and caves, but they are also regarded as intelligent and kind animals willing to help travellers by providing them with food (e.g., the legend of *La Grotte des Singes* among others, see Kabylie Djurdjura, 2009). As reported by Joleaud (1931: 119), the Berbers of Algeria and Morocco used many names in their native languages to refer to the Barbary Macaque, but they preferred using Islamic names (e.g., *Chadi*, pl. *Chuada*; *Querd* pl. *Querud*; *Sàdan*, pl. *Sàdin*) to avoid using their names directly, because doing so would have brought bad luck. Periphrasis was also used to define these monkeys, which testifies to the attitude toward these animals considered part of our species or our predecessors: "they who live in the mountains", or "the children of the forest." Here and there, folklore evokes the memory of populations that still had the Barbary Macaque as a totem, at least until the first half of the last century.

In the past, crop raids by monkeys were tolerated (Joleaud, 1931), but currently the situation is different, even in places where monkeys have traditionally been respected, and conflicts with local people have been reported in Algeria and Morocco (Alexander, 2018). Since the 19th century, Berbers and other traditional peoples who lived for millennia in these fragile ecosystems have been marginalized, their communities fragmented, and their territories expropriated. Berber cultural life also suffered long-term discrimination with official 'Arabization' campaigns (Ferhat, 1981; Brugnatelli, 2006). The reduction, degradation, and fragmentation of macaque habitats, and the loss of traditional practices that helped to maintain these territories, are probably key factors influencing the escalation of conflicts with monkeys, something that has happened in other regions of Africa (e.g., Lamarque et al., 2009).

2.2.2.2 Baboons and Patas Monkeys

During the mid-Holocene, c.9–6 thousand years ago, North Africa was characterized by a wetter climate, and evidence from fossil pollen, suggests what is now the Saharan desert was covered by annual grasses and low shrubs (Claussen et al., 1999: 2037). Zooarchaeological evidence is consistent with this picture, for the early Holocene mammalian fauna was mainly composed of Ethiopian elements, such as elephants, rhinos, hippos, giraffes, and other large ungulates. The zooarcheological record of southern Libya includes a few monkey bones, most of them olive baboons (cf. **Box Details 2.2**). It appears there was no connection with human settlements (cf. Masseti, 2010: 801), and Saharan art, from the Mesolithic to historical times, includes few representations of monkeys compared to the incredible richness of depictions of other fauna (e.g., giraffes, elephants, ungulates, carnivores etc.). Representations of monkeys are associated with different chronological periods, but the dating is often difficult to assess and thus any ethnozoological interpretations remain hypothetical. The first group of images are likely associated with the Large Wild Fauna Period (Bubaline Period) and Round Heads period (c.10.000 BP - or before - to 5,500 BP, cf. Soukopova, 2012: 8–10) both referring to hunter-gatherer communities (**Fig. 2.14**). The first group includes likely stylized representations of monkeys with other animals (e.g., Di Lernia & Manzi, 2002: 354; Muzzolini, 1989: Fig. 1.b; Huard & Allard, 1977: Fig. 7), monkeys which appear to be threatened by felids (e.g., Franklin & Strecker, 2008: 55), or other monkeys, similar to cercopithecids (see Frobenius, 2013: Fig. 7), but with characteristics of both felids and monkeys. In the later Middle Pastoral period (c.6,000–5,000 BP), a group of four baboons was found in a wall drawing at the site of Teshuinat III, in the Tradart Acacus (southwest Libya) (Mori, 1965; **Fig. 2.14**). Given these sparse data, it is difficult to gain an insight into the interaction between humans and baboons (or other monkeys) in the archaic period. However, Muzzolini (1989: 3) argues that in the Round Head paintings (9.500–c. 7,000 BP) of Acacus in the southwestern part of Libya, primates may have been regarded as symbolic animals. In the later pastoral period, which slowly replaced the hunter-gatherer economy, pictures show people herding and hunting animals. In this period, contact with baboon troops occupying the remaining forest patches seems to have been closer, with baboons being portrayed in more detail. With nomadism and herding, the primary method of food production for several millennia (Garcea, 2004), baboons may have been regarded as pests who preyed on sheep/goat herds, much as they do today in African pastoral societies (e.g., Butler, 2001) (see also next sections about baboons).

The Tuareg people who inhabit the Sahara today, use a variety of names for primates, which points to a long history of interaction across the Sahara and in other places (see below). An ancient reference to the Garamantes (a Berber community who developed an advanced civilization in ancient southwestern Libya, and who can be considered the ancestors of Tuaregs, cf. Gautier, 1935) is made by the Hellenized Syrian writer Lucian (c. 125–c. 180 CE), who wrote that these peoples used to hunt monkeys:

> They are "a savage nation that live generally in tents and are much addicted to hunting. Their sport is in hunting wild asses and large birds that fly close to the ground, but chiefly apes [i.e. monkeys] and sometimes elephants, for these animals alone are capable of enduring thirst and the excessive heat of the sun" (Lucianus, *Dipsas* Vol II, ed., 1710: 464).

Tuareg contemporary mythology is a complex mix of ancient Egyptian and Libyan myths, regional belief systems, rituals, cosmologies, and magical animist beliefs coming from sub-Saharan and Equatorial Africa, even though the Tuaregs have adopted the Islamic religion (El-Zein, 2015: 200–202). Tuareg myths conceived humans and animals as forming a single entity, and they considered themselves descendants of a tribe of *djinn,* who have taken the shape of animals, such as the wild sheep (El-Zein, 2015: 202). Currently, the Kel Ewey Tuaregs, a division living in the Aïr massif in the north of Niger, where the olive baboon is still found, regard it as a black spirit, a sort of *djinn* with evil characteristics. The black monkey or baboon, it is said, "rarely enters *villages* or camps but sometimes rapes women who wander alone" (Rasmussen, 1991: 758). Many of the Tuaregs' names to designate primates are the same as those used by the Berbers of the Atlas. *Ibiddauen* is the Barbary Macaque; *éwerked wa kăwălăn,* the olive baboon; *auerked* or *éwerkerd*, (adâgel or adaged), especially point to the "red *cercopithecus*", the patas monkey (Joleaud, 1931). When it comes to patas monkeys, it was challenging to find references in the traditional culture of the Sahara, although these animals feature in some traditional beliefs of Sahelian countries as discussed in the next section.

The Tuareg people also knew of the chimpanzee, *Pan troglodytes* (Blumenbach, 1775), due to the frequent contact and trade with the regions of Western Africa. The name they gave to the chimpanzee is *Kaya*, pl. *Kayatène* (Joleaud, 1931: 120), which is currently the name of a large city in Burkina Faso, where in the past chimpanzees existed (today chimpanzees are extinct in Burkina Faso, cf. Jane Goodall Institute, 2018). The myth of anthropomorphic creatures in far and unknown African regions pervaded all European antiquity and the Middle Ages until the encounter of Europeans with real great apes in the 16th century (cf. Veracini, Chapter 9, this volume). One might wonder how many of these myths could have originated from legends and stories about chimpanzees that made their way to the Mediterranean region through the narratives of nomadic peoples crossing the Sahara.

BOX DETAILS 2.2: SAHARAN BABOONS

Marco Masseti
Istituto Zooprofilattico della Sicilia

Scientific investigations have provided evidence about the ancient Saharan dispersion of baboons (genus *Papio* Erxleben, 1777). These animals inhabited several territories in the mountain ranges of the Tibesti (between Ciad and Libya), the Tassili n'Ajjer (Algeria) and the Tadrart Acacus region in south-western Fezzan (Libya) (**Fig. 2.12**).

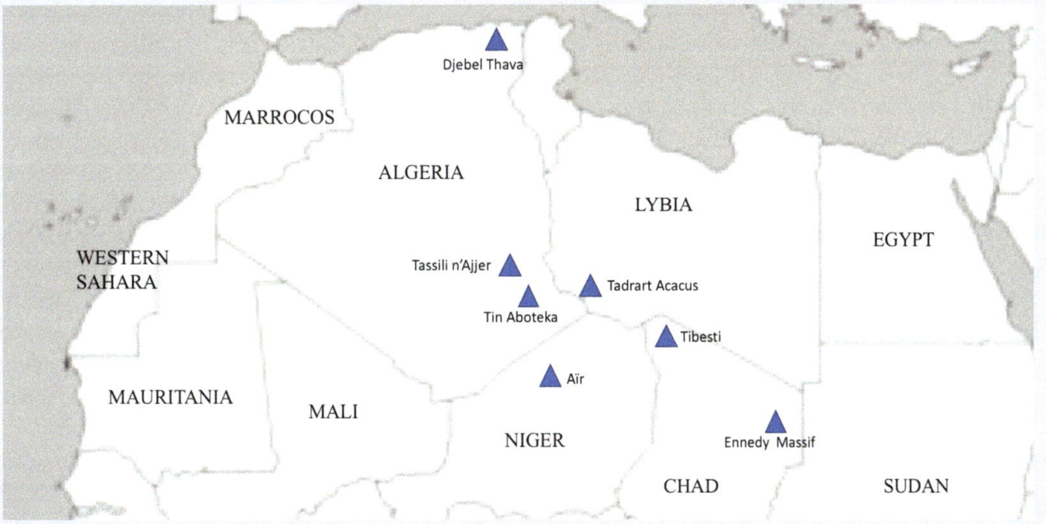

FIGURE 2.12 Map of the archaeological sites mentioned in the text.

Although these territories of the central Sahara have furnished numerous osteo-archaeological records, monkey bones are very scarce.

Remains from Uadi Ti-n-Thora (Tadrart Acacus), dating to circa. 8,000–8,500 uncalibrated BP, were classified as *Papio cynocephalus* (L., 1766) (Gautier & Van Neer, 1977) (a species now widespread in sub-Saharan Africa, from the river Zambesi to Tanzania, coastal Kenya and Somalia, Groves, 2005). Considering the extant range of the distribution of the North African representatives of the genus *Papio*, the taxonomic assessments clearly refer to the systematic single-species interpretation of the baboons, before the split into at least five different *taxa* (cf. Groves, 2001). *Papio cynocephalus* was also described by Gautier (1987) in the fauna from the site of Uan Muhuggiag (Tadrart Acacus, central Sahara, Libya) held in the Pasa collection of the museum of Verona (north-eastern Italy). A few osteological and dental fragments reported during an archaeological survey on superficial layers at the site of Uan Kasa (sample TH 131; Sennadar, 24_540N 10_280E) are described as *P. cynocephalus* by Corridi (1998) in Tadrart Acacus and Messak. The same material was referred to as a temporal phase falling between the Mesolithic period (Late Acacus, 9,000–7,000 uncalibrated BP) and the Middle Pastoral (6740-5952/5405 uncalibrated BP). Like the majority of the finds emerging from the archaeological exploration of this area, it may be attributable to the seventh–sixth millennium BP. More recently, a fragment of maxillary bone with a third molar recovered at the site of Takarkori, in southwestern Libya, was assigned to the genus *Papio* (Swindler, 2005; Masseti & Bruner, 2009). Here, several burials were identified in a rock shelter, and the specimen was dated to the Mesolithic (9,000–7,000 uncalibrated BP) or the early Pastoral (7,000–6,000 uncalibrated BP). Considering the dimensions and biogeographical ranges of currently recognized baboon *taxa* (Groves, 2001; Mittermeier et al., 2013), both the Takarkori and the Uan Kasa specimens can be assigned to the olive baboon, *Papio anubis* (Masseti & Bruner, 2009; Masseti, 2010). Although there is confirmation of the presence of baboons in these areas, it remains to be seen whether these primates were associated with human activities (hunting, pets), and were introduced from other regions or were the remnants of a prehistoric and/or protohistoric baboon distribution.

Today, the olive baboon is still dispersed in the southern mountainous massifs of the Sahara (Le Berre, 1990), from where it has been reported in the remote mountains of Aïr (Niger), Tibesti (Chad), Manakaoki (Thomas, 1925), or Wadaï and Ennedi (Chad) (cf. Bousquet, 1992; Masseti & Bruner, 2009). More specifically, according to Malbrant (in Dekeyser, 1955), in Chad, baboons are not found beyond the northern limit of diffusion represented by the 13th parallel, while they extend eastwards as far as Tibesti through the Ouadaï and the Ennedi. In Aïr, baboons are confined to the massif of Tamgak (wadies and mountains) (cf. Hall et al., 1971). The isolation of this northern population gives it great biological significance. The Aïr olive baboons mainly feed on dates from the dom palm, *Hyphaene thebaica* (L.), and, to a lesser extent, fruits of *Acacia* sp. and few other plants (Bousquet, 1992). In this regard, an interesting fact emerging from Keimer's (1939) discussion of the baboon and the dom palm is that the fruit of the tree was called "nuts" in ancient Egypt and was considered as a nourishing food.

Up to the end of the 1960s, a population of olive baboons was also reported from the mountainous massif of Tibesti. This isolated mountainous massif, about 3,500m in height (Petragnani, 1928; Brown, 1965), is located in the middle of the Sahara desert between about 19°N and 24°N, and it is now part of Chad, but previously it belonged to the Libyan State. Western scientists were not aware of these monkeys in Central Sahara until around the end of the 19th century (Hufnagl, 1972 and references therein). In the early 1950s, a short-tailed subspecies of olive baboon was described from Tibesti as *P. anubis tibestanus* Dekeyser and Derivot, 1960 (Dekeyser, 1952). A skull of this rare *taxon* was given by Mr. Kenneth Guichard to the Museum of Natural History of Tripoli (Hufnagl, 1972), where it is still on display (Masseti, 2010) (**Fig. 2.13**). The inscription on the label has faded, but it is still possible to read the date 1953, or perhaps 1958. It is possible this subspecies survived in some remote wadies in this area until the end of the 1960s (cf. Beck & Huard, 1969). In fact, up to 1969, Beck and Huard (1969) still reported that: "The Teda, who call them *dongo* or *dounbou*, sometimes claim they are derived from humans, and were transformed by God because they had shown cowardice". According to Dorst and Dandelot (1973), and Haltenorth and Diller (1977), in the 1970s baboons were still reputed to occur in the Tibesti mountains. In 1952, Dekeyser identified the only skull of a baboon then known from Tibesti (cf. Hufnagl, 1972). Its type specimen was preserved at the Institut Français d'Afrique Noire (CG. IFAN 44-24-5) and consisted of one skull with a mandible, obtained from an unnamed site located between Zouarkè and Kachem.

Indirect evidence of the former occurrence of baboons in the Saharan region comes from images, but these are scarce (cf. Masseti & Bruner, 2009). For example, the image of a primate has been found in the rock shelter of Tin Aboteka, located in south-western Tassili n'Ajjer, along the Algerian slope of Oued Edjériou (Napier & Napier, 1967) (cf. **Fig. 2.14**). Known as Round Head paintings, this image was generated within the "Neolithic Wet Phase" (c. 4500–2500 BCE), before the "Postneolithic Arid Phase" (c. 2500–1000 BCE) (Muzzolini, 1989). Among the artistic productions of both Tassili n'Ajjer and Tadrart Acacus, few other representations of monkeys related to the Round Head group are known. There are other baboons – or pseudo-baboons, for some images, have human limbs or even appear

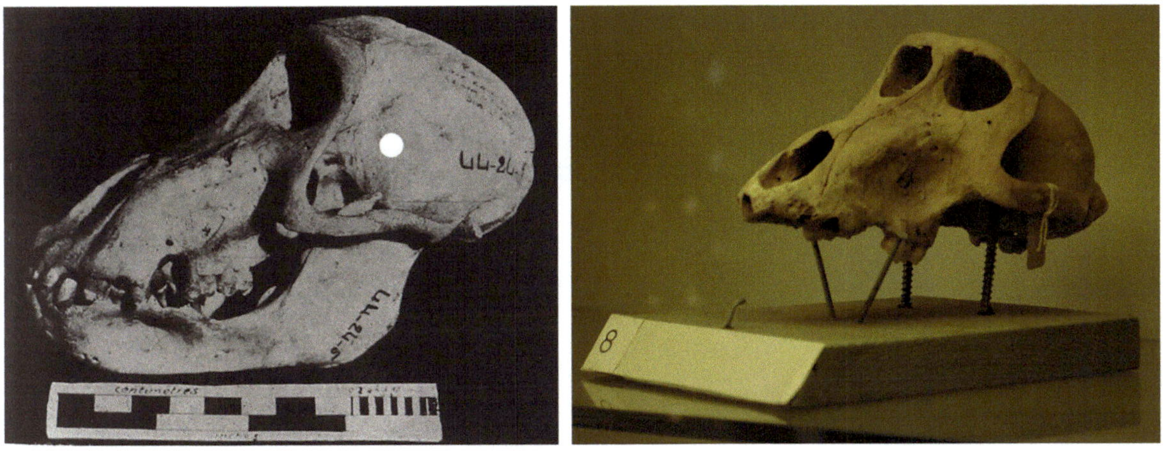

FIGURE 2.13 Left: the type specimen of the Tibesti baboon, *Papio anubis tibestianus* (Dekeyser & Derivot, 1960) was preserved at the Institut Français d'Afrique Noire (CG. IFAN 44-24-5), and consisted of a skull, obtained from an unnamed site located between Zouarkè and Kachem. (From Dekeyser, 1952 modified.) Right: the skull of the Tibesti baboon on display in the Museum of Natural History of Tripoli (Assaray Al-hambra). (Photo Marco Masseti.)

to be holding a bow – in the paintings of the shelter of Oued Tirehart, in western Tassili (Kunz, 1988). In the later Middle Pastoral period (c. 6,000–5,000 BP), a group of four baboons can easily be recognized in a wall drawing at the site of Teshuinat III, in the Tradart Acacus (southwest Libya) (Mori, 1965), and the mane and sexual characteristics suggest that the first is an adult male, followed by three females (or juveniles) (Masseti & Bruner, 2009: 84; **Fig. 2.14**).

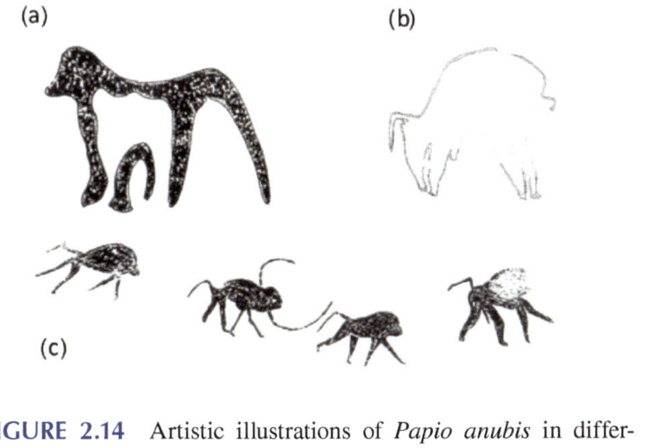

FIGURE 2.14 Artistic illustrations of *Papio anubis* in different periods: (a) Paleolithic Saharan hunters from Tibesti. (From Huard & Allard, 1977, p. 652 modified.) (b) A male baboon from the rock shelter of Tin Aboteka, located in southwestern Tassili n'Ajjer, and stylistically referred to the archaic phase of the Round Head paintings (cf. Sansoni, 1994). (From Masseti & Bruner, 2009, p. 44 modified.) (c) Adult male baboon followed by a juvenile male or a female. Teshuina III, referring to the Middle Pastoral period (6000–5000 ybb). (From Masseti & Bruner, 2009, p. 84, modified.)

2.2.3 Sub-Saharan Africa and Emblematic Primate Species

Most sub-Saharan communities included non-human primates in their cultures and spiritual beliefs. Primates affect daily life as a food resource, or as companions or pets, as well as being sacred or totemic animals playing various roles in initiation rites, ceremonies or rituals to protect tribes from evil or negative spirits. Parts of primates are still used in traditional medicine, in shamanic rituals, or as apotropaic objects, and some primate species were considered sacred and were protected by social taboos, or because they played a totemic role and thus hunting and eating them was prohibited. In many places, these traditional rules were abandoned along with the replacement of traditional cultures, but taboos and other local perceptions of primates continue to be important and are considered by scientists helpful for the protection of threatened species (Baker et al., 2018).

The symbolic value attributed to a primate species (or genus), and the attitude of people to that species, can vary depending on the geographical region, and the ethnicity and socio-economic situation of the people. Some beliefs or perceptions about primates seem to be consequences of specific historical frameworks, and can vary across time. This section will highlight some of the cultural aspects associated with the interaction between humans and primate species in sub-Saharan Africa. For the sake of brevity, only a few of the many species present in this vast territory will be discussed. They have been chosen – arbitrarily – because they are considered to be of particular cultural importance.

2.2.3.1 Baboons

Baboons belonging to the genus *Papio*, occur in most African countries, with six different species recognized (Mittermeier et al., 2013: 661–665). They are easily spotted (baboons live in large groups and are mainly terrestrial), are highly intelligent (they learn quickly to manipulate objects and perform different human tasks, cf. Cheney & Seyfarth, 2008), and their dietary and behavioral flexibility makes them highly adaptable and able to thrive in human-dominated landscapes (Mormile & Hill, 2017). The life of baboons is closely intertwined with that of people, and interactions with humans are problematic, especially because of the baboons' behavior as crop raiders and herd predators. Baboons are often cited as the most destructive for subsistence agriculture of all primate crop foraging species (cf. Findlay & Hill, 2020 and reference there in). This type of behavior must also have occurred also in the past, but competition with humans has increased in places where natural habitats have been transformed for agricultural, urban, or suburban uses, leading to a decrease in natural food sources (Lee & Priston, 2005). Currently, the perception of these animals is influenced by the socio-economic conditions and subsistence activities of local people. In general, people living in contact with baboons note their behavioral similarity to humans, and highlight their intelligence. There was a dichotomy in how traditional communities regarded baboons, with some seeing them as 'people', or human relatives, or incarnations of divinity, or as sacred and totemic animals, while others thought of them as thieves or wicked animals.

The Sahel belt, the transitional zone between the Sahara Desert to the north and the more humid savannah belt to the south is today characterized by a generally low primate diversity and endemism. This region includes the Dogon plateau, in northern Mali, inhabited by the Dogon people, who retain their animistic religion, with many totemic animals (buffalo, crocodile, leopards etc.). Among the Dogon, traditional masked dancing is of special importance and the masks are believed to possess a life force of their own. During the *Dama* masquerade, three different types of monkeys, distinguished by their colors, are invited to perform. "The black monkey is called *Dege*: he is known as the male villain of the bush, an animal that preys on the crops and is renowned for his gluttony, unpredictability and general wickedness" (Morris, 2013: 39). The Dogon believe this mask reflects the 'wild' side of mankind, uncivilized, anti-social, and dangerous. It also represents the dark forces of sorcery that are greatly feared by the tribesmen. When the dancer wearing the black monkey mask is dancing in a ceremony, he hears the words:

> Ugly male of the bush sitting at the top of a tall tree. Your stomach full of fruit, all eyes are on you…the drum plays for you. (Morris, 2013: 39)

This obscure force seems to correspond to the baboon (**Fig. 2.15**). As we have seen also among the Tuareg, baboons are regarded by some of the people living in semi-desertic or savannah habitats negatively, due to both crop raids and herd predation, and in Guinea Bissau people consider them "astute thieves" able to elude farmers (Costa et al., 2013). This association of baboons with the dark world reminds us of the Egyptian God Babi (identified with the hamadryads), a deity of the Underworld (cf. Greenlaw, this chapter). The red monkey mask of the Dogon people is known as *Ko*, and this may refer to the patas monkey, *E. patas* (see also **Fig. 2.16**). The mask known as *Omono*, the white monkey, may correspond to, *Chlorocebus sabaeus*, Linnaeus, 1758 (Blom, 2010: 365).

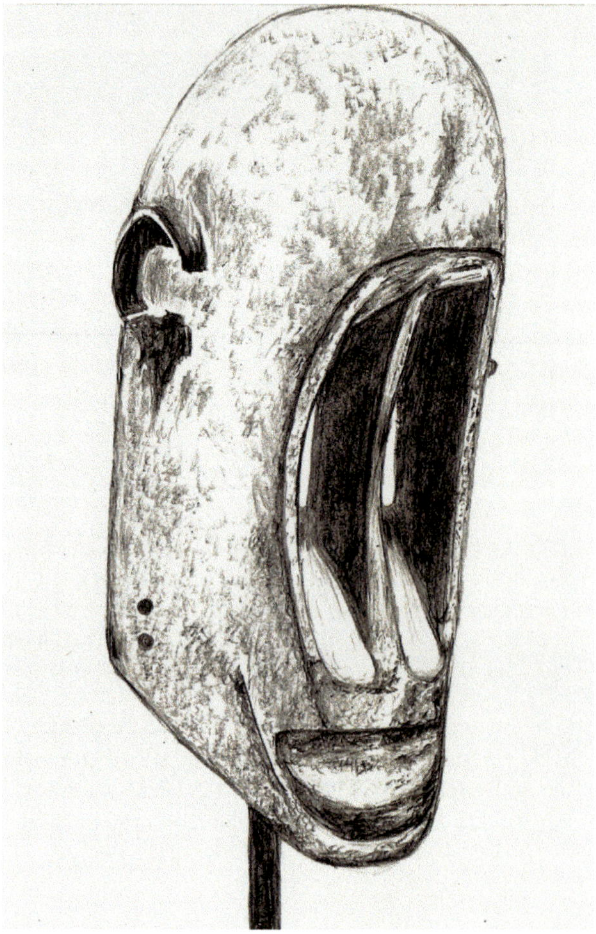

FIGURE 2.15 Masks of Dogon people: the black monkey. (Drawing Cecilia Veracini.)

FIGURE 2.16 Traditional masks of the "spirit of the bush", *Bilisi*, representing a red monkey, *Erythrocebus patas*. Bambara people. South of Macina Region, Mali. (Courtesy of the National Museum of Ethnology (MNE), Lisbon; Francisco Capelo's donation, 2004, N° MNE, BF. 863. Photo: Jean-Louis Losi, Photographic Archive of the MNE.)

The Baule people, an Akan group from the Ivory Coast of West Africa, are known for their fine art objects. Baule masks, figures and decorated objects, which have been admired and collected in Europe since the early part of the 20th century, and monkey figures and masks are considered the most powerful of their sculptural works. Monkey figures, called *gbèkrè*, are used for the divination cult called *Mbra*, which seeks to ensure success in war and may not be seen by women (Vogel, 1997a: 95). Most monkey figures do not have tails (they have a human body with the head of the monkey), but their elongated dog faces are reminiscent of the regional baboon, *P. anubis*. Delafosse (1900) noted that the animals indigenous to the Baule region were known as *Gbèkrè* or *Wotoumo*. Although all these sculptures are called *gbèkrè* in Western art, in some the monkey face seems to show cheek pouches, a feature typical of *Cercocebus atys* (Audebert, 1797), which also occurs in the Baule region. The Baule describe their figures as either spirits of nature or spirit lovers (Vogel, 1973), and they are believed to embody the animal part of the human spirit. The Baule believe this element of the human spirit exists in every person, and that it resides in the bush. The carved figures often incorporate actual skulls into wooden sculptures, and the monkey figure is usually shown holding in its paws a small bowl containing an egg during the divination rituals. Monkey figures are kept in the village, but "fed" with sacrifices in the bush (Vogel, 1997b: 238). Owned only by trance diviners and certain families, the figures remain hidden under a shelter, making it hard for outsiders to see them. Masks are hardly ever shown or handled except in preparation for performances, because the act of looking at a work of art, or at spiritually significant objects, is for the most part only for the privileged and is considered potentially dangerous (Vogel, 1997a: 95). The function of the *gbékré* figures can be better understood in terms of the centrality of agriculture in Baule life, demonstrated in the annual harvest festival in rural areas, in which the major staple food is symbolically offered to the ancestors. *Gbekre* figures are placed as guardians at the gates of villages, but they are also considered patrons of the farmers. According to Baule mythology, the baboon and other ape figures symbolize the son of the god of heaven (cf. Johnson Museum of Art, 2021). The Dan peoples, who also inhabit the Ivory Coast region, have a different approach toward monkey images. "For them the wearing of a *Koagle* monkey mask during ceremonies is intended to add a comic touch to the proceedings with the wearer playing the role of a village clown" (cf. Morris, 2013: 40).

In Kenya, elephants and the two baboon species, *P. anubis* and *Papio cynocephalus*, Linnaeus, 1776, had, and still have, a special place for people. Harry R. Tate, who travelled in Kenya in the first decades of the 20th century (Tate, 1929), said: "I cannot positively state whether all members of the monkey family come within this classification, but the *nyani* [baboon in Swahili] certainly does. Both elephants and baboons are alluded to as persons". In the Taita region he was informed that according to a native belief "baboons formerly consorted with humans and spoke their language, but that owing to misbehaviour they had been driven out and had lost their knowledge of our speech and could now jabber monkey-talk only".

Recent anthropological work (Mwangi, 2019) confirms the special consideration shown to baboons in Kenya. Among the Nthangoni people of eastern Kenya, life is entwined with the animals. For the Nthangoni, a *Mumo,* a lone elderly male baboon, is much more than an ordinary baboon. "He is god incarnate and thus has supernatural power" (Mwangi, 2019: 124); he is sacred and his presence brings good fortune to their home. Unlike other baboons living in large groups, the *mumo* normally moves and feeds alone. He is not aggressive to people, and does not kill or attack goats or chickens, and when he forages on people's crops, it is not just only tolerated, but sanctified. A *Mumo*'s prolonged absence and the misfortune that might befall the Nthongoni people, often prompts the elders to conduct a ceremony to appease the gods, because a *Mumo* is seen as a messenger from the spiritual world bringing a message of goodwill. As noted by Mwangi (2019), *Mumo* behavior demonstrates his humanity and agency, thus blurring the attributes thought to set people apart from non-human animals. Children that are born around the time a *mumo* is seen in the village are often named after him, demonstrating the distribution of personhood across humans and baboons. The other baboons are different: some residents regard baboons as 'human-like' and 'intelligent animals', while others see them as malicious and naughty. Others note their use of language to signal the peril or the arrival of the farmer when they are crop foraging, or regard their social structure as similar to that of humans. As Mwangi (2019) notes, despite the physical separation produced through the creation of the Tsavo and Chyulu National Parks, and the alienation of people from land and wildlife, there are still sociocultural bonds with baboons; they are present physically, socially, and spiritually in the lives of the Nthongoni people.

In southern Africa, different communities instinctively recognize the chacma baboon, *Papio ursinus* Kerr, 1792, as the animal closest to the human, though the significance of this perceived proximity varies (Klopper, 2010). For the Bushmen (San peoples), baboons could understand human speech and were described as people who sat on their heels. For the Boer group of Bushmen, they were part of the animal world and could speak baboon language (Klopper, 2010: 5). The Kalahari San of Botswana consider baboons to be people in another state of existence. For the ¡Xam San, baboons had similar powers to those of a shaman (Badenhorst, 2022: 383). Baboons have been protagonists of the life and cosmogony of these peoples since ancient times. Ndukuyakhe Ndlovu (2022) studied ancient rock paintings and carvings (Rock Art) in South Africa dating to different periods since approximately 40.000 years ago, and noted that baboons are not as common as other animals occurring in the same environment. The author's hypothesis is that images of baboons are incorporated into those of human figures to produce therianthropic images, representing hybrid creatures, half-human and half-animal, and that this practice reflected the importance of baboons in the cosmological vision of Bushmen. According to oral narratives recorded by well-known authors, such as Dorothy Bleek (cf. Bleek, 1923; Bleek et al., 1911), hunter-gatherer communities did not consider themselves superior to other animals, and in Bushmen folklore these animals are believed to provide protection and to be unaffected by evil and sickness. Baboons and shamans had similar spiritual powers so the incorporation of baboon elements to produce therianthropic figures is a confirmation of the closeness of baboons to people (Lewis-Williams & Dowson, 1989). Rock art depicts animals and humans believed to be medicine men or healers involved in a healing dance. Therianthropes—people with hybrid animal features are believed to be in a state of metamorphosis during trance states (Gall, 2003).

The importance of baboons in Bushmen's cosmology is reflected in them being avoided as food (Bleek, 1931). The only people in southern Africa who may have eaten baboons are the Pedi, part of the North Sotho ethnic group (Badenhorst, 2022), with consumption reserved for men and boys, revealing a ritual use of baboon flesh. Baboon skulls are impaled at the great gate of Pedi villages and other parts of the bodies of baboons are used as traditional medicine, or in material culture. Baboons, probably because of their reputation as pests and their human qualities, played a more sinister role in many farming societies in southern Africa, where skeletal parts of baboons are present in divinatory sets (Badenhorst, 2022 and references therein).

2.2.3.2 Black and White Colobuses

Black and white colobuses (*Colobus* spp.) are known for their fur (**Fig. 2.17**), especially fine in the mantled colobus, or guereza, *Colobus guereza,* Rüppell, 1835, which has at least eight subspecies in many countries of Central and Eastern Africa (Mittermeier et al., 2013: 699), but populations are threatened in parts of its range (IUCN Red List, 2021). The beauty of their fur certainly did not help these animals, and in Europe they were called *Callithrix,* meaning 'beautiful hair' in antiquity (cf. Masseti & Veracini, this volume). They have been hunted for their fur for centuries by traditional peoples, who used the fur in ritual and symbolic forms, and for commerce (cf. Veracini & Almghadi, this volume), and as late as the 1960s, in Ethiopia colobus skins were used to make cushions and carpets that could be bought without any restrictions (**Fig. 2.18**).

In the Black Bush Area of Waindow (BBAW), Menchum south, North West Region of Cameroon, between Cameroon

FIGURE 2.17 The Guereza colobus in its natural habitat (Kenya). (Photo Cecilia Veracini.)

FIGURE 2.18 Guereza skins bought in Ethiopia in 1971 by the Italian zoologist Benedetto Lanza. (Courtesy of the Museum of Zoology and Natural History, *La Specola* of Florence, N. Coll. 21748. Photo Cecilia Veracini.)

and eastern Nigeria, mantled monkeys are used in many rituals and they are reported to be sacred (cf. Angwafo & Chuo, 2016: 9). Their distinctive black and white coloration "is a symbol of certain traditional sects". If anyone hunt these monkeys "they were automatically initiated into the sect and could become seriously ill, or die, if not purified". Widely respected in the past, this taboo is now limited to families whose ancestors were servants of the gods. Unless the indigenous people are actively introduced to and sensitized on the subject of, modern-day conservation needs and benefits as a replacement, the survival of the remaining species in the study site is at risk.

At least until the 1950s–60s, colobus fur was considered by many traditional Kenyan populations to be important in various rituals, and its possession conferred superior status (cf. Adamson, 1967), and it was worn by those who had important roles, such as village chiefs, warriors or witch doctors. Splendid headdresses, coats or other traditional decorations made from colobus fur can still be seen in the National Museum of Kenya (e.g., **Fig. 2.19).** The Museum also houses a majestic and impressive cloak made of about 40 guereza furs, of the type worn by the President of Kenya Uhuru Kenyatta, on his election day, when Kenya became a republic in 1964. In the past the use of colobus furs was widespread. For example, among the Luhya people of Bantu origin of the Kakamega district of western Kenya, rulers would dress in royal robes when important occasions warranted it, and this was still the custom at the beginning of the 20th century. The Nabongo's (king's) cloak, known as *iyamwamu*, was made of several guereza skins (Openda-Omar, 1993). In Kenya in the 1950s, Joy Adamson portrayed people dressed in traditional costumes made with colobus furs, and in the Kikuyu, Tharaka, and Maragoli traditional circumcisers wore headdresses, or leggings, made of this monkey fur, and the latter were also worn by Kipsikis warriors (Adamson, 1967). Among the Embu and Idakho, warriors had headdresses of guereza fur and the Chief of the Embu

FIGURE 2.19 Traditional hat made with guereza fur. (Drawing by Cloride Baroni, from an object of the National Museum of Kenya, Nairobi.)

wore a monkey coat decorated with guereza fur, and guereza fur leggings. Among the Luo, dancing dresses were made of guereza fur, in the Kutya dancers wore ornaments made of the same monkey's fur, and the same is found among the Kissi bridegrooms. In Kenya, skins of the blue monkey (*Cercopithecus mitis,* Wolf 1822, **Fig. 2.20**) were also used for making coats or cloaks for important persons such as rulers (see Adamson, 1967), and a traditional king's cloak exhibited in the National Museum of Kenya is made of blue monkey fur with ornaments made of mantled colobus fur. The current perception of primates in Kenya is that they are crop raiders and that baboons prey on cattle. In some places (e.g., Karura Forest in Nairobi) reintroduced mantled colobus are tourist attractions (pers. obs.).

The use of colobus skin in traditional rituals, especially rituals of passage, is also evident in Ethiopia where colobus also occur (e.g., **Fig. 2.21**). Among the Daasanech, pastoralists, who live on either side of the Omo river in South West Ethiopia, and along the northeastern corner of Lake Turkana in northern Kenya, the hides of colobus monkeys (*lol*), leopards and cheetahs were (at least until the end of the 20th century) highly valued for ritual purposes, as were ostrich feathers and giraffe tails (Houtteman, 2011: 53). During the *dimi*, a complex ritual that serves to induce fertility in a man's daughter, the mothers of *dimi* girls decorate the girl's hair with colorful strings and beads, and they wear the skin of a guereza monkey, likewise kept in place by a slightly smaller belt

FIGURE 2.20 The blue monkey, *Cercopithecus mitis,* in its natural habitat, Kenya. (Photo Cecilia Veracini.)

FIGURE 2.21 Traditional mask made with guereza skins, from Ethiopia, 20th century. (Courtesy of the Museum of Anthropology and Ethnology, University of Florence. Photo Cecilia Veracini.)

of beads (Houtteman, 2011: 187). The traditional Daasanech world, and its rituals, now considered archaic by the young have been transformed during the last twenty years. However, this has not halted the increase of poverty and alcoholism among the Daasanech people. (ibid, 2011: 185).

2.2.3.3 Great Apes

The relations between indigenous people and great apes are complex and multi-faceted. "These animals have not only an important symbolical value but also an important material presence as actors who can impinge on human livelihoods and as hunted game" (e.g., in CAR, Cameroon, Giles-Vernick & Rupp, 2006: 55). The most common perception of human communities living in contact with, or in close proximity to, apes is to regard them as ancestors, descendants of humans or relatives of humans. Great apes may also have other cultural roles, as they may be associated with the world of the dead, or with ethnic groups considered inferior or hostile. The way these animals are perceived can blur and dilute the boundaries between human and non-human animals, and between the world of the living and the dead. In the past, and in the present, great apes and the humans who have lived with them in sympatry, have become deeply connected, both in cultural terms and in the sharing of common natural resources. Only recently has Western science, which for decades has conducted research on these animals that latterly has been primarily concerned with their conservation, recognized the great contribution to knowledge that can be provided by native peoples because of their profound symbolic and material relationship with the African great apes (e.g., Bortolamiol et al., 2018).

The attitude to, and perception of, great apes has adapted to the changes brought about by colonialism, contemporary conservation policies and other socio-cultural influences. For example, Giles-Vernick and Rupp (2006: 59) assume that past and contemporary narratives about gorillas of the Mpiemu people of the Central African Republic (CAR) reflect changes in their lifestyle and relationship with the forest. During the 1990s, when the CAR in conjunction with the WWF sought to protect gorillas, chimpanzees and other animals, conflict over access to the forest was expressed in Mpiemu narratives about increasingly aggressive gorilla behavior, and gorilla meat consumption began to be permitted because it was claimed gorillas were menacing women even more than before. Contemporary Mpiemu tales about the great apes reflect current debates over the competing claims of different groups to natural resources, material wealth, and development assistance.

Some insights into pre-colonial lore of the West African populations on the subject of the great apes can be found in the travel literature of the first Europeans who began to explore Africa in the 15th and 16th centuries (cf. Veracini, chapter 9, this volume). There are many accounts of the great apes, which at the time were a great novelty to European eyes. One of these accounts is that of the Jesuit priest Fr. Manuel Alvares, who belonged to a mission to Cape Verde and West Africa (Sierra Leone) in the early 17th century. He recorded much valuable information concerning Western chimpanzee (*Pan troglodytes verus Schwarz, 1934*) behavior as well as elements gathered from the oral traditions of the local people:

> There are heathens that claim to be descendants of this animal [chimpanzee], and when they see it they have great compassion: they never harm it or strike it because they consider it the soul of their forefathers, and they think themselves of high parentage. They say they are of the animal's family, and they all believe they are descendent from it and call themselves *Amienu*. (Hair, 1990: 10–12)

This, which was among the first reports of European travellers which are among the first reports to include detailed naturalistic descriptions of the chimpanzee, bore testimony to the complexity of chimpanzee-human interactions in Western Africa (e.g., Sept & Brooks, 1994: 876; Veracini & Casanova, 2016), where chimpanzee populations have lived in close proximity to humans, and have adapted to the human-influenced landscape. Contemporary respect for the *Darì* (the traditional name given to chimpanzee in 16th century Guinea-Bissau, which is still in use) is due to the belief they are the ancestors of humans (see **Box Details 2.3**).

BOX DETAILS 2.3: CHIMPANZEE – HUMAN INTERFACE IN GUINEA-BISSAU

Susana Costa
Universidade de Coimbra

In the Cantanhez Forests National Park (Tombali, Guinea-Bissau), it is common to find elderly people mentioning chimpanzees as being former humans. The legend of the origin of the chimpanzees mentions a blacksmith who was punished by God for being lazy and was consequently sent to live in the forest (Costa & Casanova, pers. obs., 2008). The difference between the species found in the wild and the ones one can find in the "civilized" part of the world (i.e., villages) is pointed out frequently by villagers who live inside the Cantanhez Forests National Park: chimpanzees (*P. t. verus*) vs. humans (*Homo sapiens*); cows (*Bos primigenious*) vs. buffalos (*Syncerus caffer*); goats (*Capra aegagrus hircus*) vs. duikers (*Cepahlophus dorsalis*); pigs (*Sus scrofa scrofa*) vs. red river hogs (*Potamochoerus porcus*) etc. The world, as in other African traditional cosmologies, is perceived as having two parallel microcosms, where chaos (the forest) and civilization (the village) coexist (Costa and Casanova, pers. obs., 2008). Nevertheless, chimpanzees are considered to be different from other animals, but this kind of taboo is not enough to protect them, because conflicts with people and the reduction of the chimpanzees' natural habitats increase day by day. In the case of Guinea-Bissau chimpanzees, similarities with humans have prevented locals from consuming the meat of the species. The species is described as human-like, and during interviews, locals appear to know a great deal about chimpanzees' physical features and behavior. Chimpanzees are described as being protective of their babies and infants, with healthy feeding habits (e.g., they peel their fruit, they pick fruit directly from trees, they do not eat fruits lying on the ground, etc.). While they do not have a tail (a fact which might have prevented them from being perceived as regular non-human animals), they do have five fingers and, possibly the most important characteristic, they are intelligent (Costa et al., 2013). Nevertheless, habitat loss and human population growth have been bringing about clashes between villagers and chimpanzees - and primates in general - and coexistence is not always easy (Lee & Priston,

2005; Alves et al., 2010; Costa et al., 2013). Chimpanzees destroy farmers' cash crops, thus they are an economic threat to these people's survival. As such, they are perceived as pests and, due to that, they are frequently and lethally shot or trapped (Lee & Priston, 2005; Costa et al., 2013, 2017). Though their meat is not seen as a source of protein, chimpanzee parts are traded to appease *Irans* (animistic entities that need to be appeased from time to time to guarantee that human needs are met) in order to prevent diseases (Alves et al., 2010) and to imrove health (Sá et al., 2013), Their parts are not bought exclusively by traditional doctors, but also by their "patients" who use them as medicines or for magical purposes. For example, in Cantanhez Forests National Park in the South of Guinea-Bissau, the skin of chimpanzees is used by pregnant women and mothers of newborn babies to prevent ugliness.

Similar narratives and beliefs occur in many traditional communities that share territories with chimpanzees, bonobos, and gorillas (e.g., Oates, 1999; Kortlandt, 1986; Köhler, 2005; Nyanganji et al., 2011; Narat et al., 2015; Bortolamiol et al., 2018). Other myths, reported by 17th-century travellers in western and central Africa, tell us that chimpanzees are able to speak as people do, but have chosen not to do so in order not to be forced to work (see Barsanti, this volume).

The Manon people, an influential family in the village of Bossou (Republic of Guinea), believe chimpanzees embody the reincarnation of their ancestors. Due to the totemic value placed on chimpanzees, they refrain from harming them, although sometimes these animals can attack humans (Hocking et al., 2010). In Uganda, near Kibale, legends and traditions related to chimpanzees are rare, but to two local communities, the Batooro and the Bakiga, chimpanzees resemble humans. According to Bortolamiol et al. (2018), the narrative of local communities depicts a close affiliation between humans and chimpanzees, which are seen as descendants of humans, from which they are distinguished by their lifestyle, with the forest providing the discontinuity between the two species. Similarly, the Mpiemu people of the middle and upper Sangha basin of the CAR, studied by the anthropologist Tamara Giles-Vernick, contend that chimpanzees (*wago*) and gorillas (*ntchilo*) share similar origins with people, although they do not live in villages, but in the forest (Giles-Vernick & Rupp, 2006). A Mpiemu tale tells that forests, chimpanzees, and gorillas were created by the deity of the forest (*Ntchambe Mekwombo*), who was the firstborn and initially commanded the other deities. Later, the deity of the fallows created the Africans and taught them to work in the fields, resulting in the people of the savannas achieving superiority over the apes in the forests (Giles-Vernick, 2002: 57).

Judging by traditional stories, local communities were well aware of the importance of the ecological role of the great apes, which resulted in taboos related to hunting and the consumption of their meat. For example, according to Giles-Vernick and Rupp (2006: 55), the oral histories of the Bangando people of South Eastern Cameroon, collected in the 19th century, spoke about the *dawa* (the broad family of monkeys, chimpanzees, and gorillas) rescuing a group of Bangando people from abduction and death. The *dawa* stood guard over the forest in the treetops and called out to warn the Banganda people as the Ndzimou [hostile people] prepared their attack. Then the *dáwá* led the Bangando through the forest to a safe place. In gratitude, this Bangando community adopted monkeys and apes as their clan totem, and they still refrain from eating their meat. "This story of mutual protection is part of an ethic that the Bangando seek to inculcate and sustain among younger generations and it has important implications for Bangando perceptions of threats to forest game populations" (Giles-Vernick & Rupp, 2006: 55).

In the already mentioned Black Bush Area of Waindow (BBAW), Menchum south, North West Region of Cameroon, historically gorillas and chimpanzees were sacred, and no one was permitted to look them directly in the face, nor talk of killing them (Oates, 1999), and the people believed that both apes were ghosts of their ancestors. Eyong (2007: 8) reports that:

> this was just a superficial explanation for those who could not understand the secret sayings of the gods. The real purpose was to prohibit the hunting of these animals to maintain their traditional values, as they were caught in secret and sacrificed to their gods for the cleansing of the village sins and for the protection of the village.

A different view of the great apes was reported by Liz Ferreira, a Portuguese anthropologist who studied gorillas in the 1930s–40s in the Maiombe Forest (Northern Cabinda), in Angola. He observed that for certain populations within Gabon, the gorilla was considered a cruel and bloodthirsty monster that silently enters huts at night and steals women, taking them to its dark hiding places in the jungle (Liz Ferreira & Magalhaes, 1945: 22). The same author noted that these legends were widespread in Central Africa, and added that, in Cameron, shamans advised people not to go to particular places at certain hours of the night, because the gorilla would be there waiting for its victims, before crushing them and listening to their cries of agony.

The body parts of apes are sometimes used in traditional medicine or, if worn, they are believed to provide protection (**Fig. 2.22**). In Sierra Leone, chimpanzee bones and teeth are used to protect infants, and in some parts of the country a piece of dried chimpanzee bone is traditionally tied around infants' bodies to make them become stronger adults, and chimpanzee central incisors are believed to protect children and give them power over others in their cohort (Alves et al., 2010). In the Maiombe Forest (Northern Cabinda), in the 1930s–40s, shamans used gorilla bones for various medicines, which they sold for high prices. One medicine was made by mixing powdered gorilla bone with snake oil; this developed strength and was given to pregnant women so that their babies would grow up strong and healthy (Liz Ferreira & Magalhaes, 1945).

FIGURE 2.22 Fetish made from wood representing a chimpanzee (Western Africa). Objects like this are believed to have magical power to protect its owner. (Photo Cecilia Veracini, private collection.) Many animals are hunted and traded specifically for use in traditional medicine, also known as wudu, juju or fetish (Adeola, 1992). They are sold at specialized fetish markets (Buij et al., 2016).

Gorillas and chimpanzees have cultural and symbolic roles for human pygmies. Pygmy populations (around 920,000 persons, Olivero, 2016) occupy a territory extending west-to-east along the central African belt from the Congo Basin to Lake Victoria. The three main groups (Bayaka, Mbuti and Twa), each contain their own ethno-linguistic groups (cf. Takeuchi, 2017; Olivero, 2016). Pygmies are the largest nomadic group of hunter-gatherers in Africa, and their life is still linked to the forest, although some groups have begun to adopt a more sedentary and agricultural lifestyle (cf. Takeuchi, 2017).

The Bayaka pygmy group (which encompasses the Aka, Luma, Mikaya, Mbendjele, Ngombe and Baka) shares the forest west of the Congo River toward the Atlantic coast with the western lowland gorilla (*Gorilla gorilla gorilla* (Savage, *1847*)) and the central African chimpanzee (*Pan troglodytes troglodytes* Blumenbach, *1775*). The essence of the Bayaka culture (but also of the other pygmy groups), which has been preserved over the millennia, is mainly expressed in music, tales, and dance that maintain bonds with their gods, spirits, and ancestors. "Tales are the explanation in poetic form of the world that came from ancient times, outside time and space, and they convey a state of affairs, an original explanation of current life, the forest and all its living beings" (Maget, 2018). Tales can include great apes, such as those of the Baka of the CAR that include gorillas in their spirit dances (Michelle Klailova pers. Comm.). Gorillas and chimpanzees also play important roles in the southeastern Cameroon Baka's traditional song stories. In these songs, apes undertake the same activities as the Baka (e.g., raising children and gathering food) and they are endowed with human emotions (Hattori, 2008: 87–89 *apud* Oishi, 2013). The Baka of Cameroon include all monkeys in the category *kema*, while the apes are either placed in an unnamed category, or are grouped in the category *so na susu* (animal with hair) (Oishi, 2013). Although *Ebobo* is the generic Baka name for gorillas, they also have at least ten terms for specific types of gorilla, depending on sex, age, and companionship status, and hunters are able to distinguish the composition of a gorilla‘s social group (Oishi, 2013). *Te-ebobo*, which can be translated as "gorilla village", indicates areas in the forest where gorilla nests are concentrated. Baka hunters' narratives refer to the intelligence and cognitive abilities of gorillas, which are "clever" and "cunning" compared with other animals, including chimpanzees (Oishi, 2013). The Baka also distinguish between the gorilla and the chimpanzee: the former is braver and fearless, and tends to defend its territory, while the latter is more cautious and runs away when it encounters people. An example of this perception is found in the Bakas' song called "*The chimpanzees and the daughter of Komba*" recorded by the anthropologist Maget:

> A chimpanzee, knowing that the daughter of Komba is very beautiful, decides to visit her and make her his bride. His whole family agrees. But once on his way through the forest he comes across a trunk of tree with many mushrooms on it. And since there were not many passers- by and there was no one there he begins to doubt and to be afraid. [The song, which is sung by the rest of the Baka group at this point, illustrates the chimpanzee's doubts as he wonders who is behind the trunk or in front of him.] The chimpanzee begins to dance and sing, afraid because he doesn't know who is there and whether the log is dangerous. After much doubt he decides, frightened, to turn back and give up marrying, and his whole family mocks him for failing to bring Komba's beautiful daughter home. [As the narrator says] the chimpanzee is the one who is always afraid, whereas the gorilla is the great ape who is the brother of the chimpanzee, and has a strong heart because when there are obstacles he always manages to face them. (From Maget, 2018. Translation and adaptation of the author)

The great apes are, for the Baka, thought to be related to humans through reincarnation (Duda et al., 2018). Some Bayaka groups traditionally regard neighboring farmers (sometimes in conflict with them, cf. Takeuchi, 2017) as apes, and they believe they will be reincarnated as gorillas. In general, the cultural antagonism between pygmies and farmer groups is reflected explicitly in the discourse of farmers, many of whom regard pygmies as quasi-human

(Takeuchi, 2005; 2017). The Baka of the CAR believe that the Bantu, the other ethnic group present in their territory, turn into gorillas at death, and the Baka turn into white people and white people turn into butterflies (Michelle Klailova pers. comm.). Among the Mbendjele Yaka Pygmies, the local name for gorilla *ebobo* is used to refer to neighboring farming villagers. According to Lewis (2002), when a Mbendjele hunter kills a gorilla, the dead gorilla was referred to as *milo*, which means "villager". Most of the Mbendjele refrain from eating gorilla meat because they consider the gorilla to be a metamorphosed farming villager. Indeed, the Baka believe that a Baka may reincarnate into a chimpanzee after death. Oishi (2013) observes that great apes, and their special meaning for traditional communities, cross both inter-ethnic boundaries between the pygmies and the farmers, and the interspecies boundaries between humans and non-human animals.

Gorillas are sometimes hunted by pygmy peoples, but this occurs occasionally and under special conditions, depending on the ethnic group. The Batwa pygmies of the Great Lake region, who inhabit the same territory as the Eastern Lowland gorilla *(Gorilla beringei graueri* (Matschie, 1914)), were traditionally associated with the help they gave to other people hunting gorillas (and often this argument was used in a discriminatory modality, Lewis, 2000: 5). The Batwa were reported by the Italian explorer Attilio Gatti in 1932 as naming each gorilla, never making the slightest mistake in the identification of an ape (cf. Gatti, 1932). The local name for gorilla is *Ngagi*: "the one who breaks, who ends the night because in the morning, in the few seconds before the first rays of the sun peep above the horizon, the gorilla stands erect, alone, beating his breast and emitting a cry" (Gatti, 1932:). In 1928, the Batwa people helped Attilio Gatti to capture an exceptionally big adult gorilla male (*G. b. graueri*), called *Moami Ngagi* (King of gorillas) in the western region of Lake Kivu (**Fig. 2.23**). The skin and skull of this specimen are preserved in the Natural Museum La Specola of Florence (cf. Ducci et al., 2015).

Gorillas are generally perceived as the most dangerous animal, after the elephant. In a study of gorilla behavior in the CAR, the trackers observing the habits of gorillas seem not to have viewed them more favorably when spending time with them. Michelle Klailova's impression (pers. comm.) is that they would have seen them as all-powerful, scary, and 'bad' in the past. In Cameroon, Baka women and children carry a special plant medicine to drive gorillas away, especially when they enter the forest unaccompanied by a male adult (Giles-Vernick & Rupp, 2006). Among the Mbuti, a group of pygmies living in the Ituri Forest of north-eastern Zaire, the gorilla is present in the culture even though it does not occur in the region (Carpaneto & Germi, 1989). The Mbuti recount tales of a mythical big ape (the *Amemangamanga*), neither a gorilla nor any other animal, living in the deep forest. They fear it as an evil spirit of the forest (Carpaneto & Germi, 1989). Some Bayaka taboos related to the consumption of gorillas and chimpanzees are due to their human-like morphology and

FIGURE 2.23 The Italian explorer Attilio Gatti and the Batwa people who helped him to capture a male of the Eastern Lowland gorilla (*Gorilla beringei graueri*) in the western region of lake Kivu (1928). (From Ducci et al., 2015.) The skin and skull of this specimen are now preserved in the Museum of Zoology and Natural History, *La Specola* of Florence (cf. Ducci et al., 2015).

because the Bayaka think that gorillas are metamorphosed farming villagers (see above). Gorilla body parts are usually named in composite forms of words referring to those of humans, such as *botekpa-ebobo* ("palm" of a gorilla) and *njo-ebobo* ("head" of a gorilla) (Oishi, 2013). In contrast, among the Mbuti, chimpanzees are treated like any other game animal, even if its behavior is similar to that of people (Carpaneto & Germi, 1989).

Native Africans understand great apes not as wild creatures living in remote forests, but through the close relationship between these animals and their own community and daily lives. But, this relationship can change according to circumstance, something that has to be taken into account by conservationists.

ACKNOWLEDGEMENTS

I would like to express my thanks to: Giulia Cavallo for her suggestions; Michelle Klailova for her comments about Baka people of CAR; Monica Zavattaro for her assistance at the Museum of Anthropology and Ethnology in Florence; the Museum of Anthropology and Ethnology for permission to publish the colobus mask; Paolo Agnelli and Laura Ducci for their help and the Museum of Zoology and Natural History, *La Specola* in Florence for permission to publish the photo of the colobus skins; the Director of the

Museum of Ethnology of Lisbon Paulo Ferreira da Costa for permission to publish the red monkey mask; Cloride Baroni for the drawing of the traditional hat; Marco Alesci for the photo of the baboon. I would also like to sincerely thank Hilary Morris for the first English revision of the text, and, last but not least, Bernard Wood for the revision of the text and for suggestions.

FUNDING OF CV

Portuguese national funds through FCT—Fundação para a Ciência e a Tecnologia in the context of the celebration of the contract under the 'Norma Transitoria', DOI 10.54499/DL57/2016/CP1456/CT0001.

REFERENCES

Adamson, J. (1967). *The Peoples of Kenya*. London: Collins & Harvill Press.

Adeola, M.O. (1992). Importance of wild animals and their parts in the culture, religious festivals, and traditional medicine of Nigeria. *Environmental Conservation*, 19, 125–134.

Alexander, S.D. (2018). *People, primates and peace, a case studies of the Barbary macaque. Ethoprimatology and interdisciplinary conservation in the Rif mountains of Morocco*. Ph.D. thesis. The University of Alabama.

Alves, R., Souto, W. & Barboza, R.R.D. (2010). Primates in traditional folk medicine: A world overview. *Mammal Review*, 40, 155–180.

Andrews, C. (1994). *Amulets of Ancient Egypt*. London: British Museum Press.

Angwafo, T.E. & Mvo Denis Chuo, M.D. (2016). Contributions of indigenous knowledge of gorilla (*Gorilla gorilla diehli*), chimpanzee (*Pan troglodytes ellioti*) and buffalo (*Syncerus caffer*) conservation, in Waindow, North West Cameroon. *Annual Research & Review in Biology*, 11(2), 1–14. doi: 10.9734/ARRB/2016/28428

Badenhorst, S. (2022). 14 citizens of the Savanna. *World Archaeoprimatology: Interconnections of Humans and Nonhuman Primates in the Past*, 90, 375.

Baker, L., Tanimola, A. & Olubode, O. (2018). Complexities of local cultural protection in conservation: The case of an endangered African primate and forest groves protected by social taboos. *Oryx*, 52(2), 262–270. doi: 10.1017/S0030605317001223

Beck, P. & Huard, P. (1969). *Tibesti carrefour de la préhistoire saharienne*. Paris: Arthaud.

Bleek, D.F. (1923). *The Mantis and His Friends*. Cape Town: Maskew Miller.

Bleek, D.F. (1931). Customs and beliefs of the ¡Xam Bushmen 1: Baboons. *Bantu Studies*, 5, 167–179. doi: 10.1080/02561751.1931.9676257

Bleek, W.H.I., Lloyd, L. & Theal, G.M. (1911). *Specimens of Bushman Folklore*. G. Allen, Limited.

Blom, H. (2010). *Dogon, Images and Tradition*. Switzerland. https://www.randafricanart.com/Dogon_masks.html.

Bodenheimer, F.A. & Rabinowitz, A. (tr., comm. and introd.) (1949). *Timotheus of Gaza. on Animals. Fragments of a Byzantine Paraphrase of an Animal-Book of the 5th century A.D.* (Collection de travaux de l'Académie internationale d'histoire des sciences, 3), Paris–Leiden.

Bortolamiol et al. (2018). Wildlife and spiritual knowledge at the edge of protected areas: Raising another voice in conservation. *Ethnobiology and Conservation*, 7, 122. https://ethnobioconservation.com/index.php/ebc/article/view/184/204

Bourriou, J. (1988). *Pharaohs and Mortels: Egyptian Art in the Middle Kingdom*. Cambridge UP.

Bousquet, B. (1992). *Guide des parcs nationaux d'Afrique*. Paris: Delachaux et Niestlé.

Boylen, P. (1922). *Thoth: The Hermes of Egypt*. Oxford: Oxford UP.

Brown L. (1965). *Africa, a Natural History*. New York: Chanticleer Press Edition.

Brugnatelli, V. (2006). *I Berberi: Elementi di Storia, Lingua e Letteratura*. Appunti per la parte generale del corso di Lingue e Letterature dell'Africa 2005-2006. Milan: Università degli Studi di Milano-Bicocca.

Buij, R., Nikolaus, G., Whytock, R., Ingram, D.J. & Ogada, D. (2016). Trade of threatened vultures and other raptors for fetish and bushmeat in West and Central Africa. *Oryx*, 50(4), 606–616.

Bulté, J. (1991). *Talismans Egyptiens d'Heureuse Maternité*. Paris: CNRS.

Butler, J.R.A. (2001). The economic costs of wildlife predation on livestock in Gokwe communal land, Zimbabwe. *African Journal of Ecology*, 38(1), 23–30.

Butynski, T.M., Cortes, J., Waters, S., Fa, J., Hobbelink, M.E., van Lavieren, E., Belbachir, F., Cuzin, F., de Smet, K., Mouna, M., de Iongh, H., Menard, N. & Camperio-Ciani, A. (2008). *Macaca sylvanus. The IUCN Red List of Threatened Species*: e.T12561A3359140. https://dx.doi.org/10.2305/IUCN.UK.2008.RLTS.T12561A3359140.en. Downloaded on 08 April 2018.

Camperio Ciani, A. (1986). Origine, evoluzione, speciazione ed ecologia del genere *Macaca* in relazione alle vicende geoclimatiche del Quaternario. *Studi per l'Ecologia del Quaternario*, 8, 9–32.

Camps, G. (1990). Des incertitudes de l'art aux "erreurs" d'Herodote. La faune des temps néolithiques et protohistorique du Nord. *Comptes rendus des séances de l'Académie des Inscriptions et Belles-Lettres*, 134-1, 35–37.

Carpaneto, G. & Germi, F. (1989). The mammals in the zoological culture of the Mbuti Pygmies in north-eastern Zaire/I mammiferi nella cultura zoologica dei Pigmei Mbuti nello Zaire nord-orientale. *Hystrix, the Italian Journal of Mammalogy*, 1(1).

Cheney, D.L. & Seyfarth, R.M. (2008). Baboon metaphysics. In *Baboon Metaphysics*. University of Chicago Press.

Claussen, M., Kubatzki, C., Brovkin, V. Ganopolsky, A., Hoelzmaan, P. & Pachur, H.J. (1999). Simulazione di un brusco cambiamento nella vegetazione sahariana nel Medio Olocene. *Geophysical Research Letters*, 26(14), 2037–2040.

Corridi, C. (1998). Faunal remains from Holocene archaeological sites of the Tadrart Acacus and surroundings (Libyan Sahara). In M. Cremaschi & S. Di Lernia (eds.): *Wadi Teshuinat. Palaeoenvironment and prehistory in south-western Fezzan (Libyan Sahara). C.I.R.S.A. Centro Interuniversitario di Ricerca per le Civiltà e l'Ambiente del Sahara Antico. C.N.R. Quaderni di Geodinamica Alpina e Quaternaria*, vol. 7, pp. 89–94.

Costa, S., Casanova, C. & Lee, P. (2017). What does conservation mean for women? The case of the Cantanhez Forest National Park. *Conservation and Society*, 15(2), 168–178.

Costa, S., Casanova, C.C., Sousa, C. & Lee, P.C. (2013). The good, the bad and the ugly: Perceptions of wildlife in Tombali (Guinea-Bissau, West Africa). *Journal of Primatology*, 2(1). doi:10.4172/2167-6801.1000110

Dahbia, A. & Chaker, S. (2004). Kabylie: Cosmogonie. *Encyclopedie Berber* 4086-93. Available online: https://journals.openedition.org/encyclopedieberbere/1443#text (accessed on 20 July 2020).

Dasen, V. (1993). *Dwarfs in Ancient Egypt and Greece*. Oxford UP.

Davis, E. (1986). Youth and age in the Theran Frescoes. *American Journal of Archaeology*, 90, 399–406.

Dekeyser, P.L. (1952). A propos de la tête osseuse d'un cinocéphale du Tibesti. *Bulletin de l'Institut français d'Afrique noire*, Dakar, 14, 537–544.

Dekeyser, P.L. (1955). *Les mammifères de l'Afrique noire française (2e edition)*. Dakar: Institut français d'Afrique noire.

Delafosse, M. (1900). Sur des traces probabile de civilisation égyptienne et d'hommes de race blanche à la Côte-d'Ivoire. Paris, *L'Anthropologie*, 9, 431–451, 543–568, 677–690.

Deputte & Anderson (2009). Baboon palm nut harvesters in ancient Egypt: New (ancient) evidence, new questions. *Folia Primatologica*, 80, 70–73.

Derchain, P. (1952). Bébon, le dieu des myths. *Révue d'Egyptologie*, 9, 23–47.

Descola, P. (2005). Ecology as cosmological analysis. *The Land Within: Indigenous Territory and the Perception of Environment*, 22–35.

Di Lernia, S. & Manzi, G. (2002). Sand, stones, and bones. The archaeology of death in the Wadi Tanezzuft Valley (5000–2000 bp). *The Archaeology of Libyan Sahara, I*. AZA Monographs 3. Florence: All'Insegna del Giglio.

Diodorus Siculus. (1954). *The Library of History*. The Loeb Classical Library Edition. Available online: https://penelope.uchicago.edu/Thayer/E/Roman/Texts/Diodorus_Siculus/20C*.html (accessed on 10 July 2020).

Dorst, J. & Dandelot, P. (1973). *Guıa de campo de los mamiferos salvajes de Africa*. Barcelona: Ediciones Omega.

Ducci, L., Veracini, C. & Agnelli, P. (2015). Note su un esemplare di Gorilla di pianura orientale, *Gorilla beringei graueri* (Matschie, 1914) conservato al Museo di Storia Naturale dell'Università di Firenze. *Museologia Scientifica Memorie*, 14, 157–162.

Duda, R., Gallois, S. & Reyes-García, V. (2018). Ethnozoology of bushmeat. Importance of wildlife in diet, food avoidances and perception of health among the Baka (Cameroon). *Revue d'ethnoécologie* [En ligne]. https://doi.org/10.4000/ethnoecologie.3976.

Dunham, S. (1985). The monkey in the middle. *Zeitschrift für Assyriologie*, 75(II), 234–264.

El-Zein, A. (2015). Mythological Tuareg Gods in Ibrahim alKoni's work. *Alif: Comparative Poetic, New Paradigms in the Study of Middle Eastern Literatures*, 200–216.

Emery, W.B. (1965). Preliminary report on the excavations at North Saqqara, 1964–5. *JEA*, 51, 3–8.

Eyong, C.T. (2007). Indigenous knowledge and sustainable development in Africa: Case study on Central Africa. *Journal of Sustainable Development in Africa*, 1, 121–139.

Fa,J.E.(1999).Macacasylvanus(L.,1758).InA.J.Mitchell-Jones,G. Amori, W. Bogdanowicz, B. Kryštufek, P.J.H. Reijnders, F. Spitzenberger, M. Stubbe, J.B.M. Thissen, V. Vohralik & J. Zima, (eds.): *The Atlas of the European mammals*. London: Academic Press, pp. 158–159.

Faulkner, R.O., Trans. (1998) [1969]. *The Ancient Egyptian Pyramid Texts*. Oxford: Oxford UP.

Ferhat, I. (1981). Berber protest songs. *Index on Censorship*, 10, 32–35. doi: 10.1080/03064228108533182.

Findlay, L.J. & Hill, R.A. (2020). Baboon and vervet monkey crop-foraging behaviours on a commercial South African farm: Preliminary implications for damage mitigation. *Human-Wildlife Interactions*, 14(3), 505–518.

Fischer, H.G. (1993). Another pithemorphic vessel of the sixth dynasty. *JARCE*, 30, 1–9.

Frankfort, H. & Pendlebury, J.D.S. (1933). *The City of Akhenaten*, vol. 2. Oxford: Oxford UP.

Franklin, N.R. & Strecker, M. (2008). *Rock Art Studies – News of the World*. Vol Oxbow Books.

Friedman, F. (1995). The underground relief panels of King Djoser. *JARCE*, 32, 1–42.

Friedman, F. (ed.) (1998). *The Gifts of the Nile: Ancient Egyptian Faience*. London: Thames and Hudson.

Frobenius, L. (2013). *Storia della Civiltà Africana*. Milano: Adelphi.

Gadgil, M., Berkes, F., & Folke, C. (1993). Indigenous knowledge for biodiversity conservation. *Ambio*, 151–156.

Gall, S. (2003). The Bushmen of the Kalahari. *Ecologist*, 33(7), 28–31.

Garcea, E.A. (2004). An alternative way towards food production: The perspective from the Libyan Sahara. *Journal of World Prehistory*, 18, 107–154.

Garcia, J. (1979). *The Famous Rock Apes of Gibraltar*. Gibraltar: Midsun Publishing LDT.

Gatti, A. (1932, May). Gorilla! By Commander Attilio Gatti. *Boy's Life*.

Gautier, A. (1987). The archaeozoological sequence of the Acacus. In B.E. Barich (ed.): *Archaeology and Environment in the Libyan Sahara. The Excavations in the Tadrart Acacus, 1978–1983*. British Archaeological Reports, International Series No. 368, pp. 283–312.

Gautier, E.F. (1935). The ancestors of the Tuaregs. *American Geographical Society*, 25(1), 12–20. https://www.jstor.org/stable/209209

Gautier, A. & van Neer, W. (1977). Prehistoric fauna from Ti-n-Torha (Tadrart Acacus, Libya). *Origini*, 82, 87–113.

Gervais, P. (1848). Sur les animaux vertébrés de l'Algérie envisagés sous le double rapport de la géographie zoologique et de la domestication *Annales des Sciences Naturelles Zoologie 3e sér* (10), 1–7.

Gervais, P. (1854). Sur les animaux vertébrés de l'Algérie, Ann. Se. Nat., 3, X. *Histoire Naturelle des Mammifères, Paris, Curmer*, 8, 98–99.

Geslin, P. (2002). Les Soussou et leur nature : La vie sociale des mangroves. *Histoire et Anthropologie*, 25, 115–130.

Giles-Vernick, T. (2002). *Cutting the Vines of the Past: Environmental History of the Central Africa Rain Forest*. Charlottersville and London: University Press, Virginia.

Giles-Vernick, T. & Rupp, S. (2006). Visions of apes, reflections on changes: Telling tales of great apes in Equatorial Africa. *African Studies Review*, 49(1), 51–73.

Gilliies, J. (1820). *The History of Ancient Greece...From the Earliest Accounts*, vol II. London: Called and Daviies.

Goudsmit, J. & Brandon-Jones, D. (2000). Evidence from the Baboon Catacomb in North Saqqara for a West Mediterranean Monkey Trade Route to Ptolemaic Alexandria. *JEA*, 86, 111–119.

Groves, C. P. (2001). *Primate Taxonomy*. Washington: Smithsonian Institution Press.

Groves C.P., (2005). Order Primates. In D.E. Wilson & D.A.M Reeder. (eds): *Mammal species of the world. A taxonomic and geographic reference*. Third edition, Vol. 1. The John Hopkins University Press, Baltimore, pp. 111–184.

Groves, C.P. (2008). *Extended Family: Long Lost Cousins. A Personal Look at the History of Primatology*. Arlington: Conservation International.

Hair, P.E.H. (translated ed.) (1990). *Manuel Álvares Etiópia Menor e descrição geográfica da província da Serra Leoa* (c. 1615). Liverpool: Department of History, University of Liverpool.

Hall, D.N., Williams, M.A.J., Desmond Clark, J., Warren, A., Bradley, P. & Beighton, P. (1971). The British expedition to the Aïr mountains. *Geographical Journal*, 137(4), 30–36.

Haltenorth, T. & Diller, H. (1977). *Säugetiere Afrikas und Madagaskars*. München: BLV Verlagsgesellschaft mb H,

Harry, R.T. (1929). Animals in African stories. *Journal of the Royal African Society*, 28(111), 323–324. Oxford University Press on behalf of the Royal African Society. https://www.jstor.org/stable/716888 (accessed on 17 April 2018).

Herodotus (1920). *The Histories. Book IV.* A.D. Godley (ed., Transl.). Cambridge. Harvard University Press. https://www.perseus.tufts.edu/hopper/text?doc=Hdt.%20 4.1&lang=original (accessed on 1 September 2018).

Hockings, K. J., Yamakoshi, G., Kabasawa, A., & Matsuzawa, T. (2010). Attacks on local persons by chimpanzees in Bossou, Republic of Guinea: long-term perspectives. *American Journal of Primatology*, 72(10), 887–896.

Houlihan, P. (1996). *The Animal World of the Pharaohs*. Cairo: American University.

Houlihan, P. (1997). Harvesters or monkey business? *Göttinger Miszelllen*, 157, 31–47.

Houtteman, Y. (2011). *Living in the navel of Waag. Ritual traditions among the Daasanech of South West Ethiopia.* Phd These, Universiteit Gent, Faculteit Letteren & Wijsbegeerte.

Huard, P. & Allard, L. (1977). Les Chasseurs anciens du Nil et leurs témoignages graves à Uweinat, au Sahara tchadien et au Fezzan oriental. *Bulletin de la Société préhistorique française*, 74, 642–660.

Hufnagl, F. (1972). *Libyan mammals*. Harrow (England): The Oleander Press,

Ikram, S. & Dodson, A. (1998). *The Mummy in Ancient Egypt: Equipping the Dead for Eternity*. London: Thames and Hudson.

Jane Goodall Institute, UK. (2018). State of Wild Chimpanzee. https://www.janegoodall.org.uk (accessed on 16 April 2018).

Jannsen, J. &Jannsen, R. (1989). *Egyptian Household Animals*. Aylesbury, UK: Shire.

Johnson Museum of Art. (2021). https://museum.cornell.edu/collections/african/sub-saharan-west-africa/baboon-gbekre (accessed on 19 June 2021).

Joleaud, L. (1931). Le rôle des singes dans les traditions populaires nord-africaines. *Journal de la Société des Africanistes*, 1(1), 117–150. doi: 10.3406/jafr.1931.1507

Kabylie Djurdjura. (2009). Available online: https://djurdjura.over-blog.net/article-28118276.html (accessed on 25 July 2020).

Karetsou, A. et al. (2000). *Kreti-Aigyptos: Politismoi Desmoi Trion Khilietion*. Heraklion: Heraklion Museum.

Keimer, L. (1939). Pavian and Dum Palme. Mitteilungen des Deutschen archäologischen Institutes, *Abtailung Kairo*, 8, 42–45.

Kessler, D. & el-H. Nur el-Din, A. (2002). Inside the Ibis Galleries of Tuna el-Gebel. *EA*, 20, 36.

Klopper, D. (2010). Boer, Bushman, and Baboon: Human and animal in nineteenth-century and early twentieth-century South African writings. *Safundi: The Journal of South African and American Studies*, *11*(1–2), 3–18.

Köhler, A. (2005). Of apes and men: Baka and Bantu attitudes to wildlife and the making of eco-goodies and baddies. *Conservation and Society*, 407–435.

Kortlandt, A. (1986). The use of stone tools by wild-living chimpanzees and earliest hominids. *Journal of Human Evolution*, 15, 77–132.

Kunz, J. (1988). Une site à peinture pariétales de l'Oued Tirenhaut-Tassili n'Ajjer. *Sahara*, 1, 95–96.

Lachaud, R. (1995). *L'Invisible Présence: Les dieux de l'Egypte pharaonique*. Paris: Champollian.

Lamarque, F., Anderson, F., Fergusson, R. Lagrange, M., Osei-Owusu, Y. & Bakker, L. (2009). *Human-Wildlife Conflict in Africa: Causes, Consequences and Management strategies*. Rome: Food and Agriculture Organization of the United Nations. Available online: https://www.fao.org/3/i1048e/i1048e00.htm (accessed on 30 July 2020).

Langdon, S. (1990). From monkey to man: The evolution of a geometric sculptural type. *AJA*, 94, 407–424.

Lasserre, F. (1959). Bonnard, A. Trans. *Archiloque: Fragments*. Paris: Société d'édition.

Le Berre, M. (1990). *Terres africaines. Faune du Sahara. 2. Mammifères*. Paris: Lechevalier - R. Chabaud,

Lee, P.C. & Priston, N.E.C. (2005). Human attributes to primates: Perception of pests, conflict and consequences for primate conservation. In J.D. Paterson & J. Wallis (eds.): *Commensalism and Conflict: The Human-Primate Interface*. Norman, OK: American Society of Primatology, pp. 1–23.

Lewis, J. (2000). *The Batwa Pygmies of the Great Lakes Region*, vol. 209. London: Minority Rights Group International.

Lewis, J. (2002). *Forest hunter-gatherers and their world: A study of the Mbendjele Yaka Pygmies of Congo-Brazzaville and their secular and religious activities and representations*. Unpublished Ph.D. dissertation, Department of Social Anthropology, London School of Economics and Political Science.

Lewis-Williams, J.D. & Dowson, T.A. (1989). *Images of Power: Understanding Bushman Rock Art*. Southern Book Publishers.

Linseele, V. (2003). Weird animals from the elite cemetery. *Archaeology's Interactive Dig*. 6 Oct.003 https://www.archaeology.org/interactive/hierakonpolis/animals.html.

Liz Ferreira, A.de & Magalhaes, H.de (1945). *Gorilas do Maiombe Portugues*. Lisboa: Imprensa Nacional.

Lucianus, S. (1710–1711). *The works of Lucian, traslated from Greek by Several Eminent Hands,* vol II. London: Ed. Samuel Briscoe and sold by J. Woodward, and J. Morphew.

Lurker, M. (1980). *The Gods and Symbols of Ancient Egypt*. London: Thames and Hudson.

Mages, L. (1997), *African Religion: The Moral Traditions of Abundant Life*. New York: Orbis Booi, Maryknoll.

Maget, L. (2018). Conte Pygmée Baka (*Likano*) (Recorded in the 2018). *Le chimpanzé et la fille de Kombe*. https://youtu.be/TJhTmX9x2CM?si=RvVwrSB65K6AjHa_ (accessed 11 April 2021).

Mannert, K. (1842). *Géographie ancienne des états Barbaresques d'après l'allemand de Mannert. France. Ministére de la guerre*. Paris: Librairies encyclopedique de Roret.

Marinatos, N. (1984). *Art and Religion in Thera*. Athens.

Marinatos, N. (1987). An offering of saffron to the Minoan Goddess of nature. *Boreas*, 15, 123–132.

Markoe, G. (1985). *Phoenician Bronze and Silver Bowls from Cyprus and the Mediterranean*, Classical Studies, vol. 26. Berkley: University of California Press.

Masseti, M. (1980). Le scimmie azzurre. La fauna etiopica degli affreschi minoici di Santorino (Thera). *Mondo Archeologico*, 51, 32–37.

Masseti, M. (2010). Holocene mammals of Libya: A biogeographical, historical and archaeozoological approach. *Journal of Arid Environments*, 74, 794–805

Masseti, M. & Bruner, E. (2009). The primates of the western Palaearctic: A biogeographical, historical, and archaeozoological review. *Jass*, 87, 33–91.

Mbiti, J.S.M. (1990). African Religion & Philosophy, 2nd ed. Johannesburg: Heinemann.

Mbiti, J.S.M. (2015). *Introduction to African Religions: Second Edition*. Waveland Press.

McDermot, W.C. (1938). *The Ape in Antiquity*. Baltimore: Johns Hopkins Press.

McDermott, W.F. (1938). *The Ape in Antiquity*. The Johns Hopkins Press.

Mittermeier, R.A., Rylands, A.B. & Wilson, D.E. (2013). *Handbook of the Mammals of the World: Volume 3, Primates*. Barcelona: Lynx Edicions.

Mori, F. (1965). *Tadrart Acacus. Arte Rupestre e Culture del Sahara preistorico*. Torino: Einaudi.

Mormile, J.E. & Hill, C.M. (2017). Living with urban baboons: Exploring attitudes and their implications for local baboon conservation and management in Knysna, South Africa. *Human Dimensions of Wildlife*, 22(2), 99–109.

Morris, D. (2013). *Monkeys*. London: Reaktion Book.

Muzzolini, A. (1989). New finds of late Round Head paintings in northern Tassili and the break of the "Postneolithic Arid Phase". Nyame Akuma. *Bulletin of the Society of Africanist Archaeologists*, 31, 2–8.

Mwangi, D. (2019). *Multi-species entanglement: Human-baboon interactions in Mthongoni, eastern Kenya*. Durham theses, Durham University. https://etheses.dur.ac.uk/13458/1/mwangi000576384.pdf

Napier, J.R. & Napier, P.H. (1967). *A Handbook of Living Primates*. London: Academic Press.

Narat, V., Pennec, F., Simmen, B., Ngawolo, J.C.B. & Krief, S. (2015). Bonobo habituation in a forest-savanna mosaic habitat: Influence of ape species, habitat type, and sociocultural context. *Primates*, 56, 339–349.

Ndlovu, N. (2022). 13 primates in South African rock art. *World Archaeoprimatology: Interconnections of Humans and Nonhuman Primates in the Past*, 90, 354.

Nyanganji, G., Fowler, A., McNamara, A. & Sommer, V. (2011). Monkeys and apes as animals and humans: Ethno-primatology in Nigeria's Taraba region. *Primates of Gashaka: Socioecology and Conservation in Nigeria's Biodiversity Hotspot*, 101–134.

Oates, J.F. (1999). *Myth and Reality in the Rain Forest: How Conservation Strategies Are Failing in West Africa*. Berkeley: University of California Press.

Ogar, E., Pecl, G., & Mustonen, T. (2020). Science must embrace traditional and indigenous knowledge to solve our biodiversity crisis. *One Earth*, *3*(2), 162–165.

Oishi, T. (2013). Human-gorilla and gorilla-human: Dynamics of human-animal boundaries and interethnic relationships in the central African rainforest. *Revue de primatologie* [En ligne], 5. https://doi.org/10.4000/primatologie.1881

Olivero, J. (2016). Distribution and numbers of pygmies in Central African Forests. *PLoS One*, 11(1), e0144499. doi: 10.1371/journal.pone.0144499

Olupona, J.K. (1993). Some notes on animal symbolism in African religion and culture. *Anthropology and Humanism*, 18, 3–12.

Openda-Omar, N.O. (1993). *A history of dress in the Kakamega district of western Kenya: Impact of socio-cultural factors*. era.library.ualberta.ca.

Osborn, D.J. & Osbornová, J. (1998). *The Mammals of Ancient Egypt*. Warminster (England): Aris & Phillips Ltd.

Petragnani, E. (1928). *Il Sahara tripolitano*. Collezione di opere e di monografie a cura del Ministero delle Colonie, 3. Roma: Sindacato Italiano Arti Grafiche.

Pinch, G. (1994). *Magic in Ancient Egypt*. London: British Museum Press.

Rasmussen, S.J. (1991). Lack of prayer: Ritual restrictions, social experience, and the anthropology of menstruation among the Tuareg. *American Ethnologist*, 18(4), 751–769.

Ryan, E.F.E. (1999). *Something about Gibraltar*. Gibraltar: Minister of Heritage.

Sá, R.M. et al. (2013). Gastrointestinal symbionts of chimpanzees in Cantanhez National Park, Guinea-Bissau with respect to habitat fragmentation. *American Journal of Primatology*, 75(10), 1032–1041.

Sansoni, U. (1994). *Le più antiche pitture del Sahara*. Milano: Jaca Book

Sax, B. (2001). *The Mythical Zoo, an Encyclopedia of Animals in world Myth, Legend & Literature*. Santa Barbara, CA: ABC-CLIO.

Sept, J.M. & Brooks, G.E. (1994). Reports of chimpanzee natural history, including tool use, in 16th-and 17th-century Sierra Leone. *International Journal of Primatology*, 15, 867–878.

Sifuna, N. (2012). The future of traditional customary uses of wildlife in modern Africa: A case study of Kenya and Botswana. *Advances in Anthropology*, 2(01), 31.

Smith, H.S. (1974). A *Visit to Ancient Egypt: Life at Memphis and Saqqara (c. 500–30BC)*. Warminster: Aris and Philips.

Soukopova, J. (2012). *Round Heads: The Earliest Rock Paintings in the Sahara*. Cambridge: Cambridge Scholar Publishing.

Spencer, A.J. (1993). *The Rise of Civilisation in the Nile Valley*. London: British Museum Press.

Swindler, D. (2005). *Primate Dentition: An Introduction to the Teeth of Non-human Primates*. (Cambridge Studies in Biological and Evolutionary Anthropology). Cambridge: Cambridge University Press.

Takeuchi, K. (2005). The ambivalent symbiosis between the Aka Hunter-gatherers and neighboring farmers. In K. Takeuchi (ed.): *Culture Conservation and Development in African Rain Forest (Research Report of MEXT Grant-in-Aid for Scientic Research)*, pp. 11–28.

Takeuchi, K. (2017). Inter-ethnic relations between Pygmies and farmers. In B.S. Hewlett (ed.) *Hunter-Gatherers of the Congo Basin: Culture, History and Biology of African Pygmies*. New Brunswick, New Jersey and London: Transaction Publishers, pp. 299–320.

Tate, H.R. (1929). Animals in African stories. *Journal of the Royal African Society*, 28(111), 323–324.

Terrace, E. (1966). Blue marble plastic vessels. *JARCE*, 5, 57–63.

Thomas, D. & Alanamu, T. (eds.). (2018). *African Religions: Beliefs and Practices through History*. abc-clio.

Thomas, O. (1925). On the mammals collected during the Whitaker expedition to Tripoli. *Proceedings of the Zoological Society of London*, 2, 2–13.

Vandier d'Abbadie, J. (1964). Les Singes Familiers dans l'Anicenne Egypte I l'Ancienne Empire. *Revue d'Egyptologie*, 16, 147–177.

Vandier d'Abbadie, J. (1965). Les Singes Familers II: le Moyen Empire. *Revue d'Egyptologie*, 17, 177–188.

Velde, G. (1988). Some remarks on the mysterious language of baboons. In *Funerary Symbols and Religion* (J.H. Kamstra et al. eds.) Kampen: Kok.

Veracini, C. (2020). A landscape without nonhuman primates? The case of the Barbary macaque, *Macaca sylvanus*, (Linnaeus, 1758) and its interaction with humans throughout recorded time. *Humanities*, 9(3), 92.

Veracini, C. & Casanova, C. (2016). How non-human primates and other terrestrial West African mammals were described in the Portuguese literary sources of the 15th and 16th centuries. *Annali Università degli Studi di Ferrara*, 12(1), 243–250.

Vogel, S. (1973). People of wood, baule figure sculpture. *The Art Journal*, 33, 23–26.

Vogel, S.M. (1997a). Baule: African art western eyes. *African Arts*, 30(4), 64–95. doi: 10.2307/3337555

Vogel, S.M. (1997b). *Baule: African Art Western Eyes*. New Haven and London: Yale University Press.

Wallis, J., Benrabah, M.E., Pilot, M., Majolo, B. & Waters, S. (2020). *Macaca sylvanus. The IUCN Red List of Threatened Species 2020*: e.T12561A50043570. https://dx.doi.org/10.2305/IUCN.UK.2020-2.RLTS.T12561A50043570.en (Accessed 15 May 2023).

Weingarten, J. (1991). *The Transformation of the Egyptian Taweret into the Minoan Genius: A Study in Cultural Transmission in the Middle Bronze Age.* SIMA 87, Partille: Paul Astrom.

Wheeler, J.T. (1854). *Geography of Herotodus.* London.

Wilkenson, T. (1999). *Early Dynastic Egypt.* London: Routledge.

Wilkinson, R. (1999). *Symbol and Magic in Egyptian Art.* London: Thames and Hudson.

Wilkinson, R. (2000). *Reading Egyptian Art: A Hieroglyphic Guide to Ancient Egyptian Painting and Sculpture.* London: Thames and Hudson.

Willis, R. (1994). *Signifying Animals: Human Meaning in the Natural World.* London and New York: Taylor & Francis.

Willis, R. (2005). *Signifying Animals.* Taylor & Francis e-library.

Zeuner, F.E. (1952). Monkeys in Europe. Past and present. *Oryx*, 1, 265–273.

3 Madagascar

Alessio Anania
Unione Bolognese Naturalisti

Giuseppe Donati
Oxford Brookes University

3.1 INTRODUCTION

Lemurs, a monophyletic group of strepsirrhine primates endemic to Madagascar, are considered the flagship group for conservation efforts on the island. The mode and timing of their origins and colonization are still to be clarified. According to the most recent hypothesis, the ancestors of modern lemurs reached Madagascar from Africa by two independent sea crossing events between the Oligocene and early Miocene (Gunnell et al. 2018), while a long-established theory considered a single dispersal in the Eocene (Yoder et al. 1996; Roos et al. 2004). Researchers proposed that the founder populations crossed from Africa by surviving on large natural rafts formed from floating vegetation that washed out to the Mozambique Channel (Ali & Huber 2010).

In contrast, humans reached the so-called "red island" relatively late in the Early Holocene. Although the timing of human colonization is debated and estimates span c.9,000 years (Dewar et al. 2013; Mitchell 2019; Douglass et al. 2019), a recent study (Hansford et al. 2018) suggests a human presence in Madagascar earlier than 10,500 cal years B.P. The first permanent settlers were likely Austronesian language-speaking people coming from Southeast Asia (Adelaar 1995; Cox et al. 2012; Crowther et al. 2016). Overall, Malagasy traditional culture shows a combination of traits typical of Austronesian, Eastern African Bantu, and Islamic cultures (Hurles et al. 2005; Blench 2008; Adelaar 2009). Modern Malagasy people are conventionally grouped into 18 major ethnic groups (Kottak 1971), embracing a diverse system of beliefs, norms, and practices. One of the common elements is a taboo system. Taboos (*fady* or *faly*) are mostly passed on orally from grandparents to grandchildren, in most cases through a story (Ruud 1960), although they are dynamic and may change over time. The tight connection with ancestors through taboos influences people's attitude toward the environment. For example, several animal species cannot be captured, killed, or consumed as this would breach a taboo and trigger dire consequences (Ruud 1960; Cinner 2007). In the case of lemurs, taboos are species-specific and not related to the life history of the species: if a given species is taboo, there is no right time and no reason to hunt it (**Fig. 3.1**). However, in rural Madagascar, most lemur species are not taboo and constitute a source of food (Jenkins et al. 2011; Borgerson et al. 2016). Additionally, a few species in the north and the east are used as part of the traditional medicine practice (*fanafody*) for various health problems (Mittermeier et al. 2010; Golden et al. 2012; Anania et al. in prep.). Lemur parts can also be used to create amulets (*ody*), including a variety that protects against fatigue during long walks (Decary 1950). Attitudes toward lemurs vary according to

FIGURE 3.1 The eastern woolly lemur (*Avahi laniger*), a nocturnal species, is considered an ancestor in animal form, so it is taboo in some areas of central-eastern Madagascar. Maromizaha Forest, Madagascar. (Photo by Alessio Anania.)

DOI: 10.1201/b21963-4

the ethnic group, the species, and the geography. The ecology of the region of the island, and the morphology and behavioral ecology of the particular species, likely played a major role in inspiring legends, beliefs, taboos, and practices that partially survive today.

3.2 STORIES OF FANTASTIC OR EXTINCT PRIMATES

The earliest written source about the lemurs of Madagascar dates back to 1625, when the English cleric, Samuel Purchas, published a monumental collection of accounts of his travels, the "Hakluytus Posthumus" or "Purchas his Pilgrimes" (Mittermeier et al. 2010). In one of those accounts (Purchas 1625), the merchant William Finch provided an accurate description of the ring-tailed lemur (*Lemur catta*; **Fig. 3.2**). Since then, observations of the then-unknown primates of the island have been brought back by European traders, explorers, missionaries, colonial administrators, professional collectors, and naturalists (Buettner-Janusch et al. 1975; Mittermeier et al. 2010). Some of these travel chronicles mention animals that no longer exist. In "Histoire de la Grande Isle Madagascar" (1658), Étienne de Flacourt, a French naturalist, explorer, and colonial governor of Madagascar, described a mysterious large-sized animal called "tretretretre" or "tratratratra":

> an animal as large as a two-year-old calf that has a round head and a human face, the front feet are monkeylike, and the rear ones as well. It has frizzy hair, a short tail, and human-like ears. [...] It is a very solitary animal, the local people fear it greatly and run away at its sight as it does from them. (translated from Flacourt 1658: 154)

It is unclear whether Flacourt ever observed the animal himself, but he could have been the first and only European to see, or to have met an informant who saw, a "giant" lemur. Madagascar was apparently home to at least 17 species of giant "subfossil" lemurs (Godfrey et al. 2010), the largest of which, *Archaeoindris,* weighed about 160–200 kg (Jungers et al. 2008; Godfrey 2017). Some of these giant species lasted until the early modern period (Crowley 2010), with *Palaeopropithecus ingens* possibly still existing in the 17th century (Godfrey et al. 2010). Their demise is assumed to be the consequence of a combination of drought and human-induced factors such as landscape modification, burning, invasive species, and, above all, overhunting (Godfrey et al. 2010; Godfrey 2017). Some scientists have interpreted Flacourt's description as an account of the koala lemur *Megaladapis* (Tattersall 1982), whereas other scholars think its description better matches the sloth lemur *Palaeopropithecus* (Godfrey & Jungers 2003; Simons 2010).

FIGURE 3.2 An illustration of the ring-tailed lemur drawn and engraved by Jean Baptiste Audebert (1799).

In Madagascar, bizarre and frightening creatures populate the collective folk imaginary (Ferrand 1893; Decary 1950; Godfrey 1986), and in some of them, there are possible traces of giant lemurs. Many years after Flacourt's report, at the end of the 19th century, the French orientalist Gabriel Ferrand recorded a Malagasy tale of a man-eating ogre with a flat human-like face and an animal-like body. This ogre, named Trimobe, which was unable to move on flat surfaces, was killed by an astute boy (Ferrand 1893). Is this a report of *Palaeopropithecus*? Raymond Decary, a French colonial administrator and scholar of Malagasy culture, reported the legend of the "tokandia", a quadrupedal animal, able to climb or leap, which lived in the forest and had a human-like call. It never showed itself, but it was feared by locals for its ferocity, and it is mentioned in a Malagasy proverb: "The *tsongomby* [TN: legendary animal] goes straight on its way, but the tokandia moves in great leaps" (translated from Decary 1950: 205). According to Decary, this legend could refer to *Megaladapis*, or to *Hadropithecus* (Decary 1950). In 1995, some interviewees from southwestern Madagascar reported eyewitness accounts of an animal with some parallels to the description of those unidentified animals from folklore. The "kidoky" or "kindoky" was described as a terrestrial lemur the size of a seven-year-old girl, with a human-like round face. Its dark coat was similar in appearance to a sifaka (*Propithecus* spp.), but its call resembled the short call of the indri (*Indri indri*).

The animal was solitary and had a baboon-like gait. The kidoky could be a misidentified sifaka, or a different extant lemur, but if it was an extinct lemur that persisted in oral traditions, the most plausible candidates would be *Archaeolemur* and *Hadropithecus* (Burney & Rimilisonina 1998). More recently, in 2011 and 2012, some villagers from northeastern Madagascar reported the presence of a "kisoala" (literally "knife of the forest"), a very large quadrupedal lemur with a blackish coat and a powerful call as loud as a thunder. Based on the similarities with variegated lemurs (*Varecia* spp.), these accounts could possibly refer to *Pachylemur, Varecia*'s closest relative extinct species (Vasey & Godfrey 2022). However, whether or not the memory of the giant lemurs is alive in people's narratives, it is apparently fixed in toponyms: in northwestern Madagascar, a locality is named after a legendary, half-human half-animal beast, the "*lavaloalika*" (literally "man with high knees") (Harpet 2011).

We will never know whether all these accounts and legends refer to extinct lemurs or not. However, it is clear that the first settlers of Madagascar had the chance to observe some, if not most, of the remarkable megafauna of the island (Godfrey et al. 2010). As some giant lemurs survived into modern times (Godfrey & Jungers 2003; Crowley 2010), and some species were probably hunted by humans (Perez et al. 2005), it is plausible that these large primates left traces in the folklore. Legends can last long after the animals on which they are based became extinct, and Madagascar is an example where folk knowledge can provide useful information about the natural history of a country.

3.3 FOLK STORIES

Malagasy traditional literature is purely oral, and one of the most common narrative forms is the *tantara*. The tantara is a story connected with events of the past that a community holds true, and although it may not necessarily be trustworthy, it conveys an important message (Jaovelo-Dzao 1996; Harpet 2011). In some cases, the story is about the origin of the ethnic group or even mankind. The tantara can also illuminate the bond between humans and animals, and norms that result from that bond. Most animal stories have a common form and recurrent themes, with some variations due to regional adaptations (Decary 1950). In the case of lemurs, the tantara legitimates a taboo that forbids killing and consuming certain species. The story-related taboos are thought to have prevented the culling of some lemur species, and by doing so they may have contributed to their preservation. Taboos based on a folk story have stronger adherence than those that lack such a link (Golden & Comaroff 2015a). Those folk stories are based on the human-lemur bond, or on the animal's singular nature. The human-lemur bond disclosed in the tantara can be a genealogical kinship. In those cases, the stories represent origin myths in which lemurs are identified as either the descendants of human ancestors that in a distant past metamorphosed and populated the forests (Abinal & La Vaissière 1885; Szumski 1968; Harpet 2005; Anania et al. 2018), or as the ancestors of humans from which people directly descend after God's intervention (Catat 1895). In one story, a man went with his wife and their child into the forest and when they were tired and hungry, they cooked some rice. Once they had finished eating, the child relieved himself. Instead of cleaning him with water or leaves, the mother used what was left of the rice. This action caused the disapproval of God who cursed the whole family and turned them into lemurs (Hildebrandt 1880; van Gennep 1904; Decary 1950; Harpet 2005). Another legend (Loudon et al. 2006) tells of a man who brought a tortoise for dinner to his first wife, making his second wife envious. The latter hit the first wife with a wooden ladle, turning her into a ring-tailed lemur. The metamorphosed wife then fought with the second wife and, beating her with the ladle, turned her into a Verreaux's sifaka (*Propithecus verreauxi*; **Fig. 3.3**).

In other stories, the human-lemur bond is based on gratitude: lemurs are portrayed as benefactors that cured or saved an ancestor from danger, or as the source of medical-botanical knowledge (Catat 1895; Varigny 1904; Decary 1950; Ruud 1960; Harpet et al. 2008; Anania et al. 2021). All descendants of those ancestors are then obliged to reciprocate by respecting lemurs. According to another story (Ruud 1960), a man who went deep into the forest to look for rubber found a giant tree and climbed up to the top of a liana that was hanging in the air and fixed to the ground.

FIGURE 3.3 An illustration of the Verreaux's sifaka by John Gerrad Keulemans in Rothschild (1894).

By cutting the liana, he collected a great amount of rubber, but then he realized that it was impossible to climb down the slippery stem and he started to shout for help. He was stuck on the tree until sunset, when a large indri lemur (**Fig. 3.4**) appeared, took the man by the hand, helped him down safely, and then disappeared. Astonished, the man ran home, told his people what happened, and they decided that from that time on no harm would be done to that type of lemur. In some variants of the story, the person is the ancestor of the Betsimisaraka people, Koto, who was looking for honey in the forest (Catat 1895; Varigny 1904; Decary 1950). This story explains the local name of the indri, *Babakoto*, which means "Koto's father". In another story, a king's ancestor was wounded in the leg while he was working in his field. A pair of crowned sifaka (*Propithecus coronatus*, **Fig. 3.5**) healed him, passed on to him the power to heal with plants, which he then passed on to his descendants (Harpet et al. 2008).

Less common stories reinforce particular beliefs by depicting the supernatural powers of one species, the aye-aye (*Daubentonia madagascariensis;* **Fig. 3.6**). Solitary and nocturnal, the aye-aye has unique characteristics: large ears, large chisel-like incisors, claws in most

FIGURE 3.4 The indri, the largest extant lemur, is featured in several legends and commonly considered taboo in eastern Madagascar. Maromizaha Forest, Madagascar. (Photo by Alessio Anania.)

FIGURE 3.5 The Coquerel's sifaka (*Propithecus coquereli*, a) and the crowned sifaka (*Propithecus coronatus*, b) are taboo in some areas of western Madagascar. (Illustrations by Marie Firmin Bocourt and Auguste Faguet, in Grandidier A. (1875), Pl. 6 and Pl. 7.)

FIGURE 3.6 The unusual appearance of the aye-aye has inspired heterogeneous beliefs, legends, and practices in Madagascar. Ankanin'ny Nofy, Ampanakary, Madagascar. (Photo by Alessio Anania.)

digits, and a highly flexible third digit used for tapping on wood prior to extracting larvae (Sterling & McCreless 2006). Naturalists and ethnographers have been recording traditions about this animal since the late 17th century. In a story from eastern Madagascar, two brothers came across an aye-aye in the forest. They hit the aye-aye three times with arrows blown through a blowpipe, then they kicked the animal to death and carried it home, where people laughed at the body, burned off the fur, and ate the meat. One of the brothers also tried to crush the skull and laughed at the effort to break it. The next day, all the people had dysentery, and when many of them died shortly afterwards, the witch doctor said their illness was the revenge of the lemur's spirit. Since then, nobody was allowed to mock, kill, or eat an aye-aye in that area (Ruud 1960). As far as we know, taboo-associated respect for the aye-aye still exists in the east and south of the country, mostly among the elders. In these areas, not only is it a taboo to kill the aye-aye (whose spirit, in turn, would kill the culprit), but in the event this happens accidentally, or if a carcass is found by chance, it is customary to prepare the body with a garment and bury it, or place it on a raised wooden platform (Shaw 1882; Lamberton 1911; Britt 2002). In some cases, a ceremony with speech, celebration, and music comparable to a funeral rite is performed (Anania et al. in prep.).

In contrast, in several areas in the north and west of the island, aye-ayes are killed on sight because when they are seen in the forest, or if they approach a house, they are considered either a harbinger of bad luck, or a direct cause of death (Goodman & Schütz 2000; Simons & Meyers 2001). In the past, these beliefs have been taken so seriously that some villages have been burned down (Glaw et al. 2008), or abandoned and relocated, following a sighting (Petter & Peyrieras 1970; Simons & Meyers 2001). Killing an aye-aye is a common way to protect family members or other villagers from the ill-fortune that could fall upon them (Simons & Meyers 2001). Following the killing, a frequently reported practice is to hang the dead body or the tail on a pole by the side of the road (Koenig 2005; Koenig & Zavasoa 2008; Glaw et al. 2008). Beliefs and practices regarding the aye-aye are so rich and varied across Madagascar that it is impossible to list them all here (for a broad overview, see Sterling & Feistner 2000; Simons & Meyers 2001).

3.4 TABOO SPECIES AND HYPOTHESIS FOR THE ORIGIN OF TABOOS

Among lemurs, genera commonly considered taboo include *Avahi*, *Daubentonia* (according to the community's culture), *Indri, Lemur,* and *Propithecus* (Britt 2002; Jones et al. 2008; Harpet 2011). However, not all lemurs are represented in the folktale repertoire. We are aware of a few stories (and associated taboos) about nocturnal species, but most folklore about human kinship involves diurnal indriids (*Propithecus* spp. and *I. indri*). This may be because indriids are the largest living lemurs, but they also have an upright posture as well as human-like hands with opposable thumbs (Napier & Walker 1967; Boyer et al. 2013). Moreover, some indriid vocalizations resemble human sounds (Wright 1998; Maretti et al. 2010). In short, more than other lemurs, they remind humans of themselves.

This cannot be said of the aye-aye, for in addition to its unique morphology, it has a mainly dark coat, and its unusual and striking appearance probably contributes to its perception as scary, ugly, or bizarre, which prompts strong feelings of fear and/or respectful reverence (Shaw 1882; Rakotomamonjy et al. 2015; Simons & Meyers 2001). Does this perception only have to do with its odd appearance? In eastern Madagascar, the Betsimisaraka(s) believe that aye-ayes are transformed or reincarnated humans and, as proof of this belief, these lemurs are encountered in inadequately closed tombs (Decary 1950). In fact, an explanation for this may lie far from the spiritual sphere. Aye-ayes prefer to eat the seeds of *Canarium madagascariensis*, and these trees are common near cemeteries because it is taboo to log trees there, and cemeteries are frequently the only remaining forested areas close to a village (Simons & Meyers 2001). The combination of their feeding ecology and human impact on their habitat mean that the aye-aye has been linked to death and dead ancestors (Simons & Meyers 2001).

The appearance and behavior of lemurs, as well as their association with deadly events, may not be the only drivers of the origin and consolidation of taboos. In recent years, several studies have demonstrated the prevalence in lemurs of potentially zoonotic ectoparasites and intestinal helminths and protozoa (Irwin & Raharison 2009; Springer et al. 2017; Klein et al. 2019), and scientists estimate that lemurs and humans share nearly 20% of their parasites (Cooper et al. 2012), so lemur taboos may have their origin in disease prevention (Golden & Comaroff 2015b). Although several lemur species are consumed in Madagascar and parasites are common among both taboo and non-taboo species (Irwin & Raharison 2009), taboos limit exposure to a broad suite of viruses and pathogens.

3.5 CONSERVATION

The 113 lemur species, which comprise around 20% of all living primate species, are an important component of Madagascar's mammalian diversity. They are also the most endangered mammal group on Earth, with 98% of species threatened with extinction and almost a third (31%) of them now classified as "Critically Endangered" (IUCN 2024). Diurnal, large-sized, and short-lived species are at higher risk of extinction (Brandon et al. 2024). While some species breed successfully in zoos elsewhere in the world, other species such as the indri cannot survive in captivity (Mittermeier et al. 2010). The lemur decline is above all caused by forest loss, fragmentation, and habitat disturbance related to traditional slash-and-burn agriculture, logging, and mining (Schwitzer et al. 2014). In addition to illegal logging of timber and gold mining within protected areas (Innes 2010; Randriamalala & Liu 2010; Cabeza et al. 2019), hunting of lemurs has also increased since the political crisis in 2009 (Jenkins et al. 2011; Jones et al. 2019a). Hunting of lemurs, which is illegal (Rakotoarivelo et al. 2011) and likely unsustainable (Golden 2009; Brook et al. 2019), is mainly driven by economic and health factors such as poverty, poor health, and malnutrition (Borgerson et al. 2016). It is primarily a means of subsistence for family consumption (Borgerson et al. 2021b), but it can also earn cash from the illegal bushmeat trade, as well as reducing the risk of crop-raiding (Lee & Priston 2005; Jenkins et al. 2011; Schwitzer 2018). There are reports that traditional taboos, which have allegedly protected some animal species, are now suffering from a process of "erosion" (Lingard et al. 2003; Jones et al. 2008; Jenkins et al. 2011; Sodikoff 2012). The decline of the power and stability of those taboos is apparently driven by the synergetic effect of increased poverty and social changes (Jones et al. 2008; Rabearivony et al. 2008; Jenkins et al. 2011; Golden & Comaroff 2015a). Although difficult to quantify, the capture of live lemurs is a growing threat linked to tourism or the pet trade (Reuter & Schaefer 2016). Finally, introduced mammalian carnivorans such as wildcats and dogs constitute an additional direct disturbance (Brockman et al. 2008), while climate change has been identified as an indirect threat to lemur survival (Brown & Yoder 2015).

The survival of lemurs is threatened by factors connected with poverty, slow socio-economic development, and low formal education levels within Madagascar (Jenkins et al. 2011; Keane et al. 2011; Borgerson et al. 2016). Malagasy people also have to struggle with the detrimental influence of local elites, the foreign exploitation of the resources, landscape erosion and drought, political crises, corruption, and insecurity. All these factors make the red island one of the poorest countries in the world, with 44% of its population undernourished (von Grebmer et al. 2019).

In the 1970s and 1980s, scientists and policy-makers raised global awareness of the importance of protecting Madagascar's biodiversity (Corson 2017). Since then, despite additional investment, more media attention, modern conservation management, and the efforts of scientists and activists to raise environmental awareness, the possible irreparable loss of Madagascar's biodiversity is both a national and a global issue (for a history of conservation in Madagascar, see Mittermeier et al. 2010; Corson 2017). Despite the efforts of governments over the last twenty years to increase the area under protection (Gardner et al. 2018), protection is insufficient, with several sites facing severe anthropic pressure, and hectares of forest are being lost at a worrying pace (Jones et al. 2019a). Researchers and conservationists have proposed new and more durable conservation strategies (Schwitzer et al. 2014; Jones et al. 2019a), but a stronger long-term commitment from the national government is urgently needed to support the network of protected areas, to guarantee equity of resource access and opportunities to local communities, and to prevent environmental crimes. Improving law enforcement and fighting against corruption would contribute to these goals (Frank & Schäffler 2019; Jones et al. 2019b). Since local communities depend mainly on wood and charcoal for heating and cooking, a well-designed reforestation plan is crucial for stemming the demand for fuelwood. In 2015, Madagascar's government committed to restoring four million hectares of forest by 2030 as part of the African Forest Landscape Restoration Initiative AFR100 (Waeber & Wilmé 2019). President Andry Rajoelina's ambitious 2019 plan to reforest 40,000 hectares per year over five years (Waeber & Wilmé 2019) was a step in the right direction. It is essential to encourage participatory management at the local community level and to provide training in resource and conflict management. Actively promoting fish, poultry, and edible insect farming would help slow the hunting of protected species (Jenkins et al. 2011; Golden et al. 2014; Borgerson et al. 2021a). In addition to these steps, the promotion and expansion of ecotourism, for which lemurs are today the main attraction, has assumed a growing importance in recent times (Schwitzer et al. 2014). Alongside ecotourism, the long-term presence of researchers has proved an effective deterrent to illegal practices (Campera et al. 2019), allowing the transfer of knowledge and conferring economic benefits on local communities.

Lemurs, unwittingly, are an "active" and "political" part of the complicated process of establishing durable protection for Malagasy nature. In Western culture, lemurs are increasingly represented in the media and have become a symbol and flag - almost a brand - of the struggle to preserve and conserve the habitats and fauna of Madagascar. In rural Madagascar, lemurs still populate legends passed down by the elders and partially known by youngsters. People's attitude to them is also influenced by a belief and taboo system that helps protect some species. Several lemur species are consumed (Jenkins et al. 2011), or culled because they are crop pests (Rakotomamonjy et al. 2015). For some people living near protected areas, lemurs are even perceived as the symbol of the influence of foreigners on Malagasy life and customs (Keller 2008; Sodikoff 2012). Beyond their perception, the reality is that most lemur species are undergoing rapid decline, running the risk they will disappear forever, along with the myths and practices that involve them. We must hope that in the coming years, an increasing commitment to an inclusive approach to conservation will mean we will have more than a handful of fascinating legends to leave to posterity.

REFERENCES

Primary Sources

Abinal A, La Vaissière C de (1885). *Vingt ans à Madagascar: colonisations, traditions historiques, mœurs et croyances.* Paris: V Lecoffre, pp. 236–238.

Audebert, JB (1799). Le Mococo. *Lemur catta.* In *Histoire naturelle des singes et des makis.* Paris: Desray.

Catat L (1895). *Voyage à Madagascar (1889-1890).* Paris: Administration de l'univers illustré.

Ferrand G (1893). *Contes populaires malgaches recueillis, traduits et annotés. No.* 19. Paris: Ernest Leroux.

Flacourt EB de (1658). *Histoire de la Grande Isle Madagascar.* Paris: Pierre l'Amy.

Grandidier A (1875). *Histoire physique, naturelle et politique de Madagascar.* Volume IX: Histoire naturelle des mammifères, Tome IV, Atlas I. Paris: Hachette. Imprimerie Nationale.

Hildebrandt JM (1880). *Ausflug zum Ambergebirge in Nord-Madagaskar. Zeitschrift der Gesellschaft für Erdkunde zu Berlin 15.* Berlin: Verlag von Dietrich Reimer, p. 278.

Purchas S (1625). Observations of William Finch, Merchant, taken out of his large Journall. In *Hakluytus Posthumus, or Purchas His Pilgrimes: Contayning a History of the World in Sea Voyages and Lande Travells by Englishmen and Others* (ed Purchas S), vol. 4. Glasgow: James MacLehose and Sons, 1905, p.11.

Rothschild W (1894). *Propithecus majori sp. nov. Novitates Zoologicae*, vol. 1. Tring: Zoological Museum. London and Aylesbury: Hazell, Watson, & Winey.

Shaw GA (1882). The Aye-Aye. *The Antananarivo Annual and Madagascar Magazine* 6. London Missionary Society Press.

Secondary Sources

Adelaar KA (1995). Asian roots of the Malagasy: A linguistic perspective. *Bijdragen tot de Taal-, Land-en Volkenkunde* (3de Afl), pp. 325–356.

Adelaar KA (2009). Towards an integrated theory about the Indonesian migrations to Madagascar. In *Ancient Human Migrations: A Multidisciplinary Approach* (eds Peregrine PN, Peiros I, Feldman MW). University of Utah Press, pp. 149–172.

Ali JR, Huber M (2010). Mammalian biodiversity on Madagascar controlled by ocean currents. *Nature* 463(7281), 653–656.

Anania A, Besnard F, Rasamisoa DC, Mosa JF, Andriantsitoherina E, Campera M, Phelps M, Gibson D, Eppley TM, Donati G (2024). Unshrouding perceptions, beliefs, and practices related to the aye-aye (*Daubentonia madagascariensis*) in eastern Madagascar. Manuscript in preparation.

Anania A, Randrianarijaona A, Kotoarivelo R, Roullet D (2021). Census of the red-bellied lemur (*Eulemur rubriventer*) in the Manirisoa-Samivar forest fragments east of Ranomafana National Park, Madagascar. *Lemur News* 23, 28–35.

Anania A, Salmona J, Rasolondraibe E, Jan F, Chikhi L, Fichtel C, Kappeler PM, Rasoloarison R (2018). Taboo adherence and presence of Perrier's sifaka (*Propithecus perrieri*) in Andrafiamena forest. *Madagascar Conservation & Development* 13(1), 6–14.

Blench R (2008). The Austronesians in Madagascar and their interaction with the Bantu of East African coast: Surveying the linguistic evidence for domestic and translocated animals. *Studies in Philippine Languages and Cultures* 18, 18–43.

Borgerson C, Fisher BL, Razafindrapaoly BN, Rasolofoniaina BJR, Randriamanetsy JM, Razafindrapaoly BL, … Aardema ML (2021a). A nutrient-rich traditional insect for improving food security and reducing biodiversity loss in Madagascar and sub-Saharan Africa. *Conservation Science and Practice* 3(9), e480.

Borgerson C, Johnson SE, Hall E, Brown KA, Narváez-Torres PR, Rasolofoniaina BJR., Razafindrapaoly BN, Merson SD, Thompson KET, Holmes SM, Louis EEJ, Golden CD (2021b). A national-level assessment of lemur hunting pressure in Madagascar. *International Journal of Primatology* 1–22.

Borgerson C, McKean MA, Sutherland MR, Godfrey LR (2016). Who hunts lemurs and why they hunt them. *Biological Conservation* 197, 124–130.

Boyer DM, Yapuncich GS, Chester SGB, Bloch JI, Godino M (2013). Hands of early primates. *American Journal of Physical Anthropology* 57, 33–78.

Brandon CJ, Pearse WD, Herrera JP (2024). The intrinsic and extrinsic drivers of extinction risk in lemurs (*Lemuroidea*). *Biological Conservation* 290, 110408.

Britt A (2002). The current status of lemurs in the Sahivo and Antanamalaza classified forests, and the forest of. Ambakaka, Toamasina province, Madagascar. *Lemur News* 7, 19–20.

Brockman DK, Godfrey LR, Dollar LJ, Ratsirarson J (2008). Evidence of invasive *Felis silvestris* predation on *Propithecus verreauxi* at Beza Mahafaly Special Reserve, Madagascar. *International Journal of Primatology* 29(1), 135–152.

Brook CE, Herrera JP, Borgerson C, Fuller EC, Andriamahazoarivosoa P, Rasolofoniaina BR, … Golden CD (2019). Population viability and harvest sustainability for Madagascar lemurs. *Conservation Biology* 33(1), 99–111.

Brown JL, Yoder AD (2015). Shifting ranges and conservation challenges for lemurs in the face of climate change. *Ecology and Evolution* 5(6), 1131–1142.

Buettner-Janusch J, Tattersall I, Sussman RW (1975). History of study of the Malagasy lemurs, with notes on major museum collections. In *Lemur Biology*. Boston, MA: Springer, pp. 3–11.

Burney DA, Ramilisonina (1998). The Kilopilopitsofy, Kidoky, and Bokyboky: Accounts of strange animals from Belo-sur-mer, Madagascar, and the Megafaunal "extinction window". *American Anthropologist* 100(4), 957–966.

Cabeza M, Terraube J, Burgas D, Temba EM, Rakoarijaoana M (2019). Gold is not green: Artisanal gold mining threatens Ranomafana National Park's biodiversity. *Animal Conservation* 22(5), 417–419.

Campera M, Phelps M, Besnard F, Balestri M, Eppley TM, Nijman V, Donati G (2019). Does forest management and researchers' presence reduce hunting and forest exploitation by local communities in Tsitongambarika, south-east Madagascar? *Oryx* 53(4), 677–686.

Cinner JE (2007). The role of taboos in conserving coastal resources in Madagascar. *Traditional Marine Resource Management and Knowledge Information Bulletin* 22, 15–23.

Cooper N, Griffin R, Franz M, Omotayo M, Nunn CL (2012). Phylogenetic host specificity and understanding parasite sharing in primates. *Ecology Letters* 15(12), 1370–1377.

Corson C (2017). A history of conservation politics in Madagascar. *Madagascar Conservation & Development* 12(1), 49–60.

Cox MP, Nelson MG, Tumonggor MK, Ricaut FX, Sudoyo H (2012). A small cohort of Island Southeast Asian women founded Madagascar. *Proceedings of the Royal Society B: Biological Sciences* 279(1739), 2761–2768.

Crowley BE (2010). A refined chronology of prehistoric Madagascar and the demise of the megafauna. *Quaternary Science Reviews* 29(19–20), 2591–2603.

Crowther A, Lucas L, Helm R, Horton M, Shipton C, Wright HT, … Picornell-Gelabert L (2016). Ancient crops provide first archaeological signature of the westward Austronesian expansion. *Proceedings of the National Academy of Sciences* 113(24), 6635–6640.

Decary R (1950). *La faune malgache, son rôle dans les croyances et les coutumes indigènes*. Paris: Éditions Payot.

Dewar RE, Radimilahy C, Wright HT, Jacobs Z, Kelly GO, Berna F (2013). Stone tools and foraging in northern Madagascar challenge Holocene extinction models. *Proceedings of the National Academy of Sciences of the United States of America* 110, 12583–12588.

Douglass K, Hixon S, Wright HT, Godfrey LR, Crowley BE, Manjakahery B, Rasolondrainy T, Crossland Z, Radimilahy C (2019). A critical review of radiocarbon dates clarifies the human settlement of Madagascar. *Quaternary Science Reviews* 221, 105878.

Frank AS, Schäffler L (2019). Identifying key knowledge gaps to better protect biodiversity and simultaneously secure livelihoods in a priority conservation area. *Sustainability* 11(20), 5695.

Gardner CJ, Nicoll ME, Birkinshaw C, Harris A, Lewis RE, Rakotomalala D, Ratsifandrihamanana AN (2018). The rapid expansion of Madagascar's protected area system. *Biological Conservation* 220, 29–36.

Glaw F, Vences M, Randrianiania RD (2008). Killed aye-aye (*Daubentonia madagascariensis*) exposed on the gallows in northeastern Madagascar. *Lemur News* 13, 6–7.

Godfrey LR (1986). The tale of the tsy-aomby-aomby: In which a legendary creature is revealed to be real. *The Sciences* 26(1), 48–51.

Godfrey LR (2017). Subfossil lemurs. In *The International Encyclopedia of Primatology* (ed Fuentes A). John Wiley & Sons, pp. 1–5.

Godfrey LR, Jungers WL (2003). The extinct sloth lemurs of Madagascar. *Evolutionary Anthropology* 12(6), 252–263.

Godfrey LR, Jungers WL, Burney, DA (2010). Subfossil lemurs of Madagascar. In *Cenozoic mammals of Africa* (eds Werdelin L, Sanders W). University of California Press, pp. 351–367.

Golden CD (2009). Bushmeat hunting and use in the Makira Forest, north-eastern Madagascar: A conservation and livelihoods issue. *Oryx* 43(3), 386–392.

Golden CD, Bonds MH, Brashares JS, Rasolofoniaina, BJ, Kremen C (2014). Economic valuation of subsistence harvest of wildlife in Madagascar. *Conservation Biology* 28(1), 234–243.

Golden CD, Comaroff J (2015a). Effects of social change on wildlife consumption taboos in northeastern Madagascar. *Ecology and Society* 20(2), 41.

Golden CD, Comaroff J (2015b). The human health and conservation relevance of food taboos in northeastern Madagascar. *Ecology and Society* 20(2), 42.

Golden CD, Rasolofoniaina BR, Anjaranirina EG, Nicolas L, Ravaoliny L, Kremen C (2012). Rainforest pharmacopeia in Madagascar provides high value for current local and prospective global uses. *PLoS One* 7(7), e41221.

Goodman S, Schütz H (2000). The lemurs of the northeastern slopes of the Réserve Spéciale de Manongarivo. *Lemur News* 5, 30–33.

Gunnell GF, Boyer DM, Friscia AR, Heritage S, Manthi FK, Miller ER, … Seiffert ER (2018). Fossil lemurs from Egypt and Kenya suggest an African origin for Madagascar's aye-aye. *Nature Communications* 9(1), 1–12.

Hansford J, Wright PC, Rasoamiaramanana A, Pérez VR, Godfrey LR, Errickson D, … Turvey ST (2018). Early Holocene human presence in Madagascar evidenced by exploitation of avian megafauna. *Science Advances* 4(9), eaat6925.

Harpet C (2005). Le lémurien dans les groupes linguistiques du nord-ouest de Madagascar et du sud de Mayotte: eléments pour une anthropologie de la biodiversité. PhD dissertation, Institut National des Langues et Civilisations Orientales - INALCO Paris - Langues O', Paris.

Harpet C (2011). Des lémuriens et des hommes: mythes, représentations et pratiques à Madagascar. *Revue de primatologie* 3(16).

Harpet C, Navarro L, Ramanankirahina R (2008). Rôle et Implications des croyances et des savoir-faire locaux dans les programmes de conservation: Exemple d'un site à lémuriens sacrés au cœur de la Station Forestière à Usages Multiples d'Antrema (pays Sakalava). *Revue Ecologie (Terre Vie)* 55, 289–292.

Hurles ME, Sykes BC, Jobling MA, Forster P (2005). The dual origin of the Malagasy in Island Southeast Asia and East Africa: Evidence from maternal and paternal lineages. *The American Journal of Human Genetics* 76(5), 894–901.

Innes JL (2010). Madagascar rosewood, illegal logging and the tropical timber trade. *Madagascar Conservation & Development* 5(1).

Irwin MT, Raharison JL (2009). A review of the endoparasites of the lemurs of Madagascar. *Malagasy Nature* 2, 66–93.

IUCN (2024). The IUCN Red List of threatened species. Version 2023-1. https://www.iucnredlist.org.

Jaovelo-Dzao R (1996). *Mythes, rites et transes à Madagascar. Angano, joro et tromba Sakalava*. Paris: Karthala.

Jenkins RK, Keane A, Rakotoarivelo AR, Rakotomboavonjy V, Randrianandrianina FH, Razafimanahaka HJ, Ralaiarimalala SR, Jones JPG (2011). Analysis of patterns of bushmeat consumption reveals extensive exploitation of protected species in eastern Madagascar. *PLoS One* 6(12), e27570.

Jones JPG, Andriamarovololona MM, Hockley N (2008). The importance of taboos and social norms to conservation in Madagascar. *Conservation Biology* 22(4), 976–986.

Jones JPG, Ratsimbazafy J, Ratsifandrihamanana AN, Watson JE, Andrianandrasana HT, Cabeza M, … Wright PC (2019a). Last chance for Madagascar's biodiversity. *Nature Sustainability* 2(5), 350.

Jones JPG, Ratsimbazafy J, Ratsifandrihamanana AN, Watson JE, Andrianandrasana HT, Cabeza M, … Wright PC (2019b). Madagascar: Crime threatens biodiversity. *Science* 363(6429), 825–825.

Jungers WL, Demes B, Godfrey LR (2008). How big were the "giant" extinct lemurs of Madagascar? In *Elwyn Simons: A Search for Origins*. New York, NY: Springer, pp. 343–360.

Keane A, Ramarolahy AA, Jones JP, Milner-Gulland EJ (2011). Evidence for the effects of environmental engagement and education on knowledge of wildlife laws in Madagascar. *Conservation Letters* 4(1), 55–63.

Keller E (2008). The banana plant and the moon: Conservation and the Malagasy ethos of life in Masoala, Madagascar. *American Ethnologist* 35(4), 650–664.

Klein A, Strube C, Radespiel U, Springer A, Zimmermann E (2019). Differences in infection patterns of vector-borne blood-stage parasites of sympatric Malagasy primate species (*Microcebus murinus, M. ravelobensis*). *International Journal for Parasitology: Parasites and Wildlife* 10, 59–70.

Koenig P (2005). Découverte d'une dépouille de Aye-aye (*Daubentonia madagascariensis*) dans le nord-ouest de Madagascar. *Lemur News* 10, 6–7.

Koenig P, Zavasoa A (2008). Et le massacre continue… Nouvelle découverte d'une dépouille d'Aye-aye (*Daubentonia madagascariensis*) dans le nord de Madagascar. *Lemur News* 13, 6–7.

Kottak CP (1971). Cultural adaptation, kinship, and descent in Madagascar. *Southwestern Journal of Anthropology* 27(2), 129–147.

Lamberton C (1911). Contributions a l'etude des mœurs du aye-aye. *Bulletin de l'Academie Malgache* 8, 129–140.

Lee PC, Priston NE (2005). Human attitudes to primates: Perceptions of pests, conflict and consequences for primate conservation. In *Commensalism and Conflict: The Human-Primate Interface* (vol. 4) (eds Paterson JD, Wallis J). American Society of Primatologists, pp. 1–23.

Lingard M, Raharison N, Rabakonandrianina E, Rakotoarisoa JA, Elmqvist T (2003). The role of local taboos in conservation and management of species: The radiated tortoise in southern Madagascar. *Conservation and Society* 1(2), 223–246.

Loudon JE, Sauther ML, Fish KD, Hunter-Ishikawa M, Ibrahim YJ (2006). One reserve, three primates: Applying a holistic approach to understand the interconnections among ring-tailed lemurs (*Lemur catta*), Verreaux's sifaka (*Propithecus verreauxi*), and humans (*Homo sapiens*) at Beza Mahafaly Special Reserve, Madagascar. Ecological and Environmental Anthropology (University of Georgia), 7.

Maretti G, Sorrentino V, Finomana A, Gamba M, Giacoma C (2010). Not just a pretty song: An overview of the vocal repertoire of *Indri indri*. *Journal of Anthropological Sciences* 88, 151–165.

Mitchell P (2019). Settling Madagascar: When did people first colonize the world's largest island? *The Journal of Island and Coastal Archaeology*, 1–20.

Mittermeier RA, Louis EE, Richardson M, Schwitzer C, Langrand O, Rylands AB, Hawkins F, Rajaobelina S, Ratsimbazafy J, Rasoloarison RM, Roos C (2010). *Lemurs of Madagascar*, Third edition. Tropical Field Guide Series. Arlington, VA: Conservation International.

Napier JR, Walker AC (1967). Vertical clinging and leaping: A newly recognized category of locomotor behaviour of primates. *Folia Primatologica* 6, 204–219.

Perez VR, Godfrey LR, Nowak-Kemp M, Burney DA, Ratsimbazafy J, Vasey N (2005). Evidence of early butchery of giant lemurs in Madagascar. *Journal of Human Evolution* 49(6), 722–742.

Petter JJ, Peyrieras A (1970). Nouvelle Contribution à l'étude d'un Lémurien malgache, le Aye-aye (*Daubentonia madagascariensis* E. Geoffroy). *Mammalia* 34(2), 167–193.

Rabearivony J, Fanameha E, Mampiandra J, Thorstrom R (2008). Taboos and social contracts: Tools for ecosystem management-lessons from the Manambolomaty Lakes RAMSAR site, western Madagascar. *Madagascar Conservation and Development* 3(1), 7–16.

Rakotoarivelo AR, Razafimanahaka JH, Rabesihanaka S, Jones JP, Jenkins RK (2011). Lois et règlements sur la faune sauvage à Madagascar: Progrès accomplis et besoins du futur. *Madagascar Conservation & Development* 6(1), 37–44.

Rakotomamonjy SN, Jones JPG, Razafimanahaka JH, Ramamonjisoa B, Williams SJ (2015). The effects of environmental education on children's and parents' knowledge and attitudes towards lemurs in rural Madagascar. *Animal Conservation* 18(2), 157–166.

Randriamalala H, Liu Z (2010). Rosewood of Madagascar: Between democracy and conservation. *Madagascar Conservation & Development* 5(1).

Reuter KE, Schaefer MS (2016). Captive conditions of pet lemurs in Madagascar. *Folia Primatologica* 87(1), 48–63.

Roos C, Schmitz J, Zischler H (2004). Primate jumping genes elucidate strepsirrhine phylogeny. *Proceedings of the National Academy of Sciences* 101(29), 10650–10654.

Ruud J (1960). *Taboo: A Study of Malagasy Customs and Beliefs*. Oslo: Oslo University Press; New York: Humanities Press.

Schwitzer C (2018). Editorial. *Lemur News* 21, 1.

Schwitzer C, Mittermeier RA, Johnson SE, Donati G, Irwin M, Peacock H, Ratsimbazafy J, Razafindramanana J, Louis EE, Chikhi L, Colquhoun IC, Tinsman J, Dolch R, LaFleur M, Nash S, Patel E, Randrianambinina B, Rasolofoharivelo T, Wright PC (2014). Averting lemur extinctions amid Madagascar's political crisis. *Science* 343(6173), 842–843.

Simons EL (2010). Inferences about the distant past in Madagascar. *Lemur News* 15, 25–27.

Simons EL, Meyers DM (2001). Folklore and beliefs about the aye-aye (*Daubentonia madagascariensis*). *Lemur News* 6, 11–16.

Sodikoff GM (2012). Totem and taboo reconsidered: Endangered species and moral practice in Madagascar. In *The Anthropology of Extinction: Essays on Culture and Species Death* (ed Sodikoff GM). Bloomington and Indianapolis: Indiana University Press, pp. 72–73.

Springer A, Fichtel C, Al-Ghalith GA, Koch F, Amato KR, Clayton JB,… Kappeler PM (2017). Patterns of seasonality and group membership characterize the gut microbiota in a longitudinal study of wild Verreaux's sifakas (*Propithecus verreauxi*). *Ecology and Evolution* 7(15), 5732–5745.

Sterling EJ, Feistner AC (2000). Aye-aye (*Daubentonia madagascariensis*). In *Endangered Animals: A Reference Guide to Conflicting Issues* (eds Reading RP, Miller B), pp. 45–48.

Sterling EJ, McCreless EE (2006). Adaptations in the aye-aye: A review. In *Lemurs. Developments in Primatology: Progress and Prospect* (eds Gould L, Sauther ML). Boston, MA: Springer, pp. 159–184.

Szumski L (1968). Extrême sud Malgache, légendes et récits. Fianarantsoa: Imprimerie catholique.

Tattersall I (1982). *The Primates of Madagascar*. New York: Columbia University Press.

van Gennep A (1904). *Tabou et totémisme à Madagascar: étude descriptive et théorique*. Paris: Ernest Leroux, pp. 214–224.

Varigny HC de (1904). *Histoire et mœurs des animaux. I, Mammifères: singes, prosimiens, chiroptères, carnivores*. Paris: J. Rueff.

Vasey N, Godfrey LR (2022). Lemur hunting in Madagascar's present and past. The case of *Pachylemur*. In *World Archaeoprimatology: Interconnections of Humans and Nonhuman Primates in the Past* (eds Urbani B, Youlatos D, Antczak AT). Cambridge: Cambridge University Press, pp. 393–416.

von Grebmer K, Bernstein J, Mukerji R, Patterson F, Wiemers M, Ní Chéilleachair R, Foley C, Gitter S, Ekstrom K, Fritschel H (2019). *2019 Global Hunger Index: The Challenge of Hunger and Climate Change*. Bonn: Welthungerhilfe; Dublin: Concern Worldwide.

Waeber PO, Wilmé L (2019). We have got to up our game substantially for forests, carbon, biodiversity, and ultimately people. *Madagascar Conservation & Development* 14(1), 3–5.

Wright PC (1998). Impact of predation risk on the behaviour of Propithecus diadema edwardsi in the rain forest of Madagascar. *Behaviour*, 483–512.

Yoder AD, Cartmill M, Ruvolo M, Smith K, Vilgalys R (1996). Ancient single origin for Malagasy primates. *Proceedings of the National Academy of Sciences* 93(10), 5122–5126.

4 The Americas

Cecilia Veracini
Universidade de Lisboa

Ana Lucia Camphora
Center for Contemporary Equine Studies - California, US

The indigenous woman would carry everything,
and in addition to the children,
she might also carry a monkey or a parrot
(Nieuhof, 1682:312)

4.1 INTRODUCTION: MAIN THEMES AND ISSUES CONCERNING INTERACTION BETWEEN MODERN HUMANS AND NEOTROPICAL PRIMATES

New World Primates (NWP) or Neotropical Primates, Suborder Platyrrhini, which occur in a wide range of ecosystems of the tropical regions of Central and South America, are a major mammalian radiation (Mittermeier et al., 2013: 202–549) with at least 218 recognized taxa (Mittermeier et al., 2022). Platyrrhine taxonomy at the family level is still disputed (Groves, 2001; Garber et al., 2009; Rylands & Mittermeier, 2009; Mittermeier et al., 2013; Schneider & Sampaio, 2015), but in this work we follow Mittermeier et al. (2013) and recognize five families of living NWP: Callitrichidae, Cebidae, Aotidae, Pitheciidae, and Atelidae. While all NWP are arboreal, they exhibit a wide range of ecology, morphology, and behavior (Fleagle, 2013). The time and place of platyrrhine origins are still controversial (Tarling, 1980; Fleagle & Kay, 1997). They are immigrant mammals from Africa whose fossil record comes from the Tertiary and Quaternary sediments of South America and the Caribbean Greater Antilles (Bond et al., 2015). They bear striking resemblances to some Eocene African anthropoids, and phylogenetic analyses suggest a close relationship with African taxa (Fleagle & Kay, 1997). The oldest fossil evidence, of the genus *Perupithecus,* discovered in Santa Rosa, Peru, is estimated to be of late Eocene age (Bond et al., 2015), but even more recently discovered scientific evidence extends the fossil record of non-human primates (primates from here onwords) in South America further back. Silvestro et al. (2019) suggest that platyrrhines originated 5–10 million years earlier than previously assumed, dating back to the Middle Eocene. For these authors the estimated ancestral platyrrhine was small—weighing 0.4 kg—matching the estimated size of their presumed African ancestors.

The indigenous people of the Americas are also immigrant mammals, but from North East Asia. According to recent data, waves of modern *Homo sapiens* immigration from East Asia may have occurred from 15 to 13,000 years (cf. Erlandson et al., 2007; Fraser, 2014) or even 22,500 years ago (Callaway, 2021). The Amerindians of South America were believed to be traditionally hunter-gatherers, but many groups also practiced aquaculture and agriculture on a large scale and others may have practiced mixed farming, hunting, and gathering (Mann, 2005). As shown by a diverse and copious body of archaeological, historical, primatological, anthropological, and ethnoprimatological studies, primates have (and had) a relevant cultural role for most of the ethnic groups with whom they share (or shared) territory. These evidences suggest primates have probably been used for millennia for consumption, companionship, cultural artifacts, and traditional medicine. The varying ways in which modern humans adapted to Neotropical habitats, and their various interactions with the diverse primate species, have given rise to an array of myths, folktales, legends, and cosmologic beliefs involving these animals (e.g., South, 2005, Cormier, 2002, 2006; Bruner & Cucina, 2005; Garcia, 2018; Urbani & Lizarralde (eds), 2020; among others).

The deconstruction of the myth of the Amazon as a pristine forest shows that American landscapes are "anthropogenic landscapes" whose habitats and ecosystems were greatly influenced or molded by human action in the course of the millennia (Balée, 1989; Kirch, 2005; Pereira, 2022). The primates' (and other animals') food preferences also play a role, and these interact with human occupation and cultivation to mold ecosystems through the selection of plant species (Bush et al., 2015). The influence of modern humans on primate density, species distribution, and even the extinction of some taxa in pre-Columbian times is shown by several studies. Subsistence hunting by native populations most likely drove several large-bodied taxa, such as the mega-*Brachyteles* of the Brazilian Atlantic forest, to extinction (Cartelle & Hartwig, 1996; Thoisy et al., 2009). The causes of extinction of the Jamaican primate *Xenothrix mcgregori* (c. 900 BP) —regarded variously as either pitheciids or a stem platyrrhine—remains a matter of debate, but the likeliest inference is that these losses were largely, if not completely, anthropogenically driven (Cooke et al., 2017). Some primate species, as well as other mammals (guinea pigs, armadillos, and opossums) and several species of birds and dogs, were introduced into the West Indian Islands in pre-Hispanic times (Hershkovitz, 1949; Newsom & Wing, 2004; Masseti, 2011). In the Antilles, a Cuban cave painting at the Cueva del Ciclon resembles a spider monkey

DOI: 10.1201/b21963-5

(Arredondo & Varona, 1983) — although this image was also interpreted as a dancing shaman — and as this genus is not part of the living and fossil Antillean fauna, that suggests evidence of early primate trade between Central and South America and the Caribbean islands (Bruner & Cucina, 2005). At the archaeological site of Bonaire, in the Netherlands Antilles, part of a skeleton of an exotic juvenile capuchin monkey (a pet, probably imported from mainland Venezuela, that died in captivity) was found (Newsom & Wing, 2004: 72). Urbani et al. (2022a) reviewing the zooarchaeological finding in the circum-Caribbean region (300–1,500 CE), show the presence of cranial and postcranial remains of the genera of *Alouatta, Ateles,* and *Cebus* in other insular archaeological sites (e.g., Colon island, northeast Panama, and Aruba island, Antilles).

With European expansion, which started at the end of the 15th century, Neotropical primate populations suffered increased pressure. They were one of the most harvested and exported "products" of the Neotropics, especially in Brazil (Teixeira & Papavero, 2010; Veracini, 2017, 2021). As shown in other chapters of this volume (Veracini, Chapter 9, this volume) the request for primates by the European market led to the removal of hundreds of thousands of animals from the 16th century onward with a probable—though difficult to evaluate—effect on forest ecosystems. European settlers, besides introducing a range of domesticated animals useful to their colonial life (Crosby, 2003), used native species for their survival. The latter played a fundamental role in the colonial societies that were consolidating in the Americas (Camphora, 2021). Europeans assimilated indigenous ways of life and consequently began to use primates for food, medicine, companion animals, and so on. Thus, the first four centuries of colonial history were not exclusively a human endeavor, but rather a construction that moved forward on the basis of interspecies relations. In this context, primates were described and thought of in often-contradictory ways not only as food, trade objects, material sources for different kinds of tools and medicine, but also as beings capable of expressing feelings and human-like behavior, sources of spiritual power, and as threats to both human survival and their property (e.g., crop raiding and pests) (cf. Veracini & Teixeira, 2017; Camphora, 2021). In different phases of the Modern period, African and Asian species of primates were introduced into some of the West Indian islands by Europeans: in the 17th and 18th centuries the green monkey, *Chlorocebus sabaeus,* and probably *Cercopithecus mona* reached the West Indies through the slave trade route (Denham, 1987; Masseti, 2011). In the early 80s of the 20th century monkeys from Asia were introduced into some West Indian islands (e.g., *Macaca mulatta* in Cayo Santiago) to create semi-free colonies used by biomedical institutions (Borroto-Páez & Woods, 2012). Recently, the primate trade in Brazil has led to the creation of stable populations of primates outside their original geographical areas, namely *Callithrix jacchus* (**Fig. 4.1**) and *Callithrix penicillata* or hybrids of these (Veracini & Ruiz-Miranda, 2004; Ruiz-Miranda et al., 2006) and *Saimiri sciureus* (Oliveira-Silva et al., 2018). The distribution and density of the primate taxa we observe today is the result of selection by, and co-evolution with, modern humans.

FIGURE 4.1 An adult and a juvenile of the common marmoset (*Callithrix jacchus*) feeding on human garbage (note the chicken bone in the juvenile's hands). This species, endemic to the northeastern Atlantic Forest of Brazil, has been introduced by illegal trade to other Brazilian regions where it forms stable wild populations due to its ability to feed on exudates, fruit, and animals. It thrives in secondary forests, forest edges, and even suburban areas. (Frame from a video by Cecilia Veracini.)

Nowadays, many Neotropical primate species are threatened with extinction as a result of past and recent commercial hunting, trade, and the various, complex and interrelated factors resulting from centuries of a generalized habitat transformation in most of the Central and South American biomes (Estrada et al., 2017; Mello et al., 2018; Fernández et al., 2022). During the last decades, the role played by indigenous communities in conservation policy has been highlighted, and international laws now recognize indigenous peoples' rights to own and manage protected areas (Colchester, 2004; Sobrevila, 2008; Chicchón, 2009). This is one of many government-level conservation approaches, but areas are also protected by civil society and individual actors (cf. Schleicher, 2018). Although conservation policy needs to develop from the community's value system and knowledge, the subsistence hunting of primates by native traditional communities is a controversial issue, and some authors have argued that biodiversity can decrease as a consequence of pressure from the local people (Peres, 2010; Luz et al., 2017). Moreover, primates have intrinsic moral value so should be conserved for their own sake, yet conservation policies should not cause harm to local human populations (Hill, 2002). Although some traditional societies of South and Central America developed taboos or other forms of social control against indiscriminate harvesting of primates, this situation is rapidly changing because of demographic, cultural, and social transformations as well as the influence of external pressures (e.g., gold mining, epidemic disease, missionary activity, contact with western societies, logging, and additive hunting by non-tribal immigrants, cf. Thoisy et al., 2009; Fernández et al., 2022). Depending on the taxon, the harvest rate in most of Amazonia does not have to reach great heights to drive the primate population into a precipitous decline (Peres, 2000; Thoisy et al., 2009).

This chapter presents a synopsis of the human–primate interface in the Neotropics with particular emphasis on the traditional myths, legends, and folk stories regarding the latter. When discussing the different periods, we rely on both primary and secondary sources (e.g., historical, archaeological, anthropological, biological, etc.). We survey the indigenous perception and traditional use of primates and discuss the appropriation of such traditions by Europeans and settlers of the colonial period, and then integrate this topic into a historical narrative that emphasizes the participation of primates in the construction of Portuguese and Spanish colonial life.

4.2 TRADITIONAL COMMUNITIES AND PERCEPTION OF NON-HUMAN PRIMATES: THEMES AND CONCEPTS

Traditional modern human communities of South and Central America have longstanding relationships with the natural world (e.g., Fernández-Llamazares & Virtanen, 2020; Pereira, 2022). Although traditional knowledge about plants and animals and their symbolic attributes has proved to be more resistant to change than other cultural domains such as kinship and politics (Balee, 2000), this knowledge is especially vulnerable to the continuous degradation of ecosystems and to the pression of exogene cultures. In some almost completely destroyed biomes, such as the Atlantic forest, ancient beliefs and traditional knowledge have generally been substituted or lost (Razera et al., 2006; Turvey et al., 2018; Silva, 2019). Regarding primates, in many of the Amerindian nations there are still complex, traditional, and multifaceted cultural interconnections with these animals and, especially in Amazonia, they are central to socio-cosmological and human nutritional realities (e.g., Cormier, 2002; Garcia, 2018; Urbani & Lizarralde (eds), 2020). Recent studies on human–primate interaction in the Americas draw on the so-called *ontological* paradigm, based on the idea that the way people interact with non-humans (i.e., the social positioning of beings) is established by both oral history and indigenous knowledge (e.g., Viveiros de Castro, 2004; Toledo & Barrera-Bassols, 2009). This approach leads to new perspectives in which all life forms (human and non-human) are interconnected and interdependent as the shaman Kopenawa tells us in his outstanding testimony (cf. Kopenawa & Albert, 2013; Cesarino, 2014). This vision overrides the classic structuralistic separation between the natural and the cultural world, the individual and society, and the human and the non-human, binary pairs that dominated the field until the late 20th century (cf. Descola, 2006; Viveiros de Castro, 2009). Although criticized by some disciplines (cf. Turner, 2021), the ontological approach, and in particular Viveiros de Castro's indigenous perspectivism, has opened new interpretative avenues for the ethnoprimatological analysis of human–primate interactions. A frequent theme in the cosmologies of Amerindians is to view animals as former human beings, and this can influence their perception and interaction with the latter. Amazonian groups perceive non-human beings as possessing a human-like soul, but differing in their physical forms and in their habitual modes of engagement with the environment. It needs to be stressed that similar myths and beliefs exist in most of the traditional communities (past and present) across the world. It is common in traditional peoples to believe in the origin of some animals from human beings or that animals mediate between the human and the divine realm (see the other chapters of this volume, Section I). Moreover, the indigenous peoples of America have complex views on the multitude of phenomena that Western views would reduce to the concept of "nature" (Kopenawa & Albert, 2013; Garcia, 2018). Recent approaches by anthropologists and thinkers in other disciplines seek to re-examine traditional ethical, political, and epistemological categories in the context of renewed attention to, and respect for, the lives of non-humans (e.g., Ogden et al., 2013; Waldau, 2013). This new attitude, which takes seriously the lives and points of view of beings other than modern humans and their agency in both historical and current narratives, is now building a useful bridge with, and a conceptual proximity to, indigenous cultures.

In this section, we summarize some of the most recent findings and useful concepts to understand the complex interaction of Amerindian traditional communities with non-human primates (see Table 4.1). While not suggesting this is a systematic or exhaustive review, we provide an overview of new findings and recent epistemologies.

As highlighted by Urbani and Cormier (2015), past cultural anthropologists produced a rich corpus of work concerning the relationship between indigenous communities and the natural world, but they focused mostly on the lifestyles and beliefs of particular cultures or ethnic groups, with most primates being assigned to inclusive categories (e.g., monkey). In recent decades, ethno-anthropological studies have been developed in part to correct this, as well as other biases that excluded human influence from research on primate behavior and ecology. Evidence points to long-term coexistence and ongoing interactions between modern humans and primates that go beyond simple ecological exploitation or competition (e.g., cf. Urbani et al. (eds), 2022b). The data demonstrate that the cultural value of primates varies with different species and with the myths and beliefs of different communities. In sum, the material and spiritual importance of a monkey species to a particular ethnicity must be understood in the context of ecological, cultural, and historical factors (Cormier, 2006).

In various ethnic groups of lowland Amazonia and other areas, there are no specific taxonomic categories for primates, although in some cases (e.g., among the Matsigenka of Peru, cf. Shepard, 2002) they stand apart from the other animals. In some communities, primates are separated from the rest of non-human animals and are often associated with the arboreal species of the raccoon family (Procyonidae), the kinkajous (*Potos flavus*) (**Fig. 4.2**) and olingo (*Bassaricyon* spp.) (cf. Lizarralde, 2002; Zent & Zent, 2020) or with sloths (*Bradypus* spp.)

TABLE 4.1
Communities Mentioned in the Text

People	Region, Country	Reference
Bari	Venezuela	Lizarralde (2002; 2020)
Guajá	Eastern Amazon	Cormier (2003)
Jotï (also Hodï)	Venezuela	Zent and Zent (2020)
Kaingang	São Paulo and Rio Grande do Sul states, Brazil	Rodrigues (2016)
Kalapalo	Upper Xingu River, Brazil	Basso (1973)
Kari'na	Guyana and Suriname	Rybka (2020)
Karitana	Rondonia state, Brazil	Velden (2016)
K'iche' (also Quiché)	Guatemala	Rosales-Meda and Hermes (2020)
Lacadon Maya	Mexico	García del Valle et al. (2020)
Lokono	Suriname and Guyana	Rybka (2020)
Matse (also Mayoruna)	Angamos, Palmeiras do Javari, Atalaia do Norte and Tabatinga, Brazil and Peru	Voss and Fleck (2011)
Matsigenka (also Machiguenga, Matsigenga)	Eastern slopes of the Andes and in the Amazon Basin jungle regions of southeastern, Peru	Shepard (2002)
Maya Q'eqchi'	Guatemala	Rosales-Meda and Hermes (2020)
Miraña	Peru, Colombia, and Brasil	Karadimas (2016)
Paraiban people	North East Atlantic Forest, Brazil	Castro Box details, 4.3, this study.
Popoluca	Los Tuxtlas region (ME)	Pinto-Marroquin and Serio Silva (2020)
Qom (Toba)	Argentina	Medrano & Suárez (2020)
Shawi	Peru	González-Saavedra (2020)
Tikuna	Amazonian region of Colômbia and Brazil	Parathian and Maldonado (2010); Maldonado & Waters (2020)
Tupinamba	Bahia state, Brazil	Soares de Souza (1587)
Xoklengs	Santa Catarina state, Brazil	Cruz (2014)
Wampis	Northern Peruvian Amazon, Peru	Świerk (2020); Gonzalez-Saavedra (2020)
Waimiri – Atroari	Amazonas, Pará, and Roraima states, Brazil	Souza-Mazurek and Bruno (2020)
Yanomami	Amazonas and Roraima states, Brazil	Kopenawa and Albert (2013), Boubli et al. (2020)

FIGURE 4.2 The kinkajou, *Potos flavus*, a rainforest mammal of the Family Procyonidae, is often considered a nocturnal primate in traditional cultures. (Source: Brehm, 1895, p. 271.)

(e.g., Gonzàlez-Saavedra, 2020). These all display several morphological traits common to Neotropical primates (e.g., grasping hands, flexible feet, long tails used for grasping or balance) and all are tree-dwelling. Among Wampis, a Jivaron people inhabiting the northern part of the Peruvian Amazon, monkeys are referred to as *ikeiñu*, a term that includes monkeys, squirrels (*Sciurus* spp.), porcupines (*Coendou* spp.), olingos, and the kinkajou, which can be translated as *arboreal mammals* (Świerk, 2020). Among the Bari of Venezuela, the kinkajou and olingo are considered to be related to the night monkey, although when questioned about the similarity with monkeys, the Bari recognized that these animals are not exactly "monkeys" (Lizarralde, 2002). The kinkajou is also known as a night monkey among the indigenous people of Guyana studied by Rybka (2020) and by the people of the Atlantic Forest (Guimarães, 2019, see below). This latter taxonomic arrangement was also found in the 16th-century Tupi-Guarani people of the Bahia region of Brazil (Tupinamba). As reported by Soares de Souza (1587), the indigenous peoples of that time mentioned a nocturnal monkey called *Saianhanga*, which means evil monkey [*bogio diabo* in Portuguese], "and it was said that when people heard them scream someone was going to die" (cf. Veracini & Teixeira, 2016: 24). Today the people of southern Bahia have resurrected a traditional myth about this animal, linking it to cacao plantations (see later). Among the Kalapalo, of the Upper Xingu River in Brazil, monkeys are classified in the category of "land mammals". They are associated with the coati (*Nasua nasua*), both called *kaiju*, which means "like human beings" (Basso, 1973: 17).

Although great caution is needed when making generalizations, despite the variation in cultures, subsistence,

FIGURE 4.3 Old drawing of *Ateles belzebul*, an endangered species endemic to northwestern Amazonia. (Source: Brehm, 1895, p. 54.)

FIGURE 4.4 Adult female of *Mico argentatus* (Family Callitrichidae), a species endemic to eastern Amazonia. (Photo Cecilia Veracini.)

and hunting behavior displayed by Amerindian traditional communities, their perceptions of primates have features in common, with the size and morphology of Neotropical primates being important factors influencing the way they are perceived and the cultural attributes they are granted. The larger species belonging to the family Atelidae (**Fig. 4.3**), which display a more "human-like" morphology, seem to elicit a greater emotional and psychological impact than the smaller Callitrichidae (**Fig. 4.4**) or squirrel monkeys (*Saimiri* spp.). In general, the similarity of animals to modern humans is frequently associated with the perception of a higher cognitive capacity and behaviors and emotions similar to our own (Urquiza-Haas et al., 2020). Atelids are used in socio-cultural practices such as hunting, feeding (see **Box Details 4.1**), and material culture, and they play a variety of symbolic roles in myths and cosmologies. Atelids, such as spider, woolly, and howler monkeys, are often recognized as ancestors, close relatives, or former humans who have undergone a metamorphosis (e.g., Urbani & Lizarralde (eds), 2020). Many of these myths and folk stories explain the similarity between human and other primates as being due to the transformation of humans into monkeys as a form of punishment for the foolish behavior of the first human or "human-monkey"

BOX DETAILS 4.1: A PERSPECTIVE ON PRIMATE HUNTING IN AMERINDIAN SOCIETIES

Cecilia Veracini

Literary sources suggest that subsistence hunting of animals, including primates, was widespread among Amerindians before and during the colonial period (see Section 4.3), and it still contributes to local food security and to the livelihoods of many forest dwellers. Primate hunting and its ecological implications (overhunting leading to population decline, alteration of forest composition, structure, and dynamics) have been addressed by many authors (Peres, 2000; Urbani, 2005; Cormier, 2006; Peres & Palacios, 2007; Thoisy et al., 2008; Peres, 2010; Luz et al., 2017 among others). In general, subsistence hunting tends to be closely correlated with body size, a key variable in determining hunter preference in terms of both prey species and sex/age classes (Alvard, 1993; Jerozolimski & Peres, 2003). There is a general tendency for large and medium-bodied primates to be exploited (this finding is confirmed by Cormier, 2006; Franzen, 2006; Peres & Palacios, 2007; Luz et al., 2017; Maldonado & Waters, 2020; Zent & Zent, 2020, among others). Atelids and capuchins and also, depending on the communities, pitheciids, and owl monkeys, are hunted more frequently than smaller callitrichids and squirrel monkeys, although smaller monkeys are sometimes hunted or captured when encountered by chance (Cormier, 2006; Luz et al., 2017). The favorite prey is *Ateles* spp. (e.g., Shepard, 2002; Lizarralde, 2002; Cormier & Urbani, 2008; Voss & Fleck, 2011), followed by *Alouatta* spp. (Cormier, 2006; Urbani & Cormier, 2015) who, for some Amazonian communities have "flesh most closely resembling the flesh of people" (Daniel, 1976 *apud* Papavero et al., 2002: 403), and then other atelids of the genera *Chiropotes* and *Lagothrix* (Peres, 1991; Shepard, 2002; Luciano, 2017).

Nevertheless, there are many exceptions to these preferences. Primates are sometimes seen as similar to humans and thus not killed, or there may be specific taboos. Spider monkeys are not eaten because they are considered to be similar to children in some communities of Brazil, Venezuela (e.g., Cormier & Urbani, 2008), and Mexico (Pinto-Marroquin & Serio-Silva, 2020; García del Valle et al., 2020). For some communities, *Alouatta* is considered less tasty (Shepard, 2002) or, due to its slothful behavior the flesh is considered taboo (Voss & Fleck, 2011; Urbani & Cormier, 2015; Boubli et al., 2020). Among the Surui of Rondonia, the meat of the woolly monkey is forbidden to young people who have to eat only spider monkeys (Coimbra Jr., 1985). The meat of the woolly monkey, in the past, was used by the Tikuna during the "Pelazon" festival, which celebrates the girls reaching puberty. A great number of these monkeys were killed and their skins were used to make drums. Due to overhunting, such traditional uses have been discontinued in most modern Tikuna society, but a few isolated communities maintain this tradition (Maldonado & Waters, 2020). Capuchin monkeys are also common game for a great number of indigenous groups of lowland Amazonia (see Peres & Palacio, 2007; Basso, 1973), as are titi and saki monkeys (Cormier, 2006).

In traditional communities, food preferences or bans on eating a particular species of monkey traditionally depended more on cultural value than on availability. Monkeys are a widely available source of food throughout South America, but some communities (e.g., the Ka'apor, cf. Balée, 1984 *apud* Cormier, 2006; the Maya – the Q'quiche, cf. Rosales-Meda & Hermes, 2020) do not use them for food, or only use them sporadically if other game is plentiful (Terra, 2007). Traditionally, different cultural mechanisms regulated primate hunting: folk tales, taboos, and the relationship between the shaman and animals. According to the Wampis, of the north Peruvian Amazon, an old spider monkey gave a poor hunter some tobacco and taught him a magical song that made him a good hunter, but he also warned the hunter not to kill the spider monkey's children and told him to tell his people not to do so. But, one day the hunter killed the children of the old spider monkey, thus provoking the spider monkey's hatred. The latter took the hunter's blowgun and inserted it into the hunter's anus, magically transforming him into a spider monkey. The hunter–monkey became the chief of the spider monkeys, helping to hide the monkeys from people (Świerk, 2020). Folk tales like this are also encountered in other ethnic groups (e.g., Cruz, 2014; García del Valle et al., 2020).

The masters of animals –"game masters" – are spiritual entities, endowed with part-ownership of the natural environment, which according to the Amerindians is something humans cannot do (e.g., Kopenawa & Albert, 2013). Game masters play a role in the ontology of many Amazonian indigenous communities, in which human and non-human life forms are considered to be interconnected (Fernández-Llamazares & Virtanen, 2020 and references therein). Game masters are spiritual beings, both visible and invisible that exist in different forms and places, typically granting hunting success to those who adhere to culturally prescribed rules and rituals established to lure, avoid, attract, and/or show respect toward these non-human agents (Fernández-Llamazares & Virtanen, 2020). The game masters protect primates from overhunting by humans and can punish the hunter and his family if he does not respect certain rules. Shamans interceded between the social/human world and the unpredictable world of the supernatural and they can contact, and intercede with, the game masters by falling into a trance, dreams, or the consumption of psychotropic substances (e.g., through dreams among the Parintintis, Kracke, 2006). Many taboos are linked to the rituals of passage from one stage of life (e.g., puberty) to another, or to age and gender. For example, among the Kalapalo of the upper region of the Xingu River in Brazil, monkey meat is prohibited only to boys entering into their period of seclusion. But, after a period in which they can eat only mandioca and fruits, monkeys are the first non-vegetable food the boys eat (Basso, 1973). In some cases, species can be eaten only after being purified, by verbal supplications addressed to the animals' respective spirit masters. Among the Jodi, capuchin, howler, and spider monkeys can only be eaten by the young and the old after verbal supplications have been addressed to the animals' respective spirit masters (Zent & Zent, 2020). These traditional forms of hunting regulation sustain the natural resources that are threatened by demographic, economic, and cultural factors that result in the exploitation of natural resources (e.g., Ferronato & Nunes, 2010; Maldonado & Waters, 2020). There is also a lack of planning due to the lack of livelihood alternatives and the limited participation of traditional populations when decisions are made (Terra, 2007). A major concern relates to large-bodied primates whose low reproductive rates and long inter-birth intervals make them particularly sensitive to these pressures (Di Fiore & Campbell, 2007; Estrada et al., 2017). Some contemporary populations have decreased their consumption of monkey meat because of prohibitive legislation (e.g., the Popoluca of Mexico, cf. Pinto-Marroquin & Serio-Silva, 2020) or for other cultural reasons (Henfrey, 2020) or because monkeys have become scarce (e.g., in the Atlantic Forest). In some cases, people only consider themselves real members of their community if they abide by the tradition of monkey consumption (cf. Lizarralde, 2002). Fernández-Llamazares and Virtanen (2020) observe that notions of Amazonian indigenous groups often differ from those of environmentalists, differences that can be explained by their peculiar ontological vision. When conservationists respond to unsustainable hunting practices, traditional communities interpret this as the result of the discontent with the animals' tutelary deities, or because a shaman's prayer to these deities was ineffective (Fernández-Llamazares et al., 2017).

beings, or a transformation useful to feed "human-people" in a world inhabited by only "human-animals". The notion of human–monkey/ape metamorphosis being a punishment is also encountered in ancient mythologies and religions of other continents where non-human primates are endemic (cf. other chapters of Section I of this volume) and also in some monotheistic religions (Veracini, chapter 7, this volume). To many Amerindians the first human-monkeys of creation were not kind to other humans, and as a consequence they were punished by being hunted and eaten. The taxonomic relationship between similar organisms is explained in popular culture in terms of kinship; thus, many narratives are formulated to justify the fact that these human-like beings, instead of being spared, are fair game for human beings.

The capacity to think, or their labeling as intelligent beings, seems unrelated to the consideration of certain animals as acceptable food sources (Urquiza-Haas et al., 2020). As argued by Shepard (2002), mythical knowledge is often the result of a cultural adaptation to the natural world, so animals could be both symbolically respected and used as food. To Amerindian indigenous peoples, the original condition common to humans and animals was not animality but humanity. Modern humans have retained this aboriginal condition: animals are ex-humans (Viveiros de Castro, 1996). Mythical narratives also use past events to explain the order of the world, and some myths current among Amerindian peoples explain the present human condition using historical facts experienced during European colonization (see below the story of the white capuchin monkey). In these myths, primates (and other animals) act as humans with agency, intention, and sociality, and these "primordial people" have different personalities whose characteristics correspond to the traditional natural history developed by indigenous communities (Zent & Zent, 2020).

4.2.1 Amazonia: Genera and Perception

4.2.1.1 Spider Monkeys

Spider monkeys (genus *Ateles*), which are distributed in Amazonia and Central America, are among the largest Neotropical primates, and display a complex social structure (Campbell (ed), 2008). Among the Bari people in Venezuela, monkeys (especially spider monkeys) were regarded as transformed humans, and according to myth they can live in the trees, eat fruit, and be hunted, and are the preferred food of the Bari (Lizarralde, 2002, 2020). Among the Matsigenka of Peru (Shepard, 2002), in ancient times only humans inhabited the earth. The first shaman used his magical breath to transform humans into the great diversity of animal species existing today. The first two animals were monkeys (in particular spider monkeys and woolly monkeys of the genus *Lagothrix*). Among the Waimiri-Atroari in the past, the black spider monkey (*Ateles paniscus*) was not hunted because it was human (*kinja*) and thus was considered a relative (*yaska*). This species, together with the golden-headed tamarin (*Saguinus midas*) and the brown capuchin (*Sapajus apella*), are humans transformed as punishment by a superior being as a consequence of their socially reprehensible attitudes (Souza-Mazurek & Bruno, 2020).

The spider monkey is also important in the Yanomami's cosmology as shaman Kopenawa tells us:

> The monkeys that we call *paxori* are humans like us. They are spider monkey humans ... but we shoot them with arrows and smoke them out. [...] Despite this, in their eyes, we are still their fellow creatures. Though we are humans, they give us the name they give themselves. This is why I think our inner part is identical to that of game and that we only attribute to ourselves the name of human beings by pretending to be so. Animals consider us their fellow creatures who live in houses while they are people of the forest. [...] In the beginning of time, there was no game. Our ancestors lived starved for meat and devoured each other. This is why they became other. They transformed themselves into game so we could eat them properly. It was so. (Kopenawa & Albert, 2013:387)

According to Yanomami beliefs, all humans have a double: image/spiritual entity called *xapiri*. The latter is an animal of the forest that is linked to an individual; if that individual is sick or dies, its counterpart will immediately share the same fate (Lizot, 1985; Cesarino, 2014). In Yanomami cosmology, the spider monkey spirits help the construction of the house of the *xapiri*: they "pull the peak of its roof to hang it on the sky's chest". Then "the spider monkey spirits grab the beams' ends and bend them to tie them together with ropes covered in celestial pitch". Moreover, the homes of the spirits of the spider monkey, the black saki, and other important animals are located not so far from this house because they overlook it and are responsible "for watching over the surroundings and protecting the house from enemy *xapiri* incursions" (Kopenawa & Albert, 2013: 99). Among the Yanomami Karohi, all the male descendants of the same ancestor have the same image animal, the emblem of the group. Emblematic animals for men were the spider monkey and the jaguar, while all the women were regarded as otters (Lizot, 1985: 175).

In other Amazonian ethnicities, the spider monkey is associated with the spirit of the dead — a dangerous creature that can kidnap women and children (Świerk, 2020). A similar thing is believed by the Takana people of the Bolivian Amazon: spider monkeys are related to negativity and evil and can steal human women from villages to take them to their forest hideouts (Townsend et al., 2020). In many indigenous communities, young spider monkeys are commonly raised as pets after the mother has been killed during hunting. The Joti say that raising these monkeys with people turns them into "a total person" (Zent & Zent, 2020).

Spider monkeys are also associated with imaginary, or past real beings. Among the Wampis (Świerk, 2020) the myth of a monkey-jaguar exists, and it is said that the Wampis formerly used to fight these animals. Other indigenous groups mention giant monkeys similar to the spider monkey: to the Bari it is gigantic with long claws and has demonic power and can rape women (Lizarralde, 2002: 90). According to the Matsigenka, a creature known as the "mother of the spider monkey" is found in the Andean foothills (Shepard, 2002). The idea of mysterious giant animals hidden in remote parts of the forest is widespread in the popular culture of lowland Amazonia. The *Mapinguari* is a monstrous being known in the Brazilian states of Acre, Rondonia, Amazonas, and Para, and it seems to correspond to that described among the Karitana people of Rondonia as a sort of giant monkey-sloth (Velden, 2016). These creatures may represent extinct animals retained in folk memory, such as the giant ground-dwelling sloths (*Mylodon* or *Megatherium* spp.) that lived on the continent until about 10,000 years ago (Ranzi, 2000).

4.2.1.2 Howler Monkeys

The adaptive radiation of howler monkeys (genus *Alouatta*) is one of the largest among New World monkeys. This genus occurs in many types of Neotropical environments and ecosystems from Central America to Argentina (Mittermeier et al., 2013; Kowalewski et al., 2015). Due to their folivorous diet, supplemented with fruit and flowers, howler monkeys are typically lethargic, needing long periods of rest. They are able to survive in small forest fragments and urban areas (Printes et al., 2010), and they share a hypertrophic hyoid bone connected to large vocal sacs, that are more marked in the larger bearded males. The vocal sacs permit the production of powerful vocalizations that can be heard across long distances. These calls are emitted, depending on the species, at different times of day, but generally show a distinct bimodality (e.g., early morning and late afternoon peaks) that may change with the seasons (e.g. *Alouatta belzebul* pers. Obs.; other species cf. Cornick & Markowitz, 2002). Howler monkey vocalizations, the ease with which these animals can be located in the forest (a great amount of time is spent resting) and their size and morphology, have probably influenced how they are perceived by native people. In particular, their ability to vocalize has resulted in them being associated with shamanistic abilities that "give them a place between the material and spiritual worlds" (Urbani & Cormier, 2015: 275). Among the Guyanian people of the Moruca River, the recognizable call in the morning signals it is time to wake up (Rybka, 2020), and howlers are used for weather forecasting in many communities of Guyana (e.g., Lokono, Kari'na, and Warao) (Rybka, 2020). Among the Shawi of Peru (González-Saavedra, 2020): when the *Alouatta* call in the morning a hot day is likely, and when they call in the late afternoon, rain is likely. They are also said to influence the weather, calling the rain with their vocalizations, and they are consequently highly respected (see also later among Mayan communities). The Paraiba in the Atlantic forest and in Rio Grande do Sul in Brazil, believe that *Alouatta* announces the rain (see later **Box Details 4.3**; Ruoso, 2012 respectively). Voss and Fleck (2011) report that the Matse people in Peru believe that when the *Alouatta* vocalize in the morning a clear day is predicted.[1] The particular morphology of the *Alouatta's* larynx has led to some forms of cultural behavior: in many Amerindian cultures, the hyoid was, and still is, used as a medicine against respiratory diseases (see Paragraph 4.3) and hairs can be used against poisonous stings and evil spirits (Rybka, 2020).

Howler monkeys play a symbolic role in many indigenous mythologies and cosmologies of South America (cf. for review Urbani & Cormier, 2015). In the Guajá community of the Eastern Amazon, studied by the anthropologist Cormier (2003: 89):

> the relationship with howler monkeys is the same as that with *'aiyã* cannibal ghosts', both of which are considered ex-humans. [...] In the Guajá creation myth, the creator hero, Mai'ira, transformed a group of Guajá into howler monkeys and instructed the Guajá to eat them. The Guajá themselves are preyed upon by the *aiyã*, who are the ghosts of their dead. The Guajá believe human death is ultimately due to the cannibalization of their spirits by the *aiyã* ex-humans. (Cormier, 2003: 89)

Only the Guajá have this relationship-siblingship with howler monkeys, which are a key game species for them, particularly in the wet season (Cormier, 2006).

Some "totemic" traditions are associated with howler monkeys. Becher (1974 *apud* Boubli et al., 2020), working among the Brazilian Yanomami at the village of Ironasitéri ("the place of the howler monkey"), noted that the howler monkeys were not hunted or eaten there, and they gave the name to one of the shamans. The people at Ironasitéri used to call themselves "howler monkey people" and stated that they have souls, and thus their forefathers were able to turn into monkeys and vice versa. Matsigenka myths, involving the transformation of howler monkeys into human beings, identify howler monkeys as lazy shamans because they rest frequently (Shepard, 2002). The mythology of the Qom of Argentina, which explains that howler monkeys used to be people because humans are ancestral to all animals (Medrano & Suárez, 2020), forbids hunting monkeys (howlers). The Jodi, who think howler monkeys are amenable to being brought up among people, consider them to be capable of affective and sensitive behavior toward those around them

[1] There is also evidence that, with changes in barometric and hydrostatic pressure, humidity, and air temperature, the animals' nervous system becomes more active than normal and they change their behavior (Acharya, S. (2011). Presage biology: Lessons from nature in weather forecasting. *Indian Journal of Traditional Knowledge*, 10(1): 114–124).

FIGURE 4.5 Tufted capuchin monkeys, *Sapajus apella*. This species is endemic to Amazonia. (Photo Arianna De Marco.)

(Zent & Zent, 2020). Traditional dances and mimic performances recall the *Alouatta* in Guyana, and the term *arawara* was used in the past as an offensive term for someone with a particularly dark face (Ahlbrinck, 1931 *apud* Rybka, 2020).

4.2.1.3 Capuchin Monkeys

Capuchin monkeys of the genera *Cebus* and *Sapajus* represent an important primate radiation in the Neotropics (Mittermeier et al., 2013: 348–389; **Fig. 4.5**). They are frugivorous and omnivorous, capable of adapting to a range of habitats, with complex behaviors, including tool-use (Visalberghi et al., 2015; Falótico et al., 2019). Many communities perceive them as intelligent and cunning (but also aggressive), with myths, cosmologies, and folk tales that reflect these perceptions. The Wampis, who inhabit the northern part of the Peruvian Amazon, tell that the white-headed fronted capuchin monkey (*Cebus albifrons*) in the past protected them against cannibals with his cunning (Świerk, 2020). The white-headed capuchin monkeys are regarded as forest people by the Shawi. This monkey is part of a mythical cosmogonic narrative that includes the disposition of their territory and their own creation as human beings: in ancient times "when animals were people and people animals" (Gonzalez-Saavedra, 2020: 337). The narrative claims that capuchin monkeys have such a noisy and restless character that they provoked the God Cunpanama, their most important deity, causing the Shawi to lose part of their territory. The white capuchin, according to them, so annoyed the wife of the god that he was turned into a white person (these are gringos, white people – foreigners). The myth also suggests that capuchin behavior caused many transformations in their society. Gonzalez-Saavedra (2020) interprets this myth as an evocation of the moment of contact with Europeans (personified as white-headed capuchin monkeys), whose arrival broke the basic norms governing the coexistence of indigenous people.

Among other cultures, cunning is associated with thieves: among the Matsigenka (Shepard, 2002) of the Arawakan cultural linguistic family, the word *Kashiri* (capuchin monkey), means thief. Children cannot eat capuchin meat, otherwise they become dishonest. For them, in the past, capuchins were humans who were transformed into monkeys because they tried to steal fire from another tribe. In Tikuna folklore, the capuchin monkey can kidnap young girls and trap them in the forest forever (Parathian & Maldonado, 2010). In popular culture, the intelligence of capuchins can pass to hunters, so when the Joti go hunting they bring along the head of a capuchin monkey for its intelligence (Zent & Zent, 2020). Capuchin teeth are used to make anklets worn in initiation rituals among the Waimiri-Atroari of Brazil (Souza-Mazurek et al., 2020).

4.2.1.4 Callitrichids and Squirrel Monkeys

Smaller monkeys such as callitrichids and squirrel monkeys are among the preferred pets. They can also be used to hunt for unpleasant insects or spiders inside houses. *Leontocebus* (former *Saguinus*) *fuscicollis* (**Fig. 4.6**) is left

FIGURE 4.6 The brown-mantled tamarin, *Leontocebus fuscicollis*, is a species endemic to western Amazonia. (Photo Montserrat Ubach.)

free in the houses and is used by the Wampis to eat the huge spiders that live in the roof (Świerk, 2020). Pygmy marmosets, *Cebuella pygmaea,* are seen by the Matsigenka as magical and somewhat dangerous creatures. They may lead a hunter astray, flitting from tree to tree until the hunter is lost, and then they vanish in a sudden gust of wind (Shepard, 2002). The Waimiri-Atroari (Souza-Mazurek & Bruno, 2020) seldom hunt the golden-headed tamarin (*Saguinus midas*), an animal that plays a central role in the myth that explains the origin of day and night. In ancient times, when the golden-headed tamarin was still human, there was no sunset, and it was never dark. One day the "tamarin/man" broke open the sun to find out what was inside it. The owner of the sun punished him and "transformed him into a monkey with golden hands because he touched the juice of the sun" (Fonseca & Fonseca *apud* Souza-Mazurek & Bruno, 2020: 231). Hence, the alternation of day and night.

Among the Jodi, squirrel monkeys are noted for the way they respond to being hunted, because when the hunters are spotted a cry of danger rings out (Zent & Zent, 2020). The alarm call is interpreted as an attempt to avoid capture, so in response squirrel monkey skulls are part of the necklace worn by the hunters.

4.2.1.5 Night Monkeys

Night monkeys of the genus *Aotus* are found in a variety of primary and secondary forest types across much of South America. Night monkeys are the only truly nocturnal primates in the New World (Mittermeier et al. 2013: 414–424), although nocturnal activity can be observed in other species (e.g., in howler monkeys). Popular culture frequently associates them with other nocturnal tree-dwelling mammals and with the malignant spirits of the night. Maldonado and Waters (2020) report that among the Tikuna in Colombia, night monkeys are linked to the spirits that harm people or transport them to another dimension. They also add that a traditional Colombian community, the Azulay Vasquez, believes night monkeys can be seen in the guise of humans, or bigger monkeys, at night. Night monkeys usually approach hunters' camps at full moon, where they harvest the hunter's blood so that the hunters die in the forest. Other versions recount that night monkeys are guardians of the moon and thus can influence female fertility. In later pre-Hispanic painted funerary textiles from the Peruvian coast, night monkeys become an illustration of a mythical episode that relies on an astronomical event and the placement of the characters on the artifact is a reflection of the disposition of stars and planets in the night sky (cf. Karadimas, 2016), and in one of the textiles stands there is a representation of an intriguing monkey, a feline creature the Karadimas interpret as a night monkey. The association between night monkeys and stars is still common today among Miraña people (who speak the Bora languages and live in Perù, Colombia, and Brazil). Thus, the representation could be a metaphorical figuration of a night monkey portrayed by the combination of a feline and a monkey.

FIGURE 4.7 Old drawing of *Cacajao melanocephalus*, endemic to Amazonia. (Image of public domain from Elliot, 1913, Vol. 1, Pl. 10. Source: https://www.biodiversitylibrary.org/bibliography/8890.)

4.2.1.6 Pitheciids

Pitheciids, which include the titi, saki, bearded saki, and uakari, in spite of their odd aspect and extensive distribution both in Amazonia and in other biomes of the South American continent (Mittermeier et al., 2013: 432–457), are the least frequently mentioned monkeys in myths and folk tales. In an extensive review of the subsistence and symbolism of indigenous lowland South American peoples, Cormier (2006) reports that the genus *Chiropotes* is mentioned only twice in myth or ritual, and uakaries (*Cacajao* spp.) (**Fig. 4.7**) are mentioned only once as an animal to be avoided, or as taboo (see **Box Details 4.2**). In Cormier's study, it also emerged that pitheciids are less often hunted than the other large, or middle-sized, genera of Platyrrhines, although many of the works cited do not specify which species are hunted (e.g., the Yanomami hunt uakari and bearded saki, Boubli et al., 2020; Kopenawa & Albert, 2013: 387). The lesser importance of pitheciids in cosmological myths is confirmed by the scarce presence of these monkeys in the collection of ethnoprimatological

BOX DETAILS 4.2: THE UACARI, THE "UNFINISHED MONKEY"

Adrian A. Barnett
Middlesex University, Hendon, London

The particular and odd appearance of the uacaris (shaggily furred, with a notably short tail) is probably at the origin of the cautious attitude that the indigenous and rural people in Amazonia show toward them. The uacari genus, *Cacajao* has two species groups; a naked red skinned head group (six subspecies of the species *Cacajao calvus*) and a black-headed group (three or four taxa, including *C. ouakary* and *melanocephalus*) in which the head has fur and the skin is dark. Part of the taboo (e.g., Peres, 1990) or antipathy to using uacari meat comes from how human the naked heads of *calvus* taxa appear. However, uacari meat has been reported as favored for use as bait in fishtraps, while the long soft fur of the tails has meant they are prized as "feather" dusters for domestic cleaning (Barnett et al., 2002, 2005). In the past, uacaris were also used as live-bait to trap big cats (which were then shot for their pelts: Mittermeier & Coimbra-Filho, 1977). Nevertheless, in some sites, uacaris have been recorded as preferentially hunted (Colombia - Defler, 1989; Peru - Aquino, 1989; Venezuela - Lehman & Robertson, 1994a, b), while at others; they are hunted only when absolutely nothing else is available (Barnett & da Cunha, 1991). Another interesting aspect that can generate a food taboo is that the uacari is regarded as an "unfinished monkey" and so dangerous to eat. This perception was reported to me at various sites on the Rio Negro, where one of the local names for the golden-backed uacari (*C. ouakary*) is "*macaco mal acabado*". The fact that it is incredibly fast, makes large leaps, and occurs high up in the canopy is probably a more practical reason to avoid trying to hunt it when other more tractable primate species are available (such as capuchin and howler monkeys). On both the Uaupes and Curicuriari rivers in the northwestern part of Brazilian Amazonia, a folk tale exists concerning why the golden-backed uacari has the appearance that it does. The story reflects this "unfinished" aspect, and centers around the time when the animals were being created by being baked from clay in a fire-pit, the kind used to bake clay pots. But, the "*bico*" [another Amazonian regional name for the golden-backed uacari, the onomatopoeia from its alarm call] was very impatient and could not wait to meet the world that was being made for him and the other animals. So, he leapt out of the fire-pit early to see it. As he did so, his legs and arms touched the hot coals of the pit and got burnt, but his back was not yet properly fired and so stayed pale. So, only some parts of the body were the finished brown color they were supposed to be. The tail of the uacari was originally made to be a very long one, but it was so long that it got burnt when the uacari was trying to get out of the fire-pit. The burnt part broke off – so now all uacaris have a very short tail, and you can still see the black tip [NB: while all golden-backs have a gold-colored back, black lower limbs and extremities, and cinnamon flanks and upper limbs, only some individuals have black tail tips]. When the uacari got to the world the trees were not ready yet, and their fruit was just forming, but he was so hungry from all the effort to escape that he ate the fruits anyway – which is why uacaris to this day still eat unripe fruits. And the uacari today is still a very impatient animal, which is why he is always the first monkey to get up in the morning [they *are* the first up in my field experience]. Outside the legend recounting it was also said that this was why uacaris leap so much because they burnt their paws and so they didn´t want to walk if they could help it. The other thing that was mentioned, outside the actual recounting of this legend, was that, because of what it did, the *bico* is always worried either that it will be captured to be taken back to be finished or that it will told off in some unspecified way — and so is always very nervous, which is why it always gives its alarm call so very easily and is so difficult to get close to. This was also given as a reason why the animals often do so poorly as pets. They are considered *manso* or *mortal* ("fragile") because they are frightened that they will be taken back and changed. Indeed, uacaris do have a reputation of doing poorly in zoological collections, while their relatives the sakis (*Pithecia*) and bearded sakis (*Chiropotes*) breed and do well (Becker et al., 2013).

works recently published in Urbani and Lizarralde (eds) (2020). For instance, some traditional groups in Guyana report that the burned hair of *Pithecia pithecia* is used as medicine, as well as their being kept as pets (Rybka, 2020). The tail of the bearded saki male is used by the Yanomami from Marari village in Brazil to make a headband (Boubli et al., 2020). In the popular culture of the Joti, the bearded saki (*Chiropotes chiropotes*) is noted for its "ceremonial feasts", that is, when different groups come together on the ground, and for its amiable attitude toward other monkeys (Zent & Zent, 2020).

4.2.2 Human Perception of Meso-American Monkeys: Past and Present

The Mayan civilization, which flourished from roughly 1500 BCE until the arrival of the Spanish conquistadors in 1521, occupied the Yucatán Peninsula, and all of the territory now incorporated into the modern countries of Guatemala and Belize, as well as the western portions of Honduras and El Salvador. In this territory, three species of atelids occur: *Alouatta pigra* (**Fig. 4.8**), *Alouatta palliata,* and *Ateles geoffroyi.* Archaeological and iconographic

FIGURE 4.8 *Alouatta pigra*, "el saraguato", endangered species endemic to Guatemala, Yucatan (ME), and Belize. (Photo Cecilia Veracini.)

records show that these primates played an important role in ancient Mayan cosmology, religion, and other cultural manifestations (Thompson, 1973; Coe, 1978; Bruner & Cucina, 2005; South, 2005; Skousen, 2009; Rice & South, 2015; South & Ford, 2022). Although the white-faced capuchin monkey (*Cebus capucinus*) is today found south of the prehistoric Mayan world, in eastern Honduras and further south and east, there are verifiable records of white-faced capuchin monkeys well within the area encompassed by both modern and ancient Maya (Baker, 2013). Even if not indigenous to the Maya region, capuchin monkeys may have been imported from further south, as was done with other mammals and birds (Masson, 1999). Monkeys such as howler monkeys and *At. geoffroyi* are central characters in numerous Maya creation myths (Rice & South, 2015), including the creation myths of the K'iche' (also spelled Quiché). The latter had an advanced civilization in pre-Columbian times and their written records of history and mythology are preserved in the 16th-century *Popol Vuh*, the sacred book of cosmogony (Tedlock, 1986). This ancient text focuses on the Hero Twins, whose intelligence, skills, and wisdom allowed them to defeat the gods of death and darkness. Before their appearance, the creator deities had tried to shape creatures from heart and mud, but these spoke nonsense, so the gods unmade them. Then the deities made humans out of wood, only to realize that they were not able to worship them, so these creatures were also destroyed, "most of them perished and their only descendants are monkeys who inhabit the forest today...This is why monkeys look like people: they are signs of a previous attempt at designing humans" (Tedlock, 1986: 34; 86). The K'iche' myth recounts that later on the Hero Twins were generated and these brought to the earth the light of wisdom and knowledge because now modern humans possessed a brain to think and reason. The Hero Twins had two elder brothers: *Hun B'atz* (a howler monkey) and *Hun Chouen* (an artisan), who, although they were successful flutists, singers, writers, and carvers, were jealous of their younger brothers. The Hero Twins took revenge on *Hun B'atz* and *Hun Chouen*, by convincing their older brothers to climb a magic tree. Once at the top, the magic tree started growing, so *Hun Batz* and *Hun Chouen* were trapped higher and higher in the canopy, and they ended up turning into monkeys: *B'atz* (howler monkey) and *Chouen* (spider monkey). The Hero Twins represent the most recent generation of creatures and *B'atz* and *Chouen* are the fallen monkeys from the original "humans-monkeys". As observed by Bruner and Cucina (2005: 113) "monkeys or monkey-like creatures are often depicted in underworld scenes even though their natural world is well above the surface. This could be linked to their ancestral state of pre- or proto-human creatures".

The presence of spider and howler monkeys in ceramic decoration and in other kinds of artworks is widespread throughout the different artistic periods of Mayan world. However, it is not always easy to recognize the species depicted from the artists' reconstruction and interpretation because these are often stylized (e.g., **Fig. 4.9**). Howler monkeys predominate in the portrayals of monkey/man scribes in Late Classic Mayan funerary pottery that represent *B'atz* and *Chowen* (Rice & South, 2015; Coe, 1978). Monkeys on pottery vessels are sometimes shown holding or writing on codices. The scribes had the shamanic ability to make contact with epochs of primordial times in order to conjure up and become their co-essence, monkey-man Gods, and the perceived connection with the creator deities, managing to recreate cosmological harmony through the divine relationship required to enter the ranks of the elite

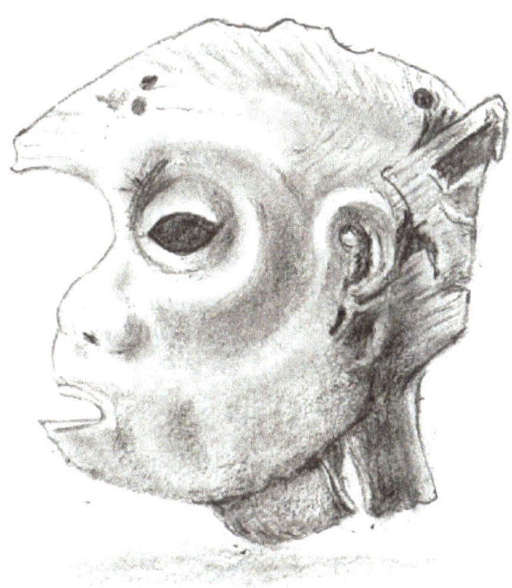

FIGURE 4.9 Pendant with monkey head (tentatively identified with a spider monkey). Aguateca, Guatemala 8th century, shell. (Drawing by Cecilia Veracini. Source: MetMuseum.org/art/collection/search/8220307.)

FIGURE 4.10 A spider monkey (right) dances in the procession with a black jaguar (in the center), while a deer plays the drum. Provenance unknown. (Source: www.mayavase.com, image by Justin Kerr (K1208). Public Domain.)

artisan (Kidder, 2009). Howlers appear more frequently in association with scribal activity and imagery such as pens, codices, and the deer's ear, while spider monkeys are found in the iconographic contexts of performance-related scenes and rituals, processions, and cacao handling (Rice & South, 2015; **Fig. 4.10**). This suggests that the Mayas saw a duality in simian roles, indicating a cultural distinction between biological roles (South & Ford, 2022). Classic Mayan art (200–900 CE) also shows an association of monkeys with sacrifice, death, and journeys to the underworld (South, 2005), having the skill to transcend natural realms. Vases adorned with monkeys probably had a special significance when placed in burials, because of the association of monkeys with death and their place as companions of the deceased on the journey to the underworld. The spider monkey is portrayed in a stylized fashion on pots of the Late and Terminal Classic Mayan periods. Skousen (2009) argues these pots had several functions: domestic, as serving vessels, or for eating solid or possibly liquid food, and in rituals associated with death and burial. As shown by García del Valle et al. (2020: 57) monkeys can evoke other meanings related to obscure forces, such as the "*Wahyis*, animated entities of the family spirits that only rulers or other important people could manipulate [...] and that can provoke diseases". Monkeys played a role in the myth of the cacao fruit, which is associated with fertility, life, and vitality. They are sometimes depicted interacting with cacao trees, or involved in the harvesting of cacao pods (Nájera Coronado, 2012). These representations may symbolize the connection between monkey behavior as seed dispersers, fertility, and agricultural practices of the Maya people. Monkeys were also related to dancing and music across Central America, and were regarded as wise and appreciated artists, as shown by the *Popol Vuh* (Coe, 1978). Moreover, iconographic studies of the Late Classical period associate monkeys with promiscuity, poetry, flowers, and procreation (Skousen, 2009). In sum, monkeys seem to have been part of many aspects of daily life and of death in ancient Mayan culture.

Among the ancient Aztecs, the spider monkey is the monkey god Ozomatli, companion spirit and servant of the god Xochipilli, the deity of music and dance (Morris, 2013: 41; **Fig. 4.11**). In the Aztec world, spider monkeys are associated with the second creation, which Ehecatl presided over as the wind deity (**Fig. 4.12**). They represented people who were turned into monkeys during the destruction of the second Sun (the mythological period of creation), and the spider monkey gives its name to the eleventh day of the calendar, and people born on this day are said to have artistic talent (Bruner & Cucina, 2005).

Recent ethnoprimatological studies carried out in Central America have found that spider and howler monkeys are still considered ancestors and close relatives of humans, originating from pre-human beings, releasing positive emotions of sympathy and curiosity and bring happiness to the forest (García del Valle et al., 2020; Rosales-Meda & Hermes, 2020). In Mayan communities, monkeys still play an important cultural role in guiding people's lives at the individual, family, and community levels, and people celebrate spider and howler monkeys with rituals, dances, festivals, and commemorative days (Rosales-Meda & Hermes, 2020; Cook & Offit, 2013: 75). As seen in other parts of tropical America, howler monkeys are highly respected for their capacity to invoke rain. Among the Popoluca, the native ethnic group of the Los Tuxtlas region (ME), when howler monkeys' vocalization is louder and more frequent than normal, they are announcing changes in the weather. "The monkey weather service" is important for

FIGURE 4.11 An Aztec stamp in ceramic representing a spider monkey dated 14th- early- 16th century. In Nahuatl, the monkey is called *ozomatli* and is the eleventh-day sign of the ancient Aztec calendar. Unlike our calendar of today, which consists of twelve months of approximately thirty days each, the Aztecs and other Mesoamerican peoples used "months" of just twenty days. Each day was given a number and named after a different animal or other natural element. (Image of public domain, source: https://www.metmuseum.org/art/collection/search/307744.)

FIGURE 4.12 This sculpture portrays a spider monkey (*Ateles geoffroyi*) wearing the regalia associated with the Aztec Wind God, Quetzalcoatl-Ehecatl. The monkey sits on its haunches with its legs folded on its sides with elongated toes projecting forward. The monkey's head is tilted back and his open mouth reveals its sharp teeth, the open mouth may have been associated with Ehecatl blowing wind. (Image of public domain. Source: https://www.metmuseum.org/art/collection/search/765334.)

these people to plan their activities (Pinto-Marroquin & Serio-Silva, 2020). Also, the Maya Q'eqchi' of Guatemala, believe that they are able to pray for rain, as they are intermediaries between humans and the gods (Rosales-Meda & Hermes, 2020). The Maya Q'eqchi' maintains the surnames *B'atz* and *Chowen* (and some of their variants): "Both names come from Mayan royal lineages which have had a long history and relationship with trade and wealth since pre-Hispanic times and are also remembered as lineages of influential families in early colonial times" (Akkeren, 2012 *apud* Rosales-Meda & Hermes, 2020). Primates are present in the sacred energies and cycles of nature guiding people's lives, so they are honored and respected, and not hunted or kept as pets.

4.2.3 The Brazilian Atlantic Forest

Centuries of human intervention in the Atlantic Forest biome were followed in the 20th century by massive and accelerated urbanization and industrialization of the Brazilian coastline. According to the Atlantic Forest Atlas, the total forest remnants correspond to 12.4% of the biome's pre-colonization coverage (cf. Fundação SOS Mata Atlântica, 2020). The increasing isolation and degradation of these unevenly distributed fragments indicate significant risk and uncertainty regarding the biome's conservation. Fragments larger than 100 ha, considered the minimum habitat for rare species, and with the largest carbon stock, represent only 8.5% of the original forest area of the biome, the largest concentrations occuring in Piauí (33.8%), Santa Catarina (22.8%), Rio de Janeiro (18.7%), and São Paulo (13.7%). The smallest concentrations of forest fragments occur in Rio Grande do Norte (3.5%), Sergipe (6.8%), Ceará (7.3%), and Rio Grande do Sul (7.8%). In this scenario, the fragmentation of officially recognized indigenous lands reflects the profound cultural transfiguration of the various native ethnic groups distributed within the Atlantic Rainforest, a scenario totally different from that of the 16th century when, according to Ribeiro (2008), around 1 million indigenous people, distributed in villages of 300–2,000 inhabitants, occupied the Brazilian coast. As examined by Jacques-Coper et al. (2019), the move from the rural to the urban world influenced all aspects of the migrants' indigenous culture including their beliefs, knowledge, and production practices. Living in cities or even in rural productive areas, part of their knowledge and beliefs can be expressed, but no longer practiced. As a consequence, the original cosmologies of the people who dwelt in the Atlantic Forest have also been deeply fragmented, and possibly lost.

In Brazil, 23 species, of which 18 are endemic, under some degree of threat occur in the Atlantic Forest (CPB, 2015), a biodiversity hot spot of global importance, despite the increasing fragmentation of the biome. The significant impacts on different species of Neotropical primates expose the fragility of some species, and the resilience of others, with a few species adapting to human-transformed environments by modifying their feeding habits and becoming closer to humans (Nunes et al., 2021 and references therein). Defined by Garcia (2018) as a vast population of non-human refugees without places of refuge, these primates establish different patterns of interactions at the boundaries between forest spaces, urban and rural areas (e.g. Box Details 4.3). The increasingly frequent encounters with humans and domestic animals provoke eventual tensions and conflicts, while revealing behavioral patterns that ensure the primates' survival in often critical socio-environmental circumstances (Costa, 2017). These events have been investigated from several perspectives, including where the ethological, anthropological, and sociological dimensions of different regional contexts converge. Historically, this proximity has been associated with economic losses occasioned by the consumption of agricultural products as described by sources of the colonial period. Nonetheless, Nunes et al. (2021) conclude that the biological characteristics of Neotropical primates, and how humans perceive them, have influenced and modulated their survival in the face of land transformation and habitat loss, which have intensified throughout the Anthropocene era.

In order to examine the state of ethnoprimatological studies of the primates of the Atlantic Forest, we verified the degree of occupation and distribution of different ethnic groups in the 13 states located in the domains of the biome. Of the total of 202 indigenous territories either legalized, or in the process of legalization, the highest concentration of indigenous lands occurs in the states of Rio Grande do Sul (42), São Paulo (30), and Bahia (28) (Funai, 2023), yet few studies have been conducted on the possible interactions

BOX DETAILS 4.3: RED-HANDED HOWLER MONKEY (*ALOUATTA BELZEBUL*) IN THE NORTH EAST ATLANTIC FOREST (STATE OF PARAIBA, BR)

Carla Soraia Soares de Castro
Universidade Federal da Paraíba

In the Brazilian Atlantic Forest, one of the most threatened tropical forests in the world, which currently covers approximately 12.5% of its original area (cf. Fundação SOS Mata Atlântica, 2015), 23 primate species occur, of which 18 are endemic (CPB, 2015). In the Pernambuco Centre of Endemism – an area that encompasses the Atlantic Forest in the states of Alagoas, Pernambuco, Paraíba, and the Rio Grande do Norte – the remnants of this forest cover less than 9% of their original area (Fundação SOS Mata Atlântica, 2015). *Alouatta belzebul* (Linnaeus 1766), endemic to Brazil, exists as two displaced populations, one in the eastern Amazon, the other in the northeast Atlantic Forest (Langguth et al., 1987; Bonvicino et al., 1989). Thirty-six years ago, the first record of this species in the Paraíba state was obtained, confirming their occurrence in the northeast Atlantic Forest (Langguth et al., 1987; Coimbra-Filho et al., 1995). "*Guariba*" – the common name for red-handed howler monkeys in Brazil (see Glossary) – was locally extinct on the north coast of Paraíba, mainly due to habitat loss and hunting pressure (IBAMA 2003). In 1990, the "Guaribas" Biological Reserve was created, covering three discontinuous areas, SEMA 01, SEMA 02, and SEMA 03. The headquarters of the Reserve – SEMA 02 – approximately 44 km northwest of the capital of Paraíba; João Pessoa, in the mesoregion of *Mata Paraibana* (Brazil IBAMA, 2003). This was repopulated by translocating individuals and, in order to protect this habitat. *A. belzebul* was designated a "flagship" species. Reintroduction efforts have successfully established a population that – although small – is able to persist and reproduce (Coimbra et al., 2022). Locals living around the Reserve mention *Alouatta belzebul* as a component of the local fauna (Castro & Casanova, 2018). The communities, that give the species the feminine name "*a guariba*", do not recognize it as a type of monkey, nor do they classify it in any animal group. They consider it unique, and they believe that when "*a guariba*" is singing (vocalizing) in the morning, "she" is calling for rain. Agriculture is the main activity and these communities have established a direct relationship between the forests and rainfall, and their narratives suggest that the red-handed howler monkey belongs in the universe of local myths, playing an important role by living in the forests and bringing rain. It is not clear whether this belief limits hunting, with c. 90.9% of the people from these communities admitting it had occurred in the past. They reported that hunting is less frequent today than in the past, stated that hunting was performed for subsistence in the past, and said that hunting is currently performed for leisure purposes so that the primate meat is consumed as a snack and for keeping as household pets (Filho et al., 2021). Currently *A. belzebul* is categorized as Vulnerable (IUCN, 2018) and is one of the primate species targeted and prioritized by the National Action Plan for the Conservation of the Primates of Northeastern Brazil (cf. Brasil 2012).

between these ethnic groups and primates. In the state of Bahia, different attitudes toward primates (and wild animals) have been found depending on the type of territory and lifestyle of the indigenous populations. Among the Tupinambá of Olivença, in the state of Bahia (BR), the perception of animal species expresses the distance from the original cultural bases of this ethnic group, which had been considered extinct until 2002. Out of the total of 55 types of animals mentioned, cats and dogs were the most frequently cited, with monkeys occupying ninth place, after other domestic animals, such as chickens, cows, oxen, and horses; 95% refer to how animals are used, and only 2.4% related to religious attributes (Razera et al., 2006). A more traditional myth about monkeys recently came to light among the communities inhabiting the lower southern part of Bahia (both indigenous and "quilombola", black slaves who resisted the slavery regime) traditionally devoted to cocoa plantations. This concerns the origin of the first plantation designated as "mixed planting" — an agricultural polyculture system that ensures income and supplies for families — which is attributed to the "monkey" called *Jupará*. This animal is identified as the "farming agent, responsible for the opening of the first plantation". Yet, the *Jupará* is not a primate, but *P. flavus*, the *Kinkajou*. It feeds on cacao fruit and distributes its seeds inside the forest, and it is often associated with primates. Incidentally, the *Jupará* was recognized and honored in the video commemorating the 50th anniversary of the federal agency to support cocoa culture (CEPLAC) (Guimarães, 2019).

In the South of Brazil, in the state of Santa Catarina, monkeys are part of the cosmology of the Kaingang, an indigenous people of the Macro-Jê language trunk. Traditionally, the Kaingang occupied territory from west of São Paulo to the northwest of the state of Rio Grande do Sul, or the Tietê River to the Ijuí River (Rodrigues, 2016). The Kaingang are classified as socio-centric societies because they recognize dualistic socio-cosmological principles. The society and the components of the natural world are divided into complementary halves (half families), called *Kamé* and *Kairu*, mythological brothers:

> A *Kamé* woman must marry a *Kairu* man, and the children follow the paternal lineage. The sun is *Kamé* and the moon *Kairu*; the pine is *Kamé* and the cedar is *Kairu*, the lizard is *Kamé* and the monkey is *Kairu* and thus distribute all beings and cosmos. (Borba, 1882 *apud* Brighenti, 2012: 18)

In his analysis of the changes that have taken place in the Kaingáng food system resulting from environmental changes occurring throughout the 20th century, Oliveira (2009) identified monkeys as one of the Kaingang community favorite "old foods". As observed by Borba at the beginning of the 20th century (1908): "The hunts they most appreciate are the tapir, the monkey (*caierê*), the bugio [monkey] (*gonguê*), the boar (*crenguê*), the collared peccary (*ókxá*), the coati (*xeê*)..." (pp. 8–11). According to the current distribution of primate species, the primates mentioned in this text were probably *Sapajus nigritus* and *Alouatta guariba*, respectively. These and other animal species used as food are no longer found in the indigenous land (called Xapecó). Monkeys play a significant role in the myth about the origin of Kaingang and the natural world. As in other traditional myths, monkeys are metamorphosed people:

> in time immemorial, there was a flood that covered the entire earth, inhabited by Kaingangs' ancestors. Only the top of the Krinjinjimbé mountain [Serra do Mar] stood out from the flood waters. Some Kaingangs and some Curutons [according to Borba, 1908, they were slaves of Kaingangs] managed to reach the mountain but were left without food for a long time. When there was no food and they were desperate the birds *saracuras* [*Aramides* spp. Rallidae] came to their aid and brought sand in order to form a spacious plain on the hill, which gave plenty of ground to the Kaingangs, with the exception of those who took refuge in the trees. These were transformed into monkeys [probably capuchin monkeys] and the Curutons into howler monkeys. (Oliveira, 2009: 140 translation from Portuguese by the authors)

The myth goes on to tell that through various actions these "fathers" of the Kaingang people gave rise to the various animal species.

In the highly modified landscape of Santa Catarina, there is also evidence of the relationship between the original inhabitants and the endemic animals of the region. The indigenous vocabulary of the Xoklengs, people of the language Family Jê, includes words describing former environments, their histories and their culture. One of their villages was called Bugio village because before the construction of the dam started in the late 1970s, they hunted monkeys (probably *A. guariba*) (Cruz, 2014). Mythical and sacred aspects are also present in memory, as in the narrative about Kámlẽn, the first Chief and spiritual leader of the Xoklengs. The following testimony was given by his grandson in the period of contact and settlement imposed by the Brazilian state in 1914:

> A group of men went hunting in Serra da Abelha and stayed there for over a month. One day Kámlẽn went hunting in the morning and came back with a fever. The boys had killed a howler monkey on the edge of a river that same day, cut it, and divided the monkey into parts. Kámlẽn felt a bad pain in his chest and said that some boys had cut him and hurt him. Kámlẽn died in the afternoon of that same day. Many believe that Kámlẽn is the spirit of the *Bugio* [the recent meaning in Portugues is howler monkey]. He talked to these animals and had special powers, as he was a healer. He was buried in Serra da Abelha, an area claimed to be traditional Xokleng territory where there are some ancient cemeteries, as the Xokleng used to perform ceremonies for the dead. There is also the belief that to get close to the place where Kámlẽn is buried and in order to kill a *Bugio* in the forest it is necessary to ask him for permission and ask this monkey for permission as well, not killing too many individuals and asking the others of the group not to leave and to stay in the forest. (Cruz, 2014: 223–224, translation from Portuguese by the authors)

Evidence of the Atlantic forest native peoples interacting with primates, and with the natural world, comes from the accounts, drawings, maps, stories and tales of the European navigators that arrived in Europe from the 16th century onwards (cf. Veracini & Teixeira, 2017). It also is attested by the various examples of pre-Columbian rock art found in the Atlantic Forest (e.g., Fonseca Souza de, 2016; de Queiroz et al., 2022).

4.3 EUROPEAN EXPLORERS, SETTLERS, AND NON-HUMAN PRIMATES IN THE NEOTROPICS SINCE THE 16TH CENTURY: A HISTORICAL OVERVIEW

After Europeans began crossing the Atlantic at the end of the 15th century, this led to a network of interactions between them and non-Europeans, and between Europeans and the animals of the Americas. Access to sources of wealth in faraway possessions was the aim of the biopolitical and geopolitical project carried out in the new territories (Gustavson & Haapoja, 2019). Over centuries, these interactions resulted in European empires wielding influence over the life and death of both human and non-human life forms (Kanji, 2017), as evidenced by reports, travel diaries, studies, and historical documents that expressed the Eurocentric values prevailing in the colonial period. Early accounts of New World nature reveal an affinity with a typical medieval culture, and with elements of exoticism, beauty, abundance, and variety of flora and fauna, reflecting both an Edenic vision and a mercantilist European logic also influenced by the dominant Catholic culture (Brennan, 2006; Seixas, 2003; Asua & French, 2005; Camphora, 2021). Nevertheless, explorers and visitors to America produced first-hand reports that contributed to the development of the earliest elements of a natural history of New World animals (Olmi, 1992; Billé, 2009; Veracini & Teixeira, 2016; Veracini, 2019). Neotropical primates, described as very abundant and always present in the colonized territories, are among the animals most widely represented in these sources. New World primates entered the life of European explorers and settlers as food, medical cures, resources for material culture, and as beings capable of expressing feelings and human attitudes (cf. for review Asua & French, 2005; Veracini & Teixeira, 2017; Camphora, 2021).

The original inhabitants of the Americas contributed to the systematic knowledge, exploration, and exploitation of the extensive territory, providing a vital understanding of its plentiful resources and wealth (Marroquin Arredondo & Bauer, 2019; Camphora, 2021). Latin American commodities became so important that they subverted the world economic order (Topik et al., 2006). In Brazil, over the first three decades of the 16th century, the pre-colonial period was characterized by the exploitation of any natural resources which could be commercialized in Europe (Dean, 1997). For example, groups of Tupinambá peoples who lived on the Atlantic coast of present-day Brazil took charge of identifying, cutting, preparing, and loading wood onto Portuguese and French ships, in exchange for iron tools such as machetes and axes, or mirrors, fishhooks, and colorful beads (Bueno, 1998). Ships carrying wood and other natural products, also carried many exotic animals, including primates, the most common animal sold in European markets (cf. Veracini, Chapter 9, this volume). Many primates were obtained by Europeans from local populations who kept them as pets, in exchange for objects of low value which, for the indigenous peoples, were a novelty (Teixeira & Papavero, 2010; Veracini & Teixeira, 2017; **Fig. 4.13**). One

FIGURE 4.13 Detail of Nicolas Vallard's Atlas (southeast America), Map 12, showing Europeans and indigenous people with their pets (parrots and monkeys, the latter similar to capuchin monkeys). Dieppe, 1547. Source: The Huntington Library Berkeley, San Marino (CA).

of the first witnesses of such exchanges was the Italian Antonio Pigafetta (1492–about 1531), who in 1519 took part in the first circumnavigation of the globe by the Spanish fleet of King Charles V under Ferdinand Magellan. When the Spanish fleet arrived at present-day Rio de Janeiro, the sailors met local people who kept animals as pets. He observes that "they have many parrots and give 8 or 10 of them for a mirror, and small sea-cats (*gati maimoni*, in this case the golden lion tamarin, *Leontopithecus rosalia*) similar to lions but yellow, what a wonderful thing!" (Pigafetta, 1984: 88). The Frenchmen André Thevet (1516–1590) and Jean de Léry (1536–1613), who participated in the French expedition to find a new settlement (the so-called *France Antartique*) on a small island in Guanabara Bay in front of present Rio de Janeiro in the mid 16th century, left eye-witness accounts of the customs of the Tupinambá people. These were said by Thevet to "hunt monkeys for food" (Thevet, 1557: 109), and Jean de Léry also mentions their hunting technique and their successful "taming":

> These savages have no other way of catching them [primates] if they do not use arrows to shoot them. They can also catch them with a sort of net. The monkeys fall down stunned and sometimes seriously wounded; after they are cured, they keep them at home for a while, and then they exchange them for any goods that foreign travellers can give, and I should say they are tamed, because at the beginning when the monkeys are caught, they are so fierce that they bite our fingers and may bite the hands of those who hold them right through. (Léry, 1578, 144–145, translation from French *apud* Veracini & Teixeira, 2017: 21-22)

In 1549, Hans Staden, a German mercenary who represented the interests of German bankers in the New World, was held prisoner in a village of the native Tupinambá people for nine months. In his book of memories of that period, he observed that the French explorers bought primates from native peoples, and he purchased three primates to take back to Germany (Staden, 1557). In the territories occupied by Spaniards, Martyr d'Anghiera (1457–1526), a Spanish historian of Italian origin, reported many accounts of the recently "discovered lands" in a series of letters written between 1488 and 1525 (*De Orbe Nouo Decades* eds. 1511, 1516, 1530). He also mentioned the local habit of the indigenous peoples of the Cumana region, present-day Venezuela, of keeping primates as pets:

> That land produces wild cats [monkeys called *gatos selvajes*]; the mother takes her young in her arms and wanders through the treetops, and it is necessary to wound her with arrows to remove her son from her; ... When the mother falls dead, the people take and breed the offspring. (Anghiera, 1989, Decade VIII, Book VII)

Many literary and iconographic sources of this, and the successive colonial period, show that primates were a common element in the daily life of the Aboriginal Americans. In addition to being a source of nourishment (see **Box Details 4.1**), primates played important cultural and symbolic roles in their lives. For example, Nieuhof Joannes, who worked for the Dutch East India Company (VOC) in Brazil between 1640 and 1649, spoke of the "Indians of Pernambuco" (Tapuia), reporting that primates, together with parrots, almost always accompanied people when they moved. They were especially appreciated by women:

> The woman would carry everything, and in addition to the children, she might also carry a monkey or a parrot. In addition, the bones of a monkey species called *namby paia*[2] were used as an embellishment for earlobes. (Nieuhof, 1682: 312, 319)

The accounts of Matías Ruiz Blanco (1643–1708) and Francisco Tauste (1625–1685) (cf. Urbani & Rodríguez, 2021) from northern Venezuela suggest that monkeys, and especially howlers, were of symbolic significance for indigenous shamans. These literary sources corroborate the corpus of evidence indicating that in the pre-colonial period primates played a significant role in the material and symbolic culture of the ancient inhabitants of Central and South America.

Among the first European reports on Neotropical mammals, there are many references to animals hunted or eaten by native peoples, or even by Europeans (e.g., "wild pigs" – peccaries – iguanas, snakes, crocodiles, fish, manatees, parrots, and other birds) (cf. Asua & French, 2005). In the 16th century, specific references to the consumption of monkey flesh, on the other hand, were less frequent, although primate hunting is referred to frequently (e.g., Gómara, 1552; Léry, 1578; Thevet, 1557; Souza, 1974). In the absence of salt, Indigenous peoples preserved mammal meat by smoking and drying it. Friar Gaspar de Carvajal, who participated in the Amazonian expedition in 1541–42 of the Spanish explorer Francisco Orellana (1511–1546), reported that the indigenous inhabitants of a village in the center of Amazonia brought them a great amount of food such as "turtles, manatees and other fish, and partridges, and smoked cats and monkeys" (Carvajal et al., 1986: 53). The use of monkey meat by the indigenous peoples and its preservation in this form, led the first Europeans who reached America to believe that cannibalism was a frequent and widespread practice throughout the continent (Armas, de, 1888). Accounts found in earlier European chronicles suggest that Europeans may have mistakenly identified the bodies of monkeys as human corpses. Here are some examples quoted by Ignacio de Armas (1888: 23): "[...] I heard they used to cook [for Montezuma] the flesh of very young boys; and as he had so many different stews and so many things, we checked to see if this was of human beings" (Diaz del Castillo, 1961); "[...] and we found on the path baskets full of corn and roasted children" (López de Gómara, 1552). Similarly, "Pizarro judged the Peruvians to be cannibals because he found feet and hands in a pot that was hanging over a fire" (*apud* Veracini & Teixeira, 2017: 17). Even the naturalist Oviedo formed the impression that

[2] Namby or nambi means ears in Tupi Guarani; paya - paia: pendulum: http://etnolinguistica.wdfiles.com/local--files/biblio%3Abarbosa-1951-pequeno/barbosa_1951_tupi-portugues.pdf

the indigenous peoples were cannibals when he saw arms, legs, and hands along with other meat on the day he arrived in Santa Marta: "These Caribe Indians are archers and eat human flesh, and that is known because in some houses there were found pots with members of men and women, including arms and legs and a hand" (Oviedo y Valdés, 1526, Lib XVII, Cap X).

In the 18th century, the Jesuit Friar Juan de Rivero noted that in the absence of salt for preservation, the people of the Orinoco basin dried and smoked monkeys to preserve them, and it was:

> horrifying to see animals like these that are so similar to human creatures, after they have passed through the fire, because both in face and hands, and the rest of the figure looks like a bunch of roasted, coal-black guys. (Rivero de, 1956: 10–11).

As observed by Humboldt, who visited the Amazon in the next century, "smoked monkey, particularly those with round heads, has an impressive resemblance to a child; wherefore when Europeans are forced to eat that flesh, they ask people to cut off their heads, and hands, and they receive only the trunk" (Armas, de, 1888: 24).

As knowledge of local customs increased, this perception and the resulting repulsion were overcome. Europeans assimilated the customs, diet, and modes of survival of native peoples, and these strategies became essential for their survival (Camphora, 2021). In rural and urban colonial environments, non-human animals were used by settlers as food and pets, and body parts were used to make utensils. For example, the Jesuit Friar Acosta, who in the second half of the 16th century lived for many years in the territory occupied by Spaniards, told of a pet monkey, probably a capuchin at the home of the Governor of Cartagena that was used to perform certain tasks normally carried out by people: it was "sent to the tavern for wine, and they put money in one of its hands and in the other put the pitcher for the wine" (Acosta, 1590: 289–290 *apud* Veracini & Teixeira, 2017: 34). Early on, Europeans began to eat wild animals and also learned some of the hunting techniques and food preferences of the indigenous peoples; they learned "how to identify animals with tasty flesh – as well as poisonous ones" (Camphora, 2021: 37). In the territory of present-day Brazil, the art of hunting, the central axis of the social and symbolic life of Amerindian peoples, became an object of systematic attention in the written accounts of Brazilian colonial society. Hunting consolidated ties and mediated relations between Amerindians, European explorers, enslaved Africans, colonists, naturalists, and travelers. Records of the period convey the idea of hunting as a sort of meeting ground, an activity that brought Amerindian groups, explorers, and settlers together.

The initial reluctance to eat the flesh of primates because of the "human" likeness to monkeys was forgotten. In 17th century French Guiana, settlers ate monkeys' brains, which, "after initial repulsion, were considered a delicacy" as reported by the French botanist Pierre Barrere, in 1741. As mentioned above, Europeans who traveled in the Orinoco basin used to cut off the heads and feet of monkeys before eating them. The Jesuit Friar Filippo Salvatore Gilij, who lived many years in the province of Caracas in the second half of the 18th century, noted that howler monkeys were highly esteemed by local people and by Spaniards, but in order to eliminate nausea caused by the human appearance of the monkeys, the Spaniards cut off the heads and feet and placed them out of sight (Gilij, 1780). The already mentioned Friar Rivero (1956) reports that monkeys of the Orinoco "were a good meal for the friars although there was no salt that would have given some seasoning to that bland meat" (1956: 236–237).

Among Brazilian settlers of this period, monkey flesh was reserved for important individuals. For example, the owner of a Rancho in the territory of São Paulo invited His Excellency the Conde de Assumar to a dinner "based on monkey meat and ants", saying that "the greatest delicacy to be found in those woods, and the ants, after being properly roasted on the fire, could be compared to the best Flanders butter" (Anonymous, 1949: 307–308). Christmas Day, which was a solemn day, was celebrated by Portuguese settlers in southern Brazil with "monkey meat cooked in four different ways with beans and *toucinho* [Brazilian bacon]" (Almeida, 1841, *apud* Papavero e Teixeira, 2007: 262).

In the city of Rio de Janeiro, as recently as the 19th century, people's eating habits followed indigenous patterns. The favored source of meat was lizards, monkeys, skunks, deer, armadillos, and pacas, which were sold door to door by slaves (De Los Rios, 1946 *apud* Camphora, 2021). The exotic taste for tropical animals became a characteristic of colonial Brazil, and the only references to any kind of food restrictions concerned the consumption of poisonous species and a few animals considered unfit for consumption (e.g., sloths). Every other kind of animal could be eaten, and the Catholic Church made no distinction between pure and impure foods (e.g., **Fig. 4.14**). In April 1832, on his first visit to Rio de Janeiro's Atlantic Forest, Charles Darwin bore witness to the hunting of howler monkeys using a variety of weapons, such as knives and thick pieces of wood, as an everyday habit:

FIGURE 4.14 Hunters with their prey in colonial Brazil, 19th century. Engraving. (Source: Maximilian Wied, Prinz Von, 1820.)

> My companion, the day before, had shot two large bearded monkeys. These animals have prehensile tails, the extremity of which, even after death, can support the whole weight of the body. One of them thus remained fast to a branch, and it was necessary to cut down a large tree to procure it. This was soon cut, and down came tree and monkey with an awful crash. Our day's sport, besides the monkey, was confined to sundry small green parrots and a few toucans. (Darwin, 1913: 29)

Over the course of the centuries, hunting acquired functions other than the supply of food. Animals were hunted for the local and European pet trade and in response to demand from European museums. European interest in Brazilian fauna was always very great and from 1770 onwards, animal shipments to Portugal were systematic. Besides monkeys of various kinds, a variety of animals, such as birds and felines, were also sent. Prior to the opening of the Brazilian ports in 1808, animals arrived in Europe via Dutch and French brokers, and from the 19th century onwards animals were sent to natural history museums all over Europe (Pereira, 2012; Camphora, 2021). Hunting was also carried out to eliminate the wild animals that threatened settlers in villages and hamlets, damaging their crops and killing their livestock (Camphora, 2021). Moreover, primate skins had an economic value as they could be used to produce a wide variety of objects, as noted by the Brazilian naturalist Alexandre Rodrigues Ferreira in his "philosophical journey" to Amazonia and Mato Grosso in the second half of the 18th century. He writes about howler monkeys:

> The skins of male guaribas [howler monkeys] are used to make excellent leather for footwear. Some of the other individuals' skins, tanned with fur, are used for holsters, chair covers, cloaks for weapon covers and hunting skins. Those of the black guariba, for the grenadiers' mitres. Guitar strings are made from their intestines. (Ferreira, 1783–1792: 140)

The French artist Debret recounted that having monkeys as pets was a habit in the 19th-century city of Rio de Janeiro, the capital of Brazil:

> Very close to her, within her reach, lies the linen basket; the lid slightly open, the end of a whip, made entirely of leather, can be glimpsed. It is used by slave owners, to intimidate their slaves. Alongside her, a small monkey, chained to one of the legs of the furniture, is a simple pastime for the lady of the house; although it is a privileged slave, with freedom of movement and manners, it is nevertheless held in check, from time to time, by the threat of the whip. (Debret, 1972: 121 *apud* Camphora, 2021)

The scarce, or non-existent, means to combat, or even minimize, the suffering produced by a wide variety of tropical diseases and epidemics were seen as justification for the controversial therapeutic practices of the colonial period. The perceived medicinal benefits derived from native formulae gave salience to the divine and miraculous powers of the elements of the natural world. In the colonial territories of present-day Brazil, horns, teeth, claws, bones, skins, hooves, shells, and tallow – all the animal products that could not serve as food – could be turned into remedies or amulets (Camphora, 2021). Settlers and missionaries learned to observe the behavior of monkeys that made use of certain plants, such as the copaíba (*Copaifera langsdorffii*), rubbing their wounds on the trunks of this healing tree. Some 16th-century accounts describe the self-medicating skills of howler monkeys (*Alouatta* spp.):

> and as they feel the arrow wounds that the Indians have inflicted on them, if they have not been felled, they take to the trees, where they settle in to chew up leaves and stick them into the wound, and in doing so drink their own blood and are healed; and on many an occasion they have been seen to remove the arrow and throw it back at its thrower, who thus may also be wounded; or on other occasions, let themselves fall, arrow in hand, on top of the Indian who has harmed them. (Souza, 1974: 138 *apud* Camphora, 2021: 58)

The Brazilian military officer and ethnologist, Couto de Magalhães (1837–1898), described another howler monkey that displayed the same behavior: when hit by a bullet, searching immediately for leaves, chewing them up and applying the paste to his wounds. In his travel diary, Couto de Magalhães wrote that, overcome by such a show of sentiment and instinct "bordering on intelligence", he put an immediate end to his hunting day, deeply moved by what then appeared to him as the extraordinary cruelty of pursuing animals "for whom death might mean more than physical destruction" (Magalhães 1975 *apud* Camphora, 2021). Muriquis were hunted for sport by explorers of the São Paulo state (Mittermeier et al., 1987; Talebi & Soares, 2005), and the demand for monkeys for medicinal purposes, illegal trade, for their skins and other body parts for ornament, and for food consumption, has resulted in the elimination of muriquis in several forest remnants in the state (Aguirre, 1971; ICMBio, 2011).

The belief in the therapeutic properties of monkey meat seems to have been common among Europeans since the 16th century in South America (Veracini & Teixeira, 2017). Juan Castellano (1522–1607), in his history of the conquest of Colombia by Spaniards, reported an episode involving a sick Spanish soldier:

> Pedro Rodrigo Pardo, desiring to know whether he was alive or dead, brushed his teeth and mouth vigorously with roasted monkey (*mico*) meat that he had brought for food. They looked to see if he was still breathing and in fact they saw he was still alive, and they asked him if he recognized his companions. (Castellano, 1942, 113, *apud* Veracini & Teixeira, 2017: 16)

In the territories occupied by the Portuguese, the supposed benefits of monkey meat were underlined by the Jesuit Father Anchieta in his work of 1560 about the province of Sao Paulo "The monkeys [in these places] are innumerable [...] and they are a very healthy and good food for the sick" (Anchieta, 1984: 123–152). Francisco Hernández (1651) also recalls the therapeutic qualities of monkeys, whose powdered bones could relieve pain and the sweat provoked by the French disease (syphilis). According to Alexandre Rodrigues Ferreira, in the second half of the 18th century:

> The country's healers [does not specify whether of indigenous or European origin] prescribe its [of howler monkeys] meat as a diet for those suffering from venereal complaints. By their advice, the hunters are careful to scrupulously collect the kneecaps of all those they kill, to be strung on strings, which serve as bracelets to those suffering from vaginal flow. The celebrated Macau beads are made from them, which they say, when worn on the left arm, cure all kinds of haemorrhoids by occult virtue. To the same bone drum [howler hyoids] extravagant virtues are attributed. (Ferreira, 1783–1792: 141)

In 1759 the Franciscan priest, Antonio Caulin, also reported that when he was close to the Island of Trinidad the use of howler hyoids was considered of great medicinal value for those suffering from asthma and other illnesses of the chest (Caulín, 1966 *apud* Urbani & Cormier, 2015: 264). Many traditional people (e.g., the Tikuna in Colombia, Maldonado & Waters, 2020; the Lacadon people of Mexico, García del Valle et al., 2020; Lokono and Kari'na in Guyana and Suriname and other ethnic groups of Arawakan and Caribbean people, Rybka, 2020 among others) still refer to the therapeutic properties of howler hyoids for some respiratory diseases (e.g., whooping cough). Recipes vary, but most commonly involve using the howler monkey's larynx as a cup.

4.4 CONCLUDING REMARKS

Modern humans and Neotropical primates have shared Central and South America for many thousands of years. The result of this interaction has produced a cultural heritage still visible today in native indigenous populations, and in historical and archaeological evidence. This ancient interaction, along with the more recent interaction resulting from the European expansion of the Modern period, has influenced the dynamics of primate populations. As a consequence, the distribution, density, and behavior of the primate taxa we see today went through both an ancient and a more recent process of selection by, and co-evolution with human beings. The future of primates can be closely linked with the future of native populations, whose contribution to the conservation of natural habitats is now well established. However, special attention must be paid to the impact of traditional hunting on primate populations, which have already been reduced, or in some cases decimated, by other factors. Traditional hunting of primates is inextricably linked to the taboos and social and religious rules that controlled the indiscriminate capture of these and other animals. The Amerindian peoples' concept of nature (flora and fauna) gives all creatures agency and intention, and the natural world is perceived as a system in which all parts are interdependent. To Amerindians the original condition common to humans and animals was not that of the animal, but that of the human. Humans have remained the same: animals are ex-humans. Non-human primates play important roles in cosmogony, rites of passage, and taboos, as well as in the origin of the world myths. Although native peoples have gone through periods of annihilation and genocide in recent centuries, many of these myths survive in contemporary societies, and their preservation and understanding are vital elements in attempts to forge a long-term future for modern humans and the natural environment.

ACKNOWLEDGMENTS

The authors would like to thank Bernard Wood for his critical reading of the text and his important suggestions; Alcide Pissinati for his help regarding the Atlantic Forest; Giulia Cavallo for her suggestions in some parts of the work; Hilary Morris for the English review of the first manuscript draft.

FUNDING OF THE FIRST AUTHOR

Portuguese national funds through FCT—Fundação para a Ciência e a Tecnologia in the context of the celebration of the contract under the 'Norma Transitoria', DOI 10.54499/DL57/2016/CP1456/CT0001.

REFERENCES

PRIMARY SOURCES

Acosta, J.de (1590) [2002]. *Historia natural y moral de las Indias*. Madrid: Dastin.

Almeida, F.J.L. (1841). *Diario da viagem do Dr. Francisco José de Lacerda e Almeida pelas capitanias do Pará, Rio Negro, Matto-Grosso, Cuyabá, e S. Paulo, nos annos de 1780 a 1790*. São Paulo: Typ. de Costa Silveira.

Alvard, M.S. (1993). Testing the "ecologically noble savage" hypothesis: Interspecific prey choice by Piro hunters of Amazonian Peru. *Human Ecology*, 21, 355–387.

Anchieta, J.de (1984). *Epistola quam plurimarum rerum naturalium quae S. Vicenti (nunc St. Pauli) provinciam incolunt*. In *Cartas, correspondência ativa e passiva*. São Paulo: Edições Loyola, pp. 123–152.

Anglería, M.de P. (Martir de Anghiera) (1989). *Décadas del Nuevo Mundo*. Madrid: Ediciones Polifemo.

Anonymous. (1949). Diario de uma Jornada que fez o Exmo. Senhor Dom Pedro desde o Rio de Janeiro athé a Cidade de Sao Paulo e desta athé Minas anno 1717. *Revta Serv. Patrim. Hist. Art. Nac, Rio de Janeiro*, 3, 307–308.

Armas y Céspedes, J.I. de (1888). *La Zoologia de Colón y de los primeros exploradores de América*. Habana: Estabelecimento Tipográfico.

Barrere, P. (1741). *Essai sur l'Histoire naturelle de la France Equinoxiale*. A Paris, chez Piget.

Cardim, F. (1997). *Tratados da terra e gente do Brasil*. Lisboa: Comissão Nacional para as Comemorações dos Descobrimentos Portugueses.

Carvajal, G.de, Rojas, A.de, Diaz, R. & Arias de Almesto, P. (1986). *La aventura del Amazonas*. Madrid: Rincón de Lectura.

Castellano, J. (1942). *Historia de la Gobernacion de Antioquia y del Choco*, vol. *I*. Bogotá.

Costa, A.K.S. (2017). *Etnoprimatologia de* Sapajus flavius *(Schreber, 1774) em fragmentos de Floresta Atlântica no Nordeste do Brasil*. Dissertação de Mestrado, PPGEC - Universidade Estadual da Paraíba.

Daniel, J.S.J. (1976). Thesouro descoberto no màximo rio Amazonas. 1°, 2°, 3° partes. *An. Biblio. Nac. Rio de Janeiro* 95 (I), 1–437.

Darwin, C. (1913). *A Naturalist's Voyage Round the World, the Voyage of the Beagle*. London: John Murray.

Debret, J.B. (1972). *Viagem pitoresca e histórica ao Brasil.* São Paulo: Livraria Martins/USP.

De Los Rios Filho, A.M. (1946). *O Rio de Janeiro Imperial.* Rio de Janeiro: A Noite.

Diaz del Castillo, B. (1961). *Historia verdadeira de la conquista de la Nueva Espana*. México DF: Fernández Editores.

Elliot, D.G. (1913). *A Review of the Primates. American Museum of Natural History, vol. 1*. New York: The American Museum of Natural History.

Ferreira, A.R. (1783–1792). *Viagem Filosófica pelas Capitanias do Grão-Pará, Rio Negro, Mato Grosso e Cuiabá. Memorias, Zoologia, Botanica.* Rio de Janeiro: Conselho Federal de Cultura (ed. 1983).

Gilij, F.S. (1780). *Saggio di storia Americana, o sia Storia Naturale, civile e sacra de regni e delle procincie spagnuole di Terra-ferma nell'America meridionale.* https://play.google.com/store/books/details?id=PQQhrv3mj4YC&rdid=book-PQQhrv3mj4YC&rdot=1.

Gómara, F.L.de (1552) [1941]. *Historia general de las Indias.* Madrid: Espasa-Calpe.

Hernández, F. (1651). *Rerum medicarum Nova Hispaniae Thesaurus seu Nova plantarum, animalium et mineralium mexicanorum historia*. Rome.

Léry, J.de (1578). *Histoire d'vn voyage fait en la terre dv Bresil, avtrement dite Amerique*. à la Rochelle.

Magalhães, Couto de J.V. (1975). *O selvagem.* São Paulo: Itatiaia/EDUSP.

Nieuhof, J. (1682). *Memorável viagem marítima e terrestre ao Brasil.* Ed Portugues (1966). São Paulo: Livraria Martins.

Oviedo y Valdés , G.F.de (1526) [1851]. *De la Natural Historia de las Indias*. Madrid Imprenta de la Real Academia de la História.

Pigafetta, A. (1984). *Antonio Pigafetta il primo viaggio intorno al mondo.* Riproduzione del Ms. Ambrosiano L103 [18]. Milano: M. Pozzi & N. Pozza.

Rivero, J. (1956). *Historia de las Misiones de los llanos de Casanare y los rios Orinoco y Meta.* Bogota, Presidencia de la Republica.

Soares de Souza, G. (1587) [1851]. *Tratado Descriptivo do Brazil.* Varhagen C di AF (ed). Rio de Janeiro: Typographia Universal de Laemmert.

Soares de Souza, G. (1974). *Notícia do Brasil*. São Paulo: MEC.

Staden, H. (1557) [1970]. *Warhaftige Historia. La mia prigionia tra I cannibali 1553-1555.* Guadagnin A (ed). Longanesi & C. Milano.

Thevet, A. (1557). *Les singvlaritez de la France Antarctiqve, avtrement nommée Amerique: & de plusieurs Terres & Isles decouuertes de nostre temps.* Chez les heritiers de Maurice de la Porte, Paris.

Secondary Sources

Aguirre, A.C. (1971). *O mono* Brachyteles arachnoides *(E. Geoffroy). Situação atual da espécie no Brasil.* Rio de Janeiro, Brazil: Academia Brasileira de Ciencias.

Ahlbrinck, W. (1931) Encyclopaedie der Karaïben. Koninklijke Akademie van Wetenschappen, Amsterdam.

Aquino, R. (1989). Preliminary survey of the population densities of Cacajao calvus uacayalii. *Primate Conservation*, 9, 24–26.

Arredondo, O. & Varona, L.S. (1983). Sobre la validez de Montaneia anthropomorpha Ameghino, 1910 (Primates: Cebidae). *Poeyana*, 255, 1–21.

Asua, M. de & French, R. (2005). *A New World of Animals. Early Modern Creature of Iberian America*. Burlington: Ashgate.

Baker, M. (2013). Revisiting capuchin monkeys (*Cebus capucinus*) and the ancient maya. *Revue de Primatologie*. Electronic document, https://primatologie.revues.org/1683, accessed April 14, 2022.

Balée, W.L. (1989). The culture of Amazonian forests. *Advances in Economic Botany*, 7, 1–21.

Balee, W.L. (2000). Antiquity of traditional ethnobiological knowledge in Amazonia: The Tupi-Guarani family and time. *Ethnohistory*, 47(2), 399–422. https://www.muse.jhu.edu/article/11649.

Barnett, A. & da Cunha, A.C. (1991). The golden-backed uacari on the upper Rio Negro, Brazil. *Oryx*, 25(2), 80–88.

Barnett, A.A., Borges, S.H., de Castilho, C.V., Neri, F.M. & Shapley, R.L. (2002). Primates of the Jaú national park, Amazonas, Brazil. *Neotropical Primates*, 10(2), 65–70.

Barnett, A.A., Castilho, C.V.D., Shapley, R.L. & Anicácio, A. (2005). Diet, habitat selection and natural history of Cacajao melanocephalus ouakary in Jaú National Park, Brazil1. *International Journal of Primatology*, 26(4), 949–969.

Basso, E.B. (1973). *The Kalapalo Indians of Central Brazil.* New York: Holt, Rinehart and Winston, Inc.

Becker, J., Baker, A.J., Frampton, T., Pullen, P.K., Bales, K.L., Mendoza, S.P. & Mason, W.A. (2013). Pitheciines in captivity: Challenges and opportunities, past, present and future. In Veiga L, Barnett A, Norconk MA & Ferrari SF (eds), *Evolutionary Biology and Conservation of Titis, Sakis, and Uacaris.* Cambridge: Cambridge University Press, pp. 344–349.

Billé, P. (2009). *La Faune bresilienne dans les ecrits documentaires du XVI siécle*. Paris: Honoré Champion.

Bond, M., Tejedor, M.F., Campbell, K.E. Jr., Chornogubsky, L., Novo, N. & Goin, F. (2015). Eocene primates of South America and the African origins of New World Monkeys. *Nature*, 520, 538–541.

Bonvicino, C.R. (1989). Ecologia e comportamento de *Alouatta belzebul* (Primates: Cebidae) na Mata Atlântica. *Revista Nordestina de Biologia*, 6(2), 149–179.

Borba, T.M. (1908). *Actualidade Indígena (Paraná, Brazil).* Curitiba: Impressora Paranaense.

Borroto-Páez, R. & Woods, C.A. (2012). Status and impact of introduced mammals in the West Indies. In Borroto-Páez R, Woods, CA & Sergile FE (eds), *Terrestrial Mammals of the West Indies.* Gainesville, FL: Wocahoota Press and Florida Museum of Natural History, pp. 241–258.

Boubli, J.P., Urbani, B., Caballero-Arias, H., Shepard, G.H. & Lizarralde, M. (2020). Primates in the lives of the Yanomami people of Brazil and Venezuela. In Urbani B & Lizarralde M (eds.), *Neotropical Ethnoprimatology: Indigenous Peoples' Perceptions of and Interactions with Nonhuman Primates*, Springer Nature, pp. 199–224.

Brasil (2012). *Portaria nº 37, de 23 de março de 2012. Aprova o Plano de Ação Nacional para a Conservação dos Primatas do Nordeste - PAN Primatas do Nordeste.* Senado. Brasília: DF.

Brehm, A.E. (1895). *Brehm's Life of Animals: A Complete Natural History for Popular Home Instruction and for the Use of Schools*. Chicago: Marquis. Collection. Biodiversity Heritage Library.

Brennan, R.E. (2006). *European Representations of the New World in Travel Narratives and Literature, Late-Fifteenth to Mid-Seventeenth Centuries*. Unpublished PhD Dissertation. Cardiff University.

Brighenti, C.A. (2012). Povos indígenas em Santa Catarina. In Vulfe Nötzold AL, Alpini Rosa H & Bringmann SF (eds), *Etnohistória, história indígena e educação: contribuições ao debate*. Porto Alegre, Pallotti, pp. 37–65.

Bruner, E. & Cucina, A. (2005). Alouatta, Ateles, and the ancient Mesoamerican cultures. *Journal of Anthropological Sciences*, 83, 111–117.

Bueno, E. (1998). *A viagem do Descobrimento: a Verdadeira História da Expedição de Cabral*. Objetiva: Rio de Janeiro.

Bush, M.B., McMichael, C.H., Piperno, D.R. et al. (2015). Anthropogenic influence on Amazonian forests in pre-history: An ecological perspective. *Journal of Biogeography*, 42, 2277–2288.

Callaway, E. (2021). Ancient footprints could be oldest traces of humans in the Americas. *Nature*, 597, 601–602. https://www.nature.com/articles/d41586-021-02597-1.

Campbell, C.J. (ed). (2008). *Spider Monkeys: Behavior, Ecology and Evolution of the Genus Ateles*, vol. *55*. Cambridge University Press.

Camphora, A.L. (2021). *Animals and Society in Brazil from the Sixteenth to Nineteenth Centuries*. Cambridgeshire, UK: The White Horse Press.

Cartelle, C. & Hartwig, W.C. (1996). A new extinct primate among the Pleistocene megafauna of Bahia, Brazil. *Proceedings of the National Academy of Sciences*, 93, 6405–6409.

Castro, C.S.S. & Casanova, C. (2018). Biodiversity conservation and local ecological knowledge from human communities in the Northeast of Brazil. *Modern Environmental Science and Engineering*, 4(9), 791–795. http://www.academicstar.us/UploadFile/Picture/2020-2/202022123643365.pdf

Caulín, A. (1966). *Historia de la Nueva Andalucía. Tomo I (1759)*. Academia Nacional de la Historia. Fuentes para la Historia Colonial de Venezuela 81, Caracas

Cesarino, P.D.N. (2014). Ontological conflicts and shamanistic speculations in Davi Kopenawas, The falling sky. *HAU: Journal of Ethnographic Theory*, 4(2), 289–295.

Chicchón, A. (2009). Working with indigenous peoples to conserve nature: Examples from Latin America. *Conservation & Society*, 7(1), 15–20.

Coe, M.D. (1978). *Supernatural Patrons of Maya Scribes and Artists*. Academic Press.

Coimbra, A., Buss, G. & de Azevedo, R.B. (2022). Abundância do Guariba-de-mãos-ruivas (*Alouatta belzebul*) na Reserva Biológica Guaribas, Paraíba, Brasil. *Biodiversidade Brasileira*, 12(1), 72–78. doi: 10.37002/biobrasil.v12i1.1842.

Coimbra-Filho, A.F., Câmara, I.G., & Rylands, A.B. (1995). On the geographic distribution of the red-handed howling monkey, *Alouatta belzebul*, in North-east Brazil. *Neotropical Primates*, 3(4), 176–179.

Colchester, M. (2004). Conservation policy and indigenous peoples. *Environmental Science & Policy*, 7(3), 145–153.

Cook, G.W. & Offit, T.A. (2013). *Indigenous Religion and Cultural Performance in the New Maya World*. UNM Press.

Cooke, S.B., Mychajliw, A.M., Southon, J. & MacPhee, R.D.E. (2017). The extinction of Xenothrix mcgregori, Jamaica's last monkey. *Journal of Mammalogy*, 98, 937–949.

Cormier, L.A. (2002). Monkey as food, monkey as child: Guaja Symbolic Cannibalism. In Fuentes A & Wolfe LD (eds), *Primates Face to Face: The Conservation Implications of Human-nonhuman Primates Interconnections*. Cambridge: Cambridge Press, pp. 63–84.

Cormier, L.A. (2003). *Kinship with Monkeys: The Guajá Foragers of Eastern Amazonia*. Columbia University Press.

Cormier, L.A. (2006). A preliminary review of Neotropical primates in subsistence and symbolism of indigenous lowland South American peoples. *Ecological and Environmental Anthropology*, 2, 14–32.

Cormier, L.A. & Urbani, B. (2008). The ethnoprimatology of the spider monkey (*Ateles* spp.): From past to present. In Campbell C (ed), *Spider Monkeys: The Biology, Behavior and Ecology of the Genus Ateles*. Cambridge: Cambridge University Press.

Cornick, L.A. & Markowitz, H. (2002). Diurnal vocal patterns of the black howler Monkey (*Alouatta pigra*) at Lamanai, Belize. *Journal of Mammalogy*, 83(1), 159–166. https://doi.org/10.1644/1545-1542(2002)083<0159:DVPOTB>2.0.CO;2

Costa, A.K.S. (2017). *Etnoprimatologia de* Sapajus flavius *(Schreber, 1774) em fragmentos de Floresta Atlântica no Nordeste do Brasil*. Dissertação de Mestrado. PPGEC - Universidade Estadual da Paraíba.

CPB – Centro Nacional de Pesquisa e Conservação de Primatas Brasileiros (2015). https://www.gov.br/icmbio/pt-br/assuntos/centros-de-pesquisa/cpb, accessed January 9, 2018.

Crosby, A.W. (2003). *The Columbian exchange: Biologia and cultural consequences of 1492*. Greenwood Publishing Group.

Cruz, T.M.S. (2014). *Etnoecologia de Paisagens na Terra Indígena Ibirama Laklãnõ, Santa Catarina, Brasil*. Dissertação. Programa de Pós Graduação em Ecologia da Universidade Federal de Santa Catarina.

Dean, W. (1997). *With Broadax and Firebrand. The Destruction of the Brazilian Atlantic Forest*. Univ of California Press.

Defler, T.R. (1989). The status and some ecology of primates in the Colombian Amazon. *Primate Conservation*, 10, 51–55.

Denham, W. (1987). *West Indian Green Monkeys: Problems in Historical Biogeography*, vol *24*. Basel, Switzerland: S. Karger.

de Queiroz, A.N., de Carvalho, O.A. & Pinto, R.R. (2022). Nonhuman primates in the archaeological record of northeastern Brazil. In *World Archaeoprimatology: Interconnections of Humans and Nonhuman Primates in the Past*, pp. 172–182.

Descola, P. (2006). Beyond nature and culture. *Proceedings of the British Academy*, 139, 137–155.

Di Fiore, A. & Campbell, C.J. (2007). The Atelines: Variation in ecology, behavior and social organization. In Campbell CJ, Fuentes A, MacKinnon KC, Panger M, Bearder SK (eds), *Primates in Perspective*. New York, Oxford University Press, pp. 155–185.

Erlandson, J.M., Graham, M.H., Bourque, B.J., Corbett, D., Estes, J.A. & Steneck, R.S. (2007). The Kelp highway hypothesis: Marine ecology, the coastal migration theory, and the peopling of the Americas. *The Journal of Island and Coastal Archaeology*, 2, 161–174. doi: 10.1080/15564890701628612.

Estrada, A., Garber, P.A. Rylands, A et al. (2017). Impending extinction crisis of the world's primates: Why primates matter. *Science Advances*, 3, e1600946. https://www.science.org/doi/10.1126/sciadv.1600946.

Falótico, T., Proffitt, T., Ottoni, E.B. et al. (2019). Three thousand years of wild capuchin stone tool use. *Nature Ecology and Evolution*, 3, 1034–1038. https://doi.org/10.1038/s41559-019-0904-4.

Fernández, D., Kerhoas, D., Dempsey, A. et al. (2022). The current status of the world's primates: Mapping threats to understand priorities for primate conservation. *International Journal of Primatology*, 43, 15–39. https://doi.org/10.1007/s10764-021-00242-2.

Fernández-Llamazares, Á, Díaz-Reviriego, I. & Reyes-García, V. (2017). Defaunation through the eyes of the Tsimane'. In Reyes García V & Pyhälä A (eds), *Hunter-Gatherers in a Changing World*. Springer, pp. 77–90.

Fernández-Llamazares, Á. & Virtanen, P.K. (2020). Game masters and Amazonian Indigenous views on sustainability. *Current Opinion in Environmental Sustainability*, 43, 21–27.

Ferronato, M.L. & Nunes, R.D.O. (2010). A exploração ilegal de madeiras na Terra Indígena Sete de Setembro, Cacoal-RO. *Revista Facimed*, 2(2), 1–12. On-line, https://ecopore.org.br/novo/wp-content/uploads/2020/05/A-EXPLORA%C3%87%C3%83O-ILEGAL-DE-MADEIRAS-NA-TERRA-IND%C3%8DGENA-SETE-DE-SETEMBRO-CACOAL-%E2%80%93-RO.pdf, accessed March 25, 2022.

Filho, R.F., de Castro, C.S.S., Casanova, C. & Bezerra, B.M. (2021). Uses of nonhuman primates by humans in northeastern Brazil. *Primates*, 62(5), 777–788. doi: 10.1007/s10329-021-00919-5.

Fleagle, J.G. (2013). *Primate Adaptation and Evolution*, 3rd Edition. Primate Adaptation and Evolution.

Fleagle, J.G. & Kay, R.F. (1997). Platyrrhines, catarrhines, and the Fossil Record. In Kinzey Warren G (ed), *New World Primates*. New York: Aldine de Gruyter, pp. 3–23.

Fonseca Souza de, T. (2016). *Pinturas rupestres e paisagem: um estudo de caso das representaçoes zoomorficas do Vale do Catimbau* - PE. Master thesis. Universidade Federal de Pernambuco, CFCH.

Franzen, M. (2006). Evaluating the sustainability of hunting: A comparison of harvest profiles across three Huaorani communities. *Environmental Conservation*, 33(1), 36–45.

Fraser, B. (2014). The first South Americans: Extreme living. *Nature*, 514, 24–26. https://doi.org/10.1038/514024a.

FUNAI. (2023). *National Foundation for Indigenous Peoples*. https://sii.funai.gov.br/funai_sii/informacoes_indigenas/visao/visao_terras_indigenas.wsp, accessed June 1, 2023.

Fundação SOS Mata Atlântica – INPE – Instituto Nacional de Pesquisas Espaciais (2015). *Atlas dos Remanescentes Florestais da Mata Atlântica – Período 2013–2014*. São Paulo: Fundação SOS Mata Atlântica.

Fundação SOS Mata Atlântica (2020). *Atlas dos Remanescentes Florestais da mata Atlântica Periodo 2018-2019. Relatorio Técnico*. São Paulo. https://cms.sosma.org.br/wp-content/uploads/2020/05/2020_Atlas_Mata_Atlantica_2018-2019_relatorio_tecnico_final.pdf

Garber, P.A., Estrada, A., Bicca-Marques, J.C., Heymann, E.W., Strier, K.B., Rylands, A.B. & Mittermeier, R.A. (2009). The diversity of the new world primates (Platyrrhini): An annotated taxonomy. In Garber PA, Estrada A, Bicca-Marques JC, Heymann EW & Strier KB (eds), *South American Primates: Comparative Perspectives in the Study of Behavior, Ecology, and Conservation*. New York: Springer, pp. 23–54.

Garcia, U. (2018). Macacos também choram, ou esboço para um conceito ameríndio de espécie. *Revista Do Instituto De Estudos Brasileiros*, (69), 179–204. https://doi.org/10.11606/issn.2316-901X.v0i69p179-204.

García del Valle, Y., Ruan-Soto, F., Guerrero-Martínez, F. & Reyes-Escutia, F. (2020). Local knowledge and cultural significance of primates (Ateles geoffroyi and Alouatta pigra) among Lacandon Maya from Chiapas, Mexico. In Urbani B & Lizarralde M (eds.), *Neotropical Ethnoprimatology: Indigenous Peoples' Perceptions of and Interactions with Nonhuman Primates*, Springer Nature, pp. 49–68.

González-Saavedra, L. (2020). The white monkey and the pelejo monkey: Primates in the social and cultural configurations of the Shawi people of Northwestern Peru. In Urbani B & Lizarralde M (eds.), *Neotropical Ethnoprimatology: Indigenous Peoples' Perceptions of and Interactions with Nonhuman Primates*, Springer Nature, pp. 331–342.

Groves, C. (2001). *Primate Taxonomy*. Smithsonian Series in Comparative Evolutionary Biology. Washington DC: Smitsonian Institution.

Guimarães, E.M. (2019). Quilombos, Japanese, and the Kinkajou "monkey" in polyculture farms of the south of Bahia, Brazil. *Antropologia Portuguesa*, 36, 191–211.

Gustavson, L. & Haapoja, T. (2019). Introduction: Museum of nonhumanity declares that dehumanization is history. In *Laura Gustavson and Terike Haapoja*. Museum of Nonhumanity, Project Muse. doi: 10.1353/book.75653.

Henfrey, T. (2020). Relationships between scientific ecology and knowledge of primate ecology of wapishana subsistence hunters in Guyana. In Urbani B & Lizarralde M (eds.), *Neotropical Ethnoprimatology: Indigenous Peoples' Perceptions of and Interactions with Nonhuman Primates*, Springer Nature, pp. 263–282.

Hershkovitz, P. (1949). Mammals of northern Colombia. Preliminary report No. 4: Monkeys (Primates) with taxonomic revisions of some forms. *Proceedings of the United States National Museum*, 98, 323–427.

Hill, C.M. (2002), Primate conservation and local communities-ethical issues and debates. *American Anthropologist*, 104, 1184–1194. https://doi.org/10.1525/aa.2002.104.4.1184.

IBAMA (2003), Instituto Brasileiro do Meio Ambiente e dos Recursos Naturais. Plano de Manejo, Fase 2, da Reserva Biológica Guaribas. *MRS Estudos Ambientais Ltda*, Brasília, pp. 1–520.

ICMbio. (2011). https://www.icmbio.gov.br/cepsul/legislacao/instrucao-normativa/349-2011.html.

IUCN, Red Book. (2018). https://www.iucnredlist.org/.

Jacques-Coper, A., Cubillos, G. & Ibarra, J.T. (2019). The Andean Condor as bird, authority, and devil: An empirical assessment of the biocultural keystone species concept in the high Andes of Chile. *Ecology and Society*, 24, 2–35.

Jerozolimski, A. & Peres, C.A. (2003). Bringing home the biggest bacon: A cross-site analysis of the structure of hunter-kill profiles in Neotropical forests. *Biological Conservation*, 111(3), 415–425.

Kanji, A. (2017). Colonial animality: Constituting Canadian settler colonialism through the human-animal relationship. In Marc W & Sebastian W (eds), *Critical Epistemologies of Global Politics*. Bristol, England: E-International Relations Publishing, pp. 63–78.

Karadimas, D. (2016). Monkeys, Wasps and Gods: Graphic perspectives on Middle Horizon and later pre-Hispanic painted funerary textiles from the Peruvian coast. *Nuevo Mundo Mundos Nuevos* (On line). https://doi.org/10.4000/nuevomundo.69281, accessed February 12, 2022.

Kidder, B.B. (2009). *Maya scribes who would be kings: Shamanism, the underworld, and artistic production in the late classic period* (Unpublished Master thesis). San Marcos, Texas. https://digital.library.txstate.edu/bitstream/handle/10877/4075/fulltext.pdf

Kirch, P.V. (2005). Archaeology and global change: The Holocene record. *Annual Review of Environment and Resources*, 30, 409–440. doi: 10.1146/annurev.energy.29.102403.140700.

Kopenawa, D. & Albert, B. (2013). *The falling sky: Words of a Yanomami shaman.* Translated by Nicholas Elliott and Alison Dundy. Cambridge, MA: Harvard University Press.

Kowalewski, M.M., Garber, A.B., Cortés-Ortiz, L., Urbani, B. & Youlatus D. (eds). (2015). *Howler Monkeys, Adaptive Radioation, Systematics, and Morphology. Developments in Primatology: Progress and Prospects.* Springer.

Kracke, W.H. (2006). To dream, perchance to cure. Dreaming and Shamanism in a Brazilian Indigenous Society. *Social Analysis*, 50(2), 106–120.

Langguth, A., Teixeira, D.M., Mittermeier, R. A., & Bonvicino, C.R. (1987). The red-handed howler monkey in northeastern Brazil. *Primate Conservation*, 8, 36–39.

Lehman, S.M. & Robertson, K.L. (1994a). Preliminary survey of Cacajao melanocephalus melanocephalus in Southern Venezuela. *International Journal of Primatology*, 15(6), 927–934.

Lehman, S.M. & Robertson, K.L. (1994b). Survey of Humboldt's uakari in south Amazonas, Venezuela. *American Journal of Primatology*, 33, 223.

Lizarralde, M. (2002). Ethnoecology of monkeys among the Bari of Venezuela: Perception, use and conservation. In Fuentes A & Wolfe LD (eds), *Primates Face to Face: The Conservation Implications of Human-nonhuman Primates Interconnections.* Cambridge: Cambridge Press, pp. 85–100.

Lizarralde, M. (2020). Frugivorous monkeys feeding in a tropical rainforest: Barí ethnobotanical ethnoprimatology in Venezuela. In Urbani B & Lizarralde, M. (Eds) *Neotropical Ethnoprimatology: Indigenous Peoples' Perceptions of and Interactions with Nonhuman Primates*, pp. 109–138.

Lizot, J. (1985). *Tales of the Yanomami.* Cambridge University Press. New York.

Luciano, G.J.dos S. (2017). Língua, educação e interculturalidade na perspectiva indígena. *Revista de Educação Pública*, 26(62/1), 295–310. doi: 10.29286/rep.v26i62/1.4996. Last accessed March 25, 2022.

Luz, A.C., Paneque-Gálveza, J., Guèzea, M.J. et al. (2017). Continuity and change in hunting behaviour among contemporary indigenous peoples. *Biological Conservation*, 209, 17–26. https://doi.org/10.1016/j.biocon.2017.02.002.

Maldonado, A.M. & Waters, S. (2020). Ethnoprimatology of the Tikuna in the southern Colombian amazon. In Urbani B & Lizarralde M (eds.), *Neotropical Ethnoprimatology: Indigenous Peoples' Perceptions of and Interactions with Nonhuman Primates*, Springer Nature, pp. 89–107.

Mann, C.C. (2005). *1491: New Revelations of the Americas before Columbus.* New York: Knopf Publishing Group. ISBN 978-1-4000-4006-3. OCLC 56632601.

Marroquin Arredondo, J. & Bauer, R. (2019). *Translating Nature: Cross-Cultural Histories of Early Modern Science.* Philadelphia: University of Pennsylvania Press.

Masseti, M. (2011). Anthropochorous mammals of the Old World in the West Indies. *Mammalia*, 75, 113–142.

Masson. (1999). Animal resource manipulation in ritual and domestic contexts at postclassic Maya communities. *World Archaeology*, 31, 93–120.

Maximilian Wied (Prinz von), (1820). *Travels in Brazil in the years 1815, 1816, 1817.* London: Sir Richard Phillips, and Company.

Medrano, C., & Suárez, V. (2020). When monkeys were humans: narratives of the relationship between primates and the Qom (Toba) people of the Gran Chaco of Argentina. In Urbani B & Lizarralde M (eds.), *Neotropical ethnoprimatology: indigenous peoples' perceptions of and interactions with nonhuman primates*, 363–376.

Mello, B., Vilela, J.F. & Schrago, C.G. (2018). Conservation phylogenetics and computational species delimitation of Neotropical primates. *Biological Conservation*, 217, 397–406. https://doi.org/10.1016/j.biocon.2017.11.017.

Mittermeier, R.A. & Coimbra-Filho, A.R. (1977). Primate conservation in Brazilian Amazonia. In Prince Rainier of Monaco & Bourne G (eds), *Primate Conservation.* New York: Academic Press, pp. 117–166.

Mittermeier, R.A., Rylands, A.B. & Wilson, D.E. (2013). *Primates, Handbook of the Mammals of the World*, vol. *3*. Barcelona: Linx.

Mittermeier, R.A., Rylands, A.B., Schwitzer, C. et al. (2022). *IUCN SSC Primate Specialist Group: Report 2018–2021.* Report to the International Primatological Society (IPS), Quito, Ecuador, 9 January 2022.

Mittermeier, R.A., Valle, C.M.C., Alves, M.C. et al. (1987). Current distribution of the muriqui in the Atlantic forest region of eastern Brazil. *Primate Conservation*, 8, 143–149.

Morris, D. (2013). *Monkeys.* London: Reaktion Books.

Nájera Coronado, M.I. (2012). El mono y el cacao: la búsqueda de un mito a través de los relieves del grupo de la serie inicial de Chichén Itzá. *Estudios de cultura Maya*, 39, 133–172.

Newsom, L.A. & Wing, E.S. (2004). *On Land and Sea: Native American Uses of Biological Resources in the West Indies.* Tuscalosa and London: The University of the Alabama Press.

Nuewied, M. (1820). *Travels in Brazil, in 1815, 1816, and 1817.* Translated from the German. London: Sir Richard Phillipe and Co.

Nunes, V.F., Lopes, P.M. & Ferreira, R.G. (2021). Monkeying around Anthropocene: Patterns of human-nonhuman primates' interactions in Brazil. *Ethnobiology and Conservation*, 10.

Ogden, L.A, Hall, B. & Tanita, K. (2013). Animals, plants, people, and things, a review of multispecies ethnography. *Environment and Society: Advances in Research*, 4(1), 5–24.

Oliveira, H.de A.P. (2009). *Comida forte e comida fraca: alimentação e fabricação dos corpos entre os Kaingáng da Terra Indígena Xapecó (Santa Catarina, Brasil).* Dissertação. Programa de PósGraduação em Antropologia Social da Universidade Federal de Santa Catarina.

Oliveira-Silva, L.R.B., Campêlo, A.C., Lima, I.M.S., Araújo, A.C.L., Bezerra, B.M. & Souza-Alves, J.P. (2018). Can a non-native primate be a potential seed disperser? A case study on *Saimiri sciureus* in Pernambuco State, Brazil. *Folia Primatologica*, 89, 138–149. doi: 10.1159/000486413.

Olmi, G. (1992). Magnus Campus: i naturalisti italiani di fronte all'America nel secolo XVI. In Olmi G (ed), *L'inventario del mondo. Catalogazione della natura e luoghi del sapere nella prima età moderna.* Bologna: Il Mulino.

Papavero, N. & Teixeira, D.M. (2007). *A fauna de São Paulo nos séculos XVI-XVII nos textos dos cronistas missionérios e relatos monçoneiros.* São Paulo: Ed. USP.

Papavero N. & Teixeira D.M. Overal W.L. (2002). *O Novo Eden.* Belém do Para: Museu Paraense Emílio Goeldi.

Parathian HE, & Maldonado AM (2010) Human-nonhuman primate interac tions amongst Tikuna people: perceptions and local initiatives for resource management in Amacayacu in the Columbian Amazon. *Am J Primatol*, 71:1–11.

Pereira, B.C. (2022). Vozes Vegetais: Um livro-planta que encanta!. *Maloca: Revista de Estudos Indígenas*, 5. e022010-e022010.

Pereira, M.R.M. (2012). O conhecimento cientifico da caatinga no século XVIII. In Kury L (Org.), *Sertões adentro: viagens nas caatinga* - seculo XVI e XIX. Rio de Janeiro, Andrea Jakobsonn Estudios.

Peres, C.A. (1990). Effects of hunting on western Amazonian primate communities. *Biological Conservation*, 54(1), 47–59.

Peres, C.A. (1991). Seed predation of Cariniana micrantha (Lecythidaceae) by brown capuchin monkeys in Central Amazonia. *Biotropica*, 262–270.

Peres, C.A. (2000). Effects of subsistence hunting on vertebrate community structure in Amazonian forests. *Conservation Biology*, 14, 240–253.

Peres, C.A. (2010). Overexploitation. In Ehrlich PR (ed), *Conservation Biology for All*. Sodhi: N.S, Oxford University Press, pp. 107–130. doi: 10.1046/j.1523-1739.2002.01635.x.

Peres, C.A. & Palacios, E. (2007). Basin-wide effects of game harvest n vertebrate population densities in Amazonian forests: Implications for animal-mediated seed dispersal. *Biotropica*, 39(3), 304–315.

Pinto-Marroquin, M., & Serio-Silva, J. C. (2020). Perception and uses of primates among Popoluca Indigenous people in Los Tuxtlas, Mexico. In Urbani B & Lizarralde M (eds.), *Neotropical ethnoprimatology: indigenous peoples' perceptions of and interactions with nonhuman primates,* Springer Nature, 3–20.

Printes, R.C., Buss, G., Jardim, M et al. (2010). The urban monkeys program: A survey of Alouatta clamitans in the South of Porto Alegre and its influence on land use policy between 1997 and 2007. *Primate Conservation*, (25), 11–19.

Ranzi, A. (2000). *Paleoecologia da Amazônia: megafauna do pleistoceno*. Florianópolis: EdUFSC.

Razera, J.C.C., Boccardo, L. & Pereira, J.P.R. (2006). Percepções sobre a fauna em estudantes indígenas de uma tribo tupinambá no Brasil: um caso de etnozoologia. *Revista Electrónica de Enseñanza de las Ciencias*, 5(3), 466–480.

Ribeiro, D. (2008). O povo brasileiro [The Brazilian People]. São Paulo: Companhia das Letras.

Rice, P. & South, K. (2015). Revisiting monkey on pots: A contextual consideration of primates imagery on classic lowland maya pottery. *Ancient Mesoamerica*, 26(2), 275–294. doi: 10.1017/S0956536115000206

Rodrigues, R. (2016). A compreensão do território Kaingang no Oeste Paulista a partir da cultura material: possibilidades interpretativas na análise arqueológica. *Revista do Museu de Arqueologia e Etnologia*, 27, 44–54.

Rosales-Meda, M. & Hermes, M.S. (2020). Representation and signification of primates in Maya-Q'eqchi'Cosmovision and Implications for their conservation in Northwestern Guatemala. In Urbani B & Lizarralde M (eds.), *Neotropical ethnoprimatology: indigenous peoples' perceptions of and interactions with nonhuman primates*, Springer Nature, pp. 69–85.

Ruiz-Miranda, C.R., Gomes, A.A., Morais, M.M., Verona, C.E., Martins A. & Beck, B. (2006). Behavioral and ecological interactions between reintroduced golden lion tamarins (*Leontopithecus rosalia* Linnaeus, 1766) and introduced marmosets (*Callithrix* spp, Linnaeus, 1758) in Brazil's atlantic coast forest fragments. *Brazilian Archives of Biology and Technology*, 1(49), 99–109.

Ruoso, D. (2012). *A percepção climática da população urbana de Santa Cruz do Sul/RS*. Raega-O Espaço Geográfico em Análise, 25.

Rybka, K. (2020). Linguistic, cultural, and environmental aspects of ethnoprimatological knowledge among the Lokono, Kari'na, and Warao of the Moruca River (Guyana). In Urbani B & Lizarralde, M. (eds) *Neotropical Ethnoprimatology: Indigenous Peoples' Perceptions of and Interactions with Nonhuman Primates*, Springer Nature, pp. 239–262.

Rylands, A.B. & Mittermeier, R.A. (2009). The diversity of the new world primates (Platyrrhini): An annotated taxonomy. In Garber PA, Estrada A, Bicca-Marques JC, Heymann EW, & Strier KB (eds), *South American Primates. Developments in Primatology: Progress and Prospects*. Springer, New York, NY. https://doi.org/10.1007/978-0-387-78705-3_2.

Schleicher, J. (2018). The environmental and social impacts of protected areas and conservation concessions in South America. *Current Opinion in Environmental Sustainability*, (32), 1–8. https://doi.org/10.1016/j.cosust.2018.01.001.

Schneider, H. & Sampaio, I. (2015). The systematics and evolution of New World primates,Äì a review. *Molecular Phylogenetics and Evolution*, 82, 348–357.

Seixas, M.L.B. (2003). *A natureza brasileira nas fontes portuguesas do século XVI*. Viseu: Passagem Editores.

Shepard. (2002). Primates in Matsigenka Subsistence and World View. *Cambridge Studies in Biological and Evolutionary Anthropology*, 101-136.

Silva,A.P.da.(2019).SaberestradicionaisTupi:etsarjunto,aprender, nhembojera. *Cadernos CEDES* [online], 39(109), 379–396. https://doi.org/10.1590/CC0101-32622019216679, Accessed Março 4. 2022.

Silvestro, D., Marcelo, F., Tejedor, M. et al. (2019). Early arrival and climatically-linked geographic expansion of new world monkeys from tiny African ancestors. *Systematic Biology*, 68(1), 78–92. https://doi.org/10.1093/sysbio/syy046.

Skousen, B.J. (2009). *Monkey pots: Inferring meaning through time and space from function, decoration, and context*. Theses and Dissertations. 1980. https://scholarsarchive.byu.edu/etd/1980

Sobrevila, C. (2008). *The Role of Indigenous Peoples in Biodiversity Conservation the Natural but Often Forgotten Partners*. Washington, D.C.: The World Bank.

South, K. (2005). *Monkening around The Maya region: A four-field look at Primate iconography and The Maya*. Southern Illinois University at Carbondale. ProQuest Dissertations Publishing.

South, K.E. & Ford, S.M. (2022). 2 Monkeys and the ancient Maya. *World Archaeoprimatology: Interconnections of Humans and Nonhuman Primates in the Past*, 90, 48.

Souza-Mazurek, R.R.D. & Bruno, A.C. (2020). Kixiri and the origin of day and night: Ethnoprimatology among the Waimiri Atroari ameindians of the Central Amazonia, Brazil. In Urbani B & Lizarralde, M. (Eds.), *Neotropical ethnoprimatology: indigenous peoples' perceptions of and interactions with nonhuman primates*, Springer Nature, pp. 225–238.

Świerk, K. (2020). Monkeys in the wampis (Huambisa) life and cosmology in the Peruvian Amazonian rainforest. In Urbani B & Lizarralde, M. (eds), *Neotropical Ethnoprimatology: Indigenous Peoples' Perceptions of and Interactions with Nonhuman Primates*, pp. 311–329.

Talebi, M. & Soares, P. (2005). Conservation research on the southern muriqui (Brachyteles arachnoides) in São Paulo State, Brazil. Neotropical Primates, 13(Suppl), 53–59.

Tarling, D.H. (1980). The geologic evolution of South America with special reference to the last 200 million years. In Ciochon RL & Chiarelli AB (eds), *Evolutionary Biology of the New World Monkeys and Continental Drift*. New York: Plenum Press, pp. 1–41.

Tedlock, D. (1986). *Popol vuh: The Mayan book of the dawn of life*. New York: Simon & Schuster.

Teixeira, D. & Papavero, N. (2010). O Trafico de Primatas brasileiros nos séculos XVI e XVII. In Pessoa ML, Correa Tavares W & Siliciano S (eds), *Mamiferos de Restingas e Manguezais do Brasil.* Rio de Janeiro: SBMz, Sociedade Brasileira de Mastozoologia. Museu Nacional, pp. 253–282.

Terra, A.K. (2007). *A caça de subsistência na Reserva de Desenvolvimento Sustentável Piagaçu-Purus e na terra indígena lago Ayapuá.* Brasil: Amazônia Central.

Thoisy, B, Richard-Hansen, C & Peres, C.A. (2009). Impacts of subsistence game hunting on Amazonian primates. In Garber PA, Estrada A, Bicca-Marques JC, Heymann EW, Strier KB (eds), *South American Primates. Developments in Primatology: Progress and Prospects.* New York, NY: Springer. https://doi.org/10.1007/978-0-387-78705-3_15.

Thompson, J. (1973). Maya ruler of the Classic Period and the divine right of kings. In *The Iconography of Middle American Sculpture.* New York: The Metropolitan Museum of Art, pp. 52–71.

Toledo, V. & Barrera-Bassols, N. (2009). A etnoecologia: uma ciência pós-normal que estuda as sabedorias tradicionais. *Desenvolvimento e Meio Ambiente, Curitiba,* 20, 31–45. https://doi.org/10.5380/dma.v20i0.14519.

Topik, S., Marichal, C. & Frank, Z. (2006). *From Silver to Cocaine: Latin American Commodity Chains and the Building of the World Economy.* Duke University.

Townsend, W.R., Wallace, R.B., Lara-Delgado, K. & Miranda-Chumacero, G. (2020). Importance of primates to tacana indigenous subsistence hunting in the Bolivian amazon. In Urbani B & Lizarralde, M. (eds), *Neotropical Ethnoprimatology: Indigenous Peoples' Perceptions of and Interactions with Nonhuman Primates,* Springer Nature, pp. 343–362.

Turner, B. (2021). The limits of culture in political theory: A critique of multiculturalism from the perspective of anthropology's ontological turn. *European Journal of Political Theory,* 20(2), 252–271.

Turvey, S.T., Bryant, J.V. & McClune, K.A. (2018). Differential loss of components of traditional ecological knowledge following a primate extinction event. *Royal Society Open Science,* 5(6), 172352.

Urbani, B. (2005). The targeted monkey: A re-evaluation of predation on New World primates. *Journal of Anthropological Sciences,* 83, 89–109.

Urbani, B. & Cormier, L.A. (2015). The ethnoprimatology of the Howler monkeys (*Alouatta* spp.): From past to present. In Kowalewski M, Garber P, Cortés-Ortiz L, Urbani B & Youlatos D (eds), *Howler Monkeys. Developments in Primatology: Progress and Prospects.* New York, NY: Springer. https://doi.org/10.1007/978-1-4939-1960-4_10.

Urbani, B. & Lizarralde, M. (eds). (2020). *Neotropical Ethnoprimatology: Indigenous Peoples' Perceptions of and Interactions with Nonhuman Primates.* Springer Nature.

Urbani, B. & Rodríguez, J. (2021). Pre-hispanic howler monkeys from two sites of northern Venezuela and the Orinocan connection: An archaeoprimatological study. *International Journal of Osteoarchaeology,* 1–14. https://doi.org/10.1002/oa.2948.

Urbani, B., Antczak, A. T., Antczak, M. M., Cannarozzi, N. R., Colten, R. H., Emery, K. F., ... & Nieweg, D. C. (2022a). 3 Monkeys on the Islands and Coasts of Paradise. *World Archaeoprimatology: Interconnections of Humans and Nonhuman Primates in the Past, 90,* 63.

Urbani, B., Youlatos, D., & Antczak, A. T. (2022b). *World Archaeoprimatology.* Cambridge University Press.

Urquiza-Haas, E.G., Martínez, R.I.O. & Kotrschal, K. (2020). Mental state attribution to nonhuman primates and other animals by rural inhabitants of the community of Conhuas near the Calakmul biosphere reserve in the Yucatan peninsula, Mexico. In Urbani B & Lizarralde, M. (eds), *Neotropical Ethnoprimatology: Indigenous Peoples' Perceptions of and Interactions with Nonhuman Primates.* Cham: Springer International Publishing, pp. 21–48.

Velden, F.F.V. (2016). Realidade, ciência e fantasia nas controvérsias sobre o Mapinguari no sudoeste amazônico. *Boletim do Museu Paraense Emílio Goeldi. Ciências Humanas,* 11, 209–224.

Veracini, C. (2017). Nonhuman primate trade in the age of discoveries: European importation and its consequences. In *Environmental History in the Making*: Volume *II: Acting,* Springer, pp. 147–171.

Veracini, C. (2019). Natural history of non-human primates in the Early Modern period: The contribution of the Italian encyclopaedist Ulisse Aldrovandi (1522–1605). In Veracini C, Casanova C & Scalfari F (eds), *History of Primatology: Yesterday and Today. The Mediterranean Tradition.* Ariccia (RM, Italy): Aracneditrice, pp. 53–91.

Veracini, C. (2021). Use and trade of non-human animals in Portuguese overseas expansion (15th and 16th centuries): Evidence from Italian travellers. *RiMe, Rivista dell'Istituto di Storia dell'Europa Mediterranea,* 8(2), 11–41. https://rime.cnr.it/index.php/rime/article/view/536/801.

Veracini, C. & Ruiz-Miranda, C.R. (2004, January). The marmosets of Rio de Janeiro: Victim or guilty? *Folia Primatologia,* 75, 347–347.

Veracini, C. & Teixeira, D., (2017). New World non-human primates in the travel literature of the 15th and 16th centuries: From myth to natural history. *Annals of Science,* 74(1), 1–39.

Visalberghi, E., Sirianni, G.k., Fragaszy, D. & Boesch, C. (2015). Percussive tool use by Taï Western chimpanzees and Fazenda Boa Vista bearded capuchin monkeys: A comparison. *Philosophical Transactions of the Royal Society* B, 370, 2014035120140351. https://doi.org/10.1098/rstb.2014.0351.

Viveiros de Castro, E. (1996). Os pronomes cosmológicos e o perspectivismo ameríndio. *Mana,* 2(2). https://doi.org/10.1590/S0104-93131996000200005.

Viveiros de Castro, E. (2004). Exchanging perspectives: The transformation of objects into subjects in Amerindian ontologies. *Common Knowledge,* 10(3), 463–484.

Viveiros de Castro, E. (2009). *Métaphisic cannibale. Lignes d'antropologie post-structurale.* Paris: Universitaires de France.

Voss, R. & Fleck, D.W. (2011). Mammalian diversity and Matses ethnomammalogy in Amazonian Peru. Part 1, Primates. *Bulletin of the American Museum of Natural History,* 351, 1–81.

Waldau, P. (2013). *Animal Studies: An Introduction.* Oxford: Oxford Unversity Press.

Zent, S. & Zent, E. (2020). Co-ecology of Jotï, primates, and other people: A multispecies ethnography in the Venezuelan Guayana. In Urbani B & Lizarralde, M. (eds), *Neotropical Ethnoprimatology: Indigenous Peoples' Perceptions of and Interactions with Nonhuman Primates.* Springer Nature, pp. 161–197.

5 Non-human Primates in the Ancient Near East, from Protohistory to the First Islamic Caliphate

Marco Masseti
Istituto Zooprofilattico Sperimentale della Sicilia, and Società italiana per la Storia della Fauna 'Giuseppe Altobello'

> [Monkey mothers] sing aloud, a youngster monkey shrieks:
> [like a band(?)] of musicians and drummers(?),
> daily they bash out a rhythm in the presence of Ḫumbaba
>
> (*Gilgamesh Tablet V.* British Museum K 8591 [MS AA]: see Al-Rawi & George, 2014: 77)

5.1 INTRODUCTION

The so-called Near East is a geographic region of south-western Asia between the Eastern Mediterranean shores to the West, the Caucasus chain and Transcaucasia to the North, and the Indian subcontinent to the East (Harrison, 1964: 5; Corbet, 1978: 2; Masseti, 2002: 47–48). As the geographical junction between Eurasia and Africa, the Near East displays a remarkable range of bioclimatic and biogeographical conditions (Zohary, 1973; Masseti, 2003a: 377–378; 2021a: 15). To the North of the Anatolian plateau, lie the mountains of Turkey where most of the fauna is Palaearctic. To the South, the steppe and the desert core of the Arabian peninsula form part of the Saharo-Sindian zone, inhabited by endemic mammals such as the Arabian oryx; *Oryx leucoryx* Pallas, 1777; the Arabian tahr; *Hemitragus jayakari* Thomas, 1894; and once, even the wild dromedary, *Camelus dromedarius* L., 1758. The southernmost territories of Arabia represent an extension of the tropical Ethiopian zoogeographical region, with species characteristic of East Africa (Harrison, 1964: 4–5; Atallah, 1977: 241–242; Harrison & Bates, 1991: 191–192; Borzatti von Löwenstern & Masseti, 1991: 145; Masseti, 2021a: 15). Fossils of modern Afrotropical ungulates are documented from several archaeological sites in the Near East (Clutton-Brock & Burleigh, 1978: 89–100; Uerpmann, 1991), and the number of Ethiopian species will probably increase as additional studies are undertaken in Arabia. With the possible exception of the tahr, no other Oriental faunal element has been identified among the Late Pleistocene and Holocene macromammal assemblage, but this probably reflects the relative lack of palaeontological, archaeozoological and zoological studies from Lower Mesopotamia, southern Iran and eastern Arabia (Uerpmann, 1991: 99–106).

Among the Afrotropical elements, the only species of monkey in the Near East is a population of hamadryas baboons or sacred baboons, *Papio hamadryas* (L., 1758), that is confined to the southeastern mountainous corner of the Arabian Peninsula up to western Yemen, in particular near Aden. This species is represented by two populations separated by the Red Sea, the second being found in the African arid zone of the Red Sea coast of Sudan, Eritrea, Ethiopia and northern Somalia (Masseti & Bruner, 2009: 48–52, and references therein). Hamadryas baboons have also been reported from the Farasan islands, in the southern Red Sea (Saudi Arabia) (Masseti, 2010: 362; Masseti et al., 2015: 109; **Fig. 5.1**). Together with the Barbary macaque; *Macaca sylvanus* (L., 1758); and the grivet, *Chlorocebus*

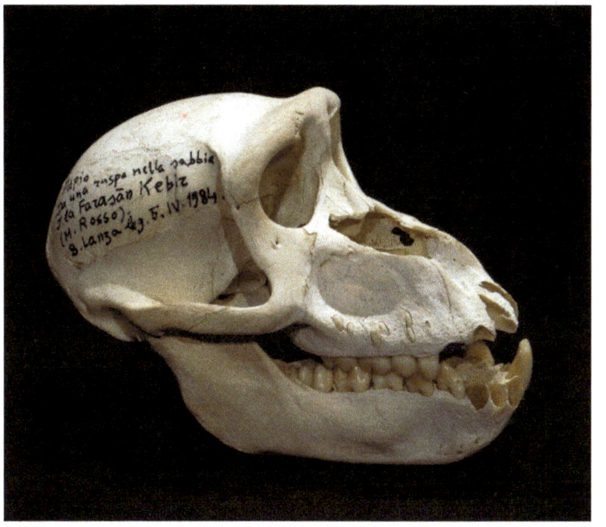

FIGURE 5.1 Sacred baboons, *Papio hamadryas* (Thomas, 1900), occur in two populations separated by the Red Sea, being found both in Africa and in Arabia. There is also evidence for the occurrence of this species on the Farasan archipelago, where a complete skull with mandible of a sub-adult female was discovered in a provisional burial on Farasan Al-Kebir, the largest island of the namesake archipelago (16°42′21″N 41°59′0″E). (Photo by Saulo Bambi; courtesy of the zoological museum 'La Specola' of the University of Florence, Italy; from Masseti & Bruner, 2009.)

DOI: 10.1201/b21963-6

aethiops (L., 1758), hamadryas baboons are one of the monkeys best known in the Western World since antiquity (Masseti & Bruner, 2009: 48–52; Morris, 2013: 11–18), and it has been suggested that in ancient times its range included Nubia and even Egypt (Osborn & Osbornova, 1998: 32–37; Dominy et al., 2020). The last country is regarded by some researchers as the 'type locality' of the species (Napier, 1981; Groves, 2001: 238; 2005: 166), even though sacred baboons have long since vanished from these parts. In any case, Linnaeus described the taxon in the first volume of the tenth edition of his *Systema Naturae* (1758: 27) on the basis of specimens from 'Egypt' and 'Upper Egypt' (cf. Groves, 2001; Masseti & Bruner, 2009: 49). Exported from its natural homeland since ancient times, the hamadryas continued to be transported to the menageries of the nobility and the princely courts of the Western and the Islamic Worlds throughout historic times.

5.2 NON-HUMAN PRIMATES IN THE ANCIENT NEAR EAST

Apart from *P. hamadryas*, monkeys are today unknown in the Near East and in the whole south-western Asia, as well as in deep time (cf. Hatt, 1959: 33; Kingdon, 1990: 15; Lion, 1992: 359; cf. Such-Gutiérrez, 2022). In fact, other representatives of the Order Primates are absent from the Upper Pleistocene and early Holocene levels of this geographical portion of the Western Palaearctic. The monkeys in Mesopotamia, during the third millennium BC, were rhesus macaques, *Macaca mulatta* (Zimmermann, 1780), introduced from northern India or western-central Asia, as reported by Such-Gutiérrez (2022). In fact, only one specimen of the latter species, from the third millennium BC levels of the Iranian site of Shahr-i Sokhta in the Iranian province of Sistan and Baluchistan (Minniti, 2018: 246; Minniti & Sajjadi, 2019: 1–11), has ever been recovered from the archaeological sites of the Near East (**Fig. 5.2**).

FIGURE 5.2 The grave (9319) of a rhesus macaque, *Macaca mulatta* (Zimmermann, 1780), from the third millennium BC levels of the Iranian site of Shahr-I Sokhta, in the province of Sistan and Baluchistan. (Photo from Minniti & Sajjadi, 2019; courtesy of Claudia Minniti.)

As observed by Barnett (1973: 3), monkeys have a long and fairly respectable history in ancient south-western Asia. In particular, various artistic representations of primates have been recovered from the archaeological exploration of Mesopotamia, the valley of the cradle of western civilization (cf. Peters et al., 2005) between the rivers Euphrates and Tigris. There are monkey amulets from archaic levels at Warka, the ancient Uruk, and a copper pin topped with a delightful gold figure of a squatting monkey was found in the grave of Meskalam-shar at Ur (Woolley, 1934: 158, pl. 165; Buren, 1939: 22). There is also a gold figure representing an unidentified primate from the Sumerian city-state of Ur, in lower Mesopotamia (Woolley, 1934: 300 and pl. 165).

A study recently carried out by Urbani and Youlatos (2021) has reviewed some statuettes of the Proto-Elamite and Elamite periods from the archaeological site of Susa, in the lower Zagros Mountains, about 250 km East of the Tigris. These figurines are among the earliest depictions of primates made by a Near Eastern civilization, or by any civilization in the world. Urbani and Youlatos (2021) have recognised two species, the olive baboon, *Papio anubis* (Lesson, 1827), and the rhesus macaque, primates characterised by ranges distributed, respectively, in sub-Saharan Africa (Groves, 2005: 166; Masseti & Bruner, 2009: 42–48) and central-eastern Asia, from Afghanistan to China, throughout Indochina (Groves, 2005: 163; Francis, 2008: 272), both very distant from the geographical location of Susa. For this reason, the opinion of Urbani and Youlatos (2021) is that Susa might have served as a Near Eastern trading post for the circulation of monkeys or primate imageries since the fourth and third millennia BCE, but in my view the representations do not provide enough evidence to justify their reliable attribution to *P. anubis* or *M. mulatta*. None of the Susa figurines shown by Urbani and Youlatos (2021: in Figure 2, p. 4) exhibit phenotypic characters that would allow them to be confidently attributed to a precise species of monkey. On the basis of the limited information provided by their depiction, it is possible to match some of them to the generic morphology of a baboon, but they could be hamadryas, as proposed for one of them by Benoit (1993: 33); or why the figurines identified as representing rhesus macaques could not depict Barbary macaques, given the consistent absence of tails in their portrayal. A limestone monkey, sitting on its hindquarters, with its back legs drawn up to its belly and its hands on its knees, has been dated to the late third millennium BCE (Paris, Louvre Museum: R. de Mecquenem excavations Sb 5884; *Room 8*- Display case 4) (**Fig. 5.3**), but according to Harper et al. (1992), the appearance of this figurine suggests perhaps an Asian macaque, but the absence of a tail in this representation is not sufficient evidence to justify its attribution to an Asian macaque rather than *M. sylvanus*.

A stone figurine representing a seated monkey – perhaps a baboon, but without a tail – came from the archaeological site of Tell Brak (northern Syria) and is dated to the early third millennium (Mallowan, 1947: pl. IX; **Fig. 5.4**). In Ur, during the Akkadian period (the second half of the third millennium BC), monkey amulets usually made of glazed frit were fairly common, with the animals always represented in the same crouching attitude (Woolley, 1934: 158; pl. XI). All these figurines depict tailless primates (Buren, 1939: 22). Mendleson (1983: pl. XXIV) refers to three clay plaques from Ur (now in the British Museum, London), each of which clearly showing a man with either one or two monkeys. Several other artistic representations of primates date to the early second millennium Iraq, among them is a curious three-inch tall figurine monkey (between 2000 and 1800 BC) carved in alabaster, from the Kititum Temple at Iščāli, Iraq (Dunham, 1985: 264; Buren, 1939: 22; Frankfort, 1943: pl. 74). This sacred building was dedicated to Ishtar, the Sumerian Inanna, which was the goddess of sexuality and war, and the most important goddess in Mesopotamian religion (Bertman, 2005: 26). A small frit figurine representing a monkey was also found in Terrace Complex II of Haft Tepe (middle of the second millennium BCE), which lies on the plain of Khuzistan, close to the ruins of the already-mentioned ancient Susa. This sculpted monkey sits with its hands towards its chest and is characterised by a rounded head "with a projecting forehead below which the eyes are large cavities" (Negahaban & Negahaban, 1991: 43).

Among the oldest artistic representations of primates from the Babylonian period, are some terracotta statuettes depicting male and female monkeys (Koldewey, 1990: 226), while an alabaster monkey has been discovered in the ruins of the Kition temple (early second millennium BCE), from Tell Ishchali, in Diyala Province in present-day Iraq (Dunham, 1985: 264 f.).

Woolley (1934: 300) noted that monkey amulets were fairly common in Mesopotamia from the Sargonid period onward, and terracotta reliefs of a man carrying a monkey, and of a man leading one by a string, which anticipate late Assyrian stone reliefs in which monkeys appear, are common in the Larsa period (circa 2000 BCE). Another artistic representation from the early second millennium BCE concerns three individuals carved in relief on a circular mould in baked clay, reportedly from the Syrian site of Mari (Tell Hariri) (cf. Basmachi, 1976: 401; **Fig. 5.5**).

From the Old Babylonian period (2000–1600 BC) onwards, small clay figurines of monkeys are present in southern Mesopotamia (i.e. Wrede, 1990: 276–277), but by

FIGURE 5.3 Limestone monkey from Susa, sitting on its hindquarters, with its back legs drawn up to its belly and its hands on its knees, dated to the late third millennium BC, (Paris, Louvre Museum: R. de Mecquenem excavations Sb 5884). It appears to be a macaque, but the taxonomic identification is uncertain.

FIGURE 5.4 The stone image of a sitting monkey from Tell Brak, dating to the early third millennium. (From Mallowan, 1947: pl. XI.)

FIGURE 5.5 An artistic representation dated to the early second millennium BCE of three primates, one long-tailed and the other two apparently not, carved in relief on a circular mould in baked clay, reportedly from the Syrian site of Mari (Tell Hariri) (Baghdad, The Iraq Museum). (From Basmachi, 1976.)

FIGURE 5.6 Ninth century fragment of wall relief from the NW Palace of Nineveh, at Nimrud, depicting a tribute of monkeys from the Phoenicians to the Assyrian king Assurnâṣirpal. (From Layard, 1849 modified.)

the Kassite period, and certainly, during the Neo-Assyrian era, it is likely that monkeys were being brought to Mesopotamia from Egypt, which had close ties to the court of the Kassite kings during the so-called Amarna period (Potts, 1997: 259–260). From the world of primate-collecting come two marble figures, allegedly from Middle Bronze Age Elam (the second half of the second millennium BCE), representing, respectively, a 'seated monkey' and a 'crouched monkey'. Urbani and Youlatos (2020) interpret them as two chimpanzees, *Pan troglodytes* Blumenbach, 1775, on the basis of arguments which are not fully convincing (see also Masseti-*Orang Utan*, in this volume). According to Urbani and Youlatos (2020: 657), they would be the oldest artistic representation of an African ape: "It would follow that chimpanzees were traded, as living animals, as artifacts or imageries, along an extended distance from the Central African forests to the east coast of Africa towards Elam, by the 2nd millennium BCE". However, Urbani and Youlatos (2020: 660–661) themselves cast doubt on the origin and dating of the two artefacts, noting also that the figures may date from the Roman period, and also observing that 'the circulation of these apes or ape representations occurred centuries later'.

Houghton (1877: 319–321), followed by Hatt (1959: 34), was among the few authors who attempted the identification of the monkeys depicted in Assyrian sculpture. A few fictile artefacts reported from Babylon, described by Koldewey (1990: 226), are claimed to be representations of male and female types of tailless primates, whereas monkeys with tails are depicted in a gypsum-wall panel relief (London, British Museum: 124562), dated to about 865 BCE (Reade, 1983: 27). This was found by Austen Henry Layard during the archaeological exploration of the Northwest palace of Nimrud, an ancient Assyrian city located South of Nineveh (Iraq, northern Mesopotamia). The panel was part of the façade of the throne room, and its reproduction can be seen in Layard (1849, vol. I: 118) (**Fig. 5.6**). One of the two human figures depicted here may be Phoenician, with a pair of monkeys, both with long tails. Layard (1849, vol. I: 118) refers to how, at the time of discovery, there were extensive traces of black pigment covering the face of the monkey bearer, and he speculated that this was either from a deliberate attempt to depict a negro, or that the paint had washed down from the man's hair. Stylistically similar monkeys are also represented in the Black Obelisk of Shalmaneser III (about 825 BCE) from Nimrud (London, British Museum: room 6 of the Ground Floor) (Reade, 1983: 17). Two phenotypic groups of primates are clearly reflected in this artefact, where an adult elephant and three monkeys are shown in the tribute to the Assyrian king of a subject prince of the enigmatic land of *Musri* (**Fig. 5.7**). One of the primates has a long tail, while the other two have no tail. In the caption, written below the representation of the bearers in the British Museum display, the English translators from the cuneiform used the terms '*monkeys*' and '*apes*' to indicate primates, respectively, with and without tails. The two tail-less primates could be Barbary macaques, while – remaining within the African biogeography – the one with a tail could be a *Chlorocebus* monkey or a young red monkey, *Erythrocebus patas* (Schreber, 1775). Nevertheless, as we shall see below (paragraph 5. Primates allegedly believed to have originated in Mesopotamia, in this chapter), it cannot be ruled out that they could also be the representations

FIGURE 5.7 Two different groups of primates are clearly reflected in the so-called Black Obelisk of Shalmaneser III (858–824 BC, from Nimrud, northern Mesopotamia, Iraq), where an adult Asian elephant and three monkeys are shown in the tribute to the Assyrian king of a subject Egyptian prince. One of the primates displays a long tail, while the other two lack tails. The legend, written below the representation of the bearers, tells: "*I received tribute from the land of Egypt: . . . cow elephants, female monkeys and apes*". The translators used the terms '*monkeys*' and '*apes*' arbitrarily, just to, respectively, indicate the primates with and without tails. ((© Trustees of the British Museum, London). https://commons.wikimedia.org/wiki/File:Salmanassar_II_Obelisk.jpg.)

of primates of Middle or Far Eastern origin. All in all, the artistic rendering of the three animals is too vague to allow their specific identification.

According to van Buren (1939), in Mesopotamian cylinder seals, a tailed monkey of the type of Hanuman from India is frequently depicted, and Budge (1914: pl. XXVIII) is of the opinion that one was brought by Phoenicians, perhaps from the fabulous land of *Ophir*, as a gift to King Ashurnasirpal II at Nimrud in the 9th century BCE. Cansdale (1970: 137) also suggests the presence of a Hanuman langur on some ancient Assyrian monuments of king Solomon's time, and later between the first half of the 10th century and the second half of the following century. The mythical land of *Ophir* has never been located in the Indian subcontinent, but some authors, such as Mills (2019: 43–50), linked this legendary land with the just as fabulous country of *Punt*, a trading partner of ancient Egypt, possibly located in the Horn of Africa on either side of the southern Red Sea. It was mentioned in the *Bible* (*Kings*, 10: 22), as the place from where King Solomon received a cargo every three years consisting of gold, silver, sandalwood, pearls, "*ivory, apes, and peacocks*", eastern spices and black human load (Schoff, 1912: 41; Ingrams, 1966: 5; Masseti, 2014: 99). *Punt* was an important emporium which provided precious and exotic goods for the ancient Egyptians, and its location has been handed down to us in ancient Egyptian records of trade expeditions. Early travellers could reach *Punt* by venturing South and East of the Nile country, suggesting that the fabulous kingdom occupied the southern Red Sea region, although its exact location is still highly debated. In agreement with other later authors, Mariette (1875) located *Punt* on the Somali coast, an area that produces premium resin of frankincense, but there are also good arguments for its location along the eastern Red Sea coast, perhaps coinciding with present-day Yemen (Meeks, 2003; Tallet, 2013). As recently observed by Dominy et al. (2020), the Egyptians also procured hamadryas baboons from *Punt*. Perhaps they could have been among the fabled 'apes' sent to king Solomon.

5.3 MONKEYS IN MESOPOTAMIAN LITERATURE. THE DIVINE CEDAR FOREST OF THE MYTHICAL GILGAMESH

Despite the almost total absence of monkeys from the biogeography of the Near East, they have featured in local literature and art since very ancient times (McDermott, 1938; Buren, 1939). The awareness of the existence of these animals by Near Eastern civilizations was clear from written historical documents dating back to the Sumerian period. Barnett (1973: 3) recalls, for example, a delightful Sumerian monkey's letter where a monkey cub complains scornfully to his mother, *Ludu-ludi* (a pure nonsense name), like any schoolboy at a boarding school, of the miserable food which is given at Ur (after Powell, 1978: 194). It was evidently given as a model for youthful scribes to copy by way of light relief from harder tasks. Barnett (1973: 4) is of the opinion that, perhaps, such ideas of animals acting like humans, which we find in Mesopotamian art, came to Sumer with monkeys from India, the classic home of the animal fable. Monkeys are preferentially associated with music in the Sumerian world, as can be understood in the following passage of the Sumerian Proverb Collection 3: 150 (after Alster, 1997: 106): "In the city Eridu, built in abundance, in a musician's house the monkey sits being gawped at". A reference to primates is also contained in a Middle Assyrian parody of legal deed, 11th century BC, about a woman who raised a monkey as her own child after rescuing it from a river (after Franke & Wilhelm, 1985: 19–26).

Despite Urbani and Youlatos (2020: 655) stating that written reports of primates in the Near East are rare and usually date from after the 5th century BCE, the epic of the hero Gilgamesh, king of Uruk, a poem from Sumerian and Babylonian Mesopotamia, dates back to the third millennium BCE (cf. Al-Rawi & George, 2014: 69–90). Monkeys are mentioned in the literary description of the *Cedar Forest*, the glorious realm of the gods of Mesopotamian mythology. It was guarded by the demigod *Ḫumbaba*, a monstrous giant of immemorial age, and was once entered by Gilgamesh who dared cut down trees from its virgin stands during his quest for fame. A recently discovered clay tablet of this poem adds 20 lines to the 'Epic of Gilgamesh'. Continuing where other sources break off, the translation of Al-Rawi and George (2014: 69–90) gives us an unexpected description of the 'forest for the gods', mythologically located in a mountainous range that may have been inspired by the Levantine Mount Lebanon, where the last centuries-old individuals of the Lebanon cedar, *Cedrus libani* A. Rich. 1823, are preserved. Providing a more detailed description of Gilgamesh *Cedar Forest*, the new section tells us this environment was no place of serene and quiet glades, but was instead full of noisy birds and cicadas, while monkeys scream and yell in the trees (Al-Rawi & George, 2014: 77). In fact, as translated by Al-Rawi and George (2014: 77), the latter animals daily "... bash out a rhythm in the presence of Ḫumbaba". In his oral paper contributed to the "Primates in Antiquity Symposium", held on August 19, 2016 at Dartmouth College (New Hampshire, USA), Andrew R. George emphasised the relationship between the musician monkeys of the Gilgamesh *Cedar Forest* and the many artistic evocations yielded by the archaeological exploration (George, 2016). In ancient Mesopotamia, the association between primates and music is also documented by artistic productions. A clay plaque coming from Ur in the early second millennium shows, for example, a man with two monkeys and a musical instrument. One of the animals is perched on his shoulder, while the other is held by a collar and plays a kind of flute (Mendleson, 1983: pl. XXIV). A monkey playing the flute is also depicted in the impression of a Neo-Babylonian cylinder seal (Spycket, 1998: 3).

For clues about the geographical location of the *Cedar Forest* of Gilgamesh, consider a relief of the 9th century BCE found far from Mesopotamia, at Karatepe, in Cilicia (see Akurgal & Hirmer, 1962: pl. 147). It shows a hideous figure derived from the Egyptian god Bes, which perhaps represents a western form of the monster *Humbaba* (cf. Kantor, 1962: 112–117), flanked by two long-tailed monkeys. In this case, it would be Mount Amanus, the present Turkish Nur Dag, that would represent the homeland of *Humbaba* and, according to Barnett (1973: 3), the presence of the monkeys would have added some local colour showing the place as a kind of jungle.

5.4 PRIMATES ALLEGEDLY BELIEVED TO HAVE ORIGINATED IN MESOPOTAMIA

Contrary to all expectations, in the ancient Near Eastern world, different varieties of monkey were regarded as originating from Upper Mesopotamia (Grayson, 1991: 150), Syria (Grayson, 1991: 218, 219, 226) and Egypt (Lion, 1992: 359, and references therein). However, as we have already observed, apart from the land of the Nile none of the latter territories is the natural homeland of these animals. Sumerians believed that monkeys came perhaps "... from the east.... In the year name of king Ibbi-Sin's twenty-third year, one hears of 'the monkeys coming from the mountains', though the interpretation is questionable" (Limet, 2005: 372). It is not easy to locate these eastern mountains from where monkeys must have come. The Zagros chain and the Iranian plateau are East of Mesopotamia, but monkeys do not live there. Otherwise, primates may have come into southern Mesopotamia from the mountains of Pakistan or India, where, for example, rhesus macaques still live. The presence of monkeys in those places had also been evidenced by the accounts of Strabo and Diodorus Siculus, after the expedition of Alexander the Great (cf. Masseti & Veracini, this volume). However, the appearance of the monkeys in the artistic representations of ancient Mesopotamia suggests they were species of a westernmost distribution, such as green monkeys of the genus *Chlorocebus* Gray, 1870, or the Maghrebi Barbary macaques.

It seems that in ancient Mesopotamia, the popular perception of primates was divided into two typological groups: primates with or without tails, respectively monkeys and apes in the modern zoological nomenclature. As we have already seen, the distinction between these two categories is reflected in the previously mentioned tribute brought from the mysterious country of *Musri* (Egypt? Kurdistan?) and illustrated in the Black Obelisk of Shalmaneser III. This monument has four sides, each of which displays five vertically arranged registers (or scenes). Each of them tells a story about a subdued king or ruler, who is paying tribute and prostrating himself before the victorious Assyrian king Shalmaneser III. Each individual scene narrates horizontally, using Akkadian cuneiform inscriptions and carved reliefs, from one side to another (in an anti-clockwise manner), wrapping around the obelisk. Scene 3 shows a parade of exotic animals brought from *Musri*. Shalmaneser comments that "*I received tribute from Muṣri, two-humped camels, a water buffalo, a rhinoceros, an antelope, female elephants, female monkeys and apes*" (http://oracc.museum.upenn.edu/nimrud/livesofobjects/blackobelisk/index.html). Side A of the register shows two camels, side B three exotic ungulates, side C an elephant and monkeys (**Fig. 5.7**), and side D two more monkeys. Looking at the individual representations in more detail, however, we note that side A depicts two-humped Bactrian camels, *Camelus bactrianus* L., 1758, while side B depicts some ungulates that could be possibly identified with (from left to right) a gaur, *Bos frontalis* Lambert, 1804; a greater one-horned rhino, *Rhinoceros unicornis* L., 1758; and perhaps a blackbuck, *Antilope cervicapra* (L., 1758), all characteristic of the Indian subcontinent (Borzatti von Löwenstern & Masseti, 1991: 147). Side 3 depicts the already-mentioned three monkeys, one of which has a long tail and the other two do not, and an Asian elephant, *Elephas maximus* L., 1758, including the small ears and convex rump (Masseti, 2002: 198; 2003a: 376). Finally, side D shows two other long-tailed monkeys, one of which is depicted with its shoulders covered with long curly hair. Such a head of hair could allow the identification of this primate with an Indian monkey, i.e., the now endangered lion-tailed macaque, *Macaca silenus* (L., 1758) (Hatt, 1959: 34; cf. Prater, 1971: 38–39). It is fairly unlikely that the animals depicted on the Black Obelisk, originated in Egypt. Instead, they likely come from the Middle East, which is within the current range of the zoological species that can be identified with some degree of certainty. Such a consideration would therefore shift their homeland from North-East Africa possibly to the Indian subcontinent, thus excluding the identification of the mysterious *Musri* country with Egypt or Kurdistan, in northern Iraq.

Although it is clear that the artist who carved the reliefs had never seen some of the animals he was called upon to represent, their images are evidence in favor of a Middle Eastern origin of the animals depicted. Since the tributes received from *Musri* include the three animal hieroglyphs which are normally shown on Indus script inscriptions (though the iconographic style may be from recollected memory), the cuneiform inscription related to these tributes (or animal hieroglyphs) as *river ox* (water buffalo?), *sakea* (translated as rhinoceros), *susu* (translated as antelope), elephants, *bazitu/uqupu* (translated as monkeys/apes) (Luckenbill, 1926: 211).

The modern translation of the cuneiform inscription of the Black Obelisk uses the terms 'monkeys' and 'apes' to indicate the primates, respectively, with and without tails (cf. Layard, 1849: Plates 53–56; Pritchard, 1954: 291), but Keller and Keller (1983: 167–168) translated instead the name of the two tailless primates with '*vervets and black baboons*', ignoring the fact that both the latter monkeys are characterised by a long tail. Also, Buren (1939) had proposed a similar identification:

> … a smooth, short-haired baboon (*Papio cynocephalus*), the other a maned ape (*Papio hamadryas*), whose head and forequarters are adorned with a thick mantle of long, coarse hair. Yet instead of the dog-faced visage of the maned ape the artist has given the animal a completely human countenance, and in other respects he has drawn largely upon the imagination in his delineation of animals which he had possibly never seen, and knew only from description.

It is clear from this that Buren (1939) should not have much confidence in the external morphology of the representatives of the genus *Papio* Erxleben, 1777.

The monkeys evoked in the Black Obelisk represent a rare case where text and picture coincide, pointing to an Indian and/or Middle Eastern source of inspiration. In view of the fact that no other representations have been found in Mesopotamian art, van Buren (1939) suggested that also the so-called 'Syrian elephants' were animals brought in tribute or used to stock royal hunting parks (Masseti, 2003a: 376).

5.5 WHERE DID 'MESOPOTAMIAN' MONKEYS COME FROM?

McDermott (1938: 15) suggests that the presence of primates in the art of the Near East indicates some exotic influence: "the main source of this influence was Egypt. The close trade connections meant that specimens of the animals and more often artistic representations would be exported from Egypt to Mesopotamia as tribute, as presents, or as objects of trade". For example, one of the tablets from Amarna lists gifts sent by Amenhotep IV (Ekhnaton, 1380–1362) to a Babylonian king. Among these was a 'silver ape', but it is not possible to tell which species it might have belonged to. However, since we can assume that its origin was Africa, it cannot be excluded that we are dealing with one of the most common species on the Egyptian market at the time: perhaps an individual green monkey of the genus *Chlorocebus* Gray, 1870, also called savanna monkeys, characterised by a silver-green coat colour, which may have inspired the metallic reflection of Ekhnaton's 'silver ape'. Not by chance, during the second millennium BCE, in the artistic evocation of grivets, *Chlorocebus aethiops* (L., 1758), both Egyptian and Minoan artists used a fascinating blue pigment to colour their coats (Masseti, 1980: 33; Masseti & Bruner, 2009: 59–62; Masseti, 2021b; see also Masseti & Veracini, in this volume). The grivet has recently been recognised as a separate guenon species, with characteristically peculiar morphological patterns (Groves, 2005; 158–159; Masseti & Bruner, 2009: 58–59, and references therein). Even though, in ancient times, its distribution extended much further North than at present, the species is still widespread from Sudan, East of the White Nile, to Eritrea and Ethiopia, East of the Rift Valley (Kingdon, 2004: 42–43; Groves, 2005: 158–159; **Fig. 5.8**). It cannot be ruled out, however, either that the Amarna tablet referred to a hamadryas baboon, characterised by a rather dull olive grey coat, tending towards silver reflections (cf. Harrison & Bates, 1991: 108). Goodman (1994: 29) is of the opinion that all the Egyptian references to the word 'primate' refer to 'small monkey', a long-tailed variety brought from abroad. If this were the case, then baboons, the ancient Greek *kinokefaloi* (*kynokephalos* = 'dog head'), would obviously have to be considered apart.

As observed by Hatt, as far as 1959 (33), the only non-human primates usually considered from ancient Mesopotamia are often representations of monkeys which appear in metal, stone, and pottery: "they were certainly not indigenous, but came in by trade as did gold, bronze, silver, copper, lapis, hardwood, and indeed, except for clay, almost everything with which Sumerian artists worked". One example is the already-mentioned monkey playing a flute in a clay plaque from Ur, early second mill. BCE (Mendleson, 1983: pl. XXIV), or several other images of musicians from Babylonia (7th century BCE) (Delaporte, 1910: pl. 382; Spycket, 1998: 1–10). This reminds us that monkeys were exotic animals that never existed in the wooded areas of the Levant, nor in the majority of the Near East. The *Cedar Forest* of Ḫumbaba may have been imaginatively placed far beyond the *eschatiai tes oikomenes*, the extreme ends of the ancient world, which coincided with the geography of Ethiopia (indicating sub-Saharan 'Black Africa' in its entirety) and India (cf. Li Causi, 2003: 62). Thus, the documentation of the occurrence of primates in ancient Mesopotamia is only possible from the evidence of their importation from foreign countries. Monkeys were exotic, held as pet animals by the upper classes and "related to sex, music, ugliness, lack of intelligence", and various other human characteristics (Kletter, 2002: 150). Their value may have been because they were curiosities and/or because of their effective economic worth (cf. Masseti, 2000: 90; 2012a: 182). In the Aegean world of the

FIGURE 5.8 Savannah monkeys of the genus *Chlorocebus* Gray, 1870, in Olduvai Gorge, Tanzania. (Photo by Marco Masseti.)

second millennium BCE, for example, African grivets were deemed so precious to be invested with the role of intermediary between the human and divine worlds (Masseti & Bruner, 2009: 61 and references therein; see also Masseti & Veracini, in this volume). As in the case of the Minoan primates, it is possible that the monkeys imported to ancient Mesopotamia could have been kept in controlled environments, in zoological parks and gardens. This is suggested by the fact that the animals had to be maintained within areas controlled by man, because of their value and rarity (cf. Masseti, 1997: 361; 2000: 90; 2003b: 275). For instance, one story on a statue from Nimrud, that the so-called ISIS/Daesh had recently tried to destroy, tells of a monkey colony bred by Ashurnasirpal II, third king of the Neo-Assyrian Empire (9th century BC). The engraving states in fact that "*herds of them* [monkeys] *in great numbers*" lived at Nimrud during the king's reign (Jarus, 2016: http://aina.org/ata/20161209174403.htm).

The figurine of a man with an antelope, a monkey and a leopard skin, now in the collections of the Metropolitan Museum of Art in New York, has been dated to the 8th century (**Fig. 5.9**). Carved in the round, this plastic group was excavated in Fort Shalmaneser, a royal building at Nimrud. It is a Phoenician product strongly influenced by the themes and styles of Egyptian art, owing to traditionally close ties between the two cultures. The depicted ungulate is almost certainly an oryx of probable African or Arabian origin, as one would think from observing its horns, which are straight and do not curve backwards, and the long-tailed monkey on the bearer's shoulders also looks like an African species, perhaps *C. aethiops*. This monkey is in many artistic representations of ancient Egypt (Masseti & Bruner, 2009: 59–60). The monkey, the leopard skin, and the oryx suggest this piece represents a Nubian bringing luxury goods in the form of live animals and animal skin as tribute to the Assyrian king. The ivory tusks that provided the raw material for this piece were almost certainly from African elephants, imported from lands South of Egypt, although Asian proboscideans did inhabit Syria before they were hunted to extinction by the end of the 8th century BCE (Masseti, 2003a: 375–379, and references therein).

Tail-less monkeys, namely Barbary macaques, may have been imported into Mesopotamia by the Egyptians who may have obtained them from the northwestern Maghreb and the wooded slopes of the Atlas mountains, in the southwestern Mediterranean, where populations of *Macaca sylvanus* still exist (Lion, 1992: 360; Masseti & Bruner, 2009: 37–42, and reference therein). Tailed monkeys came instead through Egypt from more southern African regions, such as Nubia, the extant Sudan or the so-called Horn of Africa (Eritrea, Ethiopia and Somalia), although it is very likely the Mesopotamian rulers ignored the existence of countries so far afield (Dunham, 1985: 236; Teissier, 1996: 112). After all, at the Babylonian court of Nabuchodonosor II

FIGURE 5.9 A long-tailed monkey, perhaps a grivet, *Chlorocebus aethiops* (L., 1758) is portrayed on the shoulders of a Nubian bearer in a Neo-Assyrian ivory figurine of Phoenician style, reported from Fort Shalmaneser (Nimrud, the ancient Kalhu) (New York, Metropolitan Museum of Art. Credit, Rogers Fund, 1960, https://www.metmuseum.org/art/collection/search/325089). (Image public domain.)

(*c.* 634 BC–*c.* 562 BC), the guardian of the monkeys was just an Egyptian, perhaps a deportee who was certainly able to take care of the primates better than any Babylonian could have done (Lion, 1992: 360).

5.6 BIBLICAL PRIMATES

The cultural balance of the ancient Near East was upset by the advent of the belief in a single and supreme god; initially Judaism, followed by Christianity and Islam. For the western world, the *Bible* (*Books*) is the sacred text par excellence that contains instructions, stories, poetry and prophesies, among other genres. It is a compilation of texts of a variety of forms, originally written in Hebrew, Aramaic and ancient Greek. The oldest books of the *Bible* are thought to date back to between the 8th and the 2nd century BC, but they can also refer to a much earlier tradition.

The *Bible* includes the names of many animals. According to Cansdale (1970: 13), nearly 180 Hebrew words, some still listed as 'uncertain', and just over 50 Greek words, represent a wide range of animal life. The *Bible* mentions 30 species of wild mammals, as well as 10 domestic ones (Alon, 1969: 235). Primates are only mentioned in 1 Kings 10: 22 and in 2 Chronicles 9: 21, with other precious goods imported by Solomon from the mythical land of Tharsis: "For the king's ships went to Tarshish with the servants of Huram: every three years once came the ships of Tarshish bringing gold, and silver, ivory, and apes, and peacocks" (Tristram, 1868: 37; Wood, 1875; Morris, 1896: 105–106; Cansdale, 1970: 136; Masseti, 2014: 99; Urbani & Youlatos, 2020: 654). In Hebrew, the term for monkey is *qoph* (*coph* or *koph*). *Qophim* (*kophim* or *cophim*), which is the plural of *qoph* (*kof* or *coph*) (Aldrovandi, 1637: 226 *apud* Veracini, 2019: 58), appears twice in the *Bible*. The word is from a foreign root. Groves (2008; 40) is of the opinion that it probably derives from the Egyptian *kif,* from which the Greek term *kepos* may also derive, although a possible Indian origin cannot be excluded (see Masseti & Veracini, Chapter 6, this volume), and Albright (1921: 144) and others are of the opinion that it is Egyptian in origin. Nowhere in the *Bible* is the monkey supposed to be indigenous to Israel and the Levant, instead it has an exotic origin. According to some other authors, including Tristram (1868: 37), *qoph* is derived from a Tamil word for monkey. Taylor (1814) noted that if 'monkeys' are really the creatures meant by the ancient Hebrew world *qophim*, it probably means a more uncommon species than would have come from Africa by way of Egypt. But this assumes they were seen in Judea for the first time, which something that cannot be confirmed. The same author would consequently assume an extra-African origin for some of the primates that were exported from Egypt to southwest Asia. A similar view is also held by Wood (1875) and Stiles et al. (2013).

The identification of the biblical site of *Tarshis* with a Phoenician port has been proposed (Negev & Gibson, 2001: 494). Perhaps valuable goods from markets further East, such as Indian peacocks and Middle Eastern monkeys, were brought into it.

5.7 MONKEYS FROM THE INDIAN SUBCONTINENT AND THE MIDDLE EAST

Several species of Asian primates were represented in the art of ancient Mesopotamia (cf. Such-Gutiérrez, 2022). Buren (1939: 22) suggests, for example, that some of the glazed frit monkey amulets of Akkadian Ur represent a species:

> ... of the anthropoid apes, and most likely a variety of the gibbon (*Hylobates*), an Asiatic ape whose characteristics are the excessive length of the arms, contrastingly short legs, small egg-shaped head, and human countenance; the tail is so rudimentary that it is not outwardly visible.

The same author observes that the small head and lack of tail on Sumerian monkey figurines from Ur means that a gibbon may have been the model, but these primates are currently only known from the Malay peninsula and part of the

Indonesian archipelago (cf. Corbet & Hill, 1992: 180–185; cf. Francis, 2008: 272–277). According to Houghton (1877: 329), one of the monkeys represented in Shalmaneser III Black Obelisk may be a langur of the genus *Semnopithecus* Desmarest, 1822, because the artefact – as we have already seen – also bear illustrations of other Middle Eastern animals. In favour of this hypothesis, Buren (1939: 23) notes that the African green monkeys have a simple straight tail, whereas langurs have a long tail with a tuft at the end, even though the African *Chlorocebus* also have a hairy enlargement of the distal part of the tail; the tails of the two captive primates on the stone relief from Assurnâṣirpal's palace at Niniveh are also tufted. Buren (1939: 22) suggests that these images refer to a species of langur, the Northern Plains Gray Langur, today classified as *Semnopithecus dussumieri* (Dufresne, 1797), (see Mittermeier et al., 2013) still largely dispersed in the northeastern Indian subcontinent. Thus, according to various authors, langurs would appear to be represented in some Mesopotamian art: "The apes engraved on cylinder seals, [...] almost always have very distinct, often quite long tails, and may be of the species *Semnopithecus entellus* L., the Hanuman of India" (Buren, 1939: 22).

The Asian primates most often depicted in ancient Mesopotamian art are more like macaques than langurs. Besides the already-seen lion-tailed macaque, identified by Hatt (1959: 34), among the monkeys represented in the Black Obelisk, the primates depicted on old Babylonian terracottas more closely resemble the rhesus macaque (Dunham, 1985: 261, n. 110), an animal widespread in the Indomalayan Region (cf. Corbet & Hill, 1992: 169) "... which has been, in modern times, one of the commonest monkeys kept as a pet and one which is easily carried about on the shoulder".

Another primate is represented in one of the two 'Archaic depots' dating from the Uruk period, consisting of a group of small alabaster statuettes of praying figures and animals (birds, a bear, and a pig), objects that reflect the style of sculpture in this period in Mesopotamia. A number of copper cosmetic flasks from Bactria (present-day Afghanistan) dating from the late third or early second millennium BC also represent primates and mythical animals, half-monkey half-human. Depictions of monkeys squatting and sitting on stools like humans are also to be found on metal seals from the same region. Along with other archaeological evidence, all these artifacts indicate the existence of commercial and cultural exchanges with the Indus Valley and other regions of Central Asia (such as Afghanistan), where many representations of monkeys have been found (Harper et al., 1992). As we have already noted, an articulated skeleton of a rhesus macaque was recovered during the archaeological excavations carried out by the Iranian Archaeological Mission of the necropolis of the Iranian site of Shahr-i Sokhta (Minniti, 2018: 246; Minniti & Sajjadi, 2019: 538–548). It was found in a tomb of the local necropolis and has been referred to the third millennium BC. The discovery of this species far from its natural territory casts new light on the ancient relationship between Mesopotamian people and the coeval human cultures of the Middle East.

5.8 AN ANCIENT TRADE TRADITION

The earliest contextual evidence of trade between Mesopotamia and the surrounding countries dates to the beginning of the historic period, around the early fourth millennium BCE (cf. Hallo, 1992: 353).

Various figurines of animals were among the goods imported to Mesopotamia from the Indus valley. As observed by Harper et al. (1992) although rarely depicted in the Susan region, the monkey was a common motif in many areas of eastern Asia, including the Indus civilization, notably the important Bronze Age centre of Harappa. It is now generally agreed that *Meluhha* was the name by which the Indus civilization was known to the Mesopotamians. Buren (1939: 22) noted that:

> if in very archaic times seals could be brought to Mesopotamia from the Indus Valley it is not impossible that figurines of apes were also brought, and that from these few models other figurines were copied more or less accurately.

By the third millennium, there is archaeological evidence of trade between lower Mesopotamia and the Indus region, by way of the Persian Gulf and the island of Bahrein (Leemans, 1960: 5). There is a considerable body of archaeological and textual evidence, to show that traders from Harappa were present in Sumer and Akkad. According to Sumerian and Akkadian texts, the Harappans exported timbers, gemstones and ivory; model monkeys in several materials, including cornelian and gold, have been found (McIntosh, 2008: 187).

Conversely, a number of living Indian animals were brought to Mesopotamia as gifts or exotic goods from proto-history onwards, including water buffaloes, *Bubalus arnee* (Kerr, 1792), as vividly depicted on several Sumerian cylinder seals and mentioned in a few texts of the Akkadian period (cf. Davis, 1987: 162–164; McIntosh, 2008: 187; **Fig. 5.10**). In southern Mesopotamia (Sumer), the period from approximately 2350–2150 BCE is named after the city of Agade (or Akkad), whose Semitic monarchs united the region, bringing the rival Sumerian cities under their control by conquest. Agade was probably founded before the time of Sargon (ca. 2340–2285 BCE.), the dynasty's first king. The *Curse of Agade* is a story dated to the Ur III Period (2047–1750 BC, thought to be somewhat older in origin) that tells the story of the Akkadian king Naram-Sin (reigned 2261–2224 BCE): his confrontation with the gods, particularly the god Enlil and how the city Agade was made wonderful by the goddess Inanna. It mentions monkeys in the urban streets:

> That monkeys, mighty elephants, water buffalo, exotic animals, as well as thoroughbred dogs, lions, mountain ibexes, and alum sheep with long wool would jostle each other in the public squares.
>
> *(Electronic Corpus of Sumerian Literature, "The Cursing of Agade", lines 21–24)*

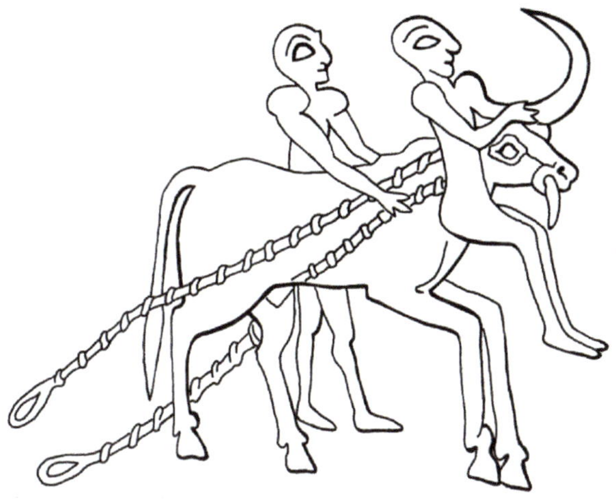

FIGURE 5.10 Impression of a cylinder seal of the Uruk period, 3,200–2,700 (the first half of the fourth millennium BCE). (From Davis, 1987 modified.)

The monkeys, elephants, buffaloes and the *alum sheep with long wool* – perhaps a domestic breed characterised by a fleece of particularly fine hair – were exotic biological elements imported from the Middle East, but a number of scholars have suggested that all of them originated in the East, probably in *Meluhha* (Dunham, 1985: 240; Potts, 1997: 259). The latter is the Sumerian name of a prominent trading partner of the ancient Mesopotamian civilization, whose precise location in the Indus valley is still debated. The site was cited in Mesopotamian texts as a source of god dust. An Ur III text describes a 'red dog' originally from *Meluhha*, which was given to king Hibbi-Sin as tribute from *Marhasi* (inland southwestern Iran) (McIntosh, 2008: 187), as probably a dhole or wild dog, *Cuon alpinus* (Pallas, 1811), a canid historically dispersed throughout south and east Asia (cf. Kamler et al., 2015). The importation to the Near East of living exotic species, and not just examples of durable parts of them, such as elephant tusks, from the Middle East reflected a practice that had been going on for millennia whenever political and economic conditions were favourable (Masseti, 2003a: 387).

One of the most interesting representations of a zoological species exotic to coeval Mesopotamia is shown on the stone reliefs from Sennacherib's palace, at Niniveh (about 700 BCE). Among the reeds of a marsh near an Assyrian town are two hornless hinds and a stag, with well-developed but scantily pointed antlers. Although this deer resembles the red deer, *Cervus elaphus* L., 1758, the shape of the antlers, the shortness of the limbs and the rather large hooves point to it being an image of a tropical Oriental swamp-dwelling species, perhaps the thamin or Eld's deer, *Rucervus eldi* (McClevelend, 1842) (Masseti, 2003a: 385), which incidentally is the most endangered now-wild ungulate in Asia (cf. Putman, 1988: 20; Corbet & Hill, 1992: 255–256; Francis, 2008: 319). Further representations of deer of probable Middle Eastern origin are documented on the stone hunting scenes of Ashurbanipal's palace, also at Ninive (about 865 BCE), where a group of these ungulates is led by the beaters against a hunting net. They resemble the Indian swamp deer or barasinga, *Rucervus duvaceli* (Cuvier, 1823) (Masseti, 2003a: 386), another endangered ungulate (Putman, 1988: 20–21; Cobet & Hill, 1992: 254–255). Like the previously mentioned thamin, this species, too, could have adapted to the extensive Middle Eastern areas of marshy grassland, prior to being imported into Mesopotamia from the Indian subcontinent by the Assyrian rulers for royal hunting parks which were located along the shores of the Euphrates and Tigris (Masseti, 2003a: 384–387; 2021a: 48–50).

5.9 A MATTER OF TAXONOMIC IDENTIFICATION

From the case studies we have considered so far, it emerges there are two main sources of the primates in the art and literature of the ancient Near East. Africa could have been the source of the Barbary macaques, *Chlorocebus* monkeys and even red monkeys, with the addition of some baboons from the Arabian Peninsula. Other species could have arrived from the Indian subcontinent, or further east. Among the latter, we could perhaps mention the image reproduced on a seal depicting a sitting monkey, dated to around the late third-early second millennium BCE and referred to the Bactria-Margiana archaeological complex, in Central Asia (New York, The Metropolitan Museum of Art; gift of Sheldon Lewis Breitbart, 1983.535.108) (**Fig. 5.11**). This metal seal, just 4.1 cm long, depicts an apparently medium-sized primate with a small, round head and a long tail. The location of the find would suggest that the artist's source of inspiration for the artefact was more likely to have been an oriental macaque than some species of African medium-sized and long-tailed primate, perhaps a long-tailed macaque, *Macaca fascicularis* (Raffles, 1821). There is no decisive evidence to link this representation with a single species. In any case, the long-tailed macaque is a widely distributed monkey in much of Indochina, Indonesia

FIGURE 5.11 The image of a sitting monkey is reproduced on a seal dated to around the late third–early second millennium BCE, and referred to the Bactria-Margiana archaeological complex, in Central Asia. (New York, The Metropolitan Museum of Art; gift of Sheldon Lewis Breitbart, 1983 1983.535.108. public domain image.)

and the Philippines (Corbet & Hill, 1992: 170–171; Groves, 2005: 161–162; Francis, 2008: 271–272).

Both of the primates represented in the already-mentioned 9th century fragment of wall relief from the NW Palace of Nineveh, at Nimrud, depicting a tribute of monkeys from the Phoenicians to the Assyrian king Assurnâṣirpal (Layard, 1849: fig. 59), have a long tail and are likely African species. These animals may show the external morphology of the so-called 'khbo' (*kepos*; *kebe*) of Aristotle (*History of animals*, VIII: 502a 17, 18 b 24–26), falling into the taxonomic category of monkeys with a tail (see McDermott, 1938: 91, 105; Masseti & Veracini, this volume). Few representations of monkeys from eastern Africa, the coastal regions of the Red Sea, or the central-western Indian subcontinent are known in the ancient world (Masseti & Bruner, 2009: 64–65). Phoenicians may have obtained them as luxuries, precious gifts, through good's exchanges with Egypt where, for example, monkeys such as grivets, *C. aethiops*, and red monkeys, *E. patas*, were particularly appreciated (cf. Masseti & Bruner, 2009: 52–54, 58–63). The already-mentioned Neo-Assyrian ivory figurine excavated at Nimrud (Metropolitan Museum of Art) (Aruz, 2010: 9) possibly represents another grivet imported into Assyria by Phoenician traders.

According to Hill (1966: 540–541), in ancient times, the range of grivets extended further north than at present. These primates have been kept as pets since time immemorial and numerous individuals have been exported out of Africa since antiquity (Masseti & Bruner, 2009: 58–63). They likely were imported into ancient Egypt from regions such as Nubia and the mythical land of *Punt*. These regions roughly correspond, respectively, to what is now northern Sudan and the territories of the Horn of Africa, where grivets are still found. According to Osborn and Osbornová (1998: 40), the importation of these and other monkeys began in the earliest Egyptian dynasties, although they did not appear in art until much later. Moreover, Houghton (1877: 319–321) noted that other primates Near-Eastern people may have been familiar with were the gelada, *Theropithecus gelada* (Rüppell, 1835), of Abyssinia, together with the sacred baboon. During the Neo-Assyrian and Neo-Babylonian periods, even the old South Arabian kingdom of Saba, located in the highlands of what is today northern Yemen, could have provided monkeys or baboons, either from the population of the latter species which inhabits certain parts of coastal Yemen, or from Ethiopia, which was colonized by the 'Sabaeans' (Potts, 1997: 259–260; cf. Veracini & Alghamdi, this volume).

5.10 'NEAR-EASTERN' MONKEYS IN ROMAN TIMES

In the centuries that saw the succession in the Near East of the Roman civilization, the Parthian and Sasanian Empires, Christianity and Islam, references to monkeys in local literature and art are always rare, as is their recovery at archaeological sites.

However, primates were quite common in the Roman world as novelties and curiosities (King, 2002: 433–434; see Masseti & Veracini this volume) and kept as pets as early as the 3rd century BC (Toynbee, 1973: 56). They were frequently dressed up as performers, with various articles as props (Toynbee, 1973: 55–60; Masseti & Bruner, 2009: 48) and this practice appears to have continued up to Medieval, Renaissance and even modern times (see Section II of this volume). An illustration of this can be found in a detail of the mosaics dating from the first half of the 6th century CE that decorate the northeastern hall of the imperial palace of Byzantium (register I, sector C, panel Co) (Istanbul, Turkey), where a clothed Barbary macaque is attempting to catch a bird from the top of a tall tree, using a lime twig (Jobst et al., 1997: 47; Masseti & Bruner, 2009: 39, 79). The Roman portrayal of monkeys is largely derived from Eastern, Greek and Hellenistic art, as demonstrated by the preponderance of catalogue entries from those sources in McDermott (1938) (see also Masseti, 2019: 21–51; Masseti & Veracini, this volume). In the Near East, the Barbary macaque was probably more accessible to Roman people than other species, but this does not rule out the importation of other primates from more far-flung areas, such as sub-Saharan Africa and/or the Indian subcontinent (Masseti & Bruner, 2009: 39, 79).

5.11 A POSSIBLE SASSANID EVIDENCE

Barnett (1973: 1) describes a white limestone slab carved in low relief from the site of Idalion on the island of Cyprus, with a curious scene (**Fig. 5.12**). It shows a man, whose head we have not been provided with, sitting turned to the

FIGURE 5.12 Limestone slab carved in low relief from the site of Idalion on the island of Cyprus, with a curious scene that has been compared to Sassanid artistic productions of the fourth and fifth centuries CE. (From Barnett, 1973.)

right, on a well-bred camel, moving to the right. At his knee, there is the top of an Iranian rider's thigh boot. The strangest feature of the whole scene is the figure of a monkey, which sits facing the rider on the camel's withers. The monkey is a prisoner, for round its neck is a collar attached by a chain to the spear shaft (upraised in the rider's left hand), which it joins just behind the blade. The rider's right hand is extended to the monkey, to steady it. According to Barnett (1973), some iconographic and stylistic elements bring this Idalion slab closer to some Sasanian artistic productions, such as three bowls of the 4th and 5th centuries CE showing a camel-riding king, allegedly the famous hero of legend, Bahram Gur (421–438 CE), famous as the hunter of onagers, *Equus hemionus* Pallas, 1775, which still exist in Iran and formerly even in Syria (Masseti, 2015: 81). The examination of these bowls suggests that the low relief could be even Sassanian in date, and probably is, and seems to have been part of a large frieze or hunting scene, since the camel's rump is missing. However, it is not a monkey that sits next to the knight, but Bahram's female companion, the beautiful but insolent Azadeh. In any case, Barnett (1973: 2) also wonders how and why the Idalion slab travelled so far from the Sassanian territory to reach Cyprus.

The morphology of the monkey depicted in the Idalion slab shows long and slender limbs, but the tail of the animal is not represented. Could this mean that it is absent? If this is the case, we may be dealing with a sub-adult Barbary macaque which, as we have already seen, was among the best-known primates of antiquity. Northwest Africa, the homeland of this monkey, was however far from the territories under the control of Sassanian rulers, as were the southernmost ranges of representatives of the genus *Chlorocebus* (see Sellier & Sellier, 2004: 21). The depiction of Idalion could also refer to one of the latter if the monkey's tail was hidden, but why would the artist deliberately conceal such an important detail? Other species could also be considered, including rhesus macaques, or even Indian langurs of the genus *Presbytis* (Müller, 1838) and *Pygathrix* (É. Geoffroy, 1812).

5.12 MONKEYS IN THE ISLAMIC WORLD

Since the 7th century of the Christian era, Muslim culture spread from the Arabian peninsula in the heart of the Near East where, as we have already seen, the primeval diffusion of non-human primates is both scarce and limited to one species. In the Islamic world, monkeys often appear as details in the description of exotic or even fantastic locations. Among the fabulous inventions and exotic legends spread by the Arabs in the ancient oecumene is, for example, that of the *Eastern Island*, a myth that perhaps spread through the world of Arabian seafaring in the wake of the tales contained in the *Periplus Maris Erythraei* ('Periplus of the Erythraean Sea'), dating to 70 CE (Ingrams, 1966: 4; Casson, 1989: 7; Masseti, 2014: 99–100), and similar accounts of exploration of exotic countries and lands (see Veracini & Alghmadi, Chapter 8, this volume). The Arabs were also responsible for the diffusion in the western world of other intriguing and long-lasting legends, such as that of the *gatto mammone*, hobgoblin, or even werecat, a fantastic character in traditional fables. Contrary to what its name would suggest, the *gattomammone* was not a true cat but a zoomorphic figure that has nothing to do with any representative of the Felidae family (Masseti, 2012b: 150). This word is found in one of the tables representing a vervet monkey (*Tavole animali*, VI, a., c. 65, see Capanna & Gippoliti, 2007: 83–87) of the 16th-century Italian naturalist Ulisse Aldrovandi'. This animal was labelled as '*Cercophetecus mas Simia caudata Gato Maimone vulgo*', which means 'a tailed monkey male also vulgarly referred to as *Gato Mammone*' (cf. Veracini, 2019). The Italian word *gattomammone* derives from the Arabic *maymūn/mīmūn*, a word that meant 'monkey' in Iran, Iraq, Turkey and Egypt (Bosworth et al., 1954: 132); the literal translation 'lucky, bringing good luck' is a euphemism, because the monkey was considered a devilish animal (Cortellazzo & Zolli, 1980). It should be noted that in the older texts, the sense is still that of 'monkey' (see Veracini & Alghmadi, Chapter 8, this volume), while the meaning of 'fantastic being' is later and probably influenced by the evangelical *Mammona*, used in the New Testament to personify profit, gain and material wealth, generally with negative connotations, i.e., quickly and dishonestly accumulated and otherwise wasted on luxuries and pleasures. *Mammona* is earthly wealth idolised, that is to say, the principle of spiritual damnation, the devil.

Among the rare artistic depictions of monkeys known from Islamic art, one that comes from the beginning of Islamic emergence in the Levant is the decoration of the 8th century Omayyad *Qasr al Amra* ('red castle', in Arabic) (**Fig. 5.13**), located in the present eastern Hashemite Kingdom of Jordan, at the southern edge of the basalt desert (also called the 'Black Desert', **Fig. 5.14**) (Mountfort, 1965: 78–80; Almagro et al., 1975: 17–23; Almagro Gorbea, 1981; Fowden, 2004: 31; Vibert-Guigue & Bisheh, 2007).

FIGURE 5.13 The extant archaeological site of Qasr al-Amra, located at the southern edge of the basalt desert (eastern Jordan), is what remains of a larger complex that included a castle, of which only the foundations have survived. The part of the palace still standing today is a small rural property, used as a royal retreat, and without any military function. (Photo by Marco Masseti.)

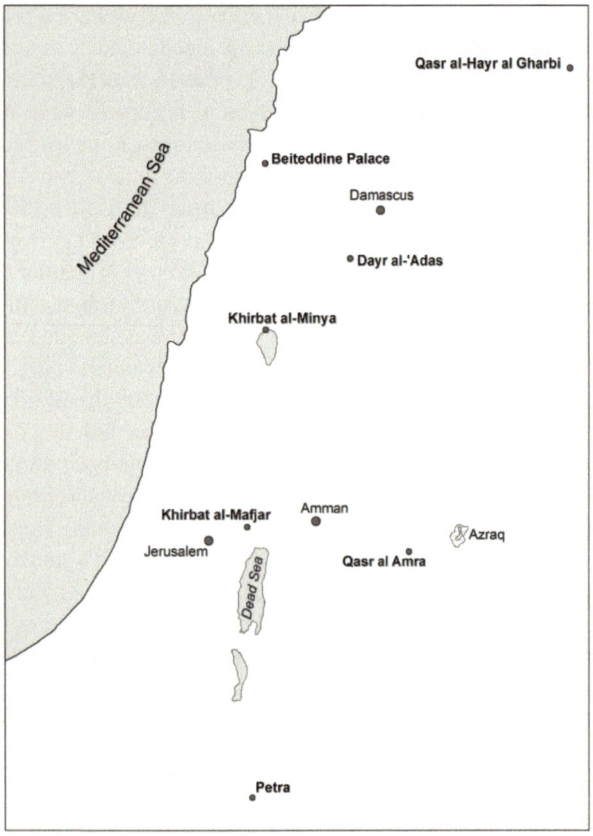

FIGURE 5.14 Geographical location of the medieval site of Qasr al-Amra, in the so-called 'Black Desert' of Jordan.

FIGURE 5.15 Detail of the frescoes on the tepidarium ceiling of Qasr al Amra. (Photo by Fabio Vianello.)

FIGURE 5.16 Olive baboon or Anubis baboon, *Papio anubis* (Lesson, 1827), Serengeti (Tanzania). (Photo by Marco Masseti.) The distinctive morphology of baboons is characterised by dog-like heads and tails relatively short, arched backward and bent at an angle above the base.

The castle was built, probably between 711 and 715, for the Umayyad caliph Al-Walid I (Mountfort, 1965: 79; Almagro et al., 1975: 23). One of the zoomorphic figures painted on the *tepidarium* ceiling of the archaeological site represents a monkey (**Fig. 5.15**). Mainly because of the prognathism of its muzzle and the mane still partially developed, this image has been interpreted as the portrait of a sub-adult hamadryas baboon (Masseti, 2015: 78–79), but it may represent a rhesus monkey because the base of tail is not baboon-like (cf. Osborn & Osbornová, 1998: 32; cf. Masseti, 2022; **Fig. 5.16**). The monkey is portrayed in one of the lozenges of the vault decoration, close to the depiction of an *oud*-playing bear (Masseti, 2015: 78–79; cf. Vibert-Guigue & Bisheh, 2007). But, unlike the carnivore, which could have come from the not-so-distant Alawit Mountains, in Syria, if it is a macaque it must have come either from India or from further afield in Asia (cf. Prater, 1971: 36–37; cf. Francis, 2008: 272). In the general context of the wall paintings of Qasr al Amra, this primate would be the only representation of a Middle-Oriental or Far Eastern zoological species, since all the other animals depicted are of African or Near Eastern origin (Masseti, 2015: 69–85). However, if it was a hamadryas, it could have reached the Black Desert of Jordan from the much closer southwestern Arabian Peninsula.

Primates continued to feature in the lives of the wealthier Islamic social classes. Elgood (1951) recalls, for instance, that "In the month of Ramaẓán in the year 221 (AD 836) there came to Samarra the Ruler of Nubia who brought to al-Mu'taṣim presents among which were some monkeys". The latter (Baghdad, 749–842) was the eighth caliph of the Abbasid dynasty. Still in the first half of the 16th century, the image of a sub-adult baboon was used to evoke in the western world the suggestive atmosphere of the Near East, as in the case of the depiction of the centre of Damascus by an anonymous Venetian painter, already attributed to the school of Gentile Bellini (Paris, Louvre) (**Fig. 5.17**). At more or less the same time, the image of a

FIGURE 5.17 Detail of an anonymous Venetian orientalist painting (already attributed to the school of Gentile Bellini), 1511, *The reception of the ambassadors in Damascus* (Paris, Louvre). (Image public domain.)

Barbary macaque smelling a Damascene rose was transformed into an amazing personification of the detested Islamic world, and its ruler was depicted in a detail of the frescoes of Buontalenti's *Grotta Grande* (1586–1587) in Boboli Gardens (Florence) Bernardino Poccetti (Masseti, 1991: 327; Masseti & Bruner, 2009: 82). The attitude in which the monkey was painted is the same as that in which Suleyman the Magnificent, extraordinary sultan of the Ottoman empire from 1520 to 1566, generally chose to be portrayed (Masseti, 1991). Suleyman, who doubled the territory of the empire during the forty-six years of his rule, wished to be painted in a pose that emphasised his sophistication and sensitivity, rather than underscoring his fame as a ruthless conqueror. His domain extended from the surroundings of Vienna far to the East and into Egypt and Persia, while his fleet dominated the Red Sea and virtually the whole of the Mediterranean basin. Consequently, it is not difficult to imagine why the artist who evoked the scene of the 'sniffing monkey' in the Boboli gardens was asked by Francesco I, or someone else at the Medici court, to paint a caricature of the Muslim enemy (Masseti & Bruner, 2009: 82; Fossi, 2021: 132).

5.13 CONCLUDING REMARKS

Despite the relative proximity to the Eastern and African areas where monkeys are still widely dispersed, it seems that from literary and archaeological evidence, primates were scarce in the ancient Near East, apart from Egypt. We can, however, say that this part of southwest Asia, although not originally populated by monkeys, has never actually been without them. In Mesopotamia, in particular, references to monkeys can be found in the pages of the world's oldest literature – the epic of the Sumerian hero Gilgamesh – and in some artistic representations, but the osteological evidence is practically non-existent other than the already-mentioned skeleton of the rhesus macaque from the archaeological exploration of Shahr-i Sokhta, referred to the third millennium BCE (Minniti, 2018: 246; Minniti & Sajjadi, 2019: 538–548). In fact, the remains of the Iranian Bronze Age urban settlement, called the 'burnt city', are located in the Sistan and Baluchistan Province in the southeastern part of Iran, on the bank of Helmand river. This archaeological site is geographically on the extreme eastern edge of the Near East, bordering the Middle East and the Indian subcontinent, and not far from the current range of the rhesus macaque, which is still widespread in the northern and central Indian subcontinent (Corbet & Hill, 1992: 169–170; Groves, 2005: 163). In spite of the importance of the human settlement of Shahr-i Sokhta between the end of the fourth and the whole third millennium BC, its geographical position places it within the biogeographical boundaries and the cultural sphere of influence of western India, rather than the Near East. Shahr-i Sokhta might ultimately provide concrete evidence of a civilization east of prehistoric Persia that was independent of ancient Mesopotamia (cf. Sajjadi, 2003: 29, 52).

ACKNOWLEDGEMENTS

I should like to express my appreciation and gratitude to Andrew R. George, SOAS, University of London, for the enjoyable and stimulating discussions on the literary occurrence of monkeys in ancient Mesopotamia; Claudia Minniti, for having put at my disposal the data and the photographs of the buried rhesus macaque recently discovered in the third millennium necropolis of the Iranian site of Shahr-i Sokhta, as well as Cecilia Veracini and Bernard Wood for scrupulous reading of the manuscript and for their precious suggestions.

REFERENCES

Akurgal E. & Hirmer M. (1962). *The Art of the Hittites*. Thames and Hudson, London.

Albright W.F. (1921). The Ivory and Apes of Ophir. *The American Journal of Semitic Languages,* 37 (2): 144–145.

Aldrovandi U. (1637). *De Quadrupedibus Digitatis Viviparis* (ed. B. Ambrosino) ibid. 1642, 1645, 1666 (nel frontespizio 1645), 1665.

Almagro M., Caballero L., Zozaya J. & Almagro A. (1975). *Qusayr 'Amra. Residencia y baños omeyas en el desierto de Jordania*. Ministerio de Asuntos Exteriores, Direccion General de Relaciones Culturales; Junta para la Proteccion de Monumentos y Bienes Culturales en el Exterior; Instituto Hispano-Arabe de Cultura, Madrid.

Almagro Gorbea A. (1981). La residencia de Qusayr 'Amra en el desierto de Jordania. In: Almagro Gorbea A. (ed.): *Tres monumentos islamicos restaurados por España en el mundo arabe*. Insitituto de España, Madrid.

Alon A. (1969). *The Natural History of the Land of the Bible*. The Jerusalem Publishing House, Jerusalem.

Al-Rawi F.N.H. & George A.R. (2014). Back to the Cedar Forest: the beginning and end of Tablet V of the Standard Babylonian Epic of Gilgameš. *Journal of Cuneiform Studies*, 66: 69–90.

Alster B. (1997). *Proverbs of Ancient Sumer: The World's Earliest Proverb Collections*. CDL Press, Bethesda.

Aruz J. (2010). Archaeology and the Department of ancient Near Eastern art. *The Metropolitan Museum of Art Bulletin*, New Series, 68 (1): 4–48.

Atallah S.I. (1977). Mammals of the Eastern Mediterranean Region; their ecology, systematics and ecological relationships. *Säugetierkundliche Mitteilungen*, 25: 241–316.

Barnett R.D. (1973). Monkey Business. *Journal of the Ancient Near Eastern Society*, 5: 1–10.

Basmachi F. (1976). *Treasures of the Iraq Museum*. Ministry of Information – Directorate General of Antiquities, Baghdad.

Benoit A. (1993). 33. Seated monkey. In: Harper P.O., Aruz J. & Tallon F. (eds.): *The Royal City of Susa. Ancient Near Eastern treasures in the Louvre*. The Metropolitan Museum of Art, New York: 33.

Bertman S. (2005). *Handbook to Life in Ancient Mesopotamia*. Oxford University Press, Oxford and New York.

Borzatti von Löwenstern E. & Masseti M. (1991). On the twisted-horn antelopes carved in the rocks of some wadies of the Jordan southern desert. *Studi per l'Ecologia del Quaternario*, 13: 143–148.

Bosworth C.E., van Donzel E., Lewis B. & Pellat C. (eds.) (1954). *The Encyclopedia of Islam*. Vol. 7. Brill, Leiden.

Budge E. (1914). *Assyrian Sculptures in the British Museum, Reign of Ashur-nasir-pal, 885–890 B.C.* British Museum Publications, London.

Buren E.D. van, (1939). *The Fauna of Ancient Mesopotamia as Represented in Art*. Pontificium Institutum Biblicum, Rome.

Cansdale G. (1970). *Animals of Bible Lands*. The Paternoster Press, Devon.

Capanna E. & Gippoliti S. (2007). I "Quadrupedi Vivipari". In: Alessandrini A. & Ceregato A. (eds.): *Natura picta. Ulisse Aldrovandi*. Editrice composizioni, Bologna: 83–87.

Casson L. (1989). *The Periplus Maris Erythraei*. Princeton University Press, Princeton.

Clutton-Brock & Burleigh R. (1978). The animal remains from Abu Salabikh: preliminary report. *Iraq*, 9: 89–100.

Corbet G.B. (1978). *The Mammals of the Palaearctic Region. A Taxonomic Review*. British Museum (Natural History)/ Cornell University Press, London and Ithaca.

Corbet G.B. & Hill J.E. (1992). *The Mammals of the Indomalayan Region*. Natural History Museum Publications, London/ Oxford University Press, Oxford.

Cortellazzo M. & Zolli P. (1980). *Dizionario etimologico della lingua italiana*. Vol. 2. Zanichelli, Bologna.

Davis S.J.M. (1987). *The Archaeology of Animals*. B.T. Batsford Ltd, London.

Delaporte L. (1910). *Catalogue des cylindres orientaux et des cachets assyro-babyloniens, perses et syo-cappadociens de la Bibliotheque Nationale*, Paris.

Dominy N.J., Ikram S., Moritz G.L., Wheatley P.V., Christensen J.N., Chipman J.W. & Koch P.L. (2020). Mummified baboons reveal the far reach of early Egyptian mariners. *eLife* 2020; 9:e60860. https://doi.org/10.7554/eLife.60860

Dunham S. (1985). The monkey in the middle. *Zeitschrift für Assyriologie*, 75: 234–264.

Elgood C. (1951). *A Medical History of Persia and the Eastern Caliphate: From the Earliest Times until the Year A.D. 1932*. Cambridge University Press, Cambridge.

Fossi G. (2021). *L'oggetto misterioso. Enigmi, segreti, viaggi nell'arte*. Giunti, Firenze.

Fowden G. (2004). *Quṣayr 'Amra. Art and the Umayyad Elite in Late Antique Syria*. University of California Press, Berkeley/ Los Angeles/London.

Francis C.M. (2008). *Mammals of the South-East Asia*. New Holland Publishers, London.

Franke S. & Wilhelm G. (1985). Eine mittelassyrische fictive Urkunde zur Wahrung des Anspruchs auf ein Findelkind. *Jahrbuch des Museums für Kunst und Gewerbe*, 4: 19–26.

Frankfort H. (1943). More sculpture from the Diyala Region. The University of Chicago Press, Chicago. *Oriental Institute Publications*, 60: xiii + 48.

George A. (2016). Monkeys in the Gilgamesh Epic. Oral paper contributed to the "*Primates in Antiquity" Symposium held at Dartmouth College*, Hanover (New Hampshire, USA), August 19, 2016.

Goodman H. (1994). *Between Jerusalem and Benares: Comparative Studies in Judaism and Hinduism*. State University of New York Press, New York.

Grayson A.K. (1991). *Assyrian Rulers of the Early First Millennium BC II (1114–859 BC)*. University of Toronto Press, Toronto.

Groves C.P. (2001). *Primate Taxonomy*. Smithsonian Institution Press, Washington DC: 350 pp.

Groves C.P. (2005). Order primates. In: Wilson D.E. & Reeder D.-A.M. (eds.): *Mammal Species of the World. A Taxonomic and Geographic Reference. Third edition, Volume* 1. The John Hopkins University Press, Baltimore: 111–184.

Groves C. (2008). *Extended Family: Long Lost Cousins. A Personal Look at the History of Primatology*. Conservation International, Arlington, Virginia, USA.

Hallo W.W. (1992). Trade and traders in the ancient Near East: some new perspectives. In: Charpin D. & Joannès F. (eds.): *La circulation des biens, des personnes et des idées dans le Proche-Orient ancien*. XXXVIIIe R.A.I. Editions Recherche sur les Civilisations, Paris: 351–356.

Harper P.O., Aruz J. & Tallon F. (1992). *The Royal City of Susa. Ancient Near Eastern Treasures in the Louvre*. The Metropolitan Museum of Art, 16 novembre 1992-7 mars 1993, New York. The Metropolitan Museum of Art/Harry N. Abrams Inc., New York.

Harrison D.L. (1964). *The Mammals of Arabia. Vol. I. Insectivora, Chiroptera, Primates*. Ernest Ben Ltd, London.

Harrison D.L. & Bates P.J.J. (1991). *The Mammals of Arabia*. Harrison Zoological Museum, Seven Oaks (Kent, UK).

Hatt R.T. (1959). Mammals of Iraq. *Miscellaneous Publications Museum of Zoology, University of Michigan*, 106: 5–113.

Hill W.C.O. (1966). *Primates. Comparative Anatomy and Taxonomy. Volume 6. Catarrhini Cercopithecoidea Cercopithecinae*. Edinburgh University Press, Edinburgh.

Houghtonk, W. (1876–1877). On the mammalia of the assyrian sculptures. Part II. Wild animals. *Transactions of the Society of Biblical Archaeology*, 5: 33–64, 319–383.

Ingrams, H. (1966). *Arabia and the Isles*. 3rd edition. John Murray, London.

Jarus, O. (2016). Inscription about ancient 'monkey colony' survives ISIS attacks. *Live Science*, December 9, 2016. https://aina.org/ata/20161209174403.htm (Accessed 15 September 2020).

Jobst, W., Erdal, B. & Gurtner C. (1997). *Istanbul. The Great Palace Mosaic*. Arkeoloji ve Sanat Yayinlari, Istanbul.

Kamler J.F., Songsasen N., Jenks K., Srivathsa A., Sheng L. & Kunkel K. (2015). *Cuon alpinus*. The IUCN Red List of Threatened Species 2015: e.T5953A72477893. https://dx.doi.org/10.2305/IUCN.UK.2015-4.RLTS.T5953A72477893.en.

Kantor, H.J. (1962). A bronze plaque with relief decoration from tell tainat. *JNES*, 21: 93–117.

Keller C. & Keller K. (1983). Die Affen des schwarzen Obelisken. *Assur*, 3: 167–168.

King A. (2002). Mammals. Evidence from wall paintings, sculpture, mosaics, faunal remains, and ancient literary sources. In: Feemster Jashemski W. & Meyer F.G. (eds.): *The Natural History of Pompeii*. Cambridge University Press, Cambridge: 401–450.

Kingdon J. (1990). *Arabian Mammals*. Academic Press, London.

Kingdon J. (2004). *African Mammals*. A&C Black, London.

Kletter R. (2002). A monkey figurine from Tel Beth Shemesh. *Oxford Journal of Archaeology*, 21 (2): 147–152.

Koldewey R. (1990). Das wieder erstehende Babylon (5th edn., Hrouda B., ed.). C.H. Beck, München.

Layard A.H. (1849). Nineveh and its Remains. Vol. I. John Murray, London:

Leemans W.F. (1960). *Foreign Trade in the Old Babylonian Period as Revealed by Texts from Southern Mesopotamia*. Brill, Leiden.

Li Causi P. (2003). *Sulle tracce del manticora. La zoologia dei confini del mondo in Grecia e a Roma*. Palumbo Editore Palermo.

Limet H. (2005). Ethnicity. In: Snell D.C. (ed.): *A Companion to the Ancient Near East*. Blackwell Publishing, Maiden (USA)/Oxford (UK): 370–383.

Linnaeus C. (1758). *Caroli Linnaei. Systema Naturae. A photographic facsimile of the first volume of the tenth edition (1758). Regnum Animale*. British Museum (Natural History), London (1956).

Lion B. (1992). La circulation des animaux exotiques au Proche-Orient antique. In: Charpin D. & Joannès F. (eds.): *La circulation des biens, des personnes et des idées dans le Proche-Orient ancien*. XXXVIIIe R.A.I. Editions Recherche sur les Civilisations, Paris: 357–365.

Luckenbill D.D. (1926). *Ancient Records of Assyria and Babylonia. Volume 1. Historical Records from Assyria from the Earliest Times to Sargon*. The University of Chicago Press, Chicago.

Mallowan M.E.L. (1947). Excavations at Brak and Chagar Bazar. *Iraq*, 9: 1–259.

Mariette A. (1875). *Les Listes Géographiques Des Pylons De Karnak Comprenant La Palestine, l'Ethiope, Le Pays De Somal*. Hinrichs, Leipzig.

Masseti M. (1980). Le scimmie azzurre. La fauna etiopica degli affreschi minoici di Santorino (Thera). *Mondo Archeologico*, 51: 32–37.

Masseti M. (1991). Dalla "turata della Gran'Bestie" allo "stanzone" degli agrumi: splendore e decadenza dei serragli faunistici del Giardino di Boboli. In: Acidini Luchinat C. & Garbero Zorzi E. (a cura di): *Boboli 90. Atti del Convegno Internazionale*. Edifir-Edizioni Firenze, Firenze: 323–337.

Masseti M. (1997). Representation of birds in Minoan art. *International Journal of Osteoarchaeology*, 7: 354–363.

Masseti M. (2000). Did the study of ethology begin in Crete 4000 years ago. *Ethology Ecology & Evolution*, 12: 89–96.

Masseti M. (2002). *Uomini e (non solo) topi. Gli animali domestici e la fauna antropocora*. Firenze University Press/Università di Firenze, Firenze.

Masseti M. (2003a). Exploitation of fauna in the Near East during ancient historical times: observations on the mammalian species represented in Assyrian art. In: Albore Livadie C. & Ortolani F. (eds.): *Variazioni climatico-ambientali e impatto sull'uomo nell'area circum-mediterranea durante l'Olocene*. Centro Universitario Europeo per i Beni Culturali, Ravello. Edipuglia, Bari: 375–390.

Masseti M. (2003b). Taxonomic and behavioural aspects of the representation of mammals in Aegean Bronze Age art) In: Kotjabopoulou E., Hamilakis Y., Halstead P., Gamble C. & Elefanti P. (eds.): *Zooarchaeology in Greece. Recent Advances*. British School at Athens Studies, 9: 267–275.

Masseti M. (2010). The mammals of the Farasan archipelago, Saudi Arabia. *Turkish Journal of Zoology*, 34: 359–365.

Masseti M. (2012a). *Atlas of Terrestrial Mammals of the Ionian and Aegean Islands*. De Gruyter, Berlin.

Masseti M. (2012b). I "boschetti con animali" e le raffigurazioni zoomorfiche. In: Hayez J. & Toccafondi D. (eds.): *Palazzo Datini a Prato. Una casa fatta per durare mille anni*. Tomo I. Edizioni Polistampa, Firenze: 145–153.

Masseti M. (2014). Herpetological enigmas from the Arabian seas, with particular reference to the Sarso island racer, *Platyceps insularis* Mertens, 1965 (Farasan archipelago, Saudi Arabia). *Scripta Herpetologica*. Studies on Amphibians and Reptiles in honour of Benedetto Lanza: 99–116.

Masseti M. (2015). The early 8th century A.D. zoomorphic iconography of the wall decorations in Qasr al-Amra, Hashemite Kingdom of Jordan. *Anthropozoologica*, 50 (2): 69–85.

Masseti M. (2019). Monkeys in the ancient Greek world (2000–500 BC). In: Veracini C., Scalfari F. & Casanova C. (eds.): *History of Primatology: Yesterday and Today*. The Mediterranean Tradition. Aracne editrice, Canterano (Rome): 21–51.

Masseti M. (2021a). Vertebrates of upper Mesopotamia: present evidence and archaeological data. In: Jawad L.A. (ed.): *Tigris and Euphrates Rivers: Their Environment from Headwaters to Mouth*. Springer Nature Switzerland AG, Cham (Switzerland): 13–72.

Masseti M. (2021b). An analysis of recent literature regarding the Minoan "blue monkeys" represented in Aegean Bronze Age art. *JASs Journal of Anthropological Sciences*, 99: 153–156.

Masseti M. (2022). Replying to Urbani, Youlatos & Binnberg (2021). *JASs Journal of Anthropological Sciences*, 100. doi: 10.4436/JASS.10004. Epub ahead of print. PMID: 35830362.

Masseti M. & Bruner E. (2009). The primates of the western Palaearctic: a biogeographical, historical, and archaeozoological review. *JASs, Journal of Anthropological Sciences*, 87: 33–91.

Masseti M., De Marchi G. & Chiozzi G. (2015). Forbidden islands. The absence of endemics among the insular non-volant terrestrial mammalian fauna of the Red Sea. *Natural History Sciences. Atti Soc. it. Sci. nat. Museo civ. Stor. nat. Milano*, 2 (2): 101–130.

McDermott W.C. (1938). *The Ape in Antiquity*. Studies in archaeology, 27. The Johns Hopkins press, Baltimore.

McIntosh J. (2008). *The Ancient Indus Valley: New Perspectives*. ABC-CLIO, Santa Barbara (California).

Meeks D. (2003). Locating punt. In: O'Connor D. & Quirke S. (eds.): *Mysterious Lands*. University College London Press, London: 53–80.

Mendleson C. (1983). More monkey business. *Anatolian Studies*, 33: 81–83.

Mills, D.K. (2019). Ophir and Punt. *SIS Chronology & Catastrophism Review*, 1: 43–50.

Minniti, C. (2018). Preliminary results from the study of animal remains at Shahr-i- Sokhta: new researches in 2017 season. *Abstract book of the 11th International Congress on the Archaeology of the Ancient Near East*, 03-07 April 2018, Ludwig-Maximilians-Universität, Munich: 246.

Minniti, C. & Sajjadi, S.M.S. (2019). New data on non-human primates from the ancient Near Est: the recent discovery of a rhesus macaque burial at Shahr-i Sokhta (Iran). *International Journal of Osteoarchaeology*, 29: 538–548.

Mittermeier R.A., Rylands A.B. & Wilson D.E. (2013). *Primates. Handbook of The mammals of the World.* Lynx Edition, Barcelona.

Morris D. (2013). *Monkeys.* Reaction book, Chicago.

Morris F.O. (1896). *Bible Natural History. A Description of the Quadrupeds, Birds, Insects, Trees, Plants, &c., &c., Mentioned in the Bible.* Christian Literature Society for India, Calcutta, Madras, and Bombay.

Mountfort G. (1965). *Portrait of a Desert. The Story of an Expedition in Jordan.* Collins, London.

Napier P.H. (1981). *Catalogue of Primates in the British Museum (Natural History) and Elsewhere in the British Isles. Part 2. Family Cercopithecidae, Subfamily Cercopithecinae.* British Museum (Natural History), London.

Negahaban E.O. & Negahaban I.A. (1991). *Excavations at Haft Tepe, Iran.* UPenn Museum of Archaeology, Philadelfia.

Negev A. & Gibson S. (2001). Tarshish. *Archaeological Encyclopedia of the Holy Land.* Continuum, New York and London.

Osborn D.J. & Osbornová J. (1998). *The Mammals of Ancient Egypt.* Aris & Phillips Ltd, Warminster (England).

Peters J., von den Driesch A. & Helmer D. (2005). The Upper Euphrates-Tigris basin: cradle of agro-pastoralism. In: Vigne J.- D., Peters J. & Helmer D. (eds.): *The First Steps of Animal Domestication. New Archaeological Approaches.* Oxbow Books, Oxford: 96–124.

Potts D.T. (1997). *Mesopotamian Civilization: The Material Foundations.* Cornell University Press/The Athlone Press, London.

Powell M.A. (1978). Ukubi to mother… The situation is desperate: a plaidoyer for methodological rigor in editing and interpreting Sumerian texts with an excursus on the verb taka: dax-dax (TAG4). *Zeitschrift für Assyriologie und Vorderasiatische Archäologie*, 68 (2): 163–195.

Prater S.H. (1971). *The Book of Indian Animals.* Bombay Natural History Society/Oxford University Press, Calcutta, Chennai, Delhi, Mumbai.

Pritchard J.S.B. (1954). *The Ancient Near East in Pictures Relating to the Old Testament.* Princeton University Press, Princeton.

Putman R. (1988). *The Natural History of Deer.* Christopher Helm, London.

Reade J. (1983). *Assyrian Sculpture.* British Museum Publications, London.

Sajjadi S.M.S. (2003). Excavations at Shahr-i Sokhta. First preliminary report on the excavations of the Graveyard 1997–2000. *Iran*, 41: 21–98.

Schoff W. (1912). *The Periplus of the Erythraean Sea: Travel and Trade in the Indian Ocean, by a Merchant of the First Century.* Longmans, Green, and Co., New York.

Sellier J. & Sellier A. (2004). *Atlas des peuples d'Orient. Moyen-Orient, Caucase, Asia centrale.* Éditions La Découverte, Paris.

Spycket A. (1998). "Le Carnaval des Animaux": on some musician monkeys from the ancient near east. *Iraq*, 60: 1–10.

Stiles D., Redmond I., Cress D., Nellemann C., & Formo R.K. (2013). *Stolen Apes. The Illicit Trade in Chimpanzees, Gorillas, Bonobos and Orangutans: A Rapid Response Assessment.* United Nations Environment Programme, GRID-Arendal, Oslo.

Such-Gutiérrez M. (2022). The Monkey in Mesopotamia during the Third Millennium BCE. In: Urbani B., Youlatos D. & Antczak A.T. (eds.): *World Archaeoprimatology Interconnections of Humans and Nonhuman Primates in the Past.* Cambridge University Press, Cambridge (UK): 419–430.

Tallet P. (2013). Deux notes sur les expéditions au Pays de Pount. *Revue d'Éyptologie*, 64: 189–209.

Taylor C. (1814). *Scripture Illustrated by Means of Natural Science: In Botany, Geology, Geography, Natural History, Natural Philosophy, Utensils, Domestic and Military. Part I. An Expository Index Referring to Subjects of Science in the Order of the Sacred Books.* Charles Taylor, London.

Teissier B. (1996). *Egyptian Iconography on Syro-Palestinian Cylinder Seals of the Middle Bronze Age.* University Press Fribourg, Fribourg and Vandenhoeck & Ruprecht, Göttingen.

Toynbee J.M.C. (1973). *Animals in Roman Life and Art.* Cornell University Press, Ithaca/New York.

Tristram H.B. (1868). *The Natural History of the Bible.* Society for Promoting Christian Knowledge, London.

Uerpmann H.P. (1991). *Equus africanus* in Arabia. In: Meadow R.H. & Uerpmann H.-P. (eds.): *Equids in Ancient World.* Vol. II. Dr. Ludwig Reichert Verlag, Wiesbaden: 12–33.

Urbani B. & Youlatos D. (2020). On the earliest representations of chimpanzees (*Pan troglodytes*): were African apes traded to bronze age elam? *International Journal of Primatology*, 41: 654–663.

Urbani B. & Youlatos D. (2021). The primates of Susa: depictions of monkeys in stone statuettes from Elam. *Ash-sharq*, 5: 1–9.

Veracini C. (2019). Natural history of non-human primates in the early modern period the contribution of the Italian encyclopaedist Ulisse Aldrovandi. In: Veracini C., Scalfari F. & Casanova C. (eds.): *History of Primatology. Yesterday and Today. The Western-Mediterranean Tradition.* Aracne editrice, Canterano (Rome): 53–91.

Vibert-Guigue C. & Bisheh G. (2007). *Les peintures de Qusayr'Amra. Un bain omeyyade dans la bâdiya jordanienne.* Bibliothèque archéologique et historique, Institut français du Proche-Orient, Beyrouth.

Wood J.G. (1875). *Wood's Bible Animals.* Philadelphia: Bradley, Garretson & Co.

Woolley C.L. (1934). *Ur excavations. Volume II. The Royal Cemetery. A Report on the Predynastic and Sargonid Graves Excavated between 1926 and 1931.* The British Museum, London, and the University Museum of Philadelphia, Philadelphia: Text/New York: Trustees of the two Museums.

Wrede N. (1990). Katalog der Terrakotten der archaeologischen Obenflächenuntersuchung (Survey) des Stadtgebietes von Uruk. *Baghdader Mitteilungen*, 21: 215–301.

Zohary M. (1973). *Geobotanical Foundations of the Middle East.* Vol. I. Stuttgard, Amsterdam.

6 Europe from the Bronze Age (Mid-Third Millennium BCE) to Greco-Roman Times

Marco Masseti
Istituto Zooprofilattico della Sicilia

Cecilia Veracini
Universidade de Lisboa

Of all the beasts the most malicious are monkeys, especially with regard to those things in which they want to imitate men

(Aelian, *De Natura Animalium,* VII, 21).

6.1 NON-HUMAN PRIMATES IN MEDITERRANEAN EUROPE: TRADE AND CULTURAL ROLE

The Mediterranean Sea, the cradle of ancient civilisations and a crossroads of people, trades, and cultures, facilitated the diffusion of many animals, which had been brought from ancient times beyond the territories of their homeland (Masseti, 2012: 21–30). Currently, non-human primates (primates from here onwards) do not figure among the mammalian fauna native to Europe, but the genus *Macaca* Lacépède, 1799, has existed in Europe since the Late Miocene (Rook et al., 2001) and inhabited this region until the Late Pleistocene (Kurtén, 2017: 58). This genus must have co-existed alongside the earlier *Homo* L., 1758, species for over one million years in the Early and Middle Pleistocene, remaining in the Late Pleistocene, widespread in central and southern Europe until they became extinct (Rosendahl et al., 2011; Schreiber & Löscher, 2011). It is still doubtful whether macaques became extinct before the arrival of modern *Homo sapiens* (L., 1758) in Europe, dated around forty-seven thousand years ago (Gibbons, 2020).

In proto-historical times, from at least the Bronze Age (mid-third millennium BCE), different species of primates, or durable parts of them, were shipped from their natural habitats to Europe. They played various cultural roles according to the different periods, being regarded as sacred and magic animals, mascots or pets, and were used for entertainment or other scopes. The best-known monkey of ancient and classical Europe was the Barbary macaque, *Macaca sylvanus* (L., 1758) (Wendt, 1956: 19; Masseti & Bruner, 2009: 39; Veracini, 2020). Individuals of this species, endemic to western North Africa, were shipped to Europe probably since the 8th century BCE (cf. Veracini, 2020 for review). They appeared to be common in this African region according to Strabo (60 BCE–24 CE) who refers, in his *Geography* (cf. Biffi, 1999, book XVII, 827: 59), to the abundance of monkeys observed along the coast of Africa by the Greek philosopher Posidonius during his journey to Rhodes.

Other monkeys, not endemic to the Mediterranean regions, would have reached the ports of Egypt from the Red Sea or would have been brought along the trans-Saharan trade routes, which connected Sudan and northern Algeria across the desert (cf. Law, 1967; Swanson, 1975). Some expeditions are known to have been sent by Egyptian rulers to the southern countries (Chami et al., 2002: 23–25) in order to collect animals and other goods. Nubia and the mythical land of Punt were places that roughly correspond to what are now, respectively, northern Sudan and some of the territories of the so-called Horn of Africa (Ethiopia, Somalia, and perhaps part of the southwestern Arabian peninsula). From there, primates of the genus *Chlorocebus* Gray, 1870, and two species of baboons, the Olive baboon, *Papio anubis* (Lesson, 1827), and the Sacred or Hamadryas baboon, *Papio hamadryas* (L., 1758), were brought to the Mediterranean shores (Masseti & Bruner, 2009: 48–52). So far, no fossil remains of Sacred baboons have been found in Egypt where there is no definite knowledge of their ancient distribution. The baboon cult, however, has appeared since Pre-dynastic times, when this primate appears to have existed in Egypt (Osborn & Osbornová, 1998: 32–37). Subsequently, it cannot be excluded that other species of primates (e.g. representatives of the genus *Erythrocebus* Trousseart, 1897), and even the African great apes, such as sub-Saharan chimpanzees and gorillas, may have arrived Egypt and from there ancient Europe.

As documented by Masseti (Chapter 5, this volume), commercial relationships and cultural exchanges between lower Mesopotamia and the Indus region were already

DOI: 10.1201/b21963-7

well-developed by the third millennium BCE; thus, also, Asiatic primates, such as the Rhesus macaque, *Macaca mulatta* (Zimmermann, 1780), might have been brought to Near East as documented by recent studies (Minniti & Sajadi, 2019). According to Warmington (1928: 146–147) followed by McDermott (1938: 73), Asian monkeys were probably only rarely imported, due to the difficulty of transporting them by sea, but in Greek classical times, Asiatic monkeys could have been brought through Persian empire. Contacts with the Middle East and the Indian subcontinent were intensified mainly with the campaigns of Alexander of Macedon (356–323 BCE). Then, the Roman conquests in the Near East gave way to a long period of political stability that favoured long-range trade, both by land and sea. Two specimens of Asian macaques (the first unidentified, *Macaca* sp.) and the other a bonnet macaque, *Macaca radiata* (E. Geoffroy, 1812) were found in a pets' necropole in the Roman Red Sea port of Berenice (Egypt) in the 1st–2nd centuries CE (Osypinska et al., 2021).

6.1.1 Bronze Age

Despite the complete absence of primates from the Mediterranean islands during the Holocene, exotic monkeys appeared during the Bronze Age in Minoan iconography. Minoan seals carved in different materials and used as a means of identifying documents and objects are important part of Bronze Age Aegean art and are the most common surviving type art after pottery from Early Minoan II (c. 2500 BCE) onwards. Seals carved in the shape of seated monkeys (e.g. **Fig. 6.1**) have been found in a number of sites on Crete, as have a few pieces of jewellery (Marinatos & Hirmer, 1960: 80; Hood, 1978: 212). Due to their stylised forms, it is generally not easy to identify the species represented, although in some cases they may resemble the morphology of baboons, Barbary macaques, or even long-tailed monkeys.

Simians were also depicted in some frescoes discovered in the palace of Knossos (Crete) (Middle Minoan III A, c. 1700–1600 BCE) (McDermott, 1938: 23), on the island of Thera, modern-day Santorini, (Late Minoan IA, about 1550 BCE, Athens, National Museum) and in the Pillar Crypt area of Milos (Marinatos, 1971: 45–46; Masseti, 1980: 33–35; 2000: 90; 2012: 179). These beautiful wall paintings have continually attracted the attention of scholars in various disciplines, especially, regarding the question of whether or not these animals were actually imported and exactly which species were depicted (see **Box Details 6.1**).

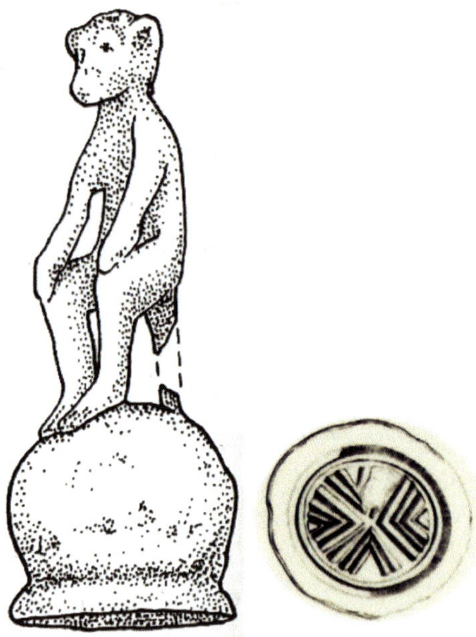

FIGURE 6.1 Knob seal surmounted by a long-tailed primate, perhaps a *Chlorocebus* monkey, from Trapeza, Lasithi (Crete), Early Minoan II–III, *c.* 2500 or later (Herakleion Archaeological Museum) (after Hood, 1978).

BOX DETAILS 6.1: THE "BLUE MONKEYS" OF THE MINOANS (SECOND MILLENNIUM BC)

Marco Masseti

Starting from the second half of the third millennium BCE, Crete and the Aegean islands were affected by a unique cultural episode which made the Eastern Mediterranean basin one of the major civil areas of the ancient oecumene: the flourishing of the Minoan civilisation (Hood, 1978; Niemeier, 1998; Stampolidis et al., 2011). Dramatically, however, when the Minoan thalassocracy was at its height, *c.* 1630 BCE, the volcanic island of Thera (Santorini) erupted with impressive consequences on the island itself and on the surrounding environment (Marinatos, 1968; Bond & Sparks, 1976). The eruption had a profound influence on the civilisations of the whole eastern Mediterranean, spreading ashes in the region and trigging tsunamis throughout the Aegean, causing the Minoan civilisation to collapse (Antonopoulos, 1992; LaMoreaux, 1995; Aydar et al., 2021).

Sumptuous relics of wall-paintings and other artistic products have come down to us partly preserved under the thick layer of eruptive materials that covered most of the Aegean islands and Crete as a consequence of the tremendous eruption. Numerous images of monkeys characterised by an intriguing blue colouration have been preserved among them at various archaeological sites in the Aegean, such as Knossos (House of the Frescoes) on Crete, Phylakopi (Pillar Crypt area) on the island of Milos, and Akrotiri on Santorini (**Fig. 6.2**). All these images have

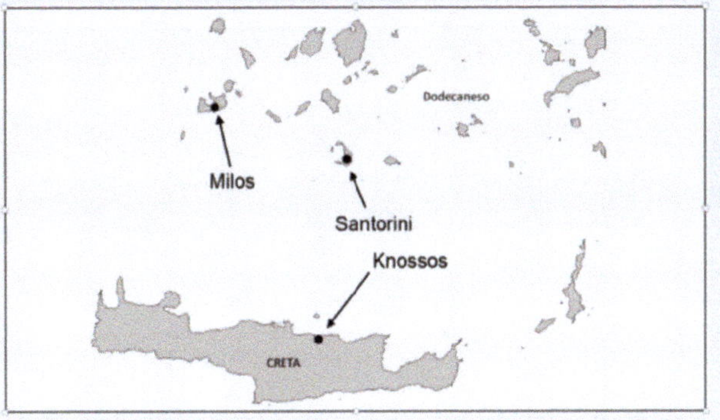

FIGURE 6.2 Archaeological location of the Minoan sites that provided painted images of "blue monkeys". (From Masseti & Bruner, 2009 modified.)

FIGURE 6.3 Detail of the "blue monkeys" wall-painting discovered in Room Berta 6 at Akrotiri (Thera - Santorini, Greece), (second Millennium BCE).

the same phenotypical characteristics, such as a slender body, black-face mask, white ruff and belly, long limbs, and tail (Masseti, 2021; **Fig. 6.3**). They feature the unmistakable morphological patterns of monkeys of the genus *Chlorocebus* (more specifically the grivet, *Chlorocebus aethiops* (L., 1758) cf. Masseti, 1980, 2003; Masseti & Bruner, 2009; or tantalus monkeys, *Chlorocebus tantalus,* Olgiby 1841 cf. Groves, 2008). The representatives of this genus are distributed in the African savannahs and steppes south of the Sahara from about 18°–15° N (Hill, 1966; Groves, 2005). In the wall-paintings of Santorini, Milos, and Crete, the images of these primates are clearly recognizable due to the peculiarity of the blue ground colour of their body (cf. Masseti, 2021). According to Bietak (2000), the use of blue instead of grey in the Aegean Bronze age settlement of Tell el-Dab'a in the Nile delta is typically Minoan. This colour convention, due to Aegean influence, was also seen in Egypt later. Blue was, in fact, the colour with which grivets were also depicted later in Egyptian art, being available to reproduce the greenish-grey coat of these monkeys and other zoological species (Masseti, 1997). It seems that among the monkeys known to the Minoans, there were almost only grivets (Masseti, 1980, 1997, 2000, 2003, 2012; Masseti & Bruner, 2009; Groves, 2008), no other primate species appearing in Late Bronze Age Aegean paintings (Masseti, 2021, 2022). It cannot, however, be ruled out that the depiction of baboons (genus *Papio* Erxleben, 1777) could be found in some ivory seals of probable Egyptian production, referring to the last Proto-Minoic phase (Marinatos & Hirmer, 1960; Masseti, 2022) and other finds of later periods recorded by Canciani (1973). "Blue monkeys" have been kept as pets since time immemorial, and numerous exemplars have been exported out of Africa since antiquity. There is considerable amount of evidence to show that grivets were traded between eastern Africa and the eastern Mediterranean basin in the course of the second millennium BCE (Masseti, 2012, and references therein). As far as is presently

FIGURE 6.4 The wall painting of the "House of the Frescoes" at Knossos, depicting a troop of "blue monkeys" in the course of a raid in a nesting area of the rock doves, *Columba livia* Gmelin, 1789, as restored by Cameron (1968). Detail of the Late Minoan IA wall-painting from the "House of the Frescoes" at Knossos (Crete, Greece), showing a "blue monkey" eating an egg from a wild pigeon nest (after Cameron, 1968).

known, *C. aethiops* was imported into the Aegean world from Egypt as a precious gift. The value of these monkeys may have been related more to their curiosity appeal than to their effective economic worth, or possibly to both (cf. Masseti, 2001, 2018). In view of their rarity, these animals had to be kept within an area controlled by humans (Masseti, 1997). Most likely for this reason, all the scenes in which the blue monkeys are depicted in Late Bronze Age Aegean art may reflect controlled environments, with those apparently set in the wild actually being located in parks (Morgan, 1988; Masseti, 1997). Grivets were to be deemed so exotic and important by the Aegean Bronze Age culture that they were not thought of simply as animals, but as creatures invested with the role of intermediary between the human and divine worlds (Marinatos, 1987; Evely, 1999; Ferrence & Bendersky, 2004). In the Aegean region, they regularly occurred even, perhaps, as a naturalised species (Masseti, in press).

On the wall-paintings of the "House of the Frescoes" at Knossos, grivets are depicted in the landscape with birds with rocks, streams, waterfalls, and many flowers and in the course of a raid in a nesting area of the rock doves, *Columba livia* Gmelin, 1789, which might explain why most of these "blue birds" were depicted in flight (Cameron, 1968; Masseti, 1997). One of the primates even seems to be eating an egg (cf. Cameron, 1968; **Fig. 6.4**). Species of the genus *Chlorocebus* are in fact often regarded as primates with a marked tendency to raid gardens, fields, and plantations, often becoming a major pest (Estes, 1992; Wafe, 2019; Veracini & Hazevoet, 2024). Studies carried out on the West African representative of this taxonomic group, the green monkey, *Chlorocebus sabaeus*, demonstrated that these primates can develop the habit of preying upon birds and small mammals, such as columbiformes and murids, respectively. Galat and Galat-Luong (1977) observed that this predation occurred in marginal habitats of the Sahelian zone, particularly in relation to some species of Columbiformes, such as the palm dove, *Streptopelia senegalensis* (L., 1766). This provides scientific verification of the behaviour of the "blue monkeys" in the Knossos wall-painting (Masseti, 2000, 2003, 2006; Masseti & Bruner, 2009).

After a careful review of the iconographic sources, it appears that monkeys of the genus *Chlorocebus* were the only primate represented in Aegean Bronze Age paintings, despite the stubborn assertions of some authors (Masseti, 2021, and references therein). The suggestion that these animals are baboons (genus *Papio* Erxleben, 1777) or other monkeys is not convincing at all and cannot be interpreted other than as a lack of specific zoological knowledge on the part of those who have attributed its identification with other species (Masseti, 2021, 2022). In the same way, the taxonomic attributions proposed by Urbani and Youlatos (2022) should be considered. There are also no definite diagnostic elements for the identification of the supposed primate rediscovered in a fragment of a Mycenaean fresco in Tiryns. Much less can we accept its attribution to a baboon-like primate, arbitrarily proposed by Youlatos et al. (2023). Nevertheless, as already observed by Trantalidou (2000), despite the frequent occurrence of grivets in the artistic production of the Aegean Bronze Age, osteological fragments of these monkeys are not reported from any sites in the Aegean. Poulianos (1972) rashly claimed the fallacious discovery of an alleged monkey skull, apparently found among sea pebbles and rocks on the east side of Santorini by the Greek archaeologist Galanopoulos, in the summer of 1966. The affair was, however, exposed as a complete fabrication, the item involved being simply a chunk of lava featuring a vague resemblance to the exterior of a monkey skull (Doumas, 2000).

In the Mycenaean period (16th–12th centuries BCE), the artistic representation of primates seems to have disappeared from Crete, mainland Greece, and the Aegean islands. Only artefacts in monkey-like form imported from Egypt or from eastern countries were found in Mycenaean contexts (Cline, 1991: 38–39). As Wolfson (2018: 44) underlines, the Mycenaeans, "did not make any simian art of their own; rather, they inherited the objects made by their Cretan predecessors without considering, or perhaps caring about, their religious implications". Two imported figurines with the cartouche of Amenhotep II (ca. 1447–1420 BCE) are among the earliest objects of this type found in a Mycenean archeological context. These were found in Mycenae and Tiryns, in mainland Greece (McDermott, 1938: 24; Cline, 1991). The Mycenaen figurine (**Fig. 6.5**) has some morphological characters (ears, eyebrowns, ridge of the nose, and the cheeks) finely detailed. The original colour in some parts is still preserved, and the final portion of the tail, visible on the rear of the figurine, shows a tuft (Cline, 1991: 31). It is likely a representation of a species of the *Chlorocebus* genus, monkey, as also Cline (1991) correctly states. Gold pendants, some in the form of squatting baboons dating between 1200 and 1000 BCE, were also found in the Greek island of Aegina (McDermott, 1938: 24). It is only during the Greek Late Geometric period (760–700 BCE, see Glossary), that primates began to emerge again in the Eastern Mediterranean in the form of recognisable simian figurines, which may have been inspired by Egyptian amulets, as well as stylised anthropomorphic bronze figurine. Many of them have been found in such diverse areas such as Elis, Laconia, Arcadia, Euboea, Rhodes, and Central Italy (Longodon, 1990: 407). The dating of these artefacts is generally difficult to ascertain, but those that have been found in recorded contexts are from sanctuaries (e.g. Olympia, Athens, Artemis Orthia at Sparta, etc.), and their generic forms are similar to those of thousands of bronze animal figurines, dating from 750 to 700 BCE, found at Olympia (Langdon, 1990: 411–412), whose provenance suggests that they were intended as votive offerings. Langdon (1990: 419) attributes the renewed popularity of images of monkeys to the Phoenicians, asserting that the distribution of monkey figurines corresponds exactly with their trade routes at the time. Different elements could have contributed to a closed link between Phoenicians and monkeys: they adopted the Egyptian dwarf god, Bes, who was closely associated with monkeys – sometimes the appearance of this god mingles with that of them (Dasen, 2013: 62); Phoenicians-Carthaginians likely adopted the traditional cult of the Barbary macaque of North African native populations (see later), and additionally in the Near East, monkeys seem to be associated with sexuality and fertility (Longdon, 1990). Greeks and Phoenicians probably exported their appreciation to these animals around the Mediterranean while trading with different civilisations.

During the first millennium BCE, the Italian peninsula and its islands were involved in the commercial network established in the central Mediterranean in the course of the 8th century by the Phoenicians and the Greeks from Euboea. Their commercial network connected the African coast with central Italy, Sicily, and southern Sardinia (D'Oriano, 2007). One monkey at least must have been seen even in Iron Age Sardinia in the course of the 9th–8th century BCE. This is attested by the bronze figure, which decorates a lamp from an unknown locality, perhaps Baunei, in the province of Ogliastra, now on display at the National Archaeological Museum of Cagliari (Masseti, 2019: 21; **Fig. 6.6**). The majority of authors agree upon the identification of this bronze figure as that of a tailless monkey, probably a Barbary macaque (D'Oriano, 2007; Masseti, 2019: 21). Barbary macaques also appear in artistic productions of the Etruscan period. Figurines and other monkey-like objects were diffused throughout the area of influence of this Italic culture for more than one century, from the 7th century BCE (Bonacelli, 1932: 372). The cultural moment that characterises this chronological period

FIGURE 6.5 Blue frit monkey figurine with cartuche of Amenothep II, found at Mycene. (Drawing by Cecilia Veracini.)

FIGURE 6.6 Bronze figurine from Sardinia, portraying a Barbary macaque. (Courtesy of the Museum of Cagliari.)

is defined as "orientalising", as it is informed on contents and styles of eastern provenance (cf. Naso, 2000). They include, among other things, the hunting use of exotic felids, such as the cheetah, *Acinonyx jubatus* (Schreber, 1775) and the caracal, *Caracal caracal* (Schreber, 1775), completely unknown previously in the western oecumene (Adinolfi et al. 2019; Masseti et al, 2022). The eastern origin of these cultural elements and traditions is essentially due to the dense network of commercial and cultural exchanges that had already been established for some time between the Near East and the central Mediterranean (Vernesi et al., 2004; Masseti, 2012). Similar sophisticated notations are destined to enhance the artistic and scenographic context that documents the persistence, still in the 5th century BCE, of those contacts with the Eastern World so evident in earlier times.

The Italian historian Bonacelli (1932: 368–369) notes that some of the Etruscan objects made in monkey shape (see **Fig. 6.7**) could have had a talismanic significance related to the sacred value of the Barbary macaque, whose worship had probably been transmitted by the Phoenicians-Carthaginians thanks to the commercial network that they had established. It seems that, at that time, the species was a venerated animal in North Africa (cf. Veracini, 2020). Depictions of monkeys in Etruscan art are known from at least three tombs of a later period (6th-5th century) (**Fig. 6.8**). Once again, it seems that we are facing the representation of Barbary macaques. This time, however, the primates appear to have lost their sacred or talismanic role. Still, according to Bonacelli (1932), the contexts in which the monkeys are presented suggest that they were probably regarded as pets since they accompany the burial of human beings. At the same time, it can't be ruled out that the image of the animals was even considered an element of the plays and games in honour of the dead.

There are still those who believe that it was the Phoenicians and/or Carthaginians who were mainly responsible for the exportation of Barbary macaques beyond their homeland in proto-historical times. This assumption has given rise to an unproven cultural model that still persisted, although not supported by any zoo-archaeological evidence. Traditionally involved in some way with the possible presence of primates is also the island of Ischia, in the Gulf of Naples, in southern Italy. First occupied by Greeks from the Euboean towns of Chalcis and Eretria (Greece) *ca.* 775 BCE, it was called *Pithecusa* or *Pithecusae*, from the ancient Greek term *Pithekoussai.* As reported by Masseti (2019: 23), this term has two contrasting believed origins: some historians claim that the name derives from the Greek *pithos*, a type of amphora, or terracotta jug, which the original settlers produced on the island, whereas others maintain that the name derives from the Greek word *pithekos*, meaning tailless monkey or prankster (Bonacelli, 1932: 356–357; Bonfante, 1992). Regarding the latter

FIGURA 6.7 Etruscan bronze Figurine. (After Bonacelli, 1932, modified.)

FIGURE 6.8 Detail of the Etruscan *Tomba della scimmia* 480–470 BCE. (Image in the Public Domain.)

derivation, it has been assumed that the first Greek settlers perhaps found a colony of monkeys on the island – although no free-ranging or wild populations of monkeys are known to have ever inhabited Ischia, either in very ancient or in more recent times. Gras (1994) argues that monkeys were transferred there by seamen within the Phoenician and Euboean circuits, and Bonacelli (1932: 358) suggests that the Phoenecians could have also introduced to the island the cult for a zoomorphic divinity with the aspect of a monkey. Thus, the toponym *Pithecusa* would mean the "isle of monkeys". This origin of the name is corroborated by the other name of the island, *inarime* (from *arim*), a word used by Tyrrhenians (Etruscans) to indicate monkeys. Bonacelli (1932: 359) is even of the opinion that the term *arim* derived from the Punic (Semitic) word *harim* which means "flat nose" and corresponds to the Latin name *simia*.

It appears that the name *Pithecusa* could even be traced back to Greek mythology (Masseti, 2019: 23; Vespa, 2021: 389–443). Ancient mythographers must have been reduced to alleging that these *pithecoi* were deceitful men whom Zeus punished by turning them into apes. Two of these were the dwarfish, larcenous sons of Oceanus (Evslin, 2006) a race of mischievous little forest creatures called *Cercopes* [=tailed ones]. According to Xenagoras, who seems to have lived in Hellenistic times, the *Cercopes* were turned into simians because of their trickery and wickedness (Jacoby, 1930: 202–203) and after the metamorphosis were banished to various volcanic areas, one of which was Ischia (cf. Graves, 1955). Either way, it, thus, appears likely that the immigrants to Ischia from the Greek island of Euboea settled in a place already known as *Pithecusa*, rather than naming it so themselves (Masseti, 2019: 22). The legend of Heracles (Hercules) defeating the *Cercopes*, which were captured by him when they played tricks on him, has been interpreted as a reminiscence of Phoenician traders bringing monkeys to Greek markets (Keller, 1887). However, as reported by Wolfson (2018), Kerkopes (*Cercopes*) that appear in Greek art as early as the Archaic Period (since 8th c. BCE) are not certainly connected with simians until the Roman era. It was Ovid (43 BCE–17 CE) in the *Methamorphoses*, who recalled in Roman times this legend linked the myth of Aeneas' journey to Italy:

> [Aeneas] with the ship deprived of his helmsman had touched Inarime and Prochite (*Prochytenque*) and the Pithecuse (*Pithecusas*) islands, arid cliffs that took their name from their inhabitants. These cercops (*cercopum*), people of cheaters, because of their fraud, perjury, and crimes had hated the Father of the Gods, who at some point decided to turn them from the men that they were, into deformed animals, in such a way that they possessed the propriety to appear at the same time similar and dissimilar to men. He shortened their limbs, flattened their nose, gouged their faces with senile wrinkles, covered their bodies with yellow hair, and sent them here not before removing from them the faculty of speech, rendering mute those tongues known only to pervert: he left them only the chance to lament with cries. (cf. Ovidio, 1992, Vol 2, book XIV: 811–812)

During the 7th and 6th centuries, different objects in the form of, or portraying monkeys (faience figurines, vases, terracotta, and ivory figurines), were discovered (and created) in the islands of Cyprus and Rhodes. These islands played a crucial role in the transmission of Egyptian and Assyrian cultures to the Greek civilisation. Additionally similarly artefacts were found also at Corinth and in Italy in Etruria (McDermott, 1838: 25; Greenlaw, 2011: 61–65). Most of those produced after the second half of the 7th century were intended for use as scent bottles (cf. **Fig. 6.9**). Baboons frequently appeared on cosmetic objects, as did other monkeys, to symbolise eroticism, fertility, and regeneration. Baboons' images also appear in different Phoenician settlements, including Sardinia where a 6th century BCE, pottery jug in the form of a squatting baboon-like monkey carrying a net on his back was discovered in Grave 5/35 at Tharros. Similarly in the Balearics a baboon amulet was found in the Punic-Phoenician burial complexes of the 6th BCE, on the island of Ibiza (it is currently housed in the Puig des Molins Museum).

6.1.2 The Classical Antiquity

In the Classical Greek period (5th–4th century BCE) and throughout the Roman empire (until *c.* 500 CE), some species of primates became well-known in Europe both through either direct contact with an improved importation of these animals and through the testimony of travellers in Africa

FIGURE 6.9 Scent bottle in form of monkey (front and profile). During the 7th and 6th centuries BCE, plastic vases, many of which were pithemorphic, began to be produced in Eastern Greek centres, particularly Rhodes, as well as Corinth and Etruria, and most of those produced after the second half of the 7th century were intended for use as scent-bottles. (Courtesy of Parco Archeologico di Cerveteri e Tarquinia – Museo Archeologico Nazionale di Tarquinia. Photo Cecilia Veracini.)

and Asia. They were depicted in different kinds of artworks and mentioned in literature works and plays. Primates morphological and behavioural characteristics were described by Greek philosophers, such as Aristotle (c. 384 BCE–322 BCE); historians, such as Strabo (c. 62 BCE–c. 24 CE); the physician Galen of Pergamum (129–216c. 201 CE); and scholars, such as Pliny the Elder (23–79 CE) and Aelian (c. 170–after 230 CE), among many others. In this historical period, according to artistic, literary, and early natural history sources, primates lost any metaphysical or divine attributes, being mainly regarded as performing animals; exotic and valuable beasts; or pets along with dogs, cats, and birds (cf. Greenlaw, 2011). It is important to stress that the term "monkey" acquired a clear negative meaning, and primates were perceived as extremely ugly and whimsical animals, showing the "unacceptable" wish to imitate human attitudes and characters.

As human companions, they could frequent kings' palaces, as well as regular houses, where monkeys moved freely enough to run away, as evidenced in the *Miles Gloriosus* (III-II century BCE). In this well-known comedy by Plautus, a case of adultery in Athens was discovered by a slave of a braggart soldier in the neighboring house. The discovery accoutred when the slaved chased an escaped monkey over the roof (cf. McDermott, 1938: 136; Cleary, 1972). Despite these examples, in the Greek archaeological records of this and in the Archaic (c. 8th–5th centuries BCE) periods (cf. Wolfson, 2018: 25), primates and other non-native animals appear only sometimes, and there is no evidence of large-scale importation, differently to what happened later in the Roman times. It seems, as reported by Calder (2011 *apud* Wolfson, 2018: 25), that only wealthy Greek people could afford such expensive pets and likely used them to underline their status.

From the 3rd century BCE, wealthy Romans had aviaries and private aquariums: monkeys, snakes, antelopes, gazelles, and other animals were kept in these menageries in order to increase the prestige of their owners (Toynbee, 1973: 17). Later, in 186 BCE, large number of animals were imported for gladiatorial games: a large number of large-sized carnivores as well as other species that came from both Africa and Asia, such as ostriches, oryxes, rhinos, tigers, hippos, and elephants. Many of these were gifts of the governors of imperial colonies or of foreign kings, but the majority were obtained from animal traders (Toynbee, 1973: 17–18). Each Roman emperor had a menagerie for triumphal processions and official celebrations, especially for those involving gladiators killing animals (Hoage & Mansour, 1996: 11). These exhibitions grew in parallel with the imperialist expansionism of Rome with the aim of showing Rome's dominion over the world (Lo Giudice, 2008: 6). Apparently, primates were not involved in the *Venationes*, staged hunts in which professional hunters attacked and slew animals held in Rome and involved many wild animals. In 55 BCE, when Pompey inaugurated his theatre, two *venationes* a day were staged for five days, during which he might show an Indian rhino; about 20 elephants; 600 lions; 410 female leopards; 1 lynx; and for the first time, an Ethiopian monkey. The latter were called *kebos* by Pliny the Elder and could be identified as a Patas monkey (cf. Lo Giudice, 2008: 7), genus *Erythrocebus* Trouessart, 1897. The Roman *kebos* reported by Pliny is the first and unique written reference to this animal in Europe until modern times. At the end of the 2nd century and during the following one, with the consolidation of the Roman expansion which gave more accessibility to animals, hunting shows became more frequent. It seems that, during the *venationes*, primates, along with elephants and camels, were trained to perform a version of the Greek dance *pyrrica* which mimicked a sort of struggle (Sabbatini Tumolesi, 1980: 97).

Among the species of primates, Barbary macaques were probably more accessible to Roman hunters and trappers than other primate species and the Barbary macaque is the only primate found to date in different epochs at European archaeological sites, having been found in Italy, England, Spain, and France (see **Table 6.1**). In Pompeii, a specimen of Barbary macaque was found under the ashes (Bailey et al., 1999). In England, many osteological remains of this species were provided by the archaeological exploration of Catterick Fort (Yorkshire, UK), dating to the Romano-British period (Lynn, 1997). Lynn (1997) recorded other British remains of the species from Dunstable, Bedforshire, dated to the late 2nd century. In Spain, too, in the old Roman city of Iulia Livica (Llívia, Cerdanya), in the vicinity of Empúries (Catalonia), a 5th–6th century tomb of a Barbary macaque was found (Guardia et al., 2005). The specimen had been buried along with a series of military decorations, including belt buckles and a number of bronze plaquettes, that were part of its personal accoutrements. The military characteristics of its apparel suggest that the Barbary macaque may have been either a pet or a military mascot (Masseti & Bruner, 2009: 40). In a tomb found in the archaeological Gallo-Roman site in France, a Barbary macaque was buried inside the tomb enclosure of the family, suggesting that this animal was not considered a mere exotic rarity but a true member of the family (cf. Gerber & Baudry-Dautry, 2012). According to Roman law, only close relatives, emancipated young people, and heirs could be buried in the family tomb; thus, the monkey could be regarded as a component of the *familia*, in the Roman sense. Yet, the role of pets in the Greco-Roman world is not yet fully understood and seems to have been very different from what we mean today (cf. Calder, 2017; Franco, 2017). Two Barbary macaques artistically evoked on a marble relief of the 2nd century, from Via della Foce at Ostia, seem to represent pets as well. They were portrayed, while they were on display in a store for public appreciation (Veracini, 2020; **Fig. 6.10**).

Representations of different species of primates as performers, pets, or exotic animals inserted in mythological or naturalistic contexts are reported from many regions of the Roman Empire. Some of them seem to be portraits of living animals; others were probably specimens depicting

TABLE 6.1

Primate Osteological Fragments Yielded by the Archaeological Exploration of European Sites from Roman Times

Species	Site	Datation	Sex	Bone Remains and Age	References
Macaca sylvanus	Navan (Northern Ireland, UK)	150±70 BP (^{14}C) 1st century AD	–	Skull	Groves (2008)
Macaca sylvanus	Pompeii (Naples, I)	50 BC–140 AD (^{14}C)	–	Juvenile-skull bones and parts of the skeleton	Henneberg (1997), Bailey et al. (1999), Ciarallo (1999), King (2002)
Macaca sylvanus	Cutry (Meurthe-et-Moselle, FR)	First half of the 2nd century AD	♀	Very old specimen	Poplin (1986), Liéger et al. (1997), Gerber and Baudry-Dautry (2012)
Macaca sylvanus	Wroxeter baths (Shropshire, UK)	*c* 150 AD	–	Phalange	Armour-Chelu (1997), Albarella (2007), Green (2018)
Macaca sylvanus	Clos de la Lombarde (Narbonne, FR)	End of the 2nd–beginning of the 3rd century AD	–	–	Sabriè and Sabriè (2011), Gerber and Baudry-Dautry (2012)
Macaca sylvanus	Llívia, Cerdanya (Empúries, SP)	515 AD	♂	Sub-adult – skeleton with the skull	Guardia and Maragall (2004), Guardia et al. (2005), Aquilué (2007), Guàrdia et al. (2007), Olesti et al. (2013)
Macaca sylvanus	Site 433, Catterick (UK)	6th–7th century AD	–	Fragment of skull	Hodgson (1990), Armour-Chelu (1997), Napier and Jenkins (1997), Hodgson (2002)
Macaca sylvanus	Dunstable (UK)	Roman levels	–	–	Hodgson (1990), Schneider (1992)
Macaca spp.	Yenikapı port quartier, European side of Istanbul	4th–15th centuries AD			Onar et al. (2013)
Chlorocebus cf. *aethiops*	Poitiers (FR)	550–620 AD	–	Adult - fragment of skull	Gerber and Baudry-Dautry (2012), Gerber (2019)

FIGURE 6.10 An image depicting two Barbary macaque pets, likely used as an attraction in a Roman shop (*taberna*) selling food. Detail from a marble relief, 2nd century CE, Parco Archeologico di Ostia Antica. (Drawing by Cecilia Veracini.)

post-mortem or images derived from the iconographic repertoire circulating among the artists of the time. Interesting representations of primates appear in the Nilotic mosaic of Palestrina of the 2nd century BCE, executed by artists probably of Alexandrian origin (see **Box Details 6.2**). The "Nilotic scenes" evoked in Roman art enhance the lush natural environment and the magnificence of Egypt's landscapes and monuments, as they were perceived in Roman times (Salari, 2006), and it is generally believed that many of them were inspired by the above-mentioned mosaic of Palestrina (Lazenby, 1949). Another evocation of Egypt is found in a fresco from the Pompeian Temple of Venus, now at the Archaeological Museum of Naples, where a Barbary macaque and a baboon along with other Egyptian animals are depict.

Monkeys feature as performers in Roman art (Toynbee, 1973: 57–58). For instance, a monkey depicted in a Pompeian painting and dressed in a sleeve coat walks on its hind legs on a lead held by its trainer, who carries a whip (see Gusman, 1899, Fig. 285 *apud* Toynbee, 1973). Sometimes performing monkeys were made to play the parts of mythological personages. One of these caricatures was Orpheus in the form of a monkey, seated with a syrinx-charming animal (see Toynbee, 1973: 58). A mosaic of the 2nd century, found in Carthage, portrays a long-tailed monkey, whose silhouette may remind that of a guenon. Considering its red colouration, it cannot be excluded that represents a Patas monkey. Very intriguing is a marble bas-relief dated 100–200 CE (see **Fig. 6.13**), where the depiction of the monkey recalls the morphology of a Bonnet macaque, *M. radiata*. Its face is not very realistic, but the hair of the head and on the body seem very similar to those of this species. Also, the general morphology and the pose remind that of a macaque. The tail could appear short, but actually, the distal part seems to be included in the sculpture, and also, it is relevant to recall that sometimes in captivity and in stressful conditions, monkeys can lose or even eat parts of their tails. New data show Romans could import this species from India (see above the Roman pet necropolis in Egypt). According to Warmington (1928: 147), two Hanuman langurs are depicted in a silvery dish found at Lampsacus (Turkey, 6th–7th centuries CE). This artifact represents India, with a woman surrounded by animals who were supposed to be typical of that region. The two monkeys evoked in the dish had long hind limbs and tails and collars round their necks. Their faces are so human-like; therefore, it is extremely difficult to figure out their taxonomic identity with certainty; plus at the time, East African regions were confused with India, and both African (Guinea Fowl) and Asiatic (tiger) animals are portrayed on the dish.

BOX DETAILS 6.2: NILE MOSAIC OF PALESTRINA (2ND C. BCE)

Cecilia Veracini and Marco Masseti

The 2nd century BCE Nile mosaic of Palestrina (the ancient Roman city Praeneste) is a late Hellenistic floor mosaic, of a width of 5.85 m and a height of 4.31 m, which was discovered at the beginning of the 16th century. This great mosaic is the depiction of a so-called "Nilotic landscape", created according to the stylistic canons drawn from tradition. The mosaic represents the flooding of the Nile over the lands of Egypt and Ethiopia (understood as Africa) and can be regarded as a demonstration of the Roman fascination with the ancient Hellenistic Egypt of the Ptolemaic period. It is considered one of the earliest Roman depictions of Nilotic scenes, of which several more were uncovered at Pompeii (Meyboom, 1995: 80; cf. Jashemski & Meyer, 2002). The mosaic portrays many of the indigenous animals which inhabited Egypt or the exotic species that were brought there by the intense trade that local rulers developed with other parts of southernmost Africa. It cannot even be excluded that some of these representations are the result of copies of the repertory images that circulated in the workshops of Hellenistic artists.

Among the primate images found in the mosaic, two are easy to identify, while the others are dubious and have given rise to speculation. We can recognise with certainty a Gelada, *T. gelada*, called here *sphingia*, and a Barbary macaque, *M. sylvanus*. Another image looks like a baboon and is called *kepos* (Salari, 2008: 53). This term was usually used to indicate a tailed primate, especially the Patas monkey. More intriguing is the depiction of an anthropomorphic figure called *sattyos* (**Fig. 6.11**). The latter creature does not have a tail, but it has long arms and a human-like face, with long hair. Salari (2008: 52–53) is of the opinion to face an erroneous wording of *satyros*; the term *satyr* refers to the comrades of Dionysios in Greek and Roman mythology, characterised by having the lower part of the body like that of a goat, with a penis permanently erect to emphasise sexual desire. The etymology of the name (like that of *tityros*) is related to the swelling of the turgid penis (Boisacq, 1923 *apud* Salari, 2008), but the term was often used to indicate anthropomorphic creatures and anthropoid primates. Salari (2008: 53) states, as does McDermott (1938: 303), that this image may refer to a chimpanzee.

Another controversial depiction is that of a quadruped with an anthropomorphic head and long blond hair, called "onocentaur" (*honokentaypa*) (**Fig. 6.12**). This word appears in various places, in the Bible (Isaiah, I, 13.22, 34, 14), and subsequently in various medieval bestiaries, in which the "onocentaur" is often a monstrous being, half woman half donkey. The biblical image of this creature is in some way connected with a demon or an unclean, negative, and destructive beast. The term could have roots in the Jewish and Christian traditions (Salari, 2006). The sole Greek description of this creature is found in the work of Aelian (1998, *De natura animalium*, book XVII, 9), where it is reminiscent of an African ape perhaps a chimpanzee (see section 6.2.1).

The anthropomorphic creatures of the mosaic of Palestrina are among the few known in the Greco-Roman world. Two images of anthropomorphic creatures, found in ancient Mediterranean artworks, had been considered by McDermott (1938: 234–238) to be representations of gorillas. These were depicted on two bowls probably of the Phoenician-Cypriot artistic school (ca. 7th c. BCE). This interpretation was later discussed by Montagu (1940), who argued that the creatures depicted were probably not gorilla. Other images considered by McDermott (1938) as chimpanzees also appear very doubtful (cf. McDermott, 1938: 248, 303).

The images described above are among the few anthropomorphic creatures found in ancient Greco-Roman art which were considered to be possibly representations of African great apes. Although it cannot rule out that these representations were actual portrayals of anthropoid apes, (however, their phenotypic characteristics are far from those of both African and Asian great apes), it still remains unclear whether living specimens of African great apes reached the Mediterranean shores in antiquity. It seems likely that depictions such as these could evoke real great apes, of which there already existed good literary descriptions at that time (e.g. Aelian, 1998, book XVII, Chapter 9; Groves, 2008: 34); otherwise, they may have been illustrations of the mythical ape-men, about whom legends were prevalent in Europe at that time (Masseti, 2019: 22). Recently, Urbani and Youlatos (2020) argued that the first images of great apes

FIGURE 6.11 The satyrs, Nilotic mosaic of Palestrina. Anthropomorphic figure in the drawing 10207 of the Dal Pozzo's Collection which reproduces the mosaic (ca. 1630), it is flanked by the inscription CATTYO (*Sattyos*) (cf. Salari, 2006, Fig. 51). (Image from Wikipedia Common.)

FIGURE 6.12 The onocentaur, *honokentaypa*. Nilotic mosaic of Palestrina. Figure who in the drawing 10207 of the Dal Pozzo's Collection which reproduces the mosaic (ca. 1630). (Image from Wikipedia Common.)

have to be referred to two Elamite marble figures of Central Asia origin and could date back to the second millennium BCE or the Roman age. The authors recognise in these artworks the morphology of the chimpanzee *Pan troglodytes* Blumenbach, 1775. However, the same authors are very doubtful about the origin and period of production of these figures, even hypothesizing that they may be fake, and conclude that further archaeological analyses are needed.

It should not be forgotten, however, that the Nilotic mosaic of Palestrina has undergone considerable and radical restoration work since its discovery in the 16th century. This fact may have irreparably altered the possibility of a correct reading of many of its parts.

FIGURE 6.13 Relief with crouching monkey and inscription, likely portraying a Bonnet macaque. Roman-Imperial Period. Thorvaldsens Museum, n° H1477. (Credit: Thorvaldsens Museum, Copenhagen, Danmark. Photographer Jo Selsing. https://kataloget.thorvaldsensmuseum.dk/en/H1477.)

6.2 BETWEEN MYTH AND REALITY: NATURAL HISTORY OF NON-HUMAN PRIMATES IN THE ANCIENT GRECO-ROMAN WORLD

Primates were studied, described, or just reported by countless Greek and Roman scholars. Aristotle and Hippocrates in the 4th and 3rd centuries BCE recognised the morphological similarity of simians with humans, while Claudius Galen showed the remarkable anatomical proximity of monkeys to human beings. None of these early scholars thought the similarity of humans to monkeys needed an explanation (McDermott, 1938: 104). Nevertheless, they did not separate human beings from the natural world, maintaining the existence of a sort of continuity among humans and the other primates at least from a morphological point of view. Many authorities commented on the vision of the natural world of Aristotle and other Greek scholars interpreting their vision as a "Chain of Being" (or *scala naturae*) (e.g. Clutton-Brock, 1995; Lovejoy, 1936; Rolfe, 1985), but this is a counterfeit (and overexploited) concept that can be regarded as a projection of modern conception of the chain of beings, precisely, to ancient representation. Aristotle just had the intuition that there are slight differences between beings that are morphologically close to each other (Arnaud, Zucker, pers. Comm.). The influence that Greco-Roman authors were to have on how primate were perceived in Western culture will be addressed and commented on in other chapters in this volume. In this section, we will merely summarise just what was known about the nature of primates in the Greco-Roman period before the advent of Christianity.

In his *Historia animalium,* Aristotle (cf. Aristotle, Thompson 2010) gave an impressive description of the primates he knew. Considering the evidence he presented, it seems unlikely that he did not rely on first-hand knowledge, although as noted by McDermott (1938: 104), his description was mainly based on the Barbary macaque:

> Some animals share the properties of man and quadrupeds, such as tailless monkeys, long tailed monkeys and baboons. [...] They have similar nostrils and ears, and teeth like those of man, both front teeth and molars. Further, whereas quadrupeds in general are not furnished with lashes on one of the two eyelids, this creature has them on both, only very thinly set, especially the under ones; in fact they are very insignificant indeed. [...] The monkey has also on his chest two teats upon poorly developed breasts. It has also arms like us, and fingers and nails like man, only that all these parts are somewhat more beast-like in appearance. The feet are like large hands. [...] This creature uses its feet either as hands or feet and doubles them up as one doubles a fist. [...] It has neither hips, inasmuch as it is a quadrupeds. [...] The genitals of the female resemble those of the female in humans; those of the male are more like those of a dog than those of a man. (*Historia animalium*, book II, 8–9 (502 a–b)

Aristotle classified primates together with other "Quadrupeds" (four-footed beasts), and although he considered primates to be an intermediate form between mankind and the other animals, he did not regard them as evidence for a transition between quadrupeds, nor did he suggest they have a common ancestor.

Claudius Galen, the Greek physician who dissected primates, pigs, bears, and other animals, assumed that the internal structure of humans was the same as that of simians. He probably mostly dissected Barbary macaques, but he used other primate species in his studies. Galen used various names for the primates he studied and dissected (e.g. *pithecus, lynx, satyrus, cynocephalus,* and *cebus*) At the beginning of Book VI of his *De anatomicis administrationibus* (cf. Galen, 1956), he includes a list of primates that he suggested are better to work on, because their anatomy is closer to that of humans than is that of other animals, and he also describes some anatomical differences among them, such as the length of jaw or the dimension of chest and the capacity to walk:

> of all animals man has the shortest jaw in proportion to his whole body. After man the ape [= Barbary macaque], than the *lynx* [unknown primate, see later], the tailed ape [= *cercopithecus*], and then the dog-faced ones [= baboon]. Their neck, too, is as long, as they all have a collar-bone like man. Some of them stand more erect than others. All [can] walk with their weight on two legs, some worse some better. (*De anatomicis administrationibus*: book VI, 97)

The zoonyms of primates used by Galen correspond to the best-known species of the time, even though the word *lynx* was never found associated with monkeys in previous authors (but see later the paragraph on tailed monkeys). Scholars after Galen argued that, among the species studied by the Greek physiologist, there could also be apes, such as gibbons, orang-utans, and chimpanzees. This is mainly because the anatomical descriptions of some parts of their bodies were more similar to some morphological characters of apes than to Barbary macaques and baboons (cf. McDermott, 1938: 97). Although it is not unlikely that he used for his studies various species of primates (cf. Galen, 1956, Book VI), some researchers maintain that the Barbary macaque was the only species used by Galen for his dissections (cf. Alghamdi et al., 2017; Diogo, 2018), but we believe this to be reductive and erroneous. Vespa (2017b: 428) observes that Galen rejects the vivisection of primates, defining it a "hideous spectacle". "Galen often uses an alternative name for monkey: 'the handsome one', which was a euphemistic and auspicious term counteraction the otherwise presumed inauspicious nature of primates, at least according to ancient Greek folklore" (Vespa, 2017b: 409). The impact of Galen's dissections, and of related topics such as the extent to which primates resemble human beings in all their internal parts, influenced Western and Islamic anatomy right up to the Renaissance, leading to many misunderstandings (cf. Diogo, 2018; Veracini, 2020).

The resemblance of monkeys to modern humans is also remarked upon by Pliny in his renowned work *Naturalis Historia* (Pliny the Elder, Ernout, (ed.) 1952, XI, 246), who states that "primates are a perfect imitation of men" and adds some anatomical details (e.g., the middle finger is the longest finger and ridged nails) primates share with modern humans. He also mentions the cheek-pouches of cercopithecids, which were not described by Aristotle.

6.2.1 Primate Taxonomy

The first known attempt to classify the primates known in the Western World at the time was carried out by Aristotle in *Historia animalium* (books II, Part 8, 502a 17-b) and in *De partibus animalium* (Aristotle, Ogle (ed.). 2008, book 689, b31). "Primates" or rather "monkeys" were divided into three typological groups: (a) πίθηκος (*pithekos*), Latinised as *pithecus*, which has no tail; (b) κῆπος (*kēpos*) or also κῆβος (*kēbos*), in the Latin form *cebus*, which referred to long-tailed (or tailed) monkeys; and (c) *kynokephalos* (κυνοκέφαλος), (pl. *kynoképhaloi*), or *cynocephalus* in the Latin form: dog's head (κυν =dog+κεφαλή=head) (cf. Rolfe & Grigson, 2006; Masseti, 2019: 26). Aristotle notes that this last form is bigger and stronger, is more savage in habits, and its face and teeth are dog-like [=baboons]. Aristotle's classification, whose criterion of the distinction of monkeys was on the basis of the possession or lack of tail, and on the form of the head, would be followed, with the addition of few other categories, until the 16th century. After the discovery of America and the Sub-Saharan Africa, the arrival of many new primate species in Europe resulted in its modifications (see Section II of this volume). We list here the description of primates and their zoonyms used in the Greco-Roman world:

6.2.1.1 Tailless Monkeys

The name *pithekos* refer mainly to the Barbary macaque (Latin authors used the term *simia*), the only well-known monkey in antiquity without a tail (i.e. with a vestigial tail) (McDermott, 1938; Rolfe & Grigson, 2006: 80; Groves, 2008; Masseti & Bruner, 2009: 39; Veracini, 2020). Sometimes, however, the word *pithekos* was also used to indicate a superordinate taxon of primates without or with tail (see for example Philostratus, Jones, (ed), 2008, book III. 3.), but it seems to refer to a monkey with a tail only when it is specified as *kerkopithekos* (from κερκος= tail+*Πιθηκος*), in Latin *cercopithecus*.

6.2.1.2 Tailed and Long-Tailed Monkeys

This category includes different taxa of primates identified with various zoonyms: *kêpos, kerkopithekos, kerkopes, callithrix, clura, sphinx* (feminine) and *sphingion* (masculine), some of them Latinised as: *cepus, cercopithecus, cercopes, cercolopes*. Usually, these names referred to the African cercopithecid species that occur in Eastern and Northern Africa, but, sometimes, they have been used to describe Indian monkeys. *Kêpos* can refer to any monkey with a tail (also baboons), but according to Groves (2008: 49), it seems "never to have meant anything other than the Patas monkey" (*Erythrocebus* spp.). Quoting Pythagoras, the already-mentioned rhetor Aelian left us a fascinating and accurate description of the external morphology of this primate:

> In his writing about the Red Sea Pythagoras says that there is an animal that lives on the shores and is called *Kêpos*. And it is well-named (*kêpos*, garden), for it is of many colours. When full-grown it is the size of an Eritrean hound. But I wish to return to the subject of its varied colouring and to describe it as Pythagoras writes. Its head, its back, and its spine down as far as the tail are pure red, though you may observe a sprinkling of golden hairs. But its face including the cheeks is white, and from there golden stripes descend as far as the neck. The lower portions down to its chest and its forefeet are all white; its two breasts, which would fill your hand, are dark, but its belly is entirely white; its hind feet are black. We would not fail to compare the shape of its face with that of a baboon. (1985, XVII: 8)

The description reported by Aelian recalls an animal of many colours, and his description is similar to that of a monkey called in Sanskrit, *kapi,* which means "little coloured animal" (see Glossary). According to Wigoder and Posner (1971), the Jewish *kof* (pl. *kofin*) derives from this Indian word, though *kof* refers generically to tailed monkeys. Others accepted that the zoonym *kēpos* derives from the Egyptian *kif* (cf. Groves, 2008: 49). As we saw in the previous paragraph, Pliny the Elder reports that an animal called *κήπουζ* arrived at the games of Pompey, in which various animals from Africa appeared: "this animal had hind legs that resemble feet, and man's legs, and those in front were similar to human hands; it was never seen again in Rome" (Pliny, *Nat. Hist.*, VIII, 70).

The term *callithrix* [=beautiful hair] was very likely used to describe the black-and-white colobus, *Colobus guereza* Rüppell, 1835, of Eastern Africa (McDermott, 1938: 105; Groves, 2008: 35), known for its fine fur. Pliny observes: "It's said that this animal can survive only in the climate of Ethiopia, where it comes from" (Pliny, *Nat. Hist.*, VIII, 216). This is in agreement with Colobinae's obligate diet of leaves, which at that time must have made their survival in captivity impossible. It cannot be excluded, however, that the term *callithrix* may have indicated other species of primates.

The word *kerkopithekos* has been known since Strabo (*Geography*, XV, 1.29; 1.37), who probably coined it to distinguish some tailed monkeys from the best-known primate of that time, *M. sylvanus*, which was tailless (cf. Salari, 2006: 33). Also, the already-mentioned *cercops* (Greek: Κέρκωπες, plural of Κέρκωψ, from κέρκος [*kerkos*="tail"]), the mythical creatures turned by Zeus into monkeys, must necessarily be identified with "long-tailed primates" (Connors, 2004), although it is impossible to match them with any extant primate species. The *sphinx* [or σφιγγίον, *sphingion* in ancient Greek] was first described in 132 BCE. One of the earliest reports is that of Agatharchides :

> the *sphingia*, the *kynokephaloi* and the *kepoi* sent to Alexandria come from the country of the *Troglodytes* [=black people] and from Ethiopia [=Sub-Saharan Africa]. They are more or less similar to the painted ones, except that they are more villous, have a docile and meek and very astute character, and when tamed reach a degree of training that can be admired in all the harmony of their movements. (Agatharchides, Henry (ed.). 1974, 250, 73)

The *sphinx* was unknown to Herodotus and Aristotle, but similar descriptions are found in Diodorus Siculus (1820–22, *Bibliotheca Historica,* III, 35.4) and Strabo (*Geography*, XVI, 4.16). Much later in the Roman world, Pliny, too, spoke about animals born in Ethiopia, among them the "*sphingas*, which have dark brown hair and pectoral mammas" (*Nat. Hist.* VIII, 30), and are "more numerous in the south of the Island of Artigala" (*Nat. Hist.* VI, 184). The presumed depiction of a *sphinx* is found in the already-mentioned Nilotic mosaic of Palestrina. According to Salari (2008: 33), this animal is a type of green monkey, but other authors are of the opinion that the *sphinx,* that appears many times in Greek literature, is the Gelada, *Theropithecus gelada* (Rüppell, 1835), (cf. Jolly & Ucko, 1969; Groves, 2008: 27), the so-called bleeding-heart monkey native to Ethiopian Highlands (see **Fig. 6.14**). The evidence tends to support this second interpretation.

The zoonym *lynx*, which Galen refers to (see the discussion on whether this is a primate in McDermott, 1938: 72, 95–96, 105), is definitely a primate, as recently demonstrated by Egyptological researches (cf. Prada, 2014). It has been demonstrated that this term is a Greek mistranslation or an error of transcription of the word *kwf,* used in a demotic text [see also Glossary *kof, kif*]. The question as to which primate species the zoonym lynx refers remains open anyway.

FIGURE 6.14 The *sphinx*, Nilotic Mosaic of Palestrina. (Image from Wikipedia Common.)

It is interesting to underline that animals called *lyncas* arrived from the remote land of Ethiopia as Pliny reported (*Nat. Hist.*, VIII, 30). Moreover, other classical authors beside Galen label the *lynx* as a monkey: in his *Oneirocritica*, (II 12), Artemidorus mentions the *lynx* in a passage dealing with dreams about monkeys (cf. Prada, 2014).

6.2.1.3 Baboons or Dog-Headed Monkeys

The dog-headed, *kynokephalos* (Latinised *cynocephalus*), also called *kynoprosopos,* was described for the first time by Ctesias of Cnidus (c. 440–397 BCE). His work, which did not survive in its original form but only in fragments quoted by later authors, continues to be of vital importance for both the study of the Near East and the western view of India before Alexander the Great. Our main source for the work of Ctesias is an epitome by Photius, the 9th-century Byzantine scholar and patriarch. Ctesias mentions a human population living in remote and mysterious "India", a place where the strangest things could happen:

> They say there are men in these mountains who have a dog's head. They wear clothes taken from the skin of wild beasts, do not speak any language but scream like dogs and understand their language. They have teeth bigger than those of a dog and claws larger than those of a dog, but longer and more curved. [...] They understand what others are saying but they express themselves by gestures and screams like deaf-mutes. Indians call them *kalystries*, which means dog-heads. (cf. Nichols, 2008, Ctesias of Cnidus, F45. Photius, 72 p. 45a21–50a4 (Sections 1–51)

Herodotus, had already spoken about dog-headed creatures when he described the fauna of Libya, regarding them as monstrous creatures (*The History*, IV.191–192, cf. Herodotus, Godey, ed., 1921). In later authors, such as Agatharchides of Cnido (220–145 BCE) they are considered similar to African monkeys, "who are too fierce to be tractable to training" (cf. Agatharchides of Cnidus, Henry, ed. 1974, Photius, 250, p. 455b). On the contrary, Pliny the

Elder does not distinguish baboons from other monkeys with tails, except by commenting on their ferocity, remarking that "those monkeys which have tails are affected by the phases of the moon", recalling the Egyptian legend about monkeys, but makes no mention of their sacredness. Other scholars confine baboons to Ptolemaic Egypt, where "the *kynokephaloi* were taught how to read, how to dance, and how to play the harp and flute" (Aelian, Scholfield, ed, 1959, VI: 1–2) referring to the Egyptian cult of baboons (Masseti, 2019: 28; Greenlaw, in this volume). Aelian also describes "dog-faces" (called *kynoprosopos*) as the inhabitants of a wild country along the road to Ethiopia (1998, VI: 10). These were able to imitate human beings showing high learning ability and could also use tools. The following excerpt from Aelian is one of the first descriptions of tool use by a non-human primate:

> It now comes to my mind to mention the following facts about the *cynocephali*. If a *cynocephalus* catches hold of something edible wrapped (for example almonds, walnuts), it pries off the skin or what surrounds it and cleans it, breaking it in a very clever way, and it knows that what is inside is edible, and that it has to remove what is outside. It drinks wine and, when shown boiled or roasted food, eats it greedily; it likes well-seasoned food, and does not like badly- cooked dishes. If it wears clothes, it takes care of them. And it does everything I have said above. And when it is still very young it will take the milk from the breast of a woman like a child. (1998, X: 30)

Among dog-headed monkeys, there is the so called "*simia porcaria*", which Aristotle called *choiropithecos* (*Hist. Anim.*, II, 503a). Its identification is controversial. McDermott (1938: 44), suggests that Aristotle's "*choiropithecos*" cannot be identified. In the majority of modern editions of the *Historia animalium* of Aristotle, the word *choiropithecos* is translated as a *baboon* or as a *pig-faced baboon* (e.g. Peck, 1965) and is considered to be the Chacma baboon, *Papio ursinus* (Kerr, 1792) (the old synonym was *Papio porcarius,* variously known in earlier literature as *Cynocephalus porcarius*, *Simia porcarius*, etc.). The Chacma baboon, found throughout the greater part of southern Africa of the river Limpopo, has a pig-like facial physiognomy, with its snout protruding above its upper lip. On the contrary, Salari (2015) cites evidences that this term refers to a small bush swine with a pointed snout, big ears, a short mane, and an upturned tail that is portrayed in the mosaic of Palestrina (see **Box Details 6.2**). This species can be identified with the bush pig, *Potamochoerus larvatus* Cuvier, 1822, dispersed in the forests and shrubland of sub-Saharan Africa. According to Salari (2015), it seems more plausible that Alexandrian Greeks were more familiar with a bush pig than with a Chacma baboon, the latter inhabiting, as it did, a more southern area.

6.2.1.4 Indian Primates

The fabulous world of India entered the western oecumene through the various stories disseminated by travellers, ambassadors, and conquerors and via those who collected and handed down their testimonies. Langurs of the genus *Semnopithecus* Desmarest, 1822, which occur with various species in the Indian sub-continent, populated western fantasies and were described by various authors, appearing beautiful in the eyes of the travellers: they were of a pale colour and were respected and loved by Indians. Ctesias of Cnidus (Photius, 45, 8), who, as early as the 5th century BCE, described the wonders of India, reported that langurs had "long tails two cubits long". At that time, much of Northwest India had been subdued by the Persians under King Darius I, who explored the entire Indus Valley. In 326 BCE, Alexander the Great, the Macedon, marched towards present-day Kabul, reaching the banks of the Indus River. He attempted to invade Punjab from North, his armies entering the region through the Hindu Kush. There, Alexander defeated the Indian king Porus, sovereign of an area between present-day Jhelum and Acesine (northwestern India) in the Punjab region. After the Indian expedition, Alexander's grandeur was exalted, as were events relating to his conquest, in the descriptions and accounts of successive authors. Thus, Indian fauna and flora, too, were often described as fabulous and with exaggerated characteristics. Megastene (c. 340–282 BCE), who visited India serving as Seleucid ambassador at the court of the Maurya king Chandragupta (Megastene: IV Indica F, 27b; Section 56; F21 Section 37), described the Indian monkeys as follows:

> In the territory of the Prasij [i.e. it extended for some distance on both sides of the Ganges] live very large tigers almost twice the size of lions. [...] The monkeys are as big as huge dogs and all white except the face, which is dark (elsewhere the opposite happens); the tail reaches 2 cubits in length. But they are very soft and do not have the wicked tendency to attack and steal. (Megastene *apud* Strabo, *Geography*, book XV, 1, 37)

Megastene's description of Indian langurs was repeated by Arrian in his work *Indiké* (cf. Biffi, 2000: XV, Chapter 61). Indikè (2nd c. CE) is indeed a summary of various Alexandrine accounts on India dating back to the 4th century BCE. Here, it is said that these monkeys were known for their size and beauty. Furthermore, Strabo (*Geography*, book, XV, 1, 29) and Diodorus Siculus (*Bibl. Hist.* book, XVII, 90), referring to Alexander's expedition, reported that in the mountains of today's Punjab: "Alexander's fellows saw many monkeys that were huge and they thought they were an army which was waiting for them". These monkeys were nothing but Rhesus macaques, *M. mulatta*, which still occurred in great quantity in those territories.

From mysterious India came many legends about animals including that which shows how people hunt monkeys using the latter imitative ability. We report here this excerpt because it had a great influence in the following centuries, lasting up to modern times:

> In addition to timber for shipbuilding, the region had many large snakes 16 cubits long and many varieties of monkeys of different sizes, and it is the nature of these animals

> themselves that suggests the technique to use for capturing them, because they imitate every action. It is not easy to catch them because of their physical strength and intelligence; thus some hunters spread honey on their eyes, others put on their shoes and others put mirrors around their necks in the presence of monkeys. Then they leave the shoes whose laces are tied together; instead of honey they put mistletoe and the mirrors attach cords. So when the animals try to imitate the actions they have seen done by men they can not do them because their eyelids are glued together, their feet are tied and it is easier to capture them. (Diodorus Siculus, Alfieri, (ed.). 1985, book, XVII, 90)

These stories and other information from India were handed down and quoted by various authors including Aelian, who played a great role in divulging them throughout the Roman world (Aelian, 1998, books XVII, 25; XVII, 39; XVI, 15). The latter scholar also reports that langurs were greatly respected and abundant in India. They were gifted with great intelligence and could dance and play the flute:

> They say that among the Prasi of India there is a kind of monkey with human intelligence. They are the same size as the dogs of Hircania and it seems that nature has provided them with a natural curl on the forehead. He who does not know the truth would say that it is artificial. A beard grows under their chins like that of the satyrs and their tail is as long as that of a lion. The rest of the body is white, except for the head and the end of the tail, which are reddish. They are wise and domestic animals by nature. They live in the forest and feed on wild fruit. They head in a group to the suburbs of Latage and eat the helpings of rice that the king has reserved for them. This food is prepared and given to them every day. And it is said that, when they have eaten until they are full, they neatly retreat to their wild habitations without giving way to anyone who meets them in their path. (book XVI, 10) [...] The Indians bring their king doves with ocher yellow plumage and also monkeys with very white fur and others with very black hair; however, they do not introduce into their cities those with reddish fur because they consider them too dangerous for women, and wherever they find these they attack them and kill them, because they hate their lecherous character .(book XV, 14)

As it is known, Indian people fed langurs because they were considered sacred (cf. Lutgendorf, this volume). In Aelian's description, several different species of monkeys can be recognised: langurs perhaps of the species *Semnopithecus dussumieri* I. Geoffroy, 1843, and Rhesus macaques, whose various behavioural traits (e.g. raids on houses) are still a matter of controversy. The black monkey that Indians brought to the king could be tentatively identified with gibbon species or with some lutungs possibly of the genus *Trachypithecus* Reichenbach, 1862, or the lion-tailed macaque, *Macaca silenus* (L. 1758).

Philostratus (170–250 CE) describes the harvest of pepper apparently collected by monkeys:

> In the fold of the mountains grow tall frank-incense trees and many other kinds too, including pepper-bearing trees that are harvested by monkeys. [...] The pepper tree looks like the Greek willow in every respect, down to the cluster of fruit, except that it grows on precipices where it is inaccessible to humans. But they say that a tribe of monkeys lives there in the clefts and hollows in the mountains, and the Indians value them so highly that when the monkeys are harvesting the peppers, they use dogs and weapons to keep the lions away from them. A lion will attack a monkey either to get a cure when it is ill, because meat arrests its diseases, or to get food when it is old. When lions are past the age for hunting deer and boar, they use their remaining strength to gobble up monkeys. People, however, feeling monkeys to be their benefactors, intervene by talking up arms to defend them against lions. The method used for taking pepper is this. The Indians make their way to the trees lower down, and after making little clearings around the trees collect the peppercorns in them, as if they were dumping them as garbage of no use to humans. The monkeys watch from their inaccessible places above, and when night comes they imitate the action of the Indians by tearing the clusters of fruit from the trees and bringing them to be dumped in the clearings. At dawn the Indians carry off the piles of spice that they have gotten with no effort at all, in fact effortlessly and during their sleep. (Philostratus, *The Life of Apollonius of Thyana,* Book III. 3)

The plant mentioned could be the long pepper, *Piper longum* L., which grew in northwestern India. It was known in Greece as early as the 4th century BCE as a widespread and expensive spice. Of course, it is impossible to verify the veracity of monkeys' behaviour described by Philostratus, but the use of other species of macaque in other parts of Asia for picking fruit is well-known (e.g. the crab-eating macaque, *Macaca fascicularis* (Raffles, 1821), or the pig-tailed macaque, *Macaca nemestrina* (L., 1766), (cf. Bertrand, 1967; Ruslin et al., 2017). The use of the Rhesus macaque as a collector of fruits for humans seems to have also existed in Medieval Islamic Persia (see Veracini & Alghamadi, in this volume). By analogy, we can also recall here the alleged ancient Egyptian custom of using green baboons, *P. anubis*, as gatherers of fruit from trees (Deputte & Anderson, 2009). This behaviour is represented in the tomb of Khnumhotep, of the 12th Dynasty (1991–1785 BCE), at Beni Hasan (Osborn & Osbornová, 1998: 38; Masseti & Bruner, 2009: 43), where dark-green-coloured baboons, with reddish-brown faces and callosities and without mantles, are depicted in a fig tree.

6.2.1.5 Great apes, Pygmaeus, Satyrs and "ape-men".

Many Greek authors wrote about mysterious anthropomorphic creatures, often living at the border of the known world. These creatures could be half humans and half animals, or monstrous people living in distant regions of Asia and Africa. Traditionally, they would have inhabited the *eschatiai tes oikomenes* – the extreme ends of the ancient world – which up to that time had been confined to the geography of Ethiopia (indicating sub-Saharan, "Black Africa" in its entirety) and India (cf. Li Causi, 2003: 63; Masseti, 2018: 64). These myths, established in both popular culture and literary tradition throughout the centuries, persisted throughout the medieval period until the 16th century and even later. Only through the modern European expansion started in the 15th century were they finally swept away by the first-hand information collected by travellers and explorers returning from all parts of the world.

How far these legends of mysterious ancient beings were based on fact (for example they could be the result of direct contact with one or more great apes) or were just a remix of mythological creatures, we do not know, but anthropomorphic creatures more or less wild, more or less mysterious and fabulous, have been part of the culture of many different people across the Old World (e.g. the Arabic *nasnas*, see Glossary). In his description of *Libya*, geographic name which indicated in ancient times North Africa west of the Nile Valley, Herodotus (5th century BCE) wrote:

> In that country are [...] horned asses, dog-headed men and headless creatures that have their eyes in their breasts, as the Libyans say, and wild men and wild women, besides many other creatures not out of the ordinary. (*The Histories*, IV: 191–192, 45)

Herodotus adds that the complete skin of one of the latter creatures was exhibited in the temple of Carthage of his time. Herodotus never reached the Punic city in his wanderings throughout the ancient world, but he was informed in the North African town of Cyrene of the feat just recently achieved by Hanno the navigator (Masseti, 1980). Hanno II was king of Carthage from 480 until 440 BCE. During his reign, he left the North African city at the head of a fleet of 60 ships to explore and colonise the northern and western coasts of Africa. It is believed that Hanno recounted the results of this expedition in his *Periplus* or *Circumnavigation of Africa* (cf. Schoff, 1912) which has, unfortunately, come down to us only in the fragmentary edition of the *Codex Palatinus Graecus* 398 (fol. 55r–56r). His fleet passed the "Pillars of Heracles" to sail into the outer Ocean reaching Western Africa. There is a lack of agreement as to precisely where to locate the farthest limit of Hanno's explorations: Sierra Leone, Cameroon, or Gabon (Harden, 1980), but the interesting thing is that in one of these geographic areas, Hanno found an island which was:

> ... filled with wild savages. The greatest number of them were females, with hairy bodies, which our Lixitael interpreters called *Gorillas*. Chasing them, we could not catch any of the males, because all of them escaped by being able to climb steep cliffs and defending themselves with whatever was available; but we caught three females who bit and scratched their captors and did not want to follow them. So we had to kill them and flayed them and we brought their skins to Carthage. (*Codex Palatinus Graecus* 398 fol. 55r–56r: 90–100, see Oikonomides & Miller, 1995)

The language of the *Periplus* is Hellenistic Greek, and, for centuries, this was the only text where one might have found Hanno's strange word *gorilla* (Masseti, 2019: 31). Today, this name is ordinarily used to indicate the largest of the three African extant great apes, which comprises two species, *Gorilla gorilla* (Savage, 1847) and *Gorilla beringei* Matschie, 1903, with both having two subspecies, respectively, *G. g. diehli* and *G. g. gorilla*, and *G. b. beringei* and *G. b. graueri*. As observed by Kitchell (2014: 77–78), debate has long existed over what animals (or native Africans) Hanno encountered (Keller, 1909, 1: 10; Lamer, 1926: 5–8; McDermott, 1938: 5155; Montagu, 1940: 89–96). If the country where Hanno encountered these intriguing creatures was really the Sierra Leone, according to what is currently the most common opinion, then the animals in question may not have been gorillas as their modern range does not extend into this area. It is possible that the western gorilla may have been present this far north, but the behaviour described by the Carthaginian navigator is not normal for the gorilla. Chimpanzees, *Pan troglodytes* Blumenbach, 1775, are a possibility: their range is suitable, their coats are sufficiently hairy, and they do throw objects when excited (Kitchell, 2014).

According to Martin-Granel (2001: 212):

> the word gorilla is not found anywhere else, except in this text. It is probably an error by the copyist, writing 'Gorillas' instead of 'Gorgadas', i.e. 'Gorgons', other monstrous creatures of Greek mythology: a very old tradition places the abode of the Gorgons and the place where Perseus met Medusa in the south west of Africa.

But the issue remains opened. Also, McDermott (1938) tried to tie the episode into versions of the Gorgon myth. However, Kitchell (2014: 78) notes also that: "certainty is impossible and we are left with a plausible choice between gorillas and chimpanzees". Historians and primatologists now agree that the hairy people of Hanno were probably native people and not gorillas or even chimpanzees (see Groves, 2008: 24; Barsanti, 2009: 18–19; Masseti, 2019: 31). Schoff (1912), for example, identifies the "gorillas" of Hanno not with anthropoid apes but with *negritos* and/or pygmies, but this hypothesis seems to be without foundation, given the current distribution of these people. Also Kitchell (2014) notices that this is unlikely, given the stress Hanno lays on the animals' hairiness.

The Latin term *pigmaeus* derives from the ancient Greek word *pygmaios* and refers to a measure. It describes a person one cubit tall, that is, the length from the elbow to the tip of the middle finger (about 50 cm or 20 inches) (cf. Thijssen, 1995: 45). Throughout history, these dwarf tailless persons have been throughout history identified with hominoid primates or with people. In fact, that they were real people seems to have been clear to the first authors who described them, such as Homer (*Iliad* III, 3–7), who actually spoke of "short men" living near the mythical river Ocean, or Ctesias of Cnidus (F45. Photius, 72 p. 45a21–50a4 (Sections 1–51) (21)), who described them as "Indian people, black and dwarfish and with long beards and long hair down to their knees" (cf. Nichols, 2008: 176). Aristotle did not believe that the accounts of pygmies were altogether fabulous; he thought that they were a tribe in Upper Egypt, or perhaps further south (Perfetti, 2020: 54). According to him, they inhabited exceedingly small caves (*Historia animalium*, VIII: 12, 597a). Herodotus reports an exploration in the interior of Africa where some members of the expedition were captured by small men, smaller than human stature (II, 32, 6, cf. Perfetti, 2020: 54). Later interpretations continued to confuse pygmies with apes or consider them as intermediate creatures between us and other animals (e.g. Albert the Great, cf. Veracini, Chapter 7, this volume). Buffon (1780–1785) observed that "the *pigmies* of Aristotle were animals perfectly similar to the *cynocephaloi*, but smaller and weaker, with shorter muzzles" (cf. Veracini, Chapter 11, this volume). However, in support of the fact that in antiquity, the pygmies were considered human beings and not apes; there is, for example, an inscription of the Ptolemaic period in the temple of Karnak regarding the people of the district of Nubia, which recounts that: "The dwarfs of the southern countries come to him, bringing their tributes to his treasury" (Schoff, 1912; Masseti, 2019: 32–33). Historical sources prove that Egyptians knew well the existence of pygmies. Egyptian documents indicate that already in 2300 BCE, the Egyptians were in contact with Aka Pygmies (Ballard, 2006: 135). The contact may have taken place in the southern part of the Republic of Sudan and most probably in a more northern location than that in which Pygmies are found today (Cavalli-Sforza et al., 1994: 180).

Another emblematic creature was the *satyrus*. As already noted, the name indicated a mythological being half man half goat, but it was also often used to indicate anthropoid apes, more often Asian apes. McDermott (1938: 79) notes that the name was given to a kind of ape because this name had already been used for mythological creatures. Pliny argues that satyrs:

> live in the mountains of Western India. They are at times quadrupeds and at times bipeds and move with such speed that only the old or sick are captured. They are the woodland race of the Chromandae, with hairy bodies, grey eyes and dog-like teeth, unable to speak speech but venting terrible shrieks. (*Nat. Hist.* VII: 24)

According to Pliny, the tribe of *khromandai* (Chromandae) was a human tribe, and this name might suggest a connection with the ancient kingdom of Mandi, located in the foothills of the Himalayas. To Aelian, "Indian satyrs have an appearance similar to mythological creatures; they live in dense forest and trees; they are also able to move at an incredible speed" (1998, XVI: 21). Both the descriptions may fit with gibbons, especially as regards how fast they can brachiate in the canopy, voice, their bipedism, and choice of habitat, although Pliny speaks of Western India, from which, in fact, they are absent. For this reason, it cannot be excluded that satirs could be other primates such as langurs, genus *Semnopithecus*, as reported by Vespa (2022). Nevertheless, Gibbons in their natural habitats were described in medieval times by Arab navigators who reached southeastern Asia in pre-Islamic times (cf. Veracini & Alghamdi, this volume) and were probably already known in classical times.

Perhaps there is no better description of a real anthropoid ape than that provided by Aelian. This creature, called *onocentaurus*, was described through quotations from Crates, a philosopher, philologist, and director of the Library of Pergamon in the 2nd century CE and Pythagoras (not the philosopher/mathematician, but the geographer), who lived at the time of Ptolemy II Philadelphus (cf. Athenaeus, IV, 183f, XIV, 634a):

> There is a certain creature which we call *Onokentauros* (*Onocentaurus*, Latin spelling), and anybody who has seen one could never doubt that the race of *Kentauroi* (Centaurs) once existed. There was a time when beings that united in themselves a mixture of dissimilar bodies were created. I want, certainly, to avoid giving an answer to the problem of whether centaurs really appeared among us in a determined and circumscribed period. [...] But I have heard this creature of which I am speaking described as follows. Its face is like that of a man and is surrounded by thick hair. Its neck below its face, and its chest are also of human form, its teats are swollen and stand out on the breast; its shoulders, arms and forearms, its hands too [...], and chest down to the waist are also those of a man. Its spine, ribs, belly and hind legs closely resemble those of an ass; likewise its colour is ashen, although beneath the flanks it inclines to white. The hands of this creature serve a double purpose, for when speed is necessary they run in front of the hind legs, and it can move quite as fast as other quadrupeds. Again, if it needs to pluck something, or to pull it down, or to seize and hold it tight, what were feet become hands; it no longer walks but sits down. The creature has a violent temper. At any rate if captured it will not endure servitude and in its yearning for freedom declines all food and dies of starvation. This also is the account given by Pythagoras and attested by Krates (Crates) of Pergamon in Mysia. (Aelian, Scholfield, ed, 1959, book XVII, 9: 333)

The initial reference to the back, and also to the hind legs similar to those of a donkey, in some ways, evokes the image of the *onocentaur*, the monstrous asinine centaur of medieval bestiaries, but immediately afterwards (when these hind legs become prehensile), it is clear that Aelian is alluding to a great ape. This description, shorn of the more fabulous elements, could refer to a chimpanzee, according Scholfield (1959: 333) followed by Groves (2008: 34). A depiction of an *onocentaurus* (*honokentaypa*) is found in the mosaic of Palestrina. It is portrayed as a quadruped with an anthropomorphic head and long blond hair, not corresponding, however, to the description of Aelian, but rather looking like the demonic donkey-centaur (see **Box Details 6.2**).

Another important account of great apes is that of Agatharchides of Cnidos, who reports about two tribes living in Ethiopia [= tropical Africa], called *spermatophagi* (seed-eaters) and *hylophagi* (wood-eaters):

> The *spermatophagi* eat seeds, fruits and herbs, the *hylophagi* the smaller branches of trees and shrubs. They climb with incredible facility, and leap with great agility. Their bodies are small and graceful and in climbing they use both feet and hands. They build tree shelters at night, they wear no clothes and have wives and children in common. When they fight they use sticks as arms, and tear the conquered to pieces. They are susceptible to cataracts and many die of hunger when their sight fails. (Agatharchides, *De mari Erythraeo* 51 copied in Diodorus, III, 24)

This description recalls many of the behavioural traits of chimpanzees. The fact that these great apes were confused or identified with people in antiquity is not surprising when we consider that even today they are considered as belonging to mankind by many traditional African communities (cf. Veracini, Chapter 2, this volume). As it is known, the modern genetic research demonstrated the strict affinities between chimpanzees and human beings.

6.3 THE MONKEY: AN AMBIVALENT ANIMAL

Some of the behavioural characteristics that were attributed to primates, or rather to a stereotyped idea of them, remained in western culture until the modern period or in some cases until today. These can be considered as the results of first-hand observations combined with element of fantasy to which moral meanings are added. It should be emphasised that some of these behavioural traits, such as imitative capacity, smartness, and parental care, were described with ambivalence. This kind of view, positive and negative at the same time, is reported in various traditional cultures (e.g. see Asian views, Lutgendorf, this volume). For instance, Aelian, whose work is an interesting mixture of scientific facts and miraculous stories from various sources (Müller-Reineke, 2010), place a positive emphasis on the imitative skills of "the monkey". However, he also reports negative aspects of this behaviour, which likely result from distorted observation mixed with fantasy:

> The monkey is a very imitative animal and all the physical activities that are taught to him he will learn point by point and can reproduce them, for example, he will dance and play the flute if he is taught. I myself saw a monkey handling and supporting the reins of a cart and waving the whip and driving the carriage. And he will not disappoint his teacher when he learns something that he has taught him. How versatile and ingenious nature is. (1998, book V, 26)
>
> Of all the beasts the most evil are monkeys, especially concerning those things in which they want to imitate men. Here is an example: a monkey saw from afar a nurse intent on giving a baby a bath in a tub; it observed that the woman, after having removed the swaddling and rewound it again, had put him [the baby] down to sleep. As the monkey saw that the child was left alone, it broke through a small, open window from which it could observe everything around it. It took the child out of bed and undressed him as it had seen the nurse do, took the tub and poured the boiling water (there was water that was heating on the fire) on the poor child, who died in the most atrocious way. (1998, book, VII, 21)

The intelligence of primates is often compared with that of their natural predators, as is especially evident in the case of African savannahs: the leopard is sometimes strangled through the intelligence and agility of the monkey, others demonstrate its superiority, deceiving monkeys in various ways. When not linked to empirical observation, the characterisation of these animals ("the leopard" and "the monkey") immediately assumes a moral and symbolic meaning, as can be seen by the comparison of the following two anecdotes of Aelian:

> In Egypt, Eudemus reports, felines were pursuing a monkey. It ran very fast and climbed up a tree; the cats were immediately behind it, because they clung to the bark and can climb trees; the monkey was about to be caught by many leopards but it jumped from the trunk to a branch at the topmost and most pendant end of the trunk, hanging on to this for a long time. The felines could not reach their prey, and thus went to look for another one, and the monkey saved itself through this great effort; and, naturally, the prize of its salvation was that it reached the ground. (1998, book V, 7)
>
> In Mauritania leopards do not attack monkeys vigorously with all the strength they possess, and the reason is that monkeys quickly flee to the trees and sit there making fun of their enemies and, crouching, spy on the insidious moves of the enemies. But leopards certainly show themselves smarter, at least judging by the tricks that they cleverly manage to play on monkeys. In fact, when monkeys stay in large numbers up a tree the leopard arrives below it and flattens itself on the ground, inflates its stomach, spreads its legs, squints, holds its breath and pretends to be dead. The monkeys, observing that, slowly descend happily and start to perform a ritual dance around the supposedly-dead leopard, but when they are tired after dancing for a long time, the leopard launches itself on them to devour them. Nature encourages the leopard to wait patiently and stoically in order to overcome the enemies' wiles. (1998, book V, 54)

Even the great care and affection that primates have for their offspring has been interpreted in an ambivalent way:

> Who doesn't hate such a race [monkeys], so ugly to look at, hateful, filthy and wicked? This animal who gave birth two filthy twins. This mother's affection was so great that the beloved one [Authors' note: it was supposed that the mother loved more one of their offspring] was killed by her hugs while the second [less loved] was carried on her back. (Oppian, *On hunting,* book II, I, 26)

This idea and image of the monkey as an incapable and merciless mother is also found in fairy tales and was later quoted by Pliny and disseminated throughout medieval literature assuming moral relevance. As for many other behavioural traits described by ancient authors, we can see a fund of truth on which popular fantasy has been superimposed. We can, in fact, suppose that monkey females, when kept in isolation and unpleasant conditions, could show typical phenomena of abusive mothers, developing aggressive behaviour towards their babies – behaviour reported in many scientific studies (e.g. Maestripieri, 1994).

Another characteristic, halfway between myth and reality, was the association of monkeys with lust and sexual perversion: "Baboons are lustful and are said to copulate with women, and become mad for damsels" (Aelian, 1998, book VII: 19). The objects of this mythicisation were above all baboons, which have conspicuous penises, a fact which had already attracted the attention of the Egyptians who, on the contrary, exalted this feature, considering it a symbol of fertility and even the key which opened the reign of Thoth, who presided over the judgement of the dead (Morris, 2013). The association of primates with sex and physical depravity could have originated either from the fact that obviously, they are not embarrassed about having intercourse or other interaction in public and because many primate species frequently use stereotyped behaviour – such as ritual mounting, touching, or sniffing genitalia, etc. – as systems for regulating social life.

Monkeys are associated to other myths such as the belief that monkey meat could be beneficial for sick people. This belief may perhaps be partially rooted on the notion that primates were used as medicine by sick lions or elderly lions unable to capture larger or faster prey. Pliny notes that the only disease of the lion is loss of appetite, and monkeys serve it as a cure (*Nat. Hist.* VIII, XIX.). The use of primate flesh or of other parts of their bodies persists in many traditional cultures across different continents (Alves et al., 2010). In the Greco-Roman world, primates are often associated with wine. Monkeys were described as fond of wine ("its masters could have given it wine to make it more amusing") and drunken primates could more easily be captured (e.g. Pliny, *Nat. Hist.* XXIII, 44). McDermott (1938: 86) reported that this method was still used at the beginning of the 20th century in the capture of chimpanzees. The imitative capacity of primates is often reported as a system for capturing them, as we have seen in Diodorus Siculus (Book, XVII, 90). This and other similar anecdotes form a considerable part of the features that will be attributed to monkeys in medieval bestiaries (cf. Janson, 1952).

The use of primates as a metaphor to ridicule other beings, and also as a symbol of bad luck is exemplified in a depiction of an ape/human hybrid which appears on a detail of a painted wall plaster from Pompeii (Naples, National Archaeological Museum) (cf. Toynbee, 1973: 58; King, 2002: 434). As reported by Ferris (2018), the scene depicted is one of the famous incidents in Virgil's Aeneid – Aeneas fleeing Troy carrying his father Anchises and his son Ascanius brought to life in dramatic fashion. However, in this interpretation the drama of the scene was totally subverted by Aeneas and Ascanius having monkey's heads and tails and being ithyphallic. Ferris (2018) interpreted this as a piece of humorous invective with political relevance that mocks part of the foundation myth of Rome about Trojan royal origins, as described by Virgil, or questions Augustus who claimed direct ancestry from Aenaes. Thus, the idea of the monkey and its interchangeable and ambivalent nature were used to create an infinite complex of interpretations and various uses of this animal as it is showed in Greek and Roman literature.

6.4 A GLANCE AT MONKEY LORE IN GREEK AND ROMAN LITERATURE

Primates are frequently evoked in Greco–Roman literature and are often invested with relevant symbolic meanings in the wide corpus of prose, theatre plays, and poetry (see among others McDermott, 1936, 1938; Connors, 2004; Ferriss-Hill, 2014; Vespa, 2017a). For Greek and Roman authors, primates, or rather the character of "the monkey", who represented simians in general and broadly identified with the Barbary macaque, were objects of parody due to their similarity with human beings. This aspect made them ridiculous and turned them into natural clowns and jesters. In a few cases the term *mimō* from the ancient Greek word *μιμώ* was used as an equivalent of *pithekos* (cf. Box details 6.3). Also, the Latin word *simia* [=monkey] was in Roman times an exact synonym of the Latin term *aemūlus* [= imitator], although its real meaning was flat nose (see Glossary). As noted by McDermott (1938), an early statement of the Scythian sage, Anacharsis (6th century BCE), quoted by the grammarian Athenaeus of Naucratis who lived in the time of the Roman emperor Marcus Aurelius (ruled from 161 to 180 CE), had a great relevance in the Greek and then Roman world:

> Anacharsis the Scythian, when on one occasion the jesters were introduced in his company, remained without moving a muscle of his countenance; but afterwards, when a monkey was brought in, he burst out laughing, and said, now this fellow is laughable by his nature, but man is only so through practice. (Athenaeus, *Deipnosophists*, XIV, 2)

As a character, the monkey could play the role of a lowborn demagogue pretending to be of higher birth than other animals, due to its intermediate nature between the latter and us. However, the human attitude normally taken towards these animals at all times was that they were ugly, evil, and unpropitious creatures, even if they could be very loved pets, as we have seen in the previous paragraphs. The idea of monkeys as evil and ugly animals is very ancient. For instance, the Greek poet Simonides, already in the 7th century BCE, identifies the very worst kind of woman as descending from the monkey:

> in appearance most ugly; when such a woman shows herself in the street, people laugh at her. She has a short neck, hardly moves, has no buttocks, is withered of limb; [...] and she knows all the intrigues and tricks, as a monkey does. (Simonides *apud* Morris, 2013: 45)

As noted by Morris (2013: 44), the ancient Greeks were "particularly offended by the monkey's lack of smoothly rounded buttocks, which we acquired during our evolution as a result of switching to upright walking" (**Fig. 6.15**). Thus, an ugly creature could not possibly have a superior mind, and the idea that primates could perform human actions without exhibiting human cleverness was widespread. The general idea of monkeys as evil, ugly imitators, and jesters, rooted in early Greek literature and in fables, pervaded European culture until medieval times.

Monkeys appear many times in Greek fables. These literary productions, to a considerable extent, derived from Near Eastern traditions, and the Greeks themselves also acknowledged their Oriental roots (Korhonen, 2019: 212). Originally, fables were considered the instruments

FIGURE 6.15 Bottom of the Barbary macaque. (Photo Cecilia Veracini.)

of persuasion established as a genre of rhetorical practice. These literary compositions were originally oral and could be told in the mists of philosophical discussions as a useful kind of story to be used in a speech, a fictitious narration that portrays reality. They are short stories in verse with animal protagonists and were "used as a means of indirect criticism, which avoids being offensive by not referring to specific persons and situations" (Zafiropoulos, 2001: 13). The iambic poet Archilochus from the 7th century BCE is one of the earliest Greek writers to use fables, and his first fable involving monkeys is one of his best-known fables *The Ape and the Fox* (Perry, 1952: n° 81). There, the character of the monkey is the dupe of the fox (Fragments 224, 233 *apud* Lasserre, 1959). The moral of the fable is that the basic nature of the monkey will ultimately betray itself.

The life of the mythical and legendary narrator of fables, Aesop, is placed in the 6th century BCE (Adrados, 1999: 274), if he existed. To him are associated a number of different sources which, belonging to the oral tradition, were not collected for some three centuries after his death. The first extant collection of fables was produced in Latin by Phaedrus a freedman of Augustus (ca. I century CE). Later on, Babrius (2nd century CE), a Hellenised Roman writer resident of Syria, produced a collection of fables in Greek. The compilation and the written form of the so-called *Augustana* collection – containing 231 Greek fables (Perry number 1–231) – may be roughly contemporary with Babrius or later. When we speak of the Aesopic fables, we mainly refer to the prose collections like the *Augustana* and some later collections, but also to Phaedrus and Babrius literary fables in verses (Korhonen, 2019: 215). Thus, what we call Aesopic fables or the Aesopic tradition contain a very variegated corpus of fables. According to Greenlaw (2011: 68), fables "are arguably the greatest indicator of attitudes towards these animals [primates] from the Archaic Period onwards".

As previously referred to in fables, the monkey is often the counterpart of the fox:

> A fox and a monkey were traveling together on the same road. As they journeyed, they passed through a cemetery full of monuments. All these monuments which you see, said the monkey, are erected in honour of my ancestors, who were in their day freedmen and citizens of great renown. The fox replied: you have chosen a most appropriate subject for your falsehoods, as I am sure none of your ancestors will be able to contradict you. (Delmont, 2014: 94)

FIGURE 6.16 The monkey and the dolphin, from the book Fontaine's fables, Engraving by Gustave Doré (1867).

The moral is that untrue stories often give themselves away, and the monkey, of course, was the falsest among animals, pretending to emulate human beings. In this category, we find one of the most famous fables containing primates: *The Monkey and the Dolphin* (Aesop, Fable 305), (see **Fig. 6.16**). This fable also contains references to myth and possible reflections of Greek daily life as people who travel by sea have the custom of taking with them Maltese

BOX DETAILS 6.3: A CULTURAL INSIGHT INTO THE ZOONYM *MIMŌ* ("THE MIMIC") IN ANCIENT GREEK AND BYZANTINE WORLD

Marco Vespa
The Hebrew University of Jerusalem

In Indo-European languages, both modern and ancient, it is difficult to pinpoint the precise meaning of the various zoonyms used to refer to non-human primates. The Latin noun *simius* is a case in point where ancient intellectuals and scholars were divided between those who wanted to see the term as of the same linguistic family as *similis* ("similar", "alike") and those who instead firmly believed that the zoonym derived from ancient Greek and in particular from the narrow and downward nose, *simos* (σιμός), that would characterize the primates among other living beings (Isidore of Seville, *Etymologies,* 12.2.30–31). The same uncertainty also surrounds the term *pithēkos* (πίθηκος), the noun by which monkeys were commonly referred to in ancient Greek (Frisk, 1970: 534; on ancient folk etymologies, see Vespa, 2021, 191–197).

The picture is not the same for another term used in the Greek language to designate a monkey, a term that is particularly transparent from the semantic point of view, because it derived from *mimos* (μῖμος), the "mime", and belonged to a group of feminine nouns with expressive doubling of the first consonant: *mimō* (μιμώ), "the mimic" (Chantraine, 1933: 115–117). The semantic transparency of the term for ancient and medieval Greek speakers was such that a distinguished intellectual and grammarian such as Manuel Moschopulus (1265–1316 CE) had even derived the verb "to imitate", *mimō* (μιμῶ) from the name of the monkey, *mimō* (μιμώ), as an animal that impersonated any action or gesture of human beings (scholia to the *Batrachomyomachia*, 7.20–21 Ludwich).

This term is never attested in the literary texts of the Archaic (VIII–VI c. BCE), Classical (V–IV c. BCE), or Hellenistic Greek (III–I c. BCE) periods. Its use is only consistently recorded since the mid-Byzantine period (VII–XII c. CE; *LBG*, 2017: 1027). Whereas in a few texts, particularly, the genre of the novel or the treatises on the interpretation of dreams, the zoonym *mimō* is employed in place of *pithēkos* (Eustathius Macrembolites, *Hysmine and Hysminias,* 9.3.10); in other cases, such as in lexicons or commentaries on literary works of Greek cultural heritage (Homer, Aristophanes, etc.), the term *mimō* is used to explain and clarify the meaning of *pithēkos* (Suda, letter *pi*, lemmas 1579–1580 Adler). This last observation could suggest a phenomenon of diachronic change in the linguistic usage where the Greek zoonym *mimō* would have gradually taken the place of *pithēkos* in the less formal contexts of oral or written communication. The zoonym *mimō* as well as another term equally used in the Byzantine period to denote the monkey, that is *arkoïzianos*, or *arkobouzianos* (ἀρκοβουζιανός), could be part of the popular or everyday register of the Greek language of the Medieval period, very often far removed from the cultivated and highly conservative language of intellectuals and the imperial court. From this perspective, we gain insight into the language of the ordinary people, *idiōtai,* who did not belong to the elites of society and who normally used similar terms to refer to non-human primates (scholia to Lycophron's *Alexandra,* v. 1000, ll. 1–5 bis, Scheer).

The modern Greek term used to refer to monkeys, that is *maïmou* (μαϊμού), is taken from the standard spoken language (Demotiki) and on the basis of the strong morpho-phonological affinities with the Greek *mimō,* a relationship of linguistic kinship should not be excluded (Andriotis, 1967: 193; Babiniotis, 2009: 806 for a possible derivation from the Turkish noun for "monkey", *maymun*; Veracini & Alghamdi, this volume).

Particularly interesting, and unique, is the mention of the zoonym in a text by Michael Psellos (1018–1078 CE) in which the term is considered a proper name, *Mimō* (Μιμώ), and is included in a list of demonic creatures of ancient Greek and Byzantine tradition, all characterised by representing malevolent figures, capable of harming the lives of mothers and their children (Johnston, 1999: 161–202; Pellizer, 1998). The list of these demonic entities is intended by Psellos to attack in a satirical poem a political opponent that he compares to a monstrous creature, capable of arousing contempt and revulsion even in God (Psellos, *Poems,* 21.285–288 Westerink; on the presence of *Gillō* and other female demonic figures in Psellos' works, see Garland, 2021). This section of the poem seems to follow the rhetorical patterns of some exorcistic formulations through which demonic creatures capable of harming the house and its inhabitants were normally expelled and chased away (Maltese, 2004).

More specifically, *Mimō* is evoked immediately after *Mormō* and before *Gillō.* Both these figures have a rich and elaborate tradition that from Archaic Greece reaches the Neo-Greek lore through the Medieval Byzantine period. The two monstrous characters embody two different manifestations of the misfortunes that can strike a young woman, marking her ruin. *Mormō,* sometimes from Late Antique sources associated with *Lamia,* would be a demonic creature hostile to infants and children because she herself would have tragically killed, devouring them, her own children (Patera, 2014: 111–129). In the case of *Gellō* (or *Gillō*), however, the danger to be averted would have come from an excessive and perilous attraction that the demon would have manifested for the children of other women since the young *Gellō* had died still a girl, a *parthenos,* without having been able to give birth (Patera, 2006–2007; cf. Sappho, fr. 186a Voigt; concerning Byzantine exorcisms against the demon *Gillō,* see recently Braccini, 2021).

The inclusion of *mimō* in the list of children's bogeymen and monsters by the Byzantine author might not only follow rhythmic-phonetic reasoning given the similarity in the morphological structure of the terms (e.g. *Mormō, Mimō*), but might also reflect certain ancient beliefs. A number of ancient texts, ranging from Aesopic fables to didactic poems of the Imperial Greco-Roman age, discuss the maternal behaviour of monkeys towards their offspring (Aesop, *Fables,* 243 Hausrath; Pliny the Elder, *Natural History,* 8.215–216; Oppianus of Apamea, *On Hunting,* 2.605–611; on monkeys and their failures to care for infants, see Ingemark & Ingemark, 2020: 109–126; Masseti & Veracini, this chapter). Ancient accounts are consistent in reporting the belief that monkeys would cause suffering and death to their cubs, in some cases, precisely, due to an excess of love and care to the point of suffocating the offspring by embracing them too tightly. A part of the Greek literary tradition on proverbs, recorded by Zenobius in the Greco-Roman Imperial age, contains a proverb, "The *Gellō* too fond of her children" (Zenobius, *Proverbs,* 3.3 Leutsch-Schneidewin), which was also used to indicate those who were capable of harming, or killing, their children due to an excess of affection and care, a situation analogous to what some ancient texts reported for monkeys. Particularly significant with regard to *mimō* and the possible links between this animal and the malevolent demons towards children is the illustration present in the illuminated codex of Oppianus of Apamea, the *Marcianus Graecus* Z 479, dating back to the XI c. CE and belonging to Cardinal Bessarion who brought it to Venice (Spatharakis, 2004: 8–13). As an illustration of the story of infanticide practiced by monkeys described by Oppianus, it is possible to observe on the right a caudate monkey holding a baby in his arms while chasing away another cub and on the left another monkey sitting on a rock seemingly holding and contemplating an object that can most likely be identified as the skull of his own killed baby (Spatharakis, 2004: 113–114; **Fig. 6.17**). In addition to the beliefs related to the

FIGURE 6.17 Monkeys. Oppian of Apamea, *Cynegetica*, Codex *Marcianus Graecus* Z 479 (= 881), folio 39r, 11th century, Biblioteca Marciana, Venice. (Photo: Courtesy of the Biblioteca Marciana. The monkey carrying the young has the appearance of a monkey of the genus *Chlorocebus* (note the beard)

negative parental behaviour of the monkey, an important role in the juxtaposition of this animal with demonic figures such as *Mormō* is undoubtedly played by the aesthetic appreciation that Greek cultures (from the VII c. BCE until the Byzantine period) had consistently expressed on the ugliness and the decrepit and monstrous features of *pithēkos*. Both the monkey and the figure of *Mormō* were readily exploited as paradigms of a physical appearance that did not correspond to the canons of youthful vigour and beauty (e.g. Michael Apostolius, *Letter to Lucius*, ll. 286–287 Stefec, where *mimō, Mormō*, and *pithēkos* are mentioned together).

dogs and monkeys for amusement during the journey (as referred in the fable cf. Zafiropoulos, 2001: 168).

After the boat was shipwrecked, the monkey that was on board was saved and picked up by a dolphin who mistook it for a man. The monkey said it was an Athenian, so the dolphin asked if it knew Piraeus. When the monkey answered that Piraeus was a good friend of it, the dolphin figured out it had been deceived and let the impostor drown.

In the Latin fables of Avianus and Phaedrus, ca. 1st century CE, (inspired by Aesop's fables), primates appear twice. The first is the story of the bad mother monkey bearing twins, of which she loves one and hates the other. The loved one is smothered by an excess of affection, while the hated one survives. The other one points out the extreme ugliness of monkeys, which is recalled in numerous fables. In this latter story, the mother monkey participates in a beauty contest for babies organised by Zeus:

> Jupiter promised a royal reward to the beast whose offspring was judged best looking. The monkey joined the rest and presented, with all tenderness, a flat-nosed, hairless young monkey as her candidate. The other beasts laughed, but she replied: I do not know if Jupiter will award the prize to my baby, but in my eyes I know him to be the most beautiful of all. (Perry, 1952: N° 364)

The moral of the story is that each person believes his own child to be the most attractive. In another fable of Phaedrus, the criticism of the simians' ugliness is aimed at the animal's bare bottom and lack of a tail (the monkey in question is perhaps the tailless Barbary macaque). The self-conscious monkey asks the fox for part of its tail to cover its bare behind. The nasty fox refuses the request with the added insult that even if its tail were longer, it would drag it through the mud rather than share a portion of it with the monkey (Perry, 1952: 533). While in early data, the monkey appears in proverbs as it appears in fable, its character is really more varied and complex than that of other animals: the monkey character could be as clownish and ugly as a donkey, wily as a dog, or greedy as a wolf, but above all, it was imitative. Ugliness and baseness are also common characteristics. Plutarch remarked that the monkey was ugly and virtueless by nature. "It has less virtue than a dog or other domestic animal because it cannot do the work they do" (Plutarch, *Moralia*, 64E).

Beside fables, other kind of Roman literature told about primates. The earliest Latin reference to monkeys is the famous sentence of Ennius, where there is an etymological pun (Zucker, pers. Comm.): "*simia quam similis turpissima bestia nobis!*" [=How ugly and, so similar to us is the monkey!], quoted by Cicero (*De natura deorum* 1.35.97). Ennius calls the monkey a very ugly beast, combining the idea of physical deformity with that of moral obloquy. An extremely interesting reference in a late poem of the *Anthologia Latina* shows the animal very low in the *scala naturae* of animals. This poem by Eucheria, written about the end of the 5th or the beginning of the 6th century CE, consists of 16 elegiac couplets expressing the indignation of the writer because of a proposal of marriage from a *servus rusticus*. The whole poem is an uninspired dilution of three of Virgil's lines in the *Eclogues* (cf. *Anthologia Latina* No. 390, 19–22, cf. Riese, 1869). The contrast is between animals of a high type (*hyaena, lynx, cerva, tigris, damma*) and those of a low type (*vulpis, simius, asinus, onager, bos*). The monkey is of course in the list of evil or base animals in contrast with the good or noble ones.

Because of its ugliness and ridiculousness, the character of the monkey was frequently used in satire and comedy. Plautus supplies us with the most striking instances of the use of this animal (cf. Wortmann, 1883, *De comparationibus* .... 5, 6, 17f). This author uses the monkey twice as a character in a dream to foreshadow the action of the play. In these two dreams, the monkey represents two different types of men: the ridiculous and the evil. Wortmann (1883) suggests that via a comparable play on *cecropius* (Athenian) and *cercops* (long-tailed monkey), Plautus' monkey business referred to the way Roman literature aped its Greek models, especially Roman comedy. When the name of the animal is used as an epithet or as a proper name, it may have either meaning, or a combination of the two.

Primates never had any religious significance for the Romans, even though they were also part of the sacred group of animals in cults of Egyptian origin, notably that of the goddess Isis (Tinh, 1964: 102–103). This aspect has even resulted in the importation of artistic conventions in the wake of the spread of Eastern religions in Italy (King, 2002: 434). Indeed the portrayal of primates is largely derived from Eastern, Greek, and Hellenistic art, as demonstrated by the preponderance of catalogue entries from those sources in McDermott (1938). Primates could, however, represent a sign of ill fortune in Roman culture. A monkey seen in a dream was a bad portent, an idea also common among Greeks, Jews, and Arabs. Nero was terrified, shortly before his death, by a dream in which the hindquarters of a favourite horse were turned into those of a monkey (McDermott, 1938: 50). The bad reputation of the monkey also extended to other social fields: the comparison of such an ugly animal to a person was meant as an offence. Moreover, there are many anecdotes which describe performing primates where a tailed monkey was very clever at dodging the spears that were thrown at it: "I am an ape, cunning in avoiding the darts hurled at me. Had I a tail, I should be a *Cercopithecus*" (Martial, 1897, *Epigrams*, Book 14, CCII. The Ape)

The worship of animals was not appreciated in Greek and in Roman times, especially after the advent of Christianity. The Greeks unfairly ridiculed the Egyptians as animal worshippers. In Greco-Roman Egypt, the worship of the baboon continued (Arnold, 1995; Osborn & Osbornová, 1998), and this practise was held in contempt because these animals themselves were considered foolish; vain; and, as we have seen, ugly. Cicero (*Tusculanae disputationes*, 5.27. 78), without mentioning monkeys, claims:

> "*Aegyptiorum morem quis ignorat? quorum imbutae mentes pravis erroribus quamvis carnifi cinam prius subierint quam ibim aut aspidem aut felem aut canem aut crocodilum violent, quorum etiam si imprudentes quid iam fecerint, poenam nullam recusant*" (Who does not know the custom of the Egyptians? Their minds are infected with degraded superstitions and they would sooner submit to any torment than injure an ibis or asp or cat or dog or crocodile, and even if they have unwittingly done anything of the kind there is no penalty from which they would recoil; translation by King, 1927).

6.5 CONCLUDING REMARKS

From the Bronze Age onwards, different species of primates were imported into Europe. They played various cultural roles according to historical periods, being regarded, as sacred and magical animals, mascots or pets, or used for entertainment and other purposes. Monkeys became integral elements in Minoan religious iconography by the time of the Late Bronze Age. Their representations became rare in Mycenaean contexts, although Aegean people were clearly familiar with these animals. Later on, the trade of Phoenician and other ancient civilisations around Mediterranean which began in the first millennium BCE contributed to a new popularity of monkeys and likely to the diffusion of a sort of worship of these animals.

In Classical times, primates were depicted in many artworks and entered literary works and plays. A highly ambivalent attitude towards these animals, considered both intelligent and imitative animals but also ugly and whimsical, is frequent in a number of works. However, one recurring and general aspect seems to be established: in the Greco-Roman world, primates had lost any metaphysical or divine attributes, being mainly regarded as performing animals, exotic and valuable beasts or as pets together with dogs, other carnivores of reduced dimensions, and birds.

The morphological and behavioural characteristics of primates were described for the first times by Greek philosophers, such as Aristotle; historians, such as Strabo; physicians, such as Galen of Pergamum; and by Roman scholars, such as Pliny the Elder among others. Apart from the first-hand scientific observations (by Aristotle and Galen), the majority of the information about primates comes from travel accounts, collected in many forms by ancient scholars. These accounts were often a mix of real observations and fantasy, and contribute to the dissemination of myths and biases about primates. It is still debated whether great apes were known or even imported into Europe at that time. According to the sources we have, the latter hypothesis should be discarded, although some correct naturalistic descriptions of African great apes (chimpanzees) exist in literature.

Primates are also frequently evoked in Greco-Roman literature and the character of "the monkey" as the representative of simians, usually synonymous with the Barbary macaque, was an object of parody due to its similarity with human beings. This ridiculous character had ambiguous and, in general, negative characteristics. It was loaded with moral meaning especially because it was considered the liar *par excellence* among animals: a low-born demagogue which pretended to high birth due to its intermediate nature between the other animals and mankind. Fables, plays, and proverbs, but also other literary works, contributed to the creation of a very negative and almost diabolical portrait of primates. This view, almost unchanged, pervaded the popular but also the scientific European culture of the Middle Ages up to the Modern Age. In particular, in Christian culture (predominant in the most recent history of the European/Mediterranean countries), the image of the monkey has always tended to be negative. Primates were frequently taken to symbolise the evil or stupid side of human beings which makes all the more intriguing the fact that Darwin's theory struck the heart of this cultural denigration of natural human origins.

ACKNOWLEDGEMENTS

We would like to dedicate this chapter to our friend and colleague Jacopo De Grossi Mazzorin (in memoriam), Dipartimento di Beni Culturali of the University of Salento (Lecce). We would like to express our thanks to Umberto Albarella, Department of Archaeology of the University of Sheffield; Chiara Assunta Corbino, Laboratorio di Scienze Applicate del Parco Archeologico di Pompei; Frédéric Gerber, Institut National de Recherches archéologiques Preventives, Poitiers; and Leonardo Salari, Department of History, Culture and Society, University of Rome Torvegata.

We would also like to thank the Parco Archeologico di Cerveteri e Tarquinia – Museo Archeologico Nazionale di Tarquinia and the Biblioteca Marciana of Venice, for permission to publish the images. We are grateful to Arnaud Zucker for the critical comments on the first draft of the manuscript and to Bernard Wood for correcting and commenting on the text. We thank Marco Vespa for his Box detail contribution.

FUNDING OF C.V.

Portuguese national funds through FCT – Fundação para a Ciência e a Tecnologia in the context of the celebration of the contract under the 'Norma Transitoria', DOI 10.54499/DL57/2016/CP1456/CT0001.

REFERENCES

Primary Sources

Aelian, (1998). *De natura animalium (La natura degli Animali)*, 2 vols. Milan: Biblioteca Universale Rizzoli.

Aelian, Scholfield, A.F. (ed) (1959). *On the Characteristics of Animals*, Volume II: Books 6–11. Loeb Classical Library 448. Cambridge, MA: Harvard University Press.

Agatharchides of Cnidus, Burstein, S.M. (ed.). (1989). *On the Erythraean Sea*. London: The Hakluyt Society.

Agatharchides of Cnidus, Henry, R. (ed.). (1974). *Photius Bibliotheque. Paris*: Les belles lettres.

Apostolius, M., Stefec, R. (eds.). (2013). *Die Briefe des Michael Apostoles. Hamburg*: Kovac Verlag.

Aristotle, Thompson, D.W. (ed.). (1910). *Historia animalium (History of Animals)*. On-line. https://classics.mit.edu//Aristotle/history_anim.html

Aristotle, Ogle, W. (ed.). (2008). *De partibus animalium (On the Parts of Animals)*, https://classics.mit.edu/Aristotle/parts_animals.html

Athenaeus of Naucratis. (1854). *The Deipnosophists. Or Banquet of the Learned of Athenaeus*. London: Henry G. Bohn, York Street, Covent Garden. https://www.perseus.tufts.edu/hopper/text?doc=Perseus:text:2013.01.0003

Batrachomyomachia, Ludwich, A. (ed.). (1896). *Die Homerische Batrachomachia des Karers Pigres: nebst Scholien und Paraphrase. Leipzig*: Teubner.

Biffi, N. (ed.). (1999). *L'Africa di Strabone. Quaderni di Invigilata Lucernis 11*. Bari: Edipuglia.

Biffi, N. (ed.). (2000). *L'Indiké di Arriano. Quaderni di Invigilata Lucernis 11*. Bari: Edipuglia.

Cicero (46 BCE). *De natura deorum*. https://la.wikisource.org/wiki/De_natura_deorum

Ctesia de Cnide. & Lenfant, D. (eds.). (2004). *La perse, L'Inde autre fragments. Paris*: Les belles lettres.

Detmold, E.J. (2014). The Fables of Aesop. New York: Calla Edition.

Diodorus Siculus (1820–1822). *Biblioteca Storica di Diodoro Siculo volgarizzata dal cav. Compagnoni, Tomi I–VII*. Milano: dalla tipografia di Giovanni Battista Sonzogno.

Diodorus Siculus, Alfieri, T. (ed.). (1985). *Biblioteca Storica, Libro XIV–XVII*. Torino: Rusconi.

Galen, Singer C. (ed) (1956). *On Anatomical Procedures: De Anatomicis Administrationibus*. Wellcome Historical Medical Museum. Available online: https://archive.org/details/b20457194/mode/2up (accessed on 25, June 2020).

Herodotus, Godey, A.D. (ed.). (1921). *The Persian war. Books III–IV*. Loeb Classical Library. Cambridge (MA) and London: Harvard University Press.

Herodotus, Colonna, A., & Bevilacqua, F. (eds) (1996). *Erodotus, Le Storie, (I-IV) (V-IX)*,2 vols. Torino: UTET.

Isidore of Sevilla, André, J. (ed.) (1989). *Isidore de Séville. Etymologies. Livre XII. Des animaux*. Paris: Les Belles Lettres.

Juvenal, S.O. (Réédition de 1921). *Satires*. Paris: Belles Lettres.

King, J.E. (1927). Cicero: Tusculan *Disputations*. Loeb Classical Library. London: William Heineman.

Lasserre, F.B.A. (ed.) (1959). *Archiloque: Fragments*. Paris: Société d'édition.

Macrembolites, E. & Marcovich, M. (eds.) (2001). *Eustathius Macrembolites De Hysmines et Hysminiae amoribus libri XI*. Munich-Leipzig: K.G. Saur.

Martial. (1897). *Epigrams. Book 14*. Bohn's Classical Library. https://www.tertullian.org/fathers/martial_epigrams_book14.htm

Oikonomides, A.N. & Miller, M.C.J. (eds.). (1995). *Hanno the Cartaginian. Periplus or Circumnavigation [Of Africa]*. Chicago: Ares Publishers.

Oppianus, Papathomopoulos, M. (eds.) (2003). *Oppianus Apameensis. Cynegetica*, Berlin: de Gruyter.

Ovidio. (1992). *Le Metamorfosi*, 2 vols. Bur: Classici Greci & Latini, Milano.

Perry, B.E. (1952). *Aesopica*. Urbana: University of Illinois Press.

Philostratus, Jones, C.P. (ed.). (2005). *The Life of Apollonius of Tyana. Books I–IV*. Cambridge, Massachussets, London: Harvard University Press.

Pliny the Elder, Ernout, A. (ed.). (1952). *Pline l'Ancien. Histoire naturelle. Livre VIII*. Paris: Les Belles Lettres.

Plutarchus, Santini, A. (eds.). (1991). *Parallel Lives, Pericle, I, 1*. Milano: Biblioteca Universale Rizzoli.

Psellos, M., Westerink, L.G. (eds.) (1992). *Michaelis Pselli poemata*. Stuttgart: Teubner.

Riese, A. (1869). *Anthologia Latina*, recensuit A. *Riese, Teubner, Lipsiae, 1870*.

Sappho, Vogt, E.M. (eds.). (1971). *Sappho and Alcaeus. Fragmenta*. Amsterdam: Polak and van Gennep.

Schoff, W.H. (1912). *The Periplus of Hanno. A Voyage of Discovery down the West African Coast, by a Carthaginian Admiral of the Fifth Century B.C.* Philadelphia: The Commercial Museum.

Scholia to Lycophron: Scheer, E. (ed.). (1958). *Lycophronis Alexandra*, vol. 2. Berlin: Weidmann,

Secondo, P.G., Barchiesi, A., Centi, R., Corsaro, M., Marcone, A., & Ranucci, G. (eds.). (1982). *Storia Naturale. I. Cosmologia e Geografia. Book 1–6*. Torino: Giulio Einaudi.

Strabo. Bohn, H.G. (eds.). (1854–1857). *Strabo's Geography in Three Volumes as Translated by H.C. Hamilton.: vols. 1–3*. Oxford.

Zenobius, Schneidewin, F.G., & von Leutsch, E.L. (eds.). (1839). *Corpus paroemiographorum Graecorum, vol. 1*. Göttingen: Vandenhoeck & Ruprecht.

Secondary Sources

Adinolfi, G., Carmagnola, R., Cataldi, M., Marras, L., Masseti, M., & Palleschi, V. (2019). Vedere l'invisibile: immagini da un "bestiario" di V sec. a.C. nella tomba dei demoni Azzurri dalle prime indagini multispettrali. *Ostraka*, 28: 5–23.

Adrados, F. R. (1999). *History of the Graeco-Latin Fable: Volume I. Introduction and from the Origins to the Hellenistic Age* (Vol. 201). Brill.

Albarella, U. (2007). Companions of our travel: the archaeological evidence of animals in exile. In: Hartmann, S. (ed.): *Fauna and Flora in the Middle Ages. Studies of the Medieval Environment and Its Impact on the Human Mind.* Verlag Peter Lang, Frankfurt am Main: pp. 133–154.

Alghamdi, M. A., Ziermann, J. M., & Diogo, R. (2017). An untold story: The important contributions of Muslim scholars for the understanding of human anatomy. *The Anatomical Record*, *300*(6): 986–1008.

Alves, R. R., Souto, W. M., & Barboza, R. R. (2010). Primates in traditional folk medicine: a world overview. *Mammal Review*, *40*(2): 155–180.

Andriotis, N. (1967). *Ετυμολογικό λεξικό της κοινής νεοελληνικής [Etymological Dictionary of Standard Modern Greek]*, Thessaloniki (1st ed. Athens, 1951), Α.Π.Θ.

Antonopoulos, J. (1992). The great Minoan eruption of Thera volcano and the ensuing tsunami in the Greek Archipelago. *Natural Hazards*, 5(2): 153–168. doi: 10.1007/BF00127003

Aquilué, X. (2007). Els animals de companyia a l'antiguitat. In: Aquilué, X. & Monturiol, J. (eds.): *Animals d'Empúries. La fauna i l'home a l'antiguitat. Exposicio, Casa dels Forestals de Sant Marti d'Empuries (juliol-setembre 2007).* Ajuntament de l'Escala/Museu d'Arqueologia de Catalunya-Empuries: pp. 32–35.

Armour-Chelu, M. (1997). Appendix 8: Faunal remains. In: Barker, P., White, R., Pretty, H., Bird, K., & Corbishley, M. (eds.): *The Bath Basilica Wroxeter. Excavations* 1966–90. *English Heritage Archaeological Report, vol. 8*. London: pp. 350–364.

Arnold, D. (1995). An Egyptian Bestiary. *Bulletin-Metropolitan Museum of Art, 52*(4).

Aydar, E., Çiner, A., Ersoy, O., Écochard, E., & Fouache, E.G. (2021). Volcanic ash and tsunami record of the Minoan Late Bronze Age Eruption (Santorini) in a distal setting, southwestern Turkey. *Journal of Quaternary Science*, 6: 586–597. https://doi.org/10.1002/jqs.3314

Babiniotis, G. (2009). *ΕΤΥΜΟΛΟΓΙΚΟ ΛΕΞΙΚΟ ΤΗΣ ΝΕΑΣ ΕΛΛΗΝΙΚΗΣ ΓΛΩΣΣΑΣ*. Athens: Lexicology Centre.

Bailey, J.F., Henneberg, M., Colson, I.B., Ciarallo, A., Hedges, H., & Sykes, B. (1999). Monkey business in Pompeii–unique find of a juvenile barbary macaque skeleton in Pompeii identified using osteology and ancient DNA techniques. *Molecular Biology and Evolution*, 16(10): 1410–1414. https://doi.org/10.1093/oxfordjournals.molbev.a026051

Ballard, C. (2006). Strange alliance: Pygmies in the colonial imaginary. *World Archaeology*, 38(1): 133–151.

Bertrand, M. (1967). Training without reward: Traditional training of Pig-tailed Macaques as coconut harvesters. *Science*, 155: 484–486.

Bietak, M. (2000). Rich beyond the dreams of Avaris: Tell el-Dab'a and the Aegean world: a guide for the perplexed: a response to Eric H. Cline. *The Annual of the British School at Athens*, 95: 185–205.

Boisacq, E. (1923). *Dictionnaire étymologique del a langue Grecque.* Heidelberg, Paris.

Bonacelli, B. (1932). La scimmia in Etruria. *Studi etruschi*, 6: 341–382.

Bond, A. & Sparks, R.S. (1976). The Minoan eruption of Santorini, Greece. *Journal of the Geological* Society, 132: 1–16. https://doi.org/10.1144/gsjgs.132.1.0001

Bonfante, G. (1992). *Il nome di Ischia. La Campania tra il VI e il III sec. a.C.* Atti del 14° Convegno di Studi Etruschi e Italici, Benevento 1981: pp. 283–284.

Braccini, T. (2021). Revisiting the 'exorcism of Gello: a new text from a Vatican manuscript, with a typological analysis of the known variants. *Medioevo Greco*, 21: 149–170.

Calder, L. (2011). Cruelty and Sentimentality: Greek Attitudes to Animals, 600-300 BC. *BAR International Series 2225.* Oxford: Archaeopress.

Calder, L. (2017). Pet and Image in the Greek World: The Use of Domesticated Animals in Human Interaction. *Interactions between Animals and Humans in Graeco-Roman Antiquity*: 61–88.

Cameron, M.A.S. (1968). Unpublished paintings from the "House of the Frescoes" at Knossos. *The Annual of the British School of Archaeology at Athens*, 63: 1–31.

Canciani, F. (1973). Scimmie a Creta. *Antichità Cretesi*, 1: 107–110.

Cavalli-Sforza, L.L., Cavalli-Sforza, L., Menozzi, P., & Piazza, A. (1994). *The History and Geography of Human Genes.* Princeton University Press.

Chami, F., Le Guennec-Coppens, F., & Mery, S. (2002). East Africa and the Middle East relationship from the first millennium BC to about 1500 AD. *Journal des Africanistes*, 7: 21–37.

Chantraine, P. (1933). *La formation des noms en grec ancien.* Paris: Klincksieck.

Ciarallo, A. (1999). Scheletro di scimmia. In: Ciarallo, A. & De Carolis, E. (eds.): *Homo Faber. Natura, Scienza e Tecnica nell'Antica Pompei.* Catalogue. Napoli, Museo archeologico Nazionale. 27 marzo -18 luglio 1999. Milan: Electa: p. 71.

Cleary, V. (1972). Se sectari simiam: Monkey Business in the Miles Gloriosus. *The Classical Journal*, 67(4): 299–305. Retrieved March 1, 2021, from https://www.jstor.org/stable/3296294

Cline, E.H. (1991). Monkey business in the Bronze Age Aegean: The Amenhotep II Faience Figurines at Mycenae and Tiryns. *The Annual of the British School at Athens*, 86: 29–42.

Clutton-Brock, J. (1995). Aristotle, the scale of nature, and modern attitudes to animals. *Social Research*, 62(3): 421–440.

Connors, C. (2004). Monkey business: imitation, authenticity, and identity from *Pithekoussai* to Plautus. *Classical Antiquity*, 23(2): 179–207.

Dasen, V. (2013). *Dwarfs in Ancient Egypt and Greece.* Oxford: Clarendon Press.

Deputte, B.L. & Anderson, J.R. (2009). Baboon palm nut harvesters in ancient Egypt: New (ancient) evidence, new questions. *Folia Primatologica*, 80: 70–73. https://doi.org/10.1159/000214225

Diogo, R. (2018). Links between the discovery of primates and anatomical comparisons with humans, the chain of being, our place in nature, and racism. *Journal of Morphology*: 472–493. https://doi.org/10.1002/jmor.20783

D'Oriano, R., (2007). *Un pithekos d'Ichnoussa tra pithekoi e Pithekoussai d'Ischia, nord Africa e Gibilterra.* In: Angiolillo, S., Giuman, M., & Pasolini, A. (eds.): Ricerca e confronti 2006. Giornate di Studio di Archeologia e Storia dell'arte. Quaderni di Aristeo. Università degli Studi di Cagliari: pp. 183–198.

Doré G. (1867). *Fable de la Fontaine illustré par Gustave Doré.* Edition du Layeur.

Doumas, C.G. (2000). Akrothiri, Thera. *To Ergon, 1999* (in Greek): pp. 77–81.

Estes, R.D. (1992). *The Behavior Guide to African Mammals.* Berkeley, CA: University of California Press.

Evely, D. (1999). *Fresco: A Passport into the Past. Minoan Crete through the eyes of Mark Cameron*. Athens: British School at Athens/N.P. Goulandris Foundation.

Evslin, B. (2006). *Gods, Demigods and Demons*. London and New York: I.B. Tauris.

Ferrence, S.C. & Bendersky, G. (2004). Therapy with saffron and the goddess at Thera. *Perspectives in Biology and Medicine*, 47: 199–226.

Ferris, I. (2018). *Cave Canem: Animals and Roman Society*. Amberling Publishing.

Ferriss–Hill, J. (2004). Horace, Catullus, and the old comic monkeys. *Paideia : rivista di filologia, ermeneutica e critica letteraria, online*: 343–350. https://doi.org/10.1400/227194

Franco, C. (2017). Greek and Latin words for human-animal Bonds: metaphors and taboos. *Interactions between Animals and Humans in Graeco-Roman Antiquity*, 39–60.

Frisk, H. (1970). *Griechisches Etymologisches Wörterbuch*. Κρ-Ω, Heidelberg, Carl Winter: Band II.

Galat, G. & Galat-Luong, A. (1977). Demographie et regime alimentaire d'une troupe de *Cercopithecus aethiops sabaeus* en habitat marginal au nord Senegal. *La Terre et la Vie*, 31: 557–577.

Garland, L. (2021). Infant mortality, Michael Psellos and the Byzantine demon Gillo. In: Beaumont, L.A., Dillon, M., Harrington, N. (eds.): *Children in Antiquity. Perspectives and Experiences of Childhood in the Ancient Mediterranean*. London & New York: Routledge: pp. 545–564.

Gerber, F. (ed.) (2019). *De la nécropole antique aux jardins des Filles de Notre-Dame*. Inrap NA&OM, Bègles.

Gerber, F. & Baudry-Dautry, A. (2012). La mode de l'animal exotique dans la haute sociéte gallo-romaine. Sépulture d'un singe dans la nécropole de la rue des Caillons à Poitiers. *Archéopages*, 35: 42–47.

Gibbons, A. (2020). Oldest *Homo sapiens* bones found in Europe. *Science*, 368(6492): 697. doi: 10.1126/science.368.6492.697

Gras, M. (1994). Pithécusses. De l'étymologie à l'histoire. APOIKIA, Scritti in onore di Giorgio Buchner. *Annali Ist. Orientale Napoli*, n.s., 1: 127–131.

Graves, R. (1955). The Greek Myths. Harmondsworth (UK): Penguin Books Inc.

Green, R.C. (2018). *A North African Barbary ape in fifth- to sixth-century Britain? A short note on the significance and context of the Wroxeter macaque remains*. https://www.caitlingreen.org/2018/02/ barbary-ape-wroxeter.html

Greenlaw, C. (2011). *The Representation of Monkeys in the Art and Thought of Mediterranean Cultures: A New Perspective on Ancient Primates*. BAR Publishing.

Groves, C.P. (2005). Order primates. In: Wilson, D.E. & Reeder, D.A.M. (eds.): *Mammal Species of the World. A Taxonomic and Geographic Reference, Third edition, vol. 1*. Baltimore: The John Hopkins University Press: pp. 111–184.

Groves, C.P. (2008). *Extended Family: Long Lost Cousins. A Personal Look at the History of Primatology*. Arlington, (USA): Conservation International.

Guardia, J. & Maragall, M. (2004). Periodització del jaciment de les Colomines, Zona A (Llívia, Cerdanya). Fase constructives i seqüencia cronologica. Setenes Jornades d'Arqueologia de les comarques de Girona (La Bisbal d'Empordà, 4 i 5 de juny de 2004). *Setenes Jornades d'Arqueologia de les comarques de Girona:* 247–252.

Guàrdia, J., Maragall, M., i Fernàndez, O.M., i Vila, O.O., Galbany, J., & Nadal, J. (2007). Enterrament d'epoca tardoromana d'un macaco Amb" Aixovar" al jaciment de Les Colomines (Llivia, La Cerdanya). *Empúries: revista de món clàssic i antiguitat tardana*, 55: 199–227.

Guardia, J., Maragall, M., Mercadal, O., Olesti, O., Galbany, J., & Nadal, J. (2005). Enterrament d'època tardoromana corresponnent a un macaco amb aixovar al jaciment de les Colomines (Llívia). *Ceretania*, 4: 66–106.

Gusman, P. (1899). *Pompei: la ville, le moeurs, les arts / Pierre Gusman; préface de M. Max. Collignon*. Paris: L. Henry May.

Harden, D.E. (1980). *The Phoenicians (revised edition)*. Harmondsworth (UK): Penguin Books.

Henneberg, M. (1997). *Skeleton of a monkey from ancient Pompeii*. Unpublished manuscript.

Hill, W.C.O. (1966). Primates. Comparative anatomy and taxonomy. In: Volume *6. Catarrhini Cercopithecoidea Cercopithecinae*. Edinburgh: Edinburgh University Press: xxiii–757 pp.

Hoage, R.J. & Mansour, J. (1996). Menagerie and Zoo. In J. Hoage & W.Deiss (eds). *New worlds, new animals: from menagerie to zoological park in the nineteenth century*. JHU Press, pp. 8–18.

Hodgson, G.W.I. (1990). *Catterick 433: animal bone report*. AMI. rep. no 2404.

Hodgson, G.W.I. (2002). A barbary ape from the 1958-9 bypass excavations (Site 433). In: Wilson, P.R. (ed.): *Cataractonium*: Roman Catterick and Its Hinterland. Excavations and Research, 1958–1997. Part II. *Council for British Archaeology Research Report*, 129: 415.

Hood, S. (1978). *The Arts in Prehistoric Greece*. Penguin (Penguin/Yale History of Art).

Ingemark, A. & Ingemark, D. (2020). *Representations of Fear. Verbalising Emotion in Ancient Roman Folk Narrative*. Helsinki: Academia Scientiarum Fennica.

Jacoby, F. (1930). *Die fragmente der griechischen Historiker, II, D, 2*. Berlin: pp. 702–703.

Jashemski, W. F., & Meyer, F. G. (Eds.). (2002). *The natural history of Pompeii*. Cambridge University Press.

Janson, H.W. (1952). *Ape and Ape Lore in the Middle Ages and the Renaissance*. The Warburg Institute, University of London.

Johnston, S.I. (1999). *Restless Dead. Encounters between the Living and the Dead in Ancient Greece*. Berkely-Los Angeles-London: University of California Press.

Jolly, C.J. & Ucko, P.J. (1969). The riddle of the sphinx-monkey. In: Douglas, M. & Kaberry, P.M. (eds.): *Man in Africa*. London: Tavistock Publications.

Keller, O. (1887). *Thiere des classischen Alterthums in kulturgeschichtlicher Beziehung*. Innsbruck: Innsbruck Verlag der Wagner'schen Universitäts-Buchhandlung.

Keller, O. (1909). *Die Antike Tierwelt*, 2 vols. Leipzig: W. Engelmann.

King, A. (2002). Mammals. Evidence from wall paintings, sculpture, mosaics, faunal remains, and ancient literary sources. In: Feemster Jashemski, W. & Meyer, F.G. (eds.): *The Natural History of Pompeii*. Cambridge: Cambridge University Press: pp. 401–450.

Kitchell, K.F. Jr. (2014). *Animals in the Ancient World from A to Z*. London and New York: Routledge.

Korhonen, T. (2019), Anthroporphism and the Aeopic Animal Fables. In: Mattila, R., Sanae Ito S., & Fink, S. (2019). *Animals and Their Relation to Gods, Humans and Things in the Ancient World*. Springer: pp. 211–234.

Kurtén, B. (2017). *Pleistocene Mammals of Europe*. London and New York: Routledge.

Lamer, H. (1926). Woher Stammt das Wort Gorilla? *Wiener Blätter für die Freunde der Antike*, 4: 5–8.

LaMoreaux, P.E. (1995). Worldwide environmental impacts from the eruption of Thera. *Geo*, 26: 172–181. https://doi.org/10.1007/BF00768739

Langdon, S. (1990). From monkey to man: the evolution of a Geometric sculptural type. *American Journal of Archaeology*, 94(3): 407–424.

Law, R. (1967). The Garamantes and trans-Saharan enterprise in classical times. *The Journal of African History*, 8(2), 181–200. doi: 10.1017/S0021853700007015

Lazenby, F.D. (1949). Greek and Roman household pets. *The Classical Journal*, 44(4): 245–252 and 44(4) (Feb. 1949), 299–307. https://penelope.uchicago.edu/Thayer/E/Journals/CJ/44/4/Household_Pets*.html

LBG. (2017). *Lexikon zur byzantinischen Gräzität, 2. Band (Λ-Ω)*, Trapp, E. (ed.). Wien: Verlag der Österreichischen Akademie der Wissenschaften: pp. 197-192.

Liéger, A., Cussenot, O., Poulain, Th., & Steinbach, D. (1997). La nécropole gallo-romaine de Cutry (Meurthe-et-Moselle). *Études Lorraines d'Archéologie Nationale*, 3 (Presses Universitaires de Nancy).

Li Causi, P. (2003). *Sulle Tracce del Manticora: La Zoologia dei Confini del Mondo in Grecia e a Roma*. Palermo: Palumbo.

Lo Giudice, C. (2008). L'impiego degli animali negli spettacoli romani: venatio damnatio ad bestias. Italies *Littérature-Civilisation-Société*, (12),: 361–395. https://italies.revues.org/1374.

Lovejoy, A.O. (1936). *The Great Chain of Being: A Study of the History of an Idea*. Cambridge: Harvard University Press.

Lynn, C.J. (1997). Excavations at Navan Fort 1961-71, County Armagh. In: Waterman, D.M. (ed.): *Northern Ireland Archaeological Monographs, vol. 3*. Belfast: Stationery Office.

Maestripieri, D. (1994). Infant abuse associated with psychosocial stress in a group-living pigtail macaque (*Macaca nemestrina*) mother. *American Journal of Primatology*, 32: 41–49.

Maltese, E.V. (2004). Osservazioni sul carme Contro il Sabbaita di Michele Psello. In: Taragna, A.M. (ed.): *La poesia tardoantica e medieval*. Alessandria, Edizioni dell'Orso: pp. 207–214.

Marinatos, S. (1968). *Excavations at Thera I. First preliminary report (1967 season)*. Athens: General Direction of Antiquities and Restoration: 58 pp.

Marinatos, S. (1971). *Excavation at Thera IV*. 1970 *season*. Athens: Hellenic Archaeological Service: 54 pp, 126 pl.

Marinatos, N. (1987). An offering of saffron to the Minoan goddess of nature: The role of the monkey and the importance of saffron. In: Linders, T. & Nordquist, G. (eds.): *Gifts to the God. Proceedings of Uppsala Symposium 1985 (Boreas 15)*. Acta Universitatis Uppsaliensis: pp. 123–132.

Marinatos, S., & Hirmer, M. (1960). Crete and Mycenae. London: Thames & Hudson.

Martin-Granel, N. (2001). Monkey business in the Congo. In: Palmberg, M. (ed.): *Encounter Images in the Meetings between Africa and Europe*. Uppsala: Nordiska Afrikainstitutet: pp. 206–220.

Masseti, M. (1980). Le scimmie azzurre. La fauna etiopica degli affreschi minoici di Santorino (Thera). *Mondo archeologico*, 51: 32–37.

Masseti, M. (1997). Representation of birds in Minoan art. *International Journal of Osteoarchaeology*, 7: 354–363.

Masseti, M. (2000). Did the study of ethology begin in Crete 2000 years ago? *Ethology Ecology & Evolution*, 12: 89–96.

Masseti, M. (2001). Did endemic dwarf elephants survive on Mediterranean islands up to protohistorical times? In: Caravetta, G., Gioia, P., Mussi, M., & Palombo, M.R. (eds.): La terra degli elefanti/The world of elephants. *Proceedings of the 1st International Congress*. Rome, 16-20 October 2001. Comune di Roma, Consiglio Nazionale delle Ricerche, Ministero per i Beni e le Attività Culturali, Università di Roma "La Sapienza": pp. 402–406.

Masseti, M. (2003). Taxonomic and behavioural aspects of the representation of mammals in Aegean Bronze Age. In: Kotjabopoulou, E., Hamilakis, Y., Halstead, P., Gamble, C., & Elefanti, P. (eds.): *Zooarchaeology in Greece. Recent Advances, vol. 9*. British School at Athens Studies: pp. 273–281.

Masseti, M. (2006). Did the study of primatology begin on the island of Crete (Greece) 4000 years ago? *Folia Primatologica*, 77(4): 296–297.

Masseti, M. (2012). *Atlas of Terrestrial Mammals of the Ionian and Aegean Islands*. Berlin: De Gruyter.

Masseti, M. (2018). The long-term redefinition of the non-volant vertebrate horizons of the Eastern Mediterranean islands. In: Sfenthourakis, S., Pafilis, P., Parmakelis, A., Poulakakis, N., & Triantis, K.A. (eds.): *Biogeography and Biodiversity of the Aegean*. Nicosia (Cyprus): Broken Hill Publishers Ltd.: pp. 191–222.

Masseti, M. (2019). Monkeys in the ancient Greek world (2000-500 b.C.). In: Veracini, C., Casanova, C., & Scalfari, F. (eds.): *History of Primatology. Yesterday and Today. The Western-Mediterranean Tradition*. Canterano (Rome): Aracne editrice: pp. 19–50.

Masseti, M. (2021). An analysis of recent literature regarding the Minoan "blue monkeys" represented in Aegean Bronze Age art. *JASs Journal of Anthropological Sciences*, 99: 153–156. doi: 10.4436/JASS.99002

Masseti, M. (2022). Replying to Urbani, Youlatos & Binnberg. *Journal of Anthropological Sciences*, 100. doi: 10.4436/JASS.10004. Epub ahead of print. PMID: 35830362.

Masseti, M. (in press). The representation of nature in Minoan art (2nd millennium BC) with particular reference to the wall paintings of Akrotiri. In: The island of Santorini (Greece). *Proceedings of ICONOGRAFIA, Convegno Internazionale di Studi* (Padova-Venezia, 12–14 December 2022).

Masseti, M. & Bruner, E. (2009). The primates of the western Palaearctic. A biogeographical, historical, ethnozoological and archaeozoological review. *Journal of Anthropological Sciences*, 87: 33–91.

Masseti, M., Adinolfi, G., Carmagnola, R., Cataldi, M., & Marras, L. (2022). Predatori esotici e prede nostrane nell'invenzione pittorica della Tomba dei Demoni Azzurri (V secolo) della necropoli di Monterozzi, a Tarquinia. *Aristonothos. Rivista di Studi sul Mediterraneo Antico*, (18), 263–283.

McDermott, W.C. (1936). The ape in Roman literature. *Transaction and proceeding of the American Philological Association*, 67: 148–167.

McDermott, W.C. (1938). *The Ape in Antiquity*. The Johns Hopkins Press.

Meyboom, P. G. (1994). The Function of the Nile Mosaic. In *The Nile Mosaic of Palestrina*, Brill: pp. 80–90

Minniti, C. & Sajadi, S.M.S. (2019). New data on non-human primates from the ancient Near Est: the recent discovery of a rhesus macaque burial at Shahr-i Sokhta (Iran). *International Journal of Osteoarchaeology*. https://doi.org/10.1002/oa.2750

Montagu, A.M.F. (1940). Knowledge of the ape in antiquity. *Isis*, 32(1): 87–102.

Morgan, L. (1988). *The Miniature Wall-Paintings from Thera: A Study in Aegean Culture and Iconography*. Cambridge: Cambridge University Press.

Morris, D. (2013). *Monkeys*. Reaction Books.

Müller-Reineke, H. (2010). Oriental animals as moral examples in Aelian's De natura animalium. *Graeco-Latina Brunensia*, 15(2): 117–126. https://digilib.phil.muni.cz/bitstream/handle/11222.digilib/115007/1_GraecoLatinaBrunensia_15-2010-2_11.pdf?sequence=1

Napier, P.H. & Jenkins, P.D. (1997). Part 4: monkey skull and mandible of the Iron Age found at site B. In: Waterman, D. M. & Lynn, C.J. (eds.): *Excavations at Navan Fort 1961-71, County Armagh. Belfast*: Depart. Environ. Northern Ireland-The Stationery Off: pp. 120–121.

Naso, A. (2000). *The Etruscans*. Bompiani Milan.

Nichols, A. (2008). *The complete fragments of Ctesias of Cnidus: translation and commentary with an introduction*. PhD These, University of Florida.

Niemeier, W.-D. (1998). The Minoans in the South-Eastern Aegean and Cyprus. In: Karageorghis, V. & Stampolidis, N.C. (eds.): *Eastern Mediterranean. Cyprus, Dodecanese, Crete 16th-6th cent. B.C.* Athens: University of Crerte and the A,G, Leventis Foundation: pp. 29–47.

Olesti, O., Guàrdia, J., Maragall, M., Mercadal, O., Galbany, J., & Nadal, J. (2013). Controlling the pyrenees: A macaque's burial from late antique Iulia Libica (Llívia, La Cerdanya, Spain). In: Sarantis, A. & Christie, N. (eds.): *War and Warfare in Late Antiquity. Current perspectives*. Leiden: Brill: pp. 703–734.

Onar, V., Alpak, H., Pazvant, G., Armutak, A., Gezer-Ince, N., & Kiziltan, Z. (2013). A bridge from Byzantium to modern day Istanbul: An overview of animal skeleton remains found during metro and Marmaray excavations. *Journal of The Faculty of Veterinary Medicine Istanbul University*, 39(1): 1–8.

Osborn, D. J., & Osbornová, J. (1998). *The mammals of ancient Egypt*. Warminster: Aris & Phillips: pp.121–123.

Osypinska, M., Skibniewski M., & Osypinski, P. (2021). Ancient pets. The health, diet and diversity of cats, dogs and monkeys from the Red Sea port of Berenice (Egypt) in the 1st-2nd centuries AD. *World Archaeology*, *52*(4): 639–653. doi: 10.1080/00438243.2020.1870545

Patera, M. (2006–2007). Gylou, démon et sorcière du monde byzantin au monde néo-grec. *Revue des études byzantines*, 64–65, 311–327.

Patera, M. (2014). *Figures grecques de l'épouvante de l'antiquité au présent. Peurs enfantines et adultes*. Leiden-Boston: Brill.

Pellizer, E. (1998). Figures de croquemitaines féminins en Grèce antique. In: Berlioz, J. & Alexandre-Bidon, D. (eds.): *Les croquemitaines. Faire peur et éduquer*. Grenoble: Jérôme Millon: pp. 141–151.

Perfetti, S. (2020). *Nature imperfette. Umano, subumano e animale nel pensiero di Alberto Magno*. Pisa: Edizioni ETS.

Poplin, F. (1986). Introduction aux animaux et aux végétaux du séjour des mortes. In: Duday, H. & Masset, C. (eds.): *Actes du colloque Anthropologie physique et Archéologie, méthodes d'étude des sepultures*. Toulouse, 4–6 November 1982. Paris: Éditions du CNRS: pp. 281–287.

Poulianos, A.N. (1972). The discovery of the first victim of Thera's Bronze Age eruption. *Archaeology*, 25(3): 229–230.

Prada, L. (2014). Translating monkeys between Demotic and Greek, or why a Lynx is not always a wild cat: (ΛΥΚΟ) ΛΥΓΞ=(WNŠ-) KWF. *Zeitschrift für Papyrologie und Epigraphik*, 189: 111–114.

Rolfe, W.D.I. (1985). William and John Hunter: breaking the Great Chain of Being. In: Bynum, W.F. & Porter, R. (eds.): *William Hunter and the Eighteenth-Century Medical World*. Cambridge: University Press: pp. 297–319.

Rolfe, W.D. & Grigson, C. (2006). Stubbs's "Drill and albino hamadryas baboon" in conjectural historical context. *Archives of Natural History*, 33: 18–41.

Rook, L., Mottura, A., & Gentili, S. (2001). Fossil Macaca remains from RDB quarry (Villafranca d'Asti, Italy): New data and overview. *Journal of Human Evolution*, 40: 187–202.

Rosendahl, W., Ambros, D., Hilpert, B., Hambach, U., Alt, K.W., Knipping, M., Reisch, L., & Kaulich, B. (2011). Neanderthals and Monkeys in the Würmian of Central Europe: The Middle Paleolithic Site of Hunas, Southern Germany. In: Conard, N.J. & Richter, J. (eds.): *Neanderthal Lifeways, Subsistence and Technology*. Vertebrate Paleobiology and Paleoanthropology Series. Dordrecht: Springer. https://doi.org/10.1007/978-94-007-0415-2_3

Ruslin, F.M.A., Azmi, M.A., Matsuda, I., Ami, Munir, R., & Md-Zain, B. (2017). Monkey school: Training phases for coconut-picking macaques (*Macaca nemestrina*). *Malayan Nature Journal*, 69(4): 301–306.

Sabbatini Tumolesi, P. (1980). *Gladiatorum paria. Annunci di spettacoli gladiatorii a Pompei*. Roma: Ed. Storia e Letteratura.

Sabriè, M. & Sabriè, R. (2011). La Maison au Grand Triclinium du Clos de la Lombarde de à Narbonne. *Archéologie et Histoire Romaine*, 19: 1–396.

Salari, L. (2006). *Mosaico Nilotico di Palestrina: animali reali o fantastici. Archeozoologia di una produzione artistica di età ellenistica. Circolo Culturale Prenestino "R. Simeoni"*. Palestrina.

Salari, L. (2015). Aristotle, the Nile mosaic of Palestrina and the *choiropithecos*. *Proceedings of the'8th Conference of the AIAZ, Lecce (I)*: pp. 193–199. doi: 10.1285/i9788883051487p193.

Schneider, J. (1992). The manshead archaeological society 1951–1991. *Bedfordshire Archaeology*, 20: 96–104.

Schreiber, H.D. & Löscher, M. (2011). The second find of a primate from the early Middle Pleistocene locality of Mauer (SW Germany): a molar of Macaca (Mammalia, Cercopithecidae). *Neues Jahrbuch für Geologie und Paläontologie - Abhandlungen*, 260(3): 297–304(8). https://doi.org/10.1127/0077-7749/2011/0137

Spatharakis, I, (2004). *The Illustrations of the Cynegetica in Venice. Codex Marcianus Graecus Z 139*. Leiden: Alexandros Press.

Stampolidis, N.C., Tassoulas, Y., & Filimonos-Tsopotou, M. (eds.). (2011). *Islands Off the Beaten Track...: An Archaeological Journey to the Greek Islands of Kastellorizo, Symi, Halki, Tilos, and Nisyros*. Museum of Cycladic Art.

Swanson, J. (1975). The myth of trans-Saharan trade during the Roman era. *The International Journal of African Historical Studies*, 8(4): 582–600. doi: 10.2307/216697

Thijssen, J.M.M.H. (1995). Reforging the Great chain of being: The medieval discussion of the human status of pygmies and its influence on Edward Tyson. In: Corbey, R. & Theunissen, B. (eds.): *Ape, Man, Apeman: Changing Views since 1600*. Leiden: Department of Prehistory of Leiden University: pp. 44–50.

Tinh, T. T. (1964). Vincent. *"Isis." In Lexicon iconographicum mythologiae classicae (LIMC) V, 1*, 761–796.

Toynbee, J.M.C. (1973). *Animals in Roman Life and Art*. London: Thames and Hudson.

Trantalidou, K. (2000). Animal bones and animal representations at Late Bronze Age Akrotiri. In: *The Wall-Paintings of Thera, International Symposium*. Athens: pp. 709–735.

Urbani, B., & Youlatos, D. (2020). On the earliest representations of chimpanzees (Pan troglodytes): Were African apes traded to Bronze Age Elam? *International Journal of Primatology*, *41*(4): 654–663.

Urbani, B. & Youlatos, D. (2022). Minoan monkeys: re-examining the archaeoprimatological evidence. In: Urbani, B., Youlatos, D., & Antczak, A.T. (eds.): *World Archaeoprimatology Interconnections of Humans and Nonhuman Primates in the Past*. Cambridge and New York: Cambridge University Press: pp. 225–279.

Veracini, C. (2020). A landscape without nonhuman primates? The case of the Barbary Macaque, *Macaca sylvanus*, (Linnaeus, 1758) and its interaction with humans throughout recorded time. *Humanities* 9(3): 92. https://doi.org/10.3390/h9030092

Veracini, C. & Hazevoet, C.J. (2024). *Homo sapiens e Chlorocebus sabaeus*, dois primatas exóticos em Cabo Verde: notas preliminares sobre suas interações e conflitos. In: F. Amante, M. Pignatelli, I. Rodrigues, & Sá, R.M. (eds.): *Latitudes em transcurso. Estudos em homenagem a Celeste Quintino por ocasião do seu 70º aniversário*. Lisboa: Colecção de Estudos Políticos e Sociais. Edições ISCSP: pp. 75–92.

Vernesi, C., Caramelli, D., Dupanloup, I. et al. (2004). The etruscans: A population-genetic study. *The American Journal of Human Genetics*, 74(4): 694–704.

Vespa, M. (2017a). A voice without a muse. Primates and the ancient phonosphere in Greek and Roman cultures. *Greek and Roman Musical Studies*, 5(2): 159–177.

Vespa, M. (2017b). Why avoid a monkey: The refusal of interaction in Galen's Epideixis. In: Fögen, T. & Thomas E. (eds.): *Interactions between Animals and Humans in Greco-Roman Antiquity*. Berlin and Boston: De Gruyter: pp: 434–449.

Vespa, M. (2021). *Geloion mimēma. Studi sulla rappresentazione culturale della scimmia nei testi greci e greco-romani*. Turnhout: Brepols.

Vespa, M. (2022). The place of nonhuman primates in ancient Roman culture. In Urbani et al. (eds.), *World Archaeoprimatology: Interconnections of Humans and Nonhuman Primates in the Past*, pp. 201–224.

Wiafe, E. D. (2019). Primates crop raiding situation on farmlands adjacent to South-West of Mole National Park, Ghana. *Ghana Journal of Agricultural Science*, *54*(2): 58–67.

Warmington, E.H. (1928). *The Commerce between the Roman Empire and India*. Cambridge.

Wendt, H. (1956). *Auf Noahs Spuren*. Halbleder: C.A. Koch's Verlag.

Wigoder, C.R.G. & Posner, R. (1971). *Encyclopedia judaica*, vols. 1–16. Jerusalem.

Wolfson, E. G. (2018). *Pictorial representations of monkeys and simianesque creatures in Greek art*. University of Missouri-Columbia.

Wortmann, E.F. (1883). *De comparationibus Plautinis et Terentianis ad animalia spectantibus* (Dissertation. Marburg).

Youlatos, D., Urbani, B., & Binnberg, J. (2023). A primate on a fresco from the Mycenaean acropolis of Tiryns. *European Journal of Archaeology*: 1–19. doi: 10.1017/eaa.2023.1

Zafiropoulos, C.A. (2001). *Ethics in Aesop's Fables: The Augustana Collection*. Leiden, Boston, Koln: Brill.

Section II

The Middle Ages and the Age of Exploration in Europe and in the Arab World

7 Non-human Primates in Medieval Europe

Cecilia Veracini
Universidade de Lisboa

The ape [...] is an untamed beast, and naturally malicious and is tamed and chastised by violence by beating, and with chains, and is restrained with an iron ball, so that he may not run about freely at his own will, to abate his fierceness and outrage

(*Bartholomaeus, De proprietatibus rerum, De simea,* c.1240).

7.1 THE PRESENCE OF NON-HUMAN PRIMATES IN EUROPE DURING THE MIDDLE AGES[1]

Since ancient times, seeking and collecting non-human animals has been a common practice among rulers all over the world (Loisel, 1912). The desire and passion for owning wild non-human animals (animals from here onward) from distant lands have always outweighed the practical difficulties of catching, transporting, and maintaining them (Kisling, 2000). The influx of exotic animals used for amphitheater games (Hughes, 2003; Mackinnon, 2006) diminished after the fall of the Roman Empire in the 5th century, and the political instability lasted until the formation of the Carolingian Empire by Charlemagne in the 9th century. The subsequent period of relative social calm resulted in an acceleration of trade (Le Goff, 1991), and exotic animals, including non-human primates (primates hereafter), began to be found more frequently in European medieval courts from the 11th century onward.

As in previous periods, in the Middle Ages charismatic carnivores (e.g., bears, wolves, and lynxes) and exotic creatures (e.g., lions, cheetahs, leopards, camels, elephants, zebras, giraffes, monkeys, and ostriches) symbolized the power of princes and rulers who displayed them to increase their influence (Loisel, 1912; Pluskowski, 2004; Salisbury, 2022). These animals, which represented the virtues related to the exercise of power, such as majesty, ferocity, and magnanimity, were displayed in parades and in more permanent venues; non-European animals (and also other exotic 'goods') were especially prized for these purposes (Simões, 2014: 524). Gifts and exchanges of animals among rulers were frequent, and such exchanges could involve a range of animals, of which primates appeared as particularly high-status gifts. For example, one or more monkeys figured in the well-known embassies sent by Harun al-Rashid, the seventh Abbassid Caliph (Viré, 1986), to Charlemagne in honor of his coronation in the 9th century (Nickel, 1991: 129). The first elephant to appear in Europa after the Roman period was also included. Menageries of exotic animals were common in the Middle Ages (e.g., see that of the King of England Henry I, 1100–1135), and some rulers, such as the Holy Roman Emperor Frederick II, 13th century, took their animals with them when they traveled (Loisel, 1912: 146). It was not just monarchs who collected animals; they were also collected by ecclesiastics, monks, aristocrats, the bourgeoisie, and even municipalities. During the Middle Ages, Italian merchants dominated the animal trade in the Mediterranean basin, trading in Syria, Palestine, North Africa, the Middle East, India, and China. The Venetian trade network stretched from the Mediterranean to the East, and Venetian merchants sold animals from these territories throughout Europe, including the main centers of trade such as Flanders and London (Howard, 2007; Molà, 2009). The activities of Venetian and other Italian traders resulted in African and Asiatic animals reaching many of the European cities. Spain's rulers enjoyed a privileged position because they could obtain exotic animals through their Arabian connections (Muñiz, 2000). The Islamic empire, which stretched, depending on the historical period and different caliphates, from Spain to the borders of China in inner Asia, created an economic network linking the maritime trade of the Mediterranean and the Indian Ocean (the largest hubs of economic activity in the medieval world). The lack of political barriers in the realm of the Islamic empire helped develop trade in goods and animals and small mammals, birds, and monkeys, could easily enter Europe at this time (Muñiz, 2000).

The interest shown by the European élite in primates is evident when we examine the inventories of contemporary menageries. Buquet (2013b: 101) found that among the 20 menageries in England, France, Italy, Germany, and Spain owned by emperors, kings, princes, and cities from the Carolingian era until the end of the 15th century, 11 included primates. Monkeys were expensive, high-status

[1] The term Middle Ages, is adopted here as encompassing the period between the fall of Rome in 476 CE until the end of the 15th century.

DOI: 10.1201/b21963-9

pets, quite common among rulers, aristocrats, and high prelates and, according to iconographic sources, they lived in houses and castles as pets as well as in menageries. Domestic monkeys were kept on leashes like dogs, or they were chained to walls, or their movements were restrained using a heavy block or bowl, like human prisoners (**Fig. 7.1**). Especially prized monkeys wore expensive collars. A collar made of red leather, with iron fittings and a brass strap, is mentioned in accounts from the court of Isabel of Bavaria, Queen of France, in the 15th century. The same monkey was provided with a "wooden bowl that turned within an iron circle with a long rope to hang on the collar of the monkeys", and Isabel ordered the monkey "a padded dress" made by a probably famous tailor, Jehan Le Lorrain (Chartier, 1858: 283 *apud* Buquet, 2013a). According to Walker-Meikle (2012: 24), in medieval society, monkeys were among the preferred species of pets, along with dogs, rare breeds of cats, birds, squirrels, ferrets, and rabbits. The use of these animals as pets and their exchange among wealthy people was defined by conventions of gender and status. The exchange of animals kept exclusively as pets was generally restricted to women and clergymen. For example, in 1413, the Duke of Burgundy gave four small monkeys to Isabella of Bavaria, Queen of France. In contrast, if a woman gave an animal to a secular man, the gift had to be one that reflected the 'right' animal for his gender, such as animals for hunting or exotic animals that symbolized the power of princes to keep a menagerie (Walker-Meikle, 2012: 25). Lions and other felids, such as cheetahs, used for hunting, were a frequent gift among rulers (Cuttler, 1991: 161). Among clerics, the use of monkeys seems to have been very common, such as in the case of the Bishop of Durham, who "kept in his court, after the custom of modern prelates, as some relief from their cares, a couple of monkeys—an old and a young one" (Maxwell, 1913: 37). The custom was criticized, with Hugh of Saint-Victor suggesting in his sermon that though the monkey was a detestable and ugly animal, "the clerics like to keep monkeys in their houses and to display them in their windows so as to impress the passing rabble with the glory of their possessions" (Ms. Latin 14934, f. 82, BnF *apud* Janson, 1952: 30, 57). The archbishop of Canterbury, Thomas Becket (c. 1119–1170), when travelling in France as Henry II's ambassador, had an entourage that included many horses and a menagerie with tame monkeys. A chronicler of the time referring to the clerics' habit of keeping monkeys described it as "the custom of modern prelates for occasionally dispelling their anxieties" (cf. Serpell, 1996: 48). Monkeys were kept on display in the cloisters of Notre Dame in Paris, although this was discouraged by the papal legate and it seems that due to the popularity of monkeys among citizens in Paris, a tax was imposed on every monkey that entered the city to be sold (Nijman & Healy, 2016).

FIGURE 7.1 Domestic monkeys were kept on leashes or they were chained to walls, or their movements were restrained using a heavy block or bowl. (a) Barbary macaque (cf. also Buquet, 2013a); (b) *Chlorocebus* sp. Details of *La Dame à la licorne, Le toucher*. Musée de Cluny Paris. Tapestry, end of the 15th century. (Images of the Public domain.)

Remains of monkeys and other exotic animals used as pets are occasionally discovered in the ruins of castles or other residences. The remains of a small Barbary macaque were found during the excavation of a stone tenement house in Southampton, in England, belonging to Richard of Southwick, burgess of the town (c. 1290) (Tomarken, 1990 *apud* Walker-Meikle, 2012: 31). Monkeys' remains are also found in other European countries, such as Russia, Germany, and Italy (cf. Walker Vadillo, 2013b).

Although these animals, given their rarity, were given close attention, their level of stress due to captivity and unsuitable nutrition is likely to have been very high. In the Greater London area, a large-scale study of animal remains from archaeological sites conducted by the Museum of London revealed fragments of the skull, particularly the maxilla, of a Barbary macaque from a site in Friars' street. Some of the cheek teeth show small pits, indicating enamel hypoplasia, a condition that may have arisen because of malnutrition and lack of exposure to the sun, such as vitamin A or D deficiency (Pipe, 1992: 190). There are examples of monkeys being fed expensive imported food; two monkeys kept by the Bishop of Durham (1274–83) were fed with "whitened almonds on a silver spoon" (cf. Maxwell, 1913: 32).

Primates, social animals *par excellence*, groom conspecifics in nature, and in captivity, they groomed other pets (cats and dogs) and even human heads when they were carried on their shoulders; as reported by the 13th century philosopher and naturalist Albertus Magnus "they collect vermin on heads and clothing and eat them" (*On animals:* 1534–1536). They also frequently hugged cats or dogs as substitutes for their need for contact with their conspecifics, as revealed by many iconographic sources. In medieval society, individuals took care of monkeys and trained them in exercises and games to entertain the public. In 13th century England, there are reports of "the Master of parrots and monkeys of the King" (Janson, 1952: 49; see **Fig. 7.2**). Monkey trainers and performing monkeys were common in the countries of Arabic influence in the Near East and North Africa since ancient times (Veracini & Alghamdi, this volume), and this tradition probably reached Europe as early as the 12th century, as shown by a bas-relief of a monkey and his trainer in Bayeux Cathedral in France (see Janson, 1952: 49).

FIGURE 7.2 A monkey and his master. (Sources: Yates Thompson MS. 'The Breviary of Renaud de Bar' or 'The Breviary of Marguerite de Bar'. (Images of the Public domain).)

Monkeys were less often used as pets on the Iberian Peninsula, Muñiz (2000: 267) arguing that because of the presence of Jewish and Muslim ethnicities, monkeys were not a favored animal. Monkeys enjoyed a bad reputation, as did pigs, among Muslims, and 'monkey' was the commonest insult offered to Jews (see below). The North-African Aghlabid dynasty, which controlled Western Libya, Tunisia, and Algeria, seems to have made the Jews wear a distinctive patch with a picture of a monkey on the shoulder of their garb. This followed the decree by Ibn Ţālib, judge of Qayrawān, that Jews should wear patches in the shape of monkeys and Christians in the shape of pigs, to distinguish them from true believers (Domínguez, 2006: 12). Chroniclers of Spain almost always refer to monkeys pejoratively. Almanzor was called by his enemies "marvelous monkey" (*quird ashab*), and Alfonso VI of Castile, the conqueror of Toledo, gave a monkey to his friend Yihya ben Abd Malik, who was proud of the animal, to indicate the latter's mental deficiency. Despite their bad reputation, monkeys were present in the daily life of Al-Andalus (the Iberian territory ruled by the Arabs in Medieval times), where it was common to see monkeys with puppeteers and snake charmers (Muñiz, 2000: 268). A study by Shemesh (2018), using a *responsum* of R. Shimon ben Zemah Duran, a Jewish halakhic adjudicator who lived in Rashbatz, Algeria, in the 15th century, shows that the trade in monkeys practiced by Algerian Jews was a common practice and that it had deep roots in the past. Trade in monkeys had been forbidden in ancient times by the Mishna sages, who were against trading in non-kosher animals, but R. Shimon ben Zemah Duran permitted monkeytrading, basing his ruling on ancient sources in rabbinical literature.

Before the 15th century, the monkey species found in Medieval Europe, which can be identified through iconographic sources, were mostly the same as those known during the Greco-Roman period.

FIGURE 7.3 The Barbary macaque was probably the best-known monkey in Europe during the Middle Ages. Here are shown two examples of the many representations found in this period. (a) Sculpture in wood, Choir of Ulm Cathedral, 15th century. (Photo Cecilia Veracini.) (b) *Symia*: Manuscript. *Der Naturen Bloeme* (The Flower of Nature), Jacob Van Maerlant, ca. 1350. p. 71 National Library of the Netherlands. (Image of the Public domain: https://publicdomainreview.org/collection/jacob-van-maerlant-der-naturen-bloeme.)

The Barbary Macaque (*Macaca sylvanus*), endemic to North Africa, was probably the best-known monkey in Europe during the high Middle Ages. It was depicted either in a naturalistic manner or in the form of a stylized model, often representing non-human primates as a whole, as is the case with allegorical images found in bestiaries, illuminated manuscripts, and other types of artwork (**Fig. 7.3**). Iconographic evidence of other species was scarce in the early Middle Ages but became more frequent from the 12th century onward. Evidence of the early importation of specimens of the genus *Chlorocebus* in Europe is found in southern Italy in the '*Porta dei Leoni*' of the Cathedral of Salerno (Campania), built at the end of the 11th century (**Fig. 7.4**). Janson (1952: 44) classifies the style of this monkey as Italo – Byzantine. Walker Wadillo (2013b) reported that remains of *Chlorocebus* sp. were found in another southern Italian locality (Otranto, Puglia). A specimen of the *Chlorocebus* cf. *aethiops*, together with a Barbary Macaque, is on one of the capitals of the Ducal Palace in Venice (1341–1350) (**Fig. 7.5**). These primates of the genus *Chlorocebus* probably originate from eastern Africa. Most could be grivet monkeys *C. aethiops*, but other species such as *Chlorocebus tantalus* and *Chlrocebus pygerythrus* may not be excluded from this list.

From the 14th century onwards, one can observe increasingly naturalistic representations of the genus *Chlorocebus*. Examples include (a) the famous studio of animals by Pisanello, '*Sturgeon and six monkeys*' (Museum of Louvre) (1430), where this animal appears in different positions, (b) the '*Journey of the Magi*' from the Dijon Museum (c. 1450–1460) (**Fig. 7.6**), (c) a miniature by Attavente Missale, at the end of the 15th century, in Lyon (Bibliothèque municipale, Ms. 512, f. 6v, cf. Buquet, 2013b), d) the famous tapestry cycle from the late 15th century called '*The Lady and the Unicorn*' now on display in the Musée de Cluny in Paris (**Fig. 7.7**) and two 14th-15th century manuscripts, defined as European arquetypes of the *Tacuinum Sanitatis* (cf. Urbani & Youlatos, 2023). A young specimen of *Chlorocebus* sp. is portrayed in the famous fresco by Gentile da Fabriano,

FIGURE 7.4 *Chlorocebus* sp. Lintel of the *'Porta dei Leoni'* of Salerno Cathedral, southern Italy, end of the 11th century. (Photo Claudia Massei.)

FIGURE 7.5 A specimen of *Chlorocebus* cf. *aethiops* together with a Barbary macaque, is a motive of one of the capitals of the Ducal Palace in Venice (1341–1350). (Photo Cecilia Veracini.)

FIGURE 7.6 *Chlorocebus* sp. in the *'Journey of the Magi'*. Dijon Museum (c. 1450–1460) (cf. also Janson, 1952). (Images of the Public domain.)

FIGURE 7.7 *Chlorocebus* cf. *aethiops*. From: *La Dame à la licorne. Le goût.* Musée de Cluny Paris. Tapestry, end of the 15th century. (Images of the Public domain.)

the *'Adoration of the Magi'* (1423), now at the Uffizi in Florence, where, together with a Barbary Macaque, it is depicted riding a horse. The Italian artist Giovannino de Grassi, in around 1390, included a depiction of a monkey in his famous sketches, where he portrayed both exotic and local animals. According to Groves (2008: 49), this portrayal is considered the earliest known image of the green monkey (*Chlorocebus sabaeus*), a species endemic to Western Africa. It is suggested that this species, and other cercopithecids, may have entered Europe with the onset of explorations in Western Africa during the late Middle Ages (cf. Veracini, Chapter 9, this volume). Baboons (*Papio anubis*), which occur in a vast area of Africa from Mali to Tanzania, are also found in a few iconographic records of the Late Middle Ages (15th c.). This species was portrayed in the *'Adoration of the Kings'* by Sandro Botticelli (now displayed in the National Gallery, London) dated between 1470 and 1475 (cf. Masseti & Bruner, 2009, Pl. 1).

From an examination of the available iconographic sources, it does not seem likely that Asian primates, such as the rhesus macaque (*Macaca mulatta*), for example, were frequently imported to Europe during the Middle Ages, although there is evidence that this species and other Asian monkeys were already traded during antiquity (cf. Masseti & Veracini, this volume). Anthropoid apes also do not seem to have reached medieval Europe, although the existence of these animals was known since antiquity (e.g., McDermott, 1938; Groves, 2008; Masseti & Veracini, this volume) and also in the Arab world (cf. Veracini & Alghamdi, this volume). The hypothesis of Urbani and Youlatus (2023), of a supposed arrival of chimpanzees in northern Italy in the late Middle Ages, based on an image found in a 15th century manuscript, does not seem supported by evidence (cf. Veracini, Chapter 9, this volume).

7.2 NON-HUMAN PRIMATES IN MEDIEVAL EUROPEAN CULTURE, FIGURATIVE ARTS, AND LITERATURE

Although not native to Europe, in the Medieval imagination, primates were common animals found in both literature and the figurative arts of the period, having various symbolic meanings. As already noted, African monkeys, and in particular the Barbary macaque, were the most familiar species in medieval Europe, and the latter is the most commonly represented monkey in the figurative arts and in popular and sacred literature. The Barbary macaque had the 'privilege', or heavy burden, of symbolically and allegorically representing all of the non-human primates known at that time. Monkeys played an emblematic role in Medieval culture, expressed in the literary and figurative arts with an impressive range of interpretations, ranging from the divine to the ridiculous, while also touching on many of the worst human vices and weaknesses. The importance of symbolic representation in medieval art was that it could convey messages even to illiterate people and thus evoke doctrinal content (Herrero Marcos, 2012: 9). Non-human animal representations were especially important because animal behavior was considered to be a critical allegorical system constructed by God to instruct men (Rogers & Holmes, 2018). The way of representing monkeys evolved along with the cultural and religious changes in Medieval society, especially after direct contact with these animals, which had become more common from the late Middle Ages onward. The symbology of primates and their iconography in figurative arts in the Middle Ages has been investigated by various authors who have produced important reference texts (e.g., Janson, 1952; Camille, 1992; Walker Vadillo, 2013a; Rogers & Holmes, 2018, among others). Below, we briefly summarize recent interpretations presented in this impressive artistic and literary corpus.

Christianity was the dominant religion in medieval European societies, and according to Christian doctrine, it was necessary to make a clear distinction between human beings created in the 'likeness of God', and all other creatures. Thus, monkeys, animals of an ambiguous nature and with a disturbing similarity to human beings, and already considered by the Greeks to be hypocritical and false creatures who tried in vain to imitate humans (cf. Masseti & Veracini, this volume), were the focus of this contrast. The association of monkeys with evil attributes, already present in classical texts, was emphasized in the early Christian period, when the Hellenistic texts were reworked in light of the Christian faith. The term *simia*[2] became pejorative and was applied to pagans, apostates, heretics, and infidels (McDermott, 1938: 63) to the point of being identified as, or associated with, the devil. As Janson (1952: 17–19) exhaustively explains, the monkey (identified with the Barbary Macaque that has a vestigial tail, thus being effectively tail-less) "had a beginning, but no end (a tail)", "the ape, not having a tail is without a species, and his rear, without a tail, is vile; like the devil, he does not have a good end". This symbolic representation is present in the *Physiologus graecus*, but not in other bestiaries (cf. Zambon et al., 2018). The shape and purpose of animals were determined by God, so it was against nature for any animal to be without a tail. The Lord, according to *Leviticus* XXII, 23, had declared the tail to be a necessary part of every animal, so animals without a tail were unfit for sacrifice. Only human beings were free to decide between good and evil, and whether their own ends were legitimately caudate or not. A Talmudic tradition says that even Adam was created with a tail, but it was removed by God (Janson, 1952: 19). Added to the idea of monkeys as evil beasts was the negative image many early Christian authors

[2] The Latin term "*simia*", with its plural form "*simiae*", refers here to non-human primates known during that time. Both medieval and classical authors used this term to denote not only the Barbary macaque specifically, but also primates in general. The use of the term ape is adopted to refer just to anthropoid apes (Family Hominidae, Subfamilies Homininae and Ponginae) and to gibbons (Family Hylobatidae). Where no specific reference is made to the original authors' words, the term "man" or "men" referring to mankind in general is reported as humans/human beings. To simplify the text, in some places primates is used in place of non-human primates. Non-human primates is only used when it is necessary to differentiate human primates from non-human primates.

had of Egypt, for it was both the land of darkness in the Old Testament, and a place where people worshiped plants and animals, with the worship of monkeys usually singled out as the most revolting of these practices. As Walker Vadillo points out (2013a: 70), Babi, the Egyptian sacred baboon and one of the gods of the underworld, could reinforce the association of primates with the underworld. During the anti-pagan riots at Alexandria in 391 CE, the Christians, led by Bishop Theophilus, destroyed the pagan temples, preserved a statue of a monkey, and set it up in a public place as a monument to heathen depravity (Socrates, *Historia ecclesiastica*, V. 16, 439 CE, *apud* Hopfner, 1922–1925: 658).

According to patristic literature (i.e., the works of the Church Fathers), one of the most important qualities of the devil is his incessant ambition to imitate God, which resulted in the devil being known as *Simia Dei* (cf. Rudwin & Rudwin, 1931: 120). The devil mimics the Creator, but with miserable and horrible results, hence the saying "*Simia Dei diabolus*", ("the devil is an imitator/the ape of God"), and this concept is found in many Christian authors, including Luther. This theme finds its expression in the art of the early Christian period until the 1000–1200 (cf. Walter Vadillo, 2013a), although, as Janson (1952) observes, there are very few artistic representations of the monkey-devil in early Medieval Romanesque art (10th–11th centuries) (see **Figs. 7.8** and **7.9**). The forbidden image of the monkey as a devilish figure persists in the textual tradition of bestiaries until 1300 and beyond. In the *Physiologus* (see **Box Details 7.1**), the devilish nature of this animal is emphasized: the naked ape with its dark, wrinkled skin, whose posterior parts are hideous (*posteriora eius magis turpia et horribilia sunt*), referring to the fact that the Barbary Macaque was not only despised for its repugnant appearance, but also for the treachery it shows to its offspring. This refers to the tale, reported by Aelian and Solinus (see later), of the monkey that has twins, but because she only loved one of her offspring, she let the other one die (Masseti & Veracini, this volume).

FIGURE 7.8 The monkey (*simia*) appears next to the satyr (upper right), representing the malevolent forces of Egypt, blocking the Jews' path to salvation (cf. Janson, 1952: Pl. I). Stuttgarter Psalter - Cod.bibl.fol.23, 191, 93r (IX), early Christian period 9th century. © 2023 Zeutschel GmbH, Tübingen | Bildrechte: Württembergische Landesbibliothek. (Image of the Public domain.)

FIGURE 7.9 Rare representations of monkeys symbolizing the Devil in the Romanisque art: (a) *Puerta de las Platerias.* Cathedral of Santiago de Compostela, beginning of the 12th century. Ape – Devil. (Image of the public domain.) (b) Baptismal Font of Master Roberto, Basilic of San Frediano, Lucca, Italy (1150–1173). Here the infant Moses presented to the Pharaoh's Daughter (cf. Janson, 1952: Pl. I). (Photo Cecilia Veracini.) Both the images represent a barbary macaque.

BOX DETAILS 7.1: SOME INFLUENTIAL TEXTS FOR MIDDLE AGES ZOOLOGY

Cecilia Veracini

Various literary works had a great influence on medieval zoology. These works gave origin to a form of medieval literature that developed, especially in the 12th century, that considered animals both as 'natural beings' and as allegorical creatures, i.e., entities that had spiritual and symbolic characteristics with intrinsic flaws and merits. The primate identification reported here is based on studies of the original sources and previous studies, among which there are: (McDermott, 1938; Groves, 2008; Vespa, 2022; cf. also Masseti & Veracini, this volume). One of them was the work of Solinus (about 210–after 258?), the *Collectanea rerum memorabilium*, which draws plentifully from the *Naturalis historia* of Pliny the Elder and from the *Chorographia* of Pomponio Mela, among other texts (cf. Masseti & Veracini, this volume; Vespa, 2022). Solinus writes down the strangest and most wonderful things inherent to people, customs, animals, and plants, illustrating them within a geographical framework. He enumerates five kinds of simians, *genus simiarum*, and identifies some of their characteristic traits:

> And the whole vast expanse that stretches between Egypt and the Ethiopians of Libya, and is dark with light, is full of various species of simians. [...] They like to imitate humans, so they can easily be caught by hunters pretending to smear their eyes with lime to induce the monkey to blind itself; they are overly affectionate with their offspring, carrying their favourite

young on their chest and having to abandon them when they jump, while the neglected ones manage to survive by clinging to their mother's neck [quoting the tale of the mother monkeys and twins from Aelian]; they rejoice when the moon is full and grieve when it is waning [quoting the *Hieroglyphica* of Horapollo that speaks of baboons]; the *cercopitheci*, who have tails but otherwise resemble those mentioned before [he probably intended the Barbary macaque]; the *cynocephali*, the most numerous of the monkeys in these parts of Ethiopia, are violent and bite fiercely and cannot be fully tamed; among simians we have the *sphinges*, very hairy, with prominent udders, docile and they learn easily and forget their native ferocity [*Theropithecus gelada*]; there the *satyros* with a good-looking face, restless in their movements [gibbons or langurs]; the *callithriches* look different from the rest in almost everything: they have beards and broad tails; they are not difficult to catch but rarely survive outside the land that gave birth to them [*Colobus guereza*]. (Solinus, *Collectanea Rerum Memorabilium*, 27: 55–60)

The *Physiologus*, a manuscript rooted in Hellenistic and Roman cultures and written in Greek circa the 2nd–3rd century by an unknown Egyptian, was fundamental in establishing the vision and perception of animals in the Middle Ages. It is a compilation of legends originating in the Indian, Egyptian, and Jewish traditions, to which are added Greek contributions from classical texts about the natural world. It has a mystic romantic tendency, and every naturalistic trait of the animals is transfigured into a religious allegory. Translations and adaptations from Latin introduced the *Physiologus* into almost all the languages of Western Europe. It retained its influence over ideas of the "meaning" of animals in Europe for over a thousand years. Each animal was described along with an anecdote, together with their moral and symbolic qualities. Medieval poetical literature is full of allusions to the *Physiologus*, and it also greatly influenced the symbolism of medieval ecclesiastical art and on medieval zoology, as well as being the predecessor of bestiaries (books of beasts). In the *Physiologus*, the monkey is a naked 'ape' with dark-brown, wrinkled skin, whose posteriors are hideous (*posteriora eius magis turpia et horribilia sunt*); its behavior has a connection with the phases of the moon; and its lack of a tail makes it resemble the devil. From the 12th century onwards, this was supplemented by a number of other tales suggestive of the new role of the monkey as the sinner, the victim of the devil rather than the devil himself. One tale, which became very popular, tells of a monkey and her twins (see also above Solinus). The twin on her back is hated, while the other, carried in front, is loved and hugged until it dies. The origin of the theme can be traced back to one of Aesop's fables from classical Greece, where its purpose was simply to demonstrate the unfortunate effect of excessive affection (cf. Walker Vadillo, 2013a: 64; Masseti & Veracini, this volume). In the text of Oppianus, *Cynegetica* the monkey kills his favorite son out of too much love while paradoxically saving the life of his unloved son by neglecting him (**Box Details 6.3**, Chapter 6, this volume).

Another important work was *Etymologiae*, produced in the 7th century by Isidore of Seville (Cartagena c. 560 – Seville, 636 CE), who based his research on the *Physiologus* and on the classical authors among Pliny and Solinus, but also on the *Hexameron* (VI, 2,4) of Ambrosius who interpreted, in his comments on biblical exegesis, animals as real species and not in an allegorical sense (Perfetti, 2012: 21). St. Isidore keeps out of the allegorization of natural entities and, regarded as primates, he incorporates the list of Solinus, stating that there are five *genera* covered by the term *simia*:

Cercopitheques, which have a tail are called *clura* by some [long tailed monkeys, guenons: as reported by Vespa, 2022 the zoonyms *clura* was taken from Paulus-Festus 48.11]; *sphinges*, are equipped with prominent nipples, docile, and learn to forget their original state [*T. gelada*]; *Cynocéphales* are similar to monkey but have doglike faces, whence their name [baboons]; *Satyrus* have a very agreeable shape, they are agile and gesticulate [gibbons or langurs]; *Callitriches* differ from the others in almost every way. In particular they have a long facial beard, and a wide tail [*C. guereza*]. (Isidore of Seville, *Etym.* Book XII, Ch. 30 – 33)

The first monkeys in Christian art were malevolent creatures, mostly depicted to look like the Barbary Macaque (e.g., **Fig. 7.10**). However, the Barbary Macaque/devil association did not persist (especially after the growing visual familiarity with the animal itself after the 11th century) probably because it was observed that these animals rarely exhibit devilish behavior. They are one of the few primate species to exhibit male parental care, and they appear amiable and friendly to each other, so in the late Middle Ages there was a shift from the 'monkey devil' to the 'monkey sinner', a sinful animal that is a symbol of human degeneration (Janson, 1952: 30). This view of the monkey as a symbol of sin persisted even after contact with the other primate species that began to arrive in Europe from the 12th century onward. Bernardus Silvestris, a philosopher and poet, explains this vision of the monkey in the *De Mundi Universitate sive Megacosmos et Microcosmos* (1170): "The monkey, a deformed image of man, come on to bring about human laughter" (Silvestris, 1978, *Cosmographia,* 225: 36). It represented, according to Bernardus, God's last effort before the creation of Adam; "if we let our animal impulses get the better of us, then we sink to the level of the ape" (Janson, 1952: 29). This vision of the monkey is reflected in two main artistic themes: the monkey on the road to sin and excess, and the punished and fettered monkey, and many representations on these themes portray the Barbary Macaque with human attributes.

FIGURE 7.10 The monkey and other demonic creatures symbolically dominated by the apostles standing above them. Cathedral *da Sé*, Evora (PT). *Porta do Sol*, sculptures of the Apostles (*Apostolado*), 14th century. Uncertain attribution of Mestre Pero. (Photo Cecilia Veracini.)

In the late Middle Ages, images of monkeys symbolizing sin and conspiracy were included in representations of Adam and Eve and their expulsion from paradise. Monkeys were placed at the edge of Eden, symbolizing the evil temptation that led Adam and Eve to sin, or were depicted on the tree of knowledge, or on the ground with half an apple or a walnut in their hand, symbolizing carnal pleasure, the real chain that binds human beings (Janson, 1952: 241). Nor did primates' apparent worship of fruit (e.g., apples) work in their favor, and the allegory of the monkey in the context of the earthly paradise is found up to the end of the 16th century. The Christian culture of this period regarded the Incarnation as the humiliation of God. The body was seen as the prison of the soul, with the horror of the body reaching its climax in its sex-related anatomy. Original Sin was interpreted as sexual sin, and the body of the woman was the place of election of the devil, according to St. Thomas Aquinas, St. Augustine, and Gregory the Great, some of the main medieval thinkers of Christian doctrine (Le Goff, 1985: 39). Eve, coming from the rib of man, was a secondary creation, not made in the image of God, and therefore of inferior substance, less rational more prone to succumbing to the serpent's temptations: in practice, she was a morally inferior human being. Thus, Thomas Aquinas echoes Paulus: "In a secondary sense, the image of God is found in man and not in woman: man is the beginning and end of woman, just as God is the beginning and end of every creature" (*Summa Theologiae,* I, 93, 4 ad.1 *apud* Warner, 2013: 179). The monkey becomes like the woman; 'by nature' an appendage of man, and in addition, they menstruate and are unstable and moody animals as Saint Hildegard noted (see later). The "woman" also used to be mocked like a monkey and compared to Eve.

In the early Middle Ages, many considered gestures to be suspect and theater was stopped because it welcomed those specializing in mime, who were regarded as the victims or servants of Satan. It was only a short step to associate mime with monkey since the latter had always been considered to be skilled at miming (cf. **Box Details 6.3**, Chapter 6, this volume). Moreover, the body was the incarnation of sin, and the idea of nature *vs* culture contrasted the savage animal world with 'man' whose role it was to dominate the animal world. This led to the second important theme, the symbol of the fettered monkey. This was evoked in the 13th century by the encyclopedist scholar Bartholomew Anglicus, who said that the monkey, which is a fierce and malicious beast, could be tamed thoroughly by violence using whips and chains. He added that wooden blocks (*trunco sive blocco*) were attached to monkeys (see **Fig. 7.1**) to prevent them from going anywhere they wanted and to suppress their insolence (*De proprietatibus rerum, Liber de animalibus, apud* Janson, 1952: 82). The image of the crouching monkey in chains made its appearance for the first time in the Romanesque capitals of churches of 11th–12th century northern Spain, such as those located in Loarre (Aragon), the Cathedral of Huesca, San Quirce, Silos, and Jaca (Olivares, 2012: 143–176), which influenced those of St. Gilles-du-Gard and others in France and Italy (**Fig. 7.11**). The chained monkey can be connected not only to Christian thought, but also to the Islamic Eschatological tradition that also relates the monkey with sinners and the devil—a tradition that could have been brought to the north of Spain by Muslim artists working in Christian territories (Monteira, 2005: 106 *apud* Walker Vadillo, 2013a). Even in this context, these figures are interpreted as symbols of vice, especially usury or lust. In addition, these fettered monkeys have also been identified with the figures of the damned in hell. In the late Middle Ages and during the

FIGURE 7.11 Two chained monkeys are eating fruit. Capital, *Torre di Pisa*, Italy, 12th–14th century. (Image of the public domain.)

Renaissance, monkeys were also often depicted fettered near women, or by the side of the Virgin Mary. In this case, they represented the repression of instincts and temptations in virtuous women, who, by chaining the monkey, managed to escape sin.

In the Gothic period, monkeys frequently appeared in manuscript marginalia, such as Books of Hours (devotional books). In these depictions, monkeys are parodying human beings and other animals, and they are monkeys dressed as Cardinals, Friars, or ordinary people. They are engaged in parties, battles or acrobatic activities or riding horses, or performing human customs, such as hunting, looking in a mirror, playing instruments, reading books, teaching, writing, and so on (cf. Janson, 1952; Camille, 1992; Walker Vadillo, 2013a, b among others; e.g., **Fig. 7.12**). Monkeys' actions are sometimes very realistic, and their behavior clearly seems to represent all the aspects of human behavior, with its vices and virtues (Rogers & Holmes, 2018). Many of the marginal illustrations in medieval manuscripts may represent traces of lost proverbs or beast fables, or they could be examples of the real behavior of Barbary Macaques or other monkeys (Domínguez, 1996: 6). Some depictions of monkeys hugging domestic animals probably reflect real behaviors observed in captive monkeys who used a cat or a dog to compensate for being unable to have contact with their own species. In Janson's (1952) interpretation, monkeys appear at the margins of full-page illuminations of devotional books, where their drolleries, including playing musical instruments such as lyres and flutes, contrast with the peace and serenity of the main characters at the center of the illuminations. According to Rogers and Holmes (2018), "in their mimicry of humanity, primates are a debased version (to the medieval mind) of humanity's imitation of the divine mind, including our imitative creative capacity in text and bookmaking." Our creative work might be laughable or evil, with a pretension to divine knowledge, including "our uniquely human knowledge of book-learning made corrupt by our base animalism". In these *marginalia* "monkeys mock our supposed 'special status' in the scale of nature pulling us down to the level of animals". In the end, monkeys are intermediaries and hybrid creatures between us and God, as we are.

FIGURE 7.12 Allegorical monkeys in combat. Marginalia. Manuscript, Royal MS 14 E IV f121r. c 1470–c 1480, British Library. (Image of the Public domain.)

7.3 DEGENERATE HUMANITY AND ETHNIC DISCRIMINATION IN THE MONOTHEISTIC RELIGIONS

The idea of non-human primates being degenerate and inferior versions of the human condition has ancient roots in Western tradition. It can be discerned in ancient Greek and Jewish myths and cosmologies, which suggest that if the supreme laws of a deity are not respected, people are turned into apes/monkeys. The collection of ancient Rabbinic writings, dating from around the 3rd century BCE, tells the story of how God punished the people who built the Tower of Babel in order to wage war by turning them into monkeys (*qofim*) and impure spirits (*Gemara,* Babylonian Talmud, Sanhedrin, f. 109a.27). Noegel (2019) argues that the concept of primates as fallen humans was adopted by Jewish culture after the exodus, when the Jewish people began to live a more urban life (changing many of their ancient traditions), and this was the opposite of the various animal cults of ancient Egypt that included the worship of monkeys. Simians were considered impure. The Sifra (51:4), following Leviticus 11:27, lists monkeys among forbidden foods. Initially, wild animals, as in other religions of the ancient Middle East and Mesopotamia, were considered sacred and were placed between divinities and human beings. The Israelites perceived domesticated animals as unintelligent beasts subject to human control, but wild animals were embodiments of divine wisdom and were subject only to Yahweh. The most convincing biblical evidence for the superiority of humans over all animals comes from the period of exile or later (Noegel, 2019). The ancient Hebrew name for primates was *Kof* (*qof,* plural *qofim*) (monkeys and apes, from the Sanscrit *kapi*), which referred to the animals Solomon imported from the Kindom of Saba (Bible, Book I Kings 10:22. II Chron. 9:21). As in the ancient Arabic world, primates were used to keep houses free of creeping things, and they were even trained for domestic use (cf. Wigoder & Posner, 1971, vol 2; 176). In the Mishnah (part of the Talmud) (Kil. 8:5), there is a discussion about whether the laws of ritual uncleanness applying to a human corpse also apply to creatures called *adonei ha-sadeh* (the lords of the field), which some scholars have identified with chimpanzees (cf. Wigoder & Posner, 1971, vol. 2: 176).

In the Greco-Roman world, the idea of the punitive transformation of the Cercopes into monkeys (Ovid, 8 CE, *Metamorphosis*, cf. Masseti & Veracini, this volume) is a late addition to the original myth. It was probably grafted on during the Hellenistic period, when Greek civilization, expanding throughout the Near East, became increasingly susceptible to Oriental influences. In the Christian context, this theme also relates to the biblical story of Adam's creation and the Fall (Janson, 1952: 90–99). The inferiority of primates is the confirmation of the Christian theological doctrine: as a consequence of his own act of disobedience, Adam suddenly found his animal instincts revolting against the control of mind, and the eating of the apple represented above all the victory of sensual appetite over law. The main body of late medieval tales explains the origin of non-human primates, following the pattern of earlier 'punitive transformation' legends (Janson, 1952: 90–99). The myth of the punitive transformation is also found in the Islamic tradition. Monkeys were mentioned in the Qu'ran three times (II, 61/65; VII, 166; V, 65/60), when God (Allah) turned some people into monkeys and pigs as a punishment for disobeying His commandments (cf. Veracini & Alghamdi, this volume). This theme is present in the myths of regions where primates are endemic (cf. the chapters of this volume about Asia, Africa, and the Americas). In many traditional cosmologies, the justification for the existence of animals so similar to us is that they are human beings transformed by a divine act following some sort of transgression. Unlike in Western cultures, these creatures are seen as at the product of a 'second' creation, and instead of being denigrated, they are treated as sacred.

Considering primates sinful and symbols of a degraded and inferior humanity meant that these animals were often identified with enemy populations or with ethnic groups that were despised and deemed to be inferior. This occurred in various social and religious contexts. According to medieval Christian thought, those who were abandoned by God could show signs of Satan's work, whether they were poor, sick (e.g., lepers), or heretics; many of these categories were associated with monkeys. Other animals (e.g., pigs, donkeys, spiders, etc.,) were considered impure, but because they looked like 'deformed human beings, monkeys lent themselves to these comparisons. The idea that monkeys were laughable and ridiculous imitations of humankind dates back to Heraclitus, whose claim that "the most beautiful of apes is ugly compared to man" and that "the wisest of men is an ape next to God" was transmitted to the early Christians through Plato (*Hippias maior*, 289 A–B) and Plotinus (*Enneads*, VI, III, 11) (cf. McDermott, 1938: 144, note 187). The negative view of primates has its origins in Judaic literature, where we find the concept that "an ape is to man, as man is to God." (Jacobs, 1901: 663, cf. McDermott, 1938). The enemies of Christ were compared to monkeys and their actions to ridiculous 'monkey business'. In the 4th century CE, the last pagan Roman emperor Julian (*Flavius Claudius Iulianus*), called the Apostate, sought to revive the dying pagan worship by imitating the Christian church. Julian conceived of a complete re-organization of the pagan church, causing Gregory Nazianzen, Archbishop of Constantinople and theologian, to sarcastically describe this imitation as "the mimicry of apes" and "parodying as if on the stage" (*Contra Iulianum*, First Invective 113, cf. McKenzie, 1959: 182).

In the Islamic world, primates were frequently associated with Jews. It was the *Tafsir* (exegesis or commentary) by al-Ṭabarī (838–923 CE), the traditional and oldest gloss of the Kur'ān, which made the connection between monkeys and Jews explicit (Domínguez, 1996). According to Domínguez (1996: 12), the monkey "never lost its association with the Jews, even as it widened its application to cover the spread of Judeo-Christian and other heresies in Islamic societies". This concept is well illustrated by a work, which, although later than the Middle Ages, clearly represents this vision. The *Carajicomedia*, an anonymous parodic poem published in 1519, tells the story of Diego Fajardo, an old man who has lost his sexual potency and seeks to have it restored. The fictional author who describes this is called Fray Bugeo Montesino. According to Domínguez (2006: 1), the name Bugeo derives from *Bugio/bugia* and means monkey (this name was used by the Spanish and Portuguese in the Modern Period to designate primates and originated in North Africa, where there exists a city called Bujia, an important trade center in Algeria since ancient times, see Glossary). Here, this term is used in a derogatory way to indicate that Fray Bugeo was probably a Jewish convert to Christianity and therefore allegorically described as a monkey. At that time, Spanish society was composed of people of three different confessional faiths, all of whom had particular ways of characterizing persons belonging to a different religion. The Arabs considered the monkey to be a symbol of the Jews, and as Domínguez states, probably the author of Carajicomedia knew this. The name of the fictional author of Carajicomedia, Fray "Bugeo" is intended to recall the Franciscan involvement in the African campaign and, more specifically, the taking of the city/kingdom of Bujía (moreover, north Africa was also known in antiquity for the worship of monkeys, cf. Veracini, 2020).

The conception and birth of simians to women who indulged in illicit unions with these animals or who were punished for other transgressions is another frequent theme. Their punishment was to give birth to fabulous hybrid creatures. The idea of interbreeding between humans and other animals was frequent in the Greek (e.g., Ovid, *The Metamorphoses*; Apuleius, *The Golden Ass*) and Arabic traditions, and we also find this motive in Christian legends widespread in medieval Europe (e.g., *Arabian Nights*, where the princess gives birth to a monkey after her relationship with a baboon). Janson (1952: 268) recalls a specific tale about a monkey/human hybrid recorded in the 11th century by the Italian St. Peter Damian, a Benedictine monk, Cardinal and Doctor of the Church, who in a pamphlet (LII *De bono religiosi status et variorum animantium tropologia*, 1503, cf. Capecelatro, 1862: 578) narrates the properties of animals that are useful to know for Christian life. The example tells the story of Count Guglielmus of Liguria, who is said to have owned a monkey that had sexual relations with his wife. One day, when the animal saw the count with his wife, it attacked and killed him in a jealous rage. Damian adds that Pope Alexander II kept the offspring of the monkey and the woman, a child called *Maimo* (the Arabian name for monkey, see Glossary), in a cage.

7.4 NATURAL HISTORY OF PRIMATES IN THE MIDDLE AGES (UNTIL THE 14TH CENTURY)

The scientific study of primates (human and non-human) in Europe began in ancient Greece with Aristotle (*Historia animalium*) and continued with Galen's anatomical studies (*De anatomicis administrationibus*) in the 1st century CE (cf. Masseti & Veracini, this volume). It also developed in the Arabic world, following the translation of Greek texts in the 9th century (cf. Veracini & Alghamdi, this volume). In the Latin Medieval world, few original contributions regarding the natural history of primates were produced until the 11th and 12th centuries.

To understand the natural history of primates in the Medieval period, we need to understand zoology as a whole at this time. In the early centuries of the current epoch, zoology, as well as the other natural disciplines, adapted to the dominant Christian way of thought which, after having spread slowly for three centuries throughout the Roman Empire, was declared a state religion by Emperor Theodosius in 382 CE. For many early Christian scholars, Christianity alone was thought to be "the sublime philosophy", although many recognized that Greek philosophy and more generally pagan knowledge, could still play a role, especially in civil life (Grant, 2002: 2). By the end of the 2nd century CE and int the 3rd century, many Christian apologists claimed that Greek philosophy was useful if domesticated by Christianity as the 'servant of theology', which was the sole, supreme science. In relation to texts dealing with animals during the Roman period, many compilations of Greek authors, already in use in the previous Hellenistic world, had become commonplace. They tended to simplify the ancient texts, making science more accessible, and along with other works of popular literature, they were translated into Latin for the parts of the Western Roman Empire where Greek was no longer spoken (Grant, 2002). This scientific simplification transmitted to early Christian scholars, together with the lack of first-hand experience with primates at least until the 11th and 12th centuries, contributed to the dissemination of simple and anecdotal notions about simians, which greatly influenced related scholarship. This approach will change around the 12th century, when Aristotle's *corpus* was 're-discovered' (see below) and began what Oelze calls the "late-medieval precursor of the modern animal turn because animals became the subject of various debates not only in zoology but also in psychology and philosophy" (2018, XII, Preface).

Alongside knowledge of the nature of animals conveyed by ancient and late antique works, "a veritable 'sacred zoology' developed, ready to translate every aspect of the animal, even fictitious animals, into allegories laden with biblical, spiritual, and moral resonances" (Perfetti, 2012: 11). The progenitor of this trend was the *Physiologus* (see **Box Details 7.1**), a collection of heterogeneous information about animals that makes invisible theological truth concrete, a perception that contrasts with ancient fables where animals embodied the dilemmas and ethical problems recurring in human experience (Perfetti, 2012: 15). This text had a great influence on medieval culture and gave rise to bestiaries (books of animals), treatises in which animal features symbolized central aspects of the Christian faith. In the *Physiologus*, and in the subsequent medieval bestiaries, the symbolic representation of primates (called the *simia* and mainly depicted as the Barbary macaque) was simple and stereotyped, recalling Greco-Roman anecdotes and tales that would have much resonance in the medieval natural history of primates.

Other 'theological' animals are encountered in biblical exegesis (the study and meaning of biblical texts, cf. Ciccarese, 1990) in the commentaries on the six days of creation, where the nature of the biblical animals is illustrated. As noted by Perfetti (2012: 18–23), at this stage, animals begin to be interpreted according to two schools of thought, which in varying degrees recur in the later medieval philosophical-theological literature dealing with primates. For authors of the Alexandrian tradition (e.g., in Philo of Alexandria, c.20 BCE–40 CE), the biblical animals constituted a representation of the passions of the soul that can both help and harm us, and they became exclusively symbolic elements unrelated to reality (Perfetti, 2012: 18). By contrast, in the naturalistic approach (e.g., the work Basil of Caesarea 4th century CE, cf. Basilius 1990), they are no longer allegorical elements, but are often described according to Aristotle's zoological *corpus*. With Basil's work, a continuity emerges between the original biblical natural *datum* and everyday observation of nature, although it should be noted that the detailed consideration of creation was always aimed at "the edification and perfecting of our souls" (Perfetti, 2012: 22) and everything else in the world, "was taken to be a sign of God's intelligence" (Oelze, 2018: 21).

Bearing these premises in mind, it is easy to understand why one of the aspects that characterized primate studies in the Middle Ages was the attempt to establish their 'true' nature as well as that of the 'anthropoid' creatures and hybrids that were taken for granted to exist in the Greco-Roman world (cf. Masseti & Veracini, this volume). "Since Christian thought was based on the scriptural revelation of one all-powerful Divine Creature, who had made humans in His own image, it became relevant to find criteria to distinguish humans from these other beings" (Janson, 1952: 74). Monsters, hybrids, and other strange creatures were considered part of nature by common people and by classical authorities such as Pliny and Aelian. Folk creatures, hybrid and metamorphosed forms, abounded in this vision of the natural world: they lived in distant countries or in the dense forests of Europe. Wild men, satyrs, dog-headed, pygmies (dwarf beings), and other anthropomorphic creatures, half monkey/ape-half human, populated the real world together with people. Early patristic literature tried to resolve the problem of their nature, declaring satyrs, centaurs, and their like, to be demons, along with the entire pantheon of strange pagan creatures, but monkeys, pygmies, and the monstrous races mentioned by Pliny (**Fig. 7.13**)

FIGURE 7.13 Head-less, one of the Plinian races. Manuscript: National Library of Netherlands, *Der Naturen Bloeme* (The Flower of Nature), Jacob Van Maerlant, ca. 1350. p. 44, https://galerij.kb.nl/kb.html#/en/dernaturenbloeme/page/11/zoom/2/lat/-61.522694945983595/lng/3.1640625. (Image of public domain: https://publicdomainreview.org/collection/jacob-van-maerlant-der-naturen-bloeme.)

presented other problems. In the *City of God* (413–426, *De civitate Dei contra Paganos*), the work of Saint Augustine (354–430), the major Christian thinker of the first millennium (cf. Livi, 2004: Vol. 1, 242), Augustine reinforced the difference between humans and demons, claiming it was mortality that distinguished humans from these creatures whom he considered evil, envious, and cunning (*De civitate Dei,* B. 8: 22). The other freakish or abnormal creatures, such as pygmies and the monstrous races mentioned by Pliny, were to Augustine sons of Adams, like normal humans (*De civitate Dei,* B. 16: 8). He claimed physical resemblance was an irrelevant criterion, and only their power of reason (*ratio*) set humans apart from the rest of the animal world (Janson, 1952: 75). Thus, Augustine differentiated these 'rational creatures' from the non-human primates, considered brutes ('irrational beasts') such as the other animals. Other early Christian authorities, such as Isidore of Seville (c.560–636), thought the freakish Plinian creatures of the Far East existed in those distant and marvellous regions (*Etymologia,* Book XII), but he included satyrs (which could be described by classical authors as simian species – perhaps gibbons or langurs, cf. also Vespa, 2022) among simians in line with classical authors (cf. Masseti & Veracini, this volume). Isidore's natural history of primates, which included five kinds of animals covered by the term *simia*, closely follows the classification of Solinus (*Collectanea*, 27.55–60, cf. **Box Details 7.1**). However, the *cynocephali* were described as similar to monkeys but with canine heads, casting doubt on whether the *cynocephali* were a genus *simiarum* and not instead a fabulous race of dog-headed people (*cynocephali*), as described by Pliny. Isidore's statements created a good deal of confusion in establishing the difference between non-human primates, humans, and human-like animals. For example, Saint Christopher, a 3rd century saint often represented in the

FIGURE 7.14 A mythical medieval satyr, here represented as hybrid creature. Manuscript. *Der Naturen Bloeme* (The Flower of Nature), Jacob Van Maerlant, ca. 1350. P. 70, National Library of Netherlands. https://galerij.kb.nl/kb.html#/en/dernaturenbloeme/page/11/zoom/2/lat/-61.522694945983595/lng/3.1640625. (Image of the Public domain: https://publicdomainreview.org/collection/jacob-van-maerlant-der-naturen-bloeme.)

form of *cynocephalus*, was a normal human, though distinct from real baboons. The belief that the saint was one of the mythical dog-heads spread in Byzantium from the 5th century onward, even though the saint's depiction as a dog-head was not dominant in Byzantine art, because the Byzantine Church disapproved of linking one of its saints with cynocephals. Isidore's medieval successors tended to equate these two kinds of *cynocephali* (human and simian ones) plus the two forms of *satyri*, real apes and mythical goat-footed creatures (on the representation of the word 'cynocephalus' by the Greeks of the classical era see also Vespa, 2019; **Fig. 7.14**).

One of Isidore's statements confused the matter still further:

> the name *simia* is Greek and they have flat noses and hideous faces covered with frightful wrinkles, just like goats which also have flat noses. Others maintain that the name *simiae* is Latin, because we can find great similarity with human mental faculties (*multa similitudo rationis humana*), but this is wrong. (Isidore of Seville, *Etym.* Book XII, Ch. 30 – 33)

This concept of similarity between *simiae* and humans, though criticized by Isidore, would have a great impact on medieval thought. It was reported and amplified in other texts, resulting in ideas and comments on the comparative psychology of monkeys and humans to help resolve their relationship (Janson, 1952: 77; cf. Oelze, 2018).

The stories about monkeys that Isidore handed down directly from Solinus (in turn drawn on classical literature, see **Box Details 7.1**) become stereotypical descriptions of

FIGURE 7.15 The mother monkey with her twins chased by a dog. British library. South-eastern England? (London?), second quarter of the 14th century, *Yates Thompson 13, f. 184v.* (Image of the Public domain.)

the behavior of monkeys, greatly influencing popular literature, bestiaries, and art. They included (a) their relationship with the moon: "the new moon delights them, the full moon and waning moon make them sad"; (b) the relationship of the mother monkey with their babies: the mother monkey bears two twins, but she loves one, which she carries in front of her, and dislikes the other, which she bears on her back (**Fig. 7.15**); and (c) the capacity of monkeys for mimic, by which hunters could easily catch monkeys that tried to imitate them.

The question of the *similitudo hominis* and the mental skills of primates was addressed by various medieval authorities throughout the entire Middle Ages, and the changes in the intellectual environment of Europe resulted in a gradual shift toward a naturalistic answer to this question. From the 11th century onward, the cessation of the invasions that had lasted since the fall of the Roman Empire resulted in relative social calm that encouraged the revival of commerce in Europe (Le Goff, 2007: 153). Also, during this period, scholars translated Greek and Greek/Arabic manuscripts, many of which had entered the Christian world after the fall of Toledo in 1085, Spain, and the conquest of Sicily in 1091 (Grant, 2002: 200–201). With these translations, which were mainly made in Spain and Sicily, Europe came into contact with ancient Arabic knowledge and with the original Greek texts. The *corpus* of Aristotle's work (of which we recall the important Latin translation of Liber de *animalibus,* which included the *Historia, De partibus*, and *De generatione animalium* - by the Scottish scholar Michael Scot, c.1175–c.1236, but also many Arabic translators and commentators of his work, such as those of the Andalusian polymath Averroes, 1126–1198) was to become dominant in philosophy and animal studies from the 12th century onward (Perfetti, 2012: 47).

Research on the nature and characteristics of primates began to flourish in this period under the influence of the original Aristotelian texts that emphasized morphological and anatomical characters instead of the previous emphasis on folkloric, allegorical, or symbolic aspects. Aristotle's biological treatises provided an alternative to the patristic and early medieval philosophical-theological culture (Perfetti, 2012: 53). The nature and origin of primates and their uncanny resemblance to humans were addressed in various ways. Interpretations of primates' nature oscillated between naturalism and symbolism, but all approaches drew a conceptual and/or biological boundary between human and non-human primates, even if, from anatomical and cognitive (e.g., Albert the Great) points of view, a clear similarity between us and the other primates, as well as their complex behavior, could not be denied. Some issues were of utmost importance: (a) whether these animals are rational creatures (in Aristotle's view, rationality is what sets humans apart from all other animals). In particular, the philosophers of the 13th and 14th centuries discussed what it means to be rational or, more precisely, what kinds of behaviors require rational cognitive processes, such as concept formation or reasoning (Oelze, 2018: 48–49); (b) whether or not they have a soul (according to Aristotle, every psychological process requires a particular type of soul or faculty of the soul). Animals have sensory souls; thus, one of the questions was: is there a difference between the sensory souls of humans and the sensory souls of non-human animals?; and (c) whether they have the ability to speak (skill that is directly linked to complex intellectual faculties).

The German Benedictine abbess, visionary, mystic, and composer, Saint Hildegard (also called Hildegard of the Rhine or Hildegarde of Bingen) (1098–1179), deserves special mention because she drew attention to the physical nature of monkeys and their similarity to women with regard to menstruation. In *Liber simplicis medicinae*, commonly known as *Physica* (a part of the *Liber subtilitatum diversarum naturarum creaturarum*, composed between 1151 and 1158), she observes that:

> The *simia* is warm because it is similar to man, behaves similarly to him, but also has something of the nature of wild beasts, but is incomplete both as a beast and as a man. Because of its humanity, it menstruates following the lunar month. It is not useful in medicine. (*Physica*, Ch. 24: 1865–1866)

Pliny had taken the opposite view, saying that "among the whole range of animated beings, the human female is the only one that has the monthly discharge" (Pliny VII, 54). Before St. Hildergard, monkey menstruation had only been noted by the Egyptians, whose priests lived in close contact with sacred baboons in temples. Horapollo had observed that baboons depended on the phases of the moon and at a particular lunar period, 'the female bleeds from her private parts' (*Hieroglyphica*, I, 14, *apud* McDermott, 1938: 43). St. Hildegard described the *cercopithecus* as cold and hot, having "something of the nature of the wolf and the cat. It contains a toxin, which it spits out and hides under the earth like its excrement, so that worms and snakes do not grow from it" (Ch. 25, 1865–1866). As Janson (1952: 77) noted, the abbess was influenced by Hermeticism and Gnosticism (Greek-Egyptian religions of the Ptolemaic and Roman periods), part of medieval culture, whereby the nature of an animal could be identified with its mood, referring to the Galenic theory of Humors (chemical systems regulating human behavior). The monkey's simultaneously warm and cold constitution makes its nature unstable and weak. Early Christian cosmological writings applied this theory to the 'fall of man': man's natural stability after sin gave rise to instability, bodily illness, and melancholy.

The English scholar, theologian and abbot, Alexander Neckam (1157–1217), presented what was known about primates at the time in *De naturis rerum*, the most important encyclopedic work at the turn of the 13th century (cf. Hunt, 1984). Neckham's compilation was based on comparisons of ancient and contemporary texts (cf. Gauvin et al., 2013). Neckam mostly openly argues against the fabulous zoology of *Physiologus*, instead proposing a naturalistic interpretation of the physical universe (Perfetti, 2012: 27). Nevertheless, with regards to primates, he adopts a more theological framework by defining the nature of primates in moralistic terms, suggesting that their "evil disposition and lack of *ratio*" demonstrated their inferiority to humans (Janson, 1952: 79). As well as citing secondary sources on primates, Neckham reported interesting facts about the monkeys he likely saw at the English court where he was raised. He writes that at court there were "noble animals such as birds, but also monkeys and bears", and he tells of a monkey who, having had a baby, as it was her habit, showed it off joyfully; but the baby was mauled by a bear tied to the same pole. The mother's sad face showed her anxiety, and she murmured instead of using words. Because of this, the local lord punished the bear by instructing the servants, happy to avenge the monkey's death, to kill it. However, Neckam concludes by saying that monkeys, though very clever, naturally cannot have a *ratio* equal to that of humans (*De naturis rerum*, 1858: 207–211).

Bartholomew Anglicus (before 1203–1272), a scholar in Paris and a member of the Franciscan order, took a different approach to primates. His work, *De proprietatibus rerum*, draws on Pliny, Solinus, and Isidore of Seville, as well as Aristotle and Avicenna, and stresses the morphological features of monkeys and compares them to human beings "it has some likeness of mankind", while recalling that traditionally, the *simia* is a "naturally malicious" beast that needs to be fettered "to abate his fierceness and outrage". The tension between the naturalistic and moralistic visions of the author is clear in the following excerpt:

> The Ape [*simia*] has the shape of man and the hair of a wolf and some apes have evil manners and thatches of hair and their teeth be as it were hounds teeth, and give malicious biting, and namely those that have tails; and some be rough and all hairy before, except the face, and such have teeth as a man, and have other things as a man, and reddish eyen and sharp, and paps and teats, in the breast, and hands, feet, and fingers, and toes, and may go and steppe on two feet…[…], as *Aristotle* said. And the female Ape is like to a woman in the private parts, and the males yard is like to an hounds yard, and his entrails be like to a mans entrails. […] The Ape is a beast wonderfully shaped, but it has some likeness of mankind, and is learned and taught, and so he is taught to leap and play in divers manner wise, and is an untamed beast, and naturally malicious and is tamed and chastised by violence by beating, and with chains, and is refrained with an iron ball, so that he may not run about freely at his own will, to abate his fierceness and outrage. The Ape eats with safety all manner of meats & unclean things, and therefore he seeks and looks at worms in men's heads, and throws them into his mouth, and eats them. (Bartholomaeus, *De simea,* Ch. 96, Trevisan (ed) 1538)

Bartholomew Anglicus refers to the five kinds of primates reported by Isidore of Seville, but he changes the spelling of three names (*clura* instead of *cluna*; *Cenophe* instead of *cynocéphales*; *Calatrices* instead of *callitriches*).

The need to reconcile opposing views of primates is evident also in the encyclopedic work *De natura rerum* of the Dominican Thomas of Cantimpré (1201–1272), whose natural history of animals brought together all the written sources of knowledge. *De natura rerum* explains both nature and the Bible, that is, the will of the Creator (Cipriani, 2014). The chapter devoted to primates is in the part of the text dedicated to quadrupeds (Ch. XCVI. *De syimia et de diversi generibus symiarum, De quadrupedibus:* IV,1: 106). Discussing the anatomy of monkeys, he notes that all quadrupeds, whose souls are based on the cardinal points, have a correct end-part, with a tail: *"quia anima fundata in quator virtibus cardinalibus bono fine concluditur"*. Only the monkey, the bear, and man are without them, with man having hips and buttocks where other animals have a tail; the same applies to monkeys and bears, which have similar hips. A comparable comparison would be repeated many times in the "*Anatomia, De anchis*" where Aristotle is cited as an authority. Thomas lists various characteristics that monkeys share with us:

> they have vulva like females and penises like dogs. Their legs and feet have the same joints as in man. They have heels on their feet, and for this reason they can stand up like a man and run when they want.

But in conclusion, Thomas comments on Aristotle's discourse by writing:

> although the external aspect is similar to man, the internal one is like other animals. Inside they have nothing of what appears, because they have a hypocritical soul and full of malice and of all the deceptions, but cannot be calmed except by human coercion, because of their nature.

Monkeys do not reach the perfection of human beings, the only creature who "lifts up his face to heaven, so that he may clearly perceive the source of his salvation". The general concept that emerges from Thomas' characterization of the *simia* is that they are hypocrites, who imitate humans without succeeding, essentially using Aristotle's assumptions to confirm the established Christian view of the monkey as a "degenerate man" (Janson, 1952: 81). Thomas quotes and confuses well-known anecdotal facts about monkeys taken from Pliny, Solinus, and Isidore of Seville:

> monkeys with tails are called *cicholiticas*, while there are also other ones, tailed, with a more handsome face and with a beard; these genera live in Ethiopia and are called *galetrices*, as Isidore says [probably the white and black colobus]. (*De syimia,* 162. 4, 96, 1)

Albert the Great (Albertus Magnus, 1200–1280), the foremost medieval scholar of nature (Vost, 2011), is an intellectual bridge between ancient and early modern natural history, and his work is an approximation to the state of scientific zoology in the Middle Ages. Albert's work re-established Aristotle's conception of zoological research after a fifteen hundred year gap (Tkacz, 2007; Krause, 2015). As well as his deep study of Aristotle, he added the works of Greek, Latin, Arabic, and other Medieval authors. Albert's monumental zoological *summa*, *De animalibus* (*On animals*) (1295), was the first Latin commentary to the *Historia animalium* of Aristotle. In addition to his commentary material, Albert included two books of comparative anatomy, with the final five books of *De animalibus* containing the results of Albert's own zoological research, much of it based on his personal observations and arranged in the manner of a traditional bestiary. *De animalibus* includes the causes that lead to animal diversity as well as a consideration of the animal's habitat (Krause, 2015: 7). The human being is put at the center of Albert's philosophical interest as "the most perfect of all living beings", and it is compared to all the rest of the animal world in an effort to explain the ultimate causes of the observed differences (Janson, 1952: 83; Miteva, 2022).

The primate nomenclature used by Albert was in accordance with the previous tradition and suggests he had direct experience with some of the primates. He described the Barbary macaque (*simia*), the *sphinga (finga),* two-tailed monkeys (*cytosica, kyboz),* the *canina simia* (baboon) and the anthropomorphic creature called *pilosus* (see **Box Details 7.2**). He may have owned a *spinga*, which he defined "as a composite of a *simia* and a wild cat, with two black spots on its cheeks and a very long grey tail with a black stripe" [probably a *Cercopithecus* sp.]. Albert, like his contemporaries, did not doubt the existence of monstrous races and composite creatures, but did not believe them to be human and described them as genera of monkeys (e.g., the *pilosus*, later recalled by Gessner, *Tierbuch*, 1551, cf. Veracini, Chapter 9, this volume). Pygmies, which held great fascination for the medieval mind, particularly attracted Albert's attention: to him, these creatures were different from simian forms and were used as a kind of boundary concept in order to define humanity vs non-humanity.

BOX DETAILS 7.2: THE DIFFERENT FORMS OF PRIMATES ACCORDING TO ALBERTUS MAGNUS' *DE ANIMALIBUS (ON ANIMALS), SYMIA. LIB XXII*

Cecilia Veracini

Some of the primates described by Albertus Magnus are easily recognizable even bearing in mind the classical tradition. Others that seem to be the result of personal observations are more difficult to identify.

Symia: the monkey is a familiar animal that has many species. The one that is found most often bears a similarity to the human in the shape of its head and ears, and in its hands, feet, and reproductive parts. It also has breasts on its chest as does a woman, but it is hairy. It has less hair on its face and has broad buttocks with a place on them prepared for sitting. It is imitative of humans and mimics human actions. For this reason, it often falls into the hands of hunters. [Here Albertus quotes the well-known stories of the shoes and the birdlime used to hunt monkeys: cf. Pliny, *Historia naturalis* 8.80.215; Solinus, *Collectanea*... 27.56; Aelian, *De Animalibus*, 17.25]. These beasts frequently bear two offspring [again, he quoted the tale of the mother monkeys and twins from Aelian]. The monkey is a tricky animal with bad habits. However much it is tamed, it is always enraged [cf. Solinus, *Collect.* 27.58], and it imitates the bad rather than the good human traits. It is playful with the small offspring of humans and dogs. But it sometimes strangles unguarded boys, sometimes hurling them off a height. It collects vermin on heads and clothing and eats them. It remembers a wrong for a long time and although it resembles a human externally, it has no resemblance with him internally, and even less than nearly all the other beasts. When it has been tamed and is living in a house and it gives birth, it shows its offspring to the people as if they were beautiful and pleasing [quoting Neckam, see Paragraph 3.]. It has breasts on its chest, as does a woman since nature gave it hands with which it can lift its young to its chest as a woman can. Other animals, however, have their breasts below them so that they can be reached by their young. As we have said repeatedly in the preceding books, there are very many genera of monkeys. One seems to be a composite of a monkey and a wild cat (*agrestis cattus*) because it has the face of a monkey, multicoloured, with two black spots on its jaws, but it has a very long grey tail that is black at the end [species of the genus *Chlorocebus* may have the tail that ends with a black tip, although the other characteristics may refer to a Patas monkey, *genus Erythrocebus*]. Some call these animals *spinga*, while others call them *cythosica symiae* [*cythosica* is evidently a corruption of *cercopithecus*, Janson, 1952: 84; see later on *finga*]. There are others, however, who have very gracious and kindly faces. They are different from other monkeys, having beards and very long tails. These are said to be Ethiopian and when they are removed from Ethiopia, they live but a short time [he is speaking of colobus monkeys, the *callithriches* of Solinus]. There are Indian ones that have completely white bodies. They are bearded and have white tails [langurs, genus *Semnopithecus*]. These are hunted with arrows by the Indians. When they are tamed, they are good at any trick, almost as if they had been created for tricks alone. The bodily parts of the monkey are said to be effective in many medications. But since this has not been personally experienced by me and since it is more fitting for another investigation, it seems better to pass it over [In India, certain monkey species are believed to have healing properties].

Kyboz: a name from Greek to define a type of tailed monkey: it has a tail, as we said, but its front part is entirely like that of woman [it recalled the idea of the gelada baboon; Avicenna used this name for baboons, also quoted as "*kiber* Avicenna" by Ulisse Aldrovandi when describing baboons, cf. Veracini, 2019].

Cefusa: as Solinus says [cf. Solinus, *Collect.* 30.20], this is a monstrous beast whose front legs bear a resemblance to the arms and hands of a human and whose rear ones resemble his legs and feet. He says that this animal was brought to Rome during the time of Caesar [a monkey called by other authors *cephos* was described by Pliny, *Hist. Nat.* 8.28.70, and Solinus, *Collect.* 30.20; it was probably a Patas monkey, genus *Erythrocebus*]. We have seen this one even in our day captured in the forest of Sclavia, a female and male, as we have said. This animal emits various calls as time passes and belongs to the monkey genus [it is difficult to interpret this part, as no primates live in Sclavia, the land of Slaves)].

Finga (or *spinga*): as Pliny recounted, it is a dark animal that has breasts on its chest as humans do; they are not so fierce that they cannot be tamed nor so docile that they do not harm those harming them. They live in peace with those who are at peace with them [the *spinga* was probably the gelada baboon, *Theropithecus gelada*, cf. Groves, 2008; Masseti & Veracini, this volume].

Mamonetus: the *mamonetus* is an animal that is smaller than a monkey. It is dusky on the back and white on the belly, with a long and shaggy tail. It has a neck that is as wide as its head, and for that reason, it is tied up at the hips rather than by its neck [this characteristic was attributed to guenons]. It has a nose that is distinct from the mouth, as does a human, and that is not continuous with it, as is a monkey's. It is called *spinga* among Italians. [This passage implies that they were kept as pets and that perhaps Albertus saw one of them]. It fights with implacable enmity with monkeys [i.e., *simia*=Barbary macaque], and although it is unequal in strength, it nevertheless wins out over them due to its boldness. It is native to the East but lives well in our climes and is often seen. It has a round head and a face that is closer to a human's than to a monkey's [This primate could be a guenon, genus *Chlorocebus*].

Pilosus: hairy-man, is a composite animal, a human above and a goat below. But it has horns on its forehead and belongs to the monkey genus. It is quite monstrous and sometimes it walks erect and is tamed. They claim that it lives in the desert regions of Ethiopia and that sometimes it is captured and led to Alexandria. A dead one was brought, preserved in salt, to Constantinople. Many such composite animals are found, as we have said in the preceding books [This could be a description of an anthropoid African ape -chimpanzee or gorilla. Nevertheless, Albertus composes this description with elements drawn from tradition. The classical figure of Pan, the satyr or Faunus, was human above and goat-like below; also, satyrs were generally, but not always, described with equine parts. Likewise, the *pilosus* belonged to the tradition of the wild man, cf. Mobley, 1997].

Papio: the *papio* is an animal that is quite numerous near the city of Caesarea [currently Israel]. It is a little bigger than a fox but has the habits of a wolf. When they eat in groups, these animals howl, with one leading and the others responding in a group after him. They are shaggy like foxes and when one of these species has died, the others howl around it as if mourning the dead. Their voices are so loud that they are reckoned to be nearby even when they are far off. This animal is said to enter the sepulchres of men when hungry and to feed on their bodies. It is composed of a wolf and a fox. [This description corresponds to a hyena, but it is included in the description of primates. Later on, it led the 16th naturalist Gessner (*Icones animalium*, 1560:76) to a misinterpretation when he described the first mandrill/drill brought to Germany (cf. Veracini, 2019; Veracini, Chapter 9, this volume].

Cynochephalus: this is a monkey with the head of a dog (*canina simia*) [baboons; it is also identified with the *canini homines* and this juxtaposition between dog-headed monkeys and dog-headed men is found in a chapter devoted to the anatomical comparison between non-human primates and humans].

The pygmies were anthropomorphic creatures first described by Homer (*Iliad*, Book III, lines 4–10). Later, their description was enriched with legendary elements that led to a lack of an unambiguous position regarding their geographical location, morphology, and customs (Perfetti, 2020: 53–54). To Albert, pygmies and monkeys were thought to share a similar anatomical form to that of human beings, but pygmies were the creatures closest to humans (Spencer, 1995: 14; Miteva, 2022). Some classical sources reported that pygmies lived in social groups, were capable of communicating with each other, and were able to use their hands to perform diverse tasks. Based on this, Albert attributes to them some rational capacity (called *umbra rationis*, the shadow of reason), but that does not reach the full light of reason that is exclusive to human beings (Perfetti, 2020: 48). Pygmies only have something human, some capacity for choice, and language that differentiates them from simians (see later). Thus, the pygmy is the creature that comes closest to human perfection despite having an imperfect *anima* (souls) and, along with the rest of animate nature, it is below humans on a descending scale of degrees of imperfection. According to Albert, humans are the only creatures that are both perfect in body and soul and have the use of reason. *Ratio* is, as for St. Augustine and Aristotle, a character that distinguishes humans and beasts.

On Albert's scale of nature, non-human primates are below pygmies. The *genus simiarum* (Albert speaks generically of primates without referring to a particular species) are human-like creatures, but they are not humans because they lack reason. Nevertheless, they are a step above brutes (all the other beasts) because they share human features such as round heads, dexterous hands, breasts on their chests, and the lack of a tail (here referring obviously to the Barbary macaque) (*De animalibus* includes a long list of anatomical features shared between humans and non-human primates); plus, they have small mouths, cannot waggle their ears, and have imagination and a sense of the value of objects. In Albert's vision, they could represent the 'missing link' between humans and other animals (Janson, 1952: 85), which is an intermediate form:

> there are certain animals that, in the natural shape of their body, share with many species or genera of animals, doing so much the same way that the middle shares in the extremes. Thus the genus of monkeys seems to be like that of the quadrupeds and the human [...]. It is a trait of almost every monkey genus that the monkeys sometimes rise up and go about on two feet, although not for long. Insofar as this genus of monkeys is hairy to the front and rear and insofar as their hair is thick and hard, they are like quadrupeds. In many of their facial members they seem to resemble humans, specifically in the nose, ears, and teeth. [...] They have both the front teeth called incisors and the rear teeth called molars, as does a human. The monkey resembles a human in its eyelids, for it has hair on both of them, upper and lower, as does a human. For other quadrupeds only have hair on the upper lid. The monkey seems to have eyebrows like a human's, but they are short and fine and are thus not noticed by some. On its chest there hang two breasts with two nipple-like cones, just as on the breasts of men and women. Its two arms are like those of humans, but they are hairy. Monkeys bend their hands and feet like a human and they have the hands we called "lesser" above, that is to say, the hands connected to the reeds of the arms (radius and ulna). They have fingers on these hands and nails on the finger as does a human. [...] The vulva of the female monkeys is like that of a woman, just as the penis of the male is like that of a man [quotations from Aristotle]. (*On animals*, Book I, Ch. IV: *On similarity and Differences among Animals in Their upper Parts....*)

As well as emphasizing that the possession of reason and soul is unique to humans, Albert also discusses how the inability of primates to speak distinguishes them from pygmies, and what this implies about their cognition:

> Speech (*sermo*) [...] is "articulated and lettered voice" (*vox articulata et literata*), making known a person's mental conception. This belongs only to humans and to the few that have an affinity to humans, such as the pygmy, the one which we named *marytonmorion* above [n.d.r. this creature is a garbled version of the mantichore cited by Aristotle, but sometimes it is said to be a monkey cf. Albertus, 2001: note p. 477], and the like. [...]. A quadruped that is viviparous and has a womb, has a simple, unlettered voice and does not speak at all. For speaking with intellect is proper to humans alone. All that speak, have a voice but the converse is not true. Thus, if some genus of manlike monkey speaks, it is from simple imagination and not from intellect. This is the case in the pygmy, and perhaps in the *marytonmorion*. Certain others of this sort may even be found (*On animals*, Book 4, Ch. II, *On the type of voice and sound of the animals*: 476–477).

The monkey has a superiority over other animals with a more complex estimative capacity, because it can estimate the useful and the harmful more than other brute animals. Albert claims that when primates search for food, they frequently err because they are driven by fantasy and not intellect; therefore, "there is no practical syllogism but an imperfect argumentation" (Albertus Magnus, 1920: 1331, *apud* Tellkamp, 2016: 121). As noted by Tellkamp, Albert "expands the horizons of medieval cognitive psychology, arguing that animals, especially primates, have the ability to bring about and act on them in a quasi-syllogistic way". To demonstrate this, Albert uses the idea that "from a physiological point of view, humans and higher animals have strong structural similarities, which entail a similarity regarding their capacities to process sensible information and transform it into meaningful content (*intentiones*)" (2016: 122). This means that while a rational human can have conscious convictions about a mental premise, animals, such as non-human primates, are moved by intentions produced by imagination or estimation, which, as mental representations of particular instances, lead to flawed outcomes. The boundary between animality and humanity thus lies in the ability or otherwise to bring the particular and fragmentary experience of animals (recognized in their various actions performed to survive) to the universality of the concept. Albert, in order to explain defence strategies and animal ways of life otherwise inexplicable in beings without reason, resorts to the expression 'natural instinct', and to this he adds other expressions that end up giving irrational beings such as pygmies a kind of natural counterpart to human rationality as 'shadow of reason' (*umbra rationis*) (Cerrito, 2020: 102).

The question of the soul and cognition in non-human primates was addressed by other late medieval scholars after Albert the Great. John Buridan (before 1300–c. 1361) asserts that highly developed animals, such as simians are capable of reasoning to some extent (*Quaestiones de anima*, cf. Oelze, 2018: 159). As Oelze states, Buridan suggests that instead of denying the capacity of reasoning to other animals, "one could speak of different degrees of reasoning" according to the kind of animal and to their soul's complexity. "Monkeys, for instance, are even smarter than dogs, whereas shellfish are comparatively dumb" (Buridan, cf. Oelze, 2018: 159). Like Buridan, Nicole Oresme (1323–1382) also states that some of the higher animals "seem to think and seem to have cognitions similar to human kinds of cognition". Because of their human-like "physiognomy", he says, their psychology is very likely to be human-like, too (*Expositio et quaestiones in Aristotelis De anima*, p. 335, cf. Oelze, 2018: 159). Nevertheless, although Albert, Buridan, and Oresme stressed the particular position of primates (*simiae*), the exceptional position of human beings continued to be claimed and defined by them.

7.5 CONCLUDING REMARKS

Primates were fairly common animals in medieval imagination and culture, and were kept as pets at court or in the homes of wealthy people as symbols of prestige. The only species with a significant presence in the early Middle Ages and then into the following centuries was the Barbary macaque, *M. sylvanus*. Otherwise, the primate species encountered most frequently from the 12th onward is the grivet monkey, *C. aethiops* endemic to Central East Africa, where trade networks to Europe remained active throughout the Middle Ages.

The symbolic and allegorical significance attributed by medieval culture to primates was primarily moral and rooted in Greco-Roman culture and in the three monotheistic religions: Judaic, Christian, and Islamic. Primates were considered to more closely resemble human beings than any other animal in appearance and bearing, but they were devoid of reason and proverbially lascivious (*turpissima bestia, simillima nostri*). In Christian culture initially, primates (the tailless Barbary macaques) were associated with the devil, then, after the sinful acts of Adam and Eve, with fallen humanity. They came to be seen as symbols of degraded and sub-human creatures, useful to establish a caesura and a clear boundary between human nature and that of the brutes (the other animals). This reinforced the superiority of the human being, created in the image of God, over non-humans and nature in general. This view is also found in Judaic texts, in which people were punished for idolatry and turned into monkeys. Similarly, the Qur'an associates primates with human beings transformed for not having respected God's laws. In late medieval Christian culture, monkeys came to be identified with sins such as lust, immoderation, malice, and envy. Thus, usually accompanied by a goat, another symbol of lasciviousness, this disparaging connotation of primates entered the imagery of the Renaissance.

One of the major interests of medieval thinkers, especially after the translation of Aristotle's *corpus*, which introduced the naturalistic paradigm for the study of animals, was to establish the nature of non-human primates and, in particular, to understand whether monkeys and other non-human creatures possess rational souls and the faculties of intellect and reason (Köhler, 2014; Oelze, 2018; Miteva, 2022). Although many medieval scholars reproduced the information found in classical religious texts, giving priority to a moralistic and unscientific view of primates, others, such as Albert the Great, by combining the information in Aristotle's texts with empirical observation, opened new fields of inquiry (Perfetti, 2020). Albert the Great, speaking about the difference between monkeys and other animals, introduced one of the earliest conceptions of the 'missing link' which was the earliest attempt to bridge the chasm between mankind and the rest of the animal world (Janson, 1952). Other late medieval philosophers spoke about different degrees of reasoning among animals, where primates have a particular, higher position among animals, although the exceptional position of human beings continued to be claimed and defined by their ability to communicate through language. These thinkers do not consider the human's special position to depend upon the immateriality of the human soul given by God, but they admit that cognitive superiority is a complex matter where materiality (the different nature of animals) has a relevant role. As Oelze argues, "this argument seems to blur the lines between humans and non-human animals. Reasoning is no longer a peculiar human capacity or reasoning is at least no longer what makes humans a peculiar species of animals" (2018: 159).

ACKNOWLEDGEMENTS

I would like to sincerely thank Marco Vespa for his important comments and suggestions to the manuscript; Hilary Morris for the first revision of English; Bernard Wood for his meticulous revision and suggestions; Claudia Massei for her photo of the Salerno Cathedral; the *I Tatti* Library of Harvard University (Florence) for years of support in my search for sources.

FUNDING

Portuguese national funds through FCT—Fundação para a Ciência e a Tecnologia in the context of the celebration of the contract under the 'Norma Transitoria', DOI 10.54499/DL57/2016/CP1456/CT0001.

REFERENCES

PRIMARY SOURCES

Aberdeen Bestiary (on-line). https://www.abdn.ac.uk/bestiary/fol.12. *Simia/Ape* Accessed 17 February 2023.

Aelian, Maspero, F. (ed). (1998). *De natura animalium (La natura degli Animali)*. Milan: Biblioteca Universale Rizzoli. 2 vols.

al-Ṭabarī, Madelung, W.F. & Jones, A. (eds) (1987). *Tafsir, The Commentary of the Qur'an Book.* Oxford University Press.

Albertus Magnus, Kitchell K.F. Jr. & Resnick I.M. (eds) (1999). *On animals, A Medieval Summa, Zoologica*. The Johns Hopkins University Press. 2 vols.
Albertus Magnus H. Stadler (ed.), (1920). *De animalibus*, vol. 2. Münster: Aschendorff.
Ambrosius, Coppa , G. (ed). (1995). *Exameron. Commento ai sei giorni della creazione.* Milan: TEA.
Apuleius, P. Gerard Walsh (ed) (1999). *The Golden Ass.* Oxford University Press.
Arabian Nights (2001). *Le Mille e una Notte*, Italian Version. Milan: Mondadori.
Aristotle, D'Arcy Wentworth Thompson (ed). *Historia animalium (History of Animals).* On-line. http://classics.mit.edu// Aristotle/history_anim.html
Basilius Magnus, M. Naldini (ed). (1990). *Homiliae in Hexamaeron* IX. 1,7. *Sulla genesi Omelie sull'Examerone.* Milano: Mondadori.
Bartholomaeus Anglicus, Trevisan (ed.) (1538) *De proprietatibus rerum.*
Bartholomew Anglicus (1483). *De proprietatibus rerum.* Johannes Koelhoff.
Bersuire, P. (1290–1362). *Reductiorum morale.* Lyon (1520); Venice (1583).
Buridan, J. (1996). *Quaestiones de anima* (2nd compilation). Paris: Bibl. Nat.
Cantimpré, T.de (1973). *De Natura Rerum.* Berlin and New York: W. de Gruyeter
Cartari, V. (1626). *Seconda nouissima editione delle imagini de gli dei delli antichi di Vicenzo Cartari Reggiano.* Padova: Stamparia Pietro Paolo Tozzi.
Chartier, J. (1858). *Chronique de Charles VII, roi de France*, III ed. Paris, Jannet: Vallet de Viriville. https://gallica.bnf.fr/ ark:/12148/bpt6k27714p
Galen C., Singer C. (ed) (1956). *On Anatomical Procedures: De Anatomicis Administrationibus.* Wellcome Historical Medical Museum. https://archive.org/details/b20457194/mode/2up
Gessner, K. (1551). *Tierbuch.* Zürich.
Homer, Rieu, E. V. (1953). *The Iliad.* London: Methuen & CO.
Isidore of Seville. Angelo Valastro (ed) (2004). *Etymologia. Etimologie o Origini.* 2 vol. Torino: UTET.
Maxwell, H. (1913). *The Chronicle of Lanercost, 1272–1346: Transalted with notes.* Glasgow: J. Maclehose. https:// archive.org/details/chronicleoflaner00maxwuoft/page/36/ mode/2up
Neckam, A. (1863). *De naturis rerum, Libri duo, with the Poem of the same author, De laudibus divinae sapientiae.* London: Thomas Wright. Longman R. and Green.
Oppianus, Papathomopoulos, M. (ed.) (2003). *Oppianus Apameensis. Cynegetica,* Berlin, de Gruyter 2003.
Oresme, N. (1995). *Expositio et quaestiones in Aristotelis De anima*, lib. iii, q. 4. Patar.
Ovid, (Ovidio) (1992). *Le Metamorfosi.* Bur, Classici Greci e Latini, Milano, 2 Vols.
Philo of Alexandria, Whitaker, G.H (ed). ((1929–1962). *Opera quae supersunt.* Vol I. *Philo in Ten Volumes, and Two Supplementary Volumes.* Cambridge, MA: Harvard University Press.
Plato, L.W.R.M. (ed). (1925). *Hippias Major. Plato in Twelve Volumes.* Cambridge, MA, Harvard University Press; London, William Heinemann Ltd.
Pliny the Elder, Ernout, A. (ed.), (1952). *Pline l'Ancien. Histoire naturelle. Livre VIII,* Paris, Les Belles Lettres.
Plotinus, M.S. & Page, B.S. (trans.). (1952). *The Six Enneads.* Chicago, London, Toronto William Benton Publisher. https://classics.mit.edu/Plotinus/enneads.html
Pomponio Mela (43CE). *Chorographia.* https://www.thelatinlibrary.com/pomponius.html
Physiologus, Curley, M. G. (ed) (2009). *A Medieval Book of Nature Lore.* Chicago: University of Chicago Press, 2nd ed.
Saint Agostine (413–426 CE). *De Civitate Dei contra paganos.* Italian transl.: "La Citta di Dio" Edizione Acrobat Ed. Patrizio Sanasi. http://www.bibliomania.com/ (2004).
Saint Hildegard, von Bingen's, H. (ed) (1986–1987). *Die Physica der heiligen Hildegard. Von Dr. J. Berendes in Goslar.* Vienna III: Druck von F. Brück & Söhne.
Silvestris, B. (1978). *Cosmographia. Creation of Animals*, vol. 20. Brill.
Socrates of Constantinople (439 CE). *Historia ecclesiastica*, vol. 16, Revised, with Notes, by THE REV. A. C. ZENOS, D.D.
Solinus, C.J. Mommsen, Th (ed) (1895) *Collectanea rerum memorabilium.* Toronto: Berolinum apud Weidmann. https:// archive.org/details/collectanearerum00soliuoft
Thomas Aquinas (1265-1273). *Summa Theologiae.* Magonza (1467).
Walberg, E. (1900). *Le Bestiaire de Philippe de Thaün P.* de (1120). Suède: Lund H. Möller. https://archive.org/details/ lebestiairedephi00philuoft

SECONDARY SOURCES

Blum, P.R. (2012). *Studies on Early Modern Aristotelianism.* Leiden, Boston: Brill.
Boehm, B.D. & Holcomb, M. (2001). *Animals in Medieval Art, Heilbrunn Timeline of Art History.* New York. https://www. metmuseum.org/toah/hd/best/hd_best.htm (last revised January 2012).
Buquet, T. (2013a). *Preventing "monkey business". Fettered apes in the Middle ages. Animal data network* (blog on Hypotheses.org). https://mad.hypotheses.org/37 (originally published 2013, last revised 2016).
Buquet, T. (2013b). Les animaux exotiques dans les menageries medievales. Jacques Toussaint. Fabuleuses histoires des bêtes et des hommes, Trema. *Societe archeologique de Namur*, 97–121.
Buquet, T. (2015). *Arab and Ottoman menageries.* Medieval animal data network (blog on Hypotheses.org). [On line] https://mad.hypotheses.org/620
Camille, M. (1992). *Image on the Edge: The Margins of Medieval Art.* Reaktion Books.
Capecelatro, A. (1862). *Storia di S. Pier Damiano e del suo tempo.* Vols 2. Firenze: G. Barbera, Ed.
Cerrito, A. (2020). Alberto Magno e la logica del vivente. In S, Perfetti, *Nature Imperfette.* Pisa: Nature Imperfette, pp. 101–126.
Ciccarese, M.P. (1990). Il simbolismo antropologico degli animali nell'esegesi cristiana antica: criteri e contenuti ermeneutici. *Annali di storia dell'esegesi*, 529–567.
Cipriani, M. (2014). *La place de Thomas de Cantimpré dans l'encyclopédisme médiéval: les sources du Liber de natura rerum* (Doctoral dissertation, Paris, EPHE).
Coelho, L. (1998). *A ciencia na Idade Media.* Lisbon: Guimaraes.
Corbey, R. (2008). *Metafisiche delle Scimmie. Negoziando il confine animali-uomini.* Italian translation, Cavalieri, P. (ed.). Torino: Bollati Boringhieri.
Cuttler, C.D. (1991). Exotics in post-medieval European art: Giraffes and centaurs. *Artibus et historiae*, 12(23), 161–179.
Daston, L. (1998). *Wonders and the Order of Nature, 1150–1750.* Zone Books.

Domínguez, F.A. (2006). Monkey business in Carajicomedia: The Parody of Fray Ambrosio Montesino as "Fray Bugeo". *ehumanista*, 7, 1–27.

Freedman, P. (2008). *Out of the East: Spices and the Medieval Imagination*. Yale University Press.

Gauvin, B., Jacquemard, C. & Lucas-Avenel, M.A. (2013). L'auctoritas de Thomas de Cantimpré en matière ichtyologique (Vincent de Beauvais, Albert le Grand, l'Hortus sanitatis). *Kentron. Revue pluridisciplinaire du monde antique* (29), 69–108.

George, W. & Yapp, B. (1991). The naming of the beasts: Natural history in the medieval bestiary, London, Duckworth, pp. xiv, 231. *Medical History*, 36(3), 340–341. doi:10.1017/S002572730005540X.

George, W. & Yapp, B. (1993). The naming of the beasts: Natural history in the medieval bestiary. *Journal of the History of Biology*, 26(1), 160–162.

Grant, E. (2002). *Os fundamentos da Ciencia Moderna na Idade Media* (Portugues Edition). Porto: Porto Editora.

Groves, C. (2008). *Extended Family: Long Lost Cousins. A personal look at the history of primatology*. Arlington, VA: Conservation International.

Herrero Marcos, J. (2012). *Bestiario románico en España*. Palencia: Calamo, 2nd ed.

Hopfner, T. (1922–1925). *Fontes historiae religionis Aegyptiacae collegit*, 2 vols. Bonn: A. Marci et E. Weberi.

Howard, D. (2007). Venice as an Eastern City. In S. Carboni (ed.), *Venice and the Islamic World, 828–1797*. New York, NY: Metropolitan Museum of Art, pp. 58–71.

Hughes, J.D. (2003). Europe as consumer of exotic biodiversity: Greek and Roman times. *Landscape Research*, 28(1), 21–31.

Hunt, R.W. (1984). *The Schools and the Cloister: The Life and Writings of Alexander Nequam, (1157–1217)*, Gibson, M.T. (ed.). Clarendon Press. (Zoological part: pp. 1–31).

Jacobs, J. (1901). *The Jewish encyclopedia*, I. New York and London: V. Apes, p. 663.

Janson, H.W. (1952). *Apes and Ape Lore in the Middle Ages and the Renaissance*. London: Warburg Institute.

Jordan, A. (2004). Rarities and novelties. In A. Jackson, and A. Jaffer (eds.), *Encounters. The Meeting of Asia and Europe, 1500–1800*. London, p. 34; Munger, J. & Frelinghuysen, A.C. (2000). *East and West: Chinese Export Porcelain*. New York: Heilbrunn Timeline of Art History. https://www.metmuseum.org/toah/hd/ewpor/hd_ewpor.htm (October 2003).

Kisling, V.N. (ed.). (2000). *Zoo and Aquarium History: Ancient Animal Collections to Zoological Gardens*. CRC Press.

Klingender, F. (2019). *Animals in Art and Thought: To the End of the Middle Ages*, vol. 28. Routledge.

Köhler, T.W. (2014). *Homo animal nobilissimum. Konturen des spezifisch Menschlichen in der naturphilosophischen Aristoteleskommentierung des dreizehnten Jahrhunderts*, Teilband 2.1 und 2.2. Leiden/Boston: Brill.

Krause, K. (2015). Albert the great on animal and human origin in his early works. *Lo Sguardo*, 18(2), 205–232.

Le Goff, J. (1985). *O maravilhoso e o quotidiano no Ocidente medieval*. Tradução por José Antonio Pinto Ribeiro. Lisboa: Edições, 70.

Le Goff, J. (1991). *Mercadores e banqueiros da idade media*. (Trans. Cardoso). Lisboa: Gradiva.

Le Goff, J. (2007). *Il cielo sceso a terra. Le radici medievali dell'Europa*. Ed. Laterza.

Livi, A. (2004). *Storia Sociale della Filosofia*, vol. I. Roma: Società Editrice Dante Alighieri, p. 242.

Loisel, G. (1912). *Histoire des ménageries de l'Antiquité à nos jours*, 3 vols. Paris: O. Doin et fils.

Lopes, P. (2006). Os livros de viagens medievais. *Medievalista*, 2(2), 1–32.

Mackinnon, M. (2006). Supplying exotic animals for the Roman amphitheatre games: New reconstructions combining archaeological, ancient textual, historical and ethnographic data. *Mouseion: Journal of the Classical Association of Canada*, 6(2), 137–161.

Masseti, M., & Bruner, E. (2009). The primates of the western Palaearctic: a biogeographical, historical, and archaeozoological review. *JASs*, 87, 33–91.

McDermott, W.C. (1938). *The Ape in Antiquity*. Johns Hopkins Press.

McKenzie, A.M. (1959). *The reaction in pagan thought to Christianity from Celsus to Julian* (PhD thesis). https://theses.gla.ac.uk/3069/

Miteva, E. (2022). Pygmies, twins, monsters: Human nature and its borderlines in Albert the Great. In *The Embodied Soul: Aristotelian Psychology and Physiology in Medieval Europe between 1200 and 1420*. Cham: Springer International Publishing, pp. 73–90.

Mobley, G. (1997). The wild man in the Bible and the ancient near East. *Journal of Biblical Literature*, 116(2), 217–233.

Molà, L. (2009). Traffici, mercanti, viaggiatori sullo scorcio del Medioevo. In *Italiani e stranieri nella tradizione letteraria. Atti del Convegno di Montepulciano, 8-10 ottobre 2007*, Roma.

Monteira, A. I. (2005). *La influencia islámica en la escultura románica de Soria. Una vía para el estudio de la iconografía en el románico*. Madrid: Fundacíon Universitaria Española, Seminario de Arte e Iconografía "Marqués de Lozoya".

Muñiz, D.C.M. (2000). La fauna exótica en la Península Ibérica: apuntes para el estudio del coleccionismo animal en el Medievo hispánico. *Espacio Tiempo y Forma. Serie III, Historia Medieval*, (13), 233–270.

Nickel, H. (1991). Presents to princes: A bestiary of strange and wondrous beasts, once known, for a time forgotten, and rediscovered. *Metropolitan Museum Journal*, 26, 129–138.

Nijman, V. & Healy, A. (2016). Present-day international primate trade in historical context. In S.A. Wich, and A.J. Marshall (eds.), *An Introduction to Primate Conservation*, pp. 129–142.

Noegel, S. (2019). From ape to zebra: On wild animals and taxonomy in ancient Israel. In R. Mattila, S. Ito, and S. Fink (eds.), *Animals and Their Relation to Gods, Humans and Things in the Ancient World*. Wiesbaden: Springer VS, pp. 95–133.

Oelze, A. (2018). What are and Why Study Later Medieval Theories of Animal Rationality? In *Animal Rationality* (pp. 5–11). Brill.

Olivares, D. (2012). Iconografía de la escultura románica en la provincia de Soria: las representaciones de simios como caso de estudio. *Estudios de Medievalismo Hispánico*, 1, 143–176.

Pastoureau, M. (2002). Le Moyen Âge en lumière. In *Manuscrits enluminés des Bibliothèques de France, sous la direction de Jacques Dalarun*. Paris: Fayard.

Pastoureau, M. (2008). *Les Animaux Célèbres*. Paris.

Perfetti, S. (2012). *Animali pensati*. Pisa: Edizioni ETS.

Perfetti, S. (2020). *Nature imperfette. Umano, subumano e animale nel pensiero di Alberto Magno*. Pisa: Edizioni ETS.

Pipe, A. (1992). A note on exotic animals from medieval and post-medieval London. *Anthropozoologica*, 16, 189–190.

Pluskowski, A. (2004). Narwhals or unicorns? Exotic animals as material culture in medieval Europe. *European Journal of Archaeology*, 7(3), 291–313.

Rogers, J. & Holmes, J. (2018). Monkey business: The Victorian natural history museum, evolution, and the medieval manuscript. *Romanticism on the Net* (70), 1–27.

Rudwin, M.J. & Rudwin, M. (1931). *The Devil in Legend and Literature*. Chicago: Open Court.

Salisbury, J.E. (2022). *The Beast within: Animals in the Middle Ages*. Routledge.

Serpell, J. (1996). *In the Company of Animals: A Study of Human-Animal Relationships*. Cambridge University Press.

Shemesh, A.O. (2018). 'There is no concern of prohibition against their trade'. A responsum by Rashbatz on the trade in monkeys practiced by Algerian Jews in the middle ages. *HTS Teologiese Studies/Theological Studies*, 74, 1–8. doi:10.4102/hts.v74i1.4768.

Simões, C.S. (2014). La importancia simbólica de lo" exótico" en la corte portuguesa a finales de la edad media. In *Anales de Historia del Arte*, vol. 24. Universidad Complutense de Madrid, , pp. 517–526. https://dx.doi.org/10.5209/rev_ANHA.2014.48291

Spencer, F. (1995). Pithekos to pithecanthropos: An abbrevieted review of changing scientific view on the relationship of the antropoid apes to homo. In R. Corbey, and T. Bert (eds.), *Ape, Man, Apeman: Changing Views since 1600*. Dept. of Prehistory, University of Leiden, pp. 13–22.

Tellkamp, J.A. (2016). Aping logic? Albert the great on animal mind and action. *Subjectivity and Selfhood in Medieval and Early Modern Philosophy*, 109–123.

Tkacz, M. (2007). Albert the great and the revival of Aristotle's zoological research program. *Vivarium*, 45(1), 30–68.

Tomarken, A.H. (1990). *The smile of truth: The French satirical eulogy and its antecedent*. Princepton.

Urbani, B. & Youlatos, D. (2023). *Simia langobardorum*: Were African apes traded in late medieval Lombardy? *American Journal of Primatology*, 85(2), e23462.

Veracini, C. (2019). Natural History of Non-Human Primates in the Early Modern Period. In, C. Veracini, F. Scalfari & C. Casanova, *History of Primatology yesterday and Today*. Roma: Aracne Editrice, pp. 53–91.

Veracini, C. (2020). A landscape without nonhuman primates? The case of the Barbary macaque, *Macaca sylvanus* (Linnaeus, 1758) and its interaction with humans throughout recorded time. *Humanities*, 9(3), 92.

Vespa, M. (2019). Le 'Tête-de-chien' (κυνοκέφαλος) des Grecs. L'Égypte au prisme des animaux'. In S. Porcier, S.Ikram, and S. Pasquali (eds.), *Creatures of Earth, Water and Sky*. Leiden: Brill, pp. 331–339.

Vespa, M. (2022). The place of nonhuman primates in ancient Roman culture. In Urbani et al. (eds.), *World Archaeoprimatology: Interconnections of Humans and Nonhuman Primates in the Past*, pp. 201–224.

Viré, F. (1986). Kird In: *The Encyclopaedia of Islam, new edition*. In H. A. R. Gibb, et al. (Eds). Leiden: E.J. Brill, 1986–2002, vol: V, pp. 131–134.

Vost, K. (2011). *St. Albert the Great*. Charlotte. North Carolina: TAN Books.

Walker-Meikle, K. (2012). *Medieval Pets*. Woodbridge, UK: The Boydell Press.

Walker Vadillo, M.A. (2013a). Los Simios. *Revista Digital de Iconografía Medieval*, 5(9), 63–77. https://pendientedemigracion.ucm.es/centros/cont/descargas/documento39857.pdf

Walker Vadillo, M.A. (2013b). *The representation of apes in medieval art*. https://pt.slideshare.net/uwstratford/the-representation-of-apes-in-medieval-art Last accessed May 1, 2023.

Warner, M. (2013). *Alone of All Her Sex: The Myth and the Cult of the Virgin Mary*. Oxford University Press.

Wigoder, C.R.G. & Posner, R. (1971). *Encyclopedia Judaica*, vols. 1–16. Jerusalem.

Willemsen, C.A. (1980). L'enigma di Otranto. In *Il mosaico pavimentale del presbitero Pantaleone nella Cattedrale, Ga-latina*. Galatina (Lecce): Mario Congedo editore.

Zambon, F., Capelli, R., Cocco, S., Cremonini, C., Sanson, M. & Villa, M. (2018). *Bestiari tardoantichi e medievali: i testi fondamentali della zoologia sacra cristiana*. Bompiani.

8 Perception and Description of Non-human Primates in the Arab World[1]

Cecilia Veracini
Universidade de Lisboa

Malak A. Alghamdi
King Saud bin Abdulaziz University for Health Sciences

> [the monkey] swept the house, lit the fire under the pot, fulled the fire ... and wind his master with a fan.
>
> *(Al-Rāmhurmuzī, 1883–1886: 77)*

8.1 PRE-ISLAMIC PERCEPTION OF AND CONTACT WITH NON-HUMAN PRIMATES IN ANCIENT ARABIANS AND BEDOUINS

Modern humans living in the Arabian peninsula have been in contact with non-human primates (hereafter primates) since the Neolithic period, especially with the only endemic species to Arabia, the hamadryas baboon, *Papio hamadryas,* L. 1758 (**Fig. 8.1**), which can still be found in the highlands (up to 2,600 m above the level sea) of the Republic of Yemen and the Kingdom of Saudi Arabia (Mittermeier et al., 2013: 664–665). Arabian baboon populations were believed to have been introduced from Africa in historic times, but recent molecular evidence (Wildman et al., 2004: 7) suggests that the ancestors of today's Arabian baboons crossed the Red Sea region during a low sea-level event in the Middle Pleistocene. Since the Bronze age (3200 BCE), Neolithic human groups likely occupied Yemen to take advantage of what was then the moister climate of Arabia, settling in large and elaborate villages in the Western and South highlands (Edens, 2002: 81). These communities relied on foraging, herding sheep and goats and farming for their livelihood, cultivating wheat, barley, sorghum and millet well before the Islamic era (Varisco, 2009: 14). Human interactions with baboons were probably agonistic, given that in contemporary traditional agro-pastoral societies, most interactions with baboons involve the latter raiding crops, or killing or injuring immature goats and lambs (e.g., in Awash National Park, Ethiopia, see Mesele et al., 2014; or in South Africa, Cheney & Seyfarth, 2010: 34). In ancient Yemen, baboons were likely to have been a major agricultural pest. This is also corroborated by the testimonies of some authors of the early Islamic period. According to the testimony of al-Hamdānī, (Abū Muḥammad al-Ḥasan ibn Aḥmad ibn Ya'qūb al-Hamdānī, 892–945), an author of the early Islamic period and known for being "the voice of Yemen" during the Abbasid dynasty (750–1258), baboons (*kird*) and hydrax (*wabr*) were considered pests at Jabal Tukhlā near Aden (cf. Varisco, 2009: 14). Al-Masʿūdī (ʿAlī ibn al-Ḥusayn ibn al-Masʿūdī, c. 896–956), the famous Arabian historian and geographer, reports that the Yemeni term for baboon was *rubah* (meaning fox), a usage which continues to the present day, probably referring to their role as a predator. He also describes the territories of Wādī Nakhla, a day's journey between Zabīd and al-Janad, and those of Ṣanʿāʾ and Kuḥlān, as having many baboons (Al-Masʿūdī, 1861–77, *Les Prairies d'or* vol. 9: 53).

Representations of baboons are absent in pre-Islamic times. They are absent in the ancient South Arabian graffiti and are not included among the sacred animals of the pantheon of South Arabian pre-Islamic religions, such as that of the Sheba Kingdom, where we find animals such as the bull, sacred to the God *Almaqah*; the eagle, symbol of the deity *Sayīn*; the ibex, often represented as common motif; or the camel depicted many times in a votive form (Simpson (ed.), 2002: 99). Primates appear sometimes (cf. Simpson (ed.), 2002: 177) in the form of apotropaic amulets, together with sphinxes, mice, serpents, dogs, antelopes and other animals of probably Egyptian origin, since trade and commerce of Arabs in the Red Sea was developed already since the 10th century BCE (Sidebotham, 1996: 786; Sperveslage & Eichmann, 2012: 172). However, the legends surrounding Yemenite monkeys confirm that the nature of their relationship with people was firmly routed in the pre-Islamic Semitic tradition (Kruk, 1996: 34). The pre-Islamic legend of the pact the Jewish King Solomon made with monkeys, to guard the demons vanquished by the genii, continued to be reported until the 10th century CE. For example, al-Masʿūdī wrote that among these troops of monkeys is found "one larger than the rest wearing at its neck the seal bestowed

[1] Arabic primary sources were translated in English by Alghamdi Malak. The translations in English of sources edited in French, Italian, Latin, etc., are by Cecilia Veracini.

DOI: 10.1201/b21963-10

FIGURE 8.1 Adult male hamadryas baboon. (Drawing Cecilia Veracini.)

FIGURE 8.2 Females and young of a Saudi Arabian baboon harem in the Hijaz Mountains, Western Saudi Arabia. In recent years, baboons have become increasingly dependent on human food sources. Local people distribute increasing amounts of food to baboons because some still believe that 'monkeys and pigs are people punished by God for disobeying divine law,' and, as a result, baboons are fed to earn spiritual credits and seek mercy from God (cf. Teller, 2012). In addition, feeding baboons has become an attraction for tourists. This has led to an increase in the baboon population, and because their higher numbers over the past three decades, human-baboon conflicts in the Kingdom of Saudi Arabia have escalated. The National Center for Wildlife is studying the baboon population to mitigate this conflict and promote better human-primate coexistence (Alboug & Pebsworth, 2023). (Image source: Wikimedia Commons, photo credit: Nick Shields.)

by Solomon" (Al-Masʿūdī, 1861–77, *Les Prairies d'or* vol. 9: 50–51), and many other sources have remarked on the intimate relationship between humans and monkeys. The Spanish Arabs of Middle Ages designated a monkey with the tail *maymūn*, the "Yemen Monkey" (*quérd al Iemén*) (see Pedro de Alcala, *apud* Blondheim, 1912: 262), probably because in the Arabic/Muslim world, they were especially abundant in Yemen, and this is supported by the words of an ancient proverb: *donum regionis Jemen simiae sunt* (monkeys are a gift of Yemen). Thus, the zoonym *maymūn,* (pl. *mīmūn),* may originally designate the baboon, as hamadryas are the only monkeys in Yemen although historical sources on this point are not clear (cf. Blondheim, 1912), but this word is still used in Arabic and Turkish (cf. also Masseti, this volume).

The habits and behaviour of baboons were probably well known since ancient times when contact with these animals was part of the daily life. Farmers and Bedouins must have had an intimate relationship with these animals, as they have today. Contemporary Bedouin of the Ṭāʾif region, located in the western province of Makka, have a sufficiently close relation with baboons that they share similar gut biomes (Angelakis et al., 2016), and baboons commonly consume the leftover food of Bedouins, and are fed by the Bedouin (Biquand et al., 1992). As reported by Teller (2012), growing commensalism in the Ṭāʾif region, especially related to tourism, has been a problem for a long time, but in the last years, several thousand individuals have become reliant on human food sources (see **Fig. 8.2**). As noted by Viré (1986: 131), according to Bedouin's animistic and polytheistic beliefs, pre-Islamic Arabian Bedouin viewed baboons "with respect and circumspection or with element of superstitious fear or even as taboo". Kruk (1996: 37) observed that "in Bedouin populations the views on matter such as the classification of animals and their relationship to man (as well as on many other things) often have nothing in common with that of official culture". Evidence of these different visions can be found in Arabic oral poetry transcribed and preserved since 1500 years ago. The Arabian Desert has long been the locus of an ancient history of poetry, beginning with the pre-Islamic odes and including an on-going tradition of transmitted and disseminated oral poetic forms in many different Semitic languages. They present a profound poetic relationship with animals and the pastoral landscape (e.g. Al-Ghadeer, 2009). Maʿāza Bedouin living in the northern half Eastern Desert of Egypt, who emigrated from Arabia in the 19th century, regard the human being, the "Son of Adam", as a type of swine. They recognize various categories within the "swine category"; human beings, human-like swine; the *namnam* (chimpanzee); and the *silʿawi* (gorilla), "who feed upon man, but have religious and other cultural practises", and the proper swine (*khanzir*) who include the monkeys (*gird*), domestic pig, donkey, horse and all carnivores. They believe that "God also gave man the burden of learning how to live responsibly with other living things. God gave man a soul and superior intellect" (see Hobbs, 1992: 87–102; Appendix).

Other contemporary Bedouin ethnic groups have a different view of primates, as ugly, bad and false. This more negative viewpoint may have been influenced by contact with other cultures such as the *shabāha* (similarity), a form of satirical poem in which someone compares another person to a monkey or a dog, while the other person answers back in the same vein. The *shabāha* is usually practised among circles of elderly people, and it is common among the Bedouin of the Jawāzī tribe (cf. Al-Jawazi, 2010: 216), but the majority of Bedouin consider some of its vocabulary abusive. In the Arahwat tribe of the Bedouin of Sinai, "the word *gird* (monkey), is used quite frequently always in a negative context although they did not recognize the animals when researchers showed it in a photograph during interview" (Stewart, 1990: 101).

Much of Islamic lore about baboons, or about primates in general, such as the prohibition to eat monkey flesh, comes from sources of pre-Islamic knowledge. Under the scheme of the Quranic law, eating the flesh of the monkey is forbidden, since the animals are omnivorous and have canine teeth:

> The animal should be a *Ḥalāl* (lawful) one, according to the Sharīʿah. Dogs, cats, monkeys or the meat of any such *Ḥarām* (unlawful) animal can be eaten by a Muslim only in an extreme urgency, when a person is threatened with starvation and his life has to be saved. (Samiullah, 1982: 75)

Another reason to avoid eating monkey flesh was an instinctive repulsion. "Some claimed that the pre-Islamic Arabs did not eat it, as it is unpleasant enough to forbid itself; a prohibition would thus have been otiose" (Al-Jāḥiẓ, 1938, XX 61.3 *apud* Cook, 1999). Nevertheless, ancient medical opinion held that the flesh of the monkeys protects against elephantiasis, whereas other parts of the animal, such as the teeth, the hair and the skin possessed "specific qualities" as told in the *Ḥayât al-ḥayawâne* (*A zoological lexicon*) by al-Damīrī, one of the most famous Islamic encyclopaedist of the 14th century:

> if a monkey scratches a person with its teeth then that person is not going to be overcome by sleep or fear at night. Moreover, eating its flesh prevents leprosy and its skin protects trees from cold and plants from locusts and a man will be speechless once he drinks hot blood of a monkey. It was also said that monkeys cry and panic when they see poisonous foods. (Ad-Damīrī, 1844: vol. 1: 254)

Other circumstantial evidence of frequent connections between baboons and humans is the incorporation in pre-Islamic times of aspects of baboon behaviour in elements of anthropomorphism. An example of this comes from the renowned Islamic scholar of the 9th century, Ibn Qutayba, who confronts an embarrassing tradition concerning an "adulterous monkey female," which traced back to the observation of a man named ʿAmr ibn Maymin (cf. Cook, 1999). ʿAmr testimony dates back to before his conversion to Islam, an important aspect since Islamic authors tend to differentiate what was produced before and after the Prophet's arrival. ʿAmr was an Awdi tribesman who grew up in south-western Arabia before the rise of Islam; he later moved to Kufa in Iraq, and, perhaps at that time followed Islamic religion. As reported by Cook, there is a long tradition of citation of ʿAmr's anecdote. In one of the longer forms, ʿAmr looking after his family's flocks on an eminence in Yemen observed that:

> A male and a female monkey appear; he rests his head on her hand [and falls asleep]. At this point a younger male appears and makes a sign to her. She gently withdraws her hand from under the head of the older male, and follows the younger one, who copulates with her. She then returns and gently begins to reinsert her hand under the older one's cheek. He wakes up in consternation, sniffs her, and cries out. When the other monkeys gather round, he again cries out and points to her with his hand. The monkeys then exit to the right and left, and come back with the younger male. Thereupon they dig a trench for the adulterous pair and stone them. (*apud* Cook, 1999: 47)

ʿAmr assures us that he recognized the male and female culprits. The episode illustrates a supposed natural morality that monkeys and other animals were supposed to share with human beings, and denotes the belief that monkeys have the strong propensity to immorality. From a contemporary ethological point of view, this seems unlikely, but parts of the story can be seen to be consistent with baboon behaviour (sniffing genitals, extra pair copulation, etc.), showing that part of the ethology of this species, after removing the anthropomorphic interpretations, was already well-known at the time. In fact, much of hamadryas baboon zoology we find in Medieval Arabia seems to have been produced by rural or nomadic people whose voice is found in the work of the scholar Al-Jāḥiẓ (c. 781–869). He notes, that although monkeys are "the most adulterous of animals", "the strength of marriage and the sexual jealousy is prevalent among them". "They combine both virtues in a way not found among other animals" (Al-Jāḥiẓ, 1938: 4:98.8, 98.12).

The different life styles of the many ethnic groups occupying the Arabian peninsula in pre-Islamic times resulted in a variety of attitudes and approaches to primates. Sedentary communities were reported to have domesticated young baboons (Viré, 1986; 131) to turn millstones, pick up fruits or guard goods (e.g. al-Damīrī and al-ʿUmarī) still in the 14th century:

> The people of Yemen also taught the monkeys some skills to perform and certain things to do like guarding shops. For

> example, the butcher and the grocer could teach a monkey to keep their shops safe till they came back. Moreover, the monkey can learn how to steal: Al-Sheikhan narrated that Al-Kaqdy Hussein said "if a monkey was taught how to steal and infiltrate houses, it should not be punished by law because animals are incapable of choosing". (Al-Damîrî, 1844: 247)

Other services reportedly rendered by baboons, perhaps with a dose of fantasy, were described by Buzurk ibn Shahrayār al-Rāmhurmuzī in his book, *Wonder of India* (*Kitāb ʿajāʾib al-Hind,* c. 953) where he collected witnesses of sailors and other peoples:

> XL. A blacksmith of Zhafar [Zafar, an ancient place in Yemen] had a monkey who helped him with his bellows all day long. This monkey had served him for at least five years because on various trips I made I saw this animal with him every time. XLII. A person told me that he had seen in a village, in a trader's house, a monkey who tended it; it swept the house, lit the fire under the pot, and fuelled the fire carried the necessary wood, swept the flies off of the table, and wind his master with a fan. (Al-Rāmhurmuzī, 1883–1886: 77)

This kind of exploitation is not surprising given that, until the last century, baboons were used in Africa as shepherds, or to keep track of the trains in small stations (cf. Cheney & Seyfarth, 2010: 36–37). An image reveals how even in ancient (14th century) Persia, monkeys, in this case probably a rhesus monkey (*Macaca mulatta*), were used to perform various tasks, including picking fruit (cf. Al-maḫlūqāt, 14th c.).

The long tradition of trade between Southern Arabia and Eastern Africa and elsewhere in the Indian Ocean brought the Arabian world into contact with other primate species. The Arabs began exploiting the Red Sea as a communication and commercial thoroughfare long before the Romans arrived in 30 BCE. The Kingdom of Axum and several of the South Arabian Kingdoms (e.g. Maʿīn, Saba and later Ḥimyar and the Kinda federation) dominated the trade in the southern part of the Red Sea, throughout the last the centuries BCE until the 7th century CE (Sidebotham, 1996: 786). This meant that Arabic populations living on the coast were exposed to imported primate species, particularly grivet (*Chlorocebus aethiops,* L. 1758) and perhaps vervet (*Chlorocebus pygerythrus,* F. Cuvier, 1821, **Fig. 8.3**) monkeys, both common species in the Horn of Africa, as well as the mantled guereza (*Colobus guereza*, Rüppell, 1835) and perhaps the Olive Baboon (*Papio anubis,* Lesson, 1827). Grivet (or vervet) monkeys were appreciated for their small size, agility and acrobatic behaviour, which were exploited by street entertainers who performed with animals. According to Islamic source, Yemenite people, who have a long tradition of using monkeys in this way, were referred to disparagingly as "monkey tamers" (cf. Viré, 1986: 132). Moreover, during the Neo-Assyrian and Neo-Babylonian periods, the kingdom of Saba (i.e. Yemen and its colonized Ethiopian territories) seems to have been a source of monkeys or baboons (Potts, 1997: 259–260).

FIGURE 8.3 An infant of Vervet monkeys, *Chlorocebus pygerythrus*. (Photo Marco Alesci.)

More recently, in Islamic times, hunters and seafarers from the coastal regions of the Ḥaḍramawt in Yemen went, via Socotra Island, to the country of the Sufāla and Zanj in the Swahili Coast, in search of white and black colobuses (*Colobus* spp.) considered, because of their fine black pelt, an item of great luxury. It was so sought after that pelts were exported as far as central Asia and Europe (Viré, 1986: 131; cf. also about guereza trade Veracini, Chapter 2, this volume). Al-Idrīsī (1619, 1st climate, Section 7) told of an island in the Indian Ocean full of monkeys (probably Pemba Island) where hunters from Khūriyā-Mūriyā and Socotra went to trap monkeys:

> Depending on the species, some were killed and their pelts sent to the fur market, and others, with collars on their necks, were sold at a high price either to merchants or to aristocratic collectors of rare beasts or to itinerant animal showmen. (Viré, 1986: 132)

One of these species, the vervet monkey, is still found in some of the Indian Ocean Islands, such as Pemba, Mafia and the Manda, in the Lamu Archipelago of northern Kenya (Mittermeier et al., 2013).

Vervet or grivet monkeys together with leopard skins, antelopes and birds of prey, were regarded as suitable princely gifts and were frequently used as tribute both to Egyptian Pharaohs and Ancient Assyrian rulers (cf. Moshe, 1978: 22; Aruz, 2010: 9; Masseti, this volume). A monkey was among the gifts sent by the fifth Abbassid Caliph Hārūn al-Rashīd to Charlemagne when ambassadors were exchanged between the two sovereigns in 797–801 (Viré, 1986: 132). Later, monkeys were offered by the Amir of Oman to the Abbasid Caliph al-Muqtadir bi-llāh, who reigned from 908 to 932, to add to the court menagerie in Baghdad (Murudi, I, Section 487, *apud* Viré, 1986).

Commercial sea routes to Eastern India were opened up, perhaps, by the 6th century BCE. With help from monsoon winds, boats from the Gulf of Aden could sail directly to the Indian peninsula, pass down the coast of India, then continue on to Ceylon, Malayan or Sumatra (Tibbetts, 1956: 184). In Roman times, people of present-day Oman had commercial relationships with India. According to the Schoff's comments about the *Periplus*, a famous ancient anonymous document of the 1st century, Arabic people of the vales of Oman brought back the worship of the Indian Monkey God Hanuman:

> In the rich Wadi Tyin in Oman, the trade of which passed through the port of Kalhat – that *Acila* of Pliny (VI, 32) from which persons embarked for India-- General Miles found a town called *Sibal* (a word related to monkey) [cf. Gen. S.B. Miles, Geographical Journal, VII, 336], which he observes, means "monkeys" and was the name of a famous pre-Islamic idol. No monkeys exist in Oman, but a temple stood here dedicated to that image. (Schoff, 1912: 237)

Two shrines of Hanuman statues at Surat on the Cambay coast were still venerated in Schoff's time, and the inhabitants were in regular communication with Arabia. Schoff reported that according to local tradition, this was the original capital of Dravida Desam (one of the kingdoms in the southern part of present-day mainland India) and the birthplace of the dynasties ruling in Southern India at the time of *Periplus*. These travels, and the close commercial relationships, enabled contact with many types of monkeys and perhaps even with Asiatic great apes of which the stories of legendary creatures present in the Islamic world are testimony (see below about the *nasnas* and other myths of ape-men). The long tradition of seafarers of Sīraf and Oman navigating the Indian ocean, and then on to China, was the principal source of information on Asiatic fauna and flora in ancient times.

8.2 PRIMATES IN THE ISLAMIC/ARABIC[2] WORLD: NATURALISTIC, PHILOSOPHICAL AND RELIGIOUS POINTS OF VIEW

Primates have been part of Arab and Middle Eastern world since antiquity, and many of the ideas and concepts developed in the pre-Islamic Arab world were inherited by the Islamic world, and then mixed with ideas and concepts from Greek, Indian and Persian cultures. The rise of Islam was one of the significant events in the thousand years between the fall of Roman Empire and the discovery of the New World (Al-Khalili, 2011: 4). United by the monotheistic religion revealed to the Prophet Muhammad in the early 7th century, the Arabs overthrew the Persian Empire in Iraq and Iran, and expelled the Byzantines from Jordan, Palestine, Syria and Egypt. The Muslims had replaced Sassanid power all over the Middle East by 700 CE, when Umayyad Caliphs took power. At this time, Muslim forces were advancing through North Africa and Asia. Further conquests by non-Arab Muslims carried Islam across India, where the Moghuls created one of the most sophisticated civilizations in history (Richards, 1996). Meanwhile, Muslim traders and missionaries spread the faith further east to Indonesia, the Malay Peninsula and southeast China (Burjor Avari, 2013: 131). Besides conquering territory larger than that conquered by Alexander the Great, this historical period is also remembered for its advances in culture and science. The Abbassids, of mixed Arabic and Persian blood, who claimed descent from the Prophet Al-ʿAbbās, ruled for five hundred years. Al-Ma'mūn Hārūn al-Rashīd (786–833 CE), the seventh Abbasid caliph, who reigned from 813 until his death in 833, led the Islamic Empire at the zenith of its power. He was apparently a cultured, passionate and enthusiastic man (Al-Khalili, 2011: 4), who created a unique environment that encouraged original thinking and free debate. Al-Ma'mūn's kingdom is marked by his efforts in the translation of Greek manuscripts, but also Latin, Persian and Indian ones, about philosophy and science. In this period, also called "Golden Age", we found many examples of philosophical treatises, encyclopaedic works and "Books of Animals", (*Kitāb al-Ḥayawān*) dealing with the natural world and animals (see Table 8.1). As well as being prime sources for a study of the zoological knowledge of Islamic scholars, they are also literary works, mainly concerned with philology rather than zoological detail (McDonald, 1988: 9). The origins of this type of writing are complex, with sources such as the pre-Islamic Arabian traditional knowledge, travellers' and sailors' tales (e.g. the *Livre des merveilles de l'Inde* of Buzurg Shahriyar, c. 950, already mentioned), along with the writings of Greek scholars, playing a major role. Among the latter, Aristotle is particularly significant because the *Historia Animalium, De Partibus Animalium* and *De Generatione Animalium* were translated quite early into Arabic, by Abū Yaḥyā ibn al-Bitrīq, c. 815 (cf. McDonald, 1988: 3). Islamic natural philosophy also has roots in the Islamic Mu'tazilism, a school of theology based on reason and neoplatonically oriented rational thought.

Abū ʿUthmān ʿAmr ibn Baḥr, better known by the nick name Al-Jāḥiẓ (776–869), born in Iraq and exponent of the Muʿtazilism, wrote one of the more important Book of Animals (*Kitāb al-Ḥayawān*) that was published in seven volumes (Al-Jāḥiẓ, 1938). He was a polymath, who wrote about literature, theology and politics, but he was a town-dweller and not a specialist in the study of animals. His sources were pre-Islamic and early Islamic poetry, historical accounts from the Jāhiliyya (previous period of Prophecy of Muhammad), early Islam, and the Umayyad and early Abbasid periods. He located monkeys[3], and other animals resembling them, on the penultimate step of the hierarchical "Chain of Being", (*scala naturae* of Greek

[2] Following the practice of Jim Al-Khalili (2011), we use the term Arabic in its broadest sense. Arabic science is that carried out by those who were politically under the rule of the Abbasids, whose official language was Arabic. A large part of the scientific works was initially (9th–10th c.) carried out in present-day Iraq. Many of the scientific figures were Persian and even anti-Arab, but most of the scientific work was written in Arabic.

[3] The term "monkey" (in Arabic *kird*), as used in this section, should be understood as synonymous with simians. In fact, although monkeys were the most well-known primates, we cannot exclude that Arabic authors might also include apes, Superfamily Hominoidea (see the next section).

TABLE 8.1
Arabic Sources (chronological order of authors' birth) and Their Studies Cited in the Text

Author	Source	Span Life	Studies
Al-Jāḥiẓ	*Kitāb al-Hayawān (Book of animals)*	c.791–c. 869	
Al-Mas'ūdī	*Maçoudi (Les Prairies d'or)*	c.896–956	
Al-Hamdānī	*Kitāb al-Iklīl. Al-Juz' al-Awwal (Crowns from the Accounts of al-Yemen…)*	c.892–945	Varisco, 2009
Ibn-Šhahrīyār Buzurg	*Kitāb 'Ajāib al-Hind …: (Livre des merveilles de l'Inde…)*	10th century	
Ibn al-Faqīh al-Hamadhānī	*Kitāb al-buldan (The Book of the Countries)*	869 - c. 951	
Ibn Abū Miskawayh	*Al-fawz, al-asghar (Le petit livre du Salut)*	932–1030	
Al-Marwazī	*Kitāb Ṭabā'i' al-Ḥayawān al-Baḥrī wa-al-Barrī (Nature of Animals)*	1056/57–1124/25	Iskandar, 1981
Al-Idrīsī	*Geographia Nubiensis: id est accuratissima totius orbis…*	c. 1099–1165	
Al-Qazwīnī	*Ajā'ib al-makhlūqāt wa-gharā'ib al-mawjūdāt (The Wonders of Creation)*	1203–1283	Bellino (ed.), 2008
Ibn Khaldūn	*The Muqaddimah (Prolegomena)*	1332–1406	
Al-'Umarī	*Masālik al-abṣār fī mamālik al-amṣār*. Vol XX.	1301–1349	Kruk, 1996
Ad-Damīrī	*Ad-Damîrî's Ḥayât al-ḥayawân (A Zoological Lexicon)*	1341–1405	
Ibn Qutayba	*Ta'wil mukhtalif al-hadith*	14th century	Cook, 1999
Al-Ma'lūf	*Mu'jam Al Hayawan. (Lexicon of Animals)*	Born 1949	
Muḥammad Farīd Wajdī	*Encyclopaedia of the 14th Century…. A Reference for the Arabic Language and the Universal Sciences.*	1875–1954	

origin) just below human males (Kruk, 1996: 30). In this interpretation of the world, there is progress from the simplest forms of the mineral world, through the vegetative, to the simplest animals and then up to the perfection of humans, with their rational soul and access to the spiritual world. Monkeys, because of their cleverness and perception, are the highest of the non-human animals. According to the Neoplatonist ibn Abū Miskawayh (Abū 'Alī Aḥmad ibn Muḥammad ibn Ya'qūb ibn Miskawayh) (932–1032), a Persian philosopher and historian, plant life is at the lowest stage of evolution, and in the initial stage of animality, among the senses, touch is the first to appear, and sight the last. With the development of the senses, animals, such as worms, reptiles, ants and bees, acquire freedom of movement. Animality reaches its perfection among quadrupeds, in the form of the horse, and among birds in the falcon, finally arriving at monkeys, which are just a degree below humans in the scale. The Muslim thinker stated that:

> monkeys then evolved into a lower kind of barbarian man, then becoming a superior human being, whereas man can become a saint, a prophet or evolve into a higher stage and becomes an angel. The being higher than angels is indeed none but God. Everything begins from Him and everything returns to Him. (Miskawayh, 1987: 116–173)

As observed by Kruk (1996: 30), the type of human beings above the monkey on the scale of life is the "animal-like" kind of man. The Central-Asiatic Turk, or the Negro, are still far removed from the optimal stage of humanity and were considered lower categories of human beings. Later on, the Arab historian ibn Khaldūn, (d. 1406) in his *Prolegomena*, a book written in 1377, which records an early view of universal history, asserted that humans developed from "the world of the monkeys", and noted:

> this is the case with simple material elements: with palms and vines, [which constitute] the last stage of plants, and snails and shellfish, the (lowest) stage of animals. It is also the case with monkeys, creatures combining in themselves cleverness and perception, in their relation to man who is the being who has the ability to think and to reflect. The preparedness for [transformation] that exists on either side, at each stage of the worlds, is meant when [we speak abou] their connection. (Ibn Khaldūn, 1980: 195)

Many Muslim thinkers developed an idea of transformation and change as a transition from one form to another, often in relation to processes that envisaged an organism's adaptability to the environment, and a struggle for existence, which some commentators interpreted as a conceptual corpus similar to the biological Theory of Evolution developed in the 19th century (Shanavas, 2010; Malik et al., 2017). For example, Al-Jāḥiẓ discusses the influence of environment saying that humans can revert to monkeys, if under adverse environmental conditions: "some may grow excessive hair, others a tail, still others develop different kinds of reddishness and some may start to look like monkeys" (Al-Jāḥiẓ *apud* Kruk, 1996: 31; Bayrakdar, 1983). Although the culture of Islam embraces a wide variety of approaches, this notion of "progressive transformation" may have its roots in the idea of metamorphosis (*maskh*), a concept of popular and pre-Islamic Arabic origin and Iranian tradition (Pellat, 1986: 737), as well as in the Classical civilization. In general, metamorphosis indicated a transformation into lower animals inflicted by the Gods to punish those who did not respect divine wills (e.g., the cercopes described in Ovid's book *Metamorphoses,* cf. Ovidio, 1992: Vol 2, Book XIV: 811–812, see Masseti & Veracini, this volume). The concept of metamorphosis in the Islamic world usually sees the transformation of humans into monkeys or pigs. Monkeys were mentioned in the Holy Book Kur'ān three times (2: 65–66; 7:163–166; 5:60) when God (Allah) turned some people into monkeys and pigs as a punishment for disobeying His commandments (cf. Silverstein, 2020). Also, the collection of ancient Rabbinic writings recounts a narrative about people who built the Tower of Babel in order to wage war. As punishment, God transformed them into monkeys and impure spirits (*Gemara,* Babylonian Talmud, *Sanhedrin*, f. 109b).

Obliged by the Qur'an to accept the reality of transformation (*maskh*), many jurists and theologians pondered its real meaning and wondered why when the God mentioned other animals in the Qur'an, he didn't turn His enemies into one of these animals? Or are the monkeys they saw alive actually metamorphosed humans? Al-Jāḥiẓ (1938, iv, 39) explains that God has chosen monkeys and pigs because they are more antipathetic than other animals, and monkeys are only mentioned a few times because metamorphosis into a monkey is the most severe punishment. Traditionally, elephants, hares, spiders, the eel and naturally the mouse, the monkey and the pig are transformed humans (Al-Jāḥiẓ, 1938, i, 390). However, many Muslim scholars disagreed, such as al-Damīrī (1844: 252), who wrote that after metamorphosis, animals do not live more than three days and cannot not eat, drink or have offspring. Metamorphosis, like evolutionary theory, establishes a genetic relationship between humans and monkeys that creationism precludes. The taboo on eating monkey flesh could be interpreted as support for monkeys being metamorphized humans.

Other books about animals, or later encyclopaedic works, provide more insight into monkey lore in the Islamic period. Sharaf al-Zamān Ṭāhir al-Marwazī's (fl. 11th–12th century) *Nature of Animals* (*Kitāb Ṭabā'i' al-Ḥayawān al-Baḥrī wa-al-Barrī*) (cf. Iskandar, 1981) is especially important because it covers a wider range of animals, and provides more information, than its Arabic predecessors, and while his quotes from the Ancients, which cover a period of 16 centuries, repeat common errors he made some interesting personal observations. Many passages from Timotheus of Gaza (fl. c. CE. 500), parts of whose book are now lost, are preserved in *Nature of Animals*., and in the part of the work called "*Treatise II: On domestic animals, wild beasts, animals of prey, and other quadrupeds*", Marwazi uses their behaviour, habits and some anatomy to differentiate among types of primates, but he relied on Galen's statements rather than investigating their anatomy himself:

> As to the monkeys called *pithekos* in Greek, they have tails; in any case, the bellies of monkeys are similar to those of man. Galen slew and dissected monkeys in order to learn about the anatomical condition of human organs, until he ascertained their qualities, quantities, and places, and nothing remained mysterious to him…. (Al-Marwazi *apud* Iskandar, 1981: 129)

Muḥammad ibn Mūsā al-Damīrī (1344 ca.–1405), an Egyptian writer on canon law and natural history, wrote the first systematic work on Arabic zoological knowledge. The work exists in various versions, of which the most comprehensive is *Hayāt al-Ḥayawān* (*The Life of Animals*). The book, which is mostly arranged alphabetically, includes descriptions of animals drawn entirely from literary sources (e.g., whether it is lawful to eat, medical properties, its meaning if seen in a dream, and proverbs that mention the animal). What Ad-Damīrī thought about monkeys is presented here:

> The monkey is a well-known animal, it is dubbed *Abo-Khalid*, *Abo-khabib*, *Abo-khalaf* and other different names. The monkey is viewed as an ugly creature but it is smart and educable. […] According to Ibin-Abas and ikrima (may Allah be pleased with them), monkeys are not perfect but they are finely -formed and some specimens are better than others. Allah says "Verily, we created man in the best state" [al-Teen 95:4]. A monkey can give birth to ten or even twelve babies. Male monkeys are very jealous concerning female monkeys. Unlike other animals, Monkeys are very similar to humans in most aspects because monkeys can laugh, sing, learn, socialize, and grab things with their hands. Like humans, monkeys that do not know how to swim will drown if they jump into water. Monkeys are also conservative and deal with their sexual needs in ways that are similar to human activities. Moreover, female monkeys carry their babies in a way similar to what women do with their children. […] Another unique feature of these animals is their tendency to sleep beside each other in lines and they wake up in an organized manner. Monkeys can also learn how to ride animals such as donkeys and they can even learn to race. In his book "Alkaamel", Ibn-day has narrated a story about a monkey in the Al-ramla area that could shape material with the help of a man. Imam Ahmad narrated that the Prophet said "there was a man on a ship who used to mix wine with water and sell it; the man was also accompanied by a monkey; yet the monkey took the money and threw the half of it into the water". (Al-Damîrî, 1844: 36, 68, 223)

As these excerpts from ad-Damīrī and other sources (see the important encyclopaedic author of the 14th century: Al-Umarī, 1989: *apud* Kruk, 1996: 29 or Al-Qazwīnī, 1280 see **Fig. 8.4**) show monkeys were perceived as a mix of good and bad. They could show moral sensitivity similar to that of human beings, they can learn to use tools and live in society, and female monkeys carry their babies in a way similar to what of human women. Yet, they were also considered ugly and treacherous, with an unrestrained sexuality that easily leads to treacherous immorality. Proverbs or beliefs support this duality. Ad-Damīrī (1844) quoting al-Qazwīnī (1280), said "whoever wakes up in the morning seeing a monkey for ten days will hear good news and he will be loved and get rich" and at the same time referred to proverbs that denote a negative quality: "Prostrate to the vicious ape as long as you are in his land and has the power" or another one which says: "that person is more talkative or more vicious than an ape".

FIGURE 8.4 Image of the entry of monkey, from Al-Qazwīnī's (1280) *Marvels of Things Created*...: "It is an ugly and friendly animal that laughs and winks, and understands quickly, and is able to learn precise crafts such as weaving. In fact, wide clothes are not woven by one person. The weaver teaches a monkey, the craftsman teaches the monkey, throwing the fabric to the monkey who throws it back to him. The king of Nubia gave the Caliph al-Mutawakkil two monkeys, one of which could weave (he was a tailor) and the other was a goldsmith". (Translation from Arabic by Marco Ammar.)

8.3 PRIMATE DIFFERENCES

Islamic scholars provided simple descriptions of the different types of primates as well as information about their geographical distribution based on sailors' and travellers' stories, mixed with Greek and other ancient myths. In general, they did not pay much attention to the presence or absence of the tail, a taxonomic character considered important to the classical authors (e.g. Aristotle). In the 10th century, al-Masʿūdī provides a classification of the species he was aware of:

> Monkeys live in hot countries; they are native to Nubia and the high part of Abyssinia in the direction of the upper source of the Nile. The individuals who belong to this species are called "Nubian", they are small and have small faces. They are always seen with monkey exhibitors who make them climb to the top of a spear. [...] There is also another species in the numerous creeks that form the sea of China on the Zabedj coast and the Empire of the Maharadja, king of these islands [n.d.r. Borneo, Java and Sumatra]. [...] This species of monkey is very well known in this nation; it exists in large numbers and is also a perfect type of monkey. One of these was offered to Moktadir-billah with a large chained snake. They had beards and long moustaches, and there were old and young ones. [...] The sailors of Siraf and Oman who always follow this route know this species of monkey very well. (Al-Masʿūdī, 1861–77: 51–52)

The first species likely refers to the grivet, *C. aethiops*, which still lives in the geographical regions cited by the author and was commonly used by the street entertainers, while the other species probably refers to the crab-eating macaque (*Macaca fascicularis* Raffles, 1821), which is abundant in Southeast Asia. Al-Masʿūdī also gave a detailed, almost naturalistic, description of Yemen baboons' behaviour talking about a Yemen valley where they gathered in large groups:

> The monkeys gather here in large numbers; they are divided into many groups each of which is led by a *hazar*, as we call the largest and strongest male that is in charge of others. Females can give birth to up to ten or twelve babies, like a female pig [quoting here al-Qazwīnī]. Females behave in a motherly way to their children. The monkeys hold public meetings and assemblies or gather in large numbers, and we hear the piercing and confused sound of their shouting. Females are separated from males as we separate women from men. Whoever hears these nocturnal meetings, whose participants, hidden in the mountains and among the trees, he cannot distinguish, could not doubt for a moment that there is a great gathering of men near him. This illusion also occurs during the day. There are no places where the monkeys are more beautiful nor more mischievous: we call them *er-roubbah*. The males, like the females, have abundant hair that falls on their shoulders and in some the colour is darker. When they are reunited, they sit, each according to their rank, below the chief, and they are in everything that they do very similar to men. (Al-Masʿūdī, 1861–77: 51–52)

According to al-Masʿūdī, a different kind of monkey was found in the "Slaves' countries",[4] although there are no primates known from this region. Al-Masʿūdī, treats them as a mysterious people of anthropoid apes who lived somewhere in central Asia:

> between four mountains, near the edge of the sea, there is another shallow depression which contains woods of various kinds and is inhabited by a kind of monkey of average size, which have round faces, and a human appearance, even though their bodies are covered with hair. It is very rare for these monkeys to fall into the traps that the hunters set for them; when you come face to face with one of them, you are amazed at their intelligence and their subtlety; even if they can not express themselves in words, they perfectly comprehend our language. (Al-Masʿūdī, 1836: 54)

Mention of anthropoid apes in this region is also found in naturalistic treatises of the 16th century (cf. Veracini, 2019). Perhaps this can be traced back to ancient legends concerning the presence of wild men and later re-interpreted as "ape forms". Arabic imagination was full of images of ape-men, half-men or other anthropomorphic creatures. It is difficult to say whether these creatures are the product of real contacts with anthropoid primates, whose description was then distorted by the imagination. or whether they are just mythological creatures. For example, contact with the African great apes seems to have already taken place in the Middle Ages (or before). Evidence for this is found in an Arabic iconographic source of the 15th century from Central Asia, where a gorilla and perhaps some chimpanzees are portrayed in the form of demons (Veracini, 2021). These beliefs may also have originated from ancient myths of monstrous creatures, common to many cultures,

[4] This is difficult to interpret: it could be Slavonic (of the Slav ethnicity) or Slaves' (those who act as slaves to another ethnicity).

and reinforced by the contribution of Classical authors who described borderline creatures, such as *dog-headed* or *headless* men located at the edges of the known world, or beyond. These monsters had been described as associated with the western lands of Libya by Herodotus in the *Histories* (cf. 1921 (Godey ed): IV, 191–192). Other classical and medieval authors (e.g., Ctesias of Cnidus, *Indika,* 5th century BCE; Pliny the Elder, *Historia Naturalis* of 23–79 CE; Solinus *Collectanea rerum memorabilium,* 3rd century) also reported these creatures (cf. Masseti & Veracini, 2016: 47–48).

Fabulous creatures are abundant in the typical Arabic *mirabilia* literature, such as the *The Wonders of Creation* of the medieval cosmographer, Zakariyā' al-Qazwīnī (1203–1283), who wrote the outstanding cosmography in Arabic literature. His work is divided into two unequal parts, the first dealing with the heavens and the second with earthly matters (divided in turn into Fire, Air, Water and Earth). In the part dealing with the land, there is a description of mountains, rivers, etc., of minerals and plants, and finally of animals, subdivided into humans; *jinns*; horses and similar animals, cattle, wild beasts, birds and finally vermin and creeping creatures (McDonald, 1988: 5). Among many fabulous creatures, Al-Qazwīnī (1280) reports that:

> Ibn al-Faqih claims to have seen on the island of Zābaj [likely the island of Giava] quadrupedal animals never seen elsewhere.[...] And there were monkeys similar to buffaloes or rams, and another type with white breasts and black backs. And Zakaria Ben Yahya Ben Khalliqan reports: I have seen on the island of Zābaj individuals with the body of human, who eat and drink like human, and have wings to fly from tree to another and articulates words similar to the voices of starlings, of which nothing is meant, and they are white, black and green. (Al-Qazwīnī, 1280: 59, Translation from Arabic by Marco Ammar, see **Fig. 8.5**)

Although these tales were part of an oral tradition, they managed to capture some characteristics that can be attributed to genuine primates. In Java, there are several species of primates, including *Presbytis comata* (Desmarest, 1822), which has a 'white breast and black back' (see **Fig. 8.6**). Other langurs or lutugs of the genus *Trachypithecus* Reichenbach, 1862, endemic to the island of Java, are characterized by varying colouration ranging from yellow to black (*Trachypithecus auratus* and *Trachypithecus mauritius*). As for human-like flying creatures, the observer likely encountered the Java gibbon, *Hylobates moloch* Audebert, 1798 (see also later al-Idrīsī, 1619 description).

Orang-utans and gibbons probably came into contact with Arab travellers and merchants who had been sailing the Indian Ocean since antiquity. In his only surviving work, "*The Book of the Countries*" (*Kitāb al-buldan*), written around the year 903 CE, ibn al-Faqīh al-Hamadhānī, an Iranian author of geography written in Arabic, describes his native town Hamadān and many countries including Sumatra (*al-Ramnī*), where he encountered such anthropomorphic creatures:

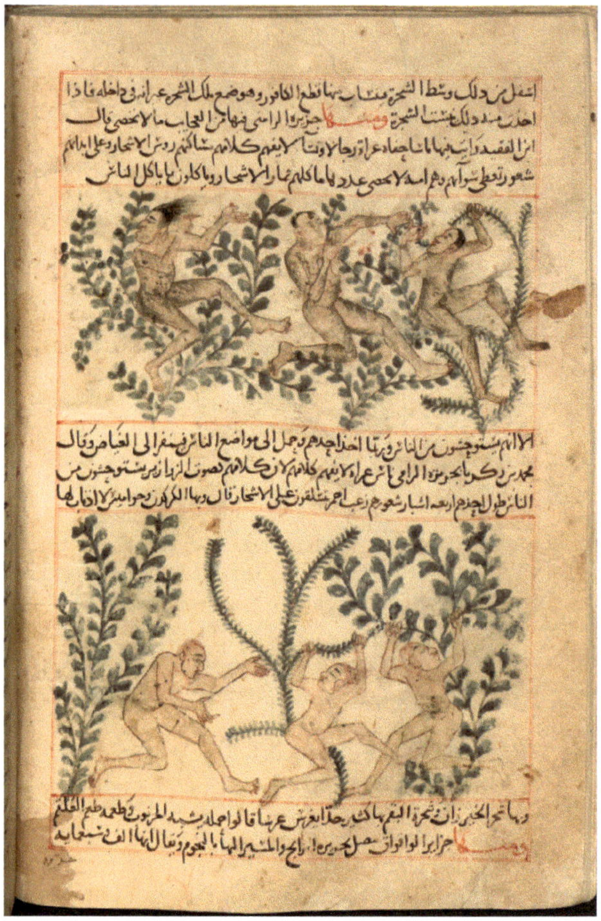

FIGURE 8.5 Image of forest dweller creatures in the Island of Java, probably referring to gibbons, according to the text of Al-Qazwīnī (1280): "...individuals with the body of human, who eat and drink like human, and have wings to fly from tree to another and articulates words similar to the voices of starlings ...". (Translation from Arabic of Marco Ammar.)

> They are human beings who go bare-footed and without clothing. Their speech is incomprehensible, and they live in the tree-tops. They are entirely covered with hair and they live on a diet of fruit; they are very numerous, and flee at the approach of man. If one of them is brought to places inhabited by man it does not stay there but runs away to the jungle. Another man, Muhammad ibn Zakariya' ar-Razhi, has stated that on the island of *al-Ramnī* there live naked people. They shun the vicinity of other people. They are four span in height, their language is a kind of whistling and their faces are covered with russet-coloured down and they climb up trees. (cf. Al-Qazwīnī, 2008: 201)

On the island of Sumatra (Indonesia), the common, or white handed, gibbon (*Hylobates lar vestitus* Miller 1942) shares the tree tops with the Sumatran orang-utan (*Pongo abelii* Lesson 1827, **Fig. 8.7**), which, like gibbons, rarely come down from the trees due to predators, such as tigers and possibly sun bears. Viré (1986: 133) thought this description best corresponded to gibbons rather than orang-utans (both russet-coloured and both living in Sumatra), especially because of the small size (four span), which refers to an animal of about 1 m. However, Sumatran orang-utan females

FIGURE 8.6 *Presbytis comata*, a langur species, endemic to Java. (Source: *Iconographia_Zoologica*, 1700–1880. Special Collections University of Amsterdam, UBA01_IZ19900062.tif, image of public domain.)

FIGURE 8.7 Orang-utan of Sumatra, adult male (*Pongo abelii*). (Photo Antonio Alexandre.)

seldom, if ever, reach a meter in height, so could also fit this description. Other reports about Sumatra island by al-Idrīsī appear to give a better description of gibbons:

> their movement through the trees is accomplished by means of the arms alone and no use is made of the feet, which gives the impression that they are flying from one branch to another. (Al-Idrīsī, 1619: 1st clime, Section 8)

But al-Qazwīnī gave more description of the reddish-coloured people, this time living on the Island of Java:

> In this Island there is a mountain in which there is a huge fire [probably a volcano]. At night it is seen at a great distance, and by day it can be seen by its smoke. No one can approach. On this island there are aloe wood, bananas, coconuts and sugar cane. Its inhabitants are reddish-colored people and have human characteristics, except that they have their faces on their chests. (Al-Qazwīnī, 2008: 210)

It is true that orang-utans seem to have no necks, and in fact, their faces seem to be on their chests, which corresponds to al-Qazwīnī's description. Questions arise about the mythical headless creatures whose origin is lost in the mists of time. Could this myth have originated from the physical characteristics of the orang-utan?

There has been much speculation about the anthropoid creatures mentioned in these travel reports, in particular, because they are reminiscent of the mysterious creature of the medieval Islamic tradition called the *nasnas* (Arabic: *nasnās* or *nisnas*, pl. *nasanis*) described as "demi-man with a human face and vertical stance, without a tail and possessing the faculty of speech, but also covered with a thick fleece, usually "russet-coloured" according to Viré (1986: 133). The *Nasnās* is associated with a wide range of popular lore about strange anthropomorphic beings supposed to live in mysterious foreign countries, and it is, at times, also described as a being, one-eyed, one-armed and one-legged – literally half-human, according to ad-Damīrī:

> It has been said in Almohakim's book that *al-nesnas* was created in the same form as people and it is a kind of creatures that can hop on one leg. [...] In the book of *Tareekh Sana* (*The History of Sana*), there is a story which indicates that a merchant travelled to their lands and saw them while they were hopping on one leg, climbing trees and running away from dogs. (Ad-Damīrī, 1844)

Sometimes they are also reported as monstrous creatures hunted for their meat in Yemen. Legend suggests they were originally ordinary people, but were transformed into half-humans as punishment for their godless behaviour. Al-Jāḥiẓ, who has described many fabulous beings of popular imagination, assimilates the *nasnās* to the *wak - wak* (one of the many Arabic demons) saying they are the Biblical people of Gog and Magog (Al-Jāḥiẓ, 1938: i, 189: VII, 178). This idea is quoted by ad-Damīrī who added "they are created in the form of people, yet they are different in certain aspects and they are not sons of Adam" (Ad-Damīrī, 1844). It is evident that these anthropomorphic creatures (and monkeys) were associated with debased populations, or with ethnicities considered culturally inferior such as Jews:

> It is said that monkeys were descendent from Iram, the son of Sam and the brother of Aad and Thamoud, and that they live near the coast the sea of India and can speak Arabic, reproduce, sound and quote poetry; and they have Arabic names. (Ad-Damīrī, 1844)

In "The Story of the Sage and the Scholar", a tale from the collection of the *Arabian Nights*, a pupil is turned into a *nasnās* after a magician applies kohl to one of his eyes (Burton, 2008, Vol. 14: 57). According to Amīn Ma'lūf (cf. Al-Maalouf, 1932: 132), the *nasnās* has also been associated with human pygmies. The confusion between anthropoid apes and human pygmies has occurred many times across history until the recognition and description of all the great apes by naturalists in the 19th century (see Veracini, Chapter 7, and Barsanti, this volume).

8.4 MONKEYS AND APES IN POPULAR ISLAMIC LITERATURE, POETRY AND ART

Because of multiple contacts between primates and humans, and their apparent presence in the everyday life of Arabic people, many literary works, popular fiction, poetry and travel books show a great variety of types of stories featuring primates. Travel literature is rich in reports concerning these animals as protagonists of the merchants' and seafarers' adventures especially on their frequent Indian Ocean journeys. In the book *Kitāb ʻağāʼib al-Hind* (*The book of Indian Wonders*, ca. 953–956 or 985), a collection of 134 stories and anecdotes of sailors in the Indian Ocean by Buzurk ibn Shahrayār al-Rāmhurmuzī (cf. Buzurk ibn Shhrayār al-Rāmhurmuzī, 1883–1886), these animals appear frequently. In one of these tales, a mariner tells the author his adventure when he embarked. After arriving at their destination, the merchants had pulled the boat into a small bay, leaving him watching the boat with many supplies while they went to sell their merchandise in the city. He says:

> After their departure there came a group of monkeys that surrounded the ship, trying to climb onto it. I pelted them with stones. A big tailed monkey managed to get inside the ship. I repulsed it and beat it up. This distracted my attention and it succeeded in getting close to me when I was eating, and so I threw it a piece of bread. Then it stayed there some time and then disappeared from my sight. In the evening it returned carrying twenty bananas in its mouth. And it put them in front of me and I took them. From that moment on, it did not leave me anymore. Every day it went away and returned bringing me bananas and other fruits of the forest and then spending the night in the ship near me. Thus this she-monkey awoke my desires and I enjoyed my passion with her. Then she started to walk heavily and the signs she showed me made me realise she was pregnant. I felt extreme anguish and I thought of the shame when the others understood what had happened. So I fled and took the ship with a mattress, water and supplies when the monkey was absent. I left the coast at my own risk and peril. After twenty days and more of navigation during which I had risked death I reached the coast of one of the Andaman Islands where I stayed some time to find comfort and look for fruit and other food. I thought that when the crew returned to the ship, they would find a monkey who had given birth to a baby or two with a human face, a hairless chest and a shorter tail. And they would not have failed to suppose that I was the father. But they might also have thought the monkey had killed me and someone had taken the ship. The sailor who made this confession, says Mohammed, son of Babichad, can't see well, and maybe that's why he had this relationship with the monkey, also due to his long stay at sea. (Buzurk ibn Shahrayār al-Rāmhurmuzī, 1883–1886: 67–68)

The sexual use of these animals (*zoophilia/zooerastia*) by both men and women, appears several times in Arabic popular literature, based on the common idea that monkeys possessed inexhaustible sexual desire and immorality. Besides this story in the *Arabian Nights* (2011: 615), we find a princess who tried to satisfy her irrepressible sexual urges with a monkey (probably a big male baboon) whom her servant bought for her. Having to flee after her father discovers this, she takes refuge in another town, but here too she is discovered, and the man who desires her consults an old woman who by a magical device removes from the woman's belly a black and a yellow worm, fruit of her relationships with the monkey and a black man, and in the end she marries that man. As Kruk (1996) observes, on this and other occasions, monkeys are associated with humans considered lower in the scale of importance. Although transfigured by fantasy, and considered deplorable, we cannot entirely rule out such occurrences, for the aforementioned al-Marwazī tells about a market place in Aden, where female primates were offered for sale to visitors who could not afford to buy slave girls, and he describes how sexually demanding females of these species were, "so that its buyer had to return it to the ape-dealer, selling it at the low price of three dirham". He also adds:

> The vagina of the female monkey resembles that of a woman, while the penis of the male resembles that of a dog [here quoting Aristotle, *De animalibus*]. Male monkeys indulge in sexual intercourse; they love women, have coitus with them, and covet adolescent girls. He who acquires a monkey male for his household should know that he has brought a stallion among his harem.... (*apud* Iskandar, 1981)

Even today in certain parts of the world (e.g., Borneo and India), as NGOs sometimes admit, female orang-utans are victims of sexual abuse, thus this kind of behaviour likely occurred in the past. There even was legislation concerning intercourse between humans and non-human primates. According to *al-Baghawī*, a famous 12th century Persian jurist, in his "Islamic punishment of Adultery Section", "if a woman gave up her virginity copulating with a monkey, she should be punished. However, there is another interpretation, which exonerates such acts" (*apud* ad-Damīrī, 1844: 36).

Other themes related to monkey's encounters with shipwrecks are mentioned in two of the voyages of Sinbad the Sailor in the book *Arabian Nights* (See **Fig. 8.8**). On one of his journeys, Sinbad meets people who gather coconuts by pelting a group of monkeys with stones, who retaliate by throwing coconuts from the treetops. In the travels described in *The book of Indian Wonders*, victims of shipwrecks seemed to have benefited from the behaviour of some monkeys wearing gold on the beach (Buzurk ibn Shahrayār al-Rāmhurmuzī, 1883–1886). It is repeatedly reported in popular fiction that troops of monkeys assault boats arriving on the beach, or caravans, to steal food. This may not be far from the truth, given reports about the

FIGURE 8.8 Apes surrounding the ship of Simba in its third travel of the sea. (From: *The Thousand and one Night*, Harvey et al., 1859: 23.) (https://archive.org/details/thearabiannightvol3/page/n31/mode/2up. Image of Public domain).)

FIGURE 8.9 The monkey and the turtle. The story is about a monkey and a turtle who stole and planted banana plants. The tortoise's plant was tall and bore fruit, while the monkey's plant was dead. The monkey offered to climb the tree to get the bananas for the tortoise, but it ate them all by itself, so the tortoise got angry and killed it. The moral of history is that gaining something is often easier than retaining it. One should not be neglectful of what he already has. (*Kalīla wa-Dimna.*Timurid school, Painted around 1410–1420 CE. National Library of Iran (Folio 61 - Verso). (Source: *Wikipedia Commons.*)

interaction between monkeys and tourists (e.g., Lee et al., 1986; Fuentes et al., 2007). Thus, in *The book of Indian Wonders*, it is narrated that in certain Indian localities:

> On the Sanfin coast, in the Lameri valley and in Qagola, monkeys of extraordinary grandeur can be found living, they walk in troops, and each has its own chief, who is the largest of the troop. Sometimes they come out of the forest on the way and invade pathways. They get between the travellers and they do not allow them to go any further, thus they have to continue their walk without touching any kind of food. One narrator was traveling in a caravan which had encountered a troop of monkeys that had blocked the passage. And there ensued a battle. The monkeys jumped from all sides and stripped their clothes, breaking the vats with water and they were in the desert, far from a spring. In the end, the travellers gave something to the monkeys so they let them pass. And because of the lack of water, most travellers died: a small number only succeeded in reaching the nearest spring. (Buzurk ibn Shahrayār al- Rāmhurmuzī, 1883–1886: 66–67)

In the famous Indian book *The Panchatantra* (1965), which was translated into Arabic early on, there are stories involving monkeys (e.g., see **Fig. 8.9**); one of these, "*The monkey and the Wedge*", was a popular theme quoted and depicted repeatedly (see **Fig. 8.10**). The story tells us that:

FIGURE 8.10 The monkey and the wedge. Folio from a *Kalīla wa-Dimna* second quarter 16th century. Attributed to India, Gujarat, probably based on an Egyptian original. (Source: https://www.metmuseum.org/art/collection/search/453080. Image of Public Domain.)

> The master-builders who were building a temple went into the city at noon-time to eat dinner. A beam of wood had been split half way through by one of the workmen, and it was left held apart by a wedge of wood which was driven

into it by a mechanical contrivance. And it chanced that a great crowd of monkeys, who dwelt in the forest, came to the spot and began playing about at random here and there among the tree-tops, the towers of the building, and the piles of wood. But in the course of this play one of the monkeys climbed upon the beam, so that his testicles hung down into the crack; and saying "who drove in this wedge which doesn't belong?" He took hold of it and began to pull it out with his hands. What happened when the wedge came out from its place, you know already without my telling you. (*The Panchatantra*, 1965: *26*)

Many other references to monkeys can be found in Arabic poetry in which, in general, the monkey is considered in a negative way and makes people laugh. Cook quotes some cases where poets compare unpleasant persons to monkeys, among whom is Ismāʿīl ibn ʻAmmār al-Asadī, a poet of the 8th century, who wrote of a slave-girl he loathed, that she had the face of a monkey (*wajhu qirdin*) when she prettified herself, and "The poet Farazdaq is referred to by his rival Jarir (729 CE) as 'the little bald monkey' (*al-qurayd al-asla*')" (Cook, 1999: 70). As noted by Kruk (1996: 36), in Arabic popular fiction, as well as among more cultured scholars, monkeys, in connection with other ethnicity, are often considered "sub-humans", or unpleasant or servile or slavish people, or women (cf. Veracini, Chapter 7, this volume), and these animals tended to be ridiculed. The Umayyad Caliph Yazīd ibn Muʿāwiya (ruled 680–683 CE) is said to have possessed a monkey, which he claimed was an old Israelite who had committed some sin and been metamorphosed. He used to get it drunk, laugh at its somersaults, and he would mount it on a wild ass and have it race against horses. According to one account, it was in the course of participating in this activity that Yazīd met his death. Given this royal depravity, "although the monkey was a clown for Yazīd, in the same way, Yazīd was a clown for the narrator" (Cook, 1999: 119).

8.5 SOME NOTES ON THE MODERN (EARLY 20TH C.) ISLAMIC VISION OF PRIMATES

Though this brief report cannot give a complete picture of the cultural differences found in the modern Islamic world regarding the perception of primates, presenting the content of two important Arab works of the 20th century may indicate how Islamic ideas about monkeys have evolved over the centuries. First, we report the content of the entry *"querd"* found in the work of Muḥammad Farīd Wajdī, or Wajdī (1875–1954), (*Encyclopaedia of the Fourteenth Century A.H., Twentieth Century A.D...., querd*: pp. 738–742), who made an effort to reconcile Islamic belief with scientific and intellectual Western currents early in the 20th century. The work is arranged alphabetically, and in long essays on important topics, the author argues that the Quran and the prophetic traditions of Islam are predictive of, or compatible with, modern science and rational (but not materialistic) philosophies. The encyclopaedia was intended as a comprehensive reference book for students on the entirety of human knowledge, and was approved for distribution to all Egyptian schools. Wajdī came to be honoured with the title *shaykh*, even though he had no academic degree or training at Islamic academies. We also report on the *Lexicon of Animals* (cf. Al-Maalouf, 1932: 23, 117, 123, 149, 155, 162, 182) by the celebrated Lebanese novelist Amīn Maʿlūf (born 1949), who wrote a dictionary of the animal world.

The monkey (*qerd*) was still defined as an animal located at the top of the scale of mammals in terms of structure and was considered the closest animal to humans in terms of physical constitution (Wajdi, 1923; Al-Maalouf, 1932), especially regarding the thumb, which, unlike that of other animals, is nearly opposite the other fingers. In addition, their skulls, eyes and foreheads are similar to those of humans (Wajdi, 1923). The *Qurduh, Qarduh, Hawdal* or *Mimoon* was described as a monkey with a giant body, a short tail and an ugly face. This entire species lives in Africa except one species called *Al-Rubh* or *Al-Rubah* (in fact, as we have seen, it is the Hamadryas baboon) (Al-Maalouf, 1932):

The *Ghorilla* or *Ghol* is a gigantic, strong and aggressive monkey with no tail (Wajdi, 1923; Al-Maalouf, 1932). Its head is bigger, its shoulders are wider, its arms are taller, and its thighs are shorter than those of humans and it walks on 4 limbs. (Wajdi, 1923)

The *Shambanzy* or *Ba'am* is smaller, thinner and shorter than the *Ghorilla* and it is able to stand on two limbs [chimpanzee] (Wajdi, 1923). The *Habbar* or *Hobr* is a small monkey with a long tail and thick hair; and it is beautiful looking (Al-Maalouf, 1932) [probably a guenon species]. The *Moroccan monkey* (Barbary macaque) has no tail. The *Nisnas, Nasnas, Shadi* or *Sa'adan* (see also Glossary) is considered a small monkey with a long tail (Al-Maalouf, 1932) and is no longer considered an anthropoid ape (in the modern sense) but a grivet or vervet monkey. In the Arabic books, there is no special word for "ape", all the types being referred to as monkeys, whether they have tails or not. As reported also by Kruk (1996: 35), "Arabs have taken for granted that monkeys have tails". From a behavioural point of view, Wajdi considered monkeys according to previous Islamic tradition as having a predisposition to be disciplined:

They are also active, have strong muscles, live in trees, and feed on fruits and birds' eggs. Most monkey types live in tribes in the forests and have a real social life, and they are mostly found in the warm areas of Africa and America. (Wajdi, 1923)

Unlike the ancient tradition (as mentioned before, ad-Damīrī, 1844, stated that a female monkey could have up to ten or even twelve babies quoting al-Masʿūdī), Wajdi (1923) writes that female monkeys give birth to only one or two infants in each pregnancy. Moreover, he reports that the lifespan of the large species of monkey is around forty years and that there are a lot of different types. Some Arab authors considered monkeys as pests for the farms and forests (Wajdi, 1923).

8.6 CONCLUDING REMARKS

This chapter analyses the interaction between humans and non-human primates in the Arabic world – the Pre-Islamic period, the Middle Ages and the early Modern period – taking into account historical, archaeological and Islamic literary sources (8th–16th centuries and 20th century). In the Ancient Arabic peninsula, modern humans have, since the Neolithic period, shared the Western and South highlands, Arabian regions of Yemen, with the only endemic species of primates: the hamadryas baboon. For these communities, which relied on farming and herding, baboons, together with the hyrax, were probably the major agricultural pests. Perhaps this was one of the reasons why baboons, unlike other native animals, did not enter the pantheon of sacred animals of ancient Yemen. The Arabic literature dating from the early Islamic period testifies to a close relationship between humans and baboons, whose habits and behaviour had probably been known since antiquity, when contact with monkeys was part of daily life. Bedouin shepherds must have had an even closer relationship with these animals, and they left some behavioural descriptions that entered Islamic literature. Sedentary Arabic people had another attitude and tamed young monkeys, using them for services such as turning millstones, picking fruit, looking after goods, etc.

The long tradition of South Arabian navigation and trade in Eastern Africa made known to the Arabian world other species of monkeys besides hamadryas (e.g., grivet and vervet monkeys, the favourite animals of street entertainers, who performed with animals). Moreover, much evidence has shown that monkeys were hunted, traded and used as gifts for rulers. Some literary travel accounts refer to different kinds of Asiatic primates which entered into contact with Arabian navigators and merchants. These descriptions include those of monkeys such as langurs and macaques, and apes such as orang-utans and gibbons. Anthropomorphic creatures, real and mythical, were part of the imaginary world of the Arabic and Islamic civilisations. Many works of literature, and also popular fiction, poetry and travel books, show a variety of stories featuring such creatures, and these stories may have been inspired by a contact with hominoids in Africa or Asia.

The concept of *marsh*, or metamorphosis, common in the Arabic-Islamic world, referred to a transformation of humans into non-human animals, (humans were turned into monkeys as consequence of their sins and their bad behaviour). This idea that took root in the Ancient classical world and probably in other popular ancient myths contemplates the possibility of a transformation from one form of life to another. Muslim thinkers and philosophers developed the idea of transformation of natural beings as a transition from one form to another, often in relation to processes that envisaged an organism's adaptability to the environment, and a struggle for existence that some commentators have considered as a simple and preliminary form of the modern theory of evolution. Some authors believed that humans may have originated from a progressive transformation of the "world of monkeys". Primates occupied the highest step in the "Chain of Being" before the one occupied by humans. In this hierarchical order of creation, progressing from the simplest to more complex forms, primates endowed with cleverness and perception, stand as the closest counterpart to humans.

Therefore, primates in the Islamic world were regarded as pleasant or detestable animals, sometimes even being used as sexual objects. While recognized for their intelligence and for being similar to us in many aspects, simultaneously they were described as physically ugly. Furthermore, they were frequently associated with considered ethnically inferior groups and also with women, who were often subject to ridicule.

ACKNOWLEDGEMENTS

We would like to sincerely thank Lucia Avallone for her work transliterating and correcting Arabic words, and for comments and suggestions; Marco Ammar for translating parts of the Al-Qazwīnī manuscript; Francesca Bellino for her advice about some primary sources; Marco Alesci for the photo of the vervet monkey; Antonio Alexandre for the photo of *Pongo abelii*; Hilary Morris for the English review; and Bernard Wood for his careful revision of the manuscript. We would also like to thank Rui Diogo for facilitating the writing of this chapter, and for his advice and help.

FUNDING OF THE FIRST AUTHOR

Portuguese national funds through FCT—Fundação para a Ciência e a Tecnologia in the context of the celebration of the contract under the 'Norma Transitoria', DOI 10.54499/DL57/2016/CP1456/CT0001.

REFERENCES

Primary Sources

Ad-Damīrī (1844). *Ad-Damîrî's Ḥayât al-ḥayawân (A Zoological Lexicon)* (A.S.G. Jayakar, Trans.). London: Luzac & co. (Original work published 14th c.).

Al-Idrīsī (1619). *Geographia Nubiensis: id est accuratissima totius orbis in septem climata diuisi descriptio, continens praesertim exactam vniuersae Asiae, & Africae, rerumq[ue] in ijs hactenus incognitarum explicationem*, Driessen, H. E. fmo RPJCB; RGabriel Sionita; Hesronita, Joannes. Paris: Typographia Hieronymi Blageart, propè Collegium Rhemense.

Al-Jāḥiẓ (1938). *Kitāb al-Hayawān* (A.S.M. Härün, ed.), 7 vols. Cairo: Maktabat Mustafā al-Bābi al-Halabi.

Al-Jāhiz (2003). *Le Livre des animaux. De l'étonnante sagesse divine dans sa Création et autres anecdotes.* (M. Mohamed, Trans.). Paris: Biblioteque Mâktâbâ, Fayard.

Al-Maalouf A. (1932). *Dar Al-Raed Al-Arabi.* (In Arabic; English translation: Lexicon of Animals). Beirut.

Al-maḫlūqāt va ġarayib al-mawǧūdāṭ' (14th c.). Image: *"Un singe ramasse dans un panier les fruits d'un arbre (lūbiyā, le cocotier) qui pousse au bord de la mer"*. Manuscript. MS. Supp Pers., BN Paris, 235°.-F. 238v.- Supplément Persan 332.

Al-Mas'ūdī 'Alī ibn al-Ḥusayn (1861–77). *Maçoudi. Les prairies d'or*, 9 vol. Paris: C. Barbier de Meynard et Pavet de Courteille (eds and trans.). https://archive.org/details/lesprairiesdor02masuuoft

Al-Qazwīnī Zakariyā' (1280). *'Ajā'ib al-makhlūqāt wa-gharā'ib al-mawjūdāt (Marvels of Things Created and Miraculous Aspects of Things Existing or The Wonders of Creation).* https://www.wdl.org/en/item/8962/

Al-Qazwīnī Zakariyā' , Bellino F. (ed.) (2008). *Le Meraviglie del Creato e le stranezze degli esseri.* Milano: Mondadori.

Al-Umarī, ibn Faḍl Allāh (1989). *Masālik al-abṣār fī mamālik al-amṣār*, vol XX. Frankfurt a/M: Institut für Geschichte der Arabisch-Islamischen Naturwissenschaften.

Arabian Nights (2001). *Le Mille e una Notte*, Italian Version. Milan: Mondadori.

Burton, R.C. (ed.) (2008). *Arabian Nights*, 16 vols. Cosimo Classics. (First edition 1885–8).

Buzurg ibn Shahrayār al-Rāmhurmuzī (1883). *Kitāb 'ajā'ib al-Hind...Livre des merveilles de l'Inde, par le capitaine Bozorg, fils de Chahriyde de Rde Riyde. Texte arabe published l'Inde, par le capitaine Schefer, collationnlationnhriy l'Inde, par le capitaine Bozorg, fils n der Lith. Traduction de l'Inde, par Devic.* Leiden: Brill.

Ibn Khaldūn (1980). *The Muqaddimah* (F. Rosenthal, Trans.), vol. 1. (First edition 1967). Princeton: Princeton University Press.

Ibn Miskawayh (1987). *Le petit livre du Salut (Al-fawz, al-asghar)* (R. Arnaldez, Trans.). Beirut, pp. 116–173.

Ovidio (1992). *Le Metamorfosi.* Milan: Bur, Classici Greci e Latini. 2 Vols.

Schoff, W.H. (1912). *The Periplus of the Erythraean Sea.* New York: Longmans, Green.

The Panchatantra (1965). (F. Edgerton Trans.). London: Allen G. & Unwin.

Wajdi, M. F. (1923). *Encyclopedia of the Fourteenth Century A.H., Twentieth Century A.D.: A Reference for the Arabic Language and the Universal Sciences.* Cairo: Maṭba'at DDṭba'a Maabaaa al-Qarn al-'Ishrīn. https://www.wdl.org/en/item/14556/

Secondary Sources

Alboug, A. & Pebsworth, P. (2023). *Reducing Human-Baboon Conflict in the Kingdom of Saudi Arabia.* Communication. Symposium Abundant Primate Populations. 23th International Primatological Society Congress, 19–24 August, Kuching Malesia.

Al-Ghadeer Moneera (2009). *Desert Voices. Bedouin Women's Poetry in Saudi Arabia.* London-New York: Tauris Academic Studies.

Al-Jawazi T. al-Kilani (2010). Bedouin poetic legacy in Egypt. Mediterranean world. *The Mediterranean Studies Group Hitotsubashi University*, 20: 209–233. https://hdl.handle.net/10086/19193 (Accessed 24 July 2018)

Angelakis, E.l et al. (2016). Gut microbiome and dietary patterns in different Saudi populations and monkeys. *Scientific Reports* 6: 32191. doi: 10.1038/srep32191

Arab News (5th February, 2023). *Saudi Arabia's 'Baboon Problem' Targeted by National Center for Wildlife.* https://www.arabnews.com/node/2244826/saudi-arabia (Accessed 1 May 2023).

Aruz, J. (2010). Discovering the art of the ancient Near East. Archaeological excavations supported by the Metropolitan Museum of Art, 1931–2010. *The Metropolitan Museum of Art Bulletin, New Series*, 68(1): 4–48.

Bayrakdar, M. (1983). *kJahiz and the Rise of Biological Evolution. tion.udi Arabia.* Tauris Academic Stu.

Biquand, S., Biquand-Guyot, V., Boug, A., & Gautier, J.P. (1992). The distribution of *Papio hamadryas* in Saudi Arabia: ecological correlates and human influence. *International Journal of Primatology*, 13(3): 223–243.

Blondheim, D.S. (1912). Maimon. *Romania*, 41(162): 260–265.

Burjor, A. (2013). *Islamic Civilization in South Asia: A History of Muslim Power and Presence in the Indian Subcontinent.* Routledge.

Cheney, D. & Seyfarth, R. (2010). *Il babbuino e la metafisica. Evoluzione di una mente sociale.* (Baboon Metaphysics: The Evolution of a Social Mind). Bologna: Saggi Zanichelli. (Original work published 2008).

Cook, M. (1999). Ibn Qutayba and the monkeys. *Studia Islamica*, 89: 43–74.

Edens, C. (2002). 81. Before Sheba. In: Simpson J. (ed.). *Queen of Sheba. Treasure from Ancient Yemen.* London: The British Museum Press

Fuentes, A., Shaw, E. & Cortes, J. (2007). Qualitative assessment of Macaque tourist sites in Padangtegal, Bali, Indonesia, and the Upper Rock Nature Reserve, Gibraltar. *International Journal of Primatology*, 28(5): 1143–1158.

Harvey, W., Lane, E.W. & Stanley, E. (1859). *The Thousand and One Nights. Vol. 3. A New Translation.* Edward Stanley Poole.

Herodotus, Godey A.D. (ed.), (1921). *The Persian war.* Books III-IV. Cambridge (Massachusetts) and London: Loeb Classical Library, Harvard University Press.

Hobbs, J. J. (1992). *Bedouin life in the Egyptian wilderness.* University of Texas Press.

Iskandar, A.Z. (1981). A doctor's book on zoology: al-Marwazī's Ṭabā'i' al-ḥayawān (Nature of animals). *Oriens*, 27/28: 266–312. https://www.jstor.org/stable/1580569.

Jim Al-Khalili (2011). *The House of Wisdom. How Arabic Science Saved Ancient Knowledge and Gave Us the Renaissance.* Penguin Book.

Kruk, R. (1996). Traditional Islamic views of apes and monkeys. In: Corbey R. & Theunissen (eds.). *Ape, Man, Apeman: Changing View since 1600.* Leiden, pp. 29–40.

Lee, P.C., Brennan, E. & Else, J.G. (1986). Ecology and behaviour of vervet monkeys in a tourist lodge habitat. In: Else J.G. & Lee P.C. (eds.). *Primate Ecology and Conservation.* Cambridge University Press, Cambridge, pp. 229–236.

Malik, A.H., Janine, M., Ziermann, J.M. & Diogo, R. (2017). An untold story in biology: the historical continuity of evolutionary ideas of Muslim scholars from the 8th century to Darwin's time. *Journal of Biological Education*, 52(1): 3–17. doi: 10.1080/00219266.2016.1268190

Masseti, M. & Veracini, C. (2016). The zoomorphic representations of the Piri Reis map (1513). *Anthropozoologica*, 51(1): 41–54. https://dx.doi.org/10.5252/az2016n1a3

McDonald, M.V. (1988). Animal-books as a genre in Arabic literature. *Bulletin (British Society for Middle Eastern Studies)*, 15(1/2): 3–10.

Mesele, A., Yosef, M. & Afework, B. (2014). Abundance of hamadryas baboon (*Papio hamadryas hamadryas*) and conflict with humans in Awash National Park, Ethiopia. *International Journal of Biodiversity and Conservation*, 6(3): 200–209.

Mittermeier, R.A., Rylands, A.B. & Wilson Don, E. (eds.) (2013). *Handbook of the Mammals of the World: 3.* Primates. Barcelona, Spain: Lynx Ediciones.

Moshe, E. (1978). The economic relations of the neo-Assyrian empire with Egypt. *Journal of the American Oriental Society*, 98(1): 20–34.

Pellat, Ch. (1986). Maskh. In: Gibb H.A.R. et al. (eds.). *The Encyclopaedia of Islam, New Edition*, vol V. Leiden: E.J. Brill, 1986–2002.

Potts, D.T. (1997). *Mesopotamian Civilization: The Material Foundations*. London: Cornell University Press/The Athlone Press, 366 pp.

Richards, J.F. (1996). *The Mughal Empire*. Cambridge: Cambridge University Press.

Samiullah, M. (1982). The meat: lawful and unlawful in Islam. *Islamic Studies,* 21(1): 75–104. Islamabad: Islamic Research Institute, International Islamic University.

Shanavas, T.O. (2010). *Islamic Theory of Evolution: The Missing Link between Darwin and the Origin of Species*. New York: Brainbrow Press.

Sidebotham, S.E. (1996). Romans and Arabs in the Red Sea. *Topoi*, 6/2: 785–797. doi: 10.3406/topoi.1996.1695

Silverstein, A. (2020). Unmasking maskh: the transformation of Jews into "apes, driven away" (Qur'ān 7: 166 in Near Easterncontext). *Jerusalem Studies in Arabic and Islam*, 49: 177–216.

Simpson J. (ed.) (2002). *Queen of Sheba. Treasure from Ancient Yemen*. London: The British Museum Press.

Sperveslage, G. & Eichmann, R. (2012). Egyptian cultural impact on north-west Arabia in the second first millennia BC. In: *Proceedings of the Seminar for Arabian Studies*, vol. 42, Papers from the forty-fifth meeting of the Seminar for Arabian Studies held at the British Museum, London, 28–30 July 2011, pp. 371–383.

Stewart, F.H. (1990). Texts in Sinai Bedouin Law, Parte 2. In: *Mediterranean Language and Culture Monograph. Serie 5*. Wiesbaden: Otto Harrassowitz.

Teller, M. (2012). The happy ones? *Archive, Aramco world*, 63: 6.

Tibbetts, G.R. (1956). Pre-Islamic Arabia and South-East Asia. *Journal of the Malayan Branch of the Royal Asiatic Society*, 29, 3(175): 182–208.

Varisco, D.M. (2009). Agriculture in al-Hamdani's Yemen: a survey from early Islamic geographical texts. *Journal of the Economic and Social History of the Orient*, 52: 382–412.

Veracini C. (2019). Natural history of non-human primates in the Early Modern period: the contribution of the Italian encyclopaedist Ulisse Aldrovandi (1522–1605). In: Veracini, C. Casanova, C. Scalfari F. (Eds), *History of Primatology: yesterday and today. The Mediterranean Tradition*. Aracneditrice: Ariccia (RM, Italy), pp. 53–91.

Veracini, C. (2021). An early representation of a gorilla from fifteenth-century Central Asia. *Primates*, 62(3): 457–462.

Viré, F. (1986). Kird. In: Gibb, H.A.R. et al. (eds.). *The Encyclopaedia of Islam, new edition*, vol V. Leiden: E.J. Brill, 1986–2002, pp. 131–134.

Wildman, D.E., Bergman, T.J., al-Aghbari, A., Sterner, K.N., Newman, T.K., Phillips-Conroy, J.E., Jolly, C.J. & Disotella, T.R. (2004). Mitochondrial evidence for the origin of hamadryas baboons. *Molecular Phylogenetics and Evolution*, 32: 287–296.

9 Non-human Primates in the 'Age of Exploration' (15th and 16th Centuries)

Trade, Species, Renaissance Courts, Iconography, and Natural History[1]

Cecilia Veracini
Universidade de Lisboa

> They have infinite parrots and give 8 or 10 of them for a mirror,
> and small sea cats similar to lions but yellow,
> what a wonderful thing!.
>
> Pigafetta, 1519, *First Journey Around the Globe.*

9.1 NON-HUMAN PRIMATES AND OTHER ANIMALS IN THE FIRST GLOBAL MARKET

The history of European expansion is intimately linked to that of non-human animals (hereafter referred to as animals), which have been a presence throughout this process (Crosby, 2003; Asúa & French, 2005), yet they have not featured in most historical accounts. Animals were an essential food source on long ocean voyages. They could be brought on board at the beginning of a voyage, or collected *en route*, and could be left in remote lands to breed, or bartered for goods, or for exotic animals or their byproducts (e.g., skins, bones, teeth, feathers, bezoars, shells, scent glands, etc.). All of these possibilities made for a very profitable global trading network, which, along with spices, timber, precious metals, and human trafficking, powered the engines of European expansion.

The Battle of Ceuta in 1415, which ended in the defeat of the Marinid Sultanate, marked the beginning of the period of European oversea expansion. In 1434, the Portuguese explorer Gil Eanes rounded Cape Bojador, a feat that was a major breakthrough for European explorers and traders *en route* for Africa and later India. At the end of the century, in 1498, Vasco da Gama reached India, thus opening the sea route to Asia, and following Columbus' journey in 1492, by the turn of the 16th century much of the South American coast, and some territories of continental America, had been reached by Europeans. By the 15th century, new maritime trading routes were opened up bringing hitherto unknown goods, plants, and animals into Europe.

This first economic globalization (Godinho, 1963; Rodrigues & Devezas, 2009) gave European courts and other collectors a unique opportunity to buy and collect an array of novel commodities (Findlen, 1994; Smith & Findle, 2013), including spices, medicinal drugs, wood, seeds, precious metals, and stones as well as art and craft work and a wide range of objects derived from the natural world. The latter included ivory, pearls, nacre shells, tortoise shells, corals, feathers, and odoriferous secretions produced by Asian and African mammals (the Siberian musk deer, *Moschus moschiferus;* African and Asian civet cats of the genus *Civettictis*), skins of various mammal species, rhinoceros horns, and bezoar stones (Kemp, 1995: 177–196; Teixeira & Papavero, 2010: 258; Veracini, 2021b). Seaborne travel also offered the possibility of transporting live animals with greater ease than overland routes. The arrival of exotic live animals from foreign lands encouraged the formation of menageries (Hallet, 2009: 9), and Renaissance princes and rulers used exotic beasts as an image of power (e.g. Pieper, 1998; Bedini, 1997; Ringmar, 2006; Groom, 2018; Simões, 2019). Control over the animal world had always figured among the paraphernalia of rulers in many societies (Sterckx, 2000: 4), but the desire to collect rare, live exotic wild beasts became especially evident in this period (Belozerskaya, 2008: 59). The rule was simple: the more exotic and rare an animal was, the more it was worth (Lloyd, 1971: 47). Although most exotic animals were housed in menageries (fenced spaces), some, especially birds and small monkeys, were kept in homes as pets.

The list of the taxa that reached Europe alive in the 15th and 16th centuries is long. Some were exclusive gifts offered by foreign sovereigns or by ambassadors to European princes, others were traded like any other type of goods, making merchants and navigators wealthy (Teixeira & Papavero, 2010; Veracini, 2017). The taxa included Indian elephants (*Elephas maximus*), tigers (*Panthera tigris*), giraffes (*Giraffa camelopardis*), Indian rhinoceroses (*Rhinoceros unicornis*), leopards (*Panthera pardus*), lions (*Panthera leo*), cheetahs (*Acinonyx jubatus*), jaguars (*Panthera onca*), civet cats (Family Viverridae), guinea pigs (*Cavia porcellus*), many species of psittaciformes (parakeets, parrots, and cockatoos), and other birds (e.g., ostriches, peacocks, cranes, and pelicans) (cf. Ferronha et al., 1993; Clarke, 1986; Lloyd, 1971; Mantua, 1992: 34–37; Masseti, 2016, 2018) and, naturally, non-human primates.

[1] The translation in English of the 16th century Latin, Portuguese, French, Spanish and Italian primary sources are by the author.

DOI: 10.1201/b21963-11

Non-human primates (hereafter referred to as primates) and psittaciformes, which were among the most sought-after taxa, were shipped in large numbers to Europe throughout the early and later phases of European expansion. Historical and iconographic evidence points to the trade in primates at this time being routine. The colorful sub-Saharan African guenons and the smaller Neotropical callitrichids and capuchin monkeys, new to European nobles, quickly became high-status pets (Veracini, 2017). Despite their initial reputation as devilish and lustful (Janson, 1952; Veracini, Chapter 7, this volume), the dissemination of primates in Europe was ubiquitous. The primate species that reached Europe in the greatest quantities (e.g., capuchin monkeys, marmosets, baboons, guenons, and macaques) were mainly omnivorous species (Veracini, 2011a) and thus stood a greater chance of surviving the long sea journeys because they could eat the same food as the sailors.

The agents and people responsible for this kind of trade were initially Portuguese and Spanish because they had acquired the monopolies of the Atlantic trade (and of the import of exotic commodities) granted by the Pope with the Treaty of Tordesillas of 1494, but soon other monarchies and private merchants contributed directly or indirectly to the animal trade (see later). As a consequence, the ports of Lisboa and Seville became key trading centers from which goods were distributed to European markets (Jiménez Montes, 2016; Gschwend, & Lowe, eds. 2015).

Before examining in detail the trade in primates by continents, it is important to recall the central role Portugal played in this process at the beginning of the 16th century. During the reign (1495–1521) of Manuel I, King of Portugal of the Avis dynasty, the commercial activity of the country was at its peak, and Lisbon became one of the main entrepots for goods coming from Asia, Africa, and America (Godinho, 1963). No other contemporary European city could compete with Lisbon and its court for the range of exotic animals (Villhena Barbosa, 1885; Gschwend, 2009). "Exotic animals and prestige-oriented menageries became a fundamental part of the image of the Portuguese court, which tried to imitate those of Indian potentates and quickly influenced all the other European courts" (Pérez de Tudela & Gschwend, 2007: 421). Elephants, an Indian rhino, big cats, and other mammals were recorded during the reign of Manuel I, The Fortunate (*O Venturoso* in Portuguese). Menageries, rides on elephants in the center of Lisbon, with a procession of Arabian horses, including one bearing a cheetah, along with aviaries full of all kinds of exotic birds, symbolized its immense power (Góis, 1566 cf. Barbosa, 1885). This tradition was followed by the sovereigns who succeeded Manuel I. During the time of Catherine of Austria (the youngest sister of the Habsburg Emperor, Charles V, and spouse of the King of Portugal, John III), Portugal was the distribution center for exotic animals for the whole of Europe. Catherine invested a great deal in procuring exotic and rare creatures for herself and her extended Habsburg family in Spain, Central Europe, and the Netherlands, helped by her global network of acquaintances and connections (Pérez de Tudela & Gschwend, 2007: 423; Gschwend, 2015).

Evidence of primates symbolizing control and dominion comes from a tapestry now housed in the Portuguese museum of Caramulo, one of a series of tapestries known since the beginning of the 16th century as *"à maneira de Portugal e da Índia"*. The series, which was commissioned by Manuel I to celebrate his successes in India (Dias, 2007; Hallet (ed), 2009) sometimes in a fanciful form, represents Vasco da Gama's voyage to India and his encounter with non-European lands and peoples, and some of the tapestries show parades, with the Portuguese bringing exotic animals from Africa and Asia (Hallet (ed), 2009). One of the tapestries in the Caramulo Museum depicts a triumphal procession with dromedaries and various African monkeys (probably of the genus *Chlorocebus*), with an oriental man handing over a monkey to Manuel I, who is receiving the wonders of the Orient (Dias, 2007; **Fig. 9.1**).

Normally, animals arriving from any of the overseas Portuguese territories were taxed by the crown, but many live animals, such as monkeys and parrots, were imported 'unofficially' by Portuguese sailors and travelers for sale as pets (Gschwend, 2009: 35; Teixeira & Papavero, 2010: 256). A network of agents, merchants, and ambassadors of the major European powers established offices in Lisbon. Merchants and traders were attracted by the Portuguese king, whose financial difficulties led him to offer generous concessions to merchant-financiers (Mathew, 1997: 3). For example, a German banking family, the Welsers of Augsburg, was directly involved in the colonization of the Americas and in the Atlantic trade and had representatives in the most

FIGURE 9.1 An oriental man hands over a monkey to Portuguese King Manuel I, who is receiving the wonders of the Orient. Detail of one of the tapestries *"à maneira de Portugal e da Índia"*. (Museu do Caramulo, Portugal. Source: Dias, 2007, modified.)

important entrepots in Europe and overseas (Roth, 2017). Lucas Rem, an agent in Lisbon at the time of the Welsers, in 1508 reported on the "strange parrots and monkeys with long and hairy tails" coming from Africa that he had purchased in Lisbon (Lach, 2010, vol II: 177). In 1523, the Welsers purchased animals from Dom Manuel's animal park and brought them to Augsburg. The Fugger family, another prominent German family of bankers and merchants, was also involved in overseas ventures (Hauptmann et al., 2016), and they, along with the Welsers, supplied merchandise to the Portuguese crown (Chalmin, 1987: 22). In 1505, Jakob Fugger the Rich participated in the expedition to the East Indies under the viceroy Francisco de Almeida, and throughout the reign of Manuel I (i.e., until 1521), the Fuggers' trading post in Lisbon provided primates, big cats, parrots, peafowl, etc., for clients (Gorgas, 1997: 219). The trade in exotic animals extended to Antwerp, which at the end of the 15th century had become the preeminent center of trade in Northern Europe:

> Between 1520 and 1530 the Fugger representative in Antwerp moved into excellently equipped business premises with a large garden next to which there were the cages to transport animals. Many of the animals were temporarily kept here and then sold to rulers, princes and rich citizens in Europe. (Gorgas, 1997: 220)

The importance of the monkey trade was reflected in the complaint by the Fuggers' agent based in Antwerp about the work involved with the care of primates in their facilities. "The head office ignored their plea because the business with live animals and their products was gaining more and more economic importance" (Gorgas, 1997: 220).

Ambassadors, such as Hans Khevenhüller de Aichelberg (1538–1606), the permanent ambassador of the Holy Roman Empire to the court of Madrid from 1572 to 1606, supplied the Habsburg emperors with exotic creatures (cf. Pérez de Tudela & Gschwend, 2007: 432; Gschwend, 2018). Animals were often used as gifts between rulers. For example, an anonymous letter written in Milan in 1578 mentions Don Remiro Nuñez de Guzmán, a Spanish nobleman, who had been sent by the Catholic king, Philip II to meet his nephew, Emperor Rudolf II of Habsburg, to bring him "22 young very beautiful horses, a mule, a lion, a leopard, seven monkeys, and six greyhounds with him as a gift from his uncle" (ASF, *MdP 3254 f. 427*, 1578; Veracini, 2011).

Exotic creatures were also sought after by the merchant classes for their own use, and as well as in Lisbon, primates and psittaciformes could be purchased in street markets in Seville, Antwerp (see **Fig. 9.2**), Bruges, and in many French ports. When the artist Albrecht Dürer visited Flanders in 1521 with the Emperor Charles V, he purchased exotic animals and objects for his personal *Wunderkammer*, writing in his diary (Lugli, 2005: 104) that he "bought a small monkey for four gulden and gave 14 stivers for five fish", a high price for that time. The small monkey was perhaps a common marmoset, *Callithrix jacchus*, one of the first Neotropical animals to reach Europe alive. As the Swiss naturalist Conrad Gessner noted, "these animals can be bought for 50 crowns in Antwerp" (Gessner, 1551: 158b).

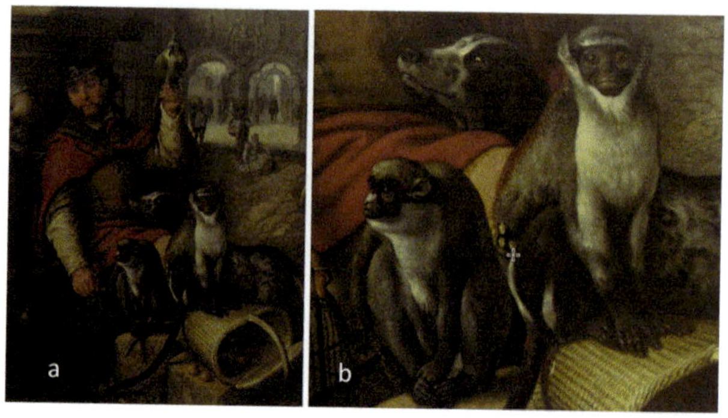

FIGURE 9.2 Seller of exotic animals in Antwerp, selling two monkeys: *Cercopithecus petaurista* and *Cercopithecus* cf. *roloway*. Joachim Beucklaer, 1566. Oil on canvas (Courtesy of the Museo di Capodimonte, Naples).

Besides the Portuguese maritime trade, African and Indian animals continued to arrive in Europe by the regular trading networks via Alexandria in Egypt, a county that had been part of a globalized economy in fauna long before the early modern period. Animals represented an important facet of the Indian Ocean trade among Mughal South Asia, Safavid Iran, and Ottoman Egypt (Mikhail, 2013: 360–362), and primates arrived in Europe this way as early as the Middle Ages, and on into the modern era.

9.1.1 Non-human Primates from Africa

During the Middle Ages, Italian merchants dominated the Mediterranean basin, developing trade in Syria, Palestine, North Africa, the Middle East, India, and China. Venetian merchants sold products from these territories throughout Europe, including the main centers of trade such as Flanders and London (Molà, 2009; Howard, 2007). There were also close ties with German traders who had established their own agency in Venice: the famous *Fondaco dei Tedeschi* (Hoffmann, 1932; Franzone, 2017). This trade also involved live animals that arrived in Venice via Alexandria from Africa and India. For example, Adam de Moleynus, in 1436 (cf. Molà, 2009: 46), in his *Libelle of Englyshe Polycye*, commented on the activities of Italian traders in London, writing:

> *The grete galees of Venees and Florence, Be wel ladene wyth thynges of complacence,*
> *All spycerye and other grocers ware, Wyth swete wynes, all manere of chaffare,*
> *Apes* [i.e. barbary macaques] *and japes and marmusettes taylede* [tailed monkeys, cercopithecids], (...).

Other sources (Nordio, 2003) describe the trade in monkeys in Bruges and London by Venetian merchants. The arrival of monkeys from Alexandria continued during the 16th century, as demonstrated by correspondence between Venice and Mantua (cf. Sermidi, 2003: 99), which mentions the importation of "*gatti mammoni*" (long-tailed monkeys, see Glossary) from Alexandria to Venice in 1589, including animals seized by the Pasha of Algiers. The Venetians' privileged relationship with Christian Ethiopia (aka the region of Priest John), which had been isolated from the 7th century after the Islamic expansion under the Caliphates and later by the Ottoman Empire, led at the beginning of the 15th century to the establishment of several embassies that included animals (Heldman, 1990). Among these was the arrival of the skin of a "big" primate [a great ape? a baboon?]. The presence of a gelada baboon (*Theropithecus gelada*), a species endemic of Ethiopia, in Verona in the mid-16th century was recorded by the Italian humanist Piero Valeriano Bolzani, who described geladas in his *Hieroglyphica* (Valeriano, 1556) and recognized this species being used for street entertainment in Verona, then part of the Venetian Republic, in 1554 (cf. Jolly, 2016) (**Fig. 9.3**).

Portuguese navigators and explorers were the first to make direct contact with animals from sub-Saharan Africa. The first Western African travel chronicles, besides describing the variety and abundance of primates, also stated that monkey species, together with parrots and civet cats, were commonly brought back alive to Europe soon after traders made contact with these territories (Veracini & Casanova, 2016). The Venetian Alvise da Mosto, also called Cadamosto, traveled in Western Africa for the King of Portugal in 1455–1456. Reaching the Gambia River, he reported (Cadamosto, 1988: 72) that there were many "*gatti mammoni*" (tailed monkeys in Italian, see Glossary) and "*babuini*" (baboons) of every kind, and the local people bartered them for a price of "*10 marchetti*" (the ancient Venetian coins), much less than the local value of an ounce of civet musk ("*50 marchetti*") (Veracini, 2021). The Gambia River was favored for the monkey trade,

FIGURE 9.3 Adult male, *Theropithecus gelada*. (Photo Ivan Norcia).

especially the Cantor riverside market. From the mid-15th century until 1586, the Portuguese frequented fairs and markets in Gambia and on the Guinean coast, including the important market of Guinalà (today in Guinee Bissau) (Godinho, 1963: 178). The Portuguese writer D'Almada (1594:186) reports that in these markets "were sold all the things that were being produced in those lands and those in the surrounding lands". In addition to the Portuguese traders, English, Flemish, French, and Spanish traders were also active in Western Africa (D'Almada, 1594). Other chronicles (Coelho, 1953: 18; Donelha, 1977: 90) reported that civet cats, parrots, and monkeys, as well as animal derivatives (ivory, skins, horns, civet musk, etc.), were traded with the local people, and Fernandes (cf. Monod et al., 1951: 34) states that monkeys were gifted to Europeans by local rulers.

By 1479, the Portuguese had a settlement in present-day Elmina (Ghana), and in 1481, they erected a fortress, *São Jorge da Mina* (St. George of the Mine), which became Portugal's West African headquarters for trade and the exploitation of African wealth (Ballong-Wen-Mewuda, 1993), but in 1637, it came under Dutch control. Iconographic sources (see below) and information gathered by Catherine of Austria suggest that many of the primates reaching Europe came from this settlement. For example, a Portuguese official, Afonso Gonçalves Botafogo, sent the queen two bearded monkeys (*bugios*) (Pérez de Tudela & Gschwend, 2007: 424), most likely *Cercopithecus roloway*, currently found in a small area of eastern Ivory Coast and Ghana.

By comparing historical and iconographic evidence about exploration and contact with local people, we know that primates were shipped to Europe from the territories of Senegal, Guinea-Bissau, Gambia, Sierra Leone, Ghana, the Ivory Coast, and later also from Central Africa. Some African primate species were traded not only with Europe, but also with other tropical territories. The islands of Cape Verde, which were an important source of shelter, water, and food for boats engaged in the slave trade, had been occupied by the Portuguese since 1462, beginning with the Island of Santiago (Barcelos, 2003). The presence of green monkeys, *Chlorocebus sabaeus*, in Santiago was reported by the Florentine Merchant Francesco Carletti in 1594. He mentions an infinite number of "*Bertucce* (monkeys) that we call *Gatti mammoni*, which have long tails, called here by the Portuguese *Bugios* [see Glossary], which they teach to dance and to play many other games" (Carletti, 1701: 12). These primates were most likely descended from animals imported by slave traders from the Gambia River and the Cacheu region in what is now Guiné-Bissau and Senegal (cf. Green, 2011; Hazevoet & Masseti, 2011: 20; Veracini & Hazevoet, 2024). In later centuries, this species, as well as another guenon, *Cercopithecus mona,* were introduced onto other islands off of continental Africa (e.g., São Tomé), as well as onto some of the West Indian Islands (Denham, 1987; Masseti, 2011). African guenons, which were among the most regularly traded primates, likely due to their small size, ease of breeding, and their "character", reached many locations in Europe, where they were among the most commonly portrayed monkeys (see later).

9.1.2 The Trade in Neotropical Primates

With the discovery of the Americas, many species of neotropical creatures began to be brought back to Europe. The first neotropical creatures landed on Spanish soil. In 1493, dozens of colorful parrots arrived in Seville and accompanied Columbus on his journey to Barcelona, then the home of the Castilian-Aragonese court, and other parrots and small rodents, all endemic to Caribbean islands, followed a year later (Berchet, 1892: 167–168; Masseti & Veracini, 2014: 134). In 1500, Vincente Yanez Pinzon reached the Amazon River and returned to Spain with an opossum (*Didelphis marsupialis*) (Anghiera, 1511, *First decade*, book VII). This may have been the first live marsupial to reach Europe, but as Anghiera explains "it died for the lack of proper food after some months".

Gaspar de Lemos, who commanded one of the ships of Pedro Alvares Cabral's fleet (that landed in the current state of Bahia on April 22nd, 1500), returned home to Lisbon from the 'new land' (aka Brazil), bringing in his ship parrots, minerals, and the wood of the red tree [dyewood, *pau-brasil* in Portuguese, *Paubrasilia echinata*] (Bueno, 1998: 110). In 1503–1504, Gonçalo Coelho brought to Manuel I, a ship loaded with red trees, monkeys (*bogios*), and parrots (Góis, 1566: 320). Since these first travels, Brazil's Atlantic coast became a rich source of dyewood, used to extract red dye used in the manufacture of cloth. The growing demand for red dye in Europe was also satisfied by an Asian wood of the same genus traded, often precariously, through the Near East (Dean, 1997: 45). Around 1503, the Portuguese crown 'leased' Brazil to a consortium of merchants led by Fernando of Noronha, who was given a monopoly on trade. They began the exploitation of what Manuel I thought were the only valuable raw materials in Brazil: red trees, cotton in a minor form, oils, seeds, and then animals, especially parrots, monkeys, and felines. On these early expeditions, sailors conducted their own private business, capturing exotic animals to sell in Europe (see Johnson, 1995; Teixeira & Papavero, 2010). In a short time, many ships crossed the Atlantic from Brazil, including those returning from India that called at Brazil (Pieper, 1998).

Portuguese and French merchants competed for these natural resourses and their ships crossed the Atlantic Ocean with cargoes of red trees and animals. In 1511, the Portuguese ship *Bretoa* brought back 5,000 red trees, along with 22 parakeets, 16 *sagoins* (likely marmosets or lion tamarins), 15 parrots, and 3 *bugios* (probably capuchin monkeys) (Anonymous, 1881), valued at 24,220 *reis* (Portuguese currency), of which a quarter was paid to the King (Bueno, 1998: 85). As Teixeira and Papavero (2010: 258) have noted, the monthly salary of a caravel captain in the 16th century corresponded to 2,000 *reis*, while a cabin boy received 933 and a half *reis*. Thus, the sale of exotic creatures represented an important supplement to the sailors' incomes. In 1531, the French ship *La Pélerine* returned from Pernambuco with 300 tons of wood; 1.8 tons of cotton, seeds, and various oils; and 3,000 skins of felids, 600 parrots, and 300 monkeys (Gaffarel, 1889; Dean, 1997: 48). Teixeira and Papavero (2010) report the value of *Pelerine*'s 3,000 animal skins as 9,000 ducats, its 600 parrots as 3,600 ducats, and the 300 monkeys as 1,800 ducats (i.e., approximately 1,000 Euros per specimen today).

The example of *La Pélerine* was not an isolated one. In 1504, the four French ships arrived at the coast of Bahia to carry out their trade (Gaffarel, 1889), and Normandy became a key gateway for goods and animals from the Americas. Captains employed in the service of the King of France, Francis I of Valois (1494–1547), imported large quantities of Brazilian dyewood for the textile manufacturers of Dieppe and Rouen (Johnson Tomlinson, 1970). For example, the ship-owner Jean Ango, a merchant, banker, and future Viscount of Dieppe (Guérin, 1900), had a large residence in Dieppe made from Brazilian precious hardwood, the courtyard of which was filled with indigenous people and creatures arriving from Brazil. Exotic animals from all over the world would arrive in Dieppe, and its citizens delighted Francis I by presenting him with the skins of various exotic animals, birds, ostrich eggs, and a few stuffed crocodiles and lizards (Lloyd, 1971: 65), and the sovereign had a building specially constructed for his collection of *curiosities* at Fontainebleau. As reported by Gaffarel (1889), travel to Brazil increased greatly between 1516 and 1550, with Norman sailors (from Dieppe, Malon, and other places) obtaining dyewoods, spices, cotton, parrots, and monkeys from the native Brazilian people. Exotic birds, such as Toucans and hummingbirds, provided feathers for the hairstyles of the ladies of the court (Denis, 1875). In the North of France, New World monkeys, in particular marmosets, were so common in the 16th century that a character used in a famous "querelle", between Marot and Sagon (Marot, 1537; Grenier, 1920; Papavero & Teixeira, 2014: 20) was based on them (**Fig. 9.4**). Dieppe retained its primacy as a center for the sale of animals during the 16th century, as evidenced by the fact that the King of France Henri III and his spouse Queen Louise traveled to Normandy in 1576, "returning to Paris with a large number of monkeys, parrots, and small dogs which they had bought in Dieppe" (L'Estoil, 1963: 62 *apud* Hazan, 2015).

At the end of the 16th century, the market for exotic animals also flourished in Marseille, in southern France. In 1597, the naturalist Thomas Platter reported that in the port of Marseille he had witnessed the disembarkation of "foreign animals and monkeys", and in various houses, he observed the presence of ostriches, leopards, and other wild felids. Finally, inside a guesthouse, he was able to admire "an animal similar to a big monkey but of a somewhat different shape. His name was Bertrand. He played many hilarious and amazing tricks, quite amazing and hilarious" [was it a chimpanzee?]. As reported by Hamy (1903: 317), the animals found in Marseille came from both the Mediterranean coasts of Africa and from the Atlantic trade.

Pre-Colombian indigenous peoples, like contemporary native Amazonian populations, kept primates as pets (**Fig. 9.5**), which were usually acquired when their mothers were killed for food, and the infants were taken and raised by the people. The demand for primates as pets among the European elite in the 16th century meant that the natives traded a great number of these animals with Europeans. Many are the accounts testify to the natives' habit of exchanging monkeys with the Europeans for any simple metallic or

FIGURE 9.4 Frontispice of the quarrel between Marot and Sagoin or Sagon. *"Le valet de Marot contre Sagon"* de Marot, Clément (1496–1544), 1537. (Source: Gallica. Bibliothèque nationale de France, https://gallica.bnf.fr/ark:/12148/btv1b8618498d.)

other small artefacts (e.g., Pigafetta, 1519; Soares de Souza, 1587: 297; Jean de Léry, 1578: 144; Veracini & Camphora, this volume). Some chronicles report the difficulties of crossing the ocean experienced by the callitrichid species (e.g., *Leontopithecus rosalia*, *Callithrix* spp.), as explained by the Portuguese chronicler Pero de Magalhães Gândavo in his *História da Província de Santa Cruz*, published in 1576.

> they are so dainty and delicate that when they are taken from their country and shipped to this kingdom [i.e. Portugal] and arrive in the colder air, almost all die at sea, and only a few manage to survive, to our great wonder. (Gândavo, 1576: 158–163: *apud* Veracini & Teixeira, 2017)

Many factors determined the fate of these creatures. Léry explains that during the return journey from Brazil, a storm delayed their return and, as they had nothing to eat, he was obliged to eat monkeys and parrots, though he did so reluctantly: "despite hunger, during which were eaten all the monkeys and parrots that we brought, I had kept a large talking bird, but such was my need that he met the same fate as the others" (Léry, 1578: 241).

After conquering the Caribbean islands, the Spanish moved on to Central and South America, reaching Peru in 1515 and Chile in 1535. The trade and exchange of goods

FIGURE 9.5 Detail of one of the many maps of the 16th century showing indigenous people of Brazil, with monkeys and parrots. Giacomo Gastaldi's map. (Source: Gian Battista Ramusio, 1556 modified.)

from the so-called New Spain, known as *La carrera de las India*, grew steadily throughout the 16th century. In addition to silver and precious stones, a vast range of animal and plant products, such as amber, pearls, Campeche wood (logwood), precious woods, indigo, and above all, cochineal, as well as live animals, were brought to Europe (Lee, 1948; Rodriguez, 2012). In 1525, a jaguar reached Toledo, and according to the Italian Andrea Navagero, ambassador of Venice in Spain, it made a great impression (Masseti & Veracini, 2014: 135). Unfortunately, no information is available about the numbers of creatures imported from these Spanish colonies, but there is circumstantial evidence about Neotropical monkeys shipped to Spain (cf. Veracini, 2011a). Gonzalo Fernández de Oviedo, historian and naturalist, the official Chronicler of the West Indies under Charles V, reports about the "wild monkeys of many kinds and varieties" and "because they are brought daily to Spain, I will not say much about them" (Fernández de Oviedo, 1526: 415). In the first half of the 16th century, Galeotto Cei, an Italian merchant traveling in what is now Venezuela, wrote: "in some parts there are certain cats as small as mice [callithichids], which are very beautiful, but they die, and it is difficult to bring them to Spain" (Cei, 1539–1553). Tomáz Lópes Mendel also commented on the trade in primates from the territories under Spanish control. Traveling to Guatemala as an official observer for the Spanish crown in 1549, he reached various locations in Central and South America and reported that "monkeys [*monos*], of which people bring back many here and to Seville", could be observed there in large quantities "because in that warmer land they can live better" (López Mendel, 1570: 176).

9.1.3 The Trade in Asiatic Primates

With the sea route to Asia open, besides the shipping of large quantities of spices and other natural products, Portuguese merchants, and later merchants from elsewhere in Europe, brought back a very different kind of natural product. As early as 1531, ships sailed to the Far East on behalf of the Portuguese queen, Catherine of Austria, wife of King John III, to find new Indian animals. Catherine's agent in Goa, Antonio Correa, "had to buy from direct sources in Asia at the cheapest price and keep her regularly informed of goods and animals for sale in these markets" (Pérez de Tudela & Gshwend, 2007: 424). In the first decades after the discovery of the Cape of Good Hope route to Asia, large mammals, such as Indian elephants (e.g., the famous Hanno, the Indian elephant sent to Pope Leo X by the Portuguese King Manuel I) or Indian Rhinos, arrived in Lisbon (Mantua, 1992; Bedini, 1997). Big cats also arrived, their skins used to make splendid garments for the royal houses and rich citizens, but live civet cats were also in demand in 16th century Europe (Gorgas, 1997: 215). Iconographic sources suggest that many species of Asian birds arrived alive in Europe (Masseti, 2016), whereas birds of paradise arrived dead and without legs (Swan, 2015; Swadling, 2019).

There is little evidence to show that in the 16th century many Asian primates reached Europe alive. Although Gorgas (1997: 214), and other authors who reported his pronouncements, claim that thousands were shipped to Europe, if so few reached Europe alive. Until now, unlike the case with African and Neotropical species, there is only one depiction of a langur (*Semnopithecus* sp., **Fig. 9.6**), which dates from the second half of the 15th century in Siena (I) and one example of a sculpture of a supposed orangutan (*Pongo* sp.) found in Italy (see **Box Details 9.1**). Lach (2010) suggested that a woodcut attributed to Nicolò Boldrini, after Titian, also depicts an orangutan, but whether it is a real or an imaginary animal is hard to say (**Fig. 9.7**). If this great ape had reached Venice, it would probably have been brought by Arabian ships via Alexandria, a marketplace still much used by Venetians in the 16th century. The final depiction

FIGURE 9.6 This artwork portrays a monkey displaying characteristics reminiscent of the Hanuman langur, *Semnopithecus* spp. It is a detail of *The Story of Griselda, Part III: Reunion*, which is the final piece of a triptych. These panels were likely commissioned around the time of the weddings of two brothers from the esteemed Sienese Spannocchi family in January 1494. Their father, Ambrogio, served as the papal banker to Pius II Piccolomini, who was also from Siena. The panel features various exotic animals, including a peacock and a Barbary macaque. (Courtesy of the National Gallery of London.)

FIGURE 9.7 Niccolò Boldrini after Titian. *Caricature of the Laocoön* scuptural group, excavated in Rome in 1506. 1540–45. (Source: Metropolitan Museum, NY: https://www.metmuseum.org/art/collection/search/360382 Image in the public domain)

of an Asiatic primate and the first naturalistic description of a prosimian, *Nycticebus coucang*, is found in a manuscript produced at the court of Emperor Rudolf II at the turn of the 16th century (Cod. Min 129, cf. Haupt et al., 1990).

The rarity of Asian primates in Europe at that time is probably due to the difficulty of providing food on the c.120day journey from Asia, even though the prevailing winds and favorable currents meant that ships returning from India had no trouble reaching the coast of Brazil and hence continuing to Europe (Alencastro, 2000). The Portuguese crown tried to systematically integrate the Brazilian coast within the Atlantic trade routes, and many vessels coming from India stopped there, where they took the opportunity to load merchandise and animals, which were often carried as ballast and provided an additional source of income (Pieper, 1998). An example of this is the French merchant Binot Paulmier, Captain of the Gonville, who tried to repeat the Portuguese route to the Indies, and because the wind from the east pushed the ship westward, they reached the southern coast of Brazil in 1504, where they filled the holds with merchandise in order to reimburse their travel expenses (Papavero & Teixeira, 2002). Many of the animals referred to generically as being "Indian" were Brazilian or African because people were unable to distinguish their places of origin. Tailed or long-tailed monkeys (guenons and other cercopithecids, but also capuchins and callitrichids) were all commonly named "sea cats" (*meerkatzlein, gatti mammoni* etc., see Glossary) because of their overseas origin. Gorgas (1997: 214) also reports that John's langur (*Trachypithecus johnii*), a species endemic to South India, was greatly valued because Indian people believed that their flesh, blood, and organs possessed healing powers against different diseases (it must be remembered that these folivore primates produce in their stomach bezoar stones, which at that time were considered a universal medicine). Gorgas claimed (but without indicating any historical source) that hundreds of these monkeys were transported by ship to Western Europe, or were exported directly via the silk route and by ships through the Red Sea route to

BOX DETAILS 9.1: ORANGUTANS, *PONGO* SP. LACÉPÈDE 1799, IN 16TH CENTURY, ITALY

Marco Masseti
Istituto Zooprofilattico della Sicilia

Barbary macaques, *Macaca sylvanus* (L., 1758), monkeys of the genus *Chlorocebus* Gray, 1870, and baboons of the genus *Papio* Erxleben, 1777, were the best-known primates in the ancient and medieval Western World (Wendt, 1956; Masseti, 1991; Rolfe & Grigson, 2006; Masseti & Bruner, 2009). This was mainly because their primeval geographical distribution was relatively close to the European borders and, therefore, it was possible to import living individuals from there (cf. Masseti & Bruner, 2009).

This was at least the case until the western oecumene managed to establish safe trade routes from the Far East for the overseas importation of other living animals. Thereafter, many species began to appear in European markets and palaces, originating from ever more distant regions, such as the Lesser Sunda Islands and archipelagos located further east in the Pacific Ocean. Already in the 13th century, in southern Italy or Sicily, for example, there is evidence for the presence of Indonesian birds, such as the sulfur-crested cockatoo, *Cacatua galerita* (Latham 1790), a large white parrot native to the tropical and subtropical forests of Australasia. This cockatoo is possibly the first species of South-Pacific parrot to have been portrayed in any European artistic production. Indeed, its image appears under the indication *psittacus* in certain illuminations contained in the pages of the 13th century manuscript of the "*De arte venandi cum avibus*" by Frederick II of Hohenstaufen in the Biblioteca Apostolica Vaticana (Codex Ms. Pal. 1071, cf. Willemsen, 1980) (Masseti, 2016, 2019a; Dalton et al., 2018). We have to wait a few more centuries (Masseti, 2016) for the importation into Europe of other wildlife from these distant archipelagos, such as birds of paradise. In fact, these animals appear not to have been imported before the beginning of the second decade of the 16th century, when, in September 1522, the wrecked craft Vittoria entered the harbor of Seville, the only survivor of five vessels that had proudly set sail more than three years earlier under the leadership of Ferdinand Magellano (Fuller, 1995).

In 16th century Europe, the arrival of another interesting animal from the remote Far-Eastern islands was documented: the orangutan, the largest contemporary tree-living animal. At the end of the Pleistocene, and in the early Holocene, orangutans were dispersed throughout continental southeastern Asia (Hertler & Rizal, 2005; Harrison et al., 2006), but today only three species: *Pongo pygmaeus* (L., 1760), *Pongo abelii* (Lesson, 1827) (Corbet & Hill, 1992; Groves, 2005), and *Pongo tapanuliensis* Nurcahyo *et al.*, 2017 survive in Borneo and Sumatra. Redmond (2011: 152) is of the opinion that orangutans were among the first apes to be shipped to Europe in the 18th century for display to an incredulous public, but Urbani and Youlatos (2020; 657) suggest the first known images of great apes date back to the second millennium BCE, but the evidence that this is an image of the chimpanzee, *Pan troglodytes* Blumenbach, 1775 is not convincing (see also Masseti, this volume). Trying to identify animals from their artistic representation is always a difficult task, but in my opinion the "Elamite Marble Figure of a Monkey; SKU: LM.61; Circa: 2100 BCE to 1500 BCE; Medium: Marble; Origin: Central Asia; Gallery Location: UK" (Urbani and Youlatos, 2020: 656) is more likely to be representations of a female, or a sub-adult, orangutan, *Pongo* sp. Lacépède 1799, on the basis of its appearance and the posture in which it is portrayed.

A few years ago, the Venetian art historian Renzo Mangili discovered an orangutan terracotta sculpture in a private Italian collection (**Fig. 9.8**). According to stylistic evaluation, even before any possible laboratory endorsement, Mangili (2015: 31) was able to recognize that the sculpture appeared to belong in the autographed bestiary by the famous Renaissance artist Jean de Boulogne or Giambologna (Douai, 1529 - Florence, 13 August 1608). Other evaluations support the suggestion that the terracotta *Orangutan with a hand on its head* dates to between 1575 and the end of the same decade (Mangili, 2015). This important discovery is evidence that one of the most important European artists of the time was aware of the existence of the orangutan, and the accuracy of the depiction suggests that Gianbologna may even have used a live specimen as a model, and the morphology of orangutans was well-known in Europe long before it was previously believed. Prior to the re-discovery of the statue by Gianbologna, it was generally believed that the first artistic depiction of this ape only appeared in 1658 in a posthumous work by the Dutch doctor Jakob de Bondt (see Barsanti, 2009: 24, and references therein).

FIGURE 9.8 Oran-gutan with a hand on its head (1575–1580 by Jean de Boulogne or Giambologna. Terracotta.

Egypt. Unfortunately, it has not been possible to find any images of this species from the 16th century.

The Portuguese trade in non-human animals in South and East Asia flourished in the second half of the 16th century. The Portuguese arrived in Japan in 1543 and were portrayed by local artists on the famous *Namban* screens (the Portuguese were known as *namban jin*, barbarians of the south), which mirrored their contact with Japan. These screens captured the moment when the Japanese met the Europeans and their culture, which they considered revolutionary and wonderful (Yoshi, 2009: 68). In the *Namban* screens, one can see the truly global character of the Portuguese empire. The Portuguese in fact brought 'globalization' to the Japanese nation and a very different kind of goods and animals, originating mainly from China but also from India and other Southern Asian countries, all linked by important trade routes (Curvelo, 2015: 28–30; Barreto, 2000: 47–50). *Namban* screens show many different animals shipped by Portuguese and Portuguese creoles to the port of Nagasaki: horses, camels, goats, dogs, leopards and other felids, peacocks and other birds, and monkeys. According to Carneiro (2009: 88), some animals would have been destined as gifts to the *daimyos*, the local feudal lords. A monkey brought on the shoulders of one of the Portuguese merchants looks like a capuchin monkey (a New World monkey), but judging from its size, it may also be a representation of a young crab eating macaque, *Macaca fascicularis*, a species endemic to Southeast Asia (**Fig. 9.9**). If it was a capuchin monkey, we can only hypothesize that this animal may have arrived on ships that had stopped in South America. Another American animal, the turkey, arrived in Goa at the beginning of the 17th century, which was portrayed by an artist at the court of the Mughal Emperor Jahangir (cf. Verma, 1999). The Mughal emperor Jahangir (ruled 1605–1627) was interested in natural science and placed Muqarrab Khan, a high-ranking noble, as his agent at Goa to purchase rare objects and exotic animals (Koch, 2009). In March 1612, he brought a turkey cock and a monkey to the court, which were described in Jahangir's memoir:

> He also brought a simian of a strange color and curious shape. Its hands, feet, ears, and head are exactly like a monkey's, but its face resembles a fox's. The color of its eyes is like a hawk's, but its eyes are larger than a hawk's. It is a cubit from its head to the base of its tail, shorter than a monkey but longer than a fox. Its fur is like the wool of a sheep, and it is grey. From its earlobe to its chin it is a wine – colored red. Its tail is longer than an half a cubit and three fingers. Unlike other monkeys this one's tail hangs down like a cat's. Sometimes it makes a noise like the cry of an antelope fawn. All in all it was extremely strange. (The *Jahangirnama*, [85b])

FIGURE 9.9 Portuguese merchant with a monkey on his back. One of the many animals brought to Nagasaki by the Portuguese. (Source: *Namban screen*, Seal of Kano Naizen (1570–1616) (Courtesy of the Museu Nacional de Arte Antiga, Lisbon.)

According to the description, in particular, the eyes and the color of the chin, this animal could be a Mongoose lemur, *Eulemur mongoz,* a species endemic to Madagascar, a region used by both Arabs and Portuguese for the slave trade in the early 1600s (cf. Armstrong, 1983). During the 17th century, the chief Malagasy port for Arab and European slave trade was Massailly (i.e., Nosy Antsoheribory, in Boina Bay, Armstrong, 1983), where this species can be found today (Mittermeier et al., 2013: 138). The Emperor also described the arrival of a dervish from Ceylon who brought an unusual animal called a *devang,* the Telugu name for the loris (*Loris* spp.) of south India and Ceylon:

> the face looks like that of a large bat, and overall, the body is similar to a monkey's but it has no tail. [...] The body is the size of a baby monkey two or three month old. [...] It drinks milk and eats bananas. It is really horrible looking. (*The Jahangirnama* [57b])

9.2 THE DIFFERENT SPECIES OF NON-HUMAN PRIMATES AT THE EUROPEAN COURTS: STATUS SYMBOLS, PETS, AND SUBJECTS OF ARTWORKS

In recent decades, the history of the interface between humans and other animals in the early Modern period has received more academic attention as demonstrated by the many studies of the perception and meaning of animals in Renaissance culture (e.g., Asúa & French, 2005; Fudge, 2002; Cuneo (ed), 2014, 2017; Drumond Braga & Drumond Braga (eds), 2015), their roles in early modern collections (Findlen, 1994; Smith & Findlen, 2013), at Renaissance courts (e.g., Tudela & Gschwend, 2007; Pieper, 2008; Gschwend, 2018; Masseti, 2018; Groom, 2018; Haag (ed), 2015; Simões, 2019), in art and natural studies (Egmond, 2017; Curry et al. (eds), 2018; MacGregor, 2018), and non-human animal agency (e.g., Rees, 2017). Due to the great variety of animals reaching Europe, the role of primates is often underestimated, and in general, little attention is paid to the species encountered. Through the depiction and identification of species, we can gain information about their place of origin, the trade routes, the agents, and intermediaries involved in the trade, and it is also possible to gain information about the original distribution and abundance of the species in question. This section focuses on the identification of primates in artworks of the late 15th and 16th centuries in Europe and on their iconography. Many of the animals mentioned here today are in danger of extinction, and it is relevant to know or estimate the impact that this first big wave of exploitation had on their natural populations.

In the Age of Discovery, anthropoid primates from Africa and tropical America were, after parrots and parakeets, most sought-after by nobles and wealthy citizens, and historical and iconographic sources suggest they were common in high-class residences and menageries (Tompkins, 1994; Zuckerman, 1998; Urbani, 2007; Pérez de Tudela & Gschwend, 2007; Teixeira & Papavero, 2010; Masseti & Veracini, 2010; Veracini, 2011a, 2017; Haag (ed), 2015; Gschwend, 2018; among others). While the transportation and housing of large exotic animals could be complicated (cf. Cockram, 2017), smaller creatures such as monkeys and parrots were more easily managed and accommodated.

Asian primates and great apes (cf. **Box Details 9.1**) are relatively rare in the iconographic record (see above), with no convincing evidence for trade in chimpanzees in the 15th century (cf. Urbani & Youlatos, 2023) (see **Box Details 9.2**).

9.2.1 New World Primates

Small Brazilian callitrichids, mostly originating in the Brazilian Atlantic Forest, capuchin monkeys, and squirrel monkeys were likely the first New World primates to reach European courts alive. The first representations that allow us to identify these primates were found mainly in Northern Europe, England, and Italy in the first half of the 16th century, but by the second half of the century, their presence in artworks was widespread.

9.2.1.1 Atlantic Forest Marmosets and Other Callitrichids

The Holy Roman Emperor Maximilian I of Habsburg (1459–1519) was one of the first rulers to acquire Atlantic forest marmosets (Groves, 2008: 53; Teixeira & Papavero, 2010; Veracini, 2011a). Collecting exotic pets was part of the tradition of the Habsburg court and of other wealthy German

BOX DETAILS 9.2: AFRICAN GREAT APES: THE FIRST CONTACT WITH EUROPEAN CULTURE IN THE 16TH CENTURY

Cecilia Veracini

Since antiquity, myths and legends have considered anthropomorphic creatures as more or less fabulous. In medieval times, some interpretative models linked these beings with the notions of the monstrous races of Pliny the Elder, or to the 'man of wood', or even hybrids between humans and monkeys (cf. Husband & Gilmore-House, 1980: 44; Veracini, Chapter 7, this volume; and Barsanti, this volume). In the natural history of medieval natural philosophers, these disturbing creatures resembled something between humans and apes, and were believed to live in very distant countries or on the edge of the known world (Mason, 2009: 98). The first explorers in Western Africa found a creature reminiscent of the mythical man of the woods: the West African chimpanzee (*Pan troglodytes verus* Schwarz, 1934). Although Portuguese exploration in Africa had begun in the previous century, it is only at the beginning of the 16th century that we find the first descriptions of these animals in navigators' manuscripts that report accurate information on the Western African coast, its natural resources (especially those that could be traded, such as ivory, animals, gold, etc.) and the peoples encountered on those shores (cf. Veracini & Casanova, 2016). These first accounts of great apes did not enter the official natural sciences of the 16th century, for many of these written documents remained in manuscript form, and were not widely circulated. In the case of Portuguese writings, it is well-known that travel reports were regarded as state secrets and hence not generally disclosed (cf. Lach, 2010; Costa, 2009).

In the earliest Portuguese accounts of Western Africa, descriptions of chimpanzees were not naturalistic, but instead continued the medieval tradition of relating tales of unusual wild men living in forests (Veracini & Casanova, 2016). This was the case for the Portuguese sea captain, explorer, and cartographer Duarte Pacheco Pereira, who described, in his work *Esmeraldo de Situ Orbis* (c. 1506–1508), chimpanzees for the first time when he outlined the characteristics of the territory now known as Sierra Leone:

> There are also wild men here, whom the ancients called satyrs. They are covered with hair, almost like that of pigs, and they resemble human creatures. They engage in intercourse with their women as we do with ours. Instead of speaking when someone harms them, they shriek. They live in the highest part of these mountains, so you can rarely catch them except when they are small boys. Many other things could be said about them, but to avoid making a lengthy sermon, I will end here (cf. Leitão (ed.), 2022: 112)

The journeys to Western Africa of the Portuguese explorer Diogo Gomes were recorded in a text known as the manuscript of Valentim Fernandes (1506–1508), and here we also find the description of the chimpanzee, considered to be an animal very similar to human beings:

> there is here an animal which seems like two different animals [n.d.r.: because it is considered false]. It has a white face and goes on four feet; it is similar to a man and it is black in the body and in the fur and spiky. (cf. Monod et al., 1951: 98)

Other navigators in successive decades reported the presence of great apes in Western and central Africa as common. This is the case of an anonymous Portuguese pilot, who traded sugar between Portugal and São Tome. His accounts were published in one of the many volumes of the famous collection of travel literature of the Venetian Giovanni Battista Ramusio (1550, vol. 1), *Libro delle navigationi e viaggi.* The above mentioned pilot makes a comment about the well-known Hanno's voyage around the coast of Western Africa and his dubious observation of anthropoid apes. According to the pilot, "Hanno would have brought a skin of a mythical gorgon; what else could those big monkeys be, that we call babuini, and that look like humans and that can be found in endless numbers along this coast" (cf. Caldeira, 2000: 79). While these first accounts of chimpanzees were short and uncommon, around the end of the 16th century, more in-depth reports on this species were produced. For example, André Alvares d'Almada (1594), a Luso-African captain who traded in Africa during this period, reported that

> In Sierra Leone there lives a kind of monkey not found elsewhere in Guinea; they are called *daris* [the current vernacular name given to chimpanzees in this region] and have no tail, and if they were not hairy it would be possible to declare that they were human like ourselves, for in other respects there is little difference. They walk on their feet, and some are so clever that if they happen to be captured when young and are brought up in a house, they go to the river to seek water and bring it back in a pot on their head. [...] They pound foods in the blacks' mortars like a man. They are short and thick in body, with good legs and arms. (D'Almada, 1594: 75)

These notes are probably a mixture of personal observation and accounts reported to the captain by native peoples, but it is relevant that chimpanzees are here considered a kind of monkey and not a human-like creature of the woods. Nevertheless, in this account, the old European myth of "Beauty and the Beast" is reaffirmed and still appears in a later writing by André Donelha, another Luso-African, who traded in Sierra Leone during the 1560s and 1570s and wrote his account during the 90s: "native people say that if this animal [a chimpanzee] meets a woman alone, it makes a match with her". Donelha described the morphology and behavior of the West African chimpanzee, noting that "*Daris* are of the size and appearance of men. The females have breasts like young women, the old ones have them hanging down like an old woman". He also reports the first scientific evidence that chimpanzees make beds to sleep in:

> this animal would not go to sleep unless it was given a mat or something on which to lie. They say that in the wild they break off branches and lay them across each other in forks of trees, and they make a bed with many leaves on which they lie. (Donelha, 1977:92)

So far, no iconographic sources indisputably testify to the trade in chimpanzees or gorillas in Europe before the 17th century (cf. Barsanti, 2009; Veracini, 2021). It is also very rare to find European historical sources that speak of the possible arrival of these great apes on the continent (see above, Section 9.1). It must also be considered that, as was the case with the arrival of the mandrill/drill which caused a great stir, if a completely different previously unkown animal had arrived, its appearance would have been recorded in some chronicle or treatise on nature. On the other hand, it is certainly established that the Arabian and Tuareg populations were familiar with chimpanzees (cf. Veracini & Almghadi, this volume) and also with gorillas (Veracini, 2021a). It is also possible that these animals could have reached the Mediterranean shores, transported via the trade routes that crossed the Sahara from time immemorial. The two 'ape-men' found in a 15th-century Lombard manuscrit ("Sloane MS 4016;", Folio 91v) and described as chimpanzees by Urbani and Youlatos (2023) cannot be considered proof of the trade of these animals in Europe, and can be hardly considered a trustworthy representation of these animals. These creatures do not show any of the morphological characteristics typical of a chimpanzee, having human hands, feet, faces, bodies, proportions, posture, and even beards. The only resemblance to the chimpanzee may be the dark coloring. In my opinion, considering the historical reconstruction set out in this and other chapters of this volume, and taking into account the notions about nature available in the 14th-15th century, the origin of the image may refer to the artist's attempt to reproduce the report, perhaps received from Arab, Venetian, or other sources, of a "black tailless monkey" with human features. The artist, using traditional conceptual models such as the myth of wild men and images found in earlier or contemporary manuscripts (one of these, called "Masson 116" depicts two Barbary macaques in the same position and acting as the supposed chimpanzees, cf. Urbani & Youlatos, 2023:6), attempted to reproduce an animal he had never seen but may have heard about.

citizens (MacDonald, 2002; Pérez de Tudela & Gschwend, 2007: 420). The close commercial and personal relationship between the Emperor Maximilian I and the King of Portugal Manuel I (Pohle et al., 2017) was probably the reason for the emperor being one of the first to receive these precious animals. Maximilian I, a cultured man, was the patron of some of the most important artists of the time, such as Albrecht Dürer, Hans Burgkmair, and Cranach the Elder. The widely known *Triumph of Maximilian I,* comprising three large groups of woodcuts: the *Triumphal Arch*, the *Triumphal Procession*, and the *Large Triumphal Carriage*, which was referred to by the art historian Hyatt (1971) as Maximilian's program of paper grandeur, portrayed marmosets, capuchin monkeys, and Barbary macaques. Live marmosets, which seem to have reached Germany by 1513, were portrayed by Dürer in the *Triumphal arch of the Emperor Maximilian I*, where two marmosets stand at the base of the portal columns of the arch, and in two "Folios" of The *Book of Hours of Maximilian I* (Teixeira & Papavero, 2010: 266). Veracini (2011a) argues that the *Triumphal arch* "chained monkeys", which traditionally symbolize the control of temptations (cf. Janson, 1952), were included to promote Habsburg universality and glorification (Panofsky, 1971; MacDonald, 2002), and were put at the entrance of the *Arch* to symbolize Maximilian I's overseas dominions.

The specimens (or the specimen) portrayed seem to be juveniles of the genus *Callithrix*. Due to their immaturity and because these works are not in color, it is not possible to establish the exact species. Considering the areas of the Brazilian Atlantic Forest explored up to that date by Europeans, these creatures could belong to either *C jacchus* or *Callithrix kulhii* (Veracini, 2011a), but the former is still the only marmoset species to appear in portraits shortly after these and the only one to be portrayed until the second half of the 16th century.

A common marmoset is also depicted in 1528 by the German painter and printmaker Hans Burgkmair (1473–1531) in his *Story of Esther*, now at the Alt Pinacothek in Munich. The monkey, which is portrayed on the porch steps where the main scene takes place, is accompanied by a dwarf (**Fig. 9.10**). This type of iconography, in which monkeys, together with other exotic animals, fill the side parts of paintings, is found in Renaissance court imagery. The dwarf's rise to pictorial fame came of age in the 15th century, when they became a symbol of the prince's magnificence, part of an impressive entourage that also included horses, dogs, exotic animals, singers, and buffoons (O'Bryan, 2012). The monkey-dwarf association is more complex, for like dwarfs, monkeys served to entertain the court with their antics, and sometimes they were dressed in the same striped garments

FIGURE 9.10 Common marmoset portrayed with a jester/dwarf in 1528 by Hans Burgkmair. (Source: detail of *Story of Esther*, Alt Pinacothek, Munich, detail) (Image in the public domain).

FIGURE 9.11 Margaret of Austria (Parma), daughter of Emperor Charles V, with a common marmoset (*Callithrix jacchus*) ca.1538–1540. (Courtesy: Maike Vogt- Lürssen.)

used for court fools. Monkeys and dwarfs were also related on a philosophical level, for both were considered degenerate forms of mankind, being at the same time both repellent and attractive (O'Bryan, 2012: 13).

Many iconographic sources from this period depict marmosets as symbols of princely authority and as beloved pets. In 1516, Cardinal Ciocchi dal Monte was portrayed in Rome with its dearest pet by Sebastiano del Piombo (Garas, 1994–95). The Cardinal, because of the presence of this monkey, was known as "*Cardinale scimmia*" (the Monkey Cardinal). A common marmoset was portrayed eating a small piece of food in the hands of Margaret of Austria of Habsburg (1522–1586), Duchess of Parma and Piacenza, and Governor of the Netherlands from 1559 to 1567 (**Fig. 9.11**). Margaret, the natural daughter of Emperor Charles V, was legitimized by her father and educated according to the dictates of her rank. The picture was painted when Margaret was 16–18 years old (i.e., between 1538 and 40) when she was in Rome, where she had arrived after becoming the widow of Alexander I de Medici, and when she was betrothed to Ottavio, the young nephew of the new Pope, Paul III, who belonged to the Farnese family. In that period (1537–38), several important Portuguese ambassadors came to Rome to negotiate with Pope Paul III the control and enforcement of the Inquisition in Portugal, and some were received at court by Emperor Charles V (cf. Barroso, 2022: 36). The common marmoset may have come with them and been given to the royal family or the curia. Her illustrious preceptor, Margaret of Austria, daughter of Maximilian I, and ruler of the Netherlands (Brandi, 2008: 152) also loved monkeys and as Dürer recalled in his memoires she had a collection of these animals (Eisler, 1991) and started a tradition that would be common to other women of the Habsburg-Spanish dynasty (see later).

The early presence of the common marmoset in Europe is testified by iconographic portraits of Tudor family members in Britain. Margaret Tudor (1489–1541), Queen of Scotland, was portrayed with a marmoset c.1515 or earlier (**Fig. 9.12a**). Hans Holbein painted the *Portrait of a Boy with a marmoset* c.1526 (**Fig. 9.12b**). Holbein was at the time at the court of Henry VIII in Britain, and the boy was likely Henry FitzRoy, the future King Edward VI (Foister, 2006: 60). A well-known painting c.1568, shows the family of William Brooke, tenth Baron Cobham, who seems to celebrate his lordship's family pride. It shows four of six children portrayed in the paintings, seated around a dining table with their pets: a parrot that was allowed to walk around the table, a goldfinch, and a marmoset (Grigson, 2016: 10; **Fig. 9.13**).

In 1548, one of the sons of the Holy Roman Emperor Charles V of Hapsburg, Prince Don Filippo, governor of Spain, and the future Philip II, who was going to Flanders to meet his father, stopped in Genoa with a flotilla of 58 ships (Saltini, 1879). Many rulers of the powerful Italian states went

FIGURE 9.12 The Tudor family with monkeys: (a) Margaret Tudor, Queen of Scottland, with a marmoset. The painting and a version by Mytens were copied from a lost original thought to be by a French artist, possibly Jean Clouet or Jean Perreal. The inscription in the Mytens copy 'Aetatis suae 26' indicates 1514/1515 as the date of the original. (b) *Portrait of a Boy with a marmoset* by Hans Holbein, c. 1526. The boy was likely Henry Fitzroy, the future king Edward VI. (c) Caterina di Aragona with a blond capuchin monkey - Miniature. There are two versions of this miniature, the other is by Lucas Horenbout, 1527. (Image in the public domain)

FIGURE 9.13 Details of the painting of the family of William Brooke, tenth Lord Cobham. It shows four of six children portrayed in the paintings, seated around a dining table with their pets: a parrot (*Amazona aestiva*), which was allowed to walk around the table, a goldfinch, and a common marmoset, c. 1567. The painting, formerly attributed to Hans Eworth, is now thought to have been painted by the Master of the Countess of Warwick (active 1567–1569). Oil on panel. (Image in the public domain)

there to pay homage to him or sent solemn embassies, but Cosimo I, distrusting the Genovese, sent his son Francesco, still a boy of seven years old, with a rich embassy (Saltini, 1879, 1883). Apparently, the young Francesco behaved like a gentleman despite his tender age (Saltini, 1879), which may be why Prince Philip gave him an important gift: a common marmoset. When Francesco succeeded his father as Grand Duke of Toscany, he cultivated his passion for the natural sciences and became a collector of novelties from other continents. His correspondence with the naturalist Ulisse Aldrovandi is well-known (Tosi, 1989); Francesco's young son was portrayed with a marmoset (**Fig. 9.14**).

FIGURE 9.14 Filippo de' Medici, son of Grand Duke of Tuscany Francesco I de' Medici, with a common marmoset. He died at four years of age (1577–1582) (unknown artist, second half of the 16th century). (Image in the public domain)

The women of the extended Habsburg court in Austria, Italy, Spain, and Portugal seem to have been inordinately fond of small monkeys. In the second half of the 16th century, common marmosets and other callitrichid species endemic to the Atlantic Forest of Brazil, as well as species from the territories ruled by Spain (Venezuela, Columbia), were often portrayed with princesses and queens of the Habsburg-Spanish family. These may have symbolized the influence this extended family acquired after the unification of the Spanish and Portuguese empires under Philip II (ruled 1581–1598). During Philip's reign, Spain reached the height of its influence and power, the so-called Spanish Golden Age, and after the unification of Spain and Portugal, the menagerie of Philip II became one of the most important in Europe, such that his gardens and aviary were seen as a "terrestrial paradise" (Pérez de Tudela & Gschwend, 2007: 443). The Italian artist Sofonisba Anguissola, who was introduced at the Spanish court by Queen Isabel de Valois, herself of Italian origin (Borghini, 2006), often painted the daughters of Philip II, the Infantas Clara Eugenia and Catarina Micaela, together with their exotic pets. The second daughter of Philip II, Catarina Micaela (and not the Infanta Clara Eugenia as commonly believed, cf. Cuadernos de Sofonisba), was portrayed in 1571 holding a common marmoset in her hand (Zuckerman, 1998); this is one of the best known portraits of this species. The same Isabella was also portrayed in 1585 by Alonso Sánchez Coello, hired by Philip II as a portrait painter for the royal family, with Magdalena Ruiz who is holding a Golden Lion tamarin (*Leontopithecus rosalia*) endemic to the state of Rio de Janeiro in Brazil, and a Cotton-Top tamarin (*Saguinus aedipus*) endemic to Colombia (Teixeira & Papavero, 2010; **Fig. 9.15**). The two animals, from two different South American territories – Portuguese Brazil and Spanish New Spain – symbolize how extensive Philip II's dominions were (Veracini, 2011a). Margaret of Austria (1584–1611), Queen of Spain, Portugal, Naples, and Sicily, grandchild of the Holy Roman Emperor Ferdinand I, was portrayed by Gonzalez (ca. 1600) while she was pregnant, with one of his sons holding a golden lion tamarin (Veracini, 2011a; **Fig. 9.16**).

FIGURE 9.15 The Cotton-Top tamarin, *Saguinus aedipus*, endemic to Colombia, was portrayed for the first time in Europe by Alonso Sánchez Coello (c.1585), a court painter at the Spanish court of Philip II, in a painting with Magdalena Rui and the Infanta Isabel Clara Eugenia (the painting is hosted at the Museum del Prado, Madrid). (Photo Cecilia Veracini)

9.2.1.2 Capuchin Monkeys and Others Neotropical Primates

Capuchin monkeys from the Brazilian Atlantic Forest and, in particular from early Portuguese settlements (e.g., Pernambuco and Bahia), reached Europe soon after Europeans made contact with Brazil. In 1508, Hans Burgkmair published six engravings illustrating the journey of the Tyrolean merchant Balthasar Springer in the East Indies. To match the traveler's description, Burgkmair introduced many elements of exoticism, including an early example of printed ethnography (Leitch, 2009) and an image of a Brazilian capuchin monkey that resembles *Sapajus xanthosternos,* endemic to Central Eastern Brazil (cf. Veracini, 2017; **Figs. 9.17a and b**). In 1512, Hans Burgkmair designed a great part of the *Triumphal Procession of the*

FIGURE 9.16 Margaret of Austria (1584–1611), Queen of Spain, Portugal, Naples, and Sicily, grandchild of the Holy Roman Emperor Ferdinand I, and wife of Phillip III of Spain, was portrayed by Bartolomé Gonzalez (c. 1600), while she was pregnant, with one of his sons holding a golden lion tamarin (Kunsthistorisches Museum, Munich). (Image in the public domain)

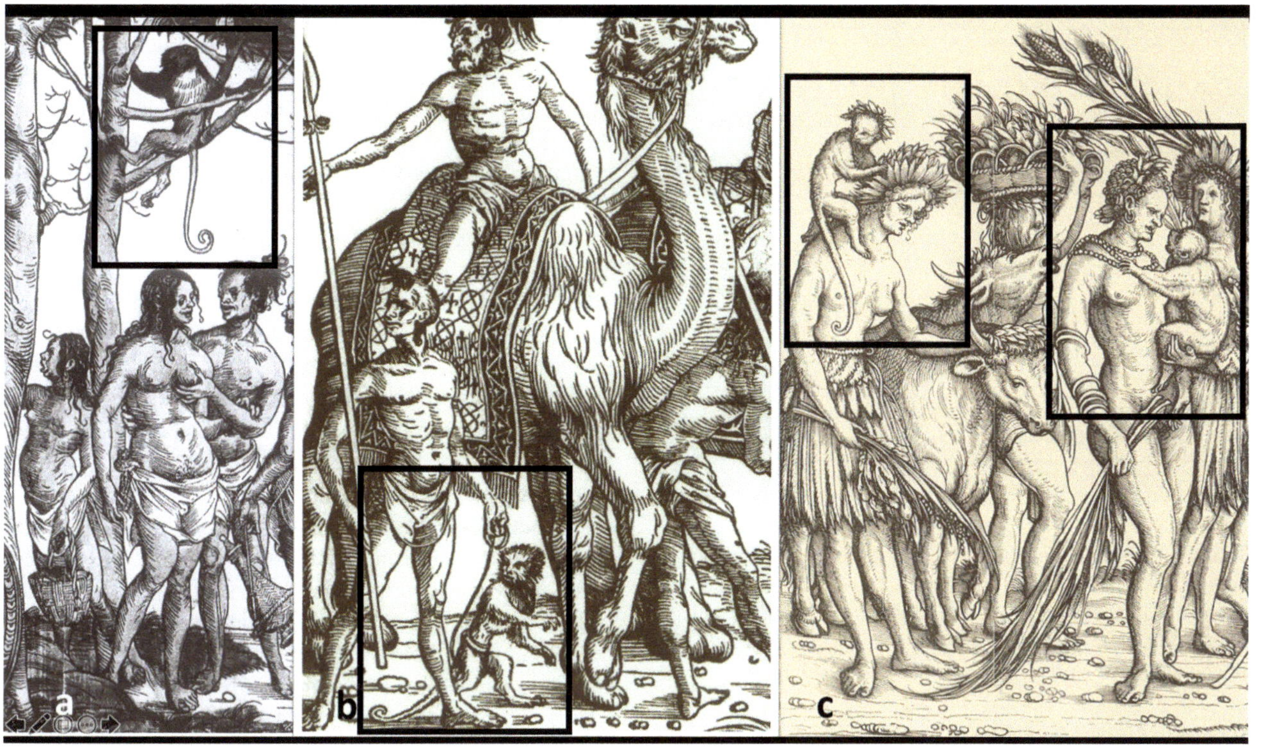

FIGURE 9.17 (a) A probable Sapajus xanthosternos in Natives with a flock of animals. Woodcut. Hans Burgkmair (1508); (b) capuchin monkey in King of Cochin, Woodcut. Hans Burgkmair (1508); (c) capuchin monkey. People of Calicut, in the Triumphal Procession of the Emperor Maximilian I. Hans Burgkmair (1518) (Images in the public domain). The first two images are likely portray animals painted from the life, in contrast to the capuchin depicted in the third work.

Emperor Maximilian I, contributing 67 woodblocks (see above). The part of the procession depicted by Burgkmair shows those following the royal carriage of the Emperor. After the Emperor comes oversea representatives, including exotic people from Calicut carrying "native" natural presents. Placed symbolically at the end, behind the European peoples of the empire, the people of the Indies are incorporated into the display of the Habsburgs' domains. An iconographic analysis of Burgkmair's depictions reveals a mix of ethnicities and elements from the New and Old Worlds. As already noted, at the beginning of the 16th century people spoke generically of the Indies and did not discriminate among the commodities and animals arriving in Europe from different places. This mixture suggests that the artist used his general knowledge to portray an imaginary exotic "Indian" place, where he inserts: an ear of corn, a large parrot (a mix of *Amazona* spp. and *Ara* spp.) held by a native dressed as a Brazilian *Tupinambà* (name used to designate the natives indigenous to the Atlantic Brazilian coast), and two monkeys: a Barbary macaque (endemic of North Africa) and a capuchin monkey carried on the shoulder by another indigenous individual (Masseti & Veracini, 2010; Veracini, 2011a; **Fig. 9.17c**). In 1518, Hans Burgkmair produced the first naturalistic painting of the squirrel monkey, *Saimiri sciureus*, in the Altarpiece of John the Evangelist (cf. Teixeira & Papavero, 2010), with another portrait of this species being produced at the turn of the 16th century at the court of Emperor Rudolf II (cf. Cod. Min 129, f. 49, Pl. 48 cf. Haupt et al., 1990, see below).

Between 1507 and 1511, a golden-bellied capuchin monkey (*S. xanthosternos*) was depicted in color by Gerolamo Genga in a fresco produced for the noble residence of the Sienese ruler Pandolfo Petrucci (Veracini, 2011a). The image of the monkey, which was probably a symbol of exoticism in the context of Greek mythology, was the first recognizable representation of this species (Veracini, 2011a; **Fig. 9.18**). Also in Italy, in 1519, the Renaissance artist

FIGURE 9.18 Golden-bellied capuchin monkey (*Sapajus xanthosternos*), depicted by Gerolamo Genga (1507–1511). Fresco, Escape from Troy, was produced for the noble residence of the Sienese ruler Pandolfo Petrucci. (Pinacoteca di Siena, image in the public domain)

FIGURE 9.19 *Sapajus flavius* depicted in two artworks: (a) a detail of Andrea del Sarto's (1519) fresco, Villa Medicea di Poggio a Caiano, Florence) (From Masseti & Veracini, 2010) and (b) detail of *Der Sundefall (Adam und Eva)*, Jan Gossaert, 1525. Gemeldegallerie, Berlin. (Image in the public domain)

Andrea del Sarto painted a fresco, commissioned by Pope Leo X, in a Medici residence in Florence, which included the first recognizable depiction of the blond capuchin monkey, *Sapajus flavius* (Masseti & Veracini, 2010), among the most traded species in the 16th century (**Fig. 9.19a**). It was probably an animal in the Pope's menagerie and was part of one of the embassies sent by the Portuguese King Manuel I to Pope Leo X (Bedini, 1997), who, like his late father Lorenzo de Medici, collected rare and exotic animals from all corners of the world (Masseti, 2018).

Catherine of Aragon was portrayed by Lucas Hombolte (or Hornebolte), the Tudor miniaturist and court painter around 1525, together with a yellow capuchin monkey (Zukemberg, 1998; Urbani, 2007). Two versions exist, the second one, from an unknown source, is the largest miniature of Henry VIII's first wife (**Fig. 9.12c**). In the second miniature, the capuchin can easily be identified as a blond capuchin, *S. flavius* (cf. Papavero & Teixeira, 2010). Monkeys were very rare in England at that time (Grigson, 2016: 8), and thus the inclusion of these animals in the portraits of Catherine must have been a mark of prestige. As Licence (2016) notes, the Queen is here portrayed as if with reference to Renaissance works such as da Vinci's Lady with an Ermine. The monkey holds flowers in one hand and appears to be playing with the queen's brooch with the other (Licence, 2016). Monkeys were not favored or loved by Catherine's rival, Anne Boleyn. A monkey, probably a capuchin monkey from Brazil, was sent in 1534 to Queen Anne Boleyn by Lady Lisle, the wife of the Governor of Calais, but the gift was not a success because the Queen "nor can scarce abide the sight of them" (Grigson, 2016: 9). Blond capuchins, especially young specimens, are portrayed in other artworks such as: (a) the painting by Jan Gossaert (1478–1532) entitled *Adam and Eve* (**Fig. 9.19b**); (b) a fresco by Giorgio Vasari and Giovanni Stradano (1561–1562), probably a copy of the del Sarto's fresco of Poggio a Caiano. The capuchin depiction is found in Palazzo Vecchio, which was the Cosimo I de Medici's residence in Florence, in the wing of the palace dedicated to the rooms of Cosimo's wife Eleonora di Toledo.

Capuchin monkeys of different species (see **Table 9.1**) were more common in Europe during the course of the 16th century and thus might have lost some of their importance as

TABLE 9.1
'New' Primates Introduced in Europe in the 15th and 16th Century – Iconographic Sources[a]

Taxon	First Known Source	Data	Studies
	New World Monkeys		
Callithrix cf. *jacchus* (Linnaeus, 1758)	*Triumphal arch of the Emperor Maximilian I. Book of Hours of Maximilian I.* Albrecht Dürer	1513	Groves (2008)
Callithrix jacchus (Linnaeus, 1758)	1. *Portrait of Cardinal Ciocchi dal Monte.* Sebastiano del Piombo.	1. 1516	1. Garas (1994–95);
	2. *Portrait of Margaret Tudor, Queen of Scotland.*	2. c. 151 5	2. This study
Callithrix geoffroyi (E. Geoffroi in Humboldt, 1812)	*Tavole animali BUB, vol. VI, a.c. 98.* Ulisse Aldrovandi	Second half, 16th c.	Capanna and Gippoliti (2007)
Saguinus oedipus Linnaeus 1758	*Infanta Isabel Clara Eugenia and Magdalena Ruiz.* Alonso Sánchez Coello	1585	Teixeira and Papavero (2010)
Leonthopithecus rosalia Linnaeus 1766	*Infanta Isabel Clara Eugenia and Magdalena Ruiz.* Alonso Sánchez Coello	1585	Teixeira and Papavero (2010)
Saimiri sciureus (Linnaeus, 1758)	*Altarpiece of John the Evangelist.* Hans Burkmair	1518	Teixeira and Papavero (2010)
Sapajus flavius Schreber, 1774	*Tributo a Cesare.* Andrea del Sarto	1519	Masseti and Veracini (2010)
Sapajus cf. *xanthosternos* (Wied -Neuwied, 1826)	*The journey of Balthasar Springer.* Hans Burkmair	1508	Veracini (2011a, 2017)

(*Continued*)

TABLE 9.1 (Continued)
'New' Primates Introduced in Europe in the 15th and 16th Century – Iconographic Sources[a]

Taxon	First Known Source	Data	Studies
Sapajus xanthosternos (Wied -Neuwied, 1826)	*Escape from Troy.* Gerolamo Genga	1507–1511	Veracini (2011a, 2017)
Sapajus cf. *nigritus*[e] (Goldfuss, 1809)	*The family of Darius before Alexander.* Paolo Veronese	1565–1570	Tompkins (1994), This study
Cebus capuchinus (Linnaeus, 1758)	*Tavole animali, vol. I. c. 84.* Ulisse Aldrovandi	Second half 16th c.	Capanna and Gippoliti (2007)
Cebus olivaceus Schomburgk, 1848	Cod. Min. 129, f. 40 r. Pl. 39. Austrian National Library	Second half 16th c.	Haupt et al. (1990)
Pithecia sp. Desmarest, 1804	Cod. Min. 129, f.27 r. Pl. 26. Austrian National Library	Second half 16th c.	Haupt et al. (1990)
Sapajus[e] sp.	*Sala degli Acrobati.* Castle of Torrechiara (Parma). Andrea Baglioni	Second half 16th c.	Bruner et al. (2008)
	Old World Monkeys and Apes		
Chlorocebus sabaeus[b] (Linnaeus, 1766)	*Tributo a Cesare.* Andrea del Sarto	1519	Masseti and Veracini (2010)
Chlorocebus tantalus[c] (Ogilby, 1841).	MS, Harley 3469. British Library	1582	This study
Cercopithecus cephus (Linnaeus, 1758)	*Expulsion from Paradise.* MIchael Coxcie	c. 1560	This study
Cercopithecus cf. *roloway*[b] (Schreber, 1774)	*Portrait of Margaret of Austria and Flanders, Duchess of Savoy.* *Seller of exotic animals in Antwerp.* Joachim Beucklaer	1. First half 16th century 2. 156 6	Veracini (2011a, 2017)
Cercopithecus petaurista (Schreber, 1774)	*Seller of exotic animals in Antwerp*, Joachim Beucklaer	1566	Veracini (2011a, 2017)
Cercopithecus mona (Schreber, 1774)	Cod. Min 129, f. 39r. Pl. 38. Austrian National Library		Haupt et al. (1990)
Semnopithecus sp. Desmarest, 1822	*The Story of Griselda, Part III: Reunion.* Master of the Story of Griselda (Siena)	c. 1494	This study
Cercocebus galeritus Peters, 1879	*Buontalenti's Grotta Grande*, the Boboli gardens. Bernardino Poccetti	1586–1587	Veracini (2017)
Colobus guereza (Rüppell, 1835)	Cod. Min. 129, f. 43 r. Pl. 42. Austrian National Library		Haupt et al. (1990)
Papio cf. *ursinus* (Kerr, 1792)	Cod. Min. 129, f. 38 r. Pl. 37. Austrian National Library		Haupt et al. (1990)
Papio hamadryas[d] (Linnaeus, 1758)	*Animals of the Brussels menagerie of Charles V.* Albrecht Dürer.	1520	Lugli (2005), Veracini (2011a)
Mandrillus sp. Ritgen, 1824	*Icone animalium*, Conrad Gessner	1560	Groves (2008)
Mandrillus leucophaeus F. Cuvier, 1807	Fresco, Sala Paolina Castel Sant'Angelo, Rome.	1520–1540	Veracini & Casanova (2015)
Pongo sp. Lacépède, 1799	*Orangutan with a hand on its head.* Gianbologna	1575	Mangili (2015)
	Prosimians		
Nycticebus cougang (Boddaert, 1785)	Cod. Min. 129, f. 38 r., Pl. 37 Austrian National Library	Second half 16th c.	Haupt et al. (1990)

[a] This table is obviously not exhaustive and relates to my current knowledge. Primates known since antiquity or in the Middle Ages are excluded from this list (e.g., the Barbary macaques, the olive baboon, and the grivet monkey).

[b] With regard to the genus *Chlorocebus*, this list is just indicative because the species of this taxon are not easily identifiable due to their similarity, especially since many specimens brought to Europe were young. Uncertain attributions to this species can be found in 15th century sources (cf. Veracini, Chapter 7, this volume); the same applies to *C. diana* and *roloway*, which are two very similar species.

[c] This species was identified by Groves (2008: 19) as the one depicted in the Bronze Age frescoes of the Minoan culture of Thera, but given the discussion regarding these frescoes (e.g., see, Box details 6.1, Chapter 6), I think it is useful to cite here a source that seems to portray this species unequivocally.

[d] This species was known since antiquity but was rarely portrayed in Europe before the 16th century.

[e] Tufted capuchin monkeys *S. nigritus, S. apella*, and *S. robustus* are very difficult to distinguish due to the great individual variation and similarity between species.

status symbols. These neotropical animals became a recurrent motif in different artistic contexts and were portrayed in many media (e.g., textiles, frescos, paintings, woodcuts, watercolors, tiles **Fig. 9.20**, etc.), and by a variety of artists such as Pieter Coecke van Aelst the Elder (1502–1550), Nicolaes de Bruyn (1571–1656), Brueghel the Elder (1525–1569), Paolo Veronese (1528–1588), and Cranach the Elder (1472–1553) (**Fig. 9.21**; Bruner et al., 2008; Veracini, 2011a, 2017). In most of these artworks, the depiction of the capuchins is accurate, suggesting the artists painted from live

FIGURE 9.20 A rare image of capuchin monkeys from Portugal: *Sapajus* cf. *flavius*. *Azulejo de Rodapé* c. 1565. Tile. Royal Palace of the Quinta do Bacalhau, Setubal. (Museu Nacional de Machado de Castro, Coimbra. Courtesy of Museu Nacional do Azulejo/ADF/DGPC, photo Francisco Matias, 1991.)

FIGURE 9.21 (a) *Book of Hour*, Cranach the Elder, 1513; (b) Detail of *The Family of Darius before Alexander*. Paolo Veronese ca. 1565–1570. (National Gallery of London; Image in the public domain.)

specimens. However, many of these renderings are clearly copies of those that circulated in the artistic milieu of that time, and in these cases it is impossible to identify the taxon. At the end of the 16th century, other New World monkey species appeared in Europe, as shown by Emperor Rudolf II's collection of animal drawings, where Amazonian capuchin species such as *Cebus olivaceus* are portrayed for the first time, along with the first image of a Saki monkey, *Pithecia* sp. (Haupt et al., 1990).

9.2.2 Old World Primates

In 1477, one elephant, two dromedaries, and many monkeys from the west coast of Africa were sent to the King of Naples, Rene d'Anjou by the Portuguese King Afonso the African. The latter is said to have maintained a menagerie with baboons in his palace in Sintra, near Lisbon (Lloyd, 1971), but none of their images have been preserved. Guenons and other cercopithecids from Western Africa could reach Northern Europe with great ease via the trade routes developed by Portuguese and other merchants. *Cercopithecus mona, Cercopithecus diana*, and *C. roloway* (**Fig. 9.2**), and *Chlorocebus sabaeus, Cercopithecus cephus,* and *Cercopithecus petaurista* (**Fig. 9.2**) appear frequently in iconographic sources of this period in many European countries. Their geographical range coincides with the first territories explored by the Portuguese in Western Africa, which were the base of extensive trade in ivory and slaves, as already noted. One of the first portraits of a Western African cercopithecid is found at the beginning of the 16th century in Northern Europe. The daughter of Emperor Maximilian I, Margaret of Austria and Flanders, Duchess of Savoy (1480–1530), was portrayed praying with a guenon (*C. diana* vel *roloway*) and a pet dog (cf. Veracini, 2011a, 2017). Was the monkey a symbol of temptation and the dog of fidelity to God, according to classical iconography, or was the monkey a pet of the princess, as reported above? A passion for exotic animals was also shown by the son of Maximilian I, the Emperor Charles V (ruled 1519–1558), whose menagerie in Brussels housed several animals portrayed by Dürer during his famous journey to Flanders (Lugli, 2005: 204). The artist expressed his impression in a lively way after observing the strength and beauty of a male hamadryas baboon (*Papio hamadryas*) (Veracini, 2017).

The Habsburg tradition of keeping and owning wild and domestic animals continued with Charles V's successors. The emperor Maximilian II (ruled 1562–1576) kept rare and foreign plants and animals in his gardens and corresponded with his ambassadors to the court of King Philip II of Spain about the acquisition of exotic flora and fauna (Pérez de Tudela & Gschwend, 2007:432). In 1583, his son and successor Rudolf II (ruled 1576–1612) transferred his court from Vienna to Prague, where he assembled an impressive collection of flora and fauna, and an extensive collection of nature drawings, objects, and specialized instruments that helped him to add to his knowledge of nature (cf. Haupt et al., 1990). Rudolf II's uncle, Archduke Ferdinand II (1529–95), ruler of Bohemia at Prague and later in charge of Tyrol and Further Austria, also kept many types of wild creatures at his castle in Prague (Haag (ed), 2015). Emperor Rudolf II invited outstanding artists from all over Europe to Prague to satisfy his longing to illustrate natural items collected during his reign. At the turn of the century, Rudolf II commissioned nature albums known as the *Museum of Rudolf II,* whose purpose was to illustrate not only the objects in his collection and the animals in his menagerie, but also other creatures he had learned about and which he possessed pictures of. In this work, produced between 1600 and 1611 (cf. Haupt et al., 1990), we find the first scientific description of the mantled colobus, *Colobus guereza*, a portrait done after the death of a young specimen. The caption accompanying the image reads: 'he died because he did not want to eat and had climbed a tree'. These monkeys were already known to the ancient Greeks as 'callitrichines' (i.e., with beautiful hair) and, despite their beauty, it was said "they could not be brought to Europe as they could only live in their own land" (cf. Masseti & Veracini, this volume). This was because their diet consisted solely of leaves, a fact that was evidently not understood in the 1600s.

The first drills and mandrills, cercopithecids of the genus *Mandrillus* native to west central Africa, appeared in Europe in the 16th century. The naturalist Conrad Gessner reported the arrival of a beast that "caused a great sensation when it arrived in Augsburg" (*Icones animalium*, 1560:76). This animal was a male drill or mandrill (*Mandrillus leucophaeus* vel *sphinx*), whose image was published by Gessner (see next section). Previously, a magnificent life-size representation of a male and a female drill (*M. leucophaeus*) had been depicted in Rome in the Sala Paolina of Sant'Angelo Castle (1520–1540) (**Fig. 9.22**; Veracini & Casanova, 2015). The Sala Paolina, which is one of the most magnificent rooms in the castle, was used for welcoming and receiving important visitors, which suggests how important these animals were. During Pope Paul III's pontificate, there was intense diplomatic activity between the Portuguese King John III (who controlled the Gulf of Guinea trade) and the Roman Curia, including gifts of animals that may have been delivered by the ambassadors themselves. Two baboons were sent by a noble Florentine, M. Olivieri, on the 12th of October, 1547, to the Duchess Eleonore of Toledo, spouse of Cosimo I de Medici Duke of Florence, as a sign of loyalty (Veracini, 2011a). The instructions accompanying the animals noted their use of tools, saying that "when given almonds or walnuts in their shells, by using a small rock, the baboons will crack them gently open, much as a person would", and it was noted one of them had "some colored stripes on its face" (likely it was a mandrill, *M. sphinx*) (ASF, Mediceo del Principato 1173 fol.).

The Medici family had a long tradition of possessing and using exotic animals for propaganda (Belozerskaya, 2006; Masseti & Veracini, 2010; Masseti, 2018; Groom, 2018), and in 1537, when the family regained power after a period of struggle and exile, Cosimo I de Medici employed important artists to create works whose allegorical meaning commemorated the greatness of his grandfather Lorenzo the Magnificent (*il Magnifico*). The latter's image was exalted by the exotic animals he had received as gifts during his reign (Lazzaro, 1995) (e.g., Vasari, *Tributo a Lorenzo de Medici,* 1556, Palazzo Medici Riccardi). At the Pitti Palace, the residence purchased in 1549 by Eleonora of Toledo to escape the unhealthy climate north of the city, a large garden, the Boboli Garden, was conceived as an extension of the Palazzo. In the first hall of Buontalenti's Grotta Grande in the Boboli gardens, frescoes painted by Bernardino Poccetti between 1586 and 1587, that is during the last years of the reign of the Grand Duke Francesco I de' Medici, son of Cosimo I and Eleonora, contain one of

FIGURE 9.22 Two drills (*Mandrillus leucophaeus*), male and female, depicted in a fresco within the Sala Paolina at Castel Sant'Angelo in Rome (c. 1540). The decorative ensemble in the Sala, attributed to Perino del Vaga and his collaborators, is dedicated to celebrating the virtues of Pope Paul III, portraying him as the new Alexander the Great, while also commemorating the martyrdom of St. Paul. The two drills, the memory of a gift brought to the Pope by foreign ambassadors, are positioned beneath the image of the Archangel Michael, possibly symbolizing the triumph over sin. The male drill presents a bunch of grapes to the female, symbolizing wealth and the essence of life. Both animals wear a red belt around their waists, which could signify dedication and loyalty. (Courtesy by the Direzione Musei Statali della città di Roma – Museo Nazionale di Castel Sant'Angelo.)

the earliest European representations of a puma (Masseti, 1991), and a representation of a white monkey, *Cercocebus galeritus*, a species endemic to Kenya (Veracini, 2017). Historical archives, such as the Universal Correspondence of the Grand Dukes of Tuscany (cf. Veracini, 2011a; Groom, 2018), suggest that some primates arrived in Florence via the Medicean port of Livorno in Tuscany.

By the 16th century, it was easy to find depictions of the species belonging to the genus *Chlorocebus* from both Eastern (*C. aethiops*, *C. tantalus* and *C. pygerythrus*) and Western Africa (*Chlorocebus sabaeus*). The Eastern African species of this genus have been known in Europe since the Bronze age. In the late Middle Ages, they could be found in noble residences, and they were the subject of artworks (cf. Veracini, Chapter 7, this volume). These monkeys were the most commonly traded, and they had been used for different purposes (e.g., entertainment, pets, and work) by a range of ethnicities and peoples ever since the Bronze Age. Despite being among the most common and well-known primates, they were treated with respect in many European courts. The King of France, Francis I, a sovereign who showed a particular interest in animals, allowed a likely vervet monkey to attend an official meeting (cf. Veracini, 2017).

The life of captivity and torture that awaited non-human primates, as well as other animals, taken from their natural habitat is well depicted by the sadness conveyed by the two red-capped mangabeys (*Cercocebus torquatus*), native to the Atlantic coast of West and Central Africa. They arrived in Antwerp, which was a great economic and commercial center in the 16th century, and were portrayed in the well-known painting *Two chained monkeys* (1562) by Pieter Brueghel the Elder, sitting chained to a ring in a deep stone window embrasure of a castle, with Antwerp harbor visible through the window.

9.3 PRIMATE NATURAL HISTORY IN THE EARLY MODERN PERIOD: CONRAD GESSNER AND ULISSE ALDROVANDI

The 16th century is considered a turning point for the natural sciences, when natural history emerged as a recognized discipline. As noted by Ogilvie, "from the 1530s through the 1630s, the task of natural history was describing nature, cataloging its marvelous and mundane products" (2006: 6). The work of inventory, and the will to collect and transmit the *summa* of natural knowledge, characterized the work of many scholars at this time. Europe in the 16th century was still steeped in the classics, which had been the cultural reference points during the previous centuries (cf. Veracini, Chapter 7, this volume). Early modern naturalists (i.e., people who concerned themselves with nature included natural philosophers, physicians, apothecaries, anatomists, etc.) wished to take stock of past knowledge and collect and compile the known facts in order to review them from a new, empirical perspective (Sabelli & Tommasini, 2004: 25; Pinon, 2005; Ogilvie, 2006). It is important to underline that among the early modern naturalists, a clear-cut boundary between the study of nature and the study of cultures did not exist, and direct observation was preceded and accompanied by laborious compilations that culled information from earlier texts (Pomata & Siraisi, 2005). Attention to reality was a novelty of the Renaissance: the empirical approach of naturalists was stimulated by the rediscovery of the biological work of Aristotle and other ancient texts (Tugnoli Pattaro, 1981: 76) and by the explorations of new continents, which contained previously unknown or undescribed plants and animals. The knowledge brought by travelers and explorers introduced new information and helped to eradicate erroneous traditional doctrines, such as the uninhabitability of the torrid zone or erroneous information on the characteristics and behavior of animals and plants (Asúa & French, 2005; Lloyd, 1971). Moreover, artists began to carefully report the world around them and produce faithful images of natural objects (Smith, 2006; Egmond, 2017). Natural history was thus a collective enterprise fuelled by the relationship between scholars and the whole of society (Findlen, 2006), including princes and patrons who were able to support their activities and diplomats and agents working in important trade centers who could provide scholars with the novelties of the newly explored lands and, not least, by emerging technologies, such as printing, which favored the rapid dissemination of texts (Eisenstein, 1979).

In this period of discovery and cultural ferment, two scholars dedicated their lives to the work of assembling, collating and cataloging the natural world: the polymath Conrad Gessner (Zurich 1516–1565), a Swiss bibliographer and apothecary, and the Italian physician Ulisse Aldrovandi (Bologna 1522–1605), born several years after Gessner. Gessner and Aldrovandi, imitating ancient writers such as Aristotle and Pliny, established the image of the naturalist as someone who aspired to describe a universal nature and continued the previous encyclopedic tradition typical of the Middle Ages (Findlen, 1994, 2006). Despite a theoretical framework still linked to the classical authorities, their conception of knowledge arose from an increased emphasis on direct experience and field observation, which often led them to amend information from authorities of the past. They assembled all the information available in ancient sources and compared it with the source of the time. Therefore, any claim to novelty in the identification of a species required a careful historical-philological investigation of the ancient texts in search of any references to the allegedly unidentified species (Pinon, 2005). The importance of verification, especially when dealing with non-European creatures such as primates, was paramount for naturalists, who depended to a great extent (because they could not verify them themselves) on second-hand sources. For instance, Aldrovandi, in a manuscript entitled *Discorso Naturale* written when he was forty-seven years old, confessed his ardent desire to visit the West Indies (Americas) and, as a "new Columbus, to discover their natural riches, animals, and plants that, due to the climate, were completely different from those in Europe". He wished to lead a great expedition to survey the natural world of the Indies with painters, scholars, and writers, and only in this way, he commented, "could those secrets of nature be unlocked" (cf. Cermentati, 1906: 361).

The new forms of non-human primates that appeared on the European scene were included in the classical Aristotelian scheme classification, which was, with some Middle Ages additions, still in use. Aristotle knew only a few species of African monkeys (and perhaps a few Asiatic species); he divided these simians into three groups (cf. Masseti & Veracini, this volume): (a) *pithekus* or *pithekos* (*πίθηκος*), which means flat nose and which is tailless (e.g., the Barbary macaque *Macaca sylvanus*); (b) *Kepos* (*κήπος*) or also *kebos* (*κῆβος*), in the Latin form *cebus*, to indicate a monkey with a tail (called also *Cercopithecus*); and (c) *kynokephalos* (*κυνοκέφαλος*) or *cynocephalus* in the Latin form: dog's head (κυν=dog+*κέφαλος*=forehead) for the baboon. Aristotle classified these creatures with other quadrupeds (four-footed beasts) and considered them as animals intermediate between humans and other animals (Aristotle, *Historia animalium*, Book II, Part 8). The early modern naturalists still took this classification for granted and treated the new primate species brought to Europe as 'varieties' of the existing species. However, in some cases, this traditional taxonomy had to be modified, adding new taxonomic groups. These scholars made a great effort to reconcile the past and the present in a plausible and coherent form. In particular, as we shall see in the case of Gessner and Aldrovandi, the type of classification they chose revealed their approach to the nature of primates at a time of transition between old and more modern ways of thought.

Conrad Gessner's monumental history of animals, *Historia animalium* (in German *Tierbuch*: Book of Animals), published in Zurich in several volumes between 1551 and 1558 and in a posthumous edition (1602), is an encyclopedic printed work whose text, enriched with illustrations, codified what Asúa and French call "the pattern of philological discourse on animals" (2009: 183). Primates were included in the first volume, which dealt with quadruped viviparous beasts (in the Latin edition, *De Quadrupedibus digitatis viviparis, liber I*). Within the quadruped category (as for birds and fish), the animals are arranged alphabetically, and there is an attempt at grouping them according to their natural affinities. The classification system adopted by Gessner was inspired by Aristotle, but it is more complex and contains much new information. For each animal, the description is divided into paragraphs devoted to its various aspects (e.g., animal philology, including etymology, names, epithets, and proverbs; morphological description and place of origin; physiology; disease and reproduction; habits and instincts; capture and domestication; and food use, moral, and literary observations). Because each animal is the sum of many concepts and meanings, its description had to include everything the authors had written about it, as stated by Gessner on the title page (Kasukawa, 2010). The first volume of the *Historia animalium* is based on 251 sources listed by Gessner, including the Old Testament, Hebrew texts, classical, and medieval authors. As far as primates are concerned, some 16th century authors are also cited, including the philosopher Gerolamo Cardano (1501–1576) and the anatomist Andreas Vesalius (1514–1564). Contemporary travel literature such as *De Orbe Novo Decades* (1511–1530) by Martir de Anghiera was also included. The *Tierbuch* (1551) came out in various editions, each update including new animals from Africa and America that were described, or brought to Europe. Some of this information was delivered to Gessner via the important Fugger Family of Augsburg, who held him in high esteem and sent him both live animals and drawings (Asúa & French, 2005: 191–95). The pictures of primates in Gessner's work came from a variety of sources, including the watercolor drawings of Felix Platter's collection (cf. Egmond, 2013), from Gessner's acquaintances, and from medieval sources (Barsanti, 2009). In addition to the *Historia animalium*, Gessner published in parallel the *Icones animalium* (1553–1560), a kind of illustrated encyclopedia that reproduced the woodcuts of the former with a shorter text.

Ulisse Aldrovandi, a scholar of the University of Bologna (Montalenti, 1960, vol.2: 118–24), actively participated in the debate on the interpretations of Aristotle's writings, persuaded by the necessity of an empirical study of a nature unhampered by influences of religion and metaphysics (Tugnoli Pattaro, 1981: 76). His museum of natural curiosities, the *Theatro di Natura*, which was one of the largest and most important of his time, contained 8,000 watercolors, 11,000 natural objects (animals, minerals, and fruits), and 7,000 dried plants collected in 15 volumes (Olmi, 1985: 5–16). Over the years, Aldrovandi wrote descriptions of his natural objects using all kinds of sources, both literary and experimental (Sabelli & Tommasini, 2004: 24). Aldrovandi's contribution to the natural history of primates consists of ten watercolors entitled *Tavole di Animali*, as well as the chapters about primates in *De Quadrupedibus digitatis viviparis* (*De Quadrupedibus* hereafter), published posthumously in 1637, which included engraved plates of primates that were printed to complement the text (Veracini, 2019), and two images copied from Gessner, whose *Historia animalium* constituted one of Aldrovandi's reference sources. In *De Quadrupedibus* Aldrovandi divided quadrupeds into three groups: *Fere* (big cats and bears), *Semifere* (other smaller animals), and domestic animals, while the ungulates were covered in a separate book. Primates were included in *Liber II* of the *De Quadrupedibus*, together with the *Semifere*, namely, foxes, otters, other small carnivores, and rodents such as rabbits, hares, etc. As with Gessner, the descriptions not only cover the physical and behavioral features of each animal, but also provide information that allows the animals to be appreciated as symbols of places, cultures, and traditions (cf. Ashworth, 1996; Veracini, 2019).

The following paragraphs briefly discuss the taxonomic groups of primates described in the works of Gessner and Aldrovandi (see **Table 9.2**). These authors' zoological nomenclature and taxonomic choice will be compared, taking into account the previous tradition, with special attention to the Aristotelian classification.

9.3.1 *De Simia*

The first primate chapter in both Gessner's and Aldrovandi's works is *De Simia* (On the primates). In both Gessner and Aldrovandi, this introductory chapter on primates is mostly a compilation of classical and medieval sources (cf. Masseti & Veracini, Chapter 6, this volume;

TABLE 9.2
Primate Taxonomy According to Gessner and Aldrovandi

Conrad Gessner
***Historia animalium* (1602, postumous edition) - *Thierbuch* (1551)**
De Simia - Ein Aff [*Macaca sylvanus* and other primate species – general introduction on primates]
De Simijs diversis - Diefe art der Affen [different kinds of primates]
De Cercopitheco - Meerkatz [guenons and other tailed primates]
De Cepo - Blumenaff [tailed monkeys and the patas monkey]
De Callitriche Simia Caudata Barbataque – Bartschwanzer [tailed monkeys and guereza monkeys]
De Simijs Sive Cercopithecis Prasianis et Alii Magnis, Caudatis – Indianischen Munaff [*Cercopithecus prasianis;* some New world monkeys; *Baaben*]
De Cynolocephalo – Hundskopff [baboons]
De Satyro – Geyssmannleinen [humanoid creatures, *e.g.*, *Pilosus*; devil of the forest]
De Sphinge, simiarum generis – Jungfrauwaff [the gelada baboon]
De Galeopithecus qui Sagoin à Bresiliensibus nominatur – Meerkatz [the common marmoset]
De Arcopithecus qui haut ab americis vocant [sloths]
De Simivulpa [opossums]
Ulisse Aldrovandi
***De Quadrupedibus Digitatis Viviparis* (1621)**
De Simia [*Macaca sylvanus* and other primate species – general introduction on primates]
De Cercopitheco [all the Old and New World tailed monkeys]
De Cynolocephalo [baboons]
De Papione [the wolf of Gessner, 1560, *Mandrillus* sp.]

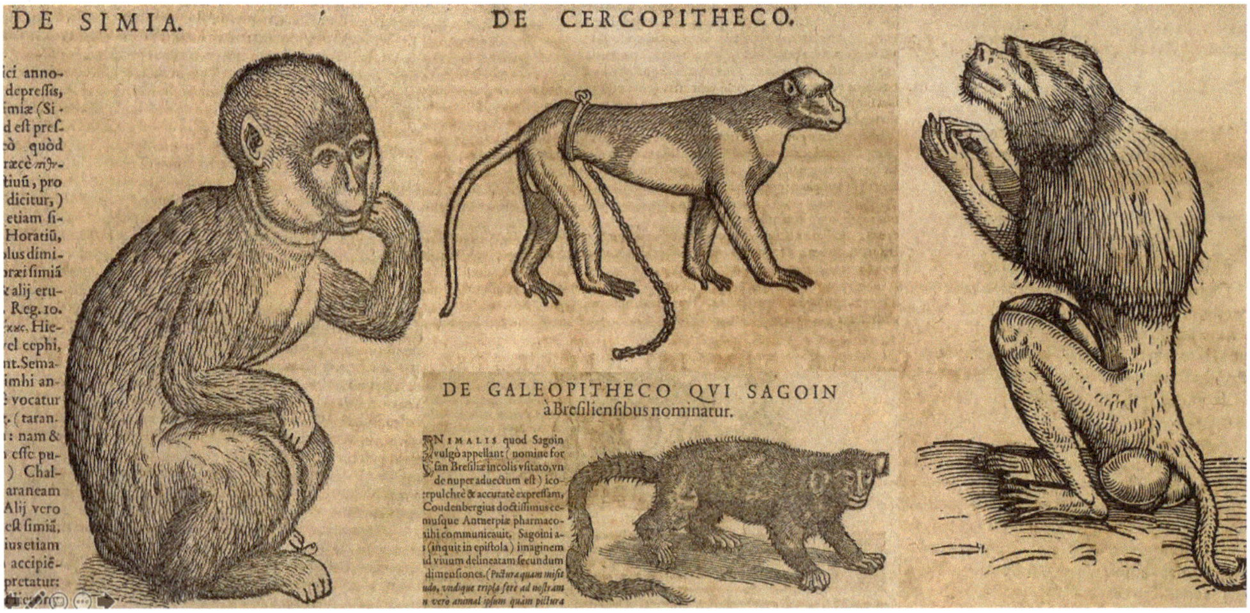

FIGURE 9.23 Some of the primates described by Conrad Gessner (*Historia animalium*). *De simia, De cercopitheco, Sagoin* and *De cynocephalo* (right side). (Images in the public domain)

Veracini, Chapter 7, this volume), where primates' main characteristics are reported, along with the traditional biogeographical, morphological, anatomical, and ethnological information. In the ancient and medieval world, *simia* was used to indicate primates as a whole (although, the only known primates were likely only some anthropoid species), but, as also noted by McDermott (1938), it was often used as a synonym for the Barbary macaque *alias* Aristotle's *pithekos* (untailed monkey). Gessner introduces primates using an image of the Barbary macaque (**Fig. 9.23**) and then discusses the etymology of the Latin name *simia,* quoting Isidore of Seville (560–636 CE) whose work, *Etymologia*, had a great influence in the medieval period: "the name *simia* Greek, *σιμος,* which means snub-nosed, that is, short- and flat-nosed". Although others maintain the name *simia* because we can find great similarity with humans, it refers to a "Greek adjective that means flat with an ugly face" (Gessner, *Historia animalium*, 1602: 847). Gessner provides a list of names in different languages of the *simia* (*Historia animalium*, 1602: 847–848), and this list was followed by Aldrovandi and reproposed as follows:

> This beast is called *Koph* by Jews and, when there are many, *Kophim*; the Chaldeans call it *Kophim*, and Aristotle *pithecos*. In the ancient Etruscan language, it was called *Arimo*; the noun *Inarime* refers to the Island of the Monkeys, a fortified island [Ischia, I]; the Moors call this monkey *Bugio* [...], the Spaniards *Mona*, or *Ximio*. In Italy it is called *Simia* or *Bertuccia*, in France *Singe*, in Germany *Aff*, in England *Ape*, and in Flanders it is called *Sime*... (Aldrovandi, *De Quadrupedibus*: 226)

In accordance with tradition, both authors report some of the negative characteristics attributed to these animals *simia quam similis turpissima bestia nobis!* (How ugly and, so similar to us is the monkey!) (Gessner, *H. animalium*, 1602: 853) from the famous sentence of Ennius, quoted by Cicero (*De natura deorum* 1.35.97), where the idea of physical deformity of monkey is combined with that of its moral obloquy: or *"nullum est animal in rerum natura, quod melees daemoni assimilate possit, quam simia"* (there is no animal in the world that can be compared to a demon, more than a monkey) (*De Quadrupedibus*: 230) where, quoting a medieval source, Aldrovandi raises the subject of the identification of the monkey with evil or its evil nature.

A long section is devoted in both Gessner and Aldrovandi to the anatomical aspects of primates, where the similarities between them and human beings listed in the works of Aristotle and Galen are given. It is worth noting that both naturalists include the work of the 16th century anatomist Andreas Vesalius, whose work, *De Humani Corporis Fabrica* (1543), proved that the writing of Galen was not absolutely correct, Vesalius made a close scientific study of primate anatomy highlighting the importance of defining the structural difference between humans and simians:

> The *simia* differs from man in many more parts than Galen thought, as Vesalius teaches (...). For example, in the thoracic muscles, in the movements of the arms, elbow and hip, in the muscles of the toes, in the movements of the foot and scapula; in the tendons of the foot .... (Gessner, *Historia animalium*, 1602: 849–850; Aldrovandi, *De Quadrupedibus:* 228)

In both Gessner and Aldrovandi, the introduction to the primates is a large fresco full of all kinds of information drawn for the most part from sources of the ancient and medieval tradition, which, however, could be refuted by contemporary works to which the authors had access, thus escaping a strict interpretation in the name of tradition.

9.3.2 Tailed Monkeys

The two authors chose differing modes of classification for tailed monkeys (the second taxonomic group of Aristotle), reflecting differences in their ideas about these animals as well as their different conceptions and interpretations of classical sources. Aldrovandi lived longer than Gessner, his life spanning all of the second half of the 16th century, and his views also likely reflect the information and images of plants and animals he was able to examine, such as those sent to him by the Archduke of Florence Francis I, who was passionate about natural history (cf. Tosi, 1989; Veracini, 2011a).

Gessner includes the various types of primates in a list that, in the Latin version of his work, is labeled *De Simijs diversis* (On the different forms of primates) (see **Table 9.2**). Here, he enumerates all the different forms of primates that he was able to know. The first chapter is dedicated to tailed monkeys, *De Cercopitheco*, whose description is accompanied by an image of a guenon with a chain around its waist. According to Gessner, tailed monkeys are called:

> *Meerkatzelein – meercatzen*, in German (sea cats), because they come from the sea (outside Europe); *gati maimoni*, or *gatti mammoni* in Italian; *marmot* in French; *mona* in Spain; *marmuset* in England but the popular word is also *munkey* a small gender with a tail and a beard. (Tierbůch, 1551: 857)

Cercopitheci "are tied by the belly because their necks are as thick as their heads and a collar would slip off them" (*H. animalium*: 1602: 856). As already observed in relation to 'sea cats', Gessner comments that most of them come from West Africa (i.e., guenons) but can also be found in India and in the New World, where very beautiful tailed monkeys live (quoting Martir de Anghiera, *De Orbe Novo*). The other tailed monkeys are listed by Gessner in chapters as follows:

De Cepo (*H. animalium*, 1602: 857). This chapter is about the *Kepos* (κηπος and also κηβος in Greek, latinized as *cebus*) or *kephos* (κηφος in Greek, Latinized as *cephus*) [the tailed monkey of Aristotle]. Gessner cites classical sources that refer to this animal as a colored monkey (quoting Diodorus Siculus, *Biblioteca Storica* and Aelian, *On animals*), saying it was also called *Kybor* by Avicenna [this term seems to have meant the common patas monkey, *Erythrocebus patas*, or other forms of this genus; cf. Masseti & Veracini, this volume; Groves, 2008: 31].

De Callitriche Simia Caudata Barbataque (*H. animalium*, 1602: 858). This section is about a monkey described by Pliny in his *Historia Naturalis* (Ch. LXXX): "it has faces with beards and tails flattened laterally in the upper part; it is said that this animal cannot live in any climate except its native Ethiopia" [this animal may correspond to *Colobus guereza,* cf. Masseti & Veracini, this volume; Groves, 2008: 31].

De Simiis Sive Cercopithecis Prasianis et Alii Magnis, Caudatis (*H. animalium*, 1602: 858). In this section, Gessner mentions: (a) the "Indian white monkey", described as *Cercopithecus prasianis*, a monkey bigger than a dog, white and with a black face [=Hanuman langur, genus *Semnopithecus*]; (b) the New World monkeys described by Martir de Anghiera in the Orbe Novo (Chapter 4), namely, howler monkeys; and (c) the *Baaben, formae rarae Cercopithecus*: *magnitude and forma hominis*, also called *hominem agreste*. Quoting Gerolamo Cardano, the Italian polymath of the 16th century, Gessner observes that the latter creature is similar to satyrs, of which it retains a similarity both in form and in lust. He shows an image of this kind of creature taken from the book *Peregrination in Terram Santa*, 1486, by Bernhard von Breydenbach (1440–1497). The same image would also be reproduced by Aldrovandi and, as noted above, labeled *Cercopithecus formae rarae* (cf. Barsanti, this volume).

Whereas Gessner mainly draws on classical and medieval authors to describe tailed monkeys (he includes in this category the monkeys known from antiquity, i.e., the African guenons, colobus monkeys, patas monkeys, some unidentified New World monkeys, Hanuman langurs, and the tailed-humanoid *Baaben*), Aldrovandi classifies them as *cercopitheci* in the chapter *De Cercopitheco (De Quadrupedibus:* 244), the taxa mentioned by Gessner, plus all the 'new' primates from the New World. Aldrovandi makes use of images (i.e., woodcuts produced from watercolors) and travel reports about America to describe these animals. For example, as well as the names attributed to tailed monkeys, mostly taken from Gessner, he also uses the indigenous names from the travel diaries of Hans Staden (1557) and Jean de Léry (1578). The former, a mercenary and representative of German bankers in the New World, was imprisoned in a Tupinambá village in the south of Brazil (today the State of São Paulo). The latter was an explorer, writer, and pastor who lived for a long time with the Indians of the region of Rio de Janeiro. Both diaries distinguish small monkeys (*Sagouin),* [marmosets and lion tamarins], medium (*Cay*) [capuchin monkeys], and large (*Aquiqui, Pricky*) [howler monkeys and muriquis] (cf. Veracini & Teixeira, 2017). Aldrovandi uses the word *Michi*, from *mico* (*De Quadripedibus*: 244), a native name from Venezuela, which according to Galeotto Cei (1513–1579), a Florentine merchant who visited and explored Colombia in the first decades of the 16th century, means small monkeys (e.g., tamarins and marmosets) (cf. Glossary, Veracini & Teixeira, 2017). The New World monkeys classified as *cercopitheci* by Aldrovrandi, *Callithrix geoffroyi, Cebus capuchins, S. xanthosternos*, and *Sapajus* sp. are portrayed in four tables (**Fig. 9.24**) (cf. Veracini, 2019 for more details).

A description of a New World primate had already been published by Gessner in the 1561 edition of *H. animalium.* In a later section of the book (page 158.b), he includes a drawing based on an image Gessner received from a German friend in Augsburg (Asúa & French, 2005: 195). Labeled *"sagoin, Ein ardt der Meerkatzen"* (a kind of seacat), it is the first printed image of a common marmoset (Veracini, 2011a), described as "animals imported from Brazil… very lovable and quick; completely covered with nice soft fur. They gobble down sultanas and white bread". In the last, posthumous edition of the *H. animalium* (Frankfurt, 1602, Vols 1: 869–871), there is some taxonomic revision, and the part entitled *De Simijs diversis* includes the chapters on the *Galeopithecus* (a mixture of weasel and monkey) or *Sagoin*, the *Arctopithecus* (the sloth, *Bradypus* spp.), and the *Simivulpa* (opossum, genus *Didelphis*). Thus, marmosets, sloths, and opossums became considered varieties of monkeys. It is interesting to note the effort Gessner made to fit these previously unknown animals, about which he had received only limited information and few visual representations, into a single scheme.

9.3.3 *Cynolocephali*, Dog-Headed Monkeys

Among the different forms of monkeys (*De Simijs diversis*) to which Gessner dedicates a chapter are the *cynocephali*, a term that had become established by tradition for the baboon group and dog-headed monkeys as a whole. A drawing of an adult male hamadryas, *P. hamadryas*, already appeared in the *Icones aimalium* (1560: 93), which illustrates the text of the chapter *De Cynocephalo* (*H. animalium*, 1602: 862). Aldrovandi also devotes a chapter to, which for him is "the third genus of monkeys, in accordance with Aristotle" (see above); adding:

FIGURE 9.24 Some of the New World monkeys portrayed in the *Tavole Animali* of Ulisse Aldrovandi. *Cebus capuchinus, Callithrix geoffroyi, Sapajus xanthosternos.* (Courtesy of the Biblioteca Universitaria di Bologna, BUB.)

FIGURE 9.25 Some of the Old World monkeys portrayed in the *Tavole Animali* of Ulisse Aldrovandi. *Macaca sylvanus,* immature male of *Papio hamadryas, Chlorocebus aethiops.* (Courtesy of the Biblioteca Universitaria di Bologna, BUB.)

> Aristotle also calls them *simia porcaria* [cf. the discussion about this term in Masseti & Veracini, this volume] and Avicenna calls them *Kiber* [=the *kyboz* of Albertus Magnus], and in Mecca they are called *Breca*, even if others call them *cercopiteci* [**Fig. 9.25**]. The French, according to Bellonium, call them *Magot*, and popularly *Babion*, and in Italy they are often called *babuinos*. (Aldrovandi, *De Quadrupedibus:* 253)

In their descriptions of baboons, Gessner and Aldrovandi collate information from ancient and medieval texts that report their role in ancient Egypt, and observe that baboons are among the best-visually represented monkeys. Aldrovandi also quotes excerpts from so-called 'emblematic literature', a literary genre flourishing in the 16th century in which the images are strictly linked to the text. For example, he quotes Joachim Camerarius (1534–1598), who, in his book on emblems (cf. Camerarius, 1654, vol. II: 78), writes about the affinity between baboons and the moon, one of the most important aspects attributed to baboons in classical and medieval tradition (Veracini, 2019).

9.3.4 *De papione* of Aldrovandi

A drill or mandrill (Groves, 2008; Veracini, 2017) of the genus *Mandrillus* is described by Gessner in the *H. animalium* (*Appendix liber I, De Quadrupedibus viviparis*, 1554) and in *Icones animalium* (1560: 76). In both of these works there is a section dealing with carnivores (*Von dem Wolff*, On the Wolf) with an image of a *cynocephalus*, the accompanying text says "that it caused a great sensation when it arrived in Germany at Augsburg. It is found in India, climbs trees and picks fruit; it respects elephants but drives away other animals; it always has two children, a male and a female" (Gessner, 1560: 76).

FIGURE 9.26 The *Papione* from Aldrovandi's *De Quadrupedibus Digitatis Viviparis* (1637: 260). Image after Gessner's picture of the "wolf" (*Icones animalium*, 1560: 76).

Based on Gessner's description, Aldrovandi dedicates the last of his primate chapters to this creature: *De papione, Synonima Descriptio et Historia Tota* (**Fig. 9.26**). He departs from the Aristotelian classification and coins a new taxon for this primate, which, he claims, differs from all other primates. Aldrovandi stresses that the affinity with wolves claimed by Gessner is incorrect. The fact that the Swiss naturalist, despite noting the similarities with monkeys, inserts this creature in a chapter with carnivores may be justified by a notion that goes back to Greco-Roman times. According to tradition, there was an animal called *arctopithecus*, a beast with the appearance of both a bear and a monkey. However, other reasons may also justify this choice. As noted by Aldrovandi (1637: 259), Albertus Magnus reported a similar animal called *Papionis*:

> This beast is bigger than a fox and with the manners of a wolf. This animal produces sounds that may be heard from afar; it feeds on corpses, and it has not the characteristics of a baboon but of a hyena. (Albertus Magnus, Libro 22, *De Animalibus*, cf. Veracini, Chapter 7, this volume).

,

Also, Iacobus de Vitriaco (Vitriacus, 1547 apud Aldrovandi, 1637:259) spoke about "dogs called *papioni,* which look like wild wolves and howl a lot". Aldrovandi admits he is ambivalent, "that is, I do not think these beasts are related to the animal brought to Germany, and he keeps the *papione* in the primate group.

9.3.5 Ape-Men, Humanoids, and Other Creatures

Aldrovandi and Gessner differ in the way they classify fabulous or 'untailed creatures' such as satyrs and pygmies (dwarf creatures), about which there was much debate in the Middle Ages (cf. Veracini, Chapter 7, this volume). These creatures often seemed intermediate between human and non-human primates, and therefore, how they were regarded was linked to the age-old question of defining human nature. Gessner, interpreting the medieval authority Albertus Magnus, who said that pygmies have an imperfect *loucutionem* and therefore cannot be considered humans (although they were different from other non-human primates and animals), said that pygmies should be regarded as a species of primate (*H. animalium*, 1602: 855). Gessner also classified satyrs as primates, writing that monkeys not only may lack tails like us, but also emulate the human figure and, having feet with heels, can stand up and run like a man. In the chapter *De Satyro* (*H. animalium*, 1602: 862), he affirms that "these are a rarer and more admirable race of monkeys, believed by some to be demons, and painted as animals with goats' feet and horns", but "they are true monkeys according to ancient authors, and the fact that some of the classical authors describe them as living around the Ocean may indicate Java". The location indicated by Gessner is in accord with the classical sources, which show that the *satyri* were Asiatic primates, probably gibbons or Indian langurs (cf. Masseti & Veracini, this volume). Similarly, another creature, the *Pilosus* (in German *Rauch Waldenmannlein,* translated into English by Groves as "little shaggy forest man", 2008: 60), with "the upper part of a man and the lower part of a goat, cuckhold in front", is considered "a monstrous kind of monkey that lives in the Ethiopian [=African] desert, and sometimes it walks erected" (*H. animalium*, 1602: 866). The chapter concerning the satyrs includes another humanoid creature: a sort of devil (in German *von dem Forsteufel*), "who is found in the Saxonian forests". This creature is illustrated by a quadrupedal devil with long ears and a forward-pointing beard (*H. animalium*, 1602: 866); the sphinx is also regarded as a true primate (*De Sphinge, simiarum generis*, *H. animalium*, 1602: 867). Gessner quotes ancient authors, among them Pliny, when he writes that "it has dark-brown hair, breasts like women, and is generated in Ethiopia". This creature, according to various ancient and modern sources (cf. Groves, 2008: 35; Masseti & Veracini, this volume), can be regarded as the gelada baboon, *T. gelada*, a real primate, but its mythical nature led to various interpretations. Gessner himself dedicates a paragraph, *De Sphinge fabulosa*, to this, stating that there are many images of beings that do not exist in nature, such as tritons, sphinxes with a half-lion, half-woman body, centaurs, hydras, chimeras, etc.

Aldrovandi removes most of these problematic creatures from the 'true primates', and places them in the book about monstrous creatures. Aldrovandi's *Monstruosis Historia* (1642: 23–34), includes satyrs, nereids and sirens, tritons, fauns and centaurs, and the *Populi carpophagi* (*Monstruosis Historia*: 74), tree-dwelling people who feed on leaves, who may be a refererence to great apes (i.e., chimpanzees or gorillas) (cf. Veracini & Alghamdi, this volume). The only creature Aldrovandi places in the primates (in the *De Cercopitheco* chapter) is the 'humanoid' *Cercopithecus formae rarae* (the *Baaben* of Gessner), regarded as a rare form of tailed monkey (Barsanti, 2009; Barsanti, this volume). The *Monstruosis Historia* also includes *Homo agrestis* (a lion's body with a man's head resembling a cat), which Aldrovandi says was captured in Peru (i.e., Gessner's *Pilosus*), and *Homo silvestris plumario* (BUB, *Tav. Animali*, vol. I, c. 74, p. 70), an indigenous Brazilian of the Tupi ethnic group, one of the many other indigenous people described in this work (Veracini, 2011a). It is important to underline that the term monster does not have a derogatory meaning, and that exotic and unknown beings were regarded as monsters. Gessner and Aldrovandi, like other naturalists of the 16th century, were likely not aware of the discoveries made in West Africa, where the first contacts with and description of chimpanzees, *Pan troglodytes*, had already occurred (see **Box Details 9.2**). The reports about West Africa, mainly of Portuguese navigators and explorers, had in fact remained in the form of manuscripts, were regarded as state secrets, and hence were not generally disclosed (Lach, 2010; Costa, 2009).

When analyzing the work of the early modern naturalists, we must not forget their relationship with the Catholic Church and the Inquisition (e.g., Findlen & Marcus, 2012). It is difficult to assess how much these factors influenced ideas on the nature of primates, in particular on issues such as their *similitude hominis*, which directly involved discussing and reviewing the place of humans in nature and their origin as narrated by the Holy Bible. Despite this, the issues of the attribution of soul and cognitive skills to animals, (which were already addressed by late medieval philosophers such as Albertus Magnus, cf. Perfetti, 2020), came to be discussed during the 16th century by a number of philosophers who did not follow the orthodox view concerning Aristotelian thought (Perfetti, 2000; Rubini, 2013; Veracini & Teixeira, 2017). In 1607, Gessner's *Historia animalium*, which remained one of the reference texts in the following century, was translated into English by the British cleric Edward Topsell, with the name *The History of Four-footed Beasts*.

The *sagoin* and the mandrill/drill, as well as other exotic animals for which there was little evidence in the literature, were often reproduced in treatises and texts of later centuries, sometimes with incorrect attributions. The fabulous monkeys of Gessner were less fortunate because the medieval tradition was abandoned very gradually. The images of primates in the *Theatro di Natura* by Aldrovandi rarely circulated outside Bologna and were not reproduced in the manuals of later centuries.

9.4 CONCLUDING REMARKS

The cultural assimilation of unusual and exotic animals brought to Europe by the opening of the new overseas trade routes was reflected in the entire cultural medium of early modern Europe, in particular in its figurative arts and literature. Among the animals in this wide physical and imaginary zoo, monkeys were always present in a great array of different artworks, iconographic meanings, and lores. The art found in noble and rich residences testifies to the presence and importance that non-European animals had in Renaissance, and later, society. Such works were, in most cases, a visual metaphor for European overseas conquests. From the *Triumph of Maximilian I* to the *Conquest of India*, tapestry series commissioned by Emperor Maximilian I and King Manuel I, respectively, both populated with exotic fauna and exotic peoples, to the Italian frescos and paintings of the Medici and other noble families, to the portraits of the many queens and princes, monkeys, and other non-European animals and people, were used as a medium for propaganda. In many cases, the iconographic meaning of monkeys is completely different from that of previous centuries, where they were metaphors of Christian morality and, in general, had a negative value. In the 16th century, the different monkey species entering Europe began to be represented accurately, demonstrating an emergent naturalistic interest and an attitude toward these animals that was different from the medieval iconographic tradition, where these animals mainly acted as metaphors of human bestiality (Veracini, 2011a). A similarly gradual shift to a more naturalistic viewpoint, still mixed with metaphysical ideas, can also be observed in the way early modern naturalists dealt with monkeys. Besides the symbolic value of many artworks representing these animals, they have zoological importance, being the first known pictorial renderings of these species. The value of naturalistic images of nature would become increasingly important in a period when texts, which tended to depend on classical authority, were more outdated than images. As Swan (2018) points out, in this period, images were used to close the gap between textual knowledge and direct experience of it.

During the 1500s, the first major global phenomenon of trade in animals and other natural resources occurred. We still do not have a precise picture of the consequences of this, but it is certain that countless animals were removed from their original ecosystems, where they fulfilled an important ecological function as seed dispersers, pollinators, and forest regenerators. Regarding primates, considering the high frequency of commercial voyages and the limited habitat range of some species, it is possible that this trade could have locally damaged some primate populations (e.g., the Brazilian species of capuchin monkeys, *S. flavius* and *S. xanthosternos*, and the golden lion tamarin *L. rosalia*). Teixeira and Papavero (2010) estimate that around two hundred monkeys were shipped each year from Brazil by the Portuguese, but this may be an underestimate. These authors and Veracini and Bezzerra (2013) also explain how this trade could have had a notable impact on the blond capuchin monkey (*S. flavius*), which originally had a small geographical distribution. The state of Pernambuco, encompassing *S. flavius*'s geographical range, was one of the three main centers for supplying dyewood (8,000 tons of dyewood from Brazil a year were imported to Portugal and France, cf. Dean, 1997). Later on, this area became one of the most important places for sugar cane production (Oliveira, 1990). Thousands of animals may have been taken from the wild over the course of about one century, and considering the destruction of the coastal Atlantic forests for sugar cane plantations and dyewood harvesting, it is likely there was a drastic decline in the numbers of this species in Pernambuco State by the end of the 16th century. The systematic trading in capuchin monkeys (and other species from the Atlantic coast of Brazil) continued to occur throughout the 17th century and in the following centuries, as revealed by many iconographic and literary sources (e.g., **Box Details 10.3,** Chapter 10).

It is also difficult to establish the cultural impact this had on those human populations that had coexisted with these species for tens of thousands of years. Many of the problems we see today - from the general loss of biodiversity to the extinction of many species, the spread of invasive ones, health problems, and the spread of pests - were probably triggered during this period and have continued at an ever-increasing rate throughout the modern and contemporary periods.

ACKNOWLEDGEMENTS

I would like to sincerely thank Ana Roque for her suggestions on some parts of the text; Bernard Wood for his careful revision of the manuscript; Dante Teixeira, for his advice and for sending various material; Stephen Nash for his helpfulness, advice, and kindness in sending material; Clifford Jolly for his sympathy and for sending his work on gelada baboons; Ivan Norcia for the photo of the gelada baboon; Mauro de Vito and Maike Vogt-Lürssen for their reports of various Renaissance paintings with monkeys, and later for providing me with the portrait of Margarita of Parma; Pedro Dias for permission to publish an image from one of his works; José Sousa e Silva Junior for his help with the identification of some New World capuchin monkeys; Luca Molà for the sources sent in; and Pedro Correia for information on the sources of the Namban screens, and also various colleagues at the Universidade Nova de Lisboa – CHAM, from whom I learned a lot about Early modern history, including Lella Alessandrini and Cristina Brito. Hilary Morris for proofreading the English drafts and her great helpfulness. I am also grateful to Marco Masseti for fruitful discussions and contributing the box details to this chapter. I sincerely thank Giulio Barsanti for his teaching. I would also like to express my gratitude to the late Professor Colin Groves for being a source of inspiration and an example in the study of the history of primatology. I would like to thank various institutions for permission to publish their art works: the Museo Nazionale di Capodimonte, Naples; the Biblioteca Universitaria, Bologna; the Museo Nazionale di Castel Sant'angelo, Rome; the Museu Nacional dos Azulejos, Lisbon; the Museu Nacional de Arte Antiga, Lisbon and the National Gallery of London. I would also like to express my gratitude to the I Tatti Berenson Library of Florence and its staff for their help over many years of research and to the National Library of Portugal, Lisbon.

FUNDING

Portuguese national funds through FCT—Fundação para a Ciência e a Tecnologia in the context of the celebration of the contract under the 'Norma Transitoria', DOI 10.54499/DL57/2016/CP1456/CT0001.

REFERENCES

Primary Sources

Aelian, Scholfield, A.F. (ed.) (1959). *On the Characteristics of Animals*, Volume II: Books 6–11. Loeb Classical Library 448. Cambridge, MA: Harvard University Press.

Aldrovandi, U. (1637). *De quadrupedibus digitatis viviparis*. Ambrosino, B. (ed.), *ibidem* 1642, 1645, 1663, 1665.

Aldrovandi, U. (2nd half 16th c.). *Tavole animali*. https://www.filosofia.unibo.it/aldrovandi, Accessed 29 December 2014.

Angleria (Anghiera), de P.M. (1511–1530). *De Orbe Nouo* Dec*ades* (*Décadas del Nuevo Mundo*).. Madrid: Ediciones Polifemo (1989).

Anonymous. (1881). *Catalogo da exposição Historia do Brasil*. Rio de Janeiro: Biblioteca Nacional de Rio de Janeiro.

Aristotle, D.W.T. (ed.). *Historia animalium* (*History of Animals*). On-line. https://classics.mit.edu//Aristotle/history_anim.html

ASF (Archivio di Stato di Firenze) (1578). Mediceo del Principato 1173 fol. *Lettere a Michele Olivieri a Pier Francesco Riccio. 12 October, 1547*.

Barbosa, I. de V. (1885). *Apontamentos para a história das coleções e dos estudos de zoologia em Portugal*. Lisboa: Typographia Christovão Augusto Rodrigues.

Berchet, G. (1892). Codice Cicogna 1985, c. 223, Museo di Venezia. *Fonti Italiane per La storia della Scoperta del Nuovo Mondo, Vol I. Carteggi diplomatici*. Roma: Ministeri Pubblica Istruzione.

Cadamosto, L. de (1988). *Viagens de Luis de Cadamosto e de Pedro de Sintra*, Lisboa: Academia Portuguesa da História.

Caldeira, A.M. (ed. and tras.) (2000). *Viagens de um piloto portugues do Século XVI à costa de Africa e a São Tomé*. Lisbon: Comissão Nacional para as Comemorações dos Descobrimentos Portugues.

Camerarius, J. (1654). *Symbolorum et emblematum ex re herbaria desumumtorum centuria una collecta*, Vol II. *Ex animalibus quadrupedibus*. Francofurti: Impensis Iohannis Ammonij.

Cardim, F.S.J. (1584). *Tratados da terra e gente do Brasil / Fernão Cardim*. de Azevedo, A.M. (ed.). Lisboa: Comissão Nacional para as Comemorações dos Descobrimentos Portugueses (1997).

Carletti, F. (1701). *Ragionamenti di Francesco Carletti fiorentino sopra le cose da lui vedute nè suoi viaggi …* (1594-1606). Firenze: Stamperia Manni.

Cei, G. (1992). *Viaggio e relazione delle Indie (1539–1553)*. F. Surdich (ed). Roma: Bulzoni.

D'Almada, A.A. (1594). *Tratado Breve dos Rios de Guiné do Cabo-Verde; Desde o Rio de Sanagá até aos Baixos de St^a Ana; de todas as Nações de Negros que há na dita Costa, e de seus Costumes, Armas, Trajes, Juramentos e Guerras*. Lisbon: Oficina Gráfica Limitada.

Donelha, A. (1977). *Descrição da Serra Leoa e Dos Rios de Guiné Do Cabo Verde, (1625)*. Lisbon: Junta de Investigações do Ultramar, Colecções Memórias do Agrupamento de Estudos de Cartografia Antiga.

Oviedo y Valdes, F. de G. (1526). *De la Natural Hystoria de las Indias* (on title page) or *Sumario de la Natural Hystoria de las Indias* (on inside cover). Toledo: Madrid Imprenta de la Real Academia de la história, (1851).

Galen (1956). *On Anatomical Procedures: De Anatomicis Administrationibus*. Singer, C. (ed.). Translated by C. Singer. Wellcome Historical Medical Museum.

Gândavo, P. de M. (1576). *História da província Santa Cruz a que vulgarmente chamamos Brasil*. Lisboa: Oficina de Antônio Gonçalves.

Gessner, C. (1551–1558). *Historia animalium*, 4 vols. Tiguri,. Froschouerus.

Gessner, C. (1560). *Icones animalium quadrupedum…* Tiguri, Excudebat C. Froschouerus.

Gessner, C. (1602). *Historiae animalium liber primus (I) de quadrupedibus viuiparis*. Francofurti In Bibliopolio Cambieriano 1585-1604.

Góis, D.De (1566). *Chronica do felicissimo rei dom Emanvel, composta por Damiam de Goes, dividida em quatro partes*. Lisbon: Francisco Correa.

Isidore of Sevillea, André, J. (ed.) (1989). *Isidore de Séville*. Etymologies. *Livre XII. Des animaux*. Paris: Les Belles Lettres.

Léry, J.de (1578). *Histoire d'vn voyage fait en la terre dv Bresil, avtrement dite Amerique*. La Rochelle.

L'Estoil (1963). *Journal de L'Estoil sur le régne d'Henri III*. Paris: Club des LIbraires de France.

Leitão, H. (ed) (2022). *Duarte Pacheco Pereira, Esmeraldo de Situ Orbis*. Lisbon: ERC Rutter Project.

Lopéz Mendel, T. (1570). *De los tre elementos. Tratado sobre la Naturaleza y el Hombre del Nuevo Mundo*. Madrid: Alianza Editorial Quinto Centenario .

Mauny, R. (ed.) (1956). *Esmeraldo de Situ Orbis, (Côte Occidentale d'Afrique du Sud Marocain au Gabon) par Duarte Pacheco Pereira (vers 1506-1508).* Bissau: Memória do Centro de Estudos da Guiné Portuguesa.

Monod, Th., Mota, A.T., Maunt, R. (1951). *Description de la Cote Occidentale d'Afrique (Sénégal au Cap sw Monte, Archipels) Par Valentim Fernandes* 1506-1510. Bissau: Memórias do Centro de Estudos da Guiné Portuguesa.

Marot, C. (1537). *Le valet de Marot contre Sagon, cum commento.* On les vend a Paris en la rue Sainct Jacques pres Sainct Benoist, en la bouticque de Jehan Morin, pres les Troys Couronnes d'argent.

Pigafetta, A. (1519). *Il primo viaggio intorno al mondo. Notizie del Mondo Nuovo con le figure de Paesi scoperti descritte da Ant. Pigafetta, vicentino, cavaliere di Rode.* Pozzi, M. (ed.), Facsimil edition from the manuscript "ms Ambrosiano L103 Sup". Milan: Neri Pozza. (1994).

Ramusio, G.B. (1550). Il Libro di Messer Alvise Ca da Mosto Gentilhuomo Venetiano" & "Navigatione del Capitano Pietro di Sintra Portoghese scritta per il medesimo M. Alvise da Ca da Mosto. In: *Primo volume delle Navigationi et viaggi.* Venetia: Stamperia de Giunti.

Ramusio, G.B. (1556). *Terzo volume delle Navigation et viaggi.* Venetia: Stamperia de Giunti.

Soares de Souza, G. (1587). *Tratado Descriptivo do Brazil.* Varhagen, A.F. (ed.) Rio de Janeiro: Typographia Universal de Laemmert, (1851).

Staden, H. (1557). *Warhaftige Historia vnd beschreibung eyner Landtschafft der Wilden, Nacketen, Grimmigen Menschfresser Leuthen, in der Newenwelt America gelegen.* Marpurg: Andress Kolben.

Thackston, W.M. (ed.) (1999). *The Jahangirnama-Memoirs of Jahangir, Emperor of India.* New York: Oxford University Press.

Thevet, A. (1558). *Les singvlaritez de la France Antarctique, avtrement nommée Amerique: & de plusieurs Terres & Isles decouuertes de nostre temps. Facsimile dell'ed. 1558, con un prologo di Jean Baudry.* Paris: Le temps, (1982).

Valeriano P. (1556). *Hieroglyphica, siue De sacris Aegyptiorum literis commentarii* Basilee: Michael Isegrini.

Vesalius, A. (1543). *De Humani Corporis Fabrica liber septem.* Basel: Ex Officinal Joannis Oporini.

Vitriacus, J de. (1597). *Historia orientalis.* Douai.

Willemsen, C.A. (1980). Postfazione. *De arte venandi cum avibus.* Codex Ms. Pal. Lat. 1071 Biblioteca Apostolica Vaticana. *Die bibliophilen Taschenbücher*, 152. Dortmund: Harenberg Kommunikation, pp. 229–244.

Secondary Sources

Albuquerque, L. de. (1985). *Os Descobrimentos Portugueses.* Lisboa: Alfa.

Alencastro, L.F. (2000). *O trato dos vivente: a formação do Brasil no Atlântico Sul.* São Paulo: Companhia das Letras.

Armstrong, J.C. (1983). Madagascar and the slave trade in the seventeenth century. *Omaly sy anio* 17: 20.

Ashworth, W. B. (1996). Emblematic natural history of the Renaissance. In N. Jardine; J. Second; E. C. Spary. *Culture of Natural History.* Cambridge; Cambridge University Press (pp. 17–37).

Asúa, M. de, French, R. (2005). *A New World of Animals. Early Modern Creature of Iberian America.* Burlington: Ashgate.

Ballong-Wen-Mewuda, B.J. (1993). *São Jorge da Mina 1482-1637.* Lisbonne, Paris: Fondation Calouste Gulbenkian.

Barcelos de Senna, C.J. (2003). *Subsídios para a História de Cabo Verde e Guiné (v. I p. 1 e 2).* Cidade da Praia: Instituto da Biblioteca Nacional e do Livro.

Barreto, L. F. (2000). *Lavrar o mar: os portugueses e a Ásia: c. 1480–c. 1630.* Lisboa: Comissão Nacional para as Comemorações dos Descobrimentos Portugueses.

Barroso, R.H.D. (2022). *Diplomacia e negócios da fé: a ação dos embaixadores portugueses na Cúria romana (1513-1557)* (Doctoral dissertation, Universidade de São Paulo).

Barsanti, G. (2009). *L'Uomo dei Boschi. Piccola storia delle grandi scimmie da Aristotele a Darwin.* Roma: Istituto Italiano di Antropologia.

Bedini, S.A. (1997). *The Pope's Elephant.* Manchester: Carcanet Press.

Belozerskaya, M. (2006). *The Medici Giraffe and Other Tales about Exotic Animals and Power.* New York: Little, Brown and Company.

Belozerskaya, M. (2008). Menageries as princely necessities and mirrors of their times. In: Bailey, C.B. et al. (eds.) *Oudry's Painted menageries. Portraits od Exotic Animals in Eighteenth-Century Europe.* Los Angeles: The J. Paul Getty Museum, pp. 59–73.

Borghini, M. (2006). *Sofonisba: una vita per la pittura e la libertà* (Vol. 231). Milano: Spirali.

Brandi, K., (2008). *Carlo V.* Torino: Einaudi.

Bruner, E., Masseti, M., Veracini, C. (2008). Historical notes on the early platyrrhine trade in Italy in 16th century: the wall-paintings from the castle of Torrechiara. *Folia Primatologica* (79): 152. doi: 10.1159/000111109

Bueno, E. (1998). *Naufragos, traficantes e degredados.* Rio de Janeiro: Objetiva.

Capanna, E., Gippoliti, S. (2007). I "Quadrupedi Vivipari". In: Alessandrini, A., Ceregato, A. (eds.) *Natura picta. Ulisse Aldrovandi.* Bologna: Editrice Compositori, pp. 83–87.

Cermentati, M. (1906). Ulisse Aldrovrandi e l'America. *Annali di Botanica* 4 fasc.: 313–366.

Chalmin, P. (1987). *Trade and Merchants: Panorama of International Commodity Trading.* London: Harwood Academic Publisher.

Clarke, T. H. (1986). *The rhinoceros from Dürer to Stubbs, 1515–1799.* London: Sotheby's Pubblication.

Cockram, S. (2017). Interspecies understanding: exotic animals and their handlers at the Italian Renaissance court. *Renaissance Studies, 31*(2), 277–296.

Corbet, G.B., Hill, J.E. (1992). *The Mammals of the Indomalayan Region. A Systematic Review.* London: Natural History Museum Publications/Oxford University Press.

Costa, P.F. da (2009). Secrecy, ostentation, and the illustration of exotic animals in sixteenth-century Portugal. *Annals of Science* 66(1): 59–82.

Crosby Jr, A. W. (2003). *The Columbian exchange: biological and cultural consequences of 1492.* Bloomsbury Publishing USA.

Cuneo, P.F. (ed.) (2014). *Animals and Early Modern Identity.* Burlington, USA: Ashgate.

Curry, H. A., Jardine, N., Secord, J. A., & Spary, E. C. (Eds.). (2018). *Worlds of natural history.* Cambridge: Cambridge Cambridge University Press.

Curvelo, A. (2015). Nanban Folding Screen Masterpieces: Japan-Portugal XVIIth Century. Paris: Chandeigne.

Dalton, H., Salo, J., Niemela, P., Orma, S. (2018). Frederick II of Hohenstaufen's Australian cockatoo: symbol of detente between Easy and West and evidence of the Ayyubids' global reach. *Paregon* 35(1): 35–60.

Denham, W. W. (1987). *West Indian green monkeys: problems in historical biogeography* (Vol. 24). Basel: Karger.

Dean, W. (1997). *With Broadax and Firebrand. The Destruction of the Brazilian Atlantic Forest.* London: University of California Press.

Denis, F. (1875). *Arte plumaria. Les plumes, leur valeur et leur emploi dans les arts au Mexique, au Pérou, au Brésil, dans les Indes et dans l'Océanie.* E. Leroux.

de Sofonisba, C. (2016). *Cuaderno de Sofonisba.* https://cuadernodesofonisba.blogspot.com/p/la-aventura-espanola.html, Last Access May 2023.

Dias, P. (2007). *À maneira de portugal e da índia: uma tapeçaria inédita.* Porto: Gráfica de Coimbra, Lda.

Drumond Braga, I. ; Drumond Braga, P. (eds) (2015). *Animais e Companhia na História de Portugal,* Lisboa: Círculo de Leitores.

Egmond, F. (2013). A collection within a collection: rediscovered animal drawings from the collections of Conrad Gessner and Felix Platter. *Journal of the History of Collections* 25(2): 149–170.

Egmond, F. (2017). *Eye for detail. Images of Plants and Animals in Art and Science* 1500 – 1630. London: Reaktion Books.

Eisenstein, L.E. (1979). *The Printing Press as an Agent of Change.* Cambridge: University Press.

Eisler C., (1991). *Dürer Animal's.* Washington and London: Smithsonian Institution Press.

Ferrão, C., Soares, J.P.M. (eds.). (1995). *Brasil Holandes.* Rio de Janeiro: Libri Principis.

Ferronha, A.L., Bettencourt, M., Loureiro, R. (1993). *A Fauna exotica do Descobrimentos.* Lisbon.

Findlen, P. (1994). *Possessing Nature: Museums, Collecting, and Scientific Culture in Early Modern Italy* (Vol. 20). Univ of California Press.

Findlen, P. (2006). Natural history. *The Cambridge history of science,* 3, 435–468.

Findlen, P., Marcus, H. (2012). Science under Inquisition: Heresy and Knowledge in Catholic Reformation Rome. *Isis, 103*(2), 376–382.

Foister, S. (2006). *Holbein in England.* Londres: Tate Publishing, 2006.

Franzone, A.C. (2017). Il Fondaco dei Tedeschi. *Economia Aziendale Online* 8(3): 159–161.

Fudge, E. (2002). Screaming monkeys: the creatures in the Bear Garden. In: Fudge, E. (ed.) *Perceiving Animals: Humans and Beasts in Early Modern English Culture.* Urbana: University of Illinois Press, pp. 11–33.

Fuller, E. (1995). *The Lost Birds of Paradise.* Shrewsbury (UK): Swan Hill Press.

Gaffarel. P. (1889). *Jean Ango.* Imprimerie de Espérance Gagniard.

Garas, K. (1994–95). Sebastiano del Piombos Kardinalbildnis in Dublin. Newe funde und Ergebnisse. *Acta Historie Artium*: 135–144.

Godinho, V.M. (1963). *Os descobrimentos e a economia mundial.* 4 vols. Lisboa: Editoria Arcadia Lisboa.

Gorgas, M. (1997). Animal trade between India and western Eurasia in the sixteenth century – the role of the Fuggers in animal trading. In: Mathew, K.S. (ed.) *Indo-Portuguese Trade and the Fuggers of Germany: Sixteenth Century.* New Delhi: Manohar Publishers, pp. 195–225.

Green, T. (2011). *The Rise of the Trans-Atlantic Slave Trade in Western Africa, 1300–1589.* Cambridge: Cambridge University Press.

Groom, A. (2018). Exotic animals in the art and culture of the Medici court in Florence. In: *Exotic Animals in the Art and Culture of the Medici Court in Florence.* Brill.

Groves, C.P. (2005). Order primates. In: Wilson, D.E., Reeder, D.-A.M. (eds.) *Mammal Species of the World. A Taxonomic and Geographic Reference,* Third edition, Volume 1. Baltimore: Thee John Hopkins University Press, pp. 111–184.

Groves, C. (2008). *Extended Family: Long Lost Cousins. A Personal Look at the History of Primatology.* Arlington, VA: Conservation International.

Gschwend, A.J. (2004). Exotic animals in sixteenth-century Europe. In: Jaffer, A., Jackson, A. (eds.) *Encounters: The Meeting of Asia and Europe 1500–1800,* pp. 42–43.

Gschwend, A.J. (2009). A procura portuguesa por animais exoticos/The Portuguese quest for exotic animals. In: Hallett, J. (ed.) *Cortejo triunfal com girafas/Triumphal procession with giraffes.* Lisbon: Fundação Ricardo do Espirito Santi Silva, pp. 32–42.

Gschwend, A.J. (2015). '[...] underlasse auch nit mich in Portugal vnnd ander orten umb frömbde sachen zu bewerben': Hans Khevenhüller and Habsburg Menageries in Vienna and Prague. In Haag, S. (ed.). *An exhibition of the Kunsthistorisches Museum Vienna: Ambras Castle Innsbruck: 18.6.-4.10.2015.* Wien: KHMMuseumsverband, 31–35.

Gschwend, A. J. (2018). The Emperor's Exotic and New World Animals: Hans Khevenhüller and Habsburg Menageries in Vienna and Prague. In *Naturalists in the Field* (pp. 76–103). Brill.

Gschwend, A..J., Lowe, K.J. (2015). *The global city: on the streets of renaissance Lisbon.* Paul Holberton Publishing.

Grenier, A. (1920). *Oeuvres completes de Clément Marot revues sur les meilleures éditions avec une notice et un glossaire.* Paris: Tome premier.

Grigson, C. (2016). *Menagerie. The History of Exotic Animals in England.* Oxford University Press.

Guérin, E. (1900). *Ango et ses pilotes.* Paris: Gallimard.

Janson, H.W. (1952). *Ape and Ape Lore in the Middle Ages and the Renaissance.* London: The Warburg Institute, University of London.

Jiménez Montes, G. (2016). Sevilla, puerto y puerta de Europa: La actividad de una compañía comercial flamenca en la segunda mitad del siglo XVI. *Studia historica. Historia moderna* 38 (2): 353–386.

Johnson, H.B. (1995). The Portuguese settlement of Brazil. In: Bethell, L. (ed.) *The Cambridge History of Latin America.* Cambridge University Press, pp. 249–286.

Johnson Tomlinson, R. (1970). *The Struggle for Brazil and the French Interlopers.* New York: Macmillian.

Jolly, C. (2016, August 19). *Tracking the sphinx monkey.* Lecture. Primates in Antiquity Symposium. New Hampshire: Dartmouth College Hanover.

Haag, S. (ed.). (2015). *An exhibition of the Kunsthistorisches Museum Vienna: Ambras Castle Innsbruck: 18.6.-4.10.2015.* Wien: KHMMuseumsverband.

Hallet, J. (ed.) (2009). *Cortejo triunfal com girafas/Triumphal procession with giraffes.* Lisbon: Fundação Ricardo do Espirito Santi Silva.

Hamy, E.T. (1903). Le commerce des animaux exotiques à Marseille à la fin du XVI° siècle. *Bulletin de Muséum d'historie naturelle*: 316–318.

Harrison, T., Krigbaum, J., Manser, J. (2006). Primate biogeography and ecology on the Sunda Shelf islands: a paleontological and zooarchaeological perspective. In: Lehman, S.M., Fleagle, J.G. (eds.) *Primate Biogeography*. New York: Springer, pp. 331–372.

Haupt, H., Vignau-Wilberg, T., Irblich, E. (1990). *Le Bestiaire de Rodolphe II*. Paris: Citadelle.

Hauptmann, A., Schneider, G., Bartels, C. (2016). The shipwreck of Bom Jesus, AD 1533: Fugger copper in Namibia. *Journal of African Archaeology* 14(2): 184–207.

Hazan, E. (2015). *A History of the Barricade*. Verso Books.

Hazevoet, C.J., Masseti, M. (2011). On the history of the green monkey *Chlorocebus sabaeus* (L., 1766) in the Cape Verde Islands, with notes on other introduced mammals. *Zoologia Caboverdiana* 2(1): 12–24.

Heldman, M. E. (1990). A chalice from Venice for Emperor Dāwit of Ethiopia. *Bulletin of the School of Oriental and African Studies*, *53*(3), 442–445.

Hertler, C., Rizal, Y. (2005). *Excursion Guide to the Pleistocene Hominid Sites in Central and East Java*. Universität Frankfurt am Mein, Germany/Institut Teknology Bandung, Indonesia.

Hoffmann, J.W. (1932). The Fondaco Dei Tedeschi: The Medium of Venetian-German Trade. *Journal of Political Economy* 40(2): 244–252.

Howard, D. (2007). Venice as an Eastern city. In: Carboni, S. (ed.) *Venice and the Islamic World, 828–1797*. New York, N.Y.: Metropolitan Museum of Art, pp. 58–71.

Husband, T., Gilmore-House, G. (1980). *The wild man: medieval myth and symbolism*. Metropolitan museum of art.

Hyatt M. (1971). *Prints and People*, New York, Metropolitan Museum of Art.

Kemp, M. (1995). "Wrought by no artist's hand": the natural, the artificial, the exotic, and the scientific in some artifacts from the renaissance. In: Claire, F. (ed.) *Reframing the Renaissance. Visual Culture in Europe and Latin America* 1450–1650. New Haven e New York: Yale Univ. Press, pp. 177–196.

Koch, E. (2009). Jahangir as Francis Bacon's Ideal of the King as an Observer and Investigator of Nature. *Journal of the Royal Asiatic Society*, *19*(3): 293–338.

Kusukawa, S. (2010). The sources of Gessner's pictures for the Historia animalium. *Annals of Science* 67(3): 303–328.

Lach, D.F. (2010). *Asia in the Making of Europe, Volume I: The Century of Discovery. Book 2*. University of Chicago Press.

Lazzaro, C. (1995). Animals as cultural signs. In: Claire, F. (ed.) *Reframing the Renaissance. Visual Culture in Europe and Latin America 1450–1650*. New Haven & London: Yale University Press.

Lee, R. L. (1948). Cochineal production and trade in New Spain to 1600. *The Americas*, *4*(4): 449–473.

Leitch, S. (2009). Burgkmair's Peoples of Africa and India (1508) and the origin of ethnography in print. *The Art Bulletin* XCI(2): 132–159.

Licence, A. (2016). *Catherine of Aragon: An Intimate Life of Henry VIII's True Wife*. Amberley Publishing Limited.

Lloyd, J.B. (1971). *African Animals in Renaissance Literature and Art*. Oxford: Clarendon Press.

Lugli, A. (ed.) (2005). *Dürer Albrecht,* 1521. *Viaggio nei Paesi Bassi*. Diabasis.

MacDonald, D. (2002). Collecting a new world: the ethnographic collections of Margaret of Austria. *Sixteenth Century Journal*: 649–663.

MacGregor, A. (ed.) (2018). *Naturalists in the Field. Collecting, Recording and Preserving the Natural World from the Fifteenth to the Twenty-First Century*. Leiden: Brill

Mangili, R. (2015). Il bestiario di Giambologna apre all'orango. *Saggi e memorie di storia dell'arte* 39: 30–45.

Mantua, A.A. (1992). O rinoceronte no imaginário ocidental. In: Orinoceronte, A.A.V.V. (ed.) *Pegadas na Torre, Exposiçao Torre de Belém*, Jan 92-Jun 92. Lisboa

Masseti, M. (1991). Dalla "turata della Gran'Bestie"allo "stanzone" degli agrumi: splendore e decadenza dei serragli faunistici del Giardino diBoboli. In: Luchinat, C.A., Zorzi, E.G. (eds.) *Boboli 90. Atti del Convegno Internazionale*. Firenze: Edifi r-Edizioni, pp. 323–337.

Masseti, M. (2008). Sculptures of mammals in the Grotta degli Animali of the Villa Medici di Castello, Florence, Italy: a stone menagerie. *Archives of Natural History* 35: 100–104.

Masseti, M. (2011). Anthropochorous mammals of the Old World in the West Indies. *Mammalia* 75(2): 113–142.

Masseti, M. (2016). Carpaccio's parrots and the early trade in exotic birds between the West Pacific islands and Europe. In: Thun Hohenstein, U., Cangemi, M., Fiore, I., De Grossi Mazzorin, J. (eds.) *Atti del 7° Convegno Nazionale di Archeozoologia, Annali dell'Università degli Studi di Ferrara. Museologia Scientifica e Naturalistica*, vol. 12, pp. 259–266.

Masseti, M. (2018). New world and other exotic animals in the Italian renaissance. The menageries of Lorenzo *Il Magnifico* and his son, Pope Leo X. In: MacGregor, A. (ed.) *Naturalists in the Field. Collecting, Recording and Preserving the Natural World from the Fifteenth to the Twenty-First Century*. Leiden: Brill, pp. 40–75.

Masseti, M. (2019a). Migrazioni umane e variazioni faunistiche nell'alto medioevo mediterraneo. In: Le migrazioni nell'alto medioevo. *Atti della LXVI Settimana di studio* (Spoleto, 5-11 aprile 2018), pp. 841–895.

Masseti, M. (2019b). Ricami e merletti di animali e dove trovarli. L'iconografia zoologica nei disegni di Giovanni Alfonso Samarco di Bari nelle collezioni di Palazzo Davanzati. In: Matucci, B., Rapino, D. (eds.) *Bellezza e nobili ornamenti nella moda e nell'arredo dei Seicento*. Florence: Museo di Palazzo Davanzati/Edifir, pp. 128–137.

Masseti, M., Bruner, E. (2009). The primates of the western Palaearctic. A biogeographical, historical, ethnozoological and archaeozoological review. *JASs – Journal of Anthropological Sciences* 87: 33–91.

Masseti, M., Veracini, C. (2010). The early importation of South American primates in 16th century in Italy: the case of the Marcgrave's capuchin monkey, *Cebus flavius* (Schreber, 1774). *Archive of Natural History* 37(1): 91–101.

Masseti, M., Veracini, C. (2014). Early European knowledge and trade of neotropical mammals: a review of literary sources between 1492 and the first two decades of the 16th century. *British Aqueological Review (Proceeding of the First Zooarqueology Conference in Portugal, Lisbon, 8th-9th* March 2012): 129–138.

Mathew, K.S. (ed.). (1997) *Indo-Portuguese Trade and the Fuggers of Germany: Sixteenth Century*. New Delhi: Manohar Publishers.

McDermott, W.C. (1938). *The Ape in Antiquity*. The Johns Hopkins University Studies in Archaeology, No. 27. Baltimore: The Johns Hopkins Press.

Mikhail, A. (2013). *The Animal in Ottoman Egypt*. Oxford University Press.

Mittermeier, R.A., Rylands, A.B., D.E. Wilson (eds.) (2013). Handbook of the Mammals of the World. Volume 3. Primates. 951.

Molà, L. (2009). Traffici, mercanti, viaggiatori sullo scorcio del Medioevo. In: *Italiani e stranieri nella tradizione letteraria. Atti del Convegno di Montepulciano, 8-10 ottobre 2007*. Roma.

Montalenti G. 1960. Aldrovandi in *Dizionario biografico degli Italiani*, vol. 2, pp. 118–24.

Mosco, M. (1985). Animal paintings in the Medici collections. In: Mosco, M. (ed.) *Natura viva in Casa Medici*. Firenze, pp. 17–22.

Nordio, G. (2003). *Il viazo de le galee di Fiandra: mercanti veneziani a Londra verso la metà del Quattrocento*. Tesi di laurea, Università di Venezia Ca' Foscari.

O'Bryan, R. (2012). Grotesque bodies, princely delight: dwarfs in Italian renaissance court imagery. *Preternature: Critical and Historical Studies on the Preternatural* 1(2): 252–288.

Ogilvie, B.W. (2006). The Science of Describing: Natural History in Renaissance Europe. Chicago: The University of Chicago Press.

Oliveira, de A. (1990). *Historia dos Descobrimentos e Expansão Portuguesa - Linha e Rumos da Colonização Portuguesa*. Lisboa: Universidade Aberta.

Olmi G. (1985). Science-Honour-Metaphor: Italian Cabinets and the Sixteenth and Seventeenth Centuries. In Impey and MacGregor (eds), *The Origins of Museums*, pp. 5–16.

Panofsky, E. (1971). *The life and art of Albrecht Dürer*. Princeton University Press.

Papavero, N., Teixeira, D. M. (2002). *Os primeiros documentos sobre a história natural do Brasil. Viagens de Pinzón, Cabral, Vespucci, Albuquerque, do Capitão Gonneville e da Nau Bretoa*. Belém: Museu Paraense Emilio Goeldi.

Papavero, N., Teixeira, D. M. (2014). Zoonímia tupi nos escritos quinhentistas europeus. *Núcleo de Apoio à Pesquisa em Etimologia e História da Língua Portuguesa, FFLCH/USP*, (3), 1–329.

Pérez de Tudela, A., Gschwend, A.J. (2007). Renaissance menageries. Exotic animals and pets at the Habsburg courts in Iberia and Central Europe. In: Enenkel Karl, A.E., Smith, P.J. (eds.) *Early Modern Zoology. The Construction of Animals in Science, Literature and Visual Arts*. Leiden-Boston: Bill, *Intersections*, 7/1, pp. 419–447.

Perfetti, S. (2000). *Aristotle's zoology and its Renaissance commentators, 1521–1601*. Leuven University Press.

Perfetti, S. (2020). *Nature Imperfette. Umano, subumano e animale nel pensiero di Alberto Magno*. Pisa: Edizioni ETS.

Pieper, R. (1998). Trading with art and curiosities in Southern Germany before the thirty years war. In: Nunez, C.E. (ed.) *Market for Art, 1400–1800*. Madrid-Sevilla: Proceedings 12th International Economic History Congress, pp. 87–100.

Pieper, R. (2008). Papagayos americanos, mediadores culturales entre dos mundos. In: Stols, E., Werner, T., Verberckmoes, J. (eds.) *Naturalia, Mirabilia & Monstrosa en los Imperios Ibéricos*. Madrid: Avisos de Flandes, pp. 123–134.

Pinon, L. (2005). Conrad Gessner and the historical depth of Renaissance natural history. *Empiricism and erudition in early modern Europe*, 241-67.

Pohle, J., Lisboa, J.L.D.C.C.V., Bernardo, L.M.A.V., Marques, C.T.E. (2017). *Os mercadores-banqueiros alemães e a Expansão Portuguesa no reinado de D. Manuel I*. (CHAM eBooks // Estudos; Vol. 2). Cham: FCSH/NOVA-UAc.

Pomata, G., Siraisi, N.G. (2005). Introduction. In: Pomata, G., Siraisi, N.G. (eds.) *Historia: Empiricism and erudition in early modern Europe*. MIT Press, pp. 1–38.

Redmond, I. (2011). *The primate family tree*. Buffalo, New York: Marshall Editions.

Rees, A. (2017). Animal agents? Historiography, theory and the history of science in the Anthropocene. *BJHS Themes* 2: 1–10.

Ringmar, E. (2006). Audience for a giraffe: European expansionism and the quest for the exotic. *Journal of World History* 17(4): 375–397.

Rodrigues, J.N., Devezas, T.C. (2009). *Portugal: o pioneiro da globalização: A herança das descobertas*. Lisboa: Centro Atlântico.

Rodríguez, F. Z. (2014). Interest and Curiosity: American Products, Information, and Exotica in Tuscany. In *Global Goods and the Spanish Empire, 1492–1824: Circulation, Resistance and Diversity* (pp. 174–193). London: Palgrave Macmillan UK.

Rolfe, W.D.I., Grigson, C. (2006). Stubbs's "Drill and albino hamadryas baboon" in conjectural historical context. *Archives of Natural History* 33: 18–41.

Roth, J. (2017). Sugar and slaves: the Augsburg Welser as conquerors of America and colonial foundational myths. *Atlantic Studies* 14(4): 436–456.

Rubini, P. (2013). Pomponazzi e l'anima degli animali. In *The Animal Soul and the Human Mind: Renaissance Debates*, In C. Muratori (ed). Pisa-Rome, pp. 75–85.

Sabelli, B., Tommasini, S. (2004). La zoologia di Ulisse Aldrovandi, In: B. Antonino (ed). *Animali e Creature Mostruose di Ulisse Aldrovandi*. Milano: Ed. Federico Motta, pp. 24–43.

Saltini, G.E. (1879). Di una visita che fece in Genoa nel 1548 il fanciullo Don Francesco di Cosimo I de' Medici al Principe Don Filippo di Spagna. *Archivio Storico Italiano, Serie Quarta* 4(112): 19–34.

Saltini, G.E. (1883). L' educazione del Principe don Francesco de' Medici. *Archivio Storico Italiano Serie Quarta, 11*(134): 157–172.

Sermidi, M. (2003). *Le collezioni Gonzaga: il carteggio tra Venezia e Mantova, 1588–1612*, Vol. 13. Silvana.

Simões, C. (2019). Non-European animals and the construction of royalty at the renaissance Portuguese court. *Animals and Courts: Europe, c. 1200–1800*, 55.

Smith, P. H. (2006). Art, science, and visual culture in early modern Europe. *Isis*, *97*(1), 83–100.

Smith, P.; Findlen, P. (Eds.). (2013). *Merchants and marvels: commerce, science, and art in early modern Europe*. Routledge.

Sterckx, R. (2000). *The Animal and the Daemon in Early China*. New York: State University of New York Press.

Swadling, P. (2019). *Plumes from Paradise*. Sydney University Press.

Swan, C. (2015). Exotica on the move: birds of paradise in early modern Holland. *Art History* 38(4): 620–635.

Swan, C. (2018). The use of realism in early modern botany. In: *Visual in Medieval Medicine and Natural History*, 1200–1550, pp. 243–244.

Teixeira, D., Papavero, N. (2010). O Trafico de Primatas brasileiros nos séculos XVI e XVII. In: Pessoa, M.L., Correa Tavares, W., Siliciano, S. (eds.) *Mamiferos de Restingas e Manguezais do Brasil*. Rio de Janeiro: SBMz, Sociedade Brasileira de Mastozoologia. Museu Nacional, pp. 253–282.

Tompkins, P. (1994). *Monkeys in Art*. New York: M.T.Tain/Scala Books.

Tosi, A. (1989). *Ulisse Aldrovandi e la Toscana, Carteggio e testimonianza documentarie*. Olschki, Firenze, L. (ed.).

Tugnoli Pattaro, S. (1981). *Metodo e sistema delle scienze nel pensiero scientifico di Ulisse Aldrovandi.* Bologna

Urbani, B. (2007). Further Information on Neotropical Monkeys Reported in the XVIth Century. Part. 2. Neotropical Primates, 14(3): 144–145. .

Urbani, B., Youlatos, D. (2020). On the earliest representations of chimpanzees (*Pan troglodytes*): were African apes traded to bronze age elam? *International Journal of Primatology* 41: 654–663.

Urbani, B., Youlatos, D. (2023). *Simia langobardorum*: were African apes traded in late medieval Lombardy? *American Journal of Primatology* 85(2): e23462.

Veracini, C. (2011a). *La conoscenza dei primati del Nuovo Mondo nell'Europa Rinascimentale* (Unpublished PhD thesis, University of Pisa).

Veracini, C. (2011b). Neotropical primates at the court of the emperor Maximilian the first. *Folia Primatologica* 82(6): 394–394.

Veracini, C. (2012). Non-human primates in the work of Ulisse Aldrovandi (1522–1605). *Journal of Biological Research* 1: 308–309.

Veracini, C. (2017). Nonhuman primate trade in the age of discovery: European importation and its consequences. *Environmental History in the Making: Volume II: Acting*: 147–171.

Veracini, C. (2019). Natural history of non-human primates in the early modern period. The contribution of the Italian encycloplaedist Ulisse Aldrovandi. In: Veracini, C., Scalfari, F., Casanova, C. (eds.) *History of Primatology Yesetrday and Today: The Western_Mediterrean Tradition.* Roma: Aracne Editrice, pp. 53–91.

Veracini, C. (2020). A landscape without nonhuman primates? The case of the Barbary Macaque, *Macaca sylvanus*, (Linnaeus, 1758) and its interaction with humans throughout recorded time. *Humanities* 9(3): 92. https://doi.org/10.3390/h9030092

Veracini, C. (2021a). An early representation of a gorilla from fifteenth-century Central Asia. *Primates* 62: 457–462. https://doi.org/10.1007/s10329-021-00906-w

Veracini, C. (2021b). Use and trade of non-human animals in Portuguese overseas expansion (15th and 16th centuries): evidence from Italian travellers. *RiMe, Rivista dell'Istituto di Storia dell'Europa Mediterranea* 8(2): 11–41. https://rime.cnr.it/index.php/rime/article/view/536/801

Veracini, C., Bezerra, B. (2013). Early trade of Capuchin monkeys in the Brazilian East coast: consequences and conservation. Abstract, *2nd Latin-American Primatological Congress,* Recife (BR).

Veracini, C., Casanova, C. (2015). Monkeys for the Pope: presence, importance and importation of non-human primates in Renaissance Italy (15th and 16th centuries). *Folia Primatologica* 86: 376. doi: 10.1159/000435825

Veracini, C., Casanova, C. (2016). How non-human primates and other terrestrial West African mammals were described in the Portuguese literary sources of the 15th and 16th centuries. *Annali Università degli Studi di Ferrara, 12*(1), 243–250.

Veracini, C., Hazevoet, C.J. (2024). *Homo sapiens* e *Chlorocebus sabaeus*, dois primatas exóticos em Cabo Verde: notas preliminares sobre suas interações e conflitos. In: Amante, F., Pignatelli, M., Rodrigues, I., Sá, R.M. (eds.) *Latitudes em transcurso. Estudos em homenagem a Celeste Quintino por ocasião do seu 70° aniversário.* Lisbon: Colecção de Estudos Políticos e Sociais. Edições ISCSP. pp. 75–92.

Veracini, C., Teixeira, D. (2017). New World non-human primates in the travel literature of the 15th and 16th centuries: from myth to natural history. *Annalas of Science, 74*(1): 1–39.

Verma, S.P. (1999). *Mughal Painter of Flora and Fauna Ustād Manṣūr.* Abhinav Publications.

Wendt, H. (1956). *Auf Noahs Spuren.* Halbleder: C.A. Koch's Verlag.

Yoshi, M. (2009). Biombos Namban no século XXI. In Curvelo et al. *Namban Screens.* Porto: Museu Nacional de Soares dos Reis, pp. 66–79.

Zuckerman, S. (1998). *The Ape in Myth and Art.* The Knowes, Scotland: Verdigris Press.

Section III

Modern Period (until Darwin)

10 Natural History of Non-human Primates in the 17th Century

Naturalists, Missionaries, Scientific Expeditions, and Trade

Cecilia Veracini
Universidade de Lisboa

> a Portuguese man, having brought with him up to thirty [marmosets], despite using all imaginable diligence could only save two. Of which one was given to the Queen of Portugal and the other sent to the Queen of France. (Cavazzi da Montecuccolo, 1687: 837. 152)

10.1 SCIENTIFIC TREATISES

The large influx of exotic animals and plants into Europe consequent to the voyages of exploration, which began in the 16th century, sparked a new interest in natural history (cf. Veracini, Chapter 9, this volume). The voyages were facilitated in the 17th century by the strengthening of the European mercantile companies, such as the Dutch East India Company, involved in an intense trade in any type of natural object from other continents (Winters, 2020). The formation of new merchant groups and the increase in the number of Atlantic merchants played a central role in the formation of Western Europe, changing political institutions and culture, and enabling the emergence of new types of organization and the development of new technologies (Acemoglu et al., 2005). The growing desire for more information about the nature of the new and distant lands acquired a particular character that was reflected in both courtly fashion and intellectual restlessness, but it also emphasized the need for a colonial strategy to acquire a better knowledge of the conquered territories (Teixeira, 1995a). The consolidation of the European powers within the occupied territories enabled an ever-deeper examination of the nature of these regions, which, in turn, led to the need for systems to describe and classify new forms hitherto unknown to scholars who studied nature. The 16th-century study of nature, defined as the "science of describing" (Olgivie, 2006), had emerged as an independent pursuit in the main European powers at the end of the 16th century, and during the 17th century, it would play an ever-increasing role. This field of enquiry fitted well with the images of the empire promoted by early modern rulers, and the majority of naturalists benefited from these ambitions and enjoyed the support of patrons who envisaged profit from collecting data from other worlds (Findlen, 1996: 73).

Around Europe, many centers for the viewing of novelties that were brought from other continents were established, and the collection and study of natural objects were undertaken as part of a widespread interest in rarities and wonders of all sorts, natural and artificial. In England, for instance, "these rarities or curiosities filled the collections that were eagerly formed by gentlemen and scholars who described themselves as *curiosi* or *virtuosi*"; the same people were regular attendees at meetings of the Royal Society (Whitaker, 1996: 75). In Italy, Federico Cesi (1586–1630), an Umbrian-Roman aristocrat and passionate natural scientist, founded together with other scholars, the *Accademia dei Lincei*, established to promote and cultivate the study of natural history. The name was derived from the exceptional acuity of vision attributed to the lynx, and the object of the *Accademia dei Lincei* was all the natural sciences, to be investigated with free experimental observation, beyond all constraints of tradition, above all without constraints with regard to the previous Aristotelian-Ptolemaic tradition, which new experimental science sometimes called into question. The academy published in 1628–1648 the "Mexican Treasure", *Rerum Medicarum Novae Hispaniae Thesaurus*, by Francisco Hernández, a unique and extraordinary documentation of the fauna and flora of the New World, with drawings of hundreds of plants and animals. This work was the result of a scientific expedition commissioned by Phillip II, the King of Spain and Portugal, in 1570, when the king appointed the court physician Francisco Hernandez to describe the animals and plants in the Spanish-American colonies. Hernandez's manuscript had a complicated fate and was never published in Spain.

The work *Exoticorum libri decem* (1605) of Charles de L'Ecluse, Latinized as Carolus Clusius, paved the way for a form of scientific treatise that sought to collect all the knowledge developed until then about the exotic flora and fauna, but also about medicinal products and other exotic peculiarities (Findlen, 2006; Mason, 2009). L'Ecluse, a French-speaking Flemish physician and botanist, was a naturalist of the 16th century (1526–1609), but his work published in 1605 became an important basis for the natural history of primates in the 1600s, which will be discussed

DOI: 10.1201/b21963-13

here. His interest in non-European natural objects was stimulated by his voyage to the Iberian Peninsula in 1564, during which he traveled to Lisbon and encountered both live animals from the Americas and botanical treatises not yet published in Latin. L'Ecluse lived in Vienna from 1573 to 1578 and again from 1581 to 1588 at the imperial courts of Maximilian II and then Rudolf II, and thus had the opportunity to visit their cabinets of curiosities, where New World artifacts were in plentiful supply. Moreover, when he moved to Leiden in the Netherlands in 1593 he had connections with local travelers such as Jan Huygen van Linschoten, who had spent a little over five years in Goa (Mason, 2009: 129). L'Ecluse, as a botanist, described the primates summarily, explaining the various doubts he encountered in classifying them and often referring the reader to other works for a more precise idea (see **Box Details 10.1**).

BOX DETAILS 10.1: THE MONKEYS DESCRIBED BY CHARLES DE L'ECLUSE (CAROLUS CLUSIUS) IN THE *EXOTICORUM LIBRI DECEM* (1605)

Cecilia Veracini

The primates are described in the chapter *Cercopitheci* [tailed monkeys, cf. Glossary] (Chap. XV, Lib. V) of the work *Exoticorum libri decem* (Ten books of exotic life forms). L'Ecluse described six species of monkey, illustrating by adding three images. The first image we find is that of *Papio hamadryas* (**Fig. 10.1**) which, although this attribution is obvious to current knowledge, left doubts at the time. As L'Ecluse himself reveals: "This animal had long hair on its body; a shorter tail than the *cercopitheci* and a long, bare rostrum with canines like a beaver's; its buttocks were hairless and glabrous, its feet included five toes armed with claws" (*Exoticorum*, p. 370). L'Ecluse notes that Adrianus Collardes had published a similar image in his book and that the author had attributed a similar animal to the class *Cinocephali* (i.e., baboons) but L'Ecluse had his doubts and therefore says that "it will be up to the reader to decide" whether it is indeed a baboon. The description of the other *cercopitheci* begins with *Cercopithecus barbatus I*, and is accompanied by an image that, due to the image and text description, can be identified as *Cercopithecus roloway*, (p. 371, **Fig. 10.2**). L'Ecluse says in fact "that the body is black as is the tail, which ends in a final knob", which he says he has "never noticed in other *Cercopithecus*". He then describes the form of *Cercopithecus barbatus II*: "it was an animal similar to the latter that he had seen in his last years in the Hague, at [the house of] Paul Choart de Buzenal (1551–1607), the ambassador of the French king Henry IV". But this:

> was a much more elegant and refined animal, with very neat and spiky hair, and much shorter than that in the picture; it had a pale chest and belly. It had a beard, but less abundant than the upper illustration and it measured a palm; the tail of the latter did not end in a tuft, but was similar to the tail of a vulgar *Cercopithecus.*

This animal can be tentatively identified as *Cercopithecus diana.*

L'Ecluse then describes the "beardless cercopitheci", (a) *Cercopithecus non barbatus I,* "an animal covered with hair of medium length and of a tawny yellow colour, as was also the whole of its head except the snout: its throat was covered with whitish hair, as was also the whole of its chest. Its front legs and feet were similar to human hands, its tail was not very bristly, but smooth in a way, as if the hairs had been turned and the furthermost part of it culminated in a long hairy tuft" [(this monkey is tentatively identified as *Chlorocebus sabaeus*]; (b) *Cercopithecus non barbatus II,* "appeared to be endowed with a larger body mass, its rostrum completely brown, its head grey, and its neck had a leonine mane, also grey; and this animal was not dissimilar to the earlier form" (this monkey is tentatively identified as *Chlorocebus* sp.).

FIGURE 10.1 Hamadryas baboon (*Papio hamadryas*), sent to Clusius by Iacobus Plateau, an acquaintance and observer of natural things, who wrote that this animal had been given to Prince Charles II Duke of Croÿ and Aarschot in France. (Source: Carolos Clusius *Exoticorum*, 1604, p. 370. Image in the public domain.)

FIGURE 10.2 *Cercopithecus barbatus I*, Carolos Clusius *Exoticorum*, (1604, p. 371). According to the text description and image, this animal may correspond to *Cercopithecus roloway*. (Image in the public domain.)

FIGURE 10.3 *Cercopithecus Brasilianus III, Cercopithecus Sagoin*, Carolos Clusius *Exoticorum*, (1604, p. 372), the common marmoset, *Callithrix jacchus*. A marmoset was seen by Clusius in Lisbon in 1564 brought from Pernambuco (BR) by Roderic Linz. (Image in the public domain.)

When in Lisbon, L'Ecluse visited Sebald Linz, whose son Roderic Linz had sailed to the Portuguese port from Pernambuco with some parrots and monkeys in 1564 (*Exoticorum*, p. 371). L'Ecluse describes one specimen as *Cercopithecus Brasilianus I*, an animal "not of large body size, smaller than those described above, with long golden hair [...] and which had a good smell". This animal is the blond capuchin, *Sapajus flavius*, one of the most commonly traded capuchin monkeys since the 16th century, (cf. Veracini, Chapter 9, this volume). L'Ecluse also defines *Cercopithecus Brasilianus* II, based on the description taken from the accounts of Léry (1578), who had spoken of a black monkey called *cay* that lived in the state of Rio de Janeiro (*Sapajus nigritus*, cf. Veracini & Teixeira, 2017). Roderic Linz also brought marmosets from Pernambuco– whom he fed on maize. Only one survived the voyage, and this individual was still alive in January 1565 when L'Ecluse left Lisbon for Spain. When L'Ecluse wrote his section on primates he obtained a colored image of the surviving animal, and this is what his woodcut is based on (**Fig. 10.3**, cf. Mason, 2009; *Exoticorum*, p. 372). He refers to the animal as *Cercopithecus Brasilianus III*, *Cercopithecus Sagoin*; here, he is quoting Gessner (*Tierbuch*, 1551, cf. Veracini, Chapter 9, this volume); it is a common marmoset.

FIGURE 10.4 *Historia Naturae maxime peregrinae* (Nieremberg, 1635). The upper image is the "sagoin" taken from Gessner (1551), the second is also a sagoin (*Callithrix jacchus*) taken from L'Ecluse (1605), and the third is the "*Cercopithecus barbatus I*" apparently a *Cercopithecus roloway*, also taken from L'Ecluse. (Retrieved from Biodiversity Library: https://www.biodiversitylibrary.org. Image in the public domain.)

Initially written in Latin, and later in the national languages, treatises like these satisfied the emerging curiosity for the *exotica* (extra European animals and plants), and attempted to bring order to the many confusing descriptions of new species (often designated by different vernacular names) that, with the exploration of new territories, was accumulating. In the wake of L'Ecluse's *Exoticorum, The Historia Naturae, Maxime Peregrinae* (1635), the work of Juan Eusebio Nieremberg (1595–1658), was printed and widely circulated in Europe. In that work, Nieremberg, a Spanish Jesuit of German origin, summarized the information available at the time about exotic species. Regarding primates, almost all the taxa were included in the chapter *De cercopitheci* (Concerning tailed monkeys) (Nieremberg, 1635, Chapter XLIV: 177), which is mostly a compilation of information gathered by 16th-century authors such as

FIGURE 10.5 *Historia Naturae maxime peregrinae* (Nieremberg, 1635). The upper illustration portrays a baboon (source unknown) and the lower is probably a tamarin (*Saguinus* sp). The second image is very similar to the images given by Father Cristovão de Lisboa (see **Fig. 10.14**).

Léry (1578), Gessner (1551), Hernández (1651), Acosta (2002), Oviedo y Valdés (1526) and Martyr de Anghiera (cf. Anglería, 1989) among others (cf. Veracini & Teixeira, 2017), as well as by L'Ecluse (1605). The depictions provided by the author are woodcuts taken from L'Ecluse and Gessner (**Figs. 10.4 and 10.5**). One of the most interesting aspects Nieremberg emphasizes is the cognitive skills of monkeys: monkeys "are even able to bargain over the selling price as a person could do" and "to go to buy the wine for their master" (quoting here the 16th-century excerpt of José de Acosta's, 1590, cf. Veracini & Teixeira, 2017). The author also addresses the cognitive aspects of primates in the chapter XLX, *De Baribus* [Concerning the *Baris*] (p. 179), where he quotes news of the performance of some "monkeys" called *baris* [=chimpanzees] encountered on the West African Coast, who were endowed with a great aptitude for housework: "these animals (*baris*) pound seeds in mortars and go to get water from the spring, carrying jugs on their heads" (Nieremberg, 1635: 179). This anecdote dealing with the cognitive capacity of chimpanzees, which is a blend of local tales and real observation, is similar to others reported by Portuguese navigators and missionaries of the 16th century. Such narratives, along with others of a comparable nature, were highlighted in the discussion about the nature of great apes and their likeness with human beings (cf. Barsanti, this volume). Nieremberg also provided data on the worship of the Hanuman langur in India (cf. Chapter 1, Section I, XLVI, *De simio albo*: 180), reporting that in India, some "white monkeys" are considered sacred and "even their teeth are kept and venerated in temples".

In 1625 Johannes De Laet, a Dutch geographer, renowned humanist, and director of the Dutch West India Company (West-Indische Company – WIC: sister of the VOC, Dutch East India Company), published the *Nieuwe Wereldt*, [*History of the New World*], which was then translated into Latin and French (De Laet, 1640). This work was the first extensive compendium dealing with the New World written in Dutch, and it is thought to have been written in order to map the territories De Laet hoped would become future WIC trading posts or colonies. It is a compilation of previous works dealing with geographical topics and incorporating fragments of texts, maps, and images of the 16th century (Dams, 2016). As Dams stresses, the book's novelty was that it omitted "fabulous animals or extraordinary people" and cast a *disenchanted* look at the new world "without myths and marvelous elements typical" of the previous century (2016: 36). In the French edition, the New World primates are described in "Chapter V, *Bestes à quatre piés du Brasil, selon un Autheur Portugais*" (*On the Quadruped Beasts of Brazil*, p. 486):

> there is an infinity of *singes* [he means the larger Neotropical monkeys] and *guenons* [monkeys with tails] of various sizes. […] Among them there is one that surpasses all others and which the savages call *Aquiaqui* [howler monkey]; they have black fur and a long beard on their chin; sometimes a male is born that is red and he is called the Monkey King [quoting Lery, 1578]. They "screams very loudly through a concave organ that nature has provided them with" [the Jesuit Father Cardim, 1997, described this organ in the 16th century]. "This monkey, according to the English Antoine Kniuet, is called by the Petiguares [native people who live in the Brazilian state of Pernambuco] *Uvariua*, but others call it *Goariba*. […] in the rest of the coast, according to Léry, there are some little black guenons called *Caï*" […] "but although Léry calls them *sagoin*, the animal described by him does not seem the same as that described by Charles de L'Eglise" [today we know, in fact, that they are two distinct animals: the first, described by L'Eglise (L'Ecluse) is the common marmoset *Callithrix jacchus* and the second one is *Sapajus nigritus*, a species of tufted capuchin monkey observed by Lery on his travels in Rio de Janeiro].

These *compendia* on Neotropical fauna based on secondary sources would later be replaced by first-hand descriptions by naturalists, missionaries, and travelers who reached South America for different reasons. After the aforementioned scientific expedition commissioned by Phillip II, another was organized by the Dutch in the first half of the 17th century to the northeast of Brazil. After it achieved independence from Philip II in 1581, Holland arose with astonishing rapidity and vigor to the status of a great European power. The 17th century saw the dominant role of the Netherlands that led to the so-called Dutch Golden Age, in which Dutch trade, science, military skills, and art became among the most celebrated in the world (Ames, 2011), and Amsterdam replaced Antwerp as the most important trade center. Also in the 17th century, Dutch seafarers began to trade with the Far East, and as the century wore on they gained an increasingly dominant position in world trade, a position previously occupied by the Portuguese and the Spanish (Ames, 2011). The VOC, the first "multinational corporation", founded in 1602 gained the Dutch monopoly on Asian trade, which it would keep for two centuries. The Dutch occupation of the Brazilian Northeast (1624–1654) was largely an initiative of the "sister" WIC. The Brazilian colony was an attempt to substitute the political and commercial power of Spain and Portugal, as well as providing a source of raw materials for the Netherlands (Parker Brienen, 2006: 15). The WIC also facilitated the connection between Africa (where the Dutch had occupied the commercial centers, *feitorias*, of Portuguese West Africa, such as the Mina de São Jorge, in present-day Ghana) and Brazil, where they cultivated sugar cane with the help of slaves brought from Africa.

During Dutch rule in northeastern Brazil, the assembly of knowledge about the natural history of the New World involved numerous individuals situated at very different levels in the hierarchy of the WIC. Brazilian fauna and flora were described and depicted both by lower-ranking soldiers (i.e., Zacharias Wagener and Caspar Schmalkalden) and more highly qualified staff members (i.e., Willem Piso and Georg Marcgraf), as well as by directors of the Company (e.g., Johannes de Laet). Local natural history also attracted the attention of Count Johann Moritz von Nassau-Siegen, the governor hired to manage the WIC's business in Brazil, who arrived in Pernambuco in 1637 and stayed until 1644. A member of the elite who claimed to be enlightened and protective of the arts and culture, von Nassau-Siegen followed the trend seen among the wealthier classes in 16th-century Europe. He built a palace surrounded by a park and created a menagerie composed of native and exotic species, as well as organizing a significant "cabinet of curiosities" with items from the New World, Africa, and the Orient (Parker Brienen, 2006: 15; Teixeira, 2006a). The Count's menagerie, as described by Calado (1648, *apud* Teixeira, 2006a: 4), contained a wide selection of birds and animals such as parrots, macaws, other colored birds, peacocks, chickens from Guinea, ducks, swans, turkeys, jaguars, pigeons, anteaters, monkeys.

Many of these were brought by local people with the intention of obtaining favors from the Count at a later time. Moreover, von Nassau-Siegen hired artists, such as Albert Eckhout (c.1610–1665) and Frans Post (c.1612–1680), to capture both nature and the local inhabitants; their extensive iconography on Brazil was not bettered until the 19th century.

Among the accounts of fauna written at this time, were those written by two WIC employees, the physician Willem Piso (1611–1678) and his assistant Georg Marcgraf (or Marcgrave) (1610–1643); the latter also dealt with astronomy and cartography. Thanks to the efforts of Johannes de Laet, in 1648 this material was prepared for printing as the "*Historia Naturalis Brasiliae*" (Piso & Marcgraf, 1648; **Fig. 10.6**). Representative of the characteristic "encyclopedism" of the 17th century, this well-known work was the

FIGURE 10.6 Frontispiece of Piso and Marcgraf (1648), *Historia Naturalis Brasiliae.* This frontispiece is an example of the change in the conception of natural history taking place in the 16th century. The New World is gradually passing from being perceived as a place of fable to a real space peopled by its animals, plants, and human beings. This bears witness to the fact that knowledge is becoming more precise and detailed, and observers are no longer content with second-hand stories or far-fetched hypotheses, nor with fanciful engravings (Palumbo, 2007). Despite this, this frontispiece testifies to the difficulty found in situating the new data, and especially the newly discovered species, within the theoretical frames of reference that still use classical themes.

only comprehensive publication on Brazilian natural history available until the 18th century, and it was a basic reference-book for more than one hundred and fifty years. The zoological section written by Marcgraf contains a description of 32 mammals, of which 15 are given scientific names – 12 from Linnaeus in 1758 and 1766 and 3 from other authors between 1776 and 1788 (Oliveira & Langguth, 2006: 2). In Book VI, dedicated to *De Quadrupedibus & Serpentibus* [Four-footed beasts and snakes], primates are dealt with in Chapter V: *"Guariba, çagui maior e menor, alii cercopitheci"*. Marcgraf describes nine varieties of monkeys, to which he gives vernacular names, and two other monkey varieties are described without being named (see **Box Details 10.2**). The list of primates also includes specimens brought from Africa and perhaps from Asia (see **Box Details 10.3**).

In the middle of the 17th century a book, *Historiae Naturalis de Quadrupedibus* published in Frankfurt and Amsterdam, written by John Johnston (or Jonston, Latinized Jonstonus) (1603–1675) would become the standard 17th-century encyclopedia of natural history. It ran through several editions throughout the century (cf. Jonstonus, 1650, 1657) and was translated into English (Johnston, 1678). Johnston, of Scottish origin, was born and raised in Poland, but after studying botany and medicine at Cambridge, he traveled extensively before settling in Leiden, where he practiced medicine. His history of animals, which relies on classical sources and on the travel literature of the 16th and 17th centuries, was intended as a comprehensive survey of different classes of animals (fish, birds, quadrupeds, reptiles, and insects). Primates are classified among quadrupeds in Chapter II of the section entitled *De Digitatis Semiferis*, together with foxes and other smaller carnivores, rabbits, rodents, armadillos, sloths, and anteaters. The organization of the chapters concerning primates is substantially the same as the scheme used by Ulisse Aldrovandi (cf. Veracini, Chapter 9, this volume); *De Simia,* [which includes the tailless monkey, *Macaca sylvanus*, and apes], *De Cercopithecus* [primates with tails], and *De Cynocephalo, Papione & Fera Upalim* [dog-headed monkeys]. In the first chapter, the author summarizes the general characteristics of *simiae* (i.e. primates, cf. the Glossary), basing his commentary largely on the works

BOX DETAILS 10.2: THE MONKEYS DESCRIBED BY MARCGRAF.

Cecilia Veracini

The English translation is from the Portuguese version (Marcgrave, 1942: 226–227), which has been compared with the original text in Latin (Piso & Marcgraf, 1648: 226–228). The species identifications were aided by the tables of the artists who worked for the Count of Nassau (cf. **Box Details 10.3**). Regarding African species, not all the animals came from Angola or Congo, as the author says, but from a wider area including Western Africa. The images of primates illustrating Marcgraf's work are taken from the artworks of artists working in Brazil at the time (**Fig. 10.7**).

Guariba [=*Alouatta belzebul*], a species of *cercopithecus* with a long tail that is hairless at the tip. This animal gets very agitated and always sits on its tail whatever is happening. [...] At the bottom of its chin, the hair is dark and this ends in a beard, similar to that of a goat. The hair on the hands and feet is less dark than that in the middle of

FIGURE 10.7 Images of monkeys by Piso and Marcgraf (1648). (a) Illustrates a blond capuchin *Sapajus flavius*, (b) a titi monkey of the genus *Callicebus*, (c) a common marmoset *Callithrix jacchus,* and (d) a black-and-red howler monkey *Alouatta belzebul.*

the tail. The males are a little larger than the females but of the same color. They are encountered in large numbers in the forests, where they utter great cries that can be heard from afar. Sometimes it is just one animal calling, and the others in large numbers form a circle as if listening to a speech. The volume of their cries can increase 100-fold when you are at a distance. This is their normal behavior, as I have observed many times every day before noon, and afterward, the others sit around on a lower floor; one sings aloud and, having received a signal, the others continue singing the song until a signal is given again. Everyone then stops in an instant and the one who started ends the song in a loud voice. [...] This animal has a habit of biting and cannot be tamed. There is also another large, black one with a long, dark, hairy beard [=perhaps *Alouatta caraya*].

***Cagui* (*pongi* is the name given in the Congo)**. Larger and smaller animals [= marmosets, capuchin monkeys, and titi monkeys] are found in this species.

1. The big animal is the size of a *cay* monkey [= capuchin monkey], which it somewhat resembles. Its hair is gray mixed with a darker color and it is slightly longer than that of the little *cagui*. Its face is rounded, leonine; its ears are almost round, dark, and hairless. Its dark muzzle is covered in dark hair mixed with gray, and the shape of the muzzle, like that of the lion, is blunt. Its eyes are dark. Like the little *cagui*, it is without dark hair near the ears. Its tail measures one foot and four toes [approx. 40 cm] and is covered with red hair [= *Callicebus melanochir* vel *barbarabrownae*].
2. *Cagui*, the smaller animal [=*Callithrix jacchus*]: a tender little animal with a lion's face [...]. The ears are rounded and surrounded by white hair, arranged in order and in braids as if elegantly combed. The hair of the whole body is red in the inner part near its origin. But in the outer part, there is a mixture of white and dark hair. The tail is composed of dark and white rings. It emits a very high-pitched sound. It is very quick at jumping and keeps the cold at bay; it feeds on almond meal and other types of meal. Note: Gessner presented an image of this animal received from Pedro Coudenberg, a learned and renowned pharmacist from Antwerp, who considered it to be the offspring of a greater monkey. Instead, it is a *sui generis* animal, which it is improper to classify as a monkey. Sometimes we obtain these animals from Brazil, but they are very delicate and averse to the cold and it is hard to keep them alive here in Europe.

Caitaia [=*Sapajus flavius*] also a *cercopithecus* with a longer coat, of a yellow and lighter-colored coat. Its head is rounded and the forehead is not salient. Its nose is small and flat. It has an arched tail and emits a musky smell. This is the only good thing about it. It must be treated with care, otherwise, it shouts very loudly: it easily becomes furious. There is another one of the same species, but larger and with darker fur similar to sable. It also has this musky smell [= probably *Sapajus xanthosternos*].

Cercopithecus* of Angola in Congo is called *macaquo [=*Papio* sp.]. Its colour is the same throughout its body as that of a wolf. The nostrils are divided into two protruding parts. The teats resemble those of a bear. The buttocks where it sits have no hair; it always carries its tail arched. The length of the body, from head to tail, is one foot or over; the length of the head is six toes, that of the tail one foot. The four legs are the same length, ending in hands/feet five toes long. The body at its largest can reach nine fingers, it emits a "hah haha" cry; it has very sharp teeth. It has a human-like penis like that of a child. [This description matches with that of a baboon, genus *Papio*; the name macaquo is found in one drawing of the *Libri principia* - cf. **Box Details 10.3**. and it seems to correspond to an adult individual of *Papio papio* whose geographical range does not include ancient Angola or Congo; the baboon species occurring in these regions mentioned by Marcgrave is *Papio kindae*. The dimensions of the animal here described are those of a young individual, as the "foot" corresponded to c. 30 cm].

Another *cercopithecus* from Angola: dark all over, but speckled with white; nostrils white. Size as the previous one. The length of the tail is 2 feet and four toes; the figure is as follows but it has a hairy forehead [=*Cercopithecus nictitans*].

Bearded *Cercopithecus* of Guineé: called *Exquima*, in the Congo [=*Cercopithecus diana* vel *roloway*]; has dark hair, but all over the back, it is as if burnt, or rust-colored…The belly is white and the underside of the chin has a very white beard with hair two fingers or more long, as if it had been combed. When this species becomes enraged and opens its mouth and moves its jaws, it attacks man rapidly, jumping in a remarkable way. It eats various fruits but especially loves Paco fruit. [*C diana* and C. *roloway* were among the most traded monkeys since the 16th century]

Another species of Guineé [= *Cercopithecus cephus*]. It is the same shape as the previous animal. The colour of the greater part of its body is dark as if mingled with shade; its belly is bluish-grey. The lower half of the tail is mixed yellow and red. Its mouth is pale blue, and on the jaw, there are yellow filaments as in a large goat's beard and the same near the ears. It plays and shouts like the previously described animal.

Another species of Guineé [= cf. *Chlorocebus sabaeus*]. Smaller than the previous one: it has mixed dark greyish and yellow fur, almost like that of the back of a hare. It has a small head, a long tail, and no beard.

by Greek and Roman (e.g., Aristotle, Avicenna, Aelian, Herodotus, Pliny among others cf. Masseti & Veracini, this volume and references therein), and medieval authors (such as Albertus Magnus and Marco Polo), as well as the Renaissance anatomist Vesalio, and the 16th-century naturalists Gessner and Aldrovandi.

The introduction reports the different names given to the *simiae*, their anatomical characteristics, geographical areas, feeding, reproduction, and general behavior. After this introduction, Johnston emphasizes recent discoveries, including (a) the "*orang-tang*" of Tulp (1641) (see Barsanti, this volume), which:

> walked oft upright, and with ease could hoyst up, and beare a burden. Being to drinke, shee with one hand took the can by the eare, and put the other under; and wiped her lip thes handsomly after. Going to ly down, she would lay her head or|derly on the pillow, and cover herself close with the coverlids

and (b) the *Baris* [=chimpanzee] quoted from Nieremberg (1635), see above. This beast in Guinee was able to "bray any thing in a morter, and carry pitchers of water on their heads, and empty them at their doors, to prevent spilling" (cf. Johnston, 1678: 76).

In the chapter *De Cercopitheco* (Jonstonus, 1650: 141–143), Johnston reports the information available about tailed monkeys from Asia, South America, and Africa. He quotes the various names used for the *cercopitheci* since the Middle Ages (including very old ones such as *mamonetus* and *sphingia* (cf. Veracini, Chapter 7, this volume), and mentions the Celtic name *abranas* quoted from Topsell, 1607). After a brief introduction about long-tailed monkeys, which relies on Greco-Roman and medieval sources, a long section is dedicated to the descriptions of the feeding habits and the geographical distribution of these primates, and these descriptions are enriched with anecdotes and information found in the works of Hernández, L'Ecluse, Marcgraf, and Nieremberg, and in the travel literature of the 16th century (e.g., Martyr de Anghiera, Oviedo y Valdés, and Acosta). In this chapter, both Old and New monkeys are described: the behavior of capuchin monkeys and their most extraordinary cognitive capacities reported by Oviedo y Valdés and Acosta (see Veracini & Teixeira, 2017); howler monkeys (called *guariba*) and their peculiar vocalizations; the various species of African monkeys described by L'Ecluse, and the Neotropical taxa described by Marcgraf (**Fig. 10.8**). Moreover, Johnston describes Indian langurs, observing that some of them are sacred, whereas others in the region of Malabar close to Goa are considered destructive and are sold for next to nothing to the Portuguese, who eat them when "they are sick and their bones brayed, and drunk" and

FIGURE 10.8 Images of primates from Jonstonus (1650). The author classifies the upper-left primate (a Barbary Macaque) as a *Cynocephalus*. The *Simia*=Affe [in German] refers to the ancient definition of an ape, which included a large number of primates but was essentially the Barbary Macaque, and then goes on to include the true apes [Family Hominidae]. The other *Cynocephalus* is a hamadryas, while the *cercopitheci* are probably a Barbary Macaque and a baboon. (Image retrieved from Biodiversity Library: https://www.biodiversitylibrary.org. Image in the public domain.)

FIGURE 10.9 Images of primates from Jonstonus (1650). The upper primate is a marmoset taken from Gessner, as is the second, called *Papio* (Aldrovandi called it *Papione*), representing a mandrill or a drill; the primate – lower left– is a *Cercopithecus diana* vel *roloway*, also called "meer kats" (cat from the sea), and the last one – lower right – called *Papio* 2, is an image of a sloth taken from Thevet. (Retrieved from Biodiversity Library: https://www.biodiversitylibrary.org. Image in the public domain.)

use them for protection against syphilis (Johnston, 1678: 76). The belief that meat, or other parts of monkeys, are useful protection against diseases was common in the 16th century-world of the Portuguese and Spanish (cf. Veracini & Camphora, this volume). This belief was probably passed on to them by the indigenous peoples with whom they came into contact in various parts of the world, many of whom still utilize primates or their parts as remedies for various diseases (cf. Section I, this volume).

Chapter IV, *De Cynocephalo, Papione & Fera Upalim*, (Jonstonus, 1650: 144) includes various descriptions of baboons with references to the works of ancient authors (e.g., Aelian, Strabo) and of the *Papione,* name given by Aldrovandi (1637) to a specimens of *Mandrillus* sp., first reported by Gessner (1551) as a genus of wolf (cf. Veracini, Chapter 9, this volume). Johnston also resurrected the *Fera Upalim;* an Egyptian *cercopithecus* described by Adrovandi (1637).

To summarize, the nomenclature of primates used by Johnston is based on classical and Renaissance texts supplemented by some 17th century sources. There are signs in the book that suggest Johnston had little or no direct experience with these animals, and in some cases, he confuses the names and images of the well-known taxa. For example, he gives the name *Papio* (a name commonly used for baboons) to an animal that is identical to Thevet's sloth (cf. Thevet, 1575; **Fig. 10.9**), and he calls the Barbary Macaque with the zoonym *Cercopithecus* [= tailed monkey] (**Fig. 10.10**), but this last choice may have been influenced by the work of Prospero Alpini (see below). In fact, although Aldrovandi

FIGURE 10.10 Images of primates from Jonstonus (1650). Compendium of Tables: LX and LXXIII. Primates taken from Marcgrave (left part) and from L'Ecluse (right part). The names are not reported correctly: e.g., the common marmoset in the upper center is given the name *Cajtaia*, which is the name of the blond capuchin. (Retrieved from Biodiversity Library: https://www.biodiversitylibrary.org. Image of the public domain.)

(cf. Veracini, 2019) had already described the Barbary Macaque, it would take more than three hundred years for this species and its geographical range to be correctly identified (cf. Fooden, 2007).

The work of the Venetian physician and botanist Prospero Alpini (1553–1616) had a significant influence on the naturalists of the 17th and 18th centuries. In his treatise, *Historia Naturalis Aegyoti*, (cf. Chapter XI, *De variis simiarum generibus in Aegypto observatis, & primo de Cynocephalis vocatis*, Alpini, 1592: 240), Alpini gave the Barbary macaque two names, not realizing that the larger one he called *Cynocephalus* was the same species as the smaller, younger individual, which he called *Callitrichis* (cf. Veracini, 2020). Although Alpini was not a zoologist, his taxonomic interpretation of *M. sylvanus* influenced naturalists such as Linnaeus and Buffon.

During the 1600s, many nobles and members of the rising merchant classes created or added to, private menageries, filling them with exotic animals from around the world. For instance, in Rome, the residences of the most important families included aviaries and menageries, and prominent architects of the time were involved in their construction (Nonaka, 2014). These facilities satisfied their owner's awareness of birds and other animals that had developed after the publication of the various natural history treatises, as well as being a source of entertainment (Nonaka, 2014). In France, King Louis XIV organized one of the most famous menageries of the 17th century at his new residence in Versailles (Del Pesco, 2011). It has been interpreted as being both a claim to absolute authority by the king (e.g., **Fig. 10.11**), as well as an example of the civilized life of the French aristocracy (Sahlins, 2012). Radical innovations in how animals were displayed and viewed at the Versailles menagerie included more focus on the grace and "civility" of birds over the brutality and violence of large, ferocious mammals that were often utilized in wild animal combat. The Versailles menagerie was a microcosm of the palace, with all the rarest and most curious birds, but also with small mammals, such as African civets, mongooses, badgers, porcupines, and foxes, and larger exotic mammals (e.g., a bear, lions, an elephant, and a camel) placed in cages (Sahlins, 2012). The primates in the royal menagerie were probably kept in the traditional manner, that is uncaged but shackled with a ball and chain (see below). The newly founded Royal Academy of Sciences of France carried out a project of comparative anatomy using the corpses of animals that died in the menagerie (Guerrini, 2015). The results are summarized in an extensive work containing the anatomical descriptions of these animals. It was at first issued anonymously (1671) and distributed as a royal gift by Louis XIV, then reissued in I676 in one volume bearing the name of Charles Perrault as editor *Mémoires pour servir àl' histoire naturelle* (Memoirs for the History of Natural Sciences) (Perrault, 1671–76). As observed by Guerrini, "the volumes bridged the gap between comparative anatomy and natural history, and between natural history and natural

FIGURE 10.11 Frontispiece of *Memoires pour servir a l'histoire naturelle des animaux et des plantes*, author Sébastien Le Clerc (Perrault, 1671). Louis XIV visiting the Académie des Sciences in Paris, with his minister Colbert in 1671 on the occasion of the inauguration of the "Jardin des Plantes". Plants and animals shown in the Academy played different roles in Louis' universe, as they represented his commitment to the sciences, but also the vastness and power of the French empire. (Retrieved from Perrault, cf. La Haye, P. Gosse & I. Neaulme, (eds.) (1731). Biodiversity Library: https://www.biodiversitylibrary.org.)

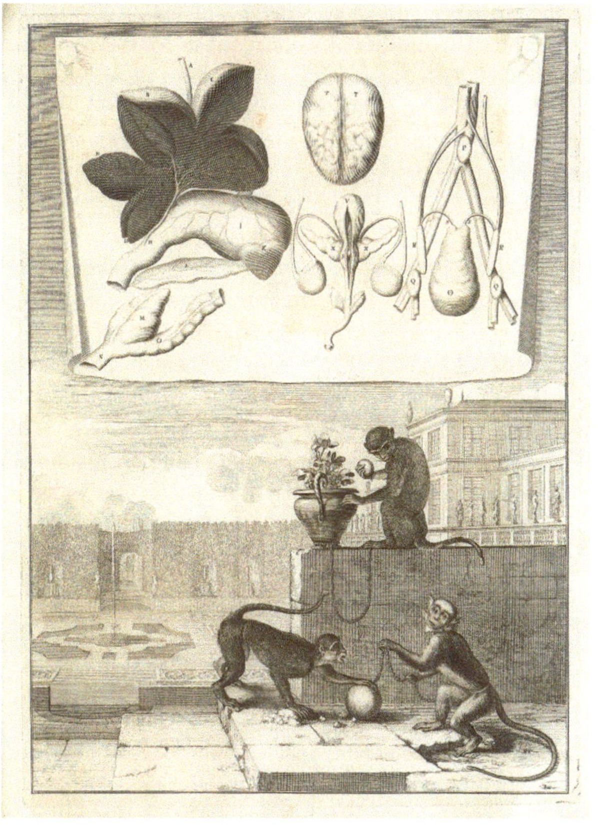

FIGURE 10.12 *Guenons et sapajous*. Three monkeys, apparently cercopithecids, are portrayed with balls and chains, a normal way of keeping monkeys, who were usually kept outside the fences. (Source: Perrault, (1676). Biodiversity Library: https://www.biodiversitylibrary.org.)

philosophy, as no other works had done and represented an unprecedented depth of collaboration between naturalists and artists" (2015: 163). It was widely influential, and an English translation of the later edition was approved for printing by a Council of the Royal Society on 17 November 1686 and published in London in 1688.

In the "*Description anatomique de Deux Sapajous et de Deux autres Guenons* (Perrault, 1671: 121–127)" an extremely detailed description of the dissection and characteristics of primates of the king's menagerie is provided. Four primates were dissected: three males and one female. The authors observed that "the four monkeys we describe are of the genus Cercopytheques [sic!] but their small size prevents them from being included among the *guenons* [this last term was used mostly for tailed monkeys of different origin]". Comparing the depiction and the anatomical descriptions, we can deduce that all the animals were African cercopithecids (although they are called *sapajous*). All the animals had cheek pods, described as "dans la Machoire d'en bas il y avoit une poche on fac de chaque costé", and according to the description, they could belong to three genera: *Chlorocebus*, *Cercopithecus,* and *Cercocebus* (see **Fig. 10.12**). There was still some confusion about the distinction between Old and New World monkeys. The text – written jointly by Perrault and Duverney – acknowledged "the close likeness between monkeys and humans while concluding in every instance that the differences between them were in fact profound" (e.g., they do not speak: although monkeys have a larynx and other anatomical parts that could form human speech, they cannot speak because they are not human) (Guerrini, 2015: 160). In the edition of 1676, the monkeys are the only animals depicted in a humanized setting, and Guerrini (2015) notes they could be on the terrace of a chateau, probably of the Palace of Versailles (**Fig. 10.12**).

The work of Perrault was included in the last important zoological work of the 17th century: the *Synopsis of Quadrupeds*, by the English Puritan clergyman John Ray (Rajus in Latin) (1627–1705). He was a renowned naturalist who studied at the University of Cambridge and traveled extensively in Europe. In 1680, he was elected a Fellow of the Royal Society, founded under the patronage of King Charles II only a few years earlier (Raven, 1986: 45). According to Groves (2008: 62) Ray was the first scholar to use the term "species" consistently. The mammals [following the previous tradition] were referred to as *Viviparous*. "He divided quadrupeds into *ungulate* and *unguiculate*, hoofed and clawed, having himself invented these Latin words. In the latter category, there are animals with broad claws such as monkeys, or others with narrow claws" (Hallam, 1839: 573). For some authors, such as the French naturalist Cuvier (18th–19th centuries), Ray's work constitutes "a clearer arrangement, more determinate than those of any of his predecessors, and applied with more consistency and precision" (Hallam, 1839: 573). Ray's description of primates does not change significantly in comparison with that of previous authors, and he omits several previously described well-known taxa of monkeys. He quotes L'Ecluse's and Marcgraf's works, reporting all the species described by these authors and the species described by the Academy of Science of Paris in the work of Perrault. Ray's classification, which is simple and recalls those of Aristotle, Aldrovandi, and Jonston, divided primates into three groups:

> The monkeys [*Simiae*] are divided into those that have no tail: called simply *simiae*, and those that are caudate: called *cercopitheci*. In English the first genus is called Apes, and the second is called Monkeys. Those which have a dog-like face are called *cynocephali*, and have a long tail, and have been called *Papiones*, in English: Baboons, and are of considerable size. (Rajus, *Simiae, Cercopitheci, Cepi & Cynocephali*, 1693: 149)

Like Perrault and colleagues, Ray provides some anatomical information, comparing these creatures, as the latter did, with human beings, also reporting data from Aristotle:

> The *Simia* is, both below and above, a very hairy animal, with a face similar to that of a man, and nostrils, teeth and ears similar to us: and it also has eyelashes in the lower part of the eyelids that other animals do not have. This was also observed by Aristotle. They have teats on their breasts, and arms like those of man, while the feet are not, but the middle toe, as in man, is the longest. As also confirmed by the Parisians [Perrault and his school] [...] even the teeth are similar to ours, except for the canines, which are longer [...] The monkey as observed by the Parisians resembles man more in the external parts than in the internal parts (Rajus: *Simiae, Cercopitheci, Cepi & Cynocephali*, 1693:149–150).

Any review of what was known of the natural history of primates at this time would not be complete without mention of the important work on Neotropical primates of Pierre Barrére. Pierre Barrére (1690–1755), who served as botanist "*du roi*" in Cayenne, lived there for three years (1722–1725) as correspondent of the French Royal Academy of Science, with instructions to prepare a detailed report on the natural history of French Guiana. His work, *Essai sur l'Histoire naturelle de la France Equinoxiale,* eventually published in 1741 as an abbreviated glossary, was divided into "*Flora*" and "*Fauna*". In the Chapter "*Quadrupèdes et Reptilles*" there are many entries about the tailed monkey, *Cercopitheci*. Unfortunately, Barrére's entries contain few descriptions of the animals, so only a few species can be reliably identified. He classifies the Neotropical monkeys as (a) *Cercopitheci,* including larger and mid-sized species of tailed monkeys; (b) *Cercopitheci minores*, which best correspond to squirrel monkeys and pithecids; and (c) *Cercopithecus minimus,* the tamarin. The descriptions of the African taxa are from Marcgraf (cf. Marcgrave, 1942) and these forms are classified as *Cercopitheci* as well. From a historical and linguistic point of view, Barrére's work is important because he provides some of the vernacular names of the primates in the Arawak language and Galibi (cf. Martius, 1863). The primates of Barrére (1741: 149–151) are listed as (the italic and the CAPS are of Barrére; the names are kept in Latin):

1. CERCOPITHECUS angolensis, major, Marcg. *Makaque*. This was the easiest monkey to tame. Indians eat them. Their flesh is as leathery as a horse. [this taxon was described by Marcgraf– probably *Papio* sp. cf. Box Details 10.2]
2. CERCOPITHECUS barbatus, maximus, ferruginoso, sterotorosus. *Cercopithecus barbatus, Guineensis Marcg. Alaouata* [sic.!]. Red monkey. A strong and big red wild animal that makes a hellish noise that can be heard from far away; this noise is produced by the hyoid bone which has a peculiar structure [*Alouatta macconelli,* Guyana howler monkey].
3. CERCOPITHECUS cinereus, cristatus .*Singe cendré* [grey monkey, possibly *Cebus olivaceus*].
4. CERCOPITHECUS cristatus, cirratus, capite nigrus [monkey with tufts and a black head, *Sapajus apella*]
5. *Kayourouré. Makaque* blanc. Rare and is found in the far south in the *terra firme* [*terra firme* is a common name used to indicate the part of Amazonia which are not subject to flooding. There are two white monkey species in Amazonia: *Pithecia albicans* and *Cacajao calvus calvus;* their distribution is far south of Guiana. *Kayourouré* is considered by Martius, 1863, a word of the Galibi-Karibe language].
6. CERCOPITHECUS major niger. With a human face. It is a *cercopithecus* of greatness and human form. It imitates humans in many things. *Qouata* [native name] [*Ateles paniscus*].
7. CERCOPITHECUS minor, albicans, brasiliensis, white *Sapajou* of Brazil [possibly *Cebus albicans albicans*]
8. CERCOPITHECUS minor albicans, maculis nigris distinctus. *Sagoin. Sapajou tigré* [perhaps *Pithecia pithecia* female]
9. CERCOPITHECUS minor, luteus, *Sapajou jaune*. [probably a misnomer for *Sapajus flavius* of Marcgraf]

10. CERCOPITHECUS minor, dilute olivaceus, parvo capite. *Cagui minor Marcg. Akarima. Sapajou.* [*Saimiri sciureus*; the term *akarima* is used today for the squirrel monkey in the Arawak language]
11. CERCOPITHECUS minimus, niger, leonthocephalus, auribus Elephantinis, *Caguis major Marcg. Galleopithecus*, the Sagoin of Gessner [Gessner Historia animalium 1602, described *Callithrix jacchus* as *Galeopithecus*, cf. Veracini, Chapter 9, this volume]. *Tamarin.* [probably *Saguinus midas,* native to French Guiana and other norther Amazonian countries].

10.2 MISSIONARIES' REPORTS

In addition to the work of Marcgraf, reports of monkeys, and their interactions with native and European people in South America, are found in the accounts of missionaries. These fathers supported the European powers who during the first half of the 17th century were attempting to establish and maintain trade centers and plantations in the Amazon region (Papavero et al., 2002:135). At the beginning of the 17th century, the fauna of the current Brazilian state of Maranhão was described by French Capuchin friars during the brief period of French occupation between 1610 and 1615, the year of the Portuguese re-conquest of the territory (Papavero et al., 2002: 96–97). The place we know today as Maranhão was of geopolitical importance, especially for delimiting the boundaries between the territories officially occupied by the Portuguese and the Spanish in South America. This second French attempt to set up in Brazil proved not as significant as the first (which occurred in 1555 a 1570 in the Rio de Janeiro territory), except that the French founded the port of Saint-Louis which became São Luis, which is still the capital of the State of Maranhão. Among the passengers on the initial voyage to Maranhão in 1612 were four Franciscan missionaries, or to be more exact Capuchins: Ambroise d'Amiens, Arsène de Paris, Claude d'Abbeville, and their superior Yves d'Evreux. They were given the task of writing about the territory, its inhabitants and their customs, and its natural history. Claude d'Abbeville and Yves d'Evreux are the authors of the two main sources of information we have about this period (the French were ruled between 1610 and 1617 by Maria de Medici regent of the king Luis XIII) in Maranhão (Ávila-Pires, 1989). As with most chroniclers of the time, in the works of the two Capuchins, we find a mixture of verifiably accurate information mixed with inaccuracies, obscurities, and sometimes even nonsense, but the quality of their work attests to the fact that they were cultured and reasonable men, who nevertheless display occasional naivety (Billé, 2015). As men of faith, missionaries, sent on voyages of exploration and colonization of the Americas were more inclined to reflect on the natural world using moral and spiritual perspectives. They sought examples of the divine secrets the Creator had left in nature, and because they tried to observe nature in its "original form" through direct experience, the information they collected was valued for its accuracy (cf. Veracini & Teixeira, 2017; Capozzi, 2021).

The work of Father Claude d'Abbeville, "*Historie de la mission des pères Capucins en l'Isle de Maragnan et terres circonvoisines*" published in 1614, contains an impressive amount of information, despite the author only spending four months in Brazil. He may have spoken the native Tupi languages, but more likely he utilized information passed on to him by two famous "tabajaras" (natives of that region) Charles des Vaux and Davi Migan, who had considerable experience in Brazilian natural history (Garcia, 2008). Claude d'Abbeville mentions some 209 Brazilian animals, including 199 named in Tupi (Billé, 2015). Primates are described in the Paragraph "*Des onces & des guenons qui sont au Bresil*" [On jaguars and monkeys of Brazil] in Chapter XXXIX. To designate the Brazilian primates, d'Abbeville used mostly the names *Monnes* (used for smaller monkeys, such as squirrel monkeys and callitrichids) and *Guenons. Cay*, which in the Tupi language means "shy and ashamed", was used as a generic name for monkeys (see also Glossary). The monkeys described in d'Abbeville's work are the following (for the identification of the species see also Billé, 2015; Ávila-Pires, 1989; d'Abbeville, 2008).

1. *Ouäriue,* they are black, as big as a dog and cry loudly and can be heard from the distance of a league [in French this corresponded to c. 4 km]. [=*Alouatta* spp., howler monkeys; in Tupi *wariwa*, today *guariba;* probably *Alouatta belzebul*];
2. *Cay-Ouässou*, they are normally brought to Europe and are seen also here [= capuchin monkeys, likely *Sapajus* spp.; literally: monkey =*Cay*+large= *Ouassou/açu*]. [The father reports that this is also a word to design a constellation of stars in the form of monkeys. This news must have come from indigenous popular culture where primates sometimes indicated an astronomical event and/or a constellation, cf. Veracini & Camphora, this volume.]
3. *Cay-ou*, they are black and have a long beard four digits or more in length; they are very beautiful and pleasant to look at [this may refer to the species of the genus *Chiropotes*];
4. *Cay-mirim* or *Sapaiou*, a small monkey with yellowish hair of a mixture of colors [=*Saimiri sciureus*; Frey Cristovão. Father Yves d'Evreux also used the word *Sapajous* to design *Saimiri sciureus,* when describing its habit of eating crickets in people's houses, cf. D'Évreux 1864 [1615], p. 187–188, *apud* Billé, 2015].
5. *Tamary*: small monkey [= probably belonging to the genus *Saguinus,* perhaps *Saguinus niger*];

6. *Mirikina*: some of these have heads in the shape of a heart [difficult to identify; it could be the night monkey, *Aotus* spp. cf. Billé, 2015; this term in the Tupi language can refer to different species of monkey: *muriki*, *buriqui*, and other variants. In Tupi *myraqui* means dirty people, d'Abbeville, 2008: 202];
7. *Ioupara*: a species of monkey. It has streaks of white over other different colors [= small arboreal mustelid carnivore, the Kinkajou, *Potos flavus;* this animal often figures in the indigenous taxonomy associated with primates, cf. Veracini & Camphora, this volume];
8. *Sagouy:* a species of monkey with silver grey hair, they are the smallest and most beautiful of all [= marmosets of Amazonia, genus *Mico,* probably *Mico argentatus*].

After the expulsion of the French, Philip IV, King of Portugal and Spain, sent Father Cristovão de Lisboa as Apostolic Inquisitor to Maranhão. When he arrived in São Luís, the Father founded the Margarida Convent and, besides acting as inquisitor, traveled to other regions of Brazil, such as Belém, Pernambuco, Amazonas, and Ceará (Papavero et al. 1999). He did not stay long in Brazil, and after returning to his country he produced a generously illustrated 198-page manuscript called "*Animaes de Maranho*" [Animals of Maranhão]. This manuscript, now held in the "Arquivo Historico Ultramarino" in Lisbon, was published in 1967 and reprinted in a new edition (Walter, (ed), 2000). It was probably a copy of an original manuscript, or part of it, written in French by the French missionaries, and then translated into Portuguese. Some of the documents making up the Lisbon book are the result of records made during a trip to the Amazonian river by Charles des Vaux and Yves d'Évreux. Des Vaux, who knew the language of the natives of the region, and who had explored at length its natural resources, was not recognized as the author of this work; instead, it was attributed to Father Cristovão de Lisboa. All this was happening during the Portuguese occupation of northern Brazil, soon after the expulsion of the French from that region, and Cristovão de Lisboa's knowledge of its natural resources had a bearing on the political and economic territorial disputes between the French and the Portuguese (Capozzi, 2021: 6). Lisboa's work, which is the first monograph on the flora and fauna of Maranhão, provides the first illustrations of the animals of the lower regions of the Amazon (Papavero et al., 2002: 102). There are two monkeys portrayed in this work (a tamarin – whose features are confused with those of a marmoset – and a squirrel monkey); they are reffered respectively as *guariba* and *bugio machacha*, with obvious references to Marcgrav's classification, albeit with misattributions (see also glossary). (see **Figs. 10.13 and 10.14**).

Other information about the monkeys of the coastal areas of Brazil colonized by the Portuguese is found in the accounts of the Italian Capuchin missionaries (cf. Papavero & Teixeira, 2015), who were carrying out apostolic work in the Kingdom of Congo. On their way to, or from, Africa they were forced to spend several months in northeastern Brazil before they could continue their journey. Many of them left vivid accounts of the lifestyle, customs, and traditions of Portuguese settlers and native people of the 17th century, including their relationship with, and use of, monkeys. It appears that the custom of using monkeys as pets or for entertainment was already widespread in the Portuguese community of the time. According to the description of Girolamo Merella, the tamed monkeys were

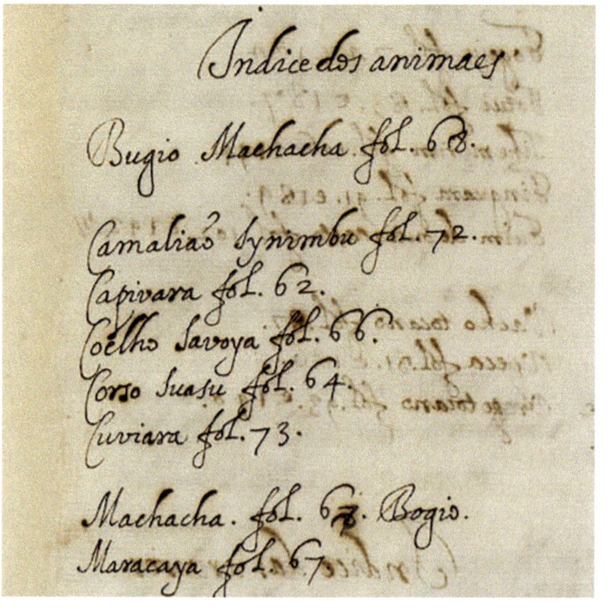
Indice dos animaes
Bugio Machacha fol. 68.
Camaliaõ Synimbu fol. 72.
Capivara fol. 62.
Coelho Savoya fol. 66.
Corso Suasu fol. 64.
Curiara fol. 73.
Machacha. fol. 68. Bogio.
Maracaya fol. 67.

FIGURE 10.13 Names of animals of the illustrated manuscript of the Frei Cristovão de Lisboa, "*Animaes de Maranho*" [Animals of Maranhão]. The primates names are: *Bugio Machacha* (upper part) and *Machacha Bogio*, penultimate line. (Courtesy of the Arquivo Histórico Ultramarino, Lisbon.)

FIGURE 10.14 Primates from the illustrated manuscript of the Frei Cristovão de Lisboa, "*Animaes de Maranho*" [Animals of Maranhão]. (Courtesy of the Arquivo Histórico Ultramarino, Lisbon.)

capuchins, and "there were many in the homes of Bahia settlers" (1692: 41). "[The] tame ones are crazy about their masters and some of the tame ones play around gracefully" (Cavazzi da Montecuccolo, 1687: 837. 152). All the Italian Capuchin fathers report that trade in these animals, especially their export to Europe (see also **Box Details 10.3**), was a common and regular practice, and monkeys figured among the most frequently traded natural resources of Brazil:

> Trade [in Brazil] consists of nothing more than wood, [the 'pau brasil', *Pau-brasilia* echinata Lam], coconut, monkeys and parrots of all kinds [...] the monkeys that are taken on board are of all manners but all with tails, one type of which is more highly esteemed than the other, so much so that it is more difficult to transport them, not to Italy, but to Spain and elsewhere because of the cold. (Merolla da Sorrento, 1692: 41)

Monkeys, along with other natural products, were used as barter currency in many commercial transactions (Bettendorf, 1698), or were used to pay taxes owed to the king of Portugal. As noted by the fathers the most highly esteemed monkeys were callitrichids (*Callithrix* spp. and *Leontopithecus rosalia*) that were given different names (e.g., *sagoris* or *sagoinos*). Among the strategies used to stop these animals from freezing to death during transportation was to keep them in sleeves filled with cotton wool (Merolla da Sorrento, 1692), but even with these precautions the mortality rate was extremely high:

> a Portuguese man, having brought with him up to thirty, despite using all imaginable diligence [in caring for them] could only save two. Of which one was given to the Queen of Portugal and the other sent to the Queen of France, but this one died soon after feeling the cold air of our climate. [...] It was easier to trade these animals in Angola where they were sold and much appreciated. (Cavazzi da Montecuccolo, 1687: 837. 152)

10.3 CONCLUDING REMARKS

The natural history of the early 17th century still drew on the descriptions of classical authors and the encyclopedic naturalists of the 16th century, although exploration, travel, and geographical expeditions led to the inclusion of information from newly explored places at the expense of tradition. Toward the middle of the 17th century, although there was a gradual move away from the use of classical authors, they continued to provide basic information about natural history (Olgivie, 2018). The tradition of directly observing animals began in the 16th century, when naturalists in France began to study physiology and anatomy (Guerrini, 2015).

At the beginning of the 17th century, works were still written in an encyclopedic style, with an anecdotal and moralistic approach to natural history typical of the medieval and, in part, the Renaissance tradition. The types of primates described are the same as those found in classical, medieval, and Renaissance texts, but their descriptions were augmented with observations drawn from the growing body of travel literature. As scientific expeditions progressed, and new animals arrived, in Europe, the types of primates included in zoological treatises increased, as did confusion about their classification. One of the major problems, as had been the case in the previous century, was that the new primates had to fit into the existing nomenclature, which, with a few exceptions, was the one used by Aristotle, which divided primates into *Simiae* (tailless monkeys), *Cercopitheci* (tailed monkeys), and *Cynocephali* (dog-headed monkeys). The taxon *Simiae* now had to accommodate the great apes, still variously described as wild men, orang-utans, baris, etc., (cf. Barsanti, this volume). All the new species described in both the Old and New Worlds were placed in the category *Cercopitheci*. The third group, the *Cynocephali*, included all the known dog-headed primates, plus sometimes the barbary macaque. The New World primates recognized in the course of scientific expeditions to South America by the European powers vying for those territories were described more accurately than those of the Old World, where interest in African and Asian primates increased, especially in relation to the great apes, as did the problems generated by their similarity to humankind. Descriptions by travelers in Africa (e.g., Dapper, 1686) provide some details of the other primate species encountered, but they were often not reported on by leading European naturalists.

Primates continued to be classified as quadrupeds, four-legged beasts, according to the classical scheme, and in some cases they are associated with the group that includes small carnivores and other medium or small mammals. In accordance with previous tradition, they are still considered "brutes" distinct from humans. For example, although their anatomy would have enabled them to speak, they did not do so because they were not endowed with this faculty in the same way as humans. Even if speech was added later, because it was unlike our language, we might not understand it!

Among the various people who gathered information on New World primates were missionaries, educated and cultured individuals sent to bring the natives of South America closer to the Catholic religion. From the perspective of the history of science, the works resulting from the French occupation of Maranhão by the Capuchin fathers were notable (Kury, 2014). The Capuchins developed a method we now call ethnobiology, in that they included Tupinambá cultural elements in their animal descriptions. Their perception of the world reflected Thomist thinking, where natural things were observed "according to Their Own Principles", thus having a more empirical approach. The same cultural paradigm had guided the Jesuit fathers in their explorations and accurate descriptions of the natural world in the 16th century (cf. Veracini & Teixeira, 2017).Although this chapter has not analyzed in detail the trade in primates in the 17th century, it is clear both from some of the sources presented here, and from analyses of the large body of travel literature from this century (not treated here), trade of these animals, and their use by wealthy Europeans, was common and widespread. These sources emphasize the commercial value of primates, which were used as trade goods between natives and Europeans in almost all parts of the world. Taking into account the subsequent loss of the original vegetation, and the large-scale trapping of these animals, which had already begun in the 16th century (cf. Veracini, chapter 9, this volume), it seems inevitable, as pointed out by Teixeira (**Box Details 10.3**), that the current patterns of distribution of the different local primates, reflects a centuries-old pattern of depredation by humankind.

BOX DETAILS 10.3: PRIMATES IN THE ICONOGRAPHY OF DUTCH BRAZIL

Dante Martins Teixeira
Museu Nacional – Universidade Federal do Rio de Janeiro

From the beginning of the discovery of America, the well-organized trade in exotic animals, such as colorfully plumaged parrots, monkeys, and felines, into Europe in the early Modern Age included the fauna of the New World. Attractive to a public avid for novelties from overseas, New World primates fetched high prices, even after travel to the New World had become established on a regular basis. In fact, although from the middle of the 16th century onwards, dozens of ships arrived annually in Europe bringing animals from the "Indies" (Monardes, 1565). In 1688 Johann Georg III, Prince-Elector of Saxony paid 80 thalers for a common marmoset (*Callithrix jacchus*), an amount equivalent to more than a year and a half of the salary of an unskilled worker (Teixeira & Papavero, 2010). It is not surprising, therefore, that primates are included in publications and illustrated manuscripts produced during the period of Dutch Rule in Brazil. For example, they are included in the works of Laet (1625), Barlaeus (1647), Piso and Marcgraf (1648), and in the original versions of Zacharias Wagener's "Thierbuch" and Caspar Schmalkalden's "Diary" (Teixeira, 1997, 1998). From the iconographic point of view, the two pre-eminent publications are those of Albert Eckhout and Frans Post, two well-known Dutch artists who came to Brazil under the patronage of Count Johann Moritz von Nassau-Siegen, the governor hired to manage the West India Company. In the case of Frans Post, this material includes several paintings executed mainly in Europe, as well as some recently discovered sketches. The most significant material comprises two volumes of illustrations, of unknown authorship, apparently annotated by Nassau – the "*Libri Principis*" – and a large series of separate paintings, mostly attributed to Albert Eckhout. Taken to Europe, the latter set was arranged in the four folios of the "*Theatrum Rerum Naturalium Brasiliae*" and in the tome of the "*Myscellanea Cleyeri*".There are three paintings of primates in the "*Theatrum Rerum Naturalium Brasiliae*" and one in the "*Myscellanea Cleyeri*", and the eight illustrations of primates in "*Libri Principis*" are one of the main iconographic sources available from this time. The six drawings in Zacharias Wagener's "Thierbuch" are of a lesser quality, and there are five more primates in Caspar Schmalkalden's "Diary" (Teixeira, 1995b, 1997, 1998). Considering other groups of mammals, primates are relatively rare in Frans Post's paintings, although isolated specimens are present in works such as the "Vision of Olinda", the "Ruins of the Vision of Olinda", and "A Sugar Mill" (Teixeira, 2006b), and images by Post discovered a few years ago include three primates (Bruin, 2016). These various images reflect the intense interest in these animals during the 17th century, but primates from the Old World are better represented than those from Brazil, with a total of at least eight Old World species, namely the collared mangabey (*Cercocebus torquatus*), mustached guenon (*Cercopithecus cephus*), greater spot-nosed monkey (*Cercopithecus nictitans*), green monkey (*Chlorocebus sabaeus*) and Guinea baboon (*Papio papio*), besides a Diana monkey (*Cercopithecus* cf. *diana*), and a possible talapoin monkey (*Miopithecus* sp.). A study by Frans Post apparently depicts a young Asiatic langur, perhaps *Presbytis* sp. (**Fig. 10.15**).

The New World was represented in the "*Libri Principis*" by six monkeys from Brazil, the red-handed howler (*Alouatta belzebul)* black-striped capuchin (*Sapajus* cf. *libidinosus*), the blond capuchin (*Sapajus flavius*), the common marmoset (*Callithrix jacchus*) and a titi monkey that has been assigned to *Callicebus coimbrai*, *Callicebus barbarabrownae*, and *Callicebus melanochir* by different authors (e.g., Kobayashi & Langguth, 1999; Jerusalinsky et al., 2006; Marques et al., 2013; Garbino et al., 2021). On the originals in the "*Libri Principis*" are based the woodcuts found in the "*Historia rerum naturalis Brasiliae*", the famous book by Willem Piso and Georg Marcgrave published in 1648 (Teixeira, 1999). Some mistakes were made, for example, the engraving of the red-handed howler in the "*Historia rerum naturalis Brasiliae*" is based on the painting of the blond capuchin in the "*Libri Principis*" (**Fig. 10.16**).

Frans Post illustrates what seems to be a spider monkey (*Ateles* sp.) in his studies, and in at least two paintings – the "See of Olinda" and the "Ruins of the See of Olinda" (**Fig. 10.17**). In two cases the specimen depicted

FIGURE 10.15 The *macacke*. A probable young Asiatic langur, perhaps *Presbytis* sp. by Frans Post.

FIGURE 10.16 The blond capuchin, *Sapajus flavius*. (Source: "*Libri Principis*".)

FIGURE 10.17 Image of a supposed spider monkey (*Ateles* sp.). Study sketch by Frans Post.

appears to be wounded or dead, or just sleeping, lying prostrate on the ground with its toes twitching and its tail twisted. As noted by certain authors, there are no other contemporary reports on the presence of spider monkeys in northeastern Brazil, and the identity of the depicted animal remains uncertain, with even speculation that a captive specimen may have been brought in by a trading vessel (Garbino et al., 2018; Mendes Pontes et al., 2019). Spider monkeys are not mentioned in the "*Historia rerum naturalis Brasiliae*" (Piso & Marcgraf, 1648), while Gaspar Barlaeus' (1647) allusion to a rufous-coated, long-tailed monkey is extremely vague. Post's own annotation is more precise, referring to monkeys "two feet long, [which] are very badly tempered, they can jump from one tree into another from a great distance, when shot they twist their tail into a knot around a branch, hanging dead" (Bruin, 2016).

Apart from the more extensive comments on the vocalizations of red-handed howlers (*A belzebul*), Moritz von Nassau's annotations accompanying the iconography of primates in the "*Libri Principis*" are generally limited to brief mentions of the size, origin, or temperament of the varieties portrayed. The more substantial commentaries by Zacharias Wagener and Caspar Schmalkalden speak of the origin, habits, and popular beliefs about certain species, as well as providing valuable details about the trade of these animals. Among the remarks, are that mustached guenons (*C cephus*) (**Fig. 10.18**) were brought from "Guinea" with some regularity and "cost little money", while common marmosets (*C jacchus*) – the most commonly depicted of all – were sent in large numbers to Holland as pets for the recreation of "idle girls" whose families were able to pay for such a distraction (**Fig. 10.19**).

Both Wagener and Schmalkalden warn that few of the African and Brazilian monkeys survived the journey to Holland, being particularly sensitive to cold, a reason why the latter author recommended that common marmosets should be carried by women capable of cradling them on their breasts. Judging by the accounts from the 16th and early 17th century, the Cebidae were more hardy, while few of the more delicate callitrichids reached Europe alive (Gandavo, 1576; Léry, 1578; L'Écluse, 1605) (cf. also Veracini, Chapter 9, this volume). According to Zacharias Wagener, of every ten common marmosets, rarely more than two or three survived the journey. Monkeys seem to have been very common in Dutch Brazil during the 17th century, something that even applies to now endangered species such as the blond capuchin (*S flavius*), which was "found in great numbers", according to Zacharias Wagener.

FIGURE 10.18 *Cercopithecus cephus.* (Source: *"Libri Principis"*.)

FIGURE 10.19 *Callithrix jacchus*, "a marmoset [called *sangwingen*] not larger than a lady fist" (Source: *"Libri Principis"*.)

Taking into account the massive loss of the original vegetation, it seems inevitable that the current patterns of distribution of the different local primates must be put down to a centuries-old pattern of depredation by mankind. In fact, even the present composition of the primate fauna in northeastern Brazil may inspire some doubts, as the region has suffered ongoing episodes of extinctions that have affected other groups of mammals and birds. In this sense, although the records of a spider-monkey should be seen with caution, it is within the bounds of possibility that an atelid did occur in the former Dutch colony and it is now part of the group of species that disappeared without trace, apart from images and reports that are often overlooked.

ACKNOWLEDGEMENTS

I would like to sincerely thank Dante Teixeira for his important suggestions, exchange of ideas, and useful materials he sent me, and for small, but important, contributions to this chapter. I would also like to thank Hilary Morris for the first English revision, Bernard Wood for meticulous editing and suggestions, and the Arquivo Historico Ultramarino of Lisbon for the permission of publication of the De Lisboa's manuscript folios.

FUNDING

Portuguese national funds through FCT—Fundação para a Ciência e a Tecnologia in the context of the celebration of the contract under the 'Norma Transitoria', DOI 10.54499/DL57/2016/CP1456/CT0001.

REFERENCES

Primary Sources

Acosta, J. de, (2002). *Historia natural y moral de las Indias.* F. J. Alcina (ed.) Cronicas de Americas, Dastin Historia (Original work published in 1590).

Aldrovandi, U. (1637). *De quadrupedibus digitatis viviparis.* B. Ambrosino (ed.) *ibidem* 1642, 1645, 1663, 1665.

Alpini, Prospero. 1592. *Historiae Aegypti Naturalis. Pars Prima.* Lugduni Batavorum: Apud Gerardum Potvliet.

Anglería (Anghiera), M. de P. (1989). *Décadas del Nuevo Mundo (De Orbe Novo Decades).* Madrid: Ediciones Polifemo (Original work published in 1511–1550).

Barlaeus, G. (1647). *Rervm per Octennivm in Brasilia. Et alibi nuper gestarum, Sub Praefectura Illustrissimi Comitis I. Mavritii, Nassoviae, &c. Comitis, Nunc Vesaliae Gubernatoris & Equitatus Foederatorum Belgii Ordd. sub Avriaco Ductoris, Historia.* Amstelodami: Ioannis Blaev.

Barrére, P. (1741). *Essai sur l'Histoire Naturelle de la France Equinoxiale ou Denombrement Des plantes, des Animaux, & des Minéraux....* Paris: Piget.

Bettendorf, J. F. (1698). *Crônica dos Padres da Companhia de Jesus no Estado do Maranhão e o Compendio da doutrina christam na Lingua Portugueza, & Brasilica.* Brasília: Edições do Senado Federal (Vol. 115). (Original work published in 1698).

Calado E. M. Frei (1648). *O Valeroso Lucideno e Triunfo da Liberdade na Restauração de Pernambuco.* Lisbon: Oficina de Paulo Crasbeeck.

Cardim, F. S. J. (1997). *Tratados da terra e gente do Brasil.* Lisboa: Comissão Nacional para as Comemorações dos Descobrimentos Portugueses.

Cavazzi da Montecuccolo, G. A. (1687). *Istorica descrizione de'tre regni Congo, Matamba ed Angola sitvati nell'Etiopia inferiore occidentale e delle missioni apostoliche esercitateui da religiosi Capuccini.* Bologna: Giacomo Monti.

d'Abbeville, C. (1614). *Historie de la Mission des Peres Capuchinas en Isle de Maragnam.* de l'impr. Paris: de F. Huby.

d'Abbeville, C. (2008). *História da Missão dos Padres Capuchinhos na Ilha do Maranhão e Terras Circunvizinhas.* Brasilia: Edições do Senado Federal - Vol. 105. (Original work published in 1615).

Dapper, O. (1686). *Description de l'Afrique.* Amsterdam: Wolfgang, Waesberge, Boom & van Someren.

d'Évreux, Y. (1864). *Voyage au nord du Brésil.* Westensee Verlag. (Original work published in 1615).

De Laet J., (1625). *Nieuwe Wereldt ofte Beschrijvinghe van West-Indien, wt veelderhande schriften ende aen-teeckeninghen van verscheyden natien.* Leiden: Isaack Elzevier.

De Laet J. (1640). *L'histoire du Nouveau-Monde, ou, Description des Indes Occidentales.* Chez Bonaventure & Abraham Elseviers.

Gandavo, P. de M. (1576). *História da prouincia sãcta Cruz a qui vulgarmente chamamos Brasil.* Lisboa: Officina de Antonio Gonsalvez.

Gessner, K. (1551). *Tierbuch. De quadrupedibus Viviparis.* Züric: Christoph Froschauer.

Hernández, F. (1651). *Rerum medicarum Nova Hispaniae Thesaurus seu Nova plantarum, animalium et mineralium mexicanorum historia.* Rome. Accademia dei Lincei.

Jonstonus, J. (1650). *Historiae naturalis de quadrupedibus libri, cum aeneis figuris, Johannes Jonstonus,... concinnavit* (Frankfurt). https://www.biodiversitylibrary.org/item/137912#page/9/mode/1up

Jonstonus, J. (1657). *Historiae naturalis de quadrupedibus libri, cum aeneis figuris, Johannes Jonstonus,... concinnavit.* Amsterdam: J. J. Schipperi.

Johnston, J. (1678). *A description of the nature of four-footed beasts with their figures en[graven in brass]* / written in Latin by Dr. John Johnston; translated into English by J.P. Amsterdam: Printed for the widow of John Jacobsen Schipper, and Stephen Swart.

La Haye, P. Gosse, I. Neaulme (1731). *Memoires de l'Academie Royale des Sciences, contenant les ouvrages adoptez par cette Academie avant son renouvellement en 1699.* Tome Prémier. Memoires pour servir a l'histoire naturelle des animaux et des plantes.

L'Écluse, C. de (1605). *Exoticorum Libri Decem: Quibus Animalium, Plantarum, Aromatum, aliorumque peregrinorum Fructuum historiae describuntur.* Antuerpiae: Ex Officina Plantiniana Raphelengii.

Léry, J. de (1578). *Histoire d'vn voyage fait en la terre dv Bresil, avtrement dite Amerique.* La Rochelle: Antoine Chuppin.

Marcgrave, J. (1942). *Historia natural do Brasil.* Portuguese version. Trad. Mons. Procopio de Magalhaes J. Museu Paulista. São Paulo: Imprensa Oficial do Estado.

Martius, K. F. P. von (1863). *Glossaria linguarum brasiliensium. Glossarios de diversas lingoas e dialectos, que fallao os Indios no imperio do Brazil. Wörtersammlung brasilianischer sprachen.* Erlangen: Druck von Junge & Sohn.

Merolla da Sorrento, G., (1692). *Breve, e succinta relazione del viaggio nel regno di Congo nell'Africa meridionale, fatto dal P. Girolamo Merolla da Sorrento, sacerdote Cappuccino missionario apostolico : continente variati clima, arie, animali, fiumi, frutti, vestimenti con proprie figure, diversità di costumi, e di viveri per l'uso umano.* Napoli: Francesco Mollo.

Monardes, N. 1565. *Dos libros. El vno trata de todas las cosas q̃ traẽ de ñras Indias Occidẽtales, que siruen al vso de Medicina, y como se ha de vsar dela rayz del Mechoacã, purga excelẽtissima. El otro libro, trata de dos medicinas marauillosas q̃ son cõtra todo Veneno, la piedra Bezaar, y la yerua Escuerçonera. Con la cura de los Venenados. Do veran muchos secretos de naturaleza y de medicina, cõ grãdes experiẽcias.* Sevilla, Casa de Sebastian Trugillo.

Nieremberg, J. E. (1635). *Historia naturae, maxime peregrinae, libris XVI. Distincta.* Antverpiae: ex officina Plantiniana B. Moreti. https://lhldigital.lindahall.org/cdm/compoundobject/collection/nat_hist/id/47961/show/47343/rec/5

Perrault, C. (1671–76). *Memoires Pour Servir A L'Histoire Naturelle Des Animaux*. Paris: De l'Imprimerie Rovale, 2 vols.

Piso, W., Marcgraf, G. (De Laet J. Org.) (1648). *Historia Naturalis Brasiliae... In qua non tantum plantae et animalia, sed et indigenarum morbi, ingenia et mores describuntur et iconibus supra quingentas illustrantur*. Lugdun. Batavorum, apud Franciscus Hackium et Amstelodami apud Lud. Elzevirium.

Rajus, J. S. R. S. (1693). *Synopsis Methodica, Animalium Quadrupedum et Serpentini Generis*. Londini: Impensis S. Smith & B. Walford. Societatis Regiae Typographorum.

Oviedo y Valdés, F.G.De (1526). *De la Natural Hystoria de las Indias* (on title page) or *Sumario de la Natural Hystoria de las Indias* (on inside cover). Toledo.

Thevet, A. (1575). *La Cosmographie Universelle*. Paris.

Topsell, E. (1607). *The historie of foure-footed beastes. Describing the true and liuely figure of euery beast, with a discourse of their seuerall names...* London: Printed by William Iaggard.

Tulp, N. (1641). *Observationum medicarum libri tres*; *editio nova, libro quarto auctior*. Amstelredami: Elzevirium, (1672).

Wagener, Z. (c. 1641). *Thier Buch*. [S.l.: s.n.]

Walter, J. (ed) (1967). *História dos animais e arvores do Maranhão*. Lisboa: Publicação do Arquivio Historico Ultramarino e Centro de Estudos Historico Ultramarinos.

Walter, J. (ed) (2000). *História dos animais e árvores do Maranhão: Frei Cristóvão de Lisboa*. Estudo de Jaime Walter. Notas e Comentários de J.Walter, F. Frade, J.E.M.Ferrão, L.F.Mendes e M.C.Liberato. Comissão Nacional para as Comemorações dos Descobrimentos Portugueses. Lisboa: Instituto de Investigação Científica Tropical, Lisboa.

Secondary Sources

Acemoglu, D., Johnson, S., & Robinson, J. (2005). The rise of Europe: Atlantic trade, institutional change, and economic growth. *American Economic Review*, *95*(3), 546–579.

Ames, G. J. (2011). *L'età delle scoperte geografiche 1500-1700* (Italian edition). Bologna: Il Mulino.

Ávila-Pires, F. D. D. (1989). Mamíferos da França Equinocial (Maranhão, Brasil). *Revista Brasileira de Zoologia,* 6, 423–442.

Billé, P. (2015). La faune brésilienne chez les chroniqueurs de la France Equinoxiale, Claude d'Abbeville et Yves d'Evreux. In: J. Penjon & C. Pereira (eds.), *L'animal dans le monde lusophone, du réel à l'imaginaire* (pp. 177–207). Presses Sorbonne Nouvelle. 978-2-87854-658-3.ffhal-02135317f

Bruin, A. de (2016). Frans Post: animals in Brazil. *Master Drawings* 54(3): 293–368.

Capozzi, R. (2021) *Experiência e descrição*: *os animais da França Equinocial entre a atividade franci cana e a filosofia natural (1612-1615)*. Master dissertation, Mestrado em História das Ciências e da Saúde, Rio de Janeiro.

Dams, B. (2016). *Comprehending the New World in the Early Modern Period: descriptions of Dutch Brazil (1624–1654)*. Doctoral dissertation, Ghent University.

Del Pesco, D. (2011). La lègation de Flavio Chigi à Paris en 1664: mémoire et documents nouveaux (avec quelques observations sur le Journal de voyage du cavalier Bernin en France de Paul de Chantelou). *La lègation de Flavio Chigi à Paris en 1664: mémoire et documents nouveaux (avec quelques observations sur le Journal de voyage du cavalier Bernin en France de Paul de Chantelou)*, 475–512.

Findlen, P. (1996). Courting nature. In: N. Jardin, J. A. Second & E. C. Spary (eds.), *Culture of Natural History* (pp. 57–74). Cambridge University Press.

Findlen, P. (2006). Natural history. *The Cambridge History of Science,* 3, 435–468.

Garbino, G.S.T., Aquino, C. C de, & Beltrão-Mendes, R. (2021). Marcgrave's red-tailed monkey: the earliest European depiction of a titi monkey. *Archives of Natural History* 48(1): 131–138.

Garbino, G. S. T., Rezende, G. C., Fernandes-Ferreira, H., & Feijò, A. (2018). Reconsidering mammal extinctions in the Pernambuco Endemism Center of the Brazilian Atlantic Forest. *Animal Biodiversity and Conservation* 41(1): 175–184.

Garcia, R. (2008). Introducão. In: C. d'Abbeville *Histoire de la mission des pères Capucins en l'Isle de Maragnan et terres circonvoisines* (1614) (pp. 15–20), vol. 105. Brasilia: Senado Federal.

Groves, C. (2008). Extended Family: Long Lost Cousins. A Personal Look at the History of Primatology. Arlington, VA: Conservation International.

Guerrini, A. (2015). *The courtiers' anatomists: animals and humans in Louis XIV's Paris*. University of Chicago Press.

Fooden, J. (2007). Systematic review of the Barbary macaque, *Macaca sylvanus* (Linnaeus, 1758). *Fieldiana Zoology*, 2007(113), 1–60.

Hallam, H. (1839). *Introduction to the Literature of Europe, in the 15th to 17th Centuries*. London.

Jerusalinsky, L., Oliveira, M. M., Pereira, R. F., Santana, V., Bastos, P. C. R., & Ferrari, S. F. (2006). Preliminary evaluation of the conservation status of *Callicebus coimbrai* Kobayashi & Langguth, 1999 in the Brazilian state of Sergipe. *Primate Conservation* 21: 25–33.

Kobayashi, S., & Langguth, A. (1999). A new species of titi monkey, *Callicebus* Thomas, from north-eastern Brazil (Primates, Cebidae). *Revista Brasileira de Zoologia* 16(2): 531–551.

Kury, L. B. (2014). *Brilhante. Representações da Fauna no Brasil, séculos XVI-XIX*. Rio de Janeiro: Andrea Jakobsson Estúdio.

Marques, E. L. N., Beltrão-Mendes, & R., Ferrari, S. F. (2013). Primates, Pitheciidae, *Callicebus barbarabrownae* Hershkovitz, 1990: new localities for the critically endangered titi monkey in the São Francisco basin, state of Sergipe, Brazil. *Check List* 9(1): 113–115.

Mason, P. (2009). *Before Disenchantment. Images of Exotic Animals and Plants in the Early Modern World*. London: Reaktion Book.

Mendes Pontes, A. R., Mariz Beltrão, A. C., & Melo Santos, A. M. (2019). Reconsidering mammal extinctions in the Pernambuco Endemism Center of the Brazilian Atlantic Forest: a critique. *Animal Biodiversity and Conservation* 42(1): 69–77.

Nonaka, N. (2014). The aviaries of the Farnese Gardens on the Palatine: Roman antiquity, the Levant, and the architecture of garden pavilions. *Memoirs of the American Academy in Rome*, 59, 361–398.

Olgivie, B. (2006). *The Science of Describing*: *Natural History in Renaissance Europe*. University of Chicago Press

Ogilvie, B. W. (2018). Visions of ancient natural history. *Worlds of Natural History*, 17–32.

Oliveira, M. M. de, & Langguth, A. (2006). Rediscovery of Marcgrave's capuchin monkey and designation of a neotype for *Simia flavia* Schreber, 1774 (Primates Cebidae). *Boletim do Museu Nacional, nova série, zoologia* 523: 1–16.

Palumbo, G. (2007). Riannodare le fola della storia della scienza tra il Vecchio Mondo e il Nuovo: la via delle immagini del XVII secolo. In F. Cantù (ed), *Scoperta e conquista di un Mondo Nuovo* (pp. 179–237). Roma:Viella.

Papavero, N., & Teixeira, M. D. (2015). *Recife e Salvador na visão dos capuchinhos italianos missionários no Reino Congo (1667-1703): Habitantes, Costumes, escravidão, comercio, matéria medica, flora e fauna do Brasil seicentista.* São Paulo: Instituto de Estudos Brasileiro, (IEB – USP).

Papavero, N., Teixeira, D. M., & Overal, W. L. (2002). *O Novo Eden.* Belém do Para: Museo Paraense Emiliao Goeldi.

Papavero, N., Teixeira, D.M. & Pujol-Luz, J.R. (1999). A fauna da Amazônia brasileira nos relatos de viajantes e cronistas dos séculos XVI ao XVIII. 11. A descrição do Estado do Maranhão, Pará, Corupá e Rio das Amazonas de Mauricio de Heriarte (1662). *Contrib. Avulsas sobre Hist. Nat. Bras. (Seropédica)* 17: 1–13.

Parker Brienen, R. (2006). *Albert Eckhout. Visões do Paraiso Selvagem. Obra completa.* J. Bandeira (ed.), (2010). Capivara Editora Ltda.

Raven, C. E. (1986). *John Ray: Naturalist: His Life and Works.* Cambridge: Cambridge Science Classic.

Sahlins, P. (2012). The royal menageries of Louis XIV and the civilizing process revisited. *French Historical Studies, 35*(2), 237–267.

Sawaya, P. (1942). Comentários sobre a parte sexta dos quadrúpedes e serpentes. In: G. Marcgrave (ed.), *História Natural do Brasil* (pp. lxxviii–lxxxviii). São Paulo: Museu Paulista.

Teixeira, D. M. (1995a). A Imagem do Paraíso: uma iconografia do Brasil Holandês (1624-1654) sobre a fauna e flora do Novo Mundo. In: C. Ferrão & J. P. M. Soares (eds.), *Brasil Holandês, T.1, Introdução e Miscelanea Cleyeri* (pp. 89–139). Rio de Janeiro: Editora Index.

Teixeira, D. M. (1995b). *Brasil Holandês: Miscellanea Cleyeri, Libri Principis & Theatrum rerum naturalium Brasiliae.* Rio de Janeiro: Editora Index.

Teixeira, D. M. (1997). *Brasil Holandês: Documentos da biblioteca universitária de Leiden, o "Thierbuch" e a "Autobiografia" de Zacharias Wagener e os quadros do "Weinbergschlösschen" de Hoflössnitz.* Rio de Janeiro: Editora Index.

Teixeira, D. M. (1998). *Brasil Holandês: o "Diário de Viagem de Caspar Schmalkalden de Amsterdã para Pernambuco no Brasil".* Rio de Janeiro: Editora Index.

Teixeira, D. M. (1999). *O mito da natureza intocada: as aves do Brasil Holandês (1624-1654) como exemplo para a história recente da fauna do Novo Mundo.* Tese (Doutorado em Zoologia) - Programa de Pós-Graduação em Zoologia, Museu Nacional, Universidade Federal do Rio de Janeiro, Rio de Janeiro.

Teixeira, D.M. (2006a). As duas tartarugas: uma pintura atribuida a Albert Eckhout (ca.1610–1666). *Publ. Avul. Mus. Nac. Rio de Janeiro,* 112, 1–24.

Teixeira, D. M. (2006b). Nature in Frans post's paintings of the new world. In: L. Krempel (ed.), *Frans Post (1612-1680): Maler des verlorenen Paradieses* (pp. 45–52). Fulda: Michael Imhof.

Teixeira, D. M., & Papavero, N. (2010). O tráfico de primatas brasileiros nos séculos XVI e XVII. In: L. M. Pessôa, W. C. Tavares & S. Siciliano (Org.), *Mamíferos de Restingas e Manguezais do Brasil* (pp. 253–282). Rio de Janeiro: Sociedade Brasileira de Mastozoologia & Museu Nacional-UFRJ.

Veracini, C. (2019). History of non-human primates in the Early Modern period: The contribution of the Italian encyclopaedist Ulisse Aldrovandi (1522–1605). In C. Veracini, C. Casanova & F. Scalfari (eds.), *History of Primatology: Yesterday and Today. The Mediterranean Tradition* (pp. 53–91). Rome: Aracneditrice

Veracini, C. (2020). A landscape without nonhuman primates? The case of the Barbary macaque, *Macaca sylvanus* (Linnaeus, 1758) and its interaction with humans throughout recorded time. *Humanities, 9*(3), 92.

Veracini, C., & Teixeira, D. M. (2017). Perception and description of New World non-human primates in the travel literature of the fifteenth and sixteenth centuries: a critical review. *Annals of Science, 74*(1), 25–63.

Whitaker, K. (1996). The culture of curiosity. In: N. Jardin, J. A. Second & E. C. Spary (eds.), *Culture of Natural History* (pp. 75–90). Cambridge University Press.

Winters, R. (2020). The Dutch East India Company and the Transport of Live Exotic Animals in the Seventeenth and Eighteenth Centuries. In: M. Chaiklin, P. Gooding & G. Campbell (eds.), *Animal Trade Histories in the Indian Ocean World* (pp. 27–65). Palgrave Macmillan, Springer.

11 Natural History of Primates in the 18th–19th Centuries, before Darwin

Cecilia Veracini
Universidade de Lisboa

If our judgment were limited to the form alone, I acknowledge that the monkey might be regarded as a variety of the human species. (Buffon (1766), *Histoire Naturelle*, Vol. XIV, p. 32).

11.1 NON-HUMAN PRIMATES IN THE "CENTURY OF THE ENLIGHTENED"

In the early modern period, naturalists were engaged in the search for adequate methods of compiling and arranging flora and fauna (Blair, 2000; Olgivie, 2006). In the 18th century, after two centuries of overseas explorations, the corpus of data arriving in Europe from the other continents (e.g., travel literature, scientific expeditions, and the great quantity of exotic animals and plants brought back from overseas) grew exponentially, as did the need to classify these different and rare natural products. It was believed that by classifying and grouping living species, humans could learn more about themselves. People thought that "if they could use observation and classification to unravel the mysteries of the natural world, the knowledge and the methods used to archive it could also be applied to the human and economic worlds for the benefit of mankind" (Huxley, 2003:79). The desire for a universal classification system and the possibility of understanding the natural relationship between living beings caused considerable debate throughout the scientific world (cf. Lefèvre, 2001). There were two main systems of classification, differing widely from each other: the natural (separation of natural objects according to their natural qualities) and the artificial (based on selected and arbitrary characteristics). These two opposing visions were identified above all with the famous naturalists Georges-Louis Leclerc Buffon (1707–1788) and Carl Linnaeus (1707–1778); the former exemplified the natural and the latter the artificial. Primates, including human beings, as well as other animals and plants, were classified differently in these two systems of classification.

With the growing arrival in Europe of specimens of the great apes - alive or dead – non-human primates, in addition to their naturalistic value, also assumed philosophical importance. These animals could discredit the idea of 'man' as a creature made in God's image, undermine his place in nature, and consequently the whole idea of the Chain of Being, opening new perspectives in the debate related to the concept of species, that until this time were seen as fixed entities determined at the moment of creation (Corbey, 2008: 65–66).

The natural sciences were also influenced by the profound philosophical and cultural changes that were taking place in the old continent. In the previous century, these disciplines had been dominated by the mathematization of the universe, and nature was imagined as a great machine, or clock composed of systems. Scientists explored all its parts to ultimately uncover its full mechanics through mathematics, but, toward the middle of the 18th century, mathematics gave way to the observational sciences (Vanommeslaeghe, 2008: 10). Scientists began to investigate nature in a more experimental and empiricist way, following, above all, the philosopher John Locke (1632–1704), who rejected any innatist views and based his theories solely on experience (Locke, 1690). He also rejected the Cartesian view that beasts were mere machines and "compared the operation of the human mind with that of the higher animals in order to ascertain the precise differences between them" (Wood, 1996: 203). The ideas of Newton and Leibniz, with their emphasis on continuity, and the philosophy of Kant and the Enlightenment, which considered reason the primary source of authority and legitimacy, had a profound effect on the natural sciences (Mayr, 1990: 140). The cultural milieu of the 18th century was infused with the ideal of advancement and progress, and scientific societies and academies grew up throughout Europe, while natural history became increasingly popular among the upper classes, as did scientific literature, a genre that took on a new importance. New and detailed images of primates, taken 'from life', together with empirical data about their anatomy and morphology, featured in treatises on nature, and scholars benefited from the proliferation of menageries, which at the beginning of the 19th century would become proper zoos (Hoage & Deiss, 1996). Since 1600, menageries have sprung up in all the most important European urban centers, and this has stimulated public interest in understanding the ordering of the natural world. The growth and improvement of menageries were facilitated by important economic and political entities, which promoted and even funded the acquisition of living animals. Exotic animals which were bought and sold like any other commodity, could be found in any of the main European ports or obtained through animal brokers. The Dutch East India Company was the major supplier of

DOI: 10.1201/b21963-14

animals to Europe between 1600 and 1700, and through it many new Asiatic species reached Europe (Winters, 2020). Travelers and traders, in the pay of this and other companies and nations, journeyed all over the world, oftentimes bringing home live animals or their remains (most of them died during those long journeys) and filling travel logs with observations on exotic animals. It should be noted that these observations gradually became more accurate and were comparable to scientific accounts, as they contained first-hand testimony about the behavior (both in nature and in captivity), diet, and habitat of the animals, as well as reports about their interaction with local populations. In addition, European naturalists could now rely on large natural history collections - both private (e.g., the cabinet of the Amsterdam-based pharmacist Albertus Seba) and state-owned (e.g., Royal Cabinets) – where they could carry out their studies. This historical moment not only fostered the discovery and description of many new forms of non-human primates, but also widened the study of primates through new viewpoints and intensified research so that by the end of the 18th century the natural history of primates would appear radically different from that of the previous centuries.

11.2 BUFFON AND LINNAEUS: PRIMATE CLASSIFICATION AND NATURAL HISTORY

Georges Buffon (Georges-Louis Leclerc, Comte de Buffon) (1707–1788), who was elected to the *Académie de Sciences* of Paris in 1734 at twenty-seven years old, was for fifty years the Regent of the King's Garden in Paris, a unique scientific institution in the world at that time. He published 36 volumes of his book, *Historie naturelle.* The original version, published by the *Imprimerie Royale* in Paris, appeared in 15 volumes between 1749 and 1767, followed by seven supplementary volumes between 1774 and 1789 (Rookmaaker, 1992: 132). Bernard de Lacépède (1756–1825), Buffon's close collaborator between 1788 and 1804, published eight posthumous volumes. The life and work of Buffon have been interpreted (or judged!) according to the different philosophical and scientific theories of the time. More recently, historians of science, by paying attention to his texts that had previously been neglected, began to consider him as a man who was one of the greatest minds of his time, transforming our understanding of nature. "He did not give new answers to traditional questions but asked new questions, opened a new field of research, and suggested a new way to conceiving natural history" (Roger, 1989, Preface). Buffon's works were at that time more widespread in France than the works of Voltaire and Rousseau, and the *Histoire naturelle* (HN) became, with the exception of *L'Encyclopedie*, the premier compendium of the culture of the Enlightenment, enjoying great influence throughout Europe (Solinas, 1965: 267). Buffon quickly became a national hero, such that Louis XVI had a statue of Buffon placed in front of the Royal Gardens (where it remains to this day) (Dugatkin, 2019). Buffon's contribution to the natural history of non-human primates[1] ranges from the classification and description of new taxonomic groups to the behavioral descriptions of some species, and finally the accurate anatomy and morphometric measurements of many taxa, carried out by his great friend and collaborator Louis Jean-Marie D'Aubenton (Daubenton). Non-human primates were relegated to the last volumes of Quadrupeds (*Histoire naturelle*, Vols 14–15), and their description was given in the last of the seven posthumous *Supplements* (1789) (S7) edited by Lacépède (cf. Contreras Martinéz, 2015: 39), whereas his rival, Linnaeus, classified them as "the first", from the Latin *primatis*, (taxon that also included human beings). Buffon was highly critical of the classification of primates as quadrupeds, stating that:

> this general description is abused when applied to certain animals. [...] Indeed man is *bimanus* and biped and the monkey is *quadrumanus.* [...] If all the animals that use the fore feet like hands are counted, we realize that half of quadrupeds do not deserve that name.
>
> (Buffon, *HN,* vol. 14: 19)

Linnaeus had already encountered the same difficulty in classifying non-human primates and other mammals as quadrupeds, and this problem forced him to significantly modify the traditional animal classification, and in the tenth edition of *Systema Naturae* he coined the word *Mammalia* [Mammals].

In Buffon's system, the Old World non-human primates (monkeys and apes) are clearly distinguished from those from the New World, while prosimians are differently classified (see below). Buffon organized primates into some 'natural groups', according to their exterior morphology, and tries to resolve the many problems generated by the classifications of ancient, medieval, and other more recent authors. He reviewed their classification in light of the new types of primates that "the Ancients could not know" because they had been discovered in the previous two centuries by European explorers. Buffon believed in the constant change of natural forms, and this belief prompted him to promote description rather than adopting what he thought was a more artificial form of classification, such as that used by Linnaeus. Buffon believed that Linnaeus' taxonomy was in danger of creating non-existent similarities among different species: "most classifications were the product of the human mind and did not take into account the continuous variation shown by living things" (Huxley, 2003: 77). Thus, it is possible to find species that can fill all

[1] Buffon uses the term *singes* (translated here as monkey) to design the non-human primate taxon, a group which in his case included apes (both great apes and gibbons) and monkeys and some of the few prosimians he knew (most of these called *makis*), although for him the "true" *singes* were only apes (*singe* is the Latin translation of *simia*, see Glossary). In this volume, we use the term apes, in the modern scientific sense: Great apes belonging to the Family Hominidae and gibbons from the Family Hylobatidae.

the possible variations between one form and another, e.g., the *magot* [=*Macaca sylvanus*]:

> occupies a place between great apes and baboons. It is the shade between these two forms. It differs from the first by having a long muzzle and large canine teeth, and from the second, because it does not have a tail.
>
> (Buffon, *HN* vol. 14: 7).

Following this principle, Buffon is certain that he is classifying animals in "their proper order, agreeable to the scale of nature and not in an arbitrary way" (Buffon, *HN*, vol. 14: 7). Buffon justifies the variation of the forms he observes, envisioning a species' transformation, "a degeneration" from their original form received at the moment of creation (cf. Buffon, *HN*, Vol. 4: *De la Dégénération des Animaux*). Thus, non-human primates are degenerative forms of human beings, who stand at the top of the Chain of Being. However, in 1749, Buffon substitutes this image of vertical progress with that of a network in which each species, through each of its common characteristics, forms different relationships with other species. The transition from a linear sequence to a network obviously had philosophical significance. The image of the great Chain of Being, a linear chain that ascended from minerals to humans, described a creation order, while:

> in contrast the image of a network is a horizontal one, which erased hierarchies, which tangled a universal order, which observed so many relationships between so many different beings that it gave rise to the feeling that the forms had happened by chance without any overall plan.
>
> (Roger, 1989: 291)

Buffon also believed that the intervention of religion in science prevented its progress. Therefore, some commentators argued that he had an atheistic and materialistic viewpoint and that the reason he often invoked the Creator was to avoid the censure of the Sorbonne and the Inquisition (Vanommeslaeghe, 2008: 15–16; see also Wood, 1996: 205).

Although Buffon described many connections among major groups (birds, mammals, and fishes), allowing that in nature there are natural families and that the members of these are 'genetically' related and have arisen from a common stock (Wilkie, 2006). He does not use the terms order, family, and class in a hierarchical and systematic manner, but uses them interchangeably, sometimes creating confusion. For instance, *singe* (the translation of the Latin word *simia*) means at the same time a specific group of tailless primates (see later) and the whole Order of Primates [*sensu* Linnaeus but without humans]. In this sense, he continued a classical and medieval tradition that used the term *simia* to refer to primates as a whole, or to just tailless monkeys, exemplified in antiquity by the Barbary macaque (see Glossary). Buffon describes more than 30 primate species: some of them, he says, are real species, whereas others can be regarded as species variations (see **Table 11.1**). His descriptions were based on a vast body of travel and scientific literature and on the examination of 40 primate specimens. Many of the latter survive in the *Muséum d'Histoire Naturelle* of Paris, where they were housed after the Royal Cabinet was taken over during the French Revolution. Buffon also bought a chimpanzee (the famous *Jocko,* see Barsanti, this volume) and a Barbary macaque. In the *Nomenclature de les Singes* (*Nomenclature of monkeys/primates*, Vol. 14: 1–17) of the *Histoire naturelle*, he describes the 'families' of primates, whose definitions are summarized here:

1. S*inge* [=ape]: "I call *singe* an animal without a tail, whose face is flat, whose teeth, hands, fingers, and nails resemble those of man, and who like him walks erect on two feet". In this group are included the "*pithecos* of the Greeks", and the "*simia* of the Romans", "and it is on this that Aristotle, Galen, and Pliny based their physical comparison between man and monkey". But "the difference between man and *singe* is in greatness, for which the ancients were right in considering it as a *homunculus*, an imperfect dwarf, *a pigmy*". This group includes the *Pongo,* the *Ourang-outang,* the pigmy, and the *gibbon*: "the latter is a monstrous race because of its long arms". According to Buffon, the *Singe* was called *pitheque* by the Ancient Greeks, whereas the *cynocephalus* corresponds to the *magot* [= *Macaca sylvanus*]. (About Buffon's classification of great apes, see Barsanti, this volume);
2. *Babouin* [=*Papio* spp.]: "this animal has a short tail, a long face, a broad high muzzle, canine teeth proportionally larger than those of man, and callosities on its buttocks". "The Italians are the first who called it *babuino*; the Germans: *bavion*; the French *babouin*; the British *baboon*; the Romans: *Papio*". Buffon includes in this group: (a) "The *papion* or *babouin* (properly called) found in Arabia, Libya, etc., and probably the *Simia porcaria* of Aristotle"; (b) "The *Mandrill* [=*Mandrillus sphinx*] found in Guinea and in other hot parts of Africa"; (c) "The *Ouanderou* [=*Macaca silenus*, see **Fig. 11.1**], found in Ceylon and Malabar and other southern regions of India"; (d) "The *cynocephalus"* [dog's head] and the *magot*, called the "fourth ape or the first baboon".
3. *Guenon* [=long-tailed monkey]: this is an animal "resembling *singes* and baboons, but they have long tails: tails as long as or longer than their bodies". Buffon states that:

> the noun *Guenon* can be derived from *kébos* or *képos*, the name given to monkeys (*singes*) with tails in antiquity. They are smaller than baboons and monkeys (*singes*), but distinct from the *makis* [= lemurs] that have an extended snout. They have four incisors like monkeys (*singes*) and baboons and not 6 like *makis*.

TABLE 11.1
The Primates According to Buffon

Histoire Naturelle and *Supplements*:	Buffon's Name	Current Taxonomy	Notes of Buffon
Singes, [apes], *HN*, Vol. 14, 1766	Les *Ourang-outangs* ou le pongo et le jocko	*Pongo* spp./*Pan troglodytes*	They are probably the same species with two distinct varieties: the first is the greater *Ourang-outang* (we have just travel reports); the other is the small *Ourang-outang* or jocko (enjocko) (we saw him alive). They have the greatest resemblance to man.
Singes, [apes], *Supplement*, Vol. 7. 1789	Le Pongo et le Jocko	*Pongo* spp./*Pan troglodytes*	Two species of orangoutan: the large pongo and the small jocko.
Singes, [apes], *HN*, Vol. 14, 1766/ *Supplement* Vol. 7, 1789	Le Pitheque	Without tail – *pithekos* of Aristotle. He misled Aristotle. In the *Supplements* is *Macaca sylvanus*, recognized by M. Desfontaines as the most similar to the Aristototle pitheque.	Animal whose nature is ambiguous, partly man, partly quadrupeds: *Simia*, Latin; *Chin chin* in Tartary; *Sinsin* in China. It is said to live in north Africa.
Singes, [apes], *HN*, Vol. 14, 1766	Le Gran gibbon/Petit gibbon	*Hylobates lar*, male/female, probably different subspecies (Cf. Groves, 2008),	Gibbon is the name under which Mr. Dupleix, Governor of French India [cf. Sidney 1886], gave us this animal. I first imagined this to be an Indian word, but I found that Strabo had denoted the *cephus* by the word *keipon*, from which *gaibon* or *gibbon* had probably been derived.
Singes, [apes], *HN*, Vol. 14, 1766/*Supplement* Vol. 7, 1789	Le Magot/Le petit Cynocéphale	*Macaca sylvanus*	This is the *Cynocephalus* of Aristotle. In the *Supplement* (1789:37), we described another varieties of the magot – who live in Spain, Gibraltar-: the small *cynocephalus* described by Prospero Alpini.
Babouin, *HN*, Vol. 14, 1766	Le Papion or Babouin	The description and engraving suggest: *Papio papio*	The baboon proper.
Babouin, *HN*, Vol. 14, 1766/*Supplement*, Vol. 7. 1789	Le Mandrill / called Le Choras in Suppl. 1789	*Mandrillus sphinx*	Mandrill is the name given by English in Guinea also called *Boggo* by the natives in Congo. It is the combination between the words man (man in general) and drill: vigorous libertine in French jargon/*Suppl.* 1789. It lives in Ceylon and India. It is the *Papio* of Gessner. The tufted ape of Pennant. (*Hist. Nat. Quad.* Vol. I, p. 174; and grand babouin p. 76 of Pennant)
Babouin, *Supplement*, Vol. 7. 1789	Le Baboin avec la face bluatre	*Papio ursinus*	Described by Pennant.
Babouin, *Supplement*, Vol. 7. 1789	Le Baboun a longues jambes	*Macaca nemestrina,* according to the Buffon's plate	It stands higher on its legs than any other baboon. Brown baboon of Pennant (*Hist Nat Quad..*, Vol. I, pag. 177) and Simia platypygos of Schreber (*Hist Nat Quad..* Vol. I , page 87).
Babouin, *Supplement*, Vol. 7. 1789	Le Babouin des Bois/Le Babouin jaune/Babouin cendré	*Papio papio/Papio cynocephalus/Papio anubis* (no plates provided)	Described by Pennant. It lives in Guinee and is called "the man of wood" (based on Pennant's description of a specimen housed in London). There are other varieties like babouin jaune.
Babouin, *Supplement*, Vol. 7. 1789	Le Babouin: A museau de chien/Le Singe de moco	*Papio hamadryas*	It occurs in Arabia, Abyssine, and Guinee to the Cape of Good Hope. The singe de moco is after Edwards, a monkey from the Persic Gulf. Simia hamadryas of Linnaeus
Babouin,, *HN*, Vol. 14, 1766/*Suppl.* Vol. 7 1789	Le Ouanderou et Le Lowando	Three species in this group: first, *Macaca silenus*; second, unknown species; third variety, *Semnopithecus* spp.	They are variety of the same species. They receive these names in Ceylon (*Ouanderou; Wanderu*) which is their proper country. It appears to exist that there are two or three varieties of this species according to their different colors of the beard and face. Le ouanderou lives in Ceylon. Lowando a variety of the ouanderou. Another variety is white.

(*Continued*)

TABLE 11.1 (*Continued*)
The Primates According to Buffon

Histoire naturelle and Supplements:	Buffon's Name	Current Taxonomy	Notes of Buffon
Babouin, *HN*, Vol. 14, 1766/*Supplement*, Vol. 7, 1789	Le Maimon	*Macaca nemestrina*	Name given to monkeys with short tail, *singe à queue de cochon*.
Guenon, *HN*, Vol. 14, 1766/*Supplement*, Vol. 7, 1789	Le Macaque and l' Aigrette	*Macaca fascicularis*	Le Macaque is the species most similar to baboons among long-tailed monkeys. *Macaquo* is its name in its native country, the Congo: it is a small species of baboon. Egrette is a guenon, a variety of the macaque. It is the aygula of Linnaeus.
Guenon, *HN*, Vol. 14, 1766	Le Patas and Patas à bandeau blanc	Two species in this group: *Erytrocebus patas* and *Chlorocebus* cf. *sabaeus*	Patas: name in its native country, Senegal (according to Brue, *Histories general des voyages*, tome II, p. 520.). "*Le singe rouge de Senegal*". It is similar to macaque but with a longer body. There are two varieties of patas in Senegal.
Guenon, *HN*, Vol. 14, 1766	Le Malbrouk et Le Bonnet-chinois	Two species in this group: *Chlorocebus cynosuros* and *Macaca sinica*	Malbrouk: name in its native country, Bengala. Bonnet-chinois is the name given for their hairs on the head top similar to a Chinese cap. It is the Faunus of Linnaeus, 10th ed. *Syst. Nat.* Le Malbrouk and bonnet chinois are probably the same species. They are similar to the macaque, egret, perhaps varieties of one species.
Guenon, *HN*, Vol. 14, 1766	Le Mangabey (Singes de Madagascar)	Two species in this group: *Cercocebus torquatus* and *Cercocebus atys* (both identified by Buffon's plates)	We received two individuals of this species, both of which were sent to us under the appellation of Madagascar monkeys. The name Mangabey is preliminary because Madagascar is near the land of Mangabey. They have a remarkable character: naked and white eye-lids.
Guenon, *HN*, Vol. 14, 1766/*Supplement*, Vol. 7, 1789	Le Mone/Le Mona:	*Cercopithecus mona*	The most common of the long-tailed monkeys. Mone, Mona, Monina, mounina … is the name of guenons or long-tailed monkeys in the Moresk, Spanish and Provençal languages. From the cost of Guinea. Le Mona: variety of the other. (To Schreber: diane; to Pennant: sponde monkey.)
Guenon, *HN*, Vol. 14, 1766	Le Callitriche	*Chlorocebus sabaeus* and other species of the genus.	*Callithrix*: beautiful color of their hairs, from Greeks. It was a long-tailed monkey, from Egypt and Aethiopie. Found in Senegal and Mauritania, Cape Verde islands.
Guenon, *HN*, Vol. 14, 1766	Le Moustac	*Cercopithecus cephus*	The name is for distinguishing him from other monkeys, from Guinea. It was called white-nose by travelers.
Guenon, *HN*, Vol. 14, 1766	Le Talapoin	*Miopithecus talapoin*, the Angolan talapoin	From Siam, and other asiatic province, as well as Africa, but there are doubts. Edward says it came from Africa. Talapoin: name given by people who gave us this monkey which we adopted.
Guenon, *HN*, Vol. 14, 1766	Le Douc	*Pygathrix nemaeus*	Name from the Cochinchine, his native country. Called *Sifac* in Madagascar. Le Douc is the last of that class of animals called singe, babouins and guenon: but it is not any of these three kinds.
Guenon, *Supplement*, Vol. 7. 1789	Le Guenon a long nez	*Nasalis larvatus*	It was sent to us from East Indies, and was unknown to any naturalist. Though very remarkable for a striking feature: a large prominent nose.
Guenon, *Supplement*, Vol. 7. 1789	Le Patas a queue courte	*Macaca mulatta*	Based on a skin we have.
Guenon, *Supplement*, Vol. 7. 1789	Le Guenon à face pourpre	*Trachypithecus vetulus*	From Ceylon, From Pennants (*Hist. Nat. Quad.*, Vol. 1, p. 184, 1771).
Guenon, *Supplement*, Vol. 7. 1789	Le roloway ou la Palatine	*Cercopithecus roloway*	From a specimen caught in Guinee and who lived in Amsterdam (source: Allamand). It cannot yet be described because it lives in Amsterdam, and is known and loved by everyone. Comes from Guinea where his name is Roloway, according to Mr. Allamand.

(*Continued*)

TABLE 11.1 (*Continued*)
The Primates According to Buffon

Histoire naturelle and Supplements:	Buffon's Name	Current Taxonomy	Notes of Buffon
Guenon, *Supplement*, Vol. 7. 1789	Le Macaque a queue corte	*Macaca assamensis*	It approaches the macaques rather more than the other guenons: but has a shorter tail and the rump nacked and of a blood-color, as well as the neighboring parts. Based on a skin we have.
Guenon, *Supplement*, Vol. 7. 1789	Le Guenon couronne	*Macaca sinica vel radiata*	Personal observation in Paris, 1774. It can be a variation of the guenon called bonnet chinois.
Guenon, *Supplement*, Vol. 7. 1789	Le Guenon a Camail	*Colobus polykomos*	Lives in Sierra Leone and Guinee, where the native give him the name King of monkeys. They use his fur to decorate various things.
Guenon, *Supplement*, Vol. 7. 1789	Le Blanc-Nez	*Cercopithecus petaurista*	Mr. Alleamand described a guenon from the Gold Cost called by Dutch: white nose. These monkeys may be the same of the mustache.
Guenon, *Supplement*, Vol. 7. 1789	Le Guenon a nez blanc prominente	*Cercopithecus nictitans*	We have the skin in the Cabinet du Roi.
Guenon, *Supplement*, Vol. 7. 1789	Le Guenon à criniere	*Macaca silenus*	We saw it. It lived many years at the manegerie of Roy at Versailles. [Figure drawn from the life, see Fig. 11.1.]
Guenon, *Supplement*, Vol. 7. 1789	Le Guenon negre	Unknown species/misled *Macaca maura*	It is a small species of guenon as big as a sagoin. The tail is long as the body.
Family Sapajous *HN*, Vol. 15, 1767/ *Supplement*, Vol. 7, 1789.	Le Ouarine et L'Alouate/ Le hurleur	Two species in this group: *Alouatta belzebul* and *Alouatta macconelli* vel *seniculus*	Ouarine (name of Maragnon) is the Guariba of Brazil; it is the same of the Alouate of Cayenne, they are different only for their color. They are the same species. (Prehensile tail and with a kind of osseous drum, in their throat.)
Family Sapajous *HN*, Vol. 15, 1767 / *Supplement*, Vol. 7, 1789.	Le Coaita et L'Exquima	Two species in this group: *Ateles paniscus* and *Cercopithecus diana*	Coaita: ou quoata, name of French Guyana; Chamek in Perù. Exquima is sinonymous of coata. Larger of the sapajous, only four fingers in the hands; very dexterous/intelligent/, wth a tail as a fifth hand.
Family Sapajous *HN*, Vol. 15, 1767/ *Supplement*, Vol. 7, 1789.	Le Sajou: Sajou brun et Sajou gris / Sajou cornu	Three species in this group: *Sapajus xanthostenos* and cf. *Cebus olivaceus*: both from the plates/Suppl. *Sapajus* cf. *nigritus*: plate XXIX.	Abridged word from Cayouassou/Sajouassou from Maragnon. The sajou cornu is called mikou in Guyana.
Family Sapajous *HN*, Vol. 15, 1767/	Le Sai: Say a gorge blanc	Two species in this group: 1) *Cebus capuchinus*; 2) the say in the plate VIII, it look likes young individuals of *Ateles fusciceps*.	The varieties of the same animal. Sai is the pronunciation of Cay, from Brazil.
Family Sapajous *HN*, Vol. 15, 1767/ *Supplement*, Vol. 7, 1789.	Le Saimiri: sapajous orange, aurore and joune	*Saimiri sciureus*	*Caymiri*, name from Maragnon/we observe some variation in the color.
Family Sagoins *HN*, Vol. 15, 1767/ *Supplement*, Vol. 7, 1789.	Le Saki: singe à queue. de renard /L'Yarqué	*Pithecia pithecia*	*Saki*, name taken from the work of Brown (Historia Natural de Jamaica, Ch. 5, sec. 5). Thevet (1558) report the name cacuein (pronunce: sacuien). It is the greatest of the sagoins/ yarqué
Family Sagoins *HN*, Vol. 15, 1767/ *Supplement*, Vol. 7. 1789	Le Tamarin/Le Tamarin négre	*Saguinus midas*	Tamarin name given in Maragnon; name at Cayenne according to Barrere and Tamary in Maragnon according to Father, d'Abbeville. The tamarins are rarer than the sapajous and live in the high parts of the forest.
Family Sagoins *Supplement*, Vol. 7. 1789	Le Tamarin négre	*Saguinus niger*	The tamarins are rarer than the sapajous and live in the high part of the canopy, while the sapajous live in the lower and wetter parts of the forest.
Family Sagoins *HN*, Vol. 15, 1767	L'Oustitì	*Callithrix jacchus*	Oustiti, a sound articulated by this animal, and which we have adopted for its name.

(*Continued*)

TABLE 11.1 (*Continued*)
The Primates According to Buffon

Histoire naturelle and Supplements:	Buffon's Name	Current Taxonomy	Notes of Buffon
Family Sagoins *HN*, Vol. 15, 1767	Le Marikina: petit singe-lion	*Leontopithecus rosalia*	Marikina is the name given to this animal in Maragnon (Mission of Father d'Abbeville).
Family Sagoins *HN*, Vol. 15, 1767	Le Pinche	*Saguinus oedipus*	Pinche is said to be "the name of this animal at Maynas" [=Manaus].
Family Sagoins *HN*, Vol. 15, 1767	Le Mico	*Mico argentatus*	Name given in the Orinoco to a small specie of monkeys, we gave it to this animal to distinguish it from the other species.
Family Sagoins *Supplement*, Vol. 7. 1789	Le Sagouin, vulgarment appelé le Singe de nuit	*Pithecia pithecia* (female)	Similiar to yarqué or saki, but different in color. The specimens came from Cayenne. Adult, female. They are very rare in that land. They make very singular motion with their head.
***Supplement* Vol. 8, Pl. 88.**	L' Indri	*Indri indri*	Description after Sonnerat 1782 (Fig. 11.12).
***HN*, Vol. 7 I, Pl. CCXXIX**	Le vari [Pl. CCXXIX]	*Varecia* sp.	
***Supplement* Vol. 7, 1789.**	Le moccoco	*Lemur catta*	More beautiful and clean than the mongous. They rest and sunbathe in front with open arms.
***Supplement*, Vol. 7. 1789**	Le grand mongous	*Eulemur* cf. *albifrons*	A variety of the one already described.
***Supplement*, Vol. 7. 1789**	Le Petit maki gris	*Hapalemur griseus*	Brought by Mr. Sonnerat from Madagascar. It seem a *moccoco*.
***Supplement*, Vol. 7. 1789**	Autre espéce de maki	*Avahi* sp.	Brought by Mr. Sonnerat from Madagascar
***Supplement*, Vol. 7. 1789**	Le Loris de Bengale	*Nycticebus bengalensis*	M. Vosmaer gave a long description of this animal when he stayed in India.
***Supplement*, Vol. 7. 1789**	Le Guenon a museau alonge	*Eulemur* cf. *coronatus*	Mr. Pennants described this animal which came from Africa.
***HN*, Vol. XIII, 1765**	Tarsier	*Tarsius* sp. (Sulawesi tarsier, cf. Groves, 2008: 79)	See comments in the text.
***HN*, Vol. VIII, 1760 not among primates**	L'Aye-aye [cf. Plate LXVIII, p. 268]	*Daubentonia madagascariensis*	See comments in the text.

This category includes many species of long-tailed monkeys (see **Table 11.1**) and also one lemur, called "guenon with a long muzzle".

Buffon differentiates clearly between Old and New World primates:

> all these monkeys [Old World primates], as well as some apes and baboons, have general and particular characteristics: naked buttocks, natural callosities peculiar to these parts. The second feature is that they have *abajoues*, or pouches, in which they can keep their food. The third one is that they have a narrow partition between the nostrils, and the apertures of the nostrils themselves are situated in the under part of the nose, like those of man. The *sapajous* and *sagoins* [New World monkeys] have none of these characteristics. The partition between their nostrils is never always placed in the under part of the nose, like those of man. The partition of their nose is always very thick, the apertures of their nostrils are situated in the sides of the nose, and not in the under part of it. They do not have buttocks, nor callosities, nor pouches. Thus they differ from the monkeys [Old World ones] not only in species, but in genus, since they possess none of the general characteristics which are common to all monkeys.
>
> (Buffon, *HN*, vol. 14: 14)

New World monkeys

1. *Sapajous* [=*Cebus*, spp., *Sapajus* spp., *Ateles*, spp., *Alouatta* spp.]: "all the *sapajous* use their tails as a finger to hang upon branches, or to lay hold of any object they cannot reach with their hands";
2. *Sagoins* [=*Callithrix* spp., *Mico* spp, *Saguinus* spp. also including *Pithecia* spp. for lacking a prehensile tail]: "they cannot employ their tails to hang by; their faces, ears, and hair are also very different, and we have to divide them into two distinct genera". (e.g., see **Fig. 11.2**).

Moreover, he suggests, in agreement with the Jesuit José de Acosta (cf. Acosta, 1590/2002, Chap. 36, v. iv, p. 33), that these New World species could have resulted from

FIGURE 11.1 Buffon's *Ouanderou* [*Macaca silenus*] "found in Ceylon and Malabar and other southern regions of India". (Courtesy of the National Library of Portugal, Lisbon.)

FIGURE 11.2 Buffon's *Sagoin* [*Pithecia*, sp.] "vulgarly called night monkey". (Courtesy of the National Library of Portugal, Lisbon.)

degeneration, environmental pressures, or the isolation of populations descended from the Old World primates. He adds that the species peculiar to the New World, or without Old World analogues, must have arisen *in situ*.

> In man air could cause changes in skin colour and other characteristics which denoted the different races, but it could not produce distinct species. By contrast, in animals the many similar species seem to have been produced by exposure to the excess of climate. Some species could exist only in a certain climate; each species has its country, its native fatherland to which each is limited by physical necessity.
>
> (Buffon, *HN*, 9:2)

The concept of degeneration as applied to animals and human beings became popular in Europe, especially the theory of New World degeneracy. This was exemplified by Abbe Cornelius de Pauw, who proposed that Europeans who migrated to the New World would also degenerate, as would their offspring (De Pauw, 1768 *apud* Dugatkin, 2019). John Ray (cf. Rajus, 1693), an English anthropologists and naturalist, strove to reconcile the idea of divine creation with that of species diversity and the more or less accidental variation observed in nature, plus those that occurred due to domestication, breeding, and agriculture (Pareti, 2016: 11). Buffon was the first naturalist to envision the mammals of the New World as a community or fauna that might well have originated independently from other regional faunas.

> Noah's ark had no place in his concept of faunal origins, and he rejected as too short the scripturally based 6,000-year estimate of the earth's age. Buffon's ideas of organic evolution and multiple centres of origin were novel and prepared the minds of his and succeeding generations for the acceptance of Darwinian evolution.
>
> (Hershkovitz, 1985: 90)

Buffon's contribution to primate natural history also included the description of the tarsius or *Wolly jerboa*: a name that referred to its elongated tarsus. It was not included among the "*Singes*", but in another section of the *Histoire naturelle* (**Fig. 11.3**). Although the origin of Buffon's tarsier specimen is not known, it is most likely a Sulawesi tarsier, *Tarsius tarsier* (cf. Groves, 2008: 79). Many new species would later be added to Buffon's list in the 1789 posthumous supplement organized by Lacépède (see **Table 11.1**). Among the new species are some new lemurs, and the Aye-aye (*Daubentonia madagascariensis*). The latter, which was placed far from monkeys and lemurs, was inserted between the flying squirrel and the Australian opossum cuscus. The description of the aye-aye was based on observations made by Pierre Sonnerat (1782), who during the 1770s had visited China and Southeast Asia as well

FIGURE 11.3 Buffon's tarsius, *Le Tarsier*, 1765, also called *Wolly jerboa*: a name which referred to its elongated tarsus. (Image of public domain, source: https://search.library.wisc.edu/digital/AOGCW4AG734I7M8Y

as Madagascar (See **Fig. 11.4**, Fig. by Sonnerat). Sonnerat's description of the aye-aye was accurate regarding both morphology and behavior; nevertheless, the unusual nature of the aye-aye - "it has features of the *maki* [=lemur], of the squirrel, and of the monkey" - led naturalists to classify it in various ways (see below). Daubenton gave a good description of a dead specimen and noted that it resembled the tarsier:

> in its flattened thumb, in its long tail covered in hair, in its upright, naked and transparent ears, in its woolly pelage, and in its very elongated digits; moreover, it was a female and had two mammas, on the lower part of the belly
>
> (*Suppl*. 1789, vol. VII, p. 268)

In the Supplement there is also a long description and an accurate illustration of the vocal organ of the Alouatta (howler monkey) ("an expansion of the hyoid bone into a large shell-like organ in the throat that gives resonance to the voice", Suppl., 1789: 93), by Camper (Fig. 11.5). Camper, intrigued by the description previously given by Buffon in his *Histoire naturelle* (*HN*, Vol. 14: X), made an accurate analysis of this organ, explaining its anatomy and function, and sent his conclusions to Buffon in recognition of his work. It is worth mentioning that the first description of the howler's vocal organ was produced in the 16th century by the Jesuit Cardim (Veracini & Teixeira, 2017).

FIGURE 11.4 Aye-aye as described by Buffon (top). (Courtesy of the National Library of Portugal, Lisbon.)

Finally, it is necessary to refer to Buffon's interest in mouers [customs] of animals. Both Daubenton and Buffon shared a common interest in investigating mouers, which is a combination of the behavior, customs, character, morality, habits, and instincts of animals. Buffon tried to describe the habits of animals either in their natural environment, replicating carefully the descriptions made by other naturalists, travelers and hunters, or by the accurate observations he made in Paris. There was no menagerie at the Royal Botanical Garden, where Daubenton could only collect stuffed animals, dissect cadavers, or measure skeletons (Contreras Martin.z, 2015: 45). In order to observe the behavior of living animals, Buffon installed a menagerie and an aviary at Montbard, his private estate, and in 1747, he bought a "*singe*" (the chimpanzee Jockò) from Angola, for which he paid 1,200 *livres* (Roger, 1989: 210). He also described the behavior of a captive female Barbary macaque kept at Montbard (Buffon, *HN*, Vol. 14: 133–134). While he affirms that "a simple comparison of behavior was enough to show man' superiority over animals and monkeys", with arguments such as "man can speak and animals cannot, although they have the vocal organs, as is shown by the monkey or the parrot"; "they

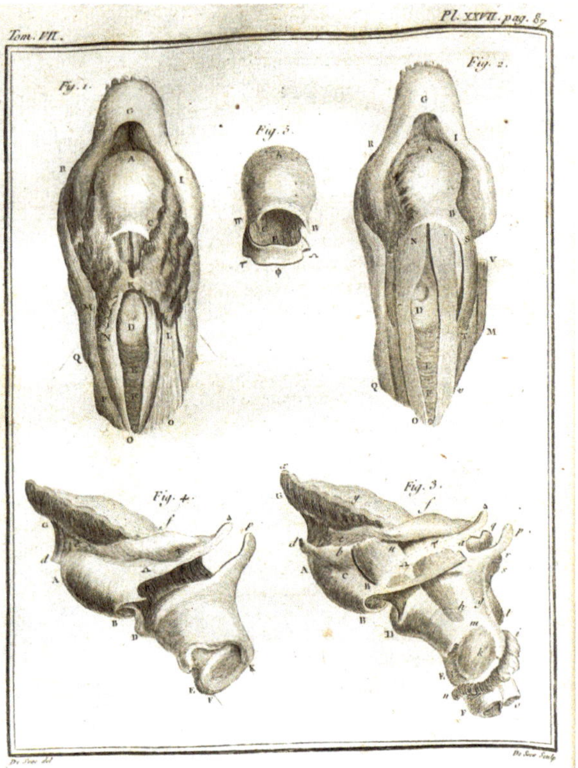

POCHE OSSEUSE DE LA GORGE DE L'ALOUATTE.

FIGURE 11.5 *"Poche Osseuse de la Gorge de l'Alouatte"*. Illustration of the vocal organ of the howler monkey, *Alouatta* spp. by Camper (*Suppl.*, 1789, Vol. VII, Pl. 27).

do not invent anything [...] they are purely material". At the same time, he notes that "animals can have a culture" and that "those who have a longer childhood also have a more developed culture" (i.e., elephants). He also presented tables of the different trends of animal organic development and made associations between behavior and physiology. Spary (2000: 111) argues that Buffon's animal description "denotes the simultaneously political, social, and scientific nature of the discourse of natural history for Buffon and Daubenton" because accounts of animal species could be converted into commentaries on human society. The intermixing of Buffon's philosophical views with his discourse on natural history is apparent, especially when he deals with great apes. On the one hand he has to defend the superiority of man, but at the same time, he conducts anatomical studies that demonstrate the opposite: that is many anatomical similarities between primates and human beings (see Barsanti, this volume). On the contrary, his contemporary and rival Linnaeus condemned rhetoric and attempted to banish from his science the use of language as a means of persuasion or for emotional effect, something for which he was highly acclaimed by his colleagues (Huxley, 2003: 78).

Carl Linnaeus (1707–1778), the Swedish physician, described as an old provincial man by his contemporary European visitors, was considered "the greatest botanist that the world ever did or probably ever will know" (Koerner, 1996: 145). In fact, although he neither identified general laws to explain life's diversity, he widened access to his classification system by providing tables that would be useful to the scientific and lay communities (Koerner, 1996: 145). Linnaeus designed a hierarchical system of classification, which he thought would help scientists group all living species in a simple and logical fashion. His global classification tree is laid out in large, inclusive divisions: the animal kingdom, the plant kingdom, and the mineral kingdom. Both animals and plants are divided into classes, and these in turn are divided into orders, which are then further divided into genera, which are then further divided into species, each identified with a binomial (Genus + Species names). This new system of two-word Latin names introduced an extremely simple form of classification and "divorced the need to put a label on an organism from the need to describe it" (Huxley, 2003: 78). As we shall see later, Linnaeus' method was highly praised by some contemporaries, but roundly criticized by others. No one before him (excluding classic authors such as Aristotle) had considered the human being an animal, and his classification system of plants based on the plant sexual system was considered immoral by some (Koerner, 1996: 154).

Many of Linnaeus' observations on animals were based on the specimens in Seba's collection in Denmark, an important cabinet of curiosities belonging to Albertus Seba (cf. Seba, 1734), and in Linnaeus' private collection (Koerner, 1996). Seba was an apothecary who had acquired his collection through the flourishing of commerce and travel during the Dutch Golden Age; some of Linnaeus' specimens are still housed in the Seba collection. Linnaeus firmly believed in first-hand observation: from 1747 on, 19 of Linnaeus' students undertook voyages of discovery to almost every part of the planet. He provided each student with "memorials" or order lists of plants and animals, and encouraged them to study the local people and their languages, cultures, and knowledge. The amount of material Linnaeus processed throughout his lifetime was huge, as shown by the records of the Linnean Society (https://www.linnean.org/). His collection of *naturalia* comprised thousands of animals and plants and 1,600 volumes; he estimated that the earth housed about 40.000 plant and animal species and a few hundred minerals. Linnaeus succeeded not only in imposing order on all these plants and animals, but he also produced manuscripts and printed output, on a scale comparable with Georges Buffon's *Histoire naturelle* (1749–1788) (cf. Müller-Wille & Charmantier, 2012: 5).

Linnaeus' Animal Kingdom comprised six classes, initially distinguished by the form of the heart and the temperature and color of the blood. Linnaeus' division of fauna (a term he coined alongside flora) was influential. The system of nature was to Linnaeus the working plan underlying creation. That is why he tried to trace a *systema naturae*, first in botany, and then later in zoology and mineralogy. He wanted to represent most of Creation in a regular sequence (chain) and noted as many details as there were objects: "one may describe some remarkable details only, but if one observed detail fails to fit the chain is incoherent" (Linné, 1735/1964, Introduction: 9). Animals are the highest and most perfect works of the Creator: "I distinguished in Tetrapodologia [study of Quadrupeds], the orders of animals according to their teeth. In Ornitologia according to their bills, in Entomologia by their antennae, wings… etc." (Linné, 1735/1964, Preface: 26). He was the first who use dentition as a taxonomic characteristic to distinguish the different groups of Quadrupeds.

In the first edition of the *Systema Naturae* (Linné, 1735/1964), human beings, together with simians and sloths, were classified in the Quadruped Order Anthropomorpha (four incisors). The Anthropomorpha were divided into: *Homo, nosce te ipsum* ["know thyself"], and into *Simia* (*Simia cauda carens, Papio, Satyrus, Cercopithecus,* and *Cynocephalus*), and *Bradypus* [sloth], (*Tardigradus*) according to the number of fingers (see **Fig. 11.6**). It was only in the tenth edition (1758) that the Swedish naturalist adopted the terms primates and *mammalia* [mammals] (cf. Schiebinger, 1993 for a critical comment about the latter word), with mammals divided into orders and genera based upon the number and structure of their teeth. The tenth edition was particularly important as it established the binomial nomenclature in zoology and it is taken as the starting point for all zoological nomenclature. "No name, however binomial, published before 1758 is available for zoological nomenclature, and no work published thereafter is available unless it consistently uses the binomial system" (Groves, 2008: 88). In the tenth edition, the Order Primates is divided into four groups:

a. *Homo* (two species): front teeth four in number, canines separated from others; two parallel mamas in the breast;
b. *Simia* (21 species): front teeth four in number, canines separated from others;
c. *Lemur* (three species): front teeth four in number; lower front teeth six in number;
d. *Vespertilio* (seven species): hands palmate, used for flying.

This division is also maintained in the 13th edition (1788), organized by Gmelin.

In the 1735 edition, Linnaeus briefly notes that the human distinction from beasts was due to "reason", while in the tenth edition, he not only invents a new term for our species, *Homo sapiens*, but also he omits the term reason, describing the "nature" of *Homo* in a more complex (and maybe less flattering) way. In fact, the human being is not only an "*animal flens, ridens, melodum, loquens, docile, judicans, admirans, and sapientissimum*" but also "fragile, in his natural state without defence, subjected to all the attacks of fate", and "endowed with a precarious and anxious spirit, persistent in his hope, and endowed with a late wisdom" (Linné, 1758: 21). The two *Homo* species of

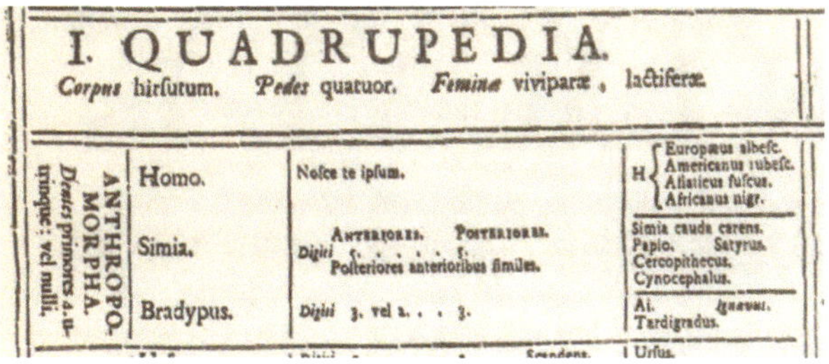
I. QUADRUPEDIA.
Corpus hirsutum. *Pedes* quatuor. *Feminæ* viviparæ, lactiferæ.

ANTHROPOMORPHA. *Dentes primores* 4. utrinque; vel nulli.	Homo.	Nosce te ipsum.	H { Europæus albesc. Americanus rubesc. Asiaticus fuscus. Africanus nigr.
	Simia.	*Digiti* Anteriores. 5. . . . . . Posteriores. 5. Posteriores anterioribus similes.	Simia cauda carens. Papio. Satyrus. Cercopithecus. Cynocephalus.
	Bradypus.	*Digiti* 3. vel 2. . . 3.	Ai. Ignavus. Tardigradus.

FIGURE 11.6 In the first edition of the *Systema Naturae* (Linnaeus, 1735), human beings together with simians and sloths were classified in the Quadruped Order Anthropomorpha (four incisors). The Anthropomorpha were divided into: *Homo, nosce te ipsum* ["know thyself"], and into *Simia* and *Bradypus* according to the number of fingers.

the tenth edition include *Homo sapiens,* defined as *diurnus* with five "*varietates*" (*Europaeus, Americanus, Asiaticus, Africanus*, and *Mostruosus*) and *Homo troglodytes* [a synonym for *Homo sylvestris,* the *Orang-outang*] which is defined by simple contrast to *Homo sapiens* as *nocturnus* (cf. Barsanti, this volume, for more details on Linnaeus' taxonomy of great apes).

The genus *Simia* includes the modern non-human Simiiformes, divided into three informal groups, in part in accordance with the earlier Aristotelian tradition based on the presence or lack of a tail (cf. Masseti & Veracini, this volume; Veracini, Chapter 9, this volume)[2] (the italics are those of Linnaeus; for the meaning of the primate names, see also Glossary):

a. *Cauda nulla* (no tail): Simiae *veterum* [the *Simia* of the Ancients].
Satyrus [=a great ape – orang-outang/chimpanzee, although for the ancient maybe refered to gibbons or langurs]; Sylvanus [=*Macaca sylvanus*].
b. *Cauda abbreviata* (short tail): Papiones.
Sphinx [=*Mandrillus* sp.]; Apedia [=*Saimiri* sp. – Linnaeus' identification based on a specimen that had lost its tail, see Groves, 2008: 94].
c. *Cauda elongata* (long tail): Cercopitheci.
Silenus [=*Macaca silenus*]; Faunus [*Cercopithecus diana* vel *roloway*]; Paniscus [=*Ateles paniscus,* confused with *Alouatta belzebul,* Marcgraf's guariba, cf. Veracini, Chapter 10, this volume]; Diana [= *Cercopithecus diana*, based on Marcgraf's *Cercopithecus barbatus guineesis,* cf. Veracini, Chapter 10, this volume]; Cephus [= *Cercopithecus cephus*, based on Marcgraf's *Cercopithecus barbatus alias guineesis* cf. Veracini, Chapter 10, this volume]; Aygula [= *Macaca fascicularis*]; Hamadryas [=*Papio hamadryas*]; Jacchus [=*Callithrix jacchus*]; Oedipus

[2] The identification of Linnaeus' taxa presented here, is based on comparison between different sources, among which those quoted by Linnaeus himself (e.g. Marcgraf and others), Buffon's work and Groves' work (2008: 92–96).

[=*Saguinus oedipus*]; Aethiops [=*Chlorocebus aethiops*]; Midas [*Saguinus midas*]; Cynomolgus [= cf. *Papio* sp., the "*cercopithecus angolensis maior*" of Marcgraf, cf. Veracini, Chapter 10, this volume]; Apella [= *Sapajus* cf. *nigritus*]; Morta [=*Saimiri* sp. according to Groves, 2008: 96]; Capucina [=*Cebus capuchinus*]; Sciurea [=*Saimiri boliviensis*, according to Groves, 2008: 96]; Syrichta [=*Carlito syrichta,* Philiphine tarsier, cf. Groves, 2008: 96].

The genus *Lemur* includes:
Tardigradus [=*Loris tardigradus,* based on one of Seba's specimens]*;* Catta [=*Lemur catta*, which has retained the same name!]; Volans [=colugos, Order Dermoptera].

The last genus, *Vespertilio*, is no longer considered a primate.

In conclusion, Linnaeus named 24 species of existing primates, 18 of which are still known by the names he gave them in 1758 (even if the genus was updated). The others are controversial or redundant: Aygula is now suppressed by the International Commission on Zoological Nomenclature (Groves, 2005: 162), and the others are not primates. Moreover, Linnaeus also combined different species into one (or overlooked the differences) (e.g., the guariba, howler monkey, of Marcgraf is not considered a separate species, but synonymous with the spider monkey, and Linnaeus' *Ateles paniscus* is described as a bearded monkey!). Some places of origin are wrong: the squirrel monkey was an Indian species, and the *Cercopithecus cephus* was believed to live in America because the source of Linnaeus was Marcgraf, who described mainly Brazilian fauna. Linnaeus was a botanist and did his best to classify exotic animals whose provenance was often misunderstood or unclear. The great German naturalist, Schreber, described graphically the difficulties encountered by naturalists of the time in doing this kind of work:

> it is very difficult for us to produce a proper classification [of primates]: most of those which arrive in Europe are mutilated, especially in the tail. [...] And, as we know, the number of monkey species according to travellers' stories is even greater.
>
> (Schreber, 1775: 63)

Linnaeus' undoubted merit, that of having found distinctive features for the class mammalia and the order primates, did not prevent him from following in his primate classification the classical division adopted by Aristotle (and then by the 16th naturalists) based mainly on the presence or absence of a tail. This led him to misunderstand many things and, for instance, to associate anthropoid apes with the Barbary macaque. In contrast, Schreber (1775) did not consider the tail as a relevant characteristic when describing cercopithecids (see later). Linnaeus acknowledged the work of the 16th century naturalists, and he especially praised the work of Aldrovandi and he might have drawn inspiration for his binomial system from Aldrovendi's nomenclature, who used a similar binomial form (Sabelli & Tommasini, 2004). For instance, the Barbary macaque, called by Aldrovandi *Simia Pithecus*, became the Linnaean *Simia Sylvanus* (see Veracini, 2019).

Linnaeus, like Buffon, updated his work regularly. In 1758, the book known as *Gleanings of Natural history* by the British naturalist George Edward (1694–1773) and the second edition of *Regnum animale* by the French zoologist Mathurin Jacques Brisson (1756–62), both important works in European natural sciences, were published. They contained descriptions of new primate species that were added to Linneaus' 12th edition of the *Systema Naturae* (Linné Part 1, 1766):

a new species without a tail: Inuus [=*Macaca sylvanus*];
a new species of "Papiones", Nemestrina [=the pig-tail macaque, *Macaca nemestrina* from Sumatra];
11 new species of Cercopitheci: Maimon [=*Mandrillus sphinx*]; Veter [=Unknown asiatic monkey]; Belzebul [=*Alouatta belzebul*]; Trepida [= *Sapajus* sp.]; Seniculus [=*Alouatta seniculus*]; Cynocephalus [=*Papio cynocephalus*]; Sabaea [=*Chlorocebus sabaeus*]; Pithecia [=*Pithecia pithecia*]; Nictitans [=*Cercopithecus nictitans*); Rosalia [=*Leontopithecus rosalia*]; Fatuellus [=*Sapajus* cf. *apella*];
Two more species of Lemur: the Edward's "mongooz" from Madagascar, Mongoz [=*Eulemur mongoz*]; the "Maucauco" from Madagascar, Macaco, [=*Eulemur macaco*].

In *Mantissa Plantarum* (Linné, 1767–1771, 2, appendix: 521), Linnaeus added some more primate species (cf. Groves, 2008: 101): Homo Lar [=*Hylobates lar*]; Sinica, [=*Macaca sinica*]; Nemaeus [=*Pygathrix nemaeus*]; and Argentata [=*Mico argentatus*]. Some wrong attributions are corrected (e.g., the confusion between the spider and howler monkeys), but still, there is no division between Old and New World monkeys, and the tarsier is still considered part of the simiiformes. Linnaeus was not able to distinguish macaques with a longer muzzle from baboons and vice versa, also because the tail morphology was considered relevant (e.g., the pig tail macaque, *Macaca nemestrina* was inserted in the Papione group – primates with short tail - and the hamadryas among long-tailed monkeys by virtue of the tail type). The 13th posthumous edition added more primate species, some taken from Buffon and Schreber and from descriptions of the travels of Bosman in Africa (1704) and Sonnerat in Madagascar (1782), for a total of 57 species (58 including the colugos; bats, still in the order primates, are not counted here). This edition included important prosimian taxa such as:

1. the potto, [=*Perodictus potto*] (already described by Pennant, 1771);
2. the mouse lemur, called *Murine maucaco* [=*Microcebus* sp.] (described as *Lemur murinus* by the English naturalist and illustrator John Frederick Miller, 1777: Plate 1325, cf. also Martin, 1972);
3. the indri, *Lemur Indri* [=*Indri indri*] is described according to Sonnerat;
4. the avahi, *Lemur Laniger* [=*Avahi laniger*] is described here for the first time.

In the 13th posthumous edition, an informal distinction between Old and New World monkeys is pointed out: the New World monkeys have "*buccis non saccatis, natibus tectis*", as highlighted by Buffon. New World monkeys were further divided into those with a prehensile tail and those who do not. As a consequence, the saki monkey is merged with the callithrichids. In Linnaeus' last edition, *Homo troglodytes* is called *Simia troglodytes*, with clear reference to the chimpanzee, and *Homo lar* is called *Simia lar* (the great gibbon of Buffon). The persistence of the redundant *S. sylvanus* and *S. inuus*, synonyms of the Barbary macaque, is interesting. It seems extremely strange that one of the best-known monkeys since antiquity, with a geographical distribution very close to Europe, was split into two different species by expert naturalists. Buffon also introduced two or three categories for the Barbary macaque, misunderstanding the word *cynocephalus* (see **Table 11.1**). One of the explanations for such uncorrected attribution could be found in the popularity enjoyed by the travel account by Prospero Alpini (1591), who talked – supposedly from first-hand experience- about various species of monkeys (*cynocephali*) he had seen in Egypt (cf. Veracini, 2020). This and other examples (e.g., the mandrill classified with three different names) help us to obtain a true picture of the difficulty at that time in classifying exotic animals with a group as diverse as primates. After all, new primate species are still being discovered today.

Both Linnaeus and Buffon have made substantial contributions to the natural history of primates. We still use the names and taxa coined and described by these two great scholars of the 18th century. Although Linnaeus' and Buffon's taxonomic systems start from opposite perspectives, their respective points of view drew closer as their work progressed. Toward the end of their lives, that is to say in their 70s, the contrast between the taxonomic methods of Linnaeus and Buffon was reduced, and their respective traditions were amalgamated in the works of their pupils. The followers of Linnaeus emphasized all those aspects of the taxonomic process that could facilitate identification, whereas Buffon and the French school focused on a better understanding of natural diversity. The Linnean school focused on discontinuity, whereas the Buffonian stressed continuity (Mayr, 1990: 121).

11.3 THE LEGACIES OF BUFFON AND LINNAEUS

In the second half of the 18th century, the publication of the works of Buffon and Linnaeus (tenth edition) provided a framework for new observations, thus stimulating the search for 'new' species (Rookmaaker, 1992: 131). Buffon's *Histoire naturelle*, which was widely translated and read throughout Europe, was particularly praised for its beautiful animal plates. Daubenton's anatomical contributions were rarely included, perhaps because, as Groves notes (2008: 78), "sales would suffer if the prospective customer encountered descriptions and engravings of livers, guts, or skeletons every dozen pages"! The organizers of these new editions, naturalists themselves, added new material in the form of personal observation in local cabinets, in markets, or from travelers (e.g., William Smellie in the 1785 English edition; Jean Nicolas Sébastien Allamand in the edition published in Holland, cf. Rookmaaker, 1992).

Linnaeus' *Systema Naturae* appeared in 12 authorized editions during his lifetime, followed by a final 13th posthumous edition, but Linnaeus' classification system was not widely used nor accepted until around the late 1760s, although his botanical work was reprinted many times and translated. So, it is important to consider Linnaeus' work concerning botany from the perspective of zoology. While the former was almost universally adopted – even in France, the country of his rival, Buffon - the classification of animals posed other questions. Two of the great issues in the zoological Linnaean system were the inclusion of humans in the order primates and the classification of Quadrupeds. Buffon's criticism of Linnaeus, especially regarding the classification of humans, is well-known (cf. Barsanti, this volume) as a threat to human dignity. Even Joahann Gmelin (1748–1804), the German naturalist who produced the 13th edition of the *Systema Naturae*, recalled Genesis (I: 26–27), where it was stated that "God created man in his own image…", and thus he could not be included among beasts. Many naturalists of the time, such as Pennant and Schreber, while admiring Linnaeus' work, declared themselves against the violation of the human-animal border (Corbey, 2008: 66). The German naturalist Philippe Louis Stace Müller, professor of philosophy at Derlang, although declaring that "man surely belongs to the animal kingdom and must place himself among the terrestrial quadrupeds", at the same time wrote that "man has an indisputable superiority over animals" and, above all, "God considered him lord of all animals" (Knorr & Müller, 1766: 93). In the 18th century, Johann Friedrich Blumenbach of Gottingen (1752–1840), a well-known naturalist considered to be the founder of physical anthropology, divided the order of primates into two parts, thus freeing the system from the unpleasant proximity of human beings with great apes. In his *De generis humani varietate nativa* (Blumenbach, 1775: 152), humans were placed in the Order Bimana, in which *Homo sapiens* was the only species, and all the other primates were in the Order Quadrumana. Blumenbach replied when someone said that even the other apes could stand upright: "it happened only occasionally and only a stick could make them stand upright".

Linnaeus' system was slowly adopted in Britain. One noteworthy aristocrat "expressed his dissatisfaction with elements of Linnaeus's system, describing it as a poor mix of natural and artificial systems" (cf. Huxley, 2003: 78). Some British naturalists of the second half of the 18th century, such as the Welsh naturalist, traveler, writer, and antiquarian Thomas Pennant (1726–1798), a fellow of the Royal Society, adopted a classification system that was a synthesis of the work of Buffon and Linnaeus. One of Linnaeus' classification issues was, as noted, the place of primates among quadrupeds. Many schemes were suggested before a common understanding about the characteristics of the different quadruped taxa was reached. As seen above, Buffon, too, had already claimed that this term was problematic. Pennant clearly states in the preface of his *History of Quadrupeds* (1793), an expanded version of his *Synopsis of Quadrupeds* published in 1771, that he would follow the general arrangement of Ray's *Synopsis of Quadrupeds* (Rajus, 1693), but he used different categories for some of the animals that Ray had classified together. He also contested the use of the number of teeth in classification, which had been used by Brisson in 1756 to separate different Quadruped categories. In the preface of the *History of Quadrupeds* (1793), Pennant stated that: "it must happen unavoidably that some quadrupeds, very distant from each other in their manner, are too closely connected in his system; a defect which, however common, should be carefully avoided by every Naturalist". He also praised Linnaeus' work regarding invertebrates but rejected his arrangement of mammals (*mammalia*) and most of its divisions (e.g., Primates, Bruta, and Faere) saying that he rejects Linnaeus' first division, "which he calls Primates or Chiefs of the Creation because my vanity will not suffer me to rank mankind with Apes, Monkies, Maucacous, and Bats, the companions Linnaeus has allotted us even in his last system". Speaking of Buffon's merits, he praises his beautiful use of language and agreeable manner and states that he had given the "amplest description of the economy of the whole four-footed creation". In fact, he says:

> we ought to drop, therefore, all thoughts of forming a system of quadrupeds from the character of a single part: but if we take combined characters, of parts, manners, and food, we will be much fairer for producing an intelligible system, which ought to be the sum of our aim.
>
> (Pennant, 1793: Preface)

In his division of quadrupeds, Pennant used both foot and tooth morphology as the main characteristics for classification. The divisions were: (a) hoofed; (b) digitated; (c) pinnated; (d) winged, and the subcategories of these were divided by tooth difference. Primates were included in the group of digitated, Section 1: Anthropomorphous. Primates [called "Apes"]:

> "have four cutting teeth in each jaw, and two canines. Each foot is formed like a hand, generally with flat nails, and, except in one instance, has four fingers and a thumb". They have Eye-brows above and below and they are: "A most numerous race. [...] Apes and parrots (the latter among birds) are more numerous in their species than any other animals"; thus, he says, "their numbers and their different appearances made it necessary to methodize and subdivide the genus".
>
> (Pennant, 1793: 94–95)

In primate classification, he makes a synthesis and elaboration of many authors' work, noting that:

> from Ray's primate division "*Linnaeus* formed his method; *M. de Buffon* followed the same; but makes a very judicious subdivision of the long-tailed apes, or the true monkies, into such which had prehensile tails, and such which had not. I shall endeavour in this genus no other reform in the system of our countryman, than what that gentleman has made; in respect to the *trivial* names of the species, I have in general invented such as I supposed congruous, or in a few instances retained those of *M. de Buffon*".
>
> (Pennant, 1793: 95–96, the italic is of the author).

Pennant's (1771) primate classification is thus the following:

DIV. II. Digitated Quadrupeds. SECT. I. Anthropomorphous

APE

1. **Without tails, the true APES.**
   - **Great [synonymous of]:** Satyrus, Pongo, Homo sylvestris, Baris, Barrys, Quojas morrou, Chimpanzee, Le jocko, Le Pongo, L'Homme de bois, Homo Troglodytes, Simia satyrus, L'orang outang.
   - **Pygmy [synonymous of]:** Pithecus (Aristotle), Simia, Ape, Le Singe, Le Pitheque, Simia sylvanus, S. ecaudatus, Le Singe commun...
   - **Long-armed:** Le grand Gibbon de Buffon [=*Hylobates lar*]
   - **Barbary** [=*Macaca sylvanus*]
   - **Tufted**: "A species of disgusting deformity...went on all fours; but would sit up on its rump, and support itself with a stick" [=*Mandrillus sphinx*, often portrayed with a stick by previous authors and associated to apes].
2. **Short tail or BABOONS**
   - **Ribbed nose** [=*Mandrillus sphinx*]
   - **Little** [= the Simia Apedia of Linnaeus]
   - **Pig-Tail** [=*Macaca nemestrina*]
3. **Long tailed or MONKEYS**
   - **A. those of the old world**, or the continents of *Asia* and *Africa*, having within each lower jaw pouches for the reception of their food.

Buttocks (generally) naked. Tails strait, not prehensile.

19 species [among which baboons, guenons, macaques, mangabeys, etc., and the *Tarsius* of Linnaeus]

B. monkeys of the new world, or the continent of *America,* having neither pouches in their jaws nor naked buttocks. Tails of many prehensiles, and naked on the under side, for a certain space next their end.

a. with prehensile tails

Seven species [among which howler, spider, and capuchin monkeys, and an uncertain species of squirrel monkey –*Morta* of Linnaeus]

b. with strait tails, not prehensile

Six species [among which the Saky monkey, tamarins, and marmosets]

MAUCAUCO

1. **Tailless** [=*Loris tardigradus*]
2. **Woolly** [=*Eulemur mongoz*]
3. **Ringtail** [=*Lemur catta*]
4. **Ruffed** o black macauco [= *Eulemur* sp.]
5. **Yellow** [= *Perodictus potto*]
6. **Flying** [= Cologus]

It is clear from Pennant's descriptions that he relied mainly on the literature at his disposal and not on any first-hand studies of his own. Nevertheless, by combining the most important aspects of Buffon's and Linnaeus' taxonomies, he produces a reasonable classification of the higher primate taxa. He also proposes an almost exhaustive description of the geographical range of non-human primates:

> almost confined to the torrid zone: fills the woods of *Africa,* from *Senegal* to the *Cape,* and from thence to *Aethiopia:* a single species is found beyond that line, in the province of *Barbary:* found in all parts of *India,* and its islands; in *Cochin-China,* in the S. of *China,* and in *Japan;* and one kind is met with in *Arabia:* they swarm in the forests of S. *America,* from the isthmus of *Darien,* as far as *Paraguay.*
>
> (Pennant, 1793: 94–95)

Nevertheless, he adopts the stereotypical traditional negative image of primates: they are "endowed with mischievous intellects: are filthy, obscene, lascivious, and thieving", moreover, they "are not carnivorous, but for mischiefs sake, will rob the nests of birds of the eggs and young". He also evokes the ancients' literature: "they are the prey of leopards and others of the feline race; of serpents, which pursue them to the summit of trees" (Pennant, 1793: 94–95).

German naturalists, while more inclined to follow the Linnean method, did not neglect the contributions of Buffon and produced various classifications of quadrupeds and primates. One of these was by the already mentioned German naturalist Johann Christian Daniel von Schreber (1739–1810), president of the German Academy Leopoldina and later a member of the Royal Swedish Academy of Science and fellow of the Royal Society. In 1774, he began writing a multivolume set of books entitled *Die Säugthiere in Abbildungen nach der Natur mit Beschreibungen* (I have consulted the French translation of 1775), which focused on mammals of the world. Many of the animals were given a scientific name for the first time, following the binomial system of Carl Linnaeus because Schreber believed that "even if Buffon did a great job, his description is too accordant to nature to be adopted".

Schreber is critical of the use of tail length or its absence, a characteristic always used for distinguishing the various groups of monkeys, because he observes:

> we have seen that in reality it cannot be adopted because in some cases a tail is not lacking, but it is short and can be covered by hairs, as can be seen even from observing the skeleton [...] and the tail increases by degrees in the various species.

Furthermore, by adopting Linnaeus' system, he states that:

> nature, as Linnaeus claims, does not recognize limits between the various forms of monkeys, baboons, guenons, sapajous, sagoin, rightly considered by Linnaeus of the same genus. However, we can use fictitious genera in order to understand what we are talking about at a first glance. Moreover, differences among them can be made in relation to their origin.
>
> (Schreber, 1775: 56)

In this sense, Schreber follows Buffon's 'biogeographic' concepts. He still uses the name quadrupeds, but divides them into various sections called genus. In the first, there are four genera of mammals, three of which are primates; the fourth is for bats (Schreber, 1775):

> *Genus I*
>
> Man (*L'Homme*), who comprises only one species, *Homo sapiens* alias *Homo diurnus.* The many varieties, and the other forms we observe are degenerations of the human race. The primitive and savage man is very different from civilized man. [...] The most adequate nutrition for man is fruit and other vegetables: that is proved by the organization and conformation of teeth and the structure of the stomach and intestine.
>
> *Genus II*
>
> Monkeys (*Les singes*): Simia (Linnaeus); Singe (Brisson); Ape (Pennant). With or without tail with: 4 incisor teeth for maxilla and mandible, of equal length and width not equal. Longer canines, 5 molars for each side. 4 legs, with 5 separate digits. These are sufficient characteristics to distinguish monkeys from the other mammals, but there are many other characteristics [...] including the resemblance to man.

This genus included 42 species, orang-utang and gibbons have two varieties, thus in total are 44 taxa, divided into the following groups:

a. without a tail [apes]
b. with a long or short tail, ischial callosities on the buttocks and cheek pouches [cercopithecids]
c. with a long prehensile tail (no ischial callosities on the buttocks and no cheek pouches) [atelids and capuchin monkeys]
d. with a long, straight tail (no ischial callosities on the buttocks and no cheek pouches) [Saki monkeys, marmosets, and tamarins]

Genus III

Le Maki: Lemur (Linnaeus); Prosimia. Maki (Brisson), Maucauco (Pennant).

4 front upper teeth, 6 lower front teeth. [...] The animals that belong to this genus resemble monkeys for their way of jumping and clinging and for their habits and feeding, but they have an elongated snout that looks more like that of a fox; [...] they have a different shape of head and a different number of teeth and mamas; they are closer to the opossums and can be seen as an intermediate genus between them.

This genus includes six species, among which one is the Colugo (Dermoptera).

Some of Schreber's primates are redundant (such as Le Cynocephale/*Simia inuus* and *Simia sylvanus,* the same mistake made by Linnaeus and Buffon), some are misinterpreted but are in accordance with the authors who first described the species. Schreber named some species using binomial nomenclature: *Cercopithecus* (*Simia*) *petaurista, Cercopithecus (Simia) mona, Cercopithecus* (*Simia*) *roloway,* and *Miopithecus (Simia) talapoin.* Schreber's primate descriptions include beautiful plates, some from Buffon, Edwards, Seba's cabinet, or his own experience, i.e., "from nature". The difficulty of understanding primate biodiversity, already noted by Schreber himself, is clearly shown in the list of synonyms (the names given to each species by different authors) provided for each taxon (see **Fig. 11.7** as an example).

Another German naturalist, Johann Christian Polycarp Erxleben (1744–1777), Professor of Philosophy at Gottingen, in his *Systema Regni Animali* (1777) adopted the Linnaeus' divisions of mammals and primates. He revised many of the known primate species, introducing some new genera to distinguish the different informal taxa that had been delineating in the works of previous naturalists (bats and colugos are still included):

Genus [the Italics follow those of Erxleben]:
Homo: sapiens
Simia: Satyrus, Silvanus, Longimana [gibbons] Inuus
Papio: Sphinx [indeterminate baboon], Maimon [*Mandrillus sphinx*], Mormon [a misunderstood mandrill, the tufted apes of Pennant], Nemestrina, Apedia
Cercopithecus: Hamadryas, Veter, Senex, Vetulus, Silenus, Faunus, Cynomolgus, Cynocephalus, Diana, Mona, Sabaea, Patas, Nictitans, Petaurista, Talapoin, Cephus, Aethiops, Aygula, Maura, Sinicus, Roloway, Nemaeus
Cebus: Belzebul, Seniculus, Paniscus, Capucinus, Apella, Trepidus, Fatuellus, Sciureus, Lugubris [the *Tarsius* of Linnaeus]
Callithrix: Pithecia, Jacchus, Oedipus, Rosalia, Argentatus, Midas
Lemur: Mongoz, Macaco, Catta, Flavus, Tarsier, Volans [not a primate]

13.

LE MACAQUE.

Pl. XIII.

Simia Cynamolgos; Simia caudata imberbis, naribus bifidis, elatis, cauda arcuata, natibus calvis. LINN. *ſyſt. p.* 38. *n.* 15. &

Simia Cynocephalus; Simia caudata, imberbis, flaveſcens, ore producto, cauda recta, natibus calvis. *ibid. n.* 16.

Cercopithecus cynocephalus, naribus bifidis, elatis, natibus calvis. BRISS. *quadr. p.* 152. *n.* 2. &

Cercopithecus cynocephalus, ex vireſcentibus & flavicantibus pilis variegatus. *ibid. p.* 151. *n.* 1.

Cercopithecus Angolenſis major, Congenſibus macaquo MARCGR. *Braſil. p.* 227. RAJ. *quadr. p.* 155.

Cercopithecus. GESN. *quadr. p* 856. avec une figure très médiocre.

Cercopithecus. JONST. *quadr. tab.* 59.

Troiſiéme ſapajou. *Mem. de l'acad. des ſciences de Paris tom.* 3. *part.* 2. *p.* 55. *Pl.* 43. La figure d'enbas à moitié érigée.

Macaque. BUFFON *hiſt. nat. tom. XIV. p.* 190. *Pl.* 20.

Hare-lipped ape. PENN. *ſyn. p.* III. *n.* 74.

Macaquo à Congo. MARCGR.

FIGURE 11.7 Example of the difficulty of understanding primate biodiversity in the 18th century, here exemplified by the list of synonyms provided for the macaque taxon by Schreber (1775).

Erxleben recognized Buffon's tarsier as a primate and placed it in the genus *Lemur* (*Lemur tarsier*). Erxleben also adopted the genus *Papio* for baboons, mandrills, and drills and the genus *Callithrix* for callitrichids – including the genus *Pithecia*, (Erxleben, 1777: 24). His "Cercopithecus Vetulus", the Ouanderou of Buffon, is *Trachypithecus vetulus*.

Many other naturalists described primates in the second half of the 18th century. These scholars added new species (or taxa) that were being discovered, or they re-described existing species ones using the Linnean binomium. Some, when translating Linnaeus' *Systema Naturae*, inserted new species. For example: the geographer and naturalist Eberhard August Wilhelm Zimmermann (1743–1815), who traveled in the Old and New World, described Cercopithecus mulatta [=*Macaca mulatta*] (cf. Groves, 2008: 103); the Italian naturalist and physician, Giovanni Antonio Scopoli, described in 1786 *Cercopithecus cynosuros* [*Chlorocebus cynosuros*] (**Fig. 11.8**); the French

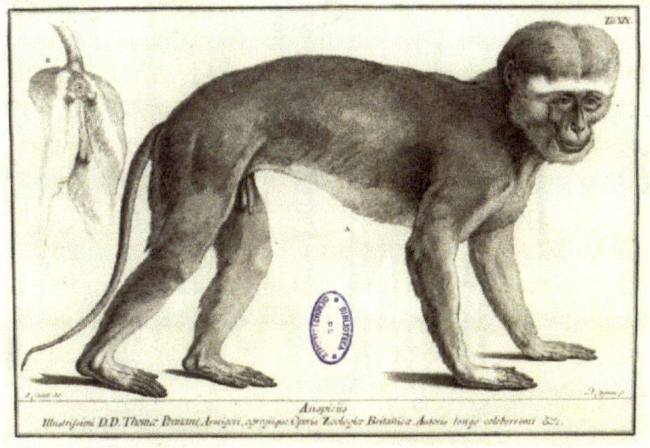

FIGURE 11.8 *Chlorocebus cynosuros*, the Malbrouck, described by the Italian naturalist and physician, Giovanni Antonio Scopoli in 1786.

naturalist, Audebert, described *Hylobates moloch* (Audebert, 1799: 1st facs sect. 2 pl.2); and the physician Pieter Boddaert (1785: 67), who owned a natural history cabinet, described the slow loris, *Nycticebus coucang* (see Buffon's plate, **Fig. 11.9**).

By the end of the 18th century, about 50 or more species of non-human primates had been described. Most important higher taxa, such as 'true apes', New and Old World monkeys, had been recognized, some genera of prosimians and the tarsier taxon had also been described. Primates were included in the Class of Mammals (though not all authors agreed), and the human being had been recognized as an animal. Linnaeus' work had introduced a simple system of classification, but it focused on the genus, and it was vague and inconsistent about intraspecific variation (Mayr, 1990: 123). In contrast, Buffon had weight to this issue, stressing both biogeographic and behavioral aspects. Other naturalists synthesized their thoughts, paving the way for 19th century primate natural history, which included morphology, physiology, anatomy, behavior, and biogeography and also laid stress on primate variation and to the phenomenon, which generate it, in other words, they paid attention to evolution.

11.4 ETIENNE GEOFFROY SAINT-HILAIRE AND OTHER NATURALISTS OF THE 19TH CENTURY (BEFORE DARWIN)

At the end of the 18th century, natural history underwent a profound transformation. While the natural history of the previous centuries had not been clearly divided into zoology and botany (although specializations could be found in some cases, e.g., Pierre Belon, 1553 *De aquatilibus*), from 1800 on, no taxonomist was able to cover both the animal and vegetable kingdoms, and thus different scientific traditions developed (Mayr, 1990: 121). Zoology slowly split into separate sub-disciplines such as physiology or palaeontology, each with its own methods, agendas, and subject matter (Outram, 1996: 249). The steady accumulation of

FIGURE 11.9 *Nycticebus coucang* as described by Buffon. Le Loris de Bengale. (Courtesy of the National Library of Portugal, Lisbon.)

animal specimens in public and private collections led to a proliferation of taxonomists who specialized in the study of only one or a few taxa. New techniques were developed that enabled the protection of animal collections from damage by moths and dermestids, and new methods of taxidermy were developed, facilitating the emergence of large museums of natural history (Mayr, 1990: 119). Interest in animal classification was also driven by new data from scientific expeditions, many of which were organized by colonial nations in order to consolidate their borders and gather information on natural resources. Many specimens were collected by field naturalists (e.g., Ferreira, Humboldt) and by the '*naturalistes-voyageurs*', individuals sent out by the Museum of Natural History of Paris to search for new and strange plants and animals (cf. Laissus, 1981). Traveling naturalists and collectors were also popular in countries such as Great Britain, where the majority of naturalists considered non-native organisms more interesting than those found at home. This and other causes led to an expansion of overseas activities concerned with natural history (Browne, 1996: 306). The scientific expeditions undertaken in South America, and especially in Brazil, whose immense territory had been closed to foreigners by Portuguese authorities for centuries, aroused much interest. The quest for a universal classification method was accompanied by another important field of inquiry, which had been latent throughout the second half of the 18th century: how the variation of species originated. In their later publications, both Buffon and Linnaeus had begun to reflect on the nature and implications of variation. The philosophical and political environments at the end of the 18th century helped this process, especially in France. The French Revolution emphasized the idea of "progress", a natural and progressive change to which humanity tended (Weckesser, 1980), paving the way for the concept of evolution. Although new knowledge about nature continued to be produced by amateurs, at the same time, "natural history was increasingly given the backing of major new state-funded and state-controlled institutions staffed by paid full-time expert researchers" (Outram, 1996: 249). These institutions were, for instance, natural history museums, the Natural History department of the British Museum and the Zoological Gardens in London, the zoological museums of the new University of Berlin, and, above all, the Muséum National d'Histoire Naturelle in Paris. Before the French Revolution, the city of Paris had become, with naturalists such as Buffon, Lacépede, Daubenton, and Brisson, an important center for the study of natural sciences, and this trend continued after the Revolution and during the Napoleonic period. After the revolution, in 1793, the *Jardin du Roi* was transformed into a *Muséum d'Histoire Naturelle* by decree of the Convention, which was fashioned into one of the leading scientific institutions in the world. It attracted researchers from outside France and was the home of many of the leading names in natural history in this period, such as Georges Cuvier, Jean-Baptiste Lamarck, and Etienne and Isidore Geoffroy Saint-Hilaire (Outram, 1996: 250). The natural history of primates benefited from this time of turmoil, and in a few decades, it would encompass a good portion of primate biodiversity, highlighting most of the primate characteristics currently recognized.

In June 1793, Etienne Geoffroy Saint-Hilaire (1772–1844), after the period of terror following the French Revolution, was appointed one of the twelve professors of the newly constituted *Muséum National d'Histoire Naturelle*, being assigned the chair of zoology. He also had the support of the anatomist Daubenton, who had worked with Buffon. Geoffroy Saint-Hilaire's scientific work was favored by the political moment. Napoleon Bonaparte, who came to power in 1799 with an army coup, through his expansionist policies enabled French science to extend beyond its border. Geoffroy was able to collect specimens during Napoleon's invasion of Egypt and when, in 1794, French forces invading the Netherlands took possession of the Prince of Orange's natural history cabinet - though it seems that Geoffroy only took with him to Paris "duplicate" specimens (cf. Groves, 2008: 113). Later on, he was sent to Lisbon by Napoleon to choose specimens from the Portuguese natural history collections on behalf of the Paris Museum. Marshal Junot, who commanded the invasion in Portugal after his entry into Lisbon in 1807, sent Geoffroy to the Royal Museum of the Ajuda, full of specimens from the Portuguese colonies, expressly in order to select and pack up everything he thought was worthy of appearing in the Paris museum. It was in this occasion that 400 animals, 3,000 mineralogical specimens, and a herbarium of 2,000 plants were brought to France (Vilhena Barbosa de, 1885; Azevedo, 1921: 14). Most of these objects were of great interest, especially those from Brazil. Likewise, it was reported that "the menagerie [of Lisbon] was completely emptied by Junot" (William Stothert *apud* Azevedo, 1921: 14). It is said that Geoffroy was a conscientious inspector of the many objects he encountered in museums, libraries, and Portuguese convents, of which, he took only away the copies (Azevedo, 1921: 20). Later, in 1860, the director of the Museum of the Polytechnic School of Lisboa, Barbosa de Bocage, negotiated the return from Paris of duplicates of the stolen specimens, and for this he warmly thanked the son of Geoffroy Saint-Hilaire, Isidore (Azevedo, 1921: 25). Many of the specimens brought to France by Geoffroy had been collected by Alexandre Rodrigues Ferreira in his famous journey, *Viagem Filosofica,* in the Amazon (Ferreira, 1783–1792, cf. Ferreira, 1971). Portuguese naturalists of the 19th century admitted the specimens taken by Geoffroy had been more valued by him than if they had remained in the Ajuda Museum. Among the 26 primate specimens brought to Paris, 25 are type specimens (cf. Ceríaco, 2014: 320). Geoffroy's great interest in primates, also helped by the opportunities referred to above, allowed him to make a significant contribution to primate natural history in terms of taxonomy, comparative anatomy, physiology, biogeography, and a subdiscipline corresponding to what today would be called eco-ethology. In a paper entitled *Histoire des Makis, ou singes de Madagascar*, written in 1795, he describes both known and new species of lemurs, lorises, galagos, and tarsiers. Most names used by him were new, while others were based on the work of Buffon and other authors. In this paper, we find for the first time the generic name *Loris*, and the author uses bushbaby when describing *Galago senegalensis,* (Groves, 2008: 113). Other works on mammals and non-human primates followed (e.g., Geoffroy, 1801, 1812; Geoffroy & Cuvier, 1798), many written with the help of George Cuvier (1769–1832), known as the founding father of palaeontology. The two authors put forward the idea of the 'subordination of characters/parts' upon which Cuvier based his zoological system. In the *Histoire des Makis, ou singes de Madagascar*, Geoffroy gave expression to Cuvier's views on the unity of organic composition, the influence of which is perceptible in his subsequent writings (see below).

Geoffroy was a comparative anatomist and he and Cuvier proposed a reform of mammal classification (Geoffroy & Cuvier, 1798). In a remarkable series of lectures delivered in the 1820s, but published in 1834 (Geoffroy, 1834, lesson IV: 26), Geoffroy explained that he and Cuvier had tried "to fill the scientific need" of classifying four-footed animals with a new approach. In this system, non-human primates were called Quadrumanes [=four-handed animals]:

> The Quadrumanes have four feet somehow metamorphosed into hands. However, the whole limbs participate

in this modification. The anterior limbs are attached to a complete shoulder, in which the front scapular bone is a strong, resistant, long and perfect clavicle as in humans. [...] A plant diet is imposed because of the existence of a very wide stomach and intestines of a medium length, and visibly because of their three types of teeth and the particular structure of their molars; the wide crown can only be used to grind grains or crush vegetables. [...] The head of the Quadrumanes is remarkable for similarities with that of man, [...] even in the forms with a very elongated jaw, all the main characteristics of the human head are preserved.

(Geoffroy, 1834, *Lesson 5:* 8–9)

In 1812, Geoffroy published the *Tableau des Quadrumanes* in which he divided non-human primates into two groups, Catarrhini (*Singes de l'Ancient continent*) plus Platyrrhini (*Singes de l'Amérique*) with the second group the *Makis* or Strepsirrhini (Lemuriens). The Strepsirrhini included all the prosimians plus the tarsier, although today we know the latter is a haplorrhine. This new division, based on the olfactory system, was truly innovative:

Primates (*singes*) have a distinct physiognomy especially for what concerns the olfactory system. The primates of the ancient world have nostrils arranged like those of man, narrow and open at the bottom, and we call them Catarrhini, therefore; the American ones on the contrary have a very wide nose, and nostrils open laterally, and are called Platyrrhini. [...] The third group, consisting of *Makis,* with their sinuous nostrils, are because of this called Strepsirrhini. The latter group are mostly found in Madagascar (with some exceptions).

(Geoffroy, 1834, *Lesson 5:* 9–10)

Geoffroy's work on primates shows how the study of mammals was evolving into a complex discipline embracing physiology, comparative anatomy, and data about structural and behavioral adaptation. All of these lines of evidence were taken into consideration when formulating the classification of primates, and in this sense Geoffroy continued the tradition of the Buffonian school by seeking an overall, not a reductionist, vision of animals. Geoffroy admits "Buffon did better than his predecessors, distinguishing important characteristics in the various groups of *singes*, such as cheek-pouches, ischial callosities [bare sitting pads] on the buttocks, and the prehensile tail in monkeys of the New World". The human being, with its structural and functional attributes, did not stay far from Geoffroy's interest and was studied in parallel with the Quadrumanes. For example, speaking of human bipedalism, Geoffroy says that this kind of locomotion:

is in fact the most convenient given the unique and peculiar position of our head with respect to the trunk. And for this reason, even if there are monkeys walking on two feet, they do not have that harmony that we have and shamble.

(Geoffroy, 1934, *Lesson 5:* 30)

Geoffroy also gives a definition of the order Quadrumanes:

Human form, a spherical skull, a frontal face. Complete orbital bursts, directed forward, and separated by temporal pits. Teeth of three types (incisors, canines and molars): flute-shaped incisors; conical canines; molars with wide crowns and tubercles. Simple stomach, medium-sized intestine, short cecum. Two pectoral breasts. Penis and testicles hanging out. Chest wide as well as deep. Perfect clavicles. Weak limbs as support, disposed to jump and hold. The two bones, both of the arm and of the leg, of the same width, articulated together but not welded; the bones of the foreparts, susceptible to pronation and supination movements. Feet composed of mobile parts, and deeply divided at the end; fingers free in their movements; the thumb, opposite to the others, enjoys independent movement.

(Geoffroy, 1834, *Lesson, 6*: 23–24)

The different manifestations of non-human primates were illustrated in the lectures given in the 1820s. Here Geoffroy explains that in addition to the shape of the nose, he also used the form of the skull and the teeth to distinguish the various taxa, and he emphasizes that many mistakes made in the past were due to the fact that the studies were carried out on young animals, in which the growth of the brain case preceded that of the muzzle.

The shape of the head changes considerably according to age, as well as in the human species, and for a long time we have paid attention to the metamorphosis it undergoes, being very round in a child. [...] But it was not known until recently that this change also exists in various monkeys. The changes in proportions are such that they naturally led naturalists to commit various errors, and thus they have increased the list of species.

(Geoffroy, 1834, *lesson, 5:* 21)

This insight leads him to reclassify the species of orang-outang, previously divided because young specimens were considered a different species from adults (see Barsanti, this volume for more details). Geoffroy's work was deeply influenced by Cuvier, who claimed that physiological systems were of such importance that they influenced a range of morphology. This was a significant conceptual innovation, for earlier anatomists thought each characteristic had a different essence (Mayr, 1990: 133). Cuvier developed this idea in the principle, or rule, he called the 'correlation of the parts', which allows you to use one aspect of an animal's morphology to predict the rest of its morphology, and thus its behavior and ecology. Geoffroy reports the description given by Cuvier of the genus *Semnopithecus*, based on its tooth shape and complex stomach:

molar teeth of *cercopitheci* [=guenons] are tubercolated, while those of *semnopitheci* are always destroyed by usage. [...] The difference between *semnopitheci* and guenons lies in the diet; guenons feed mainly on fruit and roots while the others feed mainly on leaves, and have been considered by M. Otto as ruminant monkeys, both because of their type of diet and because of their very complicated stomach; this is a hypothesis based on facts, but it is far from being admitted in science. As for guenons, their stomach is much simpler and of medium size. They also possess large cheek-pouches which they use to hold the fruit they take from gardens and trees, and then, after filling them, they eat in the trees. Guenons are well adapted to life in trees: that's there where they withdraw when they are in danger and jump from one branch to another until they disappear.

(Geoffroy, 1834: *Lesson 8:* 4)

Geoffroy also emphasized primate locomotion. He correlates the different types of primate locomotion to anatomical differences, and, above all, to adaptation to arboreal life. The "ambiguous nature of monkeys", he says:

> who can both take upright posture and walk on all four limbs is justified by the fact that they live in trees. This kind of life leads them to jump in continuous agitation from one branch to another, and for this they require the use of many muscles also, to enable them to hang in various forms to the branches.
>
> (Geoffroy, 1834, *Lesson 6:* 13)

"*Les Singes*", he says, show a certain "inflexibility in behavior (*moeurs*) that depends above all on their life in trees, a situation that is undoubtedly uncomfortable but that they have adopted because of their instinct for survival" (Geoffroy, 1834, *Lesson 6:* 17). The idea of an ecological niche also seems developed in Geoffroy, for he correlates anatomy and environment with primate behavior. For Geoffroy, in fact, the environment plays a fundamental role. As a colleague of the evolutionist Jean-Baptiste de Lamarck (1744–1829), he expanded and defended the latest evolutionary theories. Geoffroy provides an explanation for animal variability and the consequent modification over time of organisms, providing a physiological explanation related to the effect of the environment on respiratory fluids: "this phenomenon, he argues, produces an impact on the structure of the organism" - the environment thus has a direct influence on organic changes (this aspect was actually rejected by Lamarck, cf. Mayr, 1990: 307). Geoffroy does not believe in descent with modification from a common ancestor, but he believes that species started to change after the universal flood due to external influence. Useful modifications were maintained, while species with deleterious modifications were eliminated.

The notion of the physiological basis of primate behavior seems to emerge from Geoffroy's work. For example, he explains the different levels of maternal (higher) and paternal (lower) care found in a siamang living in the Paris menagerie.

> Analysing the brains of a female and a male siamang, I notice this difference: the females have their occipital lobes more developed than the male and this could perhaps distinguish the behaviour of the most active female at the time of maternity, from the male considering that this species in general is known for its indolence and stupidity.
>
> (Geoffroy, 1834, *Lesson 6:* 15)
>
> Currently we know that the main function of the occipital lobe in primates is that it contains the primary visual cortex; in any case, this example shows the sophistication of Geoffroy's analysis.

Among the many species described by Geoffroy, the aye-aye (*Daubentonia madagascariensis*) deserves a special mention. He sets the aye-aye apart in its own genus, which, in gratitude to his mentor Daubenton, who was the first to dissect this animal, called it *Daubentonia.* This species had been classified by Gmelin (1789, 1: 152) as a squirrel (*Sciurys madagascariensis*), Buffon placed it far from monkeys and lemurs while Daubenton noted its similarity to the tarsier (see above). In his anatomical lesson, Cuvier (1800) calls it *Cheiromys,* which is "rat-à- mains". In the first decades of the 19th century, the classification of the aye-aye remained uncertain because naturalists were divided in considering this species, on the one hand, more similar to quadrumanes (non-human primates) and on the other, rodents. Geoffroy supported the latter assessment, and the aye-aye did not appear in the work of 1795 nor in Geoffroy's later works on primates. On the contrary in Paul Gervais's work (*Historie Naturelle des Mammifères*, 1854: 173–174), the aye-aye is included in the Quadrumanes, Family Cheiromydés, Genus *Cheiromys* (G. Cuvier) and clearly described in anatomical and morphological terms (**Fig. 11.10**). In 1863, the British zoologist John Edward Gray, after having used Cuvier's name, coined the family name Daubentoniidae.

The extensive work of Geoffroy and Cuvier brought substantial improvements to our understanding of primate natural history. While other species and genera were being added to those already described, many, if not most, of the higher primate taxa had been established. Many books on the natural history of primates and other animals, including various encyclopedic compilations of the knowledge produced on animals and full of beautiful plates, were published. The plates illustrating primates in Buffon's publications were reproduced in many forms and with many

FIGURE 11.10 The aye-aye was included by Paul Gervais in the Quadrumanes, Family Cheiromydés, Genus *Cheiromys* (G. Cuvier). (Source: Paul Gervais, *Historie Naturelle des Mammifères*, (1854: 173–174). (Image of public domain).)

variations, emphasizing the importance of his work to both the scientific community and the general public.

To give a sense of the level of knowledge about primates around the middle of the 19th century, we conclude this pre-Darwinian section by discussing the important work of John Edward Gray (1800–1875), a fellow of the Zoological Society of London and a researcher at London's British Museum,. In 1821, at the beginning of his career, Gray proposed a new classification of vertebrates where humans were separated from other primates (i.e., Class I. Bimanes, Order Primates, with no family name, only one genus: *Homo*), but in his later publication, humans were put in the same family as great apes and Old World monkeys. This change was perhaps consequent to the observation of the many animals hosted in the London Zoo (considered "the Society's Gardens"), founded in 1826 by Sir Thomas Stamford Raffles (1782–1826), an important British functionary who during his stay in Eastern Asia collected every sort of animal (cf. Noltie, 2009). Gray inserted a new rank, the Family (Gray, 1825), between the Order and the Genus. This rank had been used before, but informally, and still today it is accepted as a formal rank between Order and Genus. As Groves (2008: 130) states "his Hominidae taxon is exactly equivalent to modern Catarrhini". Here, we present the primate classification of Gray (1825) *apud* Groves (2008: 129, modified):

Order I. Primates
Anthropomorphous
Family I. Hominidae
Tail none: (**a**) Hominina, *Homo*; (**b**) Simiina, *Troglodytes* Geoff., *Simia* Lin., *Hylobates Illiger*
Tail long or short: (**c**) Presbytina. *Presbytes* (sic) Eschy; (**d**) Cercopithecina. *Lasiopyga* Illig., *Cercopithecus* Lin., *Cercocebus* Geoff., *Macacus*. (**e**) Cynocephalina. *Cynocephalus* Brisson, *Papio* Brisson.
Family II. Sariguidae
Tail end naked: (**a**) Mycetina. *Mycetes* Illig. (**b**) Atelina. *Ateles* Geoff., *Brachyteles* Spix. *Gastrimargus* Spix, *Lagothrix* Geoff.
Tail end hairy: (**c**) Callithricina (sic). *Cebus* Erxl. (**d**) Saguinina. *Saguinus* Lacép.
Nyctipithecus Spix, *Pithecia* Geoff., *Brachypus* Spix. (**e**) Harpalina. *Jacchus* Geoff., *Midas* Geoff.
Quadrupedoid
Family III. Lemuridae
Head long; grinders blunt: (**a**) Lemurina. *Lemur* Lin. (**b**) Lichanotina. *Indris* Lacép., *Lichanotis* Illig.
Head round: (**c**) Loridina. *Loris* Geoff., *Nycticebus* (sic) Geoff. (**d**) Galagonina. *Otolienus* (sic.) Illlig., *Galago* Adams, *Cheirogallus* (sic) Geoff. (**e**) Tarsina. *Tarsius;* (**f**) Cheiromina. *Cheiromys* Cuv.
Family IV. Galeopithecidae [cologus]
Family V. Vespertilionidae [bats]

Gray outlived the publication of *The Origin of Species* and attacked Darwin for having reproduced the equivalent of Lamark's doctrine, strongly contested by English naturalists of that time. Gray continued to recognize and describe additional species, but as he worked all his life in England in the British Museum, he was not always aware of what was going on in the rest of Europe, a point emphasized by Groves, who wrote that:

> he put the siamang into its own genus, but someone else had got there first. He described the blue monkey as new, but several others had done so years before. When he realized his errors, he could often admit them, and was prepared to sink names he had coined earlier.
>
> *(2008: 132)*

11.5 CONCLUDING REMARKS

The natural history of non-human primates played an important role in 18th and 19th century western science. Indeed, it was in this period that the thorny question of *similitudo homini* with other primates and consequently, the 'place of humans in nature' was addressed in scientific terms. As argued by Raymond Corbey, metaphysical assumptions – fundamental beliefs guiding the classification and interpretation of data – about the status of non-human primates have figured as conceptual or theoretical foundations in Western scientific research (2008: 35). The nature of 'monkeys' has never been studied in a neutral way by Western science because of its philosophical, religious, and socio-political implications (cf. the other chapters in this volume). The achievements of the last decades in the philosophical, historical and anthropological fields have made it possible to highlight the key points of the western theoretical framework that nurtured a natural history of primates rooted in racist and sexist beliefs and anchored to dichotomous keypoints such as the uniqueness and superiority of 'man' *vs* other primates and the binary idea of 'nature' *vs* 'culture' (e.g., Haraway, 1989; Rodman, 1990; Descola, 2006; Marks, 2012; Wilson, 2023). We must take all this into account when we examine the natural history of primates in the 18th and 19th centuries: the nature of 'monkeys' was analyzed and weighed by naturalists responding to different philosophical approaches and social pressures that determined different methodologies of analysis (cf. Barsanti, this volume).

Different visions produced various natural histories and taxonomies, fuelled also by the many new types of primates arriving in Europe alive and by the great number of first-hand descriptions, which appeared in the rich body of travel reports of that time. This combination of factors made it difficult by the 18th and 19th centuries to use the classification of primates handed down by Aristotle and still used, albeit revised and corrected, by naturalists of the 16th and 17th centuries, and furthermore to deal with anthropoid

FIGURE 11.11 *Le guenon a crinier* [*Macaca silenus*]. Figure drawn from the life as Buffon (1789) said. He saw it because it lived many years at the menagerie of Roy at Versailles.

FIGURE 11.12 *L'indri* [*Indri indri*]. From Sonnerat (1782). *Voyage aux Indes orientales et à la Chine.*

apes. Moreover, both human and non-human primates had to be placed in a coherent classification system within the broader group of vertebrates, traditionally defined as 'four-footed beasts'. This, too, created embarrassment among naturalists and, following Linnaeus' assumption that put humans in the same taxon as other primates, the term bimanes was coined in order to separate humans from other animals. As a consequence, humans occupied a place among other animals but were distinct from them.

Although the positioning of humans appears central to the systems proposed by 18th century naturalists, the classification of other primates was also problematic. In the final years of the 18th century, we find a proliferation of names, combined with changes in the way the names were used. This created so much confusion that it became almost impossible to identify the primate being referred to, and naturalists were compelled to compile long lists of synonyms, as well as adding the names of the naturalists who had described the various forms and the titles of their publications.

At the beginning of the 19th century, the extensive work of Geoffroy Saint-Hilaire and Cuvier brought substantial improvements to our understanding of primate natural history. Most of the higher primate taxa had been established, and primates appeared in encyclopedic compilations of knowledge, which were enhanced by a variety of beautiful plates. It was with the work of Gray that human beings were placed in the family Hominidae along with the great apes and gibbons (the equivalent of the today Superfamily Hominoidea) (**Figs. 11.1.** and **11.12**).

ACKNOWLEDGEMENTS

I would like to thank Jorge Contreras Martin.z, for the precious book on Buffon, kindly donated; Giulio Barsanti for the valuable teaching; Hilary Morris for the English revision; and Bernard Wood for the important suggestions, comments, and final editing of the work. I also thank the National Library of Portugal for permission to use some of the Buffon's images.

FUNDING

Portuguese national funds through FCT—Fundação para a Ciência e a Tecnologia in the context of the celebration of the contract under the 'Norma Transitoria', DOI 10.54499/DL57/2016/CP1456/CT0001.

REFERENCES

Primary Sources

Acosta, J. de, (2002). *Historia natural y moral de las Indias.* F. J. Alcina (Ed), Cronicas de Americas, Dastin Historia (Original work published in 1590).

Alpini, P. (1591). *De medicina Aegyptiorum*. The Bavarian State Library.

Audebert, J. B. (1799). *Histoire Naturelle des Singes et Makis*. Edité par Paris: Desray.

Blumenbach, J. F. (1775). *De Generis Humani Varietate Nativa*. Gottingae: Frid. Andr. Rosenbuch.

Boddaert. (1776). Elenchus Animalium, p-66. 1785: Prosimia minima: Brown's *"little Maucauco" New illustr. Zoolo*, p. 107, pl. 44.

Boddaert, P. (1785). *Elenchus animalium, volumen I: sistens quadrupedia huc usque nota, eorumque varietates: ad ductum naturae, quantum fieri potuit disposita*. Roterodami: Apud C.R. Hake, 1784.

Bosman, W. (1704). *Nauwkeurige beschryving van de Guinese Goud-, Tand- en Slave-Kusk*. Utretch, Schouten. (English translation: London, Printed for Sir Alfred Jones by Ballantyne, 1907).

Brisson, M. J. (1756–62). *Regnum animale in classes IX distributum sive Synopsis ardencal*. Paris, Leiden: Haak.

Buffon, G. L. (1766). *Historie Naturelle*, Vol. XIV. (Quadrupèdes XI). *Nomenclature des Singes*. Paris de L'Imprimerie Royal..

Buffon, G. L. (1767a). *Historie Naturelle*, Vol. XV (Quadrupèdes XI). Paris de L'Imprimerie Royal.

Buffon, G. L. (1767b). *De la dégénération des Animaux*, (Quadrupèdes XII). Paris de L'Imprimerie Royal.

Buffon, G. L. & Lacépède, B. de (1789). *Suppléments, vol. VII : Servant de suite à l'Histoire des Animaux quadrupèdes*, (H.N. Vol. XXXVI). Paris de L'Imprimerie Royal.

Cuvier, F. (1800). Description d'un orang-outang, et observations sur les facultés intellectuelles. *Annales du Muséum*, 16: 46–65

De Pauw C. (1768). *Recherches philosophiques sur les Américains, ou Mémoires intéressants pour servir à l'Histoire de l'Espèce Humaine*. Avec une Dissertation sur l'Amérique & les Américains Berlin.

Edward, G. (1758). *Gleanings of Natural History*. London: Warwick-Lane.

Erxleben, J. C. P. (1777). *Systema Regni Animali per Classes, Ordines*, Genera, *Species. Varietates cum Synonymia et Historia Animalium*. Leipzif: Weigand.

Ferreira, A. R. (1971). *Viagem Filosófica: pelas Capitanias do Grão Pará, Rio Negro, Mato Grosso e Cuiabá: 1783-1792*. Rio de Janeiro: Conselho Federal de Cultura.

Fischer, J. B. (1829). *Synopsis mammalium*. Stuttgart.

Geoffroy Saint-Hilaire, È. (1801). *Catalogue des Mammifères du Museum National d'histoire naturelle*. Paris.

Geoffroy Sant-Hilare, È. (1795). *Histoire des Makis, ou singes de Madagascar*. Paris.

Geoffroy Sant-Hilare, E. (1812a). Tableau des quadrumanes ou des animaux composant le premier Ordre de la Classe des Mammifères. *Annales du Muséum d'Histoire Naturelle de Paris*, 19: 85–122.

Geoffroy Sant-Hilare, E. (1812b). Suite au Tableau des quadrumanes ou des animaux composant le premier Ordre de la Classe des Mammifères. *Annales du Muséum d'Histoire Naturelle de Paris*, 19: 156–170.

Geoffroy Sant-Hilare, È. (1834). *Cours de Histoire Naturelle des Mammifères*. Paris: Librairie des Sciences Médicales.

Geoffroy, E. & Cuvier, G. (1798). *mémoire sur une nouvelle division des mammiferes....par Geoffroy et Cuvier* (Man, Enclyc, t II, p. 164).

Gervais, P. (1854). *Historie Naturelle des Mammifères*. Paris: L. Curmer.

Gmelin, J. F. (1788–1790). *Systema Naturae* (Linnaeus), 13th edition. Lipsiae [Leipzig]: Georg Emanuel Beer.

Gmelin, J. F. & Kerr, R. (1792). *The Animal Kingdom, or Zoological System, of the Celebrated Sir Charles Linnæus. Containing a Complete Systematic Description, Arrangement, and Nomenclature, of all the Known Species and Varieties of the Mammalia, or Animals Which Give Suck to Their Young / Class I, Mammalia*. Edinburgh: Printed for A. Strahan, and T. Cadell, London, and W. Creech, Edinburgh. https://www.biodiversitylibrary.org/item/119041#page/9/mode/1up

Gray, J. E. (1821). On the natural arrangement of Vertebrose Animals. *London Medical Repository*, 15: 296–310.

Gray, J. E. (1825). An outline of an attempt at the disposition of Mammalia into Tribes and Families, with a list of Genera apparently appertaining to each Tribe. *Annals of Philosophy*. N. S., 10(337): 244.

Humboldt, A. V. & Bompland, A. (1812) [1811]). *Recueil d'observations de zoologie et d'anatomie comparée, faites dans l'ocean Atlantique, dans l'intérieur du nouveau continent et dans la mer du Sud pendant les années 1799, 1800, 1801, 1802 et 1803*. Paris, Vol. 1, viii, 368p, 40pls.

Knorr, G. W. & Müller, P. L. S. (1766). *Deliciae naturae selectae oder auserlesenes Naturalien-Cabinet, welches aus den drey Reichen der Natur zeiget, was von curiösen Liebhabern aufbehalten und gesammarden werden verdienet*. Nürnberg: Matthäus Berdier de la Blaquiere, viii+132 pp.

Lacépède, B. G. E. (1799). *Tableau des Divisions, Sous-Divisions, Order et Genres des Mammiferes*. Paris: Plassan.

Linné, C. (1735). *Systema Naturae*. Facsimile 1st ed. With the Introduction of Engel-ledeboer MSJ & Engel H. Nieuwkoop-De Graaf: Amsterdam (1964).

Linné, C. (1758). *Systema naturae per regna tria naturae: secundum classes, ordines, genera, species, cum characteribus, differentiis, synonymis, locis*, 10th edition. Stockholm: Laurentius Salvius.

Linné, C. (1767–1771) *Mantissa Plantarum*. Holmiae : Impensis Direct. Laurentii Salvii.

Linné, C. (1768). *Systema naturae per regna tria naturae: secundum classes, ordines, genera, species, cum characteribus, differentiis, synonymis, locis*, 12th edition. Stockholm: Laurentius Salvius.

Locke, J. (1690). *An Essay concerning Human Understanding*, 1st edition, 1 vol. London: Thomas Basset.

Miller, J. F. (1777). *Icones Animalium: [Various Subjects of Natural. History*]. London: Years: 1776–1785.

Pennant, T. (1771). *Synopsis of Quadrupeds*. Chester: J. Monk. Ann Arbor, Michigan: Michigan Library, 2011. http://name.umdl.umich.edu/004903511.0001.000

Pennant, T. (1793). *History of Quadrupeds*, 3rd edition. London: B & J White, Fleet Street. https://archive.org/details/historyofquadrup01penn. Holmiae: Impensis Direct. Laurentii Salvii. Generum editionis VI. Et specierum editionis II.

Prévost, A. F. (1746–1761). *Histories general des voyages ou Nouvelle collection de toutes les relations de voyages par Mer et par Terre, qui ont été publiées jusqu'à présent dans les différentes langues de toutes les nations connues....*, A Paris, Chez Didot, Libraire.

Rajus, J. S. R. S. (1693). *Synopsis Methodica, Animalium Quadrupedum et Serpentini Generis*. Londini: Impensis S. Smith & B. Walford. Societatis Regiae Typographorum.

Scopoli, G. A. (1786). *Deliciae florae et faunae Insubricae [...] Pars I*. Ticini : Ex Typographia Reg. & Imp. Monasterii S. Salvatoris. Praesidib. Rei litter. permittentibus.

Seba, A. (1734). *Locupletissimi rerum naturalium thesauri accurata descriptio, et iconibus artificiosissimis expressio, per universam physices historiam: Opus, cui, in hoc rerum genere, nullum par exstitit. Amstelaedami*. Apud J. Wetstenium, & Gul. Smith, & Janssonio-Waesbergios.

Shaw, G. (1792). *Musei Leveriani explicatio, anglica et latina.* London: Imprensis Jacobi Parkinson.

Schreber, F. (1775). *Histoire Naturelle des Quadrupedes represemtes d'apres Nature*, Tome I. A. Erlang, Chez Wolfgang Walther. French translation (1774).

Sonnerat, P. (1782). *Voyage aux Indes orientales et à l a Chine: Fait par ordre du Roi depuis 1774 jusqu'en 1781* … Paris: Barrois.

Thevet, A. (1558). Les singvlaritez de la France Antarctique, avtrement nomm.e Amerique: & de plusieurs Terres & Isles decouuertes de nostre temps. Facsimile dell'ed. 1558, con un prologo di Jean Baudry. Paris: Le temps, (1982).

Zimmermann, E. A. W. (1780). *Geographische Geschichte des Menschen, und der allgemein verbreiteten vierfüßigen Thiere…Enthält ein vollständiges Verzeichniß aller bekannten Quadrupeden.* Leipzig: Weygand.

Secondary Sources

Azevedo, P. de (1921). *Geoffroy Saint-Hilaire em Lisboa Academia das Sciéncias de Lisboa, "Boletim da Classe de Letras"*, Vol. XIV. Coimbra: Imprensa Portuguesa.

Blair, A. (2000). Annotating and indexing natural philosophy. In: M. Frasca-Spada, N. Jardine (Eds.). *Books and the Sciences in History*. Cambridge: Cambridge University Press, pp. 69–89.

Browne, J. (1996). Biogegraphy and empire. In: N. Jardin, A. Secord, E. C Spary (Eds.). *Cultures of Natural History*. Cambridge: Cambridge University Press, pp. 305–321.

Ceríaco, L. M. P. (2014). *A evoluçao da zoologia e dos museus de historia natural em Portugal.* PhD These, University of Evora, PT.

Contreras Martinéz, J. (2015). *Buffon's Primates*. Mexico: Siglo Ventiuno Editores.

Corbey, R. (2008). *Metafisica delle scimmie. Negoziando il confine animali-umani.* Torino: Bollati Boringhieri. Italian translation of: The Metaphysics of Apes. Negotiating the Animal-Human Boundary. New York: Cambridge University Press (2005).

Courtz, H. (2008). *A Carib grammar and dictionary*. PhD Dissertation. https://openaccess.leidenuniv.nl/bitstream/handle/1887/12578/Thesis.pdf. Accessed 18 Feb 2019.

Descola, P. (2006). Beyond nature and culture. In: *Proceedings-British Academy*, Vol. 139. Oxford University Press Inc.

Dugatkin, L. A. (2019). Buffon, Jefferson and the theory of New World degeneracy. *Evolution: Education and Outreach*, *12*(1), 1–8.

Groves C.P. (2005). Order Primates. In D.E. Wilson & D.-A.M Reeder (eds), *Mammal species of the world. A taxonomic and geographic reference. Third edition, Volume* 1. The John Hopkins University Press, Baltimore, pp. 111–184.

Groves, C. (2008). *Extended Family: Long Lost Cousins. A Personal Look at the History of Primatology*. Arlington: Conservation International.

Haraway, D. J. (1989). *Primate Visions: Gender, Race, and Nature in the World of Modern Science*. Psychology Press.

Hershkovitz, P. (1985) A history of the recent mammalogy of the neotropical region from 1492 to 1850. *Fieldiana. Zoology*; n.s. no. 39.

Hoage, R. J. & Deiss, W. A. (1996). *New Worlds, New Animals, from Menagerie to Zoological Park in the Nineteenth Century*. Baltimore and London: The Johns Hopkins University Press.

Huxley, R. (2003). "Challenging the Dogma: Classifying and describing the natural world" and "natural history collectors and their collections: Simpling Macaronis' and instruments of empire." In: *Enlightenment: Discovering the World in the Eighteenth-Century*. London: Sloan K, pp. 70–91.

Koerner, L. (1996). Carl Linnaeus in his time and place. In: N. Jardin, A. Secord, E. C. Spary (Eds.). *Cultures of Natural History*. Cambridge: Cambridge University Press, pp. 145–162.

Laissus, Y. (1981). Les voyageurs naturalistes du Jardin du roi et du M'séum d'histoire na urelle : Essai de portrait-robot. *'evue D'histoire Des Sciences*, 34(3/4): 259–317. http://www.jstor.org/stable/23632460

Lefèvre, W. (2001). Natural or artificial systems? In: Lefèvre W. (Ed.). *Between Leibniz, Newton, and Kant. Boston Studies in the Philosophy and History of Science*, Vol 220. Dordrecht: Springer.

Marks, J. (2012). Why be against Darwin? Creationism, racism, and the roots of anthropology. *American Journal of Physical Anthropology*, 149(S55): 95–104.

Martin, R. D. (1972). A preliminary field-study of the lesser mouse lemur (*Microcebus murinus* J. F. Miller 1777). *Fortschr Verhaltensforsch*, 9: 43–89.

Mayr, E. (1990). Diversità, evoluzione, eredità Ed. P. Corsi Torino: Bollati Boringhieri (original version: *The Growth of Biological Thought. Diversity, Evolution and Inheritance*. (Cambridge-London: Harvard Univ. Press, 1982).

Müller-Wille, S. & Charmantier, I. (2012). Natural history and information overload: The case of Linnaeus. *Studies in History and Philosophy of Biological and Biomedical Sciences*, 43: 4–15. doi:10.1016/j.shpsc.2011.10.02

Noltie, H. J. (2009). *'affles' Ark Redrawn: Natural History Drawings from the Collections of Sir Thomas Stamford Raffles*. British Library Publishing.

Olgivie, B. (2006). The Science of Describing: Natural History in Renaissance Europe. University of Chicago Press.

Outram, D. (1996). New spaces in natural history. In: N. Jardin, A. Secord, E. C. Spary (Eds.). *Cultures of Natural History*. Cambridge: Cambridge University Press, pp. 249–265.

Pareti, G. (2016). Prima e dopo Lamarck. Il miglioramento della specie umana tra ereditarietà e degenerazione », *Studi Francesi* [Online], 179 (LX | II), accessed 04 juin 2018. http://journals.openedition.org/studifrancesi/4263; doi: 10.4000/studifrancesi.4263.

Rodman, P. S. (1990). Flawed vision: Deconstruction of primatology and primatologists. *Current Anthropology*, 31(4), 484–486.

Roger, J. (1989). *Buffon: A Life in Natural History*. Cornell University Press, English Translation, (1997).

Rookmaaker, L. C. (1992). J.N.S. Allamand's additions (1769–1781) to the Nouvelle Edition of Buffon's Histoire Naturelle published in Holland. *Bijdragen tot de Dierkunde*, 61(3): 131–162.

Sabelli B. & Tommasini S. (2004). La zoologia di Ulisse Aldrovandi, In B.Antonino (ed), *Animali e Creature Mostruose di Ulisse Aldrovandi* (pp. 24–43). Milan: Federico Motta.

Schiebinger, L. (1993). Why mammals are called mammals: Gender politics in eighteenth-century natural history. *The American Historical Review*, 98(2): 382–411.

Sidney, J. O. (1886). Joseph François "upleix. *English Historical Review*. Oxford Journals, 1(4): 699–733. doi: 10.1093/ehr/I.IV.699

Solinas, G. (1965). Illuminismo e storia naturale in Buffon. *Rivista Critica Di Storia Della Filosofia*, 20(3): 267–312.

Spary, E. C. (2000). *Utopia's Garden: Utopia's Garden: French Natural History from Old Regime to Revolution*. University of Chicago Press.

Stearn, W T. (1959). The background of Linnaeus's contributions to the nomenclature and methods of systematic biology. *Systematic Zoology*, 8(1): 4–22.

Vanommeslaeghe, B. (2008). *L'Histoire naturelle au service de la morale naturelle. Buffon et ses Histoire générale des Animaus et Histoire naturelle de l'Homme*. Master Theses Universiteit Gent.

Veracini, C. (2019). History of non-human primates in the Early Modern period: The contribution of the Italian encyclopaedist Ulisse Aldrovandi (1522-1605). In C. Veracini, C. Casanova and F. Scalfari (eds.), *History of Primatology: Yesterday and Today. The Mediterranean Tradition*. Rome: Aracneditrice pp. 53–91

Veracini, C. (2020). A landscape without nonhuman primates? The case of the Barbary macaque, Macaca sylvanus, (Linnaeus, 1758) and its interaction with humans throughout recorded time. *Humanities*, 9(3): 92.

Veracini, C. & Teixeira, D. M. (2017). Perception and description of New World non-human primates in the travel literature of the fifteenth and sixteenth centuries: A critical review. *Annals of Science*, 74(1): 25–63.

Vilhena Barbosa de (1885). *Apontamentos para a Historia das collecções e dos Estudos de Zoologia em Portugal*. Lisboa: Sociedade do Jardim Zoologico e de Acclimaçao em Portugal.

Weckesser, T. C. (1980). *The French Revolution and the Idea of Progress*. New School for Social Research.

Wilkie, J. S. (2006). The idea of evolution in the writings of Buffon.—II. *Annals of Science*, 12(3): 212–227. doi: 10.1080/00033795600200176.

Wilson, T. C. (2023). Sincerely, a black primatologist: An open letter to biological anthropologists highlighting the experiences of a black primatologist. *American Journal of Biological Anthropology*, 180(3): 423–426.

Winters, R. (2020). The Dutch East India Company and the Transport of Live Exotic Animals in the Seventeenth and Eighteenth Centuries. In: Chaiklin, M., Gooding, P., Campbell, G. (eds.), *Animal Trade Histories in the Indian Ocean World. Palgrave Series in Indian Ocean World Studies*. Palgrave Macmillan, Cham. https://doi.org/10.1007/978-3-030-42595-1_2

Wood, B. P. (1996). The science of man. In: N. Jardin, A. Secord, E. C Spary (Eds.). *Cultures of Natural History*. Cambridge: Cambridge University Press, pp. 197–210.

12 The Natural History of Great Apes from Gesner to Huxley[1]

Giulio Barsanti
Università di Firenze

I ask you and the whole world to show me
only one characteristic which makes it possible
to distinguish man from an ape (Linné, 1747: 25).

12.1 THE "BABOON" – *CERCOPITHECUS – HOMO AGRESTIS* (1486–1637)

As reported in the other chapters of this volume, most researchers thought ancient and medieval Europe had direct contact only with non-anthropoid apes (although many legends and myths about fabulous anthropomorphic creatures had circulated since ancient times) that in size and posture looked very different from us and therefore seemed to confirm the dominant idea that our species was not part of the animal kingdom. The nature of this separation, and the problem of our place in nature, raised questions about the nature of wild beasts and how close they are to human beings. Aristotle, Pliny, and Galen had affirmed that "man is very similar to monkeys" and therefore that even anthropological discourse could be framed in naturalistic terms. Aristotle claimed that monkeys have "an intermediate nature" between man and quadrupeds (Aristotle, 347–343, II, 8), and in the pages in which he was more concerned with emphasizing differences ("man is the only being among animals to assume the upright position", Aristotle 335–330, IV, 9 [689b]) he emphasized that the human species was to be placed "among animals". Pliny also referred to some Asian monkeys (the "satyrs" which live "in India", and which some thought were orang-utans, Pliny, 77–78, *Historia Animalium*, VII.24 and VIII.216) to which he attributed sophisticated behavior, denying that man differed from them by possessing a spiritual principle. "As if our breath - he argues - had something different from that of other animals" (Pliny 77–78, *HA*, VII.188). Galen further blurred the line between humans and monkeys, stating that the organs of the latter "are arranged similarly to human organs" and according to him some monkeys also "walk and run on two legs "(Galen 165–195, I, II [222]) (see also Masseti & Veracini, this volume).

The turning point was in 1486 when a "pilgrim in the Holy Land" (Bernhard von Breydenbach, Canon of Mainz) published a travel report accompanied by depictions of various little-known, but interesting, animal species, including a large caudate monkey-man (**Fig. 12.1**) which, however, "*non constat de nomine*": no one had been able to tell him what its name was. It was, however, so intriguing that, given the authoritative first source; it was referred to shortly thereafter, in 1500, by another traveler, Noah Bianchi, a missionary friar as a "*baboino*" (baboon). The image of this fabulous being, which perhaps came from an engraving by Johann von Caub (Caub, 1485), and in turn

FIGURE 12.1 Breydenbach, 1486, *Non constat de nomine*.

[1] This chapter is a partial and revised translation of parts of the book: Barsanti G. (2008). *L'uomo dei boschi*. Roma: Università La Sapienza. The translation of the quoted primary sources (not the English ones) is from the Italian. Notes about translation: (a) although the word "man" - as human beings - is no longer in use, we used this term in the translation of the sources of the 18th and 19th centuries. Only when these sources are not cited directly, we use "human beings" or "humans"; (b) Non-human primates are referred to as 'primates'; (c) the term ape, used in the past for all monkeys, is here used only for the Family Hominidae.

DOI: 10.1201/b21963-15

was probably inspired by a medieval miniature (Niccolò fra' da Poggibonsi, 1346), reached beyond travel accounts when it was published in the first Western encyclopedia of Conrad Gesner (1551). Following Girolamo Cardano (see Cardano, 1550: 415), Gesner presented the "*baboino*" as an Asian "*cercopithecus*", "which is somewhere between an animal and a human because the *cercopithecus* has a tail", intermediate between man and animal. This creature is also included – the image was copied from Gesner – (**Fig. 12.2**) in the second Western encyclopedia of natural history, that of Ulisse Aldrovandi (1637), which gave it further credibility. While Europe was going through the scientific revolution, it was reported that this ape-man "is a ferocious animal, yes, but endowed with so much talent that many men are inferior". And so Aldrovandi attached it to *Homo* as *Homo agrestis*, or "rural man". This creature *baboino - cercopithecus - Homo agrestis* persisted up to the Enlightenment. But in the meantime, attention began to be paid to another great ape.

12.2 THE *PONGO* (1613) AND TWO *SATYRUS: HOMO SYLVESTRIS* (1641) AND THE "ORANG-UTAN" (1658)

In 1613 an English sailor and adventurer Andrew Battell, described his long, enforced stay in "Angola" (whose borders differed from today's Angola) and in particular a

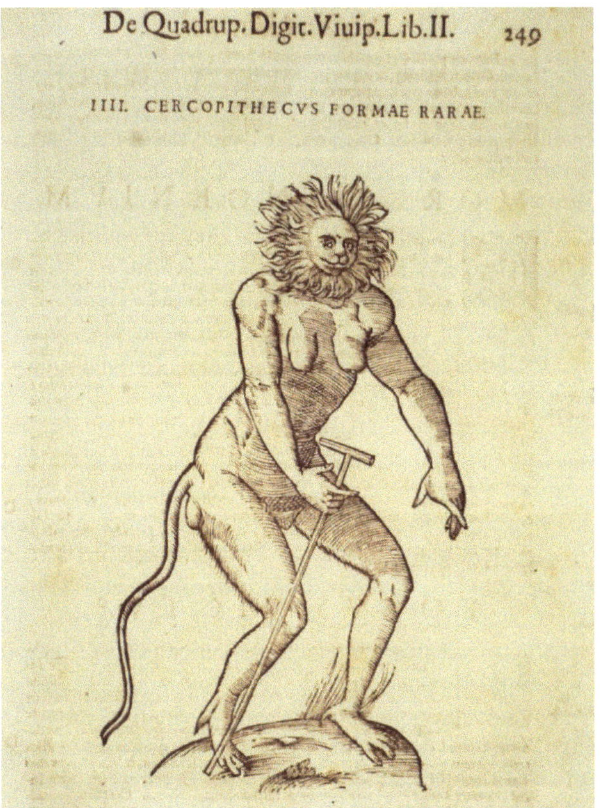

FIGURE 12.2 Aldrovandi 1637, *Cercopithecus formae rarae, Homo agrestis*, (*De quadrupedibus digitatis viviparis*: 249).

"monster of that region – the *pongo* – which deserved great attention for its resemblance to man". In that description he also wanted to reassure his readers that the resemblance did not include mental performance, for he wrote that "*pongo* is not more intelligent than other animals". But, though different cognitively, Battell admitted the creature looked very much like us:

> It is very tall and has a human face, with sunken eyes and long hair on its eyebrows. It has a hairless face and ears, as well as hands. Its body is covered with hair, but this is not very thick. [...] It is distinguished from man only by its legs, which have no calves.

Moreover, Battell admitted that the *pongo* was similar to the human species with respect to its behavior. It was not only a bipedal "monkey" ("always walking on its legs, and even holding its hands tightly behind the back of its head when walking on the ground"), but it was also nearly as tall as a man (but rather "of stature closer to that of a giant"), and ingenious ("it builds shelters against the rain and uses tools such as sticks") and it is part of a social organization (it "moves in a group"). It was even a symbolic animal, dedicated to the most sophisticated rituals: "When a *pongo* dies, its companions", the English adventurer asserts "cover the body with large heaps of wood and branches". Battell's account was credible because it was precise even with regard to the possibility of capturing the "monkey":

> adult *pongos* are never taken alive because they are so strong that ten men cannot control one; however, you can capture its babies by using poisoned arrows; in fact the young *pongo* always clings tightly to the mother's womb so that when the men of the village kill a female, they can capture the infant. (Battell, 1613: 982)

Was this the first encounter with a gorilla? Some scholars think that this primate had already been observed by Hanno in the 6th century. BCE (e.g., see Kappler, 1980: 140), but the Carthaginian admiral had reported this encounter in vague terms, writing that off the coast of:

> Ethiopia (whose word indicated Sub - Saharan Africa) there is an island that has a lake, where there is another island, full of wild men. The females were much more numerous than the males. They had hairy bodies, and our interpreters called them *gorillas*. We chased the males but we could not catch any of them because they were very agile, good climbers and defended themselves by throwing stones at us. But we captured three females who, however, biting and scratching, did not want to follow us. Then we killed them, and brought their skins to Carthage. (Hanno, 510 BCE: 13–14)

Pliny reports that their skins were still visible, in the temple of Astarte, one hundred and fifty years later (Pliny, 77–78, *NH*, VI.200). These could not have been gorillas, for morphological and behavioral reasons, plus the range of *Gorilla gorilla* is well to the south of where Hanno could have reached. Other scholars suggested the gorillas of Hanno were chimpanzees (see Yerkes & Yerkes, 1929: 16), but in my opinion, this is even more unlikely, baboons, or pygmies – since *gor-yi* means "they are men" and *gorel* is equivalent to "little men" (concerning Hanno's gorillas see also Groves, 2008: 23–24). Bernard Heuvelmans, who carried out a detailed review of the existing literature on the Periplus by Hanno, has gone so far as to suggest they were Neanderthals (see Heuvelmans, 1980: 198–201). There are also those who read this travel report in an ideological way and maintain that the label "gorillas" is only intended as a figure of speech: Hanno, according to them, stated that he had met these creatures only to emphasize that there was no longer any reason for him to continue his journey because beyond that point there were no more men to be enslaved (the true purpose of his mission, according to this interpretation) but only wild beasts (cf. Mathis, 1954: 11). Finally, there are those who think the *Periplus* is a fake, "the greatest bluff in the history of ancient navigations" (Mauny, 1971: 95). For these reasons, I firmly believe the first encounter of Western people with the gorilla must be attributed to Battell (cf. also Veracini, 2021). Out of the total of 19 anatomical, behavioral, morphological, and habitat-related characteristics Battell attributed to this creature, the majority are consistent with *G. gorilla* (cf. also Yerkes & Yerkes, 1929: 32). The behavior of these creatures, which appeared surprisingly sophisticated, gave rise to widespread discussion and helped to undermine the idea that the "monkey" in question was no more than a wild beast– consequently questioning the idea of distance between humans and the animal kingdom. But for various reasons, it would be more than a hundred years before the gorilla reappeared in literary form, and it would not be recognized by Western science until around the middle of the 19th century.

From the African regions frequented by Battell came instead news of the "Indian satyr", portrayed by Nicolaas Tulp in 1641, and from distant Borneo a report of a "wild man", which appeared in a book by Jakob de Bondt (1658). They were, respectively, the chimpanzee and (in all probability) the orang-utan. These two creatures monopolized the attention of naturalists for almost two centuries before the two species were correctly identified.

12.2.1 The Chimpanzee

It is traditionally held that the first chimpanzee (and, more generally, the first great ape) depicted in a naturalistic form is the creature that appears **(see Fig. 12.3)** in an engraving in the book "*Vera descriptio regni africani (…) quod Congo appellatur*" (…) by Filippo Pigafetta, drawn up from the notes of a Portuguese sailor, Eduardo Lopez, and published in 1598 (cf. Pigafetta, 1591, 1598). Thomas Huxley says:

> I have not met with any notice of one of these Man-like Apes of earlier date than that contained in Pigafetta's *Description of the kingdom of Congo.* […] The tenth chapter of this work…'contains a brief passage to the effect that in the Songan country, on the banks of the Zaire, there are multitudes of apes, which afford great delight to the nobles by imitating human gestures. As this might apply to almost any kind of apes, I should have thought little of it, had not the brothers De Bry, whose engravings illustrate the work, thought fit, in their eleventh "Argumentum" to figure two of these *Simiæ magnatum deliciæ*. […] It will be observed that they are tail-less, long-armed, and large-eared; and about the size of Chimpanzees. It may be that these apes are as much figments of the imagination of the ingenious brothers as the winged, two-legged, crocodile-headed dragon which adorns the same plate; or, on the other hand, it may be that the artists have constructed their drawings from some essentially faithful description of a Gorilla or a Chimpanzee. (Huxley, 1863: 10)

FIGURE 12.3 Pigafetta 1591; 1598, *Simiae magnatum deliciae*, p. 11.

I would exclude the latter explanation, but some scholars thought they could be baboons or Barbary macaques. Overall, I agree with Heuvelmans, who wrote that "indeed, all the animal figures in the work [of Pigafetta], have been taken from some medieval bestiary" (Heuvelmans, 1980: 23). As a first approximation, I suggest they are similar to the monkeys that appear, also in the act of putting on boots, in a text by Sebastian Brant. In fact, the myth of monkeys that could be caught by boots was reported in the Middle Ages (*e.g.,* the 14th century, cf. Diretano Baldo, National Library of Florence, ms. II IV 29, c. 529). Veracini (2021) showed that in the second half of the 15th century, anthropoid apes had already been represented in the form of demonic creatures by artists who lived at the Turcoman courts. This may be evidence that contact with great apes (and trade in them with the Near or Middle East), had already developed before the arrival of Europeans in sub-Saharan Africa. Moreover, an orang-utan is probably the subject of a sculpture found in 16th-century Venice (cf. **Box details 9.1, Chapter 9**, this volume).

The first report of a chimpanzee is normally attributed to Battell when he argued that in the same region as the *pongo* there also lived a smaller "monster", called *engeco* – which will reappear, as we shall see, named *jocko* in Buffon's work, but the Western chimpanzee had already been described several times by Portuguese navigators during the 16th century (e.g., see Ferdinand, 1862; cf. Veracini, **Chapter 9**, this volume). But, because these manuscripts had a very limited readership, they had little or no impact on naturalists or historians.

The first to depict the chimpanzee in Europe was Willem de Groot, brother of the famous Grotius, in a letter addressed to Pierre and Jacques Du Puy. The specimen, which was later examined by Tulp, was donated to Prince Frédéric-Henri d'Orange in 1630. The first chimpanzee to arrive alive in Europe came from Angola or Guinea (the sources differ because there wasn't a precise definition of regional borders) and was about one year old (Tulp, 1641: 240). Willem de Groot presented it as "a monster: man or beast I cannot say" adding "it is apparently able to understand our language" and in his letter of 1630, he enclosed a drawing that has unfortunately been lost (cf. Hamy, 1897; Heuvelmans, 1980). The first visual documentation of a chimpanzee is in the *Observationes medicae* (1641) of the Dutchman Nicolaas Tulp. The author portrays it alive and provides an image that in subsequent centuries became very popular, but he calls it variously: *Satyrus indicus*, *Orang-outang*, *Homo sylvestris*, *Quoias-morrou.* These names referred to different beings:

a. *Satyrus indicus,* the "satyr" was perhaps used to evoke the orang-utan or other Asian apes like gibbons (see Gesner, 1551; Pliny, 77–78; Polo, 1298; Masseti & Veracini, this volume);
b. *Orang-outang,* the orang-utan, which is the Malay name of this great ape and means "*man of the woods*" and was used by the coastal populations of Borneo and Sumatra to indicate, contemptuously, the non-civilized beings of the interior;
c. "*Homo sylvestris*" (a wild man, in the sense of sylvan, of the woods) was the scientific translation of 'orang-utan', which indicated conjectured populations of creatures intermediate between man and ape (see also Pliny, 77–78), whose existence the western world continued to believe in (cf. Aldrovandi, 1642);
d. "*Quoias-morrou*"' was, according to Tulp, the African name of the orang-utan.

This confusing nomenclature left open the problem of the identity (ape, human, mixed) of a creature that seemed to have inhabited both Africa and Asia and would lead to the conception of "monkeys" whose nature was intermediate between human beings, chimpanzees, and orang-utans. But the image provided by Tulp (**Fig. 12.4**) was a relatively good likeness of the chimpanzee, which, not by chance, became very popular in iconography, and was arguably the most successful great ape image ever. It was used by Liceti, Dapper, Zahn, Tyson, and Beeckman among others, and after being re-used by the *Histoire générale des voyages* (1748), it would survive until the time of Linnaeus, Le Cat, and White (see Liceti, 1665; Dapper, 1668; Zahn,

FIGURE 12.4 Tulp 1641, *Homo sylvestris. Orang-outang*, Pl. XIIII [sic], p. 271.

1696; Tyson, 1699; Beeckman, 1718; Aa [1720]; Linné, 1761–1785; Le Cat, 1765; Linné, 1765, 1773–1776; White, 1795–1799). In 1863, Thomas Huxley described it as "a beautiful picture, evidently taken from a live specimen [...]. It is clearly a young chimpanzee" (Huxley, 1863: 17). Tulp refers to it as a biped "neither fat nor thin; robust, yet very agile and very resistant [...], in front completely hairless, but behind covered with rough black hairs"; morphologically it is indistinguishable from man ("you could hardly find an egg more similar to another egg") even in the calves (relevant information because calves, which Battell's *pongo* was lacking, are necessary for the upright position), and incidentally, it is surprisingly educable –for example having meals at the table, or going to bed. Moreover:

> "when drinking it took the handle of the mug with one hand; with the other it helped itself by acting on the bottom of the container, and then it wiped away the liquid left on its lips, no less elegantly than you would see a courtier's end". [...] When going bed "it lowered its head onto the pillow and conveniently pulled up the blankets, it covered and settled itself just like the best-mannered of men". (Tulp, 1641: 270–272)

In the scientific literature, Tulp's contribution was treated with little respect, with one expert writing that:

> we remain appalled by the mediocrity of such a description. [...] If the burgomaster of Amsterdam had refrained from describing the chimpanzee of the Prince of Orange, [...] he would have rendered a great service to the history of zoology. (Heuvelmans, 1980: 249–250)

Perhaps one would have expected something more sophisticated from the great Dutch anatomist, but he helped refute many fanciful tales, scotching the rumors spread by Willem de Groot, according to whom the chimpanzee "is able to understand our language".

12.2.2 The Orang-Utan

The first unequivocal news of the existence of the orang-utan was contained in a letter written in 1633 by Nicolas-Claude Fabri de Peiresc in 1633 to Pierre and Jacques Du Puy. The source of the information was the poet Marc-Antoine Gérard de Saint-Amant, according to whom the orang-utans:

> constitute an intermediate genre between man and ape. They are not vicious at all and can, at home, sweep, light the fire and carry out other domestic services, which they perform with great punctuality and meekness.

More sensational still, was the depiction and description of the orang-utan in a posthumous work (1658) by Jakob de Bondt. The drawing (see **Fig. 12.5**) may not have been made by the Dutch doctor, who had died in 1631, but by his editor for the purpose of making the book more appealing. The image could have taken it from one of the many encyclopedias that still contained images from medieval bestiaries (e.g., the *Monstruorum historia* by Ulisse Aldrovandi, in which there is a similar hairy creature, named *foemina villosa*). de Bondt's picture generated confusion until Thomas Huxley suggested that the depiction "is nothing but that of a very hairy woman, rather graceful in appearance, with proportions and feet genuinely human" (Huxley, 1863: 18), and well before that Edward Tyson had written, "I confess that I completely distrust that description" (Tyson, 1699: 19). Nevertheless the "wild man" of Bondt (or of his publisher) survived, despite much criticism, up to the late 18th century, still appearing in the last works of Linnaeus (see Linné, 1760, 1761–1785, I, 1761, 1765, 1773–1776, I, 1773). Although it appeared to be unique, it was given the same names as Tulp's chimpanzee – *Satyrus, Ourang outang, and Homo silvestris* (cf. Bondt, 1658: 84). This added to the confusion between the chimpanzee and orang-utan, and for years it was thought the descriptions and depictions all referred to the same species of great ape, or ape-men. In the *Historia naturalis et medica Indiae orientalis* Bondt speaks of his *Satyr, Orang-utan, Homo silvestris* as a being not only a biped and with a human form but also modest ("the female hides herself from males which she does not know") and obviously capable of every kind of emotion:

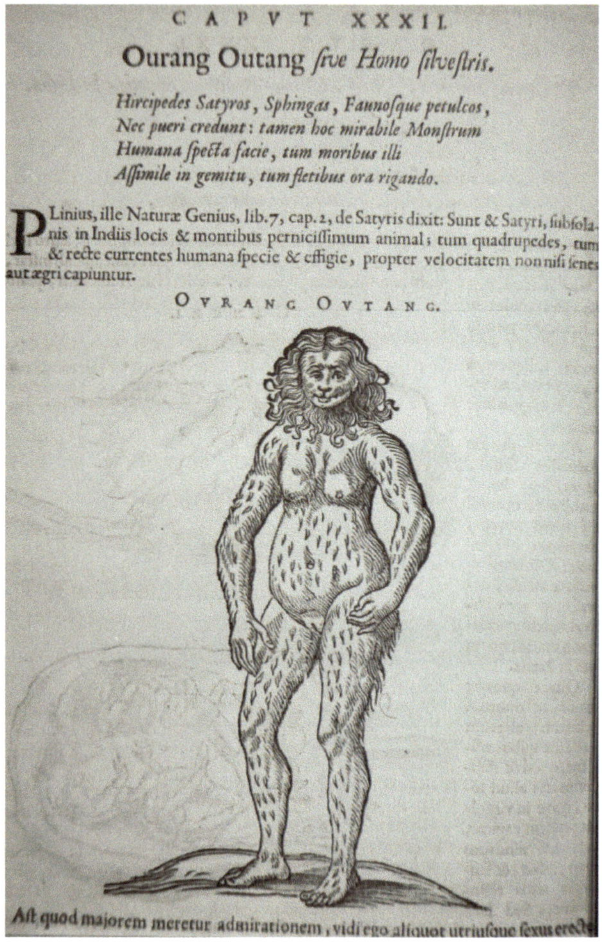

FIGURE 12.5 Bondt 1658, *Ourang Outang sive Homo silvestris*, p. 84.

> it cried, it protested and it carried out other human actions, so much so that you would have said that it lacked nothing human, except speech. [...] I saw some of both sexes walking upright. [...] The Javanese say they are able to talk but do not want to, in order not to be forced to work. [...] They were born of Indian women, who because of their abominable lust mate with apes and *cercopitheci*. (Bondt, 1658: 84–85)

Since ancient times, there have been suggestions that child-bearing couplings between humans and primates resulted in mythological animals. Pliny and Plutarch, among others, had proposed the existence of such unions, and in the Middle Ages, it was alleged Pope Alexander II used to present to his guests a "boy" of about twenty years old, who, he claimed, was the son of an ape and a countess, and unfortunately resembled the father more than the mother (see Janson, 1952: 282; Tinland, 1968: 41). At the end of the 16th century the story of Ippolita Biscontina became widespread: she was a young maid from southern Italy (Messina) who had a long affair with a quadruman (ape or monkey) and never thought she would become pregnant. "But God, just judge, not wanting such an abomination to remain hidden and unpunished, allowed her to conceive" (see Anonymous, [1600]). Martin Delrio reported in his

Disquisitiones magicae (1599–1600) that a young woman was burned alive at the stake along with the "monkey" and their baby monster, who some claimed was Satan in the form of a "monkey". Delrio also recounted the story, which was later retold by Liceti (1616) and Tyson (1699), of a young Portuguese woman who had been abandoned on a desert island after conceiving two "children" with a quadruman. When the young girl left the island, the father-monkey drowned the "children" at sea and committed suicide. In the second edition of the *Disquisitiones*, Delrio added that in Peru men take ape-females as concubines, and specified that "*natos inde foetus caput humanum et pudenda habere, caetera simiis similes esse*" (Delrio, 1599–1600, II: 151).

These reports were not confined to popular culture or to demonology. Even the great scholar and skeptic Michel de Montaigne, who believed in the idea of the contiguity between "monkeys" and humans, emphasized that "every day we see Barbary macaques (*singes*) possessed by furious love for women". Montaigne was convinced it was necessary to bring down the human being from the "ridiculous pedestal" on which he placed himself as lord and master of the universe, and he was also persuaded that animals have a different nature than that traditionally attributed to them ("beasts have many qualities that come close to ours", and indeed: "which of our faculties do we not find also in them?") (Montaigne, 1580–1588, II, 12: 580, 587, 606, 613).

Bondt's Malay legend, which suggested that some monkeys (aka orang-utans) "are able to talk but do not want to, in order to avoid being forced to work", also featured in African culture. In 1623, the English explorer Richard Jobson had already mistakenly linked the legend to baboons. His guides in Gambia had told him about "monkeys" that had set up couches in the trees (likely the nests of chimpanzees), and he – who had always seen them empty – had thought they belonged to baboons, the primates he encountered most frequently. This author also reports that they "constitute a true race of men, who refuse to speak in order to avoid being forced to work and live in slavery" (see Jobson, 1623; Heuvelmans, 1980: 237–238). For a long time, Bondt's Malay legend continued to be attributed to baboons. For example, Peter Kolb says: "The Hottentots think that baboons are able to talk, if they want to; but they remain silent, fearing that men will capture them to make them work" (Kolb, 1742, III: 64–65). Only later was the legend attributed to the correct creatures, the chimpanzees referred to by Jobson's guides. The legend was later linked to all of the apes that had some affinity with humans.

Both chimpanzees and orang-utans were referred to as *wild men*, to indicate that they are only slightly inferior to civilized humans, thus it is not surprising that those who later saw orang-utans would go so far as to include them in the human species. Wouter Schouten tells us that they have "the same form and the same size as other men" and are "intelligent". He also informs us that:

> they are strong, agile, daring, and they know how to defend themselves from armed men; [...] they have a great passion for our women and it is not safe for these to pass in the woods, because they are immediately attacked and raped by the Orang-utan. [...] They are taught to walk on their feet, or rather on their hind legs, and to use their front feet, which are roughly hands, to perform many operations, including domestic ones: how to rinse glasses, pour a drink, turn the spit etc. Many of them try to get hold of something edible, to have the pleasure of enriching the table too. (Schouten, 1676, II: 36–37)

12.3 THE MANY TERMS ATTRIBUTED TO CHIMPANZEE: *BARIS/DARIS – DRILL – SMIT* (1608–1704) AND HOW THE CHIMPANZEE WAS PERCEIVED IN TRAVEL LITERATURE

In the 17th century Europeans recognized both an African "orang-utan" and an Asian orang-utan. Both were considered inferior human beings. But it was the African "orang-utan", the chimpanzee, which dominated the literature of the 17th and 18th centuries, thanks to travelers' reports of a *baris* or *barris* or *daris* (as they were called, on the coasts of Guinea, by locals – cf. also Veracini, Section II this volume regarding the *daris*) and a *drill,* which in English meant a great ape. The derivative *mandrill* did not indicate the real mandrill, but an anthropomorphic creature – aka *smit* (see Du Jarric, 1608–1613; Nieremberg, 1635; Gassendi, 1641; Dapper, 1668; Bosman, 1704). This creature was endowed with enough intelligence to do housework (Du Jarric, 1608–1613), "the *baris* pound seeds in the mortars and go to get water at the spring, with jugs that they carry on their heads" (Nieremberg, 1635: 179), play the flute or the zither skillfully, and [...] they are monkeys with "great powers of judgment" (Gassendi, 1641: 169). In 1668 the German physician Olfert Dapper, apparently referring to first-hand experiences, but in fact reproducing the previous literature and existing iconography, summarized the performance of the *baris*:

> they are bred and tamed so well that they can do almost as many tasks as a slave. Since they usually walk upright like men, they pound the millet in the mortar. They go to draw water with a pitcher and when the pitcher falls they bear witness to their displeasure with shouts. They know how to turn the spit and play many juggling games that give a lot of amusement to their masters. (Dapper, 1668: 224)

Dapper's text (*Naukeurige beschrijvinge*) is significant because it provides a complete list of its sources (about 77 authors). The creatures' physical characteristics are reported: "They are reddish in color and become very big. I have seen one with my own eyes, which was five feet tall and therefore only slightly smaller than a man". He reports that many think the animal "is derived from a man and a monkey" but, referring to an African legend, he hypothesizes, alternatively, "that it is derived from men who became wild through having always lived in the forests" (Dapper, 1668: 257, 603–604). This idea of the autonomous degeneration of the human species, rather than its hybridization with other species, was taken up, as we shall see, also by Jean-Jacques Rousseau, who did not rely directly on Dapper's *Naukeurige beschrijvinge*, though he cited him among his sources. Dapper also describes their behavior:

> they are very fierce and very daring. What an English merchant told me seems incredible: behind the fort that the English have in Wimba [Guinea] there are an enormous number of these 'monkeys', who are so daring that they even try to attack men; what the merchant told me happened to two of his slaves. These animals grabbed them and would have gotten their eyes if they had not been rescued just in time by other Negroes; these ugly beasts had already prepared small pieces of wood to hit them with.

After Dapper, the travelers who speak about the great intelligence of chimpanzees acquire more credibility. Thus, in 1704 the Dutch merchant Willem Bosman was of the impression, among other things, that *smitten* [plural of *smit*] could listen and understand the words of natives (a circumstance already related, as we know, in a letter by Willem de Groot but later refuted by Nicolaas Tulp) and could fight with sophisticated weapons. Finally, Bosman refers, albeit indirectly, to the mental capacities of *smitten*:

> There are Negroes who firmly believe that these animals can also speak, and that they do not want to do so in order to avoid being forced to work, a thing they do not like at all. (Bosman, 1704: 279)

According to the above descriptions, the chimpanzee is the primate closest to humans, but in a very confusing way, but it is worth remembering that at this time this African great ape had not yet been identified as a species in its own right, and the term "chimpanzee" had not yet appeared. Also, individuals belonging to the same species (for example a *baris* and a *drill*, or a *smit*) were considered to belong to different species, chimpanzees were described as having characters and behavior typical of orang-utans, and the chimpanzee's "mixed" nature resulted in it being perceived as a "monster", such that it was included in works such as the *De monstris* of the Italian doctor Fortunio Liceti (1665, plate p. 399). Its monstrous nature justified the presence of "pathologies", and thus it was portrayed in an engraving published by the German physicist Johann Zahn, where the *Ouranus Outanus – alias* chimpanzee – is included with medieval mythical and ambiguous dog-headed, dog-tailed and one-eyed creatures.

In literature, and particularly in the travel literature, there were frequent inaccuracies, plagiarisms, and extrapolations about what became known as the chimpanzee. In the *Historische ende journaesche aenteyckeningh* (1646), the Belgian merchant Pieter Van Den Broecke suggests that it is inferred that the "wild man" has "a small tail of the length and thickness of an inch. [...] It does not speak and passes the nights in the trees" (Broecke, 1646 apud Heuvelmans, 1980: 251). In the *Description historique du royaume de Macacar* (1688) by Nicolas Gervaise, a French missionary, the "monkey without a tail is very daring and ferocious" and "always stands upright like a man, walking only on its two hind feet" (Gervaise, 1688: 47). In the *Voyage* by Daniel Beeckman (1718), an English explorer, there is a portrait of an anthropoid ape (see **Fig. 12.6**) looking like the *satyr - Homo sylvestris* of Tulp, but the place he visited was Borneo where the only ape he would have encountered was the modern orang-utan. The image he published, which resembles a chimpanzee, was a bad copy of the one depicted, almost eighty years earlier, in Tulp's *Observationes medicae*. The fact that Beeckman thought he could illustrate his description of the "orang-outan" with an old image of the chimpanzee – here completely hairless – is not only evidence of how unscrupulous people could be, but was also further confirmation of the lack of distinction between the two anthropoid apes and the persistent idea that the two continents were inhabited by the same great ape.

The characteristics and behaviors observed in Asian great apes were routinely attributed to chimpanzees. For example, it was written of the chimpanzee that it is "six feet

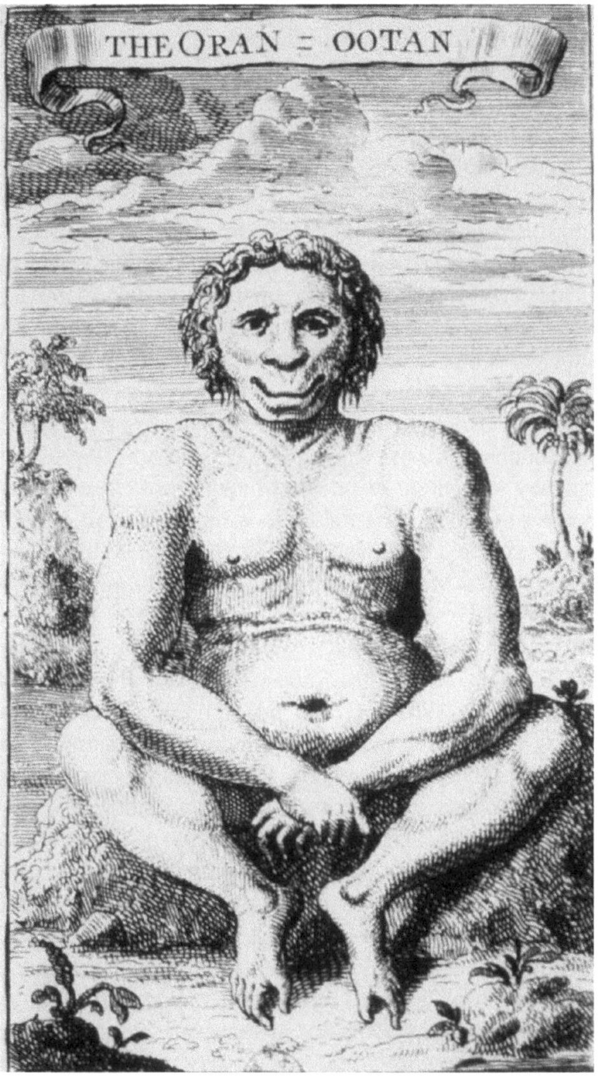

FIGURE 12.6 Beeckman 1718, *the Oran-ootan*, p. 37.

tall", "walks upright" and has "a passable face, certainly more graceful than that of many Hottentots", together with the cliché that "it throws huge stones, branches and trunks at anyone who disturbs it". But Beeckman's *Voyage* also proposed that the "*oran-ootan*" was a fallen man, as in "the natives believe that initially it was a man, which then was turned into a beast for having blasphemed" (Beeckman, 1718: 37).

In his *Description of the Coasts of North and South Guinea* (1732), Jean Barbot, a French merchant and explorer, repeats the information provided by Bosman on the *smit*, and in *Voyage to Congo river*, published the same year, John Casseneuve, adds this detail "when this animal cries it is difficult to distinguish its lament from that of a child", and John Atkins also suggests it might be "a human savage" (see Atkins, 1735). In the first publication that uses the word *chimpanzee*, in 1738 Henry Flower reported capturing a female *chimpanzee* of about eighteen months old, which he claimed "manifests a great discontent, when the dress is lifted up to ascertain its sex" (see Flower, 1738).

Flower's modest *chimpanzee* (also called – significantly – *mockman*) was portrayed in two engravings. The first was by the English merchant Thomas Boreman, in which the chimpanzee appears in motion, a perfect biped (see **Fig. 12.7**). This image was also inserted in *Histoire générale des voyages* (1748) and used by Robinet (1768). The second, drawn by Hubert-François Bourguignon Gravelot, a French draftsman, portrays the creature in the act of lifting a cup of tea to its mouth **(Fig. 12.8)**, an image re-used by Linnaeus (1760), Robinet (1768), and Schreber (1775), among others. The brief description accompanying Gravelot's image, suggests the chimpanzee is "of a very ferocious nature" and that it "is by no means lacking in intelligence, being able to reproduce the human voice" (Anonymous, 1739b: 564–565).

Other extraordinary chimpanzee behaviors were reported by a "De La Brosse", but I have been unable to locate his 1738 publication *Voyage à la côte d'Angola*. Buffon quoted an unspecified "extract" from the *Voyage* (Buffon, 1766b: 50–51, 55–56) misspelled the author's name. whom I believe to have been the philosopher and linguist Charles de Brosses de Tournay, with whom Buffon was in close contact. However, Buffon wanted to emphasize the importance of two pieces of information given by the French author. According to Buffon, De Le Brosse argued that the bipedal and 6–7 feet tall chimpanzees "try to surprise African women" not in order to rape them and immediately abandon them, as has been said so far, but because they want "to fully enjoy them"…"they take them and feed them well. I met in Loango, a woman who had stayed with these animals for three years". He also suggests that "these animals have the instinct to sit at table like men; and, to cut up and eat what is presented to them on the plate; they use fork, knife and spoon". Buffon was impressed by this last fact and reports that: "De La Brosse, who had bought from a Negro two small orang-utans [chimpanzees], does not tell us that the African had trained them; he seems to affirm, on the contrary, that they did what they did spontaneously" (Buffon, 1766b: 55). If true, this would mean that these behaviors are instinctive and not learned.

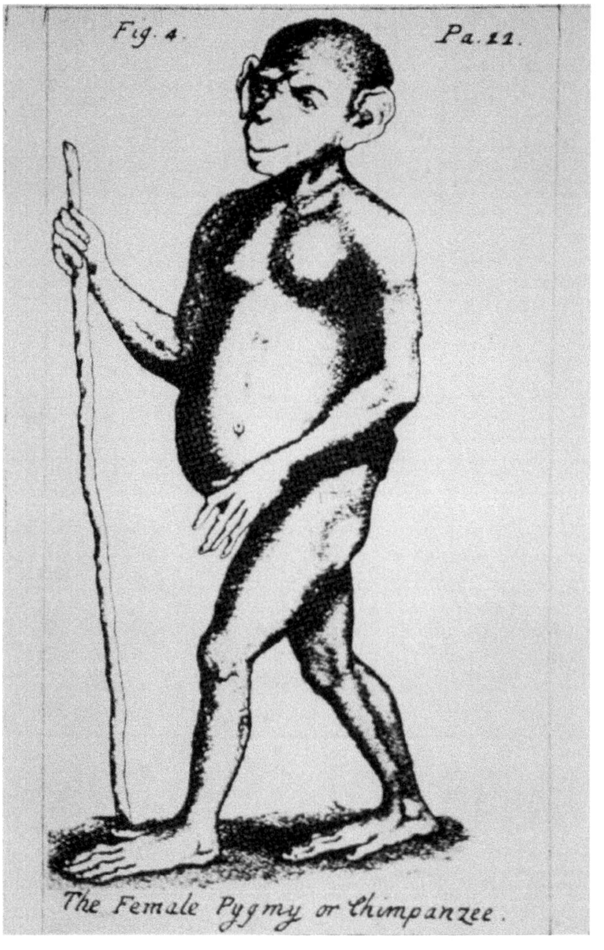

FIGURE 12.7 Chimpanzee of the English merchant Thomas Boreman (1739).

This and other (pseudo-) information is amplified in the English traveler William Smith's *New Voyage to Guinea* (1744). He reports that his chimpanzee, called *mandril,* is an animal that "has absolutely nothing in common with apes, never walks on all fours like a monkey" and, if he has to rape women, "he tries to rape white women" (Smith, 1744, I: 104, 106). Instead of approaching an objective description, the travel literature seems to progressively move away from it (e.g., the drawing in the *New Voyage* does not resemble a chimpanzee). The morphological information reported suggests that the English traveler had observed the African primate first-hand. It is possible that the drawing was carried out later on the basis of Smith's text, or that Smith, like Bondt's *Orang Outang*, never saw the image and would have disapproved.

Until the mid-18th century, the travel literature lagged far behind the scientific knowledge of the time. As early as 1699 the English physician Edward Tyson had described the chimpanzee with remarkable accuracy and precision (Wood, 2020). In order to appreciate Tyson's contribution, we can compare it with what was written about the

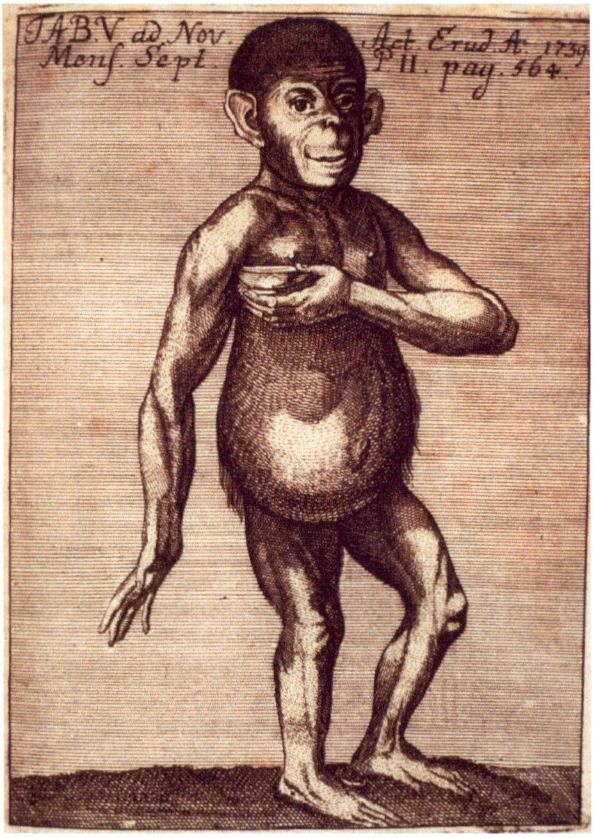

FIGURE 12.8 Chimpanzee by Hubert-François Bourguignon Gravelot (1739).

"generalized" great ape (namely orang-utan+chimpanzee) at the end of the 17th century, for example, this quote from the *Relation d'un voyage fait aux côtes d'Afrique* (1698) by François Froger, a French explorer:

> the monkeys (*singes*) living on the coast of Gambia are bigger and more vicious than in any other African region; for this reason the Negros fear them, and they cannot go alone into the countryside without running the risk of being attacked by these animals, who show them a stick and force them to fight. [...] I heard the Portuguese say that they had often seen them bring little girls between seven and eight years old up to the trees, and that it was very difficult to take them away from these animals. [...] Most of the Negroes believe that they are foreign people who have migrated to their country, and that if they do not speak it is only for fear of having to work. (Froger, 1698: 42–43)

Froger's description is a combination of all of the "information" available: Gervaise's "wicked" creatures; Battell's claim that "*pongos* attack negroes using sticks as offensive weapons"; the vast literature on the "libido" of the *satyr - baris – mandrill*; the legends by Jobson and Bondt and a new hypothesis that chimpanzees are in fact human beings who have immigrated from another country. All this is accompanied by a drawing in which there are long-tailed monkeys and baboons. The impression is that Froger, without first-hand experience, simply assembled a "chimpanzee" in his imagination.

12.4 THE *ORANG-UTAN – PYGMIE* (TYSON, 1699) AND THE FIRST OSTEOLOGICAL COMPARISONS (1721)

Tyson's contribution is different from those of previous naturalists because rather than using presumed behavioral observations, he focuses on accurate first-hand anatomical investigation. He does not distinguish the chimpanzee from the orang-utan; in fact, he refers to the chimpanzee – a young male of about 60cm – as an *Orang-utan,* a *Homo sylvestris,* or a *Pygmie* (Tyson, 1699). The confusion about names as exemplified by Nicolaas Tulp, and by Jan Jonston and Walter Charleton (see Jonston, 1657; Charleton, 1667) among others, would continue for many decades. The time was not ripe for Tyson to distinguish his *pygmie* from the orang-utan of travelers, but he could at least dissect it meticulously, with the help of the renowned anatomist William Cowper. Apparently referring to tradition, Tyson affirms that man rises above all animals and, being the only creature endowed with a soul, occupies his own place in nature, but he went beyond this principle establishing continuity between humans and other animals. Tyson rehabilitated the old idea of a *scala naturae* (for which all forms would constitute links in the same chain, which leads from the simplest of minerals, passing through plants, to the most complex of animals), but for the first time, he extends it to human beings (Tyson, 1699, *The Epistle Dedicatory*, p.n.n [3]). The careful dissection was recorded by illustrations (see **Figs. 12.9 and 12.10**), as well as by 82 defining characteristics – organs, systems, and organ systems, 34 of which are also present in monkeys, with 48 shared with the human species (see Tyson, 1699: 92–95). The fact that the chimpanzee constitutes "the link between man and apes" (Tyson, 1699: 91) was, for the first time, established empirically and quantitatively. Nor did Tyson stop there. Without even referring to travel literature, he believed the chimpanzee to be an intermediate species, not only from an anatomical point of view but also from a mental one. In the Dedication of his *Anatomy of a pygmie* Tyson put forward the hypothesis that the African primate constitutes the connecting link:

> The Animal of which I have given the Anatomy, coming nearest to Mankind; seems to be the Nexus between the Animal and Rational, as your Lordship, and those of your High Rank and Order for Knowledge and Wisdom, approaching nearest to that kind of Beings which is next above us; Connect the Visible and Invisible World. (Tyson, 1699: *The Epistle Dedicatory,* p.n.n. [3–4])

Although Tyson appealed to a figure of authority (John Sommers, Baron of Evesham, Lord Chancellor, President of the Royal Society), his investigation resulted in promoting the chimpanzee (which is not a man, "but – said Tyson – not merely an ape", Tyson, 1699: 91), and by doing so administered a severe blow to traditional metaphysics. Thanks to the *pygmie* and the calculations of the English physician, what had been conceived as a qualitative and therefore irreducible difference between "man and the ape", was transformed into an objectively defined one.

FIGURES 12.9 and 12.10 Tyson 1699, Pl. I f.t p. 95 and II f.t. p. 96.

Around the middle of the 18th century, a doctor, Julien Offray de La Mettrie, continued to see no other difference between the various *wild men* than the terminological one: "Wild men, quite common in the Indies and in Africa, are called *orang-utans* by Indians and *quroias morrou* by Africans" (La Mettrie, 1745: 156). And the naturalist, Charles Bonnet, generically (and theologically) spoke of the "monkey", without being able to distinguish a single species. As late as 1764, he rhetorically asked: "How can nature take such steps as to develop from Quadruped to Man?" It is not a question of "degree" because, between quadrupeds and man, there is a void. How is it possible "to straighten this head bent towards the earth? To change these legs into flexible arms? To turn these hooked feet into folding and elegant hands?" [...] It is not possible "although the monkey is like a sketch of t Man [...] a rough sketch, imperfect but still a similar portrait, that sheds light on the admirable progression of God's works" (Bonnet, 1764, I: 104, 120).

Tyson's conclusions were contentious, and in order to check their veracity, naturalists requested specimens of primates – of any species and in any condition – from all over the world and anatomists were urged to carry out comparative investigations. Anatomical investigations confirmed Tyson's conclusions about the relatively few anatomical differences between man and apes (e.g., the size of the thumb, the shape of the kidneys, the length of the bladder, cf. Tyson, 1699: 94–95), compared with the many similarities between our species and the chimpanzee. Many naturalists also accepted that anatomical similarities were accompanied, as reported by travelers, by behavioral resemblances. Humans are not only contiguous anatomically to the animal kingdom but also in their behavior and habits.

12.4.1 Anatomical Similarity/Mental Affinities between Humans and Animals

The type of anatomical comparison, so carefully documented by Tyson, had already been used many years earlier. In the mid-16th century, the French naturalist Pierre Belon compared two skeletons, one human and one animal (a bird), showing that twenty-six main anatomical characteristics – the cervical vertebrae, the arm bone, the two forearm bones, ribs, etc. – were present in both (see Belon, 1555: 40–42, 40–41). But man had become, in that case, slightly caudate, and the comparison was made with a gigantic bird. Although taken up by Ulisse Aldrovandi (cf. Aldrovandi, 1642: 86–98, plate p. 87), this osteological comparison did not stimulate much discussion or research. However, it did force scientists to intensify research in comparative anatomy, which led to a novel and important osteological comparison between human and non-human primates by Richard Bradley, an English naturalist, in 1721. It was still difficult to obtain specimens of great apes, so the primate comparator was a cercopithecid (**Fig. 12.11**) whose species is not specified. Bradley believed that the anatomical similarities corresponded to mental and behavioral affinities. If, on the one hand, apes are capable of the performances that had been talked about since the time of Battell, and as Bradley later affirmed, they can learn from experience and devote themselves to the same vices as the human species ("they get drunk frequently. They love tobacco and smoke abundantly"), yet they are still far from showing the rationality that distinguishes man. It is possible that this difference depends not on spiritual principles, but on the different conformation and structure of the components of the nervous system and in particular of the brain (Bradley, 1721, *Philosophical account*). His argument is linear: from a behavioral point of view there was widespread acceptance that "the ape and the *cercopithecus* [monkey] of the various species are naturally keen to imitate the actions of men"; from the anatomical one it turns out that "both the form and the disposition of their bodily parts are similar to those of man much more than to the bodily parts of any other creature"; so "some of them could be useful if educated appropriately when young", and "as the tone of the voice depends on the structure of phonic organs, so their capacity for comprehension depends on the structure of the parts that make up the brain" (Bradley, 1721: 95). This hypothesis, which would soon influence many researchers (cf. Barsanti, 1986a), was another important step toward reducing the exalted state of humans and including them in the animal kingdom, and toward the correct identification and characterization of primates.

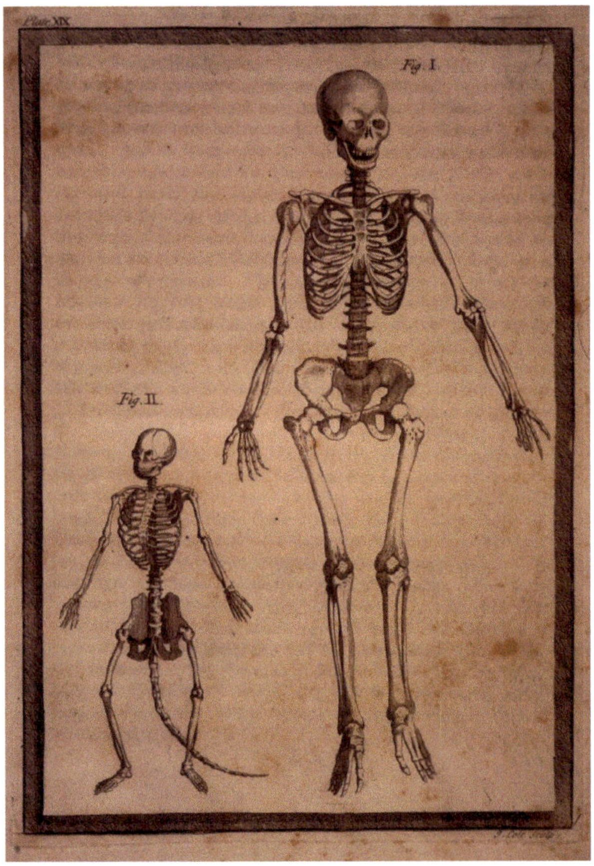

FIGURE 12.11 Bradley 1721, Pl. XIX, p. 171.

12.5 FANTASTI: THE MONKEY IS NOT A HUMAN SPECIES (1720–1732)

In 1732 the Italian physician Girolamo Cesare Fantasti opened another perspective on the same themes. Linked to traditional philosophy, he well understood the potential consequences of information about the nature of the great apes. He considered it disastrous that the likening of primates to the human species might lead to the proclamation of man's animal nature (and therefore to its inclusion in the *systema naturae*), so he published *La scimia non èspecie d'uomo* (*The monkey is not a human species*), a work dedicated to downplaying and belittling the nature of the other primates. In practice, the text, which from many points of view was well behind the knowledge of the time, had no repercussions. The "medical philosopher" (as Fantasti defines himself) does not start with the literature of the time concerning the (real or presumed) mental abilities and behavioral performance of primates, but simply from information he attributes to a conversation with a Lutheran baron. He draws on predominantly ancient and medieval sources and bases his ambitious discourse on a generic "monkey" (the Barbary macaque, the only species he knew of). He frames his treatment with Christian metaphysics, either referring to the myth of creation, or attributing human superiority to the possession of a soul, and often presents the monkey in the crudest terms (it is "a disgusting beast, a monster", "a filthy and ridiculous Brute", a "Devil's work"). Moreover, he treats those who have dared to liken the monkey to man as "crazy" "misguided", and "wicked" (Fantasti, 1732: 4, 7, 9–13, 20, 24–26, 41, 44–45). Fantasti and his editor brought out a book that, despite its limitations, deserves more attention for being the first monograph on this topic. It was not the monolithic pamphlet it might seem, because Fantasti operated on different levels, developing his own way of speaking. He emphasizes insistently that:

> in many behavioural trait these Brutes show us their little discernment, their lack of intellect, and their nonsense in wanting to imitate Man, [...] in its conduct the *Scimia* bears very little resemblance to the human species. Why do they never talk? [...] On the basis of the few pieces of information on anatomy collected by Ulysses Aldrovandi [...] we are forced to believe, that the use of speech was given by God only to the human body by imprisoning within it a soul capable of reasoning. (Fantasti, 1732: 4, 8, 18, 26, 28, 31)

This suggests that monkeys do not speak not because of any organic defect, but because they lack a spiritual principle, yet in other contexts, Fantasti insists that the monkey "is as skilful as man", and ends up admitting that "monkeys also love, dislike, rejoice and grieve", and therefore they prove to be capable of many human actions. Fantasti did not only criticize those who dared to say that the monkey was a species of Man, but he also wrote more generally "to mock the most foolish Philosophers, who ascribe the motion and the behavior [of animals] to machines only" (Fantasti, 1732: 17, 36–37, 46). He rejected the Cartesian mechanistic theory because it seemed to him to be senseless not to recognize in these animals "passions" and "actions" that machines and automata would never be able to simulate:

> All wild animals love, rejoice, become angry, fight, and when beaten cry out in pain, and among them there is sympathy, and dislike just as there is in us; [...] the artificial bodies invented by Mathematicians do not have any of these passions, nor will they ever have them. If Brute Animals were equal to a Clock, when called the Clock would come, it would answer, and if threatened it would flee, or it would confront any creature which scolded it. The Clock does not live, it does not eat, it does not drink, it does not sleep, and it does not grow, and it does not die [...]. The motion of its Wheels does not depend on the will of the Clock itself, but rather on the hands of its machinery, and then on the will of the person who possesses it. [...] In the Animal [...] one observes an immense number of movements, and actions, which depend on its willpower, such as going, staying, desire, displacement, running, resting and also breathing, [...] all things which neither the Clock nor any other machine created by Mathematicians can show us. (Fantasti, 1732: 37–39)

In his passionate anti-mechanistic manifesto, Fantasti does not mention insignificant animal skills, or anything inferior to human qualities, and he repeats the assessment that these "wild animals" also possess a "will", and certainly does not differentiate the "disgusting Beast" from humans, but rather shows its similarity to the human species. Fantasti considered the role of the soul and the brain in different ways. On the one hand, he seemed to attribute all mental behavior and performance to the soul, considering it as tripartite (vegetative, sensitive, rational) following Aristotle. In other sections, Fantasti seems to attribute the complexity of higher functions exclusively to the complexity of the brain, stating that this "is the main organ of animal actions", and that if monkeys do not speak "it is because the parts that make up their brain are less perfect and less numerous". In this section, Fantasti has significantly changed his viewpoint. While in other pages he states that it was God who denied monkeys the possession of a rational soul, here he claims that nature itself "denied the brain the nerve connections which are the cause not only of sneezing but also of smiling and sobbing". So, the ultimate cause or agent has been laicized and the question reduced to the presence of "nerve connections", and not to God. He also complained about the lack of evidence for "connections", since those who have studied the anatomy of monkeys have not described these in detail (Fantasti, 1732: 26, 37, 43, 48).

The issue remained open:

a. monkeys express sophisticated "passions", they are provided with "willpower" which promotes directed behavior, although with an elementary aim;
b. it is not certain that they have a simpler brain than humans.

Although Fantasti's views were expressed in an ambiguous framework, these were important admissions, especially since they were an attempt to demonstrate the traditional idea of the divide between humans and primates. Fantasti's contribution did not accelerate the humanization process of monkeys, but neither did it slow down the process of including humans in the natural world.

12.6 LINNAEUS' GREAT APES: FROM THE *SATYRUS*/MAN OF THE WOODS AND THE *TROGLODYTES*/*HOMO NOCTURNUS* TO THE TENTH EDITION OF THE *SYSTEMA NATURAE* (1760)

The *Systema naturae* (1735) by Carl von Linnaeus was a turning point in debates about the singularity of modern humans. Humans, apes, and monkeys are not only included in the same Class called *Quadrupedia* (later renamed *Mammalia*) but also in the same Order, *Anthropomorpha* (later on *Primates*) (cf. also Veracini, Chapter 11, this volume). Humans and the other primates/simians meet on the boundary line between two genera (*Homo* and *Simia*), which merge into one another through two species: on the human side *Homo africanus niger*, and on the simian side *Simia cauda carens* (later renamed *Satyrus*), the "tailless monkey" (i.e., the chimpanzee according to Tulp) (cf. Linné, 1735: 12; 1758–1759, I, 1758: 25). Linnaeus' classification provoked heated discussion not because Africans were considered a lower species, but because human beings were classified in the animal kingdom for the first time. Our proximity to the chimpanzee had never before been expressed with so much clarity. As a consequence, the Swedish naturalist was subjected to severe criticism, because by virtue of their intellectual abilities, human beings should be *sui generis*, philosophically, morally, and theologically, and therefore should not be included in any presentation or discussion of natural history. Linnaeus' stance surprised even Denis Diderot (who certainly could not be considered a conservative) and who complained:

> As soon as a classifier has placed, in his system, man at the top of the quadruped ladder, he conceives him as an animal. [...] In vain the sublime reason with which man is endowed protests against the designation of animal [...]; in vain, nature has directed the gaze of man towards the sky: the classification forces man to bend his body towards the earth. (Diderot, 1753: 153)

Why did Linnaeus not even consider intellectual abilities? Why did he want to "humiliate" the human species? So, for example, Buffon wrote "Why turn him [man] upside down and force us to see him only as an animal?" (Buffon, 1749b, II: 437), and:

> Let man examine, analyze, deepen, and recognize without delay the nobility of his being; he will feel the existence of his soul, he will cease to be degraded and he will suddenly see the infinite distance that God has placed between him and the beast. (Buffon, 1753: 110)

Linnaeus thought that mental characteristics should not be used in classifications (e.g., see Linnaeus, 1746: 3v), and he was also convinced that there was not a single physical characteristic capable of distinguishing humans from other primates. "It is very difficult to find the specific difference of man" (Linné, 1746: 3v). Replying to an opponent, he spoke thus: "I ask you and the whole world to show me only one characteristic which makes it possible to distinguish man from an ape. I certainly do not know of one. I would really like someone to point one out" (Linné, 1747: 25).

A few years later Linnaeus moved from questioning these views to attacking them, stating that between man and the other primates, there are not even mental differences: "The stupidest Monkey differs so little from the more knowledgeable Man, that we have still to find the geodets of nature able to draw between them a line of division" (Linné, 1766–1768, I, 1766: 34). He had observed a few years before:

> To many people it seems that between man and the monkey there is a greater difference than between day and night. [...] But they forget , that just as great or perhaps greater, is the difference between man and man: if they compared the greatest heroes of Europe to the Hottentots of the Cape of Good Hope, they would hardly think they could have had the same origin; and if they wanted to compare a noble virgin of the court, dressed and educated in the finest way, with a wild and abandoned man, it would be very difficult for them to believe them to be of the same species. (Linné, 1760: 65)

In 1758, far from renouncing his views about the inclusion of man in natural history by comparing him to the other great apes, Linnaeus doubled down. He thought it was not the chimpanzee, but the orang-utan which was the primate closest to us, and this promotion of the Asian great ape deserves attention. In the meantime (i.e., after 1735), the orang-utan had been described by travelers other than Wouter Schouten, namely the Italian Gian Francesco Gemelli Careri, an anonymous man, and the French naturalist François Le Guat. In his *Giro del mondo* (*Tour of the world*, 1699–1700), Careri had reported the orang-utan to be a "monkey" of "monstrous grandeur", a biped ("it went on two feet"), sensitive ("it cried like a child"), aggressive (it knew how to arm himself with "pieces of wood"), and very intelligent "it seems sometimes, that these monkeys surpass, in their shrewdness men themselves". These observations came from the Philippines (Zamboanga) and Borneo, and Careri focused particular attention and emphasis on observations concerning the "shrewdness" of Asian monkeys (which were not named and probably were not apes, but likely the grab-eating macaque, *Macaca fascicularis*), which were observed to behave in the following way:

> not finding fruit in the mountain, they go to collect crabs. oysters, and things like that at the sea shore. There is a species of oyster, called *Taclovo*, which has many pounds of pulp, and it is found open on the beach. Now the Monkey, fearing that when it tries to eat it, the shell could close, and its arm remain inside, first of all thrusts a stone inside the shell in order to prevent it from closing when it is devouring it and to avoid injury. (Gemelli Careri 1699–1700, V, 1700: 155–156)

This evidence was noticed by, among others, Buffon (see Buffon, 1766b: 58–59). An anonymous man, communicated from Batavia (Jakarta) that "the wild man of Borneo has a round head very similar to the human head, as well as having a wrist on its arm, exactly as we have"; it is very tall, ("the height of these animals, when they have reached full development, equals that of the tallest men"), strong ("I discovered it when I pulled it by the hand, feeling an unusually powerful resistance"), fast ("it runs faster than a deer"), libidinous ("these animals are said to be very lascivious"), aggressive ("in the woods they break tree branches, which they use to kill those who pass") and is a vampire ("when they kill someone they suck their blood, which tastes to them like the most delicious of drinks"). But the same source also suggested they are capable of human actions, for instance, when "sleeping on their sides, they rest on one hand" and perhaps they even have delicate feelings, although "their rather grim expression has a melancholy air" (Anonymous, 1700: 184–186), but this last report passed almost unnoticed. On the other hand, Le Guat, who in his *Voyages et adventures* (1720) referred to the "extraordinary monkey (*singe*) of Java Island", was widely quoted. He tells us about:

> a large female, which often walked perfectly upright on her hind feet; when she did so, with one hand (which was hairy neither above nor below) she hid the part of the body that distinguished her sex ; [...] her face was without other hairs than those of her eyebrows and she was, in general, quite like the grotesque faces of the Hottentot women I saw at the Cape; [...] every day, she made her bed, on which she lay down with her head on the pillow and covered herself with a blanket in the same way that men do; and when she had a headache, she covered her head with a large handkerchief, and it was a pleasure to see her so coiffed in her bed. (Le Guat, 1720, II: 95)

This description was accompanied by a make-believe image. The sequel is more original; in that for the first time, the hypothesis that the distribution area of the orang-utan was limited to Indonesia was put forward: "Some said it was a particular species which is only is found on the island of Java". But Le Guat himself admitted that "few thought so, and the prevailing opinion was that this beast resulted from intercourse between a monkey and a woman":

> When a slave has made a great mistake and has become aware of the dire punishment that is meted out in such

cases, it often happens that she escapes into the woods like a frightened beast, and that she lives there like a beast. And Nature, which is not opposed to the coupling of donkeys and horses, may well tolerate that of a monkey with a female animal that resembles it, if the latter is not held back by any Principle. [...] A monkey and a black slave girl, born and raised without knowing God, are just as likely to have intercourse as an ass and a horse. (Le Guat, 1720, II: 96–97)

In 1758, Linnaeus was convinced by this and other literature (e.g., Brisson, 1756: 189) that the orang-utan is not only the "highest monkey" (i.e., nearer to humans than the chimpanzee), but that it is so close to us that it can be considered *Homo* (Linné, 1758–1759, I, 1758: 24). Does not "Orang-utan" mean, in the Malay language, "Man of the woods"? Is it not known to naturalists as "*Homo sylvestris*"? Do not those who are familiar with it claim "It is able to talk but does not want to, in order to not be forced to work" (Bondt, 1658: 85)? Do travelers not report that "it is as intelligent as other men" (Schouten, 1676, II: 36) and perhaps "surpasses them, in a certain sense" (Gemelli Careri, 1699–1700, V, 1700: 156)? Did not a doctor hypothesize "that the brain of this alleged animal was originally made to feel and think like ours" (La Mettrie, 1745: 158), and has a philosopher not written "I would prefer an hour of conversation with this savage man, rather than with the finest and most ingenious man in Europe" (Maupertuis, 1752: 351)? And did not a clergyman – Melchior de Polignac – address it, saying "Speak, and I will baptize you" (cf. Paul, 1922)?

Linnaeus' chimpanzee (*Simia cauda carens* or *Simia ecaudata, Satyrus*) was a hybrid of chimpanzees, a "chimpanzee-orang-utan" which "lives in Africa and Asia" (Linné, 1758–1759, I, 1758: 25), and he had adopted the position that the orang-utan "lives in Java, in Amboina, in Ternate and in Ethiopia" (Linné, 1758–1759, I, 1758: 24). Linnaeus' main source was still Bondt, whose *ourang outang* appears, in the tenth edition of the *Systema naturae* (1758–1759), as a second human species, provided with the following characteristics:

he "has a white body, he walks upright and is less than half as tall as we are. White, frizzy hair. Round eyes: an iris and a golden pupil. Falling eyelids with nictitating membrane. Nocturnal side vision; Lifespan: 25 years. In the daytime he hides, at night he sees, he goes out and feeds. He expresses himself by whistling; he thinks, and believes - according to travelers - that the earth has been made for his benefit and that in the future he will be its master". (Linné, 1758–1759, I, 1758: 24)

Linnaeus claims he cannot distinguish either animal from *Homo sapiens*, "or only by very slightly different characteristics" (Linné, 1758–1759, I, 1758: 24n.). This shows that the two creatures belong to the same genus. Even Jean-Jacques Rousseau was heading toward this conclusion. For the philosopher, the *ourang outang* might even be a real savage, unable to civilize himself due to a lack of adequate *stimuli*:

all the observations on modifications that an infinite number of causes can produce in the human species, make me suspect that those animals similar to men, considered by travelers to be animals either because they did not thoroughly examine them, or because of some difference that they noticed in their conformation, or simply because those animals did not speak, might actually be real savages. Their race, once scattered in the woods, had not the opportunity to develop any of its potential faculties, so did not acquire any degree of perfection and still finds itself in the primitive state of nature. (Rousseau, 1755: 190)

Rousseau proposed the creatures described by Battell, Dapper, Tulp, and Merolla, all of whom he quoted from the *Histoire générale des voyages* (see Rousseau, 1755: 190–192), were also *wild men* of the type described above, writing:

From these excerpts the reasons why the authors refuse to name the animals in question wild men do not emerge; but it is easy to suppose that it is because of their stupidity, and because they do not speak. These are weak reasons for those who know that although the organ of speech is natural to man, however, speech is not, and they know to what extent perfectibility may have elevated civilized man above his original state. The few lines which contain these descriptions testify that the animals have been observed badly and with many prejudices. (Rousseau, 1755: 192)

Rousseau was unaware many scholars either called them *wild men* (men of the woods, etc.), or were unconvinced of their stupidity, and claimed they could speak. He concluded that "perhaps it will be discovered, after more accurate research, that they are men" (Rousseau, 1755: 195), but his words would have had more force if he had been familiar with the literature. It is probable that Linnaeus did not know of Rousseau's conclusion, but in the tenth, 1758, edition of the *Systema naturae*, Linnaeus goes in the direction indicated by the French philosopher and by travelers, especially Schouten. The human being and the orang-utan are so close that Linnaeus places them not only in the same class and in the same order, but also in the same genus. According to Linnaeus, they come together as *Homo sapiens* and *Homo troglodytes* – the orang-utan renamed to pay tribute to Pliny (77–78, V.45, VII.23 and 31, VIII.26 and 32, IX. 38 and XI.125) – on the blurred borderline between *Homo diurnus* and *Homo nocturnus*. The Swedish naturalist was finally able to reveal human nature, freeing it from the trappings of old metaphysics – as Michel de Montaigne, at the end of the 16th century, had hoped would happen (cf. Montaigne, 1580–1588, II, 12). Linnaeus "downgraded" human beings twice, first when he placed them in the same Order as "monkeys", and then when he unexpectedly upgraded the orang-utan, placing the latter in *Homo*.

The only great ape Linnaeus was able to examine directly was a chimpanzee sent to him in 1758 by the English naturalist George Edwards (see **Fig. 12.12**). For the rest, although Linnaeus was influenced by the reports of amateurs and sometimes of second-hand witnesses, he paid less attention to them than many of his contemporaries. This is the case, for example, with the legend reported by Bondt, according to which the orang-utan "is able to speak but does not want to, in order not to be forced to work".

FIGURE 12.12 Edwards 1758, *The man of the woods*, Pl. p. 6.

The Dutch doctor immediately added "But that is ridiculous, by Hercules!" (Bondt, 1658: 85), but when the Swedish naturalist reported this he omitted Bondt's important commentary (see Linné, 1760: 71). Thus, this myth acquired a credibility it did not have originally.

The replacement of the chimpanzee as the closest ape by the orang-utan, and the upgrading of the latter to the genus *Homo*, likely found their justification in a variety of sources, such as the now comment-less report by Bondt (which, among other things, as we know, presented the "monkey" as a demure creature: Bondt, 1658: 85), the re-elaboration of Bondt's drawing by Tyson (who, in order to assure us of her modesty, portrayed the *wild woman* in the act of hiding her genitals with a twig, and the text of a Swedish scholar, Olof von Dalin, in which he defined "*nivea*", the Asian (Afro) primate, and assured his readers that it "speaks and thinks". Linneaus describes these creatures thus:

> in the centre of Africa there exist a kind of naive men, with white, frizzy hair, long ears, drooping eyelids, round eyes, rosy irises and transparent yellow pupil membranes; side vision, simultaneously on both sides; which nevertheless see better in darkness than in full light; and they live twenty-five years. Their body is thin. They speak and they think, for example, that the earth has been created for their benefit, and they hope that one day they will dominate it. (Linné, 1760: 74, and v. Dalin, 1749: 5)

There is a noticeable paradox, for it is a presumed African sighting that brings the Asian troglodyte – alias the orang-utan of Bondt – to be considered the closest "species" to humans. Linnaeus considered every animal mentioned in the literature to be extant, and therefore he classified them in case they existed in nature. Linnaeus diligently inventoried everything, including *Homo caudatus* – which Gesner and Aldrovandi had reported (see Gessner, 1551: 970; Aldrovandi, 1637: 249–252 cf. Veracini, Chapter 9, this Volume). But he did not classify it because he could not decide "whether it belongs to the genus of man or that of monkeys" (Linné, 1758–1759, I, 1758: 24). He did so shortly afterward when he renamed it *Lucifer*.

Linnaeus' last comment concerning primates was in a 1760 monograph that was the thesis of his student C. E. Hoppius, which included a new ape-man, *Lucifer*. Anthropoid apes are described thus:

> They often walk upright, resting only on their hind limbs, they peel their food and bring it to their mouths with their hands, they drink liquids from empty coconuts, and when they are running out of water they dig springs with their fingers; they are omnivorous like us. (…) They continually groom and clean themselves, love games like children, they are exceptional tightrope walkers and so skilful in gesticulation that they are met with irrepressible laughter. They are mischievous by nature, ready for any prank, dedicated to theft, very lascivious even when they are pregnant; they have a vivid memory of the injuries suffered and it is very difficult to calm them; they worry very easily and are shy hunters; they imitate every crazy act; it is very difficult to succeed in castrating them; fathers and mothers are very fond of their children, even after having nine. They run away when they see crocodiles and snakes, and, a fact that will surprise you, they keep away from apes suffering from contagious fevers. (Linné, 1760: 67–68)

The conclusion is that "the *Anthropomorpha* are very similar to us not only in height but also for their customs" (Linné, 1760: 76). Linnaeus distinguishes four species, whose representation (**see Fig. 12.13**), verbal description, and classification reflects the knowledge at the time. The four species are aligned to form a ladder in which neither *Homo sapiens* nor *Homo diurnus* appear.

a. The first humanoid ape (the *Troglodyta*: Linné, 1760: 72–75) is a representative of the second species of man – after us – *Homo nocturnus*; this is the orang-utan of Bondt, of which a revision is presented. This ape is now corroborated, by the reports of Olof Dalin, Nils Matson Köping, and Georg Eberhard Rumpf. Köping was the first to define it as "snow-white", to affirm that it leads a nocturnal life and to attribute to it "a particular language", assigning it another name: "in Ternate, as well as on Amboina, there lives a strange people called Kakurlako. When they meet them, the natives capture and slaughter them. They are snow white, both as regards the skin and the hair". During the day they remain hidden in holes dug in the ground and when they are extracted from the earth they start to roam tentatively, "as if they eyes had been pierced". At night, on the contrary, the darker it is the better they see us:

then they abandon themselves to theft and looting, robbing the natives of all that they have planted or sown. They have a particular language, made up of flute-like sounds, which has nothing in common with the idiom of the country. (Köping, 1667: 159, and see Linné, 1760: 74–75).

Linnaeus corroborated these data with the words of Pliny: "The *Troglodytes* dig caves and use them as dwellings, they feed on snake meat and screech, instead of talking": Pliny 77–78, V.45).

b. *Lucifer* is a new species introduced by Linnaeus (Linné, 1760: 72–75). In fact, the picture recalls the old *Cercopithecus formae rarae* of Gesner, called also *Homo agrestis* by Aldrovandi. Now, it is also identified with the *Homo montanus* of Bondt:

> in the island of Borneo, in the kingdom called Succodano, frequented by our merchants near Oritia and Adamonte, we find caudate mountain men, which many have seen in the king's palace. [...] Their tail is a sort of extension of the coccyx, of four fingers or a little more, like the severed tail of the dogs called Spligioni, but hairless. (cf. Bondt, 1658: 85)

It was again reported by Köping (see Köping, 1667: 79) and Maupertuis (1752: 350–351) as "an intermediate species between us and monkeys". As far as we can deduce from their description, *Lucifer* was an Asian human being, probably a native of the Sunda Islands, with a congenital tail.

c. The third anthropoid ape, the *Satyrus*, (Linné, 1760: 69–70) is the chimpanzee depicted in 1739 by Gravelot and Scotin, whose engraving is reproduced.

d. The fourth, the *Pygmaeus* (Linné, 1760: 68–69) is the *Man of the Woods*, portrayed in 1758 by Edwards (**Fig. 12.12**) as closer to "lower" monkeys, but in reality it is a chimpanzee, which Linnaeus distinguishes and classifies differently for reasons that are unclear. Edwards himself, in fact, likened it to the ape of Tyson and Gravelot-Scotin:

> "it is supposed that this animal, which is one of the first of the genus *Simia*, is, of all, the one that comes closest, outwardly, to man. Several Englishmen, who have travelled to Africa and the Indies, report that those which have reached full development are, more or less, when they are standing, six feet tall, and that they walk on their hind limbs." [...] The specimen portrayed "is now in this city, London, in the famous British Cabinet. It was young, and when it died it was only two and a half feet tall; it was first immersed in spirit of wine, and then it was dried; finally it was disposed, before it was dried, in the position in which it has been depicted. [...] About fifty years ago Edward Tyson published an anatomical description of this same animal, which he named *Pygmie* and of which he provided two figures. Later, in 1738, another image of one of these apes was published, which came from the coast of Africa, was called Chimpanzee and was shown in London". (Edwards, 1758: 6–7)

To Linnaeus "there are no evident characteristics that distinguished the *Troglodyta* from Man" (Linné, 1760: 75). This resulted in a reversal of positions between the two creatures, because behind the poor representation of the second species of man (the *Troglodyta*) (see **Fig. 12.13**, there was a monkey, and behind the equally poor representation of the first species of monkey (*Lucifer*), there was a man.

So in Linnaeus' final scheme, the boundary between humans and other primates disappears, and people and great apes end up taking each other's place.

12.7 THE DOUBTS OF BUFFON AND OTHER SCHOLARS REGARDING THE LINNAEAN ANTHROPOMORPHIC FORMS

There was no lack of criticism of Linnaeus' classification of *Homo* in the Order Primates. Georges-Louis Leclerc de Buffon denounced Linnaeus' statements virulently and even insulted the Swedish naturalist, whom he called – albeit indirectly – a wild beast: "Man is of a nature so different, so distinct and so superior to that of wild beasts, that we would be as unintelligent as the beasts are, if we confused them" (Buffon, 1749b: 437). According to him, there are only some similarities between human beings and apes (Buffon uses the term *singes* for Simiiformes but distinguishes those without tails, and thus we use the word apes, cf. Table 1, Veracini, Chapter 11, this volume) were morphological:

> "Comparing man with the animal we will find in both a body, organized matter, senses, flesh and blood, movement and an infinity of similar things: but all these resemblances are external and not enough to make us declare that our nature is similar to that of the animal". [...] "There is an infinite distance between the faculties of man and those even of the most perfect animal, and this is clear evidence of the fact that man is of a different nature and that he constitutes alone a distinct class, from which, in order to reach the class of animals we have to descend an infinite distance". [...] It is evident that "man is of a completely different nature from that of the animal, which resembles him only outside and if we judge him according to this material resemblance it means that we are deceived by appearances and we are deliberately closing our eyes to the light which should give us the ability to distinguish appearance from reality". (Buffon, 1749b: 437, 443–444)

Therefore, any Linnaean "fusion" (inclusion in the same genus) between human beings and apes is impossible:

> to consider them as very close [or to fuse them] we would have to find intermediate bodies from every point of view (i.e. species that blur one into another, also with regard to mental performance) and this is not the case.

FIGURE 12.13 Linné 1760, *Anthropomorpha,* Pl. p. 76.

Buffon, maintaining a Cartesian view, stated:

> if men were of the order of animals there would be in nature a number of beings less perfect than man and more perfect than the animal, from which man would develop gradually with nuances between man and the monkey; but this does not happen: we pass without continuity in only one step from a thinking being to a material being; from intellectual power to mechanical force; from order and planning to blind movement; from cognition to appetite. (Buffon, 1749b: 443)

According to Buffon man was not an animal and should be clearly distinguished from the animal kingdom, so should not be placed in either the order of Primates, or in the class of Mammals. Placing him in the animal kingdom, was already, for the French naturalist, an undue "degradation" of the "masterpiece" and of the "king" of nature (Buffon, 1749a: 2). Thus, Linnaeus'attempt to include humans in the animal kingdom through the *troglodyta*, was forcefully condemned by Buffon not only for theoretical reasons and ideological concerns, but because the French naturalist was concerned about the correctness of the rare, and vague reports concerning *nocturnal man* (*Homo nocturnus*). Buffon suggests a different explanation, that *Homo nocturnus* could be an "*albino negro*", whose appearance at least partly coincides with the descriptions of nocturnal man:

> I have serious doubts about the validity of the description of this nocturnal man; in reality I also doubt its existence, and I think that it was an *albino negro*, which the travelers cited by Linnaeus have observed badly and described worse: because these *albinos negros* actually have, like the nocturnal man of our author, white bodies , red eyes, weak voices etc., but it is a man in all respects and does not whistle [...]; it thinks, speaks and acts like all other men. (Buffon, 1766b: 46)

Buffon did not deny the existence of the orang-utan, but he called it *pongo* because the *gorilla* no longer appeared in the travel literature, so *pongo*, the name used by Battell, was attributed to the orang-utan and it also became a synonym for "great ape". But before he could examine one, Buffon relied extensively on travel reports, and he did not provide any depiction, merely placing it in the large, and distinct, family of *Singes*. Regarding *Lucifer*, the other intermediate link proposed by Linnaeus in 1760, it appeared to the French naturalist so unlikely to exist as not to deserve to be referred to. Thus, *Lucifer* and *Homo nocturnus*, the two "anthropoids" validated by Linnaeus, were finally shown to have no foundation.

However, even Buffon stressed there were numerous anatomical similarities between humans and primates. When he reviewed the 48 similarities and 34 differences between humans and chimpanzees identified by Tyson (1699), Buffon suggested three additional similarities (the chimpanzee does not have cheek pads, has prominent buttocks and without callosity, and has calves) and rejected 11 other differences. Out of the remaining 23, Buffon claimed that only two were significant:

> The only essential differences between this animal and man are reduced to the shape of the pelvis bones and the length of the feet. (Buffon, 1766b: 67–70)

The Italian physician, Pietro Moscati, would soon reach similar conclusions. His *Discorso accademico delle corporee differenze essenziali che passano fra la struttura de' bruti, e la umana* (*Academic discourse of the bodily essential differences that pass between the structure of brutes, and the human one,* 1770) concludes in this way:

> no substantial bodily difference exists between them [apes], and us; because our body position is no different from theirs, nor are the structure of the viscera, of bones, nor of muscles; neither are the blood-vessels different, nor the heart, nor the sense organs, nor the nerves, nor the brain. This latter, if it is not, as a brain, less perfect than that of the brutes, certainly it is not better from an anatomical point of view. (Moscati, 1770: 59)

Like Buffon, Moscati concludes that "no substantial bodily difference exists between them, and us", and he also affirms that "human superiority to all brutes [see Glossary]" is, for this reason, disproved – thus questioning traditional anthropology. If indeed "there is not a physical difference in the organic structure which could cause, or even contribute to a cause of human superiority to all brutes", it must be due to the possession of a soul (Moscati, 1770: 14). It is indeed the lack of anatomical differences that can be cited as "proof" of the existence of a spiritual principle, since only this spiritual principle can determine intellectual differences. Thus, Moscati ended up defining the differences between man and the chimpanzee identified by Tyson's *minutiae* (small things of little value) (Moscati, 1771: 17).

Perhaps Buffon would also have referred to the differences between humans and chimpanzees identified by Tyson as *minutiae*, given his representation of the *jocko* (1766: **Fig. 12.14**), which appeared earlier as the *engeco* (see Battell, 1613). Buffon had seen a live chimpanzee or *jocko*, but in the 14th volume of his *Histoire naturelle* (1766), it appeared even more humanized than the Linnaean chimpanzees – to the point of eliciting protest from Peter Camper, who wondered why this ape was depicted with straight knees (Camper, 1779b). From an anatomical and naturalistic point of view, the drawing published by Buffon, which was made by the great Jacques-Eustache De Sève, represented a step back, both compared to the drawings approved by Tyson (see Camper, 1779b: 36) and the illustrations in the *Encyclopédie* of Diderot and d'Alembert, in which the *jocko* appeared (1768) in an even more relaxed pose. Camper may have known of the collection of images published around 1720, by a bookseller from Leiden. These prints were in the style of Dürer's image of a rhinoceros, but Pieter van der Aa had confirmed they depicted living specimens (see Aa, [1720], *Botanophilis et rerum naturalium amatoribus salutem*, p. n.n. *in principio*). Among the many reconstructions, such as another chimpanzee by Tulp (Aa, [1720] and an Orang-utang (plate 11), there was, for the first time, an image of a great ape with its arms around a tree trunk, (**Fig. 12.15**), as well as an image of a great ape offering a

FIGURE 12.14 Buffon 1766b, *Le jocko*, Pl. p. 82.

FIGURE 12.15 Aa [1720], *Satyri Sylvestris Orang Outang dicti*, Pl. 77.

female a tulip. Camper claimed that Buffon was searching for other characteristics to associate primates with us:

> so it is not only ignorant travellers and lovers of exotic legends who credit the fable of the existence of monkeys so similar to man, and indeed who wonder whether they could be men: even the director [Buffon] of the main anatomical Natural History Cabinets help to feed those superstitions. (Camper, 1779b: 36)

FIGURE 12.16 Foucher d'Obsonville 1783, Pl. p. 3.

It is true that in 1768 Buffon unified the chimpanzees of Jean-Baptiste Robinet (*orang-outang, pongo, homme des bois, satyre, barris, chimpanzé, jocko, homme de nuit* e *troglodyte*) putting them in just one category because *"all these names design the same quadruman"* (Robinet, 1768: 151). The latter amalgamated a poor version of Tyson's pygmies, with the primate illustrated in 1783 by Foucher d'Obsonville (1783; see **Fig. 12.16**), which was perhaps recalling the *jocko* of Buffon (to whom his work called the *Essais philosophiques* is dedicated). However, we know that Buffon gave no credence to the caudate, mountain, sylvan, and marine men of many of his contemporaries (defining the reports of travelers as " obscure, incorrect, exaggerated", and often resorting to expressions such as "badly observed" or "badly described": Buffon, 1766b: 44–46). He also observed that "Bondt's *Homo silvestris* is rather a woman, than a female monkey" (Buffon, 1766: 43). Buffon did not attribute any human characteristic to the behavior and mental performance of anthropoid apes. Although their behavior appeared to be complex, he emphasized that it was mechanical and automatic:

> this ape (*singe*), which philosophers and the common people considered a being difficult to define, of an equivocal and intermediate nature between that of man and that of animals, is, in fact, not a pure animal, that on the outside presents the mask of a human but inside is lacking in thought and in all that man does. [...] It is an animal that is indeed beneath several others as regards its faculties, and substantially different from man, even in instinct, temperament, amount of time necessary for education, gestation, development of the body during its life-span; that is, in all the characteristics and forms of behaviour that constitute what is regarded as the nature of a being. (Buffon, 1766: 41–42, see also Martinez Contreras, 1989)

Buffon was very far from elevating these primates to a human level, as Antoine-François Prevost, among others, had done. When in 1748, Prevost translated Battell's text for the *Histoire générale des voyages*, he changed Battell's version "it [the ape] cannot speak and has no more intelligence than animals", into "it cannot speak despite having more intelligence than animals" (*Histoire générale des voyages,* V, 1748: 89). Buffon, on the other hand, did not refer to Battell's account (Buffon, 1766b: 44–45), and Denis Diderot, perhaps inspired by Buffon's *Histoire naturelle*, wrote one of the most pessimistic accounts of the intelligence of a great ape:

> I saw an ape-man: he did not think more than an ape [*singe*] and did not speak at all, but he screamed like an ape; he moved around without stopping like an ape; he was messy like an ape; he got irritated like a ape, calmed down like a ape and shameless like an ape. (Diderot, [1778]: 48)

Buffon's classification of anthropoid apes was not only the most rigorous of its time, but it was also the most effective in showing the distance that separates these animals from the human species. The first section of the *Singe* (properly called, apes, without tails) included the *pongo*, the orang-utan of Bondt and Linnaeus, stripped of its fanciful characteristics, the *jocko* (chimpanzee), the gibbon (Buffon was the first to include it in this taxon, **Fig. 12.17**) and the Barbary macaque (see Buffon 1749–1789, XIV, 1766: 43–310; about Buffon's classification of primates see Veracini Sec. III, Ch. 3 this volume). This classification was adopted, with some variations, by Schreber, Blumenbach, and Cuvier (see Schreber, 1775–1810, I; Blumenbach 1779–1780, 1782, 1799; Cuvier, 1797, 1817). But Buffon's classification did not deter other scholars from publishing different schemes (cf. Veracini Sec. III, Ch. 3 this volume).

In 1778 Arnout Vosmaer examined young orang-utans sent to him from Borneo, but because he appreciated they were immature being half as tall as suggested by Buffon in 1778, he concluded that the great orangutan (the *pongo* of 1.6 m) was a figment of imagination, and was adamant that a primate taller than two and a half feet "certainly does not exist" (Vosmaer, 1778: 11 and n.). Anoted by the students of Georges Cuvier, the engravings of Vosmaer were "for a long time the best representations of the orang-utan" (Cuvier, [1836–1849], 108n.) and these were used by Jean-Nicolas-Sébastien Allamand (1785), in the *Nouveau dictionnaire d'histoire naturelle* (1818), and by Frédéric Cuvier with Isidore Geoffroy Saint-Hilaire (1824–1847). These images were consistent with the existence of just

FIGURE 12.17 Buffon 1766b, *Le Grand Gibbon*, Pl. II, p. 108.

one orang-utan that was "short" and "small". This theory received support from Peter Camper, who, quoting among others, James Cook ("We saw a live *Ourang Outang*: it came from Java and was only two feet six inches tall", Cook, 1777, IV: 158), argued that the short orang-utan was the only orang-utan (Camper, 1779b: 25, 37).

In the meantime, Buffon was promoting the existence of a tall orang-utan, which had been depicted in 1771 by Allamand and corroborated in 1780 by Jacob Cornelis Radermacher. Radermacher also made the error of recognizing immature orang-utans as a distinct species, but he did not reject the existence of a second taller orang-utan species, suggesting that it "is wrongly considered just a myth" (Radermacher, 1780: 140–144). Radermacher had lived for many years in Batavia, where he founded, and was the president of, the local Society for Sciences, so his report based on first-hand observation was considered authoritative. Nevertheless, Buffon considered Vosmaer and Camper to be even more authoritative than Radermacher, so he rejected the notion of a tall great ape and concluded that "*pongo*" (always considered by him a great ape) must be a chimpanzee.

12.8 PHILOSOPHICAL DEBATES AND LITERARY DIGRESSIONS ON ANTHROPOID APES AFTER LINNAEUS

The difficulty in accepting Linnaeus' new perspective, and the conflicting interpretations of the great apes and other primates in the 18th century, is evident on a social and philosophical level as well as at a scientific level. In his *Philosophie de la nature* (1770) the French philosopher Jean-Baptiste Delisle de Sales introduces two ideas that illustrate the persisting empirical difficulty of defining great apes, and the associated difficulty in defining the boundaries between humans and ape-like creatures. The first idea concerns the lack of speech in the "orang-utan", and the second is the results of their mating. Although Delisle de Sales understands that the "Orang-utan or Man of the woods, the biped that most resemble us, is a being that lives in the Torrid Zone of our hemisphere", he does not know what characteristic to attribute to it because "so far travellers have not been able to observe these when they come across this singular creature". Nevertheless, Delisle de Sales is convinced that although "the fact that the Orang-utan cannot speak, despite having speech organs, proves" that its inability to speak is because "it lives in an isolated and lonely state". He suggests that similar isolation is one of the reasons some modern humans are mute, citing the case of "the man who was found in Hanover's forests in 1724 lived in isolation and alone; that is why he did not speak" (Delisle de Sales, 1770, I: 217 and IV: 244–245, 248, 252).

Delisle de Sales also asked what would be the result of the mating of a Wild man with a Man *tout court*? He is convinced of the possibility of such a mating because "all travelers agree that the Orang-utan desires our women with a passion equal to that with which it wants its own females and that, in particular it rapes black women", and "Locke, who doubted many things, did not doubt at all that a woman could be fertilized by a monkey" (Locke, 1690, III, VI, 23), and he suggested that it is from such matings that the Albino or Negro spring. In one of the central chapters of *Philosophie de la nature* Delisle de Sales shows how difficult he found it to locate this "creature", considering it both "a kind of monkey" that "ends its unfortunate career at thirty, without having lived", while at the same time being "one of the last links in the great chain of varieties of the human species", in consideration of the fact that "the Albinos are inferior even to the Negro" (Delisle de Sales, 1770, I: 217 and IV: 181–182). The uncertainty led to Delisle de Sales persecution, and his *Philosophie de la nature* was burned at the Châtelet. His contemporaries were of the opinion that the Albino was not an accidental pathology but a population of individuals ("a distinct race, a constant variety": Delisle de Sales, 1770, IV: 186), equivalent to the Kakurlako of Köping, the snowy man of Dalin, the *Troglodyte* of Linnaeus.

Another example of the overlap between the literature of the imagination and the scientific literature was *La découverte australe* (1781) by the French philosopher Nicolas-Edme Restif de la Bretonne. In that work *ape-man* appears as both "a four-legged furry animal" that does not speak and as an animal with no tail that emits cries "that are uttered in a way that testifies to a combination of ideas and therefore constitutes a language". Even if they do not speak, the *ape-men* nevertheless "understand", and it is clear they "consult and deliberate" (Restif de la Bretonne, 1781, II: 272–274). According to Restif de la Bretonne, "they are very wicked (*ibidem*: 275, 277) but also very intelligent", so that one day, "an ape man gave checkmate to his master and, because he was given a hard slap, was careful, from then on, not to do the same thing again" (*ibidem*, IV: 171–172).

These two forms are opposed, but also complementary. One version of the *ape-man* is physically a monkey and intellectually a man (wicked). The other, which is similar to Linnaeus' *nocturnal man*, is physically a man and intellectually an ape (good). The latter has a "pale pink" skin and an "acute and strident" voice and is only active at night because it is averse to the light of day (from which it protects itself by staying in deep caves), and is "infinitely limited from the point of view of intelligence" and even seems "unable to create tools" (Restif de la Bretonne, 1781, I: 187–189; II: 264, 267). The author assures us that *nocturnal man* is not a few pathological individuals (i.e., monsters), but it is "a real race, that other men annihilated wherever they met them, and that therefore they survive only in inaccessible regions" (*ibidem*, I: 190). Restif also assures his readers that the *nocturnal man* version of *ape-man* can interbreed with the race of diurnal men (*Homo diurnus* was in Linnaeus, as we know, synonymous with *Homo sapiens*, cf. Linné, 1758–1759, I, 1758: 20), and that "their offspring are susceptible to education" and "their sole defect is squinting in the daylight" (*ibidem*, II: 267, IV: 330). Finally, in Restif de la Bretonne's *Découverte* we find an ape, in the person of Caesar of Malacca (see **Fig. 12.18**), who takes pen and paper and writes a long letter to the animals of his species (*ibidem*, III: 19–22), citing 23 varieties, including the *pongo, baris, and smit.* Caesar, the son of a son of a Malaccan woman and a baboon, who mated with a Barbary macaque, is portrayed in meditation at a desk, where he launches into a passionate defense of his species and an equally impassioned condemnation of society and of human religions. Restif invents a good half of the literature produced in support of his story, but he knows, and cites, the texts of Buffon, Bosman, Gesner, Kolb, Battell, Smith, Labat, Dapper, Tulp, Merolla, Barbot, and Gemini Careri (see Restif de la Bretonne, 1781, IV: 95–138).

Even scientists were not averse to the same kind of speculation. For example, with respect to the possibility of mating between humans and apes, and in particular about the libidinous nature of the latter, the physician, philosopher, and provocateur), Julien Offray de La Mettrie suggested that:

> often Satyrs throw themselves with fury on armed men, as well as on women and girls - to whom they actually do less harm. There is nothing more lascivious, more impudent and more inclined to fornication than these animals. Indian women do not think twice about going into the caves where they are hidden, and the Satyrs make love with them with as few scruples as dogs have. (La Mettrie, 1745: 157)

FIGURE 12.18 Restif de la Bretonne 1781, III, *Le singe. César de Malaca*, Pl. p. 18.

As regards the possibility of offspring of such a couple, the naturalist Eberhard August Wilhelm von Zimmermann had recently argued that:

> since the Orang-utan [he was reffering to the Tyson chimpanzee] looks a lot like man, from the morphological point of view, [...] it is not at all impossible that they produce an intermediate creature. It was said that such an attempt was made in London, where a young prostitute was offered to a male of Orango and was paid for this. Certainly the experiment (assuming it happened) was not successful. But how could it be? [...] Not even a man, instead of the Orang-utan, could have done better in a single meeting. Moreover the violence of the temperament of apes is known. They have rapid ejaculations from which reproduction cannot be expected. A worthwhile experiment would be one made with a man and an Orang-utan female, which had the opportunity to frequent each other for some time. (von Zimmermann, 1778–1783: 195–196)

Writers continued to explore the consequences of the 1760 ape taxonomy of Linnaeus. In the domain of natural sciences, it both confirmed the exchange of positions between man and orang-utan, and the notion that the chimpanzee was superior to some human populations. For example, Claude-Nicolas Le Cat, a French physician, declared categorically that the African "monkey" "is superior to the Moors, who are more tigers than men" (Le Cat, 1765: 35), and equally vehemently the French naturalist and philosopher Jean-Baptiste Robinet observed that "faced with the orang-utan [but he was thinking of the chimpanzee], one is tempted to ask: what is missing for him to be a man?", and with respect to certain human races Robinet asked "what kind of animals are they?" (Robinet, 1768: 155). Also Zimmermann, who conducted his research in an evolutionary context (of Buffonian influence, cf. Barsanti, 2005), suggested:

> Buffon is amazed by the fact that, although the orang-utan has a brain very similar to that of man, it does not think. Firstly I would like to observe that it is not so simple to establish whether it thinks or not; and then that the level of perfection of intellectual abilities escapes the scalpel of the anatomist. And finally that the distance between Newton's intellect and that of a New Holland native is certainly greater than the distance between the intellect of the latter and that of the orang-utan. (von Zimmermann, 1778–1783: 201)

The French mathematician and philosopher Alexandre Savérien, quoting Gassendi, Battell, Schouten, de la Brosse, Grose, Pyrard, and Brisson (cf. Savérien, 1778: 277–283), made the more frivolous observation that chimpanzees "are smarter than the Savoyards of Paris; because these are unable to make jokes in the theater, while chimpanzees can think and ratiocinate in the woods" (Savérien, 1778: 29).

12.9 NOMENCLATURAL ISSUES IN THE SECOND HALF OF THE 18TH CENTURY AND THE PERMANENCE OF THE ORANG-UTAN – CHIMPANZEE – GIBBON DISTINCTION

The many name changes in the second half of the 18th century were obstacles to the progress of understanding the nature of the extant apes. Some continued to call the "new" short orang-utan a *pongo* (that had become synonymous with a great ape) but in order to emphasize it was a "small ape" they named it *Pongo pygmaeus* (*pygmaeys* was originally the name of the chimpanzee). This label, which historically is equivalent to "*Gorilla chimpanzee*", and semantically to "*Great Small Ape*", is still used today for the scientific name of the Bornean species of orang-utan. A bizarre name, but justified by the sequence of events. Other naturalists, including Buffon, preferred to use the name *pongo* for the chimpanzee too, using the other name (*jocko*) for the orang-utan – because as *pongo* had become synonymous with a great ape, *jocko* had ended up meaning a small ape, which was a fitting name for the "small" version of the orang-utan.

The name *pongo* was used for all three great apes in the current genera *Pan, Pongo,* and *Gorilla* (cf. Battell, 1613; Buffon, 1766b, 1789). Today the chimpanzee bears the Linnaean name (*Pan troglodytes*) even though *Troglodytes* was originally used by Linnaeus for the modern orang-utan. The label *troglodyte* was used by Pliny (see Pliny, 77–78, V.34, 43, 45) for an African human population and then passed to the chimpanzee (Buffon, 1789: 1–4).

The nomenclature of the apes had always created problems, but in the final years of the 18th century the proliferation of names, combined with changes in the way the names were used, created so much confusion that naturalists were no longer able to understand each other. Because it became almost impossible to identify the primate to which people's observations referred, naturalists were compelled to compile long lists of synonyms, as well as adding the names of the naturalists who had described them, and the titles of the publications (see also Veracini, Sec. III, ch. 2). Despite these precautions, researchers engaged in time-consuming and unproductive controversies. For example, Vosmaer discovered young orang-utans do not have fingernails. This should not have created problems, but given Vosmaer's belief that the African apes were also orang-utans, the Dutch naturalist accused all his colleagues of unpardonable laxity, starting with Tyson, who claimed that the chimpanzee did have fingernails (see Vosmaer, 1778: 1, 4, 6). Likewise, Vosmaer blamed Buffon for drawing an ape with very large lips and large ears extending very far from the head (see Vosmaer, 1778: 5). These are typical characteristics of the chimpanzee, but Vosmaer had mistaken it for the orang-utan. For the same reason, Camper checked whether the chimpanzee had the vocal organs attributed to it by Tyson. So, with great care, he sectioned seven specimens of the Asian great ape (see Camper, 1779a) and discovered, among other things, organs never described (e.g., the laryngeal sacs typical of the orang-utan) and was indignant that no one had ever described them in the chimpanzee (see Camper, 1779b: 49–53), despite them not occurring in chimpanzees. Others, such as Johann Elias Ridinger, were so confused that they described a chimpanzee, depicted it as a gibbon with a human face, and attributed to it the behavior of a gorilla, as reported by Battell at the beginning of the 17th century. Ridinger, who claims to have observed these creatures, writes:

> The *Gibbon* or *wild man* deserves the first place. These beasts are very similar to men, and in particular to Negroes. They mostly live in dense forests and sleep in trees, where they erect a kind of roof with leaves and branches to protect themselves from the sun and rain. They feed on fruits and nuts. They have so much strength that they often escape from ten men who are trying to catch them and frequently kill the Negroes in the forests, where they move in groups.

Other doubts existed about some actions attributed to "gibbons", such as the fact that they capture and hold men prisoner, treating them honorably and providing them with the necessary nourishment. But there were, also many certainties:

> their height ranges from five to six feet and their homeland is Africa. When they walk they most often stand on their hind legs, although they also run on four limbs. Their arms are almost as long as their bodies and touch the ground. Since the Negroes find it difficult to catch large specimens, they first kill the mothers, whose bodies are held up by the smaller ones, and then capture the latter, which they capture to serve them. (Ridinger, 1768, II: 13)

The naturalist who was able to manage this complicated situation best was Johann Friedrich Blumenbach (see the more and more precise classifications contained in his *Handbuch*, from 1779–1780 to 1799), and the person who provided the complementary iconography was the German naturalist Johann Christian Daniel von Schreber (Schreber, 1775–1810). But it was a Scottish philosopher, James Burnett, Lord Monboddo, who did most to clarify the differences between the African apes. The report of a captain in the merchant navy distinguished clearly between the *chimpanze* (chimpanzee) and the *impungu* (gorilla). The description of the *impungu* had, however, still some imaginary elements mixed with real morphological and ecological data:

> This marvelous and formidable product of nature walks upright like a man; it is seven to nine feet tall [...] and incredibly strong; it is covered with very long hair, black as coal, on the whole body, which is even longer on the head; it has a face more like that of man than that of a the chimpanzee, but with black skin, and has no tail. [...] it lives off the fruits and roots found in its habitat, [...] and when it happens to find itself where there is no water it drinks the juice of a tree; and often it takes these trees with it when it moves, in case it does not find water along the road. I have heard that, with its prodigious strength, it can break down a palm tree, to obtain its liquid. [...] When it sees a Negro, it often chases and captures him; sometimes it kills him, sometimes it grabs him and takes him with it. Those who have escaped from it say that it does not lie down when it sleeps, but leans against a tree. In this position, when the prisoner sees it asleep, he wriggles loose and sneaks away; but sometimes he is discovered and recaptured. (Burnett, 1774: 281–2)

This, the only report of the gorilla found in 18th-century literature, was either ignored or given scant attention among naturalists. For example, Georges Cuvier denied the gorilla's existence still fifty years later (Cuvier, 1829–1830).

12.10 THE STORY OF THE ORANG-UTAN UP TO ITS DESCRIPTION IN THE WILD: 1797–1849

In 1780, Baron Friedrich von Wurmb, an amateur naturalist who lived and worked in Batavia (today's Jakarta) as an official of the Dutch East India Company (aka the United East India Company), and who was also Secretary of the local Society of Sciences, published a report that reinforced the evidence that the *pongo* was real and was the creature described by Bondt, Linnaeus, and Buffon. Friedrich von Wurmb's report concerns an orang-utan, presumably an adult, as tall as a man that had been captured and killed a short time earlier by a Dutch merchant, who sent it to Wurmb immersed in brandy. Measuring 4 Rhenish feet and 1 inch (see Wurmb, 1780: 6), which is about 1 m and 30 cm, it was the tallest ape ever described.

In 1782 Johann Friedrich Blumenbach placed the *pongo* at the top of the ape scale as regards height (see Blumenbach, 1782: 62), but he was referring to the African ape of Andrew Battell, and in 1788 when he confirmed the *pongo* classification, he eliminates any reference to Battell and to other African *pongos* (see Blumenbach, 1788: 64). In 1791, Blumenbach who referred to the *pongo* as being as tall as a man living in the Indonesian islands, as the first among the great apes. Like Jacob Radermacher, Blumenbach did not consider it "a myth" (Radermacher, 1780: 141), perhaps because he had read the account of his compatriot Wurmb. Wurmb's "Batavian *pongo*"'was met with skepticism in Europe, which was dominated by commentators who only accepted the existence of a "short" orang-utan species. Wurmb's specimen was sent to Blumenbach, but, according to rumors it was lost in a shipwreck (cf. Greene, 1959: 232), and Wurmb himself wrote that "unfortunately we heard that the ship had failed" (letter of 18 February 1781, quoted in Huxley, 1863: 30). In fact, the specimen had been stolen by the French army (which had occupied Amsterdam in 1795 and seized the entire Stathouder Natural History Cabinet) and then taken to the Museum of Natural History in Paris, where the *pongo* was examined by Étienne Geoffroy Saint-Hilaire (**see Fig. 12.19**).

Geoffroy Saint-Hilaire, who suggested the specimen was tall and could stand (see Geoffroy Saint-Hilaire, 1796: 114; 1797: 26; 1798: 345), used Camper's facial angle, which was introduced around 1768, as a sorting criterion. The facial angle is the angle between the "horizontal line", which connects the external acoustic meatus and the root of the nose, with the "facial" one, which connects the incisors and the prominence of the frontal bone (see Camper, 1791). The facial angle of the "Batavian *pongo*" was only 30°, which was low for primates because of its prognathism. The facial angle distinguished the tall orang-utan from the chimpanzee, but unfortunately it also distinguished it from the "short" orang-utan, which turned out to have a facial angle of 60°. I was able to examine Wurmb's orang-utan skeleton, and when I checked the measurements and conclusions of Cuvier and Geoffroy Saint-Hilaire, I discovered many instances of downgrading the orang-utan (see Barsanti, 1989). Therefore, to most of the scientific community, the orang-utan remained a baboon until 1817 (see for example Audebert, 1797–1800; Lacépède, 1801; Latreille, 1801; Tiedemann, 1808–1814; Illiger, 1811; Geoffroy Saint-Hilaire, 1812a, b; Cuvier, 1817), and it was not until the publication of further research by Cuvier (see Cuvier, 1829–1830) and Owen (see Owen, 1835) that the

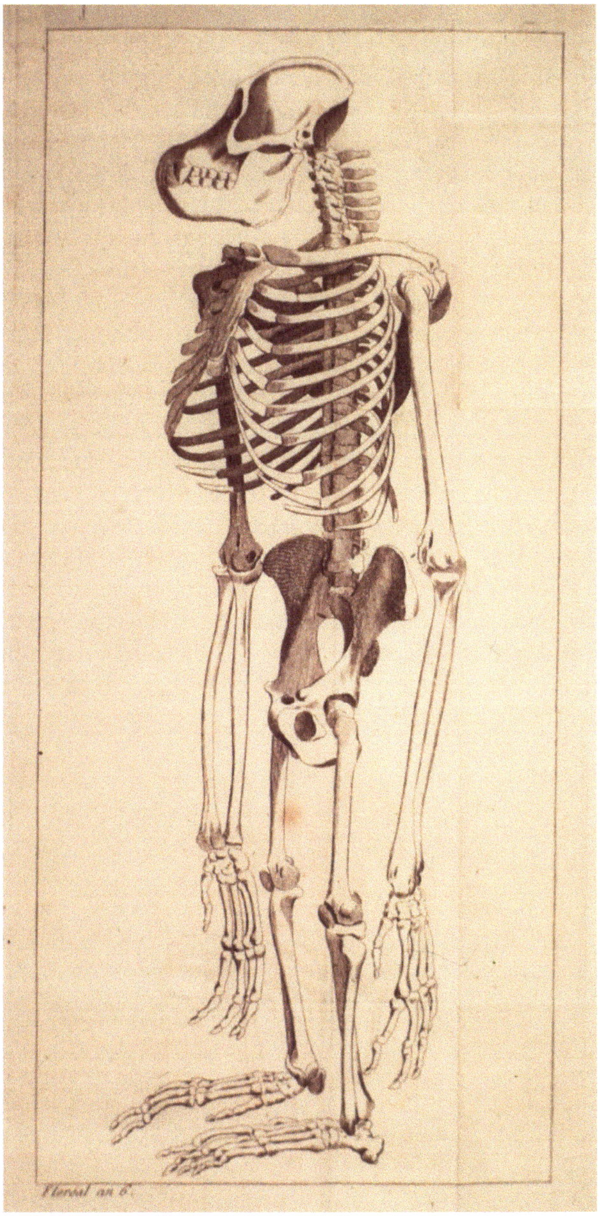

FIGURE 12.19 Geoffroy Saint-Hilaire 1798, pl. p. 400.

"Batavian *pongo*" was finally recognized as the adult form of the orang-utan. This debate, which involved some of the most famous naturalists of the time, deserves to be more widely known.

In 1797, Audebert reiterated that the "ape of Wurmb" belongs to "a species different from all those we know of", and is close to the mandrill (Audebert, 1797–1800: 21, 23–24). Four years later Lacépède suggested it belonged to a distinct genus beneath the macaques (see Lacépède, 1801: 3–4) and Latreille confirmed that the "*Pongo* of Wurmb" occupies "almost the last place in the great family of monkeys (*singes*)" suggesting "it must be placed between the mandrills and the alouattes [howler monkeys]" (Latreille, 1801: 263). Later on, Tiedemann, who coined the expression *Pongo wurmbii*, followed Lacépède in considering it a genus in its own right, and relegated it even lower, to the seventh and last place on the monkey scale (see Tiedemann, 1808–1814, I, 1808: 317–329) and Illiger placed it as a species among the *Cynocephali* (see Illiger, 1811: 67 and ss.), but, in the wake of Tiedemann, first Geoffroy Saint-Hilaire and then Cuvier, demoted it to the last place among monkeys, beneath the mandrill, (see Geoffroy Saint-Hilaire, 1812: 3–5; Cuvier, 1817, I: 102–111).

This happened as a result of extra-scientific considerations. Audebert denied that an "intermediate species between man and animals" existed and, in order to reaffirm the superiority of our species, he even denied that the great apes were able to perform elementary actions (see Audebert, 1797–1800: 12). As for Cuvier and Geoffroy Saint-Hilaire, Augustin-François Silvestre interpreted their presentation at the *Société Philomatique* as indicating that their project always intended to show that "orang-utans are real monkeys" (i.e., forms that could never be considered intermediate links). These animals, they argued, "have much more affinity with animals than with man, to whom we have the habit of comparing them" (Geoffroy Saint-Hilaire, 1798: 107). Concerning the *Pongo* of Wurmb, Geoffroy Saint-Hilaire declared that "the many errors that obscure the natural history of apes (*singes*)" are all due to the fact that "it is believed to be possible to pass from human to animal nature through almost imperceptible nuances" (Geoffroy Saint-Hilaire, 1796: 113).

The status of man as nonpareil was confirmed by Pierre André Latreille some years later, who extended his censorship to the whole Enlightenment culture. The French naturalist claimed:

> until today reason had made this cry resound at the bottom of the conscience of all men, in all ages and in all populations: they knew how to know each other and respect their own dignity. But then came men who dared to doubt this supremacy and who were not ashamed to compare themselves to brutes.

Latreille had harsh words of condemnation for men who cast doubt on human supremacy, distinguishing two kinds:

> the first ones, sad toys of the grosser error, proposing similar paradoxes, are worthy only of compassion; but there are other partisans of this error, of which we must highlight, if possible, the opprobrium and the ignominy: they are those foolish, self-styled philosophers destined to illuminate the universe, who, by denying their own reason, wish to demean man, lowering him to the condition of beasts, to prepare for him the same fate, after having extinguished in his soul any principle of justice and virtue. It is therefore important to know these strange animals [apes], because they have become, to the eyes of these self-styled sages, our rivals. (Latreille, 1801: 154–155)

The real rehabilitation of the *pongo* occurred in 1818 when Cuvier received from Nathaniel Wallich, director of the Natural Reserve of the VOC (Dutch East India Company), the skull of an orang-utan whose morphology was intermediate between that of the *orang* (the young, short, version) and the *pongo* (the adult, tall, version). All the orang skulls examined up to that moment were immature ("they had not yet lost their milk teeth"), but the Wallich specimen was "of medium age" and was surprisingly close to the form of the *pongo* (it "already has a very prominent snout and a very receding forehead; you can already see hints of temporal and occipital crests"). Cuvier finally came close to accepting that the "two" species are just one, stating that "it is not impossible that the great ape of Wurmb is simply an adult orang-utan". Although expressed in the third person, his self-criticism was severe "It was, therefore, beyond any doubt Cuvier himself, misled by the relative smallness of the skull, who relegated him [the orang] under mandrills" (Cuvier, 1818: 210–211). Henri Ducrotay de Blainville agreed with this hypothesis (see Blainville, 1818: 331), but as Anselme-Gaetan Desmarest suggests, Cuvier himself was still uncertain and had also thought, alternatively, that the three monkeys (the orang, the specimen of Wallich, and the *pongo*) were distinct species (see Desmarest, 1818: 572). Although the unification hypothesis prevailed (see Virey, 1818: 597), Desmarest still considered the *pongo* as a separate genus from the orang-utan (see Desmarest, 1820: 30–46). In 1825, Frédéric Cuvier observed that his brother's hypothesis, that the *pongo* was the adult form of the orang-utan, was only "a conjecture" (see Cuvier, 1825: 284), and in 1826 Antoine Desmoulins was in favor of leaving the classification as it was (Desmoulins, 1826: 304: the *pongo* "is of the same species neither of the orang-utan, nor of any gibbon"), and in 1827 Jean-Baptiste Bory de Saint-Vincent supported the compromise of Desmarest (see Bory de Saint-Vincent, 1827: 226). But in 1829, while René Primevère Lesson was documenting the various phases of the controversy (see Lesson, 1828–1837: 264–268, 335–348), Georges Cuvier finally eliminated the *pongo* both as a genus and as a species, and spoke about it in the appendix to his chapter on the orangutan: "we can consider it an adult, if not of the species of the orang-utan, at least of a very close species" (Cuvier, 1829–1830: 88–89).

In 1835, Richard Owen stated that the *pongo* was certainly "an adult orang-utan" and placed this species immediately below the chimpanzee (Owen, 1835: 372–373). He followed Tyson by listing and enumerating the differences, but this time including the orang-utan:

a. the chimpanzee and the orang-utan differ from us in 21 common characteristics and between the 2 in 23 specific characteristics;
b. the chimpanzee is closer to man than the orang-utan because there are 16 common characteristics, while the orang-utan and man have just three characteristics in common (see Owen, 1835: 344–370).

Owen, having taken 102 measurements of young and adult specimens of the two species (Owen, 1835: 374–375 and Fig. 54), concluded that "it follows that the *Chimpanzee* must be placed on a descending scale above the *Orang-utan*" (Owen, 1835: 369).

The reassessment of the *pongo* involved going against the evidence provided by the facial angle. When Desmarest used it as a sorting criterion, the first of the "monkeys" (the *Troglodyte*) had an angle of 50°, the second (the *Orang*) 65°, the third (the *Pongo*) 30°, the fourth (the *Colobe*) of 40°–45° (Desmarest, 1820: 30 and ss.). Evidently, the facial angle was useless if not misleading for ranking taxa, and it was soon abandoned, even by Geoffroy Saint-Hilaire. The latter was the naturalist most affected by the revaluation of the *pongo*, and he resisted it to the last by trying to have the taxonomic solution of Desmarest and Bory de Saint-Vincent (see Geoffroy Saint-Hilaire, 1828, 5e leçon and 7e *leçon*) accepted. Eventually, Geoffroy Saint-Hilaire had to concede that the facial angle was both ineffective and unreliable, and the classification used in his *Cours de l'histoire naturelle des mammifères* (1828) was devoid of this kind of quantitative support. The only facial angle he provided was for the chimpanzee, and was forced to admit that it varied between 50° and 60° depending on how you measured it (Geoffroy Saint-Hilaire, 1828, *7e leçon*: 17). The objective geometric method, which in 1795 had seemed so rigorous and promising, had resulted in misjudging the orang-utan and in an arrangement of modern humans (**Fig. 12.20**) in which the Negro was taken as being transitional between the chimp and the European, both physically and intellectually.

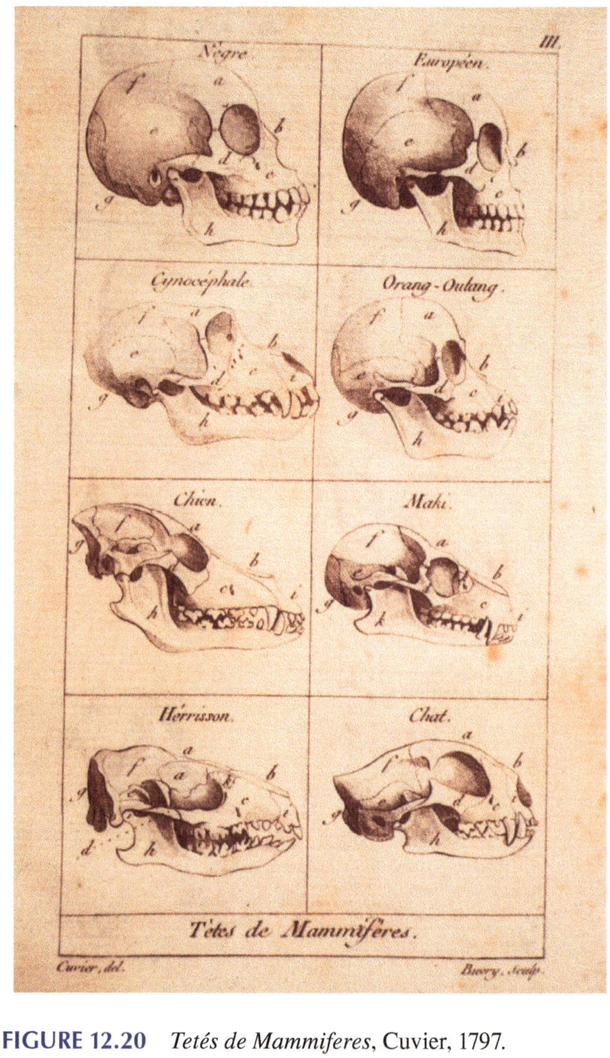

FIGURE 12.20 *Tetés de Mammiferes*, Cuvier, 1797.

When Frédéric Cuvier and Isidore Geoffroy Saint-Hilaire finally had the opportunity to examine a *pongo* specimen ten to eleven months old, which arrived in Paris in 1808 (see Chapter 1854: 144), they correctly stated that it "belongs to the same species as the apes described and depicted by Tulp, Edwards, Vosmaer, Allamand and Buffon: the *Simia satyrus* of Linnaeus" (Cuvier and Geoffroy Saint-Hilaire, 1819–1842, VI, 1824: 1). In their *Histoire naturelle des mammifères* they even provide accurate and detailed morphological and behavioral information, writing about its "reddish coat" (Cuvier and Geoffroy Saint-Hilaire, 1819–1842, VI, 1824: 2), but their depictions did not live up to what they promised in their subtitle ("original drawings, made from live specimens"). In fact, they published a version of the orang-utan of Vosmaer portrayed by Allamand in 1785 and presented it again in Vosmaer's version of 1778). Vedere se mettere questo nella nota, that they defined "exact, absolutely correct" (Cuvier, 1829: 3), and allowed – moreover – that it was colored with delightful bluish hair tending to turquoise.

In England, the recognition of the orang-utan was even more delayed. A seated mandrill was mistaken for an orang-utan (*Simia satyrus*) that was reminiscent of Edwards's chimpanzee, and it was claimed that the "wild man of the woods" inhabited not only Borneo and other regions of the East Indies, but also the interior of Africa and Madagascar.

As for the understanding in popular literature, in 1818 the *Nouveau dictionnaire d'histoire naturel* presented a caricature of Vosmaer's ape, and in 1831 the *Orang roux* that was shown in the *Dictionnaire classique d'histoire naturelle* was inaccurate. But these representations were soon superceded by the images of the chimpanzee and the orang-utan that appear in the fourth edition of Cuvier's *Règne animal*, published between 1836 and 1849, in 20 volumes, edited by his pupils – Jean-Victor Audoin, Charles-Émile Blanchard, Gérard-Paul Deshayes, Alcide d'Orbigny, Louis-Michel Doyère, Antoine-Louis Duges, Georges-Louis Duvernoy, Charles-Léopold Laurillard, Henry Milne-Edwards, François-Désiré Roulin, and Achille Valenciennes. These images, created by Jacques-Christophe Werner, show the *Simia troglodytes* which lived in the menagerie of the Parisian museum between 1837 and 1838, and the *Simia satyrus* which was housed there between 1836 and 1837 (see Cuvier, Cap, 1854: 144–145). The chimpanzee was not brownish, as it had been as a *jocko*, instead, it was "covered with black hairs", with the appropriate posture and locomotion **(see Fig. 12.21)**, and the young Sumatran male orang-utan that replaced the bizarre representation in the previous edition (Cuvier [1836–1849], I, [1836]: 109) was also taken up by the "popular" press **(Fig. 12.22)**. The *Dictionnaire pittoresque d'histoire naturelle* had already published the entry for chimpanzee under the letter C in 1835, but the Dictionnaire entry for the orang-utan (see **Fig. 12.23**) showed it correctly in two new postures. This individual portrayed was called Jack, and according to Paul-Antoine

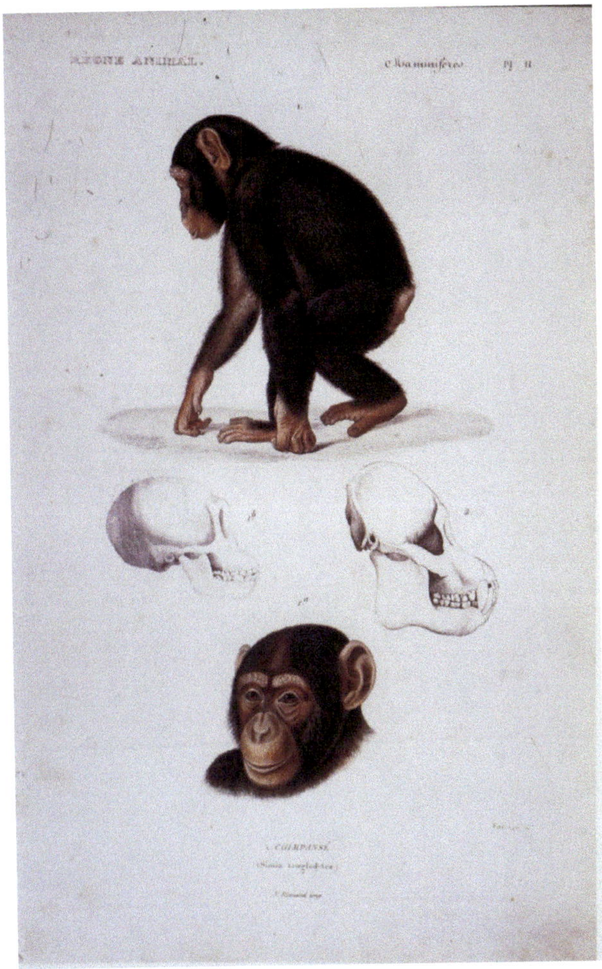

FIGURE 12.21 Cuvier [1836–1849], I, [1836], p. 110; [1837], *Chimpansé (Simia troglodytes)*, pl. II.

Cap "it was characterized by its sweetness, its likeableness and its behaviour, sometimes appropriate, sometimes awkward". "It climbed with ease along a rope placed in its quarters and, when it sat down, crossed its legs as the Turks do". But what had caught the author's attention was above all the fact that:

> it could not bear to be alone. The proximity of a dog at first made its isolation less painful, but it soon got tired of it. It needed the company of people [...] it loved to play, especially with children, and with its guardian it lived, so to speak, familiarly. (Cap, 1854: 144–145)

Coenraad Jacob Temminck, the director of the Museum of Natural History of Leiden, described and depicted the orang-utan in the wild (see **Fig. 12.24**). He made the point that it only lived in Borneo and Sumatra because "volcanic Java cannot. and could never feed the Orang-utan", because "it lacks wide extents of alluvial soil" (Temminck, 1846–1849, I: 322, II, p.87), He reports:

> they "live in these [of Borneo] humid forests, in these inaccessible retreats in which the dense foliage intercepts the rays of the sun. They slowly cross the majestic summit of these great regions and feed on the numerous fruits that the aquatic forests produce. It rarely happens

FIGURE 12.22 Cuvier [1836–1849], II, [1837], *Orang-outang (Simia satyrus)*, tav. 10.

FIGURE 12.23 *Dictionnaire pittoresque d'histoire naturelle*, VI, 1838, *Orang-Outang*, tav. 428 f.t. *in fine*.

> that they are caught on the ground, where they reveal they have few means of defense and escape; they show all their higher powers when they pass from branch to branch or take advantage of the tops of the trees to avoid danger". (Temminck 1846–1849, II, 407)

12.11 THE ANCESTOR OF HUMANS: THE CHIMPANZEE, *TROGLODYTES*, 1797–1809 (LAMARCK AND VIREY)

After its recognition by James Burnett, the gorilla was not mentioned again until 1847, and it did not receive its modern classification until 1852 (see Savage and Wyman, 1847; I. Geoffroy Saint-Hilaire, 1852a, b, see next paragraph). At the beginning of the 19th century, those in search of the ape closest to human beings began to consider the chimpanzee. It was chosen by both Julien and Joseph Virey as the creature closest to humans from the cognitive point of view, and when Jean-Baptiste de Lamarck considered the possibility of an ape-like origin for our own species, he chose the chimpanzee. There was no lack of opponents to the theory that primates and humans were close relatives. Virey declared the chimpanzee was next to Bushmen, which degraded the latter rather than enhancing the former. His comparative analyses, which were based on a racial perspective "founded" on questionable comparative evaluations and an imperfect knowledge of cerebral physiology, ended in this way:

> Do you believe that there is a great distance between the Hottentots and the chimpanzee, who lives almost in the same regions? The triangular snout, the ignoble and shady appearance, the intolerable roughness of its voice and behaviour, the excessive narrowness of the intellect, the completely animal habits and completely concentrated on the physical appetites: all this denounces their animal stupidity and their excessive hebetude; they barely ply in a disconnected way. I do not think that between Bushmen and chimpanzees there is the place for another intermediate creature.

And regarding his sources:

> Bryan Edwards writes that black people have such a prominent snout, so ugly, that they exactly resemble the

FIGURE 12.24 Temminck 1839–1844, *Simia satyrus*, II.1 (n.d.), Pl. I f.t. p. 228.

> physiognomy of the baboons. If it is correct to say, with Winkelmann and Lavater, that our way of thinking is generally analogous to the shape of our body, I do not see a great difference between the intelligence of apes, and those of Lapps, Hottentots, idiots, Papuans and Omagua people. These stupid people are certainly closer to the apes than to Newton, Montesquieu, Buffon. [...] It can be shown through physiology that the chimpanzee brain is as similar to that of the wild Hottentot as the brain of these is to that of a European: and we know that intelligence is proportional to brain mass. (Virey, 1801, I: 428–429)

Lamarck adopted an evolutionary perspective stating that "the particular structure of man has been acquired gradually over a long period of time, with the help of favorable circumstances", later (perhaps influenced by Lacépède: cf. Lacépède, 1798) choosing the chimpanzee as the candidate able to acquire the erect posture and to become a biped:

> the longer the *jocko* continues to walk erect, the less mobility it will have in the toes; so that the toes, which are already much shorter than the other fingers, will gradually cease to be comparable to the other fingers and fit to grasp objects. (Lamarck, 1802: 135; cf. Barsanti, 1990, 1995)

Lamarck's decision was reiterated, with even greater emphasis, in a famous excerpt from the *Philosophie zoologique*, in which, after having said that "the orang-utan of Angola (*Simia troglodytes*) is certainly higher than the orangs of the Indies (*Simia satyrus*), called orang-utan", he states that:

> if any population of quadrumans and in particular the most clever, lost, obliged by circumstances, the habit of climbing trees and grasping the branches with their feet as well as with their hands, to hold themselves up, and if the individuals of that population were obliged, for several generations, to use their feet only to walk, ceasing to use their hands as feet, there is no doubt [...] that they would eventually be transformed into bimans. (Lamarck, 1809, I: 349)

Lamarck "saw" the chimpanzee with the eyes of Jean-Baptiste Audebert (see **Fig. 12.25,** his orang-utan) and Pierre-André Latreille (see **Fig. 12.26** his orang-utan), and his conception of it was influenced by the verbal description from a naval officer, Louis-Marie-Joseph de Grandpré, which he quoted in full. In his *Voyage à la côte occidentale d'Afrique* (1801), Grandpré begins by stating that:

> the intelligence of this animal is truly extraordinary; it generally walks upright, leaning on the branch of a tree used as a walking-stick; Negroes fear it and not without reason, because it maltreats them when it meets them. They say that if it does not speak it is out of laziness; they think that this [speech] would identify it as a man, and that it would then be obliged to work, but that it could do both, if only it wanted to. This prejudice is so rooted in them that when they meet one they talk to it.

FIGURE 12.25 Audebert 1797–1800, *Le Pongo. Simia Troglodites*, tav. I f.t. p. 15.

FIGURE 12.26 Latreille 1801, *Jeune Pongo ou Orang-outang de la grande espèce*, Pl. II f.t. p. 79.

The officer goes on to describe how, despite failing to capture an individual, he was able to collect information about the chimpanzee:

> despite all the vain efforts I made to get a specimen of this species there, I managed to collect some data about this beast as I saw one on a vessel in port: she was a female, subject to the same inconveniences and accompanied by the same symptoms and by the same circumstances as women. I examined and measured her carefully, and she let us measure her with great pleasure. Erect, with her heels on the ground, she was four feet tall, two inches and eight lines [sic!]. Her arms extended to one inch above the knee; she was covered with hair, with a reddish back, grey legs and arms, a white belly, and with reddish head hair on her head, shorter than that on her body; the breast was hairless around the nipples; the buttocks were fleshy, yet less so than they are in the human species; these were covered with hairs and, instead of the callosity common to all monkeys, these had only a small horny part, which the animal had contracted while remaining seated; the legs were thin, but the thighs were very muscular; she had no evidence of a tail; and the surgeon of the vessel assured me, after a long examination, that it was possible that the spine ended with an internal curvature, as in the human species, and that it was certainly because of this circumstance that the animal could walk upright.

With respect to the mental abilities of the chimpanzee, Grandpré continued:

> it would be too long to cite all the evidence of intelligence that this animal gave us, so I put together only the most surprising; she had learned to heat the oven; carefully she did not let any charcoal escape that could set fire to the vessel, she knew perfectly when it was sufficiently warm and never failed to warn the baker, who, for his part, sure of the animal's sagacity, trusted her and hastened to bring the dough when the monkey went to look for it: and she had never made a mistake. (see Grandpré, 1801: 26–29; Lamarck, 1802: 136–138)

Since the chimpanzee, along with the bonobo, are currently accepted as the apes closest to humans, all in all, Lamarck made a good choice, especially if we consider the widespread confusion about evidence concerning the orangutan (see Barsanti, 1989, 1995).

12.12 THE REAPPEARANCE OF THE GORILLA: FROM A SPECIES OF CHIMPANZEE (*TROGLODYTES GORILLA*) TO THE GENUS *GORILLA* (1847–1861)

In a letter to Richard Owen dated 24 April 1847, Thomas Savage, an American missionary recently settled in Gabon, wrote "I have discovered an extraordinary animal, which I have reason to believe it is unknown to naturalists". He managed to obtain four skulls of the "extraordinary animal" (two males and two females, all adults), as well as various other parts of the skeleton, as well as collecting a great deal of evidence from the natives. These allowed him to write a long account, which was read at the *Boston Society of Natural History* on August 18th, and published in the *Boston Journal* in December of that year. It is, according to Savage, "a new species of *Orango*" (in the generic sense: an anthropoid ape), different both "from the *Troglodytes niger* [the chimpanzee] and from the orangutans of Borneo and Sumatra". Savage thinks it could have been Battell's "monster", because of both the tribe from which his behavioral descriptions came, and the region it occupied (along both banks of the river Gabon, from its mouth up to 60 miles inland). The beast was known as *Mpongwe*, of which the word *pongo* could be a corrupt form, but the natives, who called chimpanzees' *enché-eko* (from which the *jocko* of Buffon probably derived), called this "pongo" *engé-ena* (Savage and Wyman, 1847: 417, 419, 422). Thus, the almost unnoticed testimony of Thomas Edward Bowdich, who almost thirty years earlier had reported the existence of an *intchego* and an *ingena* in that region (Bowdich, 1819), was shown to be credible. And yet Savage, still influenced by what was said about the *Periplus* of Annone, called the animal *gorilla* in honor of the Carthaginian admiral (see Savage and Wyman, 1847: 419).

Savage admitted that the information about the habits of the gorilla was based solely on the tales told by local natives. Most of them were convinced that both the *enché-eko* and the *engé-ena* were human beings, even if in a degenerate form, while others thought, in a slightly more refined way, of a sort of "migration of souls" (see Chapter 2, Section I on this topic in contemporary Africa). Chimpanzees, the more riparian of the two animals, would have inherited "the spirit, less violent and more intelligent, of the inhabitants of the coast" (coast-dwellers) whereas gorillas, who inhabit the interior, would have inherited the characteristics of "men of the bush" (bushmen). But, Savage assured the reader that he had learned to distinguish beliefs from true knowledge, and so felt able to offer the assurance that:

> Its height is above five feet, it is disproportionally broad across the shoulders, thickly covered with coarse black hair. (...). With age it become gray (...) . Their gait is shuffing, the motion of the body which is never upright as in man, but bent forward, is somewhat rolling, or from side to side. The arms being longer than those of the Chimpanzée it does not stoop as much in walking; like that animal it makes progression by thrusting its arms forward, resting the hands on the ground and the giving the body a half jumping, half swinging motions between them. (...) They live in bands but are not so numerous than as the Chimpanzée's; (...) The silly stories about their carrying off women from the native towns, and vanquishing the elephants, related by voyagers and widely copied into books, are unhesitating denied. (Savage and Wyman, 1847: 422–424)

Although this was second-hand information, there was no doubt such an animal existed. The anatomist Jeffries Wyman, a professor at Harvard, contributed to the osteological part of the account and paid particular attention to identifying and highlighting the specific characteristics of this great ape. The results attested the animal was distinct from the other African "orangutan" in 13 different ways (starting from the median sagittal crest and the nuchal crest), and the two authors decided to place the new species in a lower position on the scale of the anthropoid apes. "Certainly, the *engé-ena* occupies an inferior position compared to chimpanzees [...] and it is, as some characteristics show, further from human beings than the chimpanzee and also the orang-utan" (Savage and Wyman, 1847: 427, 429, 434). Their conclusion was, as they hoped to clarify also in the title of the account, that they considered the new species a *Troglodytes gorilla*; that is, a species of the chimpanzee genus (see Joleaud, 1936; Appel, 1988).

In 1848, the English anatomist Richard Owen argued that the two great apes were species of the same genus, and in 1851 he disputed Wyman's conclusions, instead arguing that "the *Troglodytes gorilla* was nearer in structure to the human being than the *Troglodytes niger*" (Owen, 1851: 171). Thinking they were "two Chimpanzees", that is "two species of *Troglodytes*", for a long time one was known as the "small chimpanzee" and the new one identified by Savage was known as the "great chimpanzee" (Owen, 1848–1849: 381, 390–391, 416–417, 421). Isidore Geoffroy Saint-Hilaire, son of Étienne, was given access to two gorilla specimens, a young and an adult individual, preserved in alcohol. He reported on them on 19th January 1852 to the *Académie des Sciences*, when with understandable excitement ("the Museum of Natural History received a gift of extraordinary interest three days ago") he claimed that unlike the skulls and the parts of the skeleton available to Savage, Wyman, and Owen, the cadavers provided "the definitive demonstration of the existence of a second African species of anthropoid ape". Finally, Isidore Geoffroy Saint-Hilaire said it was possible "to fully understand the gorilla, in its external form and its internal structure, and to determine exactly its characteristics as well as its relationship both to chimpanzees and to orangutans". The animal was as tall as a man of average height (measuring 1 m and 67 cm), and was therefore "the greatest among the known Primates". Geoffroy Saint-Hilaire proposed that this great ape, to which he gave the name *Gorilla ngina*, was the one "most notable for its similarities with

FIGURE 12.27 *Dictionnaire universel d'histoire naturelle, Atlas, Gorille de Savage (Gorilla savagei), Mammifères*, tav. 26.

the physical structure of man", and he also ceased to consider it a species of chimpanzee, describing it as "a distinct genus in most of its characteristics inferior to the genus *Troglodytes*, in others superior" (I. Geoffroy Saint-Hilaire, 1852b: 81–84).

Sarah Wallis, the artist and writer wife of Bowdich, drew the gorilla, but her depiction of the *ingena* of 1847 owed much to her imagination (see Lee, 1847, *The Ingena:* 208). A faithful depiction of both the immature and adult gorilla by Werner appeared in the second contribution by I. Geoffroy Saint-Hilaire on the *Troglodytes gorilla*, and it was confirmed as a distinct species (*Gorilla savagei*) in the *Dictionnaire universel d'histoire naturelle* organized by Charles d'Orbigny, where its depiction (**see Fig. 12.27**) was widely acclaimed.

Paul Belloni Du Chaillu, a Franco-American naturalist (born in either Paris or New Orleans), who explored equatorial Africa (*Explorations and adventures in equatorial Africa*, 1861), saw a gorilla first-hand. He describes coming across:

> a huge male gorilla which had crossed the forest on four legs but when it saw us stood up to its full height and looked us boldly in the face. [...] It seemed to be almost six feet tall, and had an immense body, a monstrous chest, and arms of an incredible muscular power. Its large, gray, sunken eyes shone with a savage light and its face had a diabolical expression. This was how the king of the African forests appeared[...] it remained there, on the spot, beating its chest with its huge fists, which made it resound like an immense drum. This is its way of challenging the enemy; and it simultaneously roared repeatedly.

The latter behaviour made a particular impression on Du Chaillu, who supplied this vivid description:

> the roar of the gorilla is the strangest and most terrifying sound that can be heard in those forests. It starts with a kind of irregular bark, like that of an angry dog, then turns into a dull growl that resembles the distant rumbling of thunder [...]. While we remained motionless on the defensive, its eyes lit up with an ever brighter flame until the hairs on its head stood on end and it began to move fast, baring its powerful canines and emitting new rumbles of thunder. (Du Chaillu, 186: 145–146)

With the inclusion of some drawings of the "king of the African forests", Du Chaillu's book was published in London by John Murray, Charles Darwin's publisher, two years after the *Origin of species*. Du Chaillu also almost certainly encountered the animal we now know as the bonobo, which he thought was a different form of chimpanzee "because the *nshiégo-mbouwé* is smaller than the chimpanzee" as well as being bald. He called it "*Troglodytes calvus*" (Du Chaillu, 1861: 260–262, 318, 324).

Thus, by the 1860s there were good enough descriptions of the great apes to tackle the problem of how they are related to humans. What is the primate most closely related to us, or in other words, which of the great apes do we share our most recent common ancestor with? Darwin contributed to this discussion in 1871, but two Italian naturalists, Filippo De Filippi in 1864 and Giovanni Canestrini in 1866, had already made their own contributions. Filippo De Filippi argued that in the *scala naturae*:

> men are at the head of the line, true, but lined up with monkeys and that the results of the most advanced research "lead us to consider that all the monkeys derived from a common ancestor", we should not look for our primitive form in the living great apes but in a lost form of prehuman times. (...) In conclusion we have to ask ourselves: what is our closest relative among the living anthropoid apes? (De Filippi, 1864: 16–17, 44–45)

This was confirmed by Canestrini, who wrote that "we are no longer discussing whether man descends from primates or not; the question has become more demanding, because we want to know from which founder we descend and in which way" (Canestrini, 1866: 20). In 1863, Thomas Henry Huxley, citing morphological evidence, placed gorillas in a "higher" position than chimpanzees (see **Fig. 12.28**). In 1871, Charles Darwin came to a different conclusion.

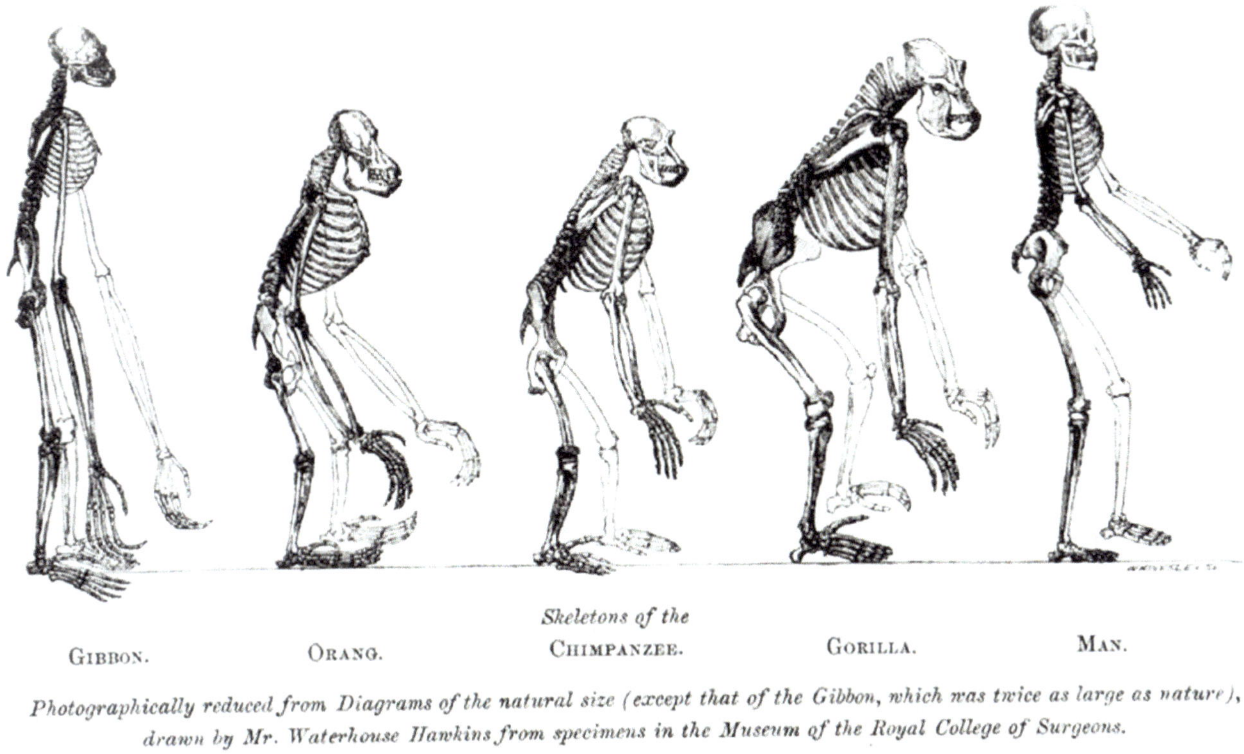

FIGURE 12.28 Huxley 1863, *Skeletons of the Gibbon, Orang, Chimpanzee, Gorilla, Man*, frontispiece.

REFERENCES

Primary Sources

Aa Peter van der (1720). *Icones arborum, fruticum et herbarum exoticarum (...) ut et animalium peregrinorum rarissimorum*, Lugduni.

Aldrovandi Ulisse (1637). *De quadrupedibus digitatis viviparis libri III et de quadrupedibus digitatis oviparis libri II.* Bononiae: Tebaldinum.

Aldrovandi Ulisse (1642). *Monstruorum historia.* Bononiae: Tebaldini.

Allamand Jean-Nicolas-Sébastien (1771). *Addition à l'article des Orangs-outangs*; In Buffon 1766–1799, V, pp. 71–76.

Allamand Jean-Nicolas-Sébastien (1785). *Addition à l'article des Orangs-outangs*; In Buffon 1766–1799, Supplément, V, pp. 45–48.

Alletz Pons-Augustin see Anonymous 1752.

Anonymous (1600). *Discours prodigieux et véritable d'une fille de chambre, laquelle a produit un monstre après avoir eu la compagnie d'un singe.* Paris: Bourriquant, s.d.

Anonymous (1700). Extrait d'une lettre écrite des Indes le 10 de janvier 1700. *Mémoires pour l'Histoire des Sciences et des Beaux-Arts*, I, 1701, 1: 184–186.

Anonymous (Thomas Boreman) (1739a). *A Description of Some Curious and Uncommom Creatures.* London: Ware.

Anonymous (Hubert Gravelot & Gerard Scotin) (1739b). Animalis rarioris, chimpanzee dicti, ex regno Angola Londinum advecti, brevior description. *Nova Acta Eruditorum*, VIII: 564–565

Anonymous (1739c). *An Essay towards the Character of the Late Chimpanzee Who Died Feb. 23, 1739.* London: Gilliver e Clarke.

Anonymous (Pons-Augustin Alletz) (1752). *Histoire des singes et autres animaux curieux.* Paris: Duchesne.

Anonymous (1772). *Tintinnabulum naturae*, a cura di S. Matton. Paris: Séha, 2002.

Anonymous (Foucher d'Obsonville) (1783). *Essais philosophiques sur les moeurs de divers animaux étrangers.* Paris: Coutourier fils.

Anonymous (1836). Delle scimie, ed in ispezialità del troglodite nero, e piteco satiro ossia urang-utano. *Teatro Universale*, III: 91–93.

Anonymous (1869). I liberali e le scimmie: confronti scientifici. *La Civiltà Cattolica*, XX: 257–266.

Aristotle (335–330 AD). *De partibus animalium.* In: Opere *biologiche*, D. Lanza & M. Vegetti (eds.). Torino: Utet, 1971, pp. 555–736.

Aristotle (347–343 AD). Historia animalium. In: *Opere biologiche*, D. Lanza & M. Vegetti (eds.). Torino: Utet, 1971, pp. 129–473.

Atkins John (1735). *A Voyage to Guinea, Brasil and the West Indies.* London: Ward and Chandler, 1737.

Audebert Jean-Baptiste (1797–1800). *Histoire naturelle des singes et des makis.* Paris: Desray, an VI – an VIII, 10 fascicoli.

Barbot Jean (1732). *Description of the Coasts of North and South Guinea; and of Ethiopia Inferior, Vurgarly Angola.* London: Churchill.

Barbot Jean & Casseneuve John (1732). *A Voyage to Congo River*; In Churchill e Churchill 1732, pp. 497–522.

Batavorum, Vander, s.d. (1800). *A General History of Quadrupeds*. Newcastle, Tyne.
Battell Andrew (1613). *The Strange Adventures of Andrew Battell*; In Purchas 1613, II, pp. 970–985.
Beeckman Daniel (1718). *A Voyage to and from the Island of Borneo*. London: Warner.
Belon du Mans Pierre (1555). *L'histoire de la nature des oyseaux*. Paris: Cauellat.
Blainville Henri-Marie Ducrotay de (1818). *Sur l'orang-outang. Journal de Physique, de Chimie, d'Histoire Naturelle et des Arts*, LXXXVI(1): 311–313.
Blainville Henri-Marie Ducrotay de (1836). Sur quelques espèces de singes confondues sous le nom d'Orang-outang. Comptes Rendus Hébdomadaires des Séances de l'Académie des Sciences, II: 73–76.
Blumenbach Johann Friedrich (1779–1780). *Handbuch der Naturgeschichte*. Göttingen: Dieterich, 2 voll.
Blumenbach Johann Friedrich (1782). *Handbuch der Naturgeschichte*, Zwente Ausgabe. Goettingen: Dieterich.
Blumenbach Johann Friedrich (1788). *Handbuch der Naturgeschichte*, Dritte Ausgabe. Goettingen: Dieterich.
Blumenbach Johann Friedrich (1791). *Handbuch der Naturgeschichte*, Vierte Ausgabe. Goettingen: Dieterich.
Blumenbach Johann Friedrich (1797). *Handbuch der Naturgeschichte*, Fünfte Ausgabe. Goettingen: Dieterich
Blumenbach Johann Friedrich (1799). *Handbuch der Naturgeschichte*, Sechste Ausgabe. Goettingen: Dieterich.
Bondt Jakob de (1658). *Historiae naturalis et medicae Indiae orientalis libri sex*; In Piso 1658, pp. 50–86.
Bonnet Charles (1764). *Contemplation de la nature (Contemplazione della natura)*. Modena: Montanari, 1769–1770, 2 voll.
Boreman Thomas see. Anonymous 1739a.
Bory de Saint-Vincent Jean-Baptiste (1827). Orang. In: *Dictionnaire classique d'histoire naturelle*, XII, pp. 261–285.
Bosman Willem (1704). *Nauwkeurige beschryving van de Guinese goud; Voyage de Guinée*. Utrecht: Schouten, 1705
Bowdich Thomas Edward (1819). *Mission from Cape Coast Castle to Ashantee*. London: Murray.
Bradley Richard (1721). *A Philosophical Account of the Works of Nature*. London: Mears.
Brant Sebastian (1501). *Mythologi Esopi*. Basileae: de Phortzheim.
Breydenbach Bernhard von (1486). *Opusculum sanctarum peregrinationum ad sepulchrum Christi venerandum*, Moguntiae, Reüwich de Trajecto; *De peregrinationibus ad sepulchrum Domini atque Sanctae Catharinae*, s.l., s.e., 1502.
Brisson Mathurin-Jacques (1756). *Le règne animal divisé en IX classes*. Paris: Bauche.
Buffon Georges-Louis Leclerc de (1749–1789). *Histoire naturelle, générale et particulière*. Paris: Imprimerie Royale, 1749–1767, 15 voll.; *Histoire naturelle des oiseaux*, 1770–1783, 9 voll.; *Supplément à l'Histoire naturelle*, 1774–1789, 7 voll.; *Histoire naturelle des minéraux*, 1783–1788, 5 voll.
Buffon Georges-Louis Leclerc de (1749a). *Histoire générale des animaux*; In Buffon 1749–1789, II, pp. 1–426.
Buffon Georges-Louis Leclerc de (1749b). *Histoire naturelle de l'homme*; In Buffon 1749–1789, II, pp. 429–603; III, pp. 305–530.
Buffon Georges-Louis Leclerc de (1753). *Discours sur la nature des animaux*; In Buffon 1749–1789, IV, pp. 1–110.
Buffon Georges-Louis Leclerc de (1766a). *Nomenclature des singes*; In Buffon 1749–1789, XIV, pp. 1–42.
Buffon Georges-Louis Leclerc de (1766b). *Les orangs-outangs, ou le Pongo et le Jocko*; In Buffon 1749–1789, XIV, pp. 43–71.
Buffon Georges-Louis Leclerc de (1766–1799). *Histoire naturelle, générale et particulière (...) Nouvelle edition*. Amsterdam: Schneider, 38 voll.
Buffon Georges-Louis Leclerc de (1789). *Addition à l'article des Orangs-outans*; In Buffon 1749–1789, Supplément, VII, pp. 1–29.
Buffon Georges-Louis Leclerc de (1799–1808). *Histoire naturelle générale et particulière (...) Nouvelle edition*. Paris: Dufart, 127 voll.
Burnett James lord Monboddo (1774). *Of the Origin and Progress of Language. Vol. I. Second edition*. Edinburgh: J. Balfour
Camper Peter (1779a). An account of the organs of speech of the Orang-outang. *Philosophical Transactions of the Royal Society of London*, LXIX: 139–159.
Camper Peter (1779b). *Histoire naturelle de l'orang-outang, et de quelques autres singes*. Harlingae: van der Plaats.
Camper Peter (1791). *Dissertation sur les variétés naturelles qui caractérisent la physionomie des hommes*. Paris: Jansen.
Canestrini Giovanni (1866). *L'antichità dell'uomo*. Lezione popolare letta in Modena li 22 marzo 1866. Modena: Vincenzi.
Cardano Girolamo (1550). *De subtilitate libri XXI*. Lugduni: Rouillium, 1559.
Charleton Walter (1667). *Exercitationes de differentiis et nominibus animalium*. Oxoniae: Sheldoniano,
Churchill Awnsham & Churchill John (1732). *A Collection of Voyages and Travels*. London: Walthoe, 6 voll.
Cook James (1777). *A voyage round the vorld*; *Voyage dans l'hémisphère austral, et autour du monde*. Paris: Hôtel de Thou, 1778, 5 voll.
Cuvier Frédéric (1810). Description d'un orang-outang, et observations sur ses facultés intellectuelles. *Annales du Muséum National d'Histoire Naturelle*, XVI: 46–65
Cuvier Frédéric & Geoffroy Saint-Hilaire Isidore (1819–1842). *Histoire naturelle des Mammifères*. Paris: Lasteyrie, 2 voll.
Cuvier Frédéric & Geoffroy Saint-Hilaire Isidore (1824–1847). *Histoire naturelle des Mammifères*. Paris: Belin, 5 voll.
Cuvier Georges-Léopold-Chrétien (1797). *Tableau élémentaire de l'histoire naturelle des animaux*. Paris: Baudoin.
Cuvier Georges-Léopold-Chrétien (1817). (con Latreille) *Le règne animal distribué d'après son organization*. Paris: Deterville, 4 voll.
Cuvier Georges-Léopold-Chrétien (1818). Analyse des travaux de l'Académie Royale des Sciences, pendant l'année 1818. Partie physique. *Mémoires de l'Académie Royale des Sciences*, III: 179–230.
Cuvier Georges-Léopold-Chrétien (1825). *Recherches sur les ossemens fossiles (...) Troisième edition*. Paris: Dufour e d'Ocagne, 7 voll.
Cuvier Georges-Léopold-Chrétien (1829–1830). *Le règne animal distribué d'après son organisation (...) Nouvelle edition*. Paris: Deterville, 5 voll.
Cuvier Georges-Léopold-Chrétien (1836). *Le règne animal distribué d'après son organisation (...) Troisième edition*. Bruxelles: Hausman, 3 voll.
Cuvier Georges-Léopold-Chrétien (1836–1849). *Le règne animal distribué d'après son organisation*. Paris: Fortin, 10 voll.
Cuvier Georges & Geoffroy Saint-Hilaire Étienne (1795). Histoire naturelle des orangs-outangs. *Magasin Encyclopédique*, I, 3: 451–463; extracted: 1–12
Cuvier Georges & Geoffroy Saint-Hilaire Étienne (1798). Mémoires sur les orangs-outangs. *Journal de Physique, de Chimie et d'Histoire Naturelle*, III, 46: 185–191.
Dalin Olof (1749). *Tal vid Praesidii afläggande om Sverige i sit ämne och Sverige i sin upodling*. Stockholm: Salvius

Dampier Guillaume (1715). *Nouveau voyage autour du monde*. Rouen: Machuel le jeune, 3 voll.

Dapper Olfert (1668). *Naukeurige Beschrijvinge der Afrikaensche Geweste*; *Description de l'Afrique*. Amsterdam: Wolfgang, 1686.

Daubenton Louis-Jean-Marie (n.d.). *Discours sur la conformation des singes* (Paris, Bibliothèque Centrale du Muséum National d'Histoire Naturelle, ms.870).

Daubenton Louis-Jean-Marie (1766). *Description du jocko*; In Buffon 1749–1789, XIV, pp. 72–83.

de Brosses Charles (1756). *Histoire des navigations aux terres australes*. Paris, 2 voll.

De Filippi Filippo (1864). *L'uomo e le scimie. Lezione pubblica detta in Torino la sera dell'11 gennaio 1864. Terza edizione*. Milano: G. Daelli e Comp., 1865.

Delisle de Sales Jean-Baptiste-Claude (1770). *De la philosophie de la nature*. Arkstée e Merkus, 3 voll.

Delrio Martin (1599–1600). *Disquisitionum magicarum libri sex*. Lugduni: Pillehotte, 1604.

Desmarest Anselme-Gaetan (1818). Pongo. *Nouveau dictionnaired'histoire naturelle*, XXVII: 571–573.

Desmarest Anselme-Gaetan (1820). *Mammalogie*. Paris: Agasse.

Desmoulins Antoine (1826). *Histoire naturelle des races humaines*. Paris: Méquignon-Marvis.

Dictionnaire classique d'histoire naturelle. Paris: Rey e Gravier, (1822–1831), 16 voll.

Dictionnaire classique des sciences naturelles. Bruxelles: Meline, (1837–1845), 10 voll.

Dictionnaire des sciences naturelles. Paris: Levrault, (1804–1806), 3 voll.; Strasbourg: Levrault, 1816–1830, 60 voll.

Dictionnaire pittoresque d'histoire naturelle. Paris: Bureau de Souscription, (1834–1840), 9 voll.

Dictionnaire universel d'histoire naturelle. Paris: Masson, (1867–1869), 14 voll.

Diderot Denis (P. Rossi ed.) (1753). *Interprétation de la nature*; *Interpretazione della natura*. In: *Opere filosofiche*. Milano: Feltrinelli, 1963, pp. 119–169.

Diderot Denis (1778). *Élements de physiologie*, par J. Mayer. Paris: Didier, 1964.

Diretano bando, Biblioteca Nazionale Centrale di Firenze, ms. II IV 29.

d'Obsonville Foucher see Anonymous (1783).

Du Chaillu Paul Belloni (1860a). Descriptions of five new species of mammals discovered in western equatorial Africa. *Proceedings of the Boston Society of Natural History*, VII: 296–304.

Du Chaillu Paul Belloni (1860b). Descriptions of mammals from equatorial Africa. *Proceedings of the Boston Society of Natural History*, VII: 358–367.

Du Chaillu Paul Belloni (1861). *Explorations and adventures in Equatorial Africa*; *Voyages et aventures dans l'Afrique équatorial*. Paris: Lévy frères, 1863.

Du Jarric père Pierre (1608–1613). *Histoires des choses les plus mémorables advenues tant ez Indes Orientales que autres païs de la descouverte des Portugais*, Bourdenaus: Millanges, 3 voll.

Dureau de Lamalle Adolphe (1852). Mémoire sur le Grand Gorille du Gabon, déterminant la limite de la navigation d'Hannon. *Annales des Sciences Naturelles*, III s., Zoologie, XVI: 183–192.

Edwards George (1758). *Gleanings of Natural History*. London: The Author, 3 voll.

Encyclopédie, ou Dictionnaire raisonné des sciences, des arts et des métiers. Paris: Briasson, 1751–1772, 17 voll.; *Recueil des planches*. Paris: Briasson, 1762–1777, voll. 18–28.

Fantasti Girolamo Cesare (1732). *La scimia non è specie d'uomo*. Verona: Ramanzini

Ferdinand Valentin (1862). Beschreibung der Serra Leoa. *Abhandlungen Der Historischen Classe der Königlich Bayerischen Akademie der Wissenschaften*, IX, 111–142.

Flower Henry (1738). London magazine and monthly chronologer, VII, September.

frate Noè (Bianchi) (1500). *Viaggio da Venitia al Sancto Sepulchro et al Mote Sinai*. Vinegia: Zoppino.

Froger François (1698). *Relation d'un voyage fait en 1695, 1696 et 1697 aux côtes d'Afrique, detroit de Magellan, Brézil, Cayenne et isles Antilles*. Paris: Saugrain.

Galeno di Pergamo 165-195 I procedimenti anatomici. In: *Opere scelte*, I. Garofalo & M. Vegetti. Torino: Utet, 1978, pp. 147–289.

Gassendi Pierre (1641). *Viri illustris Nicolai Claudij Fabricij de Peiresc, senatoris aquisextiensis*, Vita; *editio tertia*. Hagae Comitum: Vlacq, 1655.

Gemelli Careri Gian Francesco (1699–1700). *Giro del mondo*. Napoli: Roselli, 6 voll.

Geoffroy Saint-Hilaire Étienne (1795). Mémoire sur une nouvelle division des Mammifères. Magasin Encyclopédique, II, 6: 164–190

Geoffroy Saint-Hilaire Étienne (1796). Extrait d'un mémoire à la Société Philomatique. *L'Ami des Arts. Journal de la Société Philotechnique*, 15 frimaire an 5: 113–115.

Geoffroy Saint-Hilaire Étienne (1797). Extrait d'un mémoire sur les orangs-outangs. *Bulletin des Sciences par la Société Philomatique*, II, 4: 25–26.

Geoffroy Saint-Hilaire Étienne (1798). Note sur un prétendu orang-outang des Indes. *Journal de Physique*, III, 46: 342–346.

Geoffroy Saint-Hilaire Étienne (1812a). Sur trois dessins de Commerçon, représentant des Quadrumanes d'un genre inconnu. *Annales du Muséum d'Histoire Naturelle*, XIX: 171–175.

Geoffroy Saint-Hilaire Étienne (1812b). Tableau des Quadrumanes. *Annales du Muséum d'Histoire Naturelle*, XIX: 85–122; 156–170; estratto: 1–58.

Geoffroy Saint-Hilaire Étienne (1828). *Léçons sur l'histoire naturelle des Mammifères*. Paris: Pichon e Didier.

Geoffroy Saint-Hilaire Étienne (1836a). Considérations sur les singes les plus voisins de l'homme. *Comptes Rendus Hebdomadaires des Séances de l'Académie des Sciences*, II: 92–95.

Geoffroy Saint-Hilaire Étienne (1836b). Études sur l'Orang-outang de la Ménagerie. *Comptes Rendus Hebdomadaires des Séances de l'Académie des Sciences*, III: 1–8.

Geoffroy Saint-Hilaire Isidore (1852a). Note sur le gorille. *Annales des Sciences Naturelles*, III s., XVI: 154–158.

Geoffroy Saint-Hilaire Isidore (1852b). Sur le gorille. Comptes *Rendus de l'Académie des Sciences*, XXXIV: 81–84.

Gervaise Nicolas (1688). *Description historique du royaume de Macacar*. Paris: Foucault.

Gesner Conrad (1551). *Historiae animalium lib. I de Quadrupedibus viviparis*; In 1551–1558, I 1551–1558, *Historiae animalium*. Tiguri: Froschoverum, 4 voll.

Grandpré Louis-Marie-Joseph comte O'Hier de (1801). *Voyage à la côte occidentale d'Afrique*. Paris: Dentu, an XI, 2 voll.

Gravelot Hubert Bourguignon (delineavit) and Scotin Gérard (insculpsit) see Anonymous, (1739b).

Gray John Edward (1861). Observations on Mr. Du Chaillu's papers on 'The new species of mammals' discovered by him in western equatorial Africa. *Proceedings of the Zoological Society of London*: 273–278

Grose Henry (1757). *Voyage to the East-Indies*; *Voyage aux Indes orientales*. Londres: Panckoucke, 1758.

Guazzo Francesco Maria (1608). *Compendium maleficarum*. Mediolani: Lantonus.

Hanno (510 AD). *Periplus, in Geographi graeci minores*, a cura di C. Mullerus, Parisiis, Didot, 1853–1861, I, pp. 1–14.

Histoire générale des voyages. Paris: Didot, 1746–1761, 16 voll.; *Suite*. Amsterdam: Arkstée, 1761; *Continuation*. Paris: Rozet, 1768–1789, 3 voll.

Huxley Thomas, E. (1863, 1869). *Evidences as to man's place in nature; Prove di fatto intorno al posto che tiene l'uomo nella natura* (Milano: Treves) / *Il posto dell'uomo nella natura*, G. Giacobini (ed.). Torino: Utet, 2005.

Huxley Thomas, E. (1863). *Evidences as to Man's Place in Nature*. London: William and Norgate.

Illiger Johann Carl Wilhelm (1811). *Prodromus systematis Mammalium et Avium*. Berolini: Salfeld.

Jobson Richard (1623). *The Golden Trade, or a Discovery of the River Gambra*. London: Okes.

Jonston Jan (1657). *Historiae naturalis de Quadrupedibus libri*. Amstelodami: Schipper.

Kankel Johann (1667). *Een kort beskriffning vppå trenne resor och peregrinationer*. Wisingsborgh, s.e.

Kolb Peter (1742). *Description du Cap de Bonne-Esperance*. Amsterdam: Catuffe, 3 voll.

Köping Nils Matson (1667). Reesa genom Asia, Africa och många andra Hedniska Konungaryken. In: *Beskrifning om en Resa genom Asia, Africa och andra hendra länder*, Kankel (ed.). Stockholm: Salvius, 1743.

Lacépède Bernard-Germain-Étienne de (1798). *Discours d'ouverture et de clôture du cours d'histoire naturelle des animaux vertebrés et à sang rouge*. Paris: Plassan, an VI.

Lacépède Bernard-Germain-Étienne de (1801). *Tableau des divisions, sous-divisions, ordres et genres des Mammifères*. Paris: Plassan, an IX.

Lacroix A. Phérotée de (1686). *Relation universelle de l'Afrique ancienne et modern*. Lyon: Amauby.

Lamarck Jean-Baptiste de Monet de (1802). *Recherches sur l'organisation des corps vivans. Paris: Maillard.*

Lamarck Jean-Baptiste de Monet de (1809). *Philosophie zoologique*. Paris: Dentu, 2 voll.

La Mettrie Julien Offray de (1745). Histoire naturelle de l'âme; Storia naturale dell'anima. In: *Opere filosofiche*, S. Moravia (ed.). Roma: Laterza, 1974, pp. 49–162

Latreille Pierre-André (1801). *Le singe de Wurmb*; In Buffon 1799–1808, XXXV, an IX, pp. 262–269.

Lavater Johann Caspar (1775–1778). *Physiognomische Fragmente zur Beförderung der Menschenkenntnis und Menschenliebe*; *L'art de connaître les hommes par la physionomie*. Paris: Prudhomme, 1806–1809, 10 voll.

Le Cat Claude-Nicolas (1765). *Traité de l'existence, de la nature et des propriétés du fluide des nerfs*. Berlin, s.e.

L'Ecluse Charles de (1605). *Exoricorum libri decem*, Antverpiae: Plantiniana Raphelengii, 3 voll.

Lee R. (1847). *The African Wanderers, or the Adventures of Carlos and Antonio. Embracing Interesting Descriptions of the Manners and Customs of the Western Tribes, and the Natural Productions of the Country.* London: Grant and Griffith, 1850

Le Guat François (1720). *Voyages et aventures en deux isles desertes des Indes orientales*. Londres: Mortier, 2 voll.

Lesson René-Primevère (1828–1837). *Histoire naturelle générale et particulière des Mammifères et des Oiseaux découverts depuis 1788 jusqu'à nos jours*. Paris: Baudoin, 11 voll.

Liceti Fortunio (1616). *De monstrorum natura caussis et differentiis*. Paravii: Crivellarium

Liceti Fortunio (1665). *De monstris (...) Editio novissima*. Amstelodami: Frisii Licostene v. Wolffhart C.

Linné Carl (1735). *Systema naturae*. Lugduni Batavorum: Haak.

Linné Carl (1747). *Fauna svecica*. Stockholmiae: Salvii lettera a J.G. Gmelin, cit. in Greene 1909, pp. 25–26.

Linné Carl (1749–1790). *Amoenitates academicae*. Lugduni Batavorum: Haak, 10 voll.

Linné Carl (1758–1759). *Systema naturae (...) Editio* decimal. Holmiae: Salvii, 2 voll.

Linné Carl (1760). *Anthropomorpha*; In 1749–1790, VI, 1763, pp. 63–76.

Linné Carl (1761–1785). *Natuurlyke historiae of uitvoerige beschryvind der dieren, planten en mineraalen*. Amsterdam: Houttuyn, 37 voll.

Linné Carl (1765). *Sistema der natuurlyke historie*. Hagae-Comitum: Staatman

Linné Carl (1766–1768). *Systema naturae (...) Editio duodecima*. Holmiae: Salvii, 4 voll.

Linné Carl (1773–1776). *Vollständiges natursystem*. Nürnberg: Raspe, 10 voll.

Locke John (1690). Essay Concerning Human Understanding; *Saggio sull'intelligenza umana*. Bari: Laterza, 1972, 4 voll.

lord Kames Henry Home (1774). *Sketches of the History of Man*. London: Strahan, 2 voll.

Lord Monboddo see Burnett James

Magitot L.-F.-E. (1869). L'homme et les singes anthropomorphes. *Bulletin de la Société d'Anthropologie de Paris*, II s., IV: 113–142

Maillet Benoît de (1748). *Telliamed*. Basle: Libraires Associés, 1749

Mandeville Jean de (1355). *Voyage à Jerusalem*; *The voiage and Travaile of Sir J. M.* London: Lumley, 1839

Maupertuis Pierre-Louis Moreau de (1752). *Lettre sur le progrès des sciences*; In 1756, II, pp. 343–399

Maupertuis Pierre-Louis Moreau de (1756). *Oeuvres*. Lyon: Bruyset, 4 voll.

Merolla padre Girolamo (1692). Breve e succinta relazione del viaggio nel regno del Congo. Napoli: Mollo

Montaigne Michel de (1580–1588). *Essais*; *Saggi*, F. Garavini (ed.). Milano: Fabbri, 2001, 4 voll.

Moscati P. (1770). *Delle corporee differenze essenziali che passano fra la struttura de' bruti, e la umana*. Milano: Galeazzi.

Moscati P. (1771). *Appendice al discorso accademico delle corporee differenze essenziali che passano fra la struttura de' bruti, e la umana*. Brescia: Rizzardi

Niccolò fra' da Poggibonsi (1346). *Libro d'oltramare* (Biblioteca Nazionale Centrale di Firenze: *Libro d'oltramare*, Palat. 685; *Santuario d'oltramare*, Panciatich. 78–79; *Viaggio d'oltremare*, II, IV, 119)

Nieremberg Juan Eusebio (1635). *Historia naturae, maxime peregrinae*. Antverpiae: Moreti

Nouveau dictionnaire d'histoire naturelle. Paris: Deterville, (1803–1804), 24 voll.; 1816–1819, 36 voll.

Owen Richard (1835). On the osteology of the chimpanzee and orang utan. *Transactions of the Zoological Society of London*, I: 343–379.

Owen Richard (1848–1849). Osteological contributions to the natural history of the chimpanzees, including the description of the skull of a large species discovered by T. Savage in the Gaboon country. *Transactions of the Zoological Society of London*, III: 381–422; Observations ostéologiques pour servir à l'histoire naturelle des Chimpanzés. *Annales des Sciences Naturelles*, III s., Zoologie, XVI: 161–171.

Owen Richard (1851). Mémoire sur le squelette du grand Chimpanzé. *Annales des Sciences Naturelles*, III s., Zoologie, XVI: 171–172.

Owen Richard (1857). *Of the characters, principles of division, and primary groups of the class Mammalia*. Journal of the Proceedings of the Linnean Society, II: 1–37.

Pigafetta Filippo (1591). *Relatione del reame di Congo e delle circonvicine contrade; Vera descriptio regni africani, quod tam ab incolis quam Lusitanis Congus appellatur*. Francofurti: Richter, 1598.

Pigafetta Filippo (1598). *Icones ad primam partem Indiae orientalis*. Francofurti: Richter.

Piso Gulielmus (1658). *De Indiae utriusque re naturali et medica*. Amstelaedami: Elzevirios.

Plinio Caio Secondo (77–78). *Naturalis historia*; *Storia naturale*, G.B. Conte (ed.). Torino: Einaudi, II, 1983

Polo Marco (1298). *Le livre de M. P. citoyen de Venise, dit Million*; *Il libro di Marco Polo detto Milione*. Torino: Einaudi, 1954.

Prévost Antoine-François (1746–1789). *Histoire générale des voyages, ou Nouvelle collection de toutes les relations de voyages par mer et par terre*. Paris, 20 voll.

Pulci Luigi (1478–1483). *Morgante*. Firenze: Le Monnier, 1855.

Purchas Samuel (1613). *Purchas his pilgrimage*; *Hakluytus postumus, or Purchas his pilgrims*. London: Stansby, 1625–1626, 5 voll.

Pyrard François (1615). *Voyage de F. P. de Laval, contenant sa navigation aux Indes orientales, aux Moluques et au Bresil*. Paris: Thiboust

Quatrefages Armand de (1866). Sur le gorille. *Bulletin de la Société d'Anthropologie de Paris*, II s., I: 648–649

Radermacher Jacob Cornelis Mattheus (1780). Beschrijving van het eiland Borneo, voor zoo verre hetz elve, tot nu toe, bekend is. *Verhandelingen van het Bataviaasch Genootschap der Kunsten en Wetenschappen*, II, I: 107–148

Reade William Winwood (1863a). *Savage Africa: Being the Narrative of a Tour in Equatorial, South-Western, and North-Western Africa, with Notes on the Habits of the Gorilla*. London: Smith, Elder and Co.

Reade William Winwood (1863b). *Notes on the Derbyan eland, the African elephant, and the gorilla*. Proceedings of the Zoological Society of London: 169–173

Restif De La Bretonne Nicolas-Edme (1781). *La décoverte australe par un homme Volant*. Leïpsick: s.e., s.d., 4 voll.

Ridinger Johann Elias (1768). *Das in seiner grossen Mannig-faltigkeit und in seinen schönen farben / Représentations des animaux selon leur grande variété et leurs belles couleurs*. Augsburg: s.e., s.d.

Robinet Jean-Baptiste-René (1768). *Considérations philosophiques de la gradation naturelle des formes de l'être*. Paris: Saillant

Rousseau Jean-Jacques (1755). *Discours sur l'origine et les fondemens de l'inégalité parmi les hommes; Discorso sull'origine e i fondamenti dell'ineguaglianza fra gli uomini*. Roma: Editori Riuniti, 1968

Savage Thomas S. & Wyman Jeffries (1847). Notice of the external characters and habits of Troglodytes gorilla (...); Osteology of the same. *Boston Journal of Natural History*, V, IV: 417–443

Savérien Alexandre (1778). *Histoire des progrès de l'esprit humain dans les sciences et dans les arts qui en dépendent. Histoire naturelle*. Paris: Humblot

Schott Gaspar (1697). *Physica curiosa, sive Mirabilia naturae et artis*. Herbipoli: Endteri

Schouten Wouter (1676). *Oost-Indische voyagie*; *Voyage aux Indes orientales*. Amsterdam: Roger, 1707, 2 voll.

Schreber Johann Christian Daniel von (1775–1810). *Die Säugthiere in Abbildungen nach der Natur mit Beschreibung*; *Histoire naturelle des Quadrupèdes représentés d'après nature*. Erlangen: Walther, 3 voll.

Smith William (1744). *A new voyage to Guinea*; *Nouveau voyage en Guinée*. Paris: Durand, 1751

Temminck Coenraad Jacob (1839–1844). *Verhandelingen over De Natuurlijke geschiedenis der Nederlandsche overzeensche bezittingen*. Leiden: Luchtmans, 7 voll.

Temminck Coenraad Jacob (1846–1849). *Coup-d'oeil général sur les possessions néerlandaises dans l'Inde archipélagique*. Leide: Arnz, 3 voll.

Tiedemann Friedrich (1808–1814). *Zoologie*. Landshut: Webersch, 3 voll.

Tiedemann Friedrich (1826). Hirn des Orang-outangs mit dem des Menschen verglichen. *Zeitschrift für Physiologie*, II: 17–28

Tiedemann Friedrich (1836). On the brain of the Negro, compared with that of the European and the Orang-outang. *Philosophical Transactions of the Royal Society of London*, CXXVI: 497–527

Tulp Nicolaas (1641). *Observationum medicarum libri tres*; *editio nova, libro quarto auctior*. Amstelredami: Elzevirium, 1672

Tyson Edward (1699). *Orang-utan, sive Homo sylvestris; the anatomy of a Pygmie*. London: Osborne, 1751

Vallisnieri Antonio (1721). *Istoria della generazione dell'Uomo, e degli Animali*; In 1733, II, pp. 97–304

Vallisnieri Antonio (1726–1728). *Saggio d'istoria medica, e naturale*; In 1733, III, p. 363481

Vallisnieri Antonio (1733). *Opere fisico-chimiche*. Venezia: Coleti, 3 voll.

van der Broecke Pieter (1646). *Historische ende journaesche aenteyckeningh, in begin ende voortgangh der Vereenighde Oost-Indische Compagnie*. Amsterdam, s.e.

Virey Julien-Joseph (1801). *Histoire naturelle du genre humain*. Paris: Dufart, an IX, 2 voll.

Virey Julien-Joseph (1818). Orang-outang. In: *Nouveau dictionnaire d'histoire naturelle* (v.), XXIII, pp. 587–607.

von Caub Johann Wonnecke (1485). *Gart der Gesundheit*; *Ortus sanitatis major*. Mainz, 1490.

von Wurmb Friedrich (1780). Beschrijving van de Groote Borneoosche Orang Outang of the Oost-Indische Pongo. *Verhandelingen van het Bataviaasch Genootschap der Konsten en Wetenschappen*, II, 1: 245–261; Description du grand Orang-outang e Bornéo, ou Pongo des Indes orientales. *La Décade Philosophique, Littéraire et Politique*, IV, 79: 1–8.

von Zimmermann Eberhard August Wilhelm (1778–1783). *Geographische Geschichte des Menschen*. Leipzig: Weygand, 3 voll.

Vosmaer Arnout (1778). *Description de l'espèce de singe aussi singulière que très rare, nommée Ourang-outang, de l'Isle de Borneo*. Amsterdam: Meijer.

Vrolik Willem (1842). *Recherches d'anatomie comparée sur le chimpansée*. Amsterdam: Müller.

Wallis Sarah see Lee R.

White Charles (1795–1799). *An Account of the Regular Gradation in Man, and in Different Animals and Vegetables*. London: Dilly, 1799.

Wolffhart Conrad (1557). *Prodigiorum ac ostentorum chronicon*. Basileae: Petri.

Zahn dom Johann (1696). *Specula physico-mathematico-historica notabilium ac mirabilium sciendorum*. Norimbergae: Lochneri, 3 voll.

Secondary Sources

Appel Toby Anita (1988). Jeffries Wyman, philosophical anatomy, and the scientific reception of Darwin in America. *Journal of the History of Biology*, XXI(1): 69–94.

Barnard Alan (1995). *Monboddo's Orang outang and the definition of man*. In Corbey & Theunissen, pp. 71–85.

Barsanti Giulio (ed.) (1986b). *Misura d'uomo. Strumenti, teorie e pratiche dell'antropometria e della psicologia sperimentale fra Otto e Novecento*. Firenze: Giunti.

Barsanti Giulio (1986a). *L'uomo fra storia naturale e medicina, 1700-1850. Gli strumenti geometrici e la localizzazione delle funzioni cerebrali*, pp. 11–49.

Barsanti Giulio (1989). L'orang-outang déclassé. Histoire du premier singe à hauteur d'homme (1780-1801), in Histoire de l'anthropologie: hommes, idées, moments, C. Blanckaert et al. (eds.). *Bulletins et Mémoires de la Société d'Anthropologie de Paris*", n.s., I, 3–4: 67–104.

Barsanti Giulio (1990). Storia naturale delle scimmie 1600-1800. *Nuncius*, 2: 99–165.

Barsanti Giulio (1992). *La scala, la mappa, l'albero. Immagini e classificazioni della natura fra Sei e Ottocento*. Firenze: Sansoni.

Barsanti Giulio (1995). Les singes de Lamarck. In R. Corbey & Theunissen (eds.) 1995, pp. 101–115.

Barsanti Giulio (2005). *Una lunga pazienza cieca. Storia dell'evoluzionismo*. Torino: Einaudi.

Barsanti Giulio (ed.) (2008). *In camicia. L'uomo e altre scimmie nelle riflessioni, gli sberleffi e le allucinazioni di letterati, filosofi e naturalisti*. Firenze: Polistampa.

Barsanti Giulio (2009a). *Fra scienze della terra e scienze della vita. Storia dell'evoluzionismo italiano*. In: *La cultura italiana*, L. Cavalli Sforza (ed.), Torino: Utet, VIII, *Scienza e tecnologia*, a T. Pievani, (ed.), pp. 406–443.

Barsanti Giulio (2009b). *L'uomo dei boschi. Piccola storia delle Grandi scimmie*. Roma: La Sapienza.

Bénichou Claude & Blanckaert Claude (1984). L'anello mancato. *Kos*, I, ottobre: 20–32

Benitez Miguel (2002). *Une 'méthaphysique africaine' ou le panthéisme évolutionniste d'un orang-outang*; In Anonymous 1772, pp. 59–77.

Blanckaert Claude (1987). Les vicissitudes de l'angle facial et les débuts de la craniométrie (1765-1875). *Revue de Synthèse*, IV s., 3–4: 417–453.

Blanckaert Claude (1991). "Premier des singes, dernier des hommes"? Les métamorphoses de l'homme-singe aux XVII-XVIII siècles. *Alliage*, 7–8: 113–129.

Blanckaert Claude (1992). Une anthropologie de transiction. Lacépède et l'histoire naturelle de l'homme (1795-1830). *Annales Benjamin Constant*, 13: 95–111.

Blanckaert Claude (1993). Buffon and the natural history of man. *History of the Human Sciences*, VI(1): 13–50.

Blanckaert Claude (1995). La question du singe et l'ordre des Primates à la Société d'Anthropologie de Paris (1865-1870). In: Corbey R. & Theunissen (eds.), pp. 117–137.

Bourne Geoffroy H. (eds.) (1969). *The Chimpanzee: Anatomy, Behavior, and Diseases*. Basel: Karger.

Bouveignes Olivier de (1950). Le périple d'Hannon et les gorilles. *Zooléo*, n.s., 8: 7–18.

Bowler Peter J. (1995). *The Geography of Extinction: Biogeography and the Expulsion of 'Ape Men' from Human Ancestry in the Early Twentieth Century*. In: Corbey & Theunissen (eds.) 1995, pp. 185–193

Brentjes Burchard (1965). Die älteste Darstellung des Gorillas. *Säugetierk Mitteilungen*, XIII: 14–22

Broberg Gunnar (1983). Homo sapiens. Linnaeus's classification of man. In: *Linnaeus: The Man and His Work*, T. Frängsmyr (ed.). Berkeley: University of California Press, pp. 156–194

Cap Paul-Antoine (1854). *Le Muséum d'histoire naturelle*. Paris: Curmer.

Céard Jean (1977). *La nature et les prodiges. L'insolite au XVI siècle en France*. Genève: Droz.

Corbey Raymond & Theunissen Bert (eds.) (1995). *Ape, Man, Apeman: Changing Views since 1600*. Leiden: Leiden University

de Roy Pieter (1995). *In Search of Perfection: The Creation of a Missing Link*; In R. Corbey &Theunissen (eds.) 1995, pp. 195–207

Dobson Jessie (1952). John Hunter and the early knowledge of the anthropoid apes. Proceedings of the Zoological Society of London, CXXIII: 1–12.

Dougherty Frank (1995). *Missing Link, Chain of Being, Ape and Man in the Enlightenment: The Argument of the Naturalists*. In: Corbey & Theunissen (eds.) 1995, pp. 63–70.

Germain Gabriel (1957). Qu'est-ce que le périple d'Hannon? Document, amplification ou faux intégral? *Hespéris*: 205–248

Giacobini Giacomo & Giraudi Riccarda (1986). E l'uomo incontrò la scimmia. Kos: III, giugno, pp. 14–37.

Greene Edward Lee (1909). Linnaeus as an evolutionist. *Proceedings of the Washington Academy of Sciences*, XI: 17–26

Greene John C. (1959). *The Death of Adam. Evolution and Its Impact on Western Thought; La morte di Adamo. L'evoluzionismo e la sua influenza sul pensiero occidentale*. Milano: Feltrinelli, 1971

Groves Colin (2008). *Extended Family: Long Lost Cousins. A Personal Look at the History of Primatology*. Arlington: Conservation International

Hamy Ernest-Théodore. (1897) Documents inédits sur l'Homo sylvestris rapporté d'Angola en 1630. *Bulletin du Muséum National d'Histoire Naturelle*, III: 277–282

Hamy Ernest-Théodore. (1906). La légende du singe chaussé. *La Nature*, XXXIV: 113–114

Heuvelmans Bernard (1980). *Les bêtes humaines d'Afrique*. Paris: Plon.

Hill William C. Osman (1969a). *The Discovery of the Chimpanzee*; In Bourne 1969, pp. 22–49.

Hill William C. Osman (1969b). *The Nomenclature, Taxonomy and Distribution of Chimpanzees*; In Bourne 1969, pp. 22–49.

Janson Horst Woldemar (1952). *Apes and Ape Lore in the Middle Ages and the Renaissance*. London: Warburg Institute.

Joleaud Léonce (1931). Le rôle des singes dans les traditions populaires nord-africaines. *Journal des Sociétés Africanistes*, I: 117–150

Joleaud Léonce (1936) *L'origine du nom du gorille*. L'Anthropologie, XLVI: 525–528

Kappler Claude (1980). *Demoni, mostri e meraviglie alla fine del Medioevo*. Firenze: Sansoni, 1983.

Keller Otto (1887). *Thiere des classischer Alterthums in culturgeschichtlicher beziehung. Innsbruck*: Wagner

Laurent Goulven (1995). *Idées sur l'origine animale de l'homme en France au XIX siècle*; in Corbey & Theunissen (eds.) 1995, pp. 157–171

La Vergata Antonello (1993). *Il diavolo e il babbuino. L'uomo, la bestia e il darwinismo*. In: *Lo specchio oscuro. Gli animali nell'immaginario degli uomini*, L. Battaglia (ed.). Torino: Satyagraha, pp. 119–134

Leakey Richard e Slikkerveer (Jan 1993). *Man-Ape Ape-Man: The Quest for Human's Place in Nature and Dubois' 'Missing Link'*. Leiden: Netherlands Foundation for Kenya Wildlife Service

Mahodeau Pierre G. (1915). Le pongo, d'après le récit d'André Battell. *Revue d'Anthropologie*, XXV: 165–170.

Mahodeau Pierre G. (1917). Les moeurs du pongo, d'après André Battell. *Revue d'Anthropologie*, XXVII: 113–122

Martinez Contreras Jorge (1989). Las costumbres de los monos según Buffon. *Arbor*, CXXXII(517): 41–62

Martinez Contreras Jorge (1992). L'émergence scientifique du gorille. *Revue de Synthèse*, CXIII(3–4): 399–421

Martucci Vittorio (1983). Uno dei primi scritti darwiniani in Italia: 'L'uomo e le scimie' di F. De Filippi. *Physis*, XXV: 177–182

Martucci Vittorio (2008). *Scimmioni e vecchi merletti. Le scimmie antropomorfe e i dibattiti ottocenteschi sull'origine dell'uomo*. Napoli: Dante & Descartes

Mathis Maurice (1954). *Vie et moeurs des anthropoids*. Paris: Payot.

Matton Sylvain (2002). *Introduction* a Anonymous 1772, pp. 5–39

Mauny Raymond (1971). *Les siècles obscurs de l'Afrique noire*. Paris: Fayard

McDermott William Coffmann (1938). *The Ape in Antiquity*. Baltimore: J. Hopkins

Montagu M.F. Ashley (1940). Knowledge of the ape in Antiquity. *Isis*, XXXII: 87–102.

Montagu M.F. Ashley (1943). *Edward Tyson and the Rise of Human and Comparative Anatomy in England*. Philadelphia: American Philosophical Society

Moran III Francis (1993). Between primates and primitives: natural man as missing link in Rousseau's Second discourse. *Journal of the History of Ideas*, LIV: 37–58

Morris Ramona & Morris Desmond (1966). *Men and Apes*. London: Hutchinson

Mothu Alain (2002). *Rêves de singes au XVIII siècle*; In Anonymous 1772, pp. 79–156

Nash Richard (1995). *Tyson's Pygmie: the Orang-outang and Augustan 'Satyr'*; In Corbey & Theunissen 1995, pp. 51–62

Niderst Alain (2002). *Les rêveries méthaphysiques d'un lecteur de Fontenelle*, in Anonymous, 1772, pp. 49–57

Paul, P. (1922). Le Cardinal Melchior De Polignac - 1661–1741. Paris: Plon.

Penel Jean-Dominique (1982). *Homo caudatus. Les hommes à queue d'Afrique Centrale: un avatar de l'imaginaire occidental*. Paris: SELAF

Porset Charles (2002). *Le gai savoir d'un orang-outang*; In Anonymous 1772, pp. 41–47

Reynolds Vernon (1967). On the identity of the ape described by Tulp in 1641. *Folia Primatologica*, V: 80–87.

Reynolds Vernon (1968). *The Apes*. London: Cassell

Riese Alexander (1881a). Der namen des gorilla. *Der Zoologische garten*, XXII: 52–53

Riese Alexander (1881b). Gorillas bei Hanno. *Rheinisches Museum für Philologie*, XXXVI: 209–211

Roulin (1837). Les orangs. *Revue des Deux Mondes*, IX

Sanderson Ivan T. (1961). *Abominable Snowmen: Legend Come to Life*. Philadelphia: Chilton

Schmid Georg (1913). Die angeblichen Gorillas in Hannos bericht. *Zoological Annalen*, V: 67–71

Shea Brian T. (1984). Between the gorilla and the chimpanzee: a history of debate concerning the existence of the kooloo-kamba or gorilla-like chimpanzee. *Journal of Ethnobiology*, IV: 1–13

Slotkin James S. (1965). *Readings in Early Anthropology*. Chicago: Aldine

Spencer Frank (1995). *Pithekos to Pithecanthropus: An Abbreviated Review of Changing Scientific Views on the Relationship of the Anthropoid Apes to Homo*; In R. Corbey & Theunissen (eds.) 1995, pp. 13–27

Stechow Eberhard (1948). *Die Gorillas im 'Periplus Hannonis'*. *Forschungen und Fortschritte*, XXIV: 148–149

Stoczkowski Wiktor (1995). *Portrait de l'ancêtre en singe: l'hominisation sans évolutionnisme dans la pensée naturaliste du XVIII siècle*; In R. Corbey & Theunissen (eds.) 1995, pp. 141–155

Theunissen Bert (1989). *Eugène Dubois and the Ape-Man from Java: The History of the First 'Missing Link' and Its Discoverer*. Dordrecht: Kluwer

Thijssen J.M.M.H. (1995). *Reforging the Great Chain of Being: The Medieval Discussion of the Human Status of "Pygmies' and Its Influence on Edward Tyson*; In R. Corbey & Theunissen (eds.) 1995, pp. 43–50

Tinland Franck (1968). *L'homme sauvage. Homo ferus et homo sylvestris. De l'animal à l'homme*. Paris: Payot

Veracini C. (2021). An early representation of a gorilla from fifteenth-century Central Asia. *Primates*, 62: 457–462. https://doi.org/10.1007/s10329-021-00906-w

Wokler Robert L. (1976) Tyson and Buffon on the orang-utan. *Studies on Voltaire and Eighteenth Century*, 155: 2301–2319.

Wokler Robert L. (1978). Perfectible Apes in Decadent Cultures: Rousseau's Anthropology Revisited. *Daedalus*, pp. 107–134

Wokler Robert L. (1980). The ape debates in Enlightenment anthropology. *Studies on Voltaire and Eighteenth Century*, 192: 1164–1175.

Wokler Robert L. (1989). *Apes and Races in the Scottish Enlightenment: Monboddo and Kames on the Nature of Man*; In Jones 1989, pp. 145–168.

Wokler Robert L. (1993). From l'homme physique to l'homme moral and back: towards a history of Enlightenment anthropology. *History of the Human Sciences*, VI(1): 121–138.

Wokler Robert L. (1995). *Enlightenment Apes: Eighteenth-Century Speculation and Current Experiments on Linguistic Competence*; In R. Corbey & Theunissen (eds.) 1995, pp. 87–100.

Wood, B. 2020. Birth of primate anatomy. *Evolutionary Anthropology* 29, no. 1: 9–13.

Yerkes Robert M. & Yerkes Ada W. (1929). *The Great Apes*. New Haven: Yale University Press.

Section IV

Our Place in Nature

13 The Contribution of Morphology to Darwin's Understanding of the Genealogy of Modern Humans

Bernard Wood and Ryan McRae
George Washington University

Rowan M. Sherwood
University of Michigan

> Nor shall I have occasion to do more than to allude to the amount of difference between man and the anthropomorphous apes; for Prof. Huxley, in the opinion of most competent judges, has conclusively shown that in every visible character man differs less from the higher apes, than these do from the lower members of the same order of Primates.
>
> *(Darwin 1871, p. 18)*

13.1 INTRODUCTION

Darwin (1859) made a deliberate decision not to discuss the origin and relationships of modern humans in the *Origin of Species*. Browne (2021) points to a letter written by Darwin to Alfred Russel Wallace in 1857 in response to the latter's enquiry as to whether Darwin was going to discuss 'man' in his forthcoming book. In his letter, Darwin explained to Wallace "I shall avoid the whole subject" because it was "so surrounded with prejudices," yet Darwin conceded that the topic of human origins "is the highest & most interesting problem for the naturalist."

Darwin continued to ponder the origins and relationships of modern humans, eventually publishing his conclusions in his *The Descent of Man, and Selection in Relation to Sex* (Darwin, 1871). Because Darwin tackled the "highest & most interesting problem" in *The Descent of Man*, its contents and messages were at least as controversial—and arguably more controversial—than those in the *Origin of Species*. Darwin continued to develop his ideas, and in the second edition of *The Descent of Man*, published in 1874, he added 40 pages to the text and reorganized the book into three parts, 'The Descent or Origin of Man', 'Sexual Selection', and 'Sexual Selection in Relation to Man and Apes.'

In the first part of the book, Darwin draws on several lines of evidence to argue that modern humans should not be set apart from the natural world. One of those lines of evidence was comparative anatomy as explored and revealed by dissection, and in this chapter, we examine how anatomical research contributed to the developing understanding of the relationships between modern humans and the great apes. Specifically, we explore how information about the anatomy of the great apes, generated from dissections and the analysis of skeletal morphology, influenced that understanding.

Most of the apes reaching Europe alive in the 18th century and 19th century were infants, for only infants were tractable enough to be cared for during a long voyage. However, the external appearance of infants is paedomorphic, which made them seem more similar to modern humans than would have been the case if the comparisons had been made with adults. But, paedomorphosis has less effect on soft-tissue morphology, so information obtained from dissections was especially important evidence about the closeness, or otherwise, of the relationship between great apes and modern humans. We focus on reports by anglophone researchers, and we acknowledge we are likely to have omitted relevant reports published in other languages.

We reviewed investigations conducted up to, and including, 1871, the year the first edition of *The Descent of Man* was published. We used 1871 as the cut-off for two reasons. First, we wanted to capture the information that would have been available to Huxley before the publication of his *Evidence as to Man's Place in Nature* (Huxley, 1863). In this publication, Huxley reviewed the many resemblances between the African apes and modern humans and summarized the reasons why he became convinced the African apes were more closely related to modern humans than orangutans or the lesser apes. It was Huxley's evidence and conviction that convinced Darwin that Africa, not Asia, was probably the birthplace of humankind. Second, we wanted to make sure we covered any information about great ape morphology that could conceivably have been available to Charles Darwin when he was preparing *The Descent of Man*.

The publications we reviewed are listed chronologically in **Table 13.1**. Columns in the table specify which of the

DOI: 10.1201/b21963-17

TABLE 13.1
Publications and Reviews of Dissections of *Pan, Gorilla* and *Pongo* Specimens (1699–1871)

Publication	Author	Year	Troglodytes	Gorilla	Pongo	Hard Tissue	Soft Tissue	New Data
Orang-Outang, *sive Homo Sylvestris:* or, the Anatomy of a Pygmie Compared with that of a *Monkey, Ape* and a *Man*	Tyson, Edward	1699	Yes	No	No	Cranium Mandible Dentition Axial skeleton Upper limb/girdle Lower limb/girdle	Muscles Viscera Blood vessels Nerves Brain	Yes
Oeuvres de Pierre Camper, qui ont pour objet L'Histoire Naturelle, La Physiologie et L'anatomie Comparee	Camper, Pierre (aka Petrus)	1803	No	No	Yes	Cranium Axial skeleton Upper limb/girdle Lower limb/girdle	Viscera	Yes
Observations on the Anatomy of the Orang Outang	Traill, Thomas Stewart	1821	Yes	No	No	Cranium Upper limb/girdle Lower limb/girdle	Muscles Viscera Blood vessels Nerves Brain	Yes
Some Account of the Dissection of a Simia Satyrus, Ourang-Utang or Wild Man in the Wood	Jeffries, John	1824–5	No	No	Yes	Cranium Dentition Upper limb/girdle Lower limb/girdle	Viscera Brain	Yes
Des Dents des Mammifères, Considérées Comme Caractères Zoologiques	Cuvier, M.F.	1825	No	No	Yes	Dentition		Yes
Das Hirn des Orang-Outangs mit dem des Menschen Verglichen	Tiedemann F.	1827	No	No	Yes		Brain	No
On the Anatomy of the *Orang Utan* (*Simia Satyrus*, L).	Owen, Richard	1830	Yes	No	Yes	Cranium Axial skeleton Upper limb/girdle Lower limb/girdle	Muscles Viscera	Yes
On the Osteology of the Chimpanzee and Orang Utan	Owen, Richard	1835	Yes	No	Yes	Cranium Dentition Axial skeleton Upper limb/girdle Lower limb/girdle		Yes
XIII. On the Brain of the Negro, compared with that of the European and the Orang-Outang	Tiedemann, F.	1836	Yes	No	Yes		Nerves (spinal cord) Brain	Yes
Du Crane en Général, et de ses Proportions avec la Face. Leçons D'Anatomie Comparée (2nd edition)	Cuvier, Georges	1837	Yes	No	Yes	Cranium		Yes
Inquiry into the Present State of our Knowledge Respecting the Orang-Outang & Chimpanzee	Knox, Robert	1840	Yes	No	Yes	'Skeleton'		No
Recherches D'Anatomie Comparée sur le Chimpansé	Vrolik, W.	1841	Yes	No	No	'Skeleton'	Muscles Viscera Blood Vessels Nerves Brain	Yes
Art. XXXIV. Notice of the external characters and habits of Troglodytes gorilla. A new species of orang from the Gaboon River by Thomas S. Savage and osteology of the same by Jeffries Wyman.	Savage, T.S. and Wyman, J.	1847	No	Yes	No	Cranium Axial skeleton Upper limb/ Lower limb/girdle		Yes

(*Continued*)

TABLE 13.1 (*Continued*)

Publications and Reviews of Dissections of *Pan, Gorilla* and *Pongo* Specimens (1699–1871)

Publication	Author	Year	Troglodytes	Gorilla	Pongo	Hard Tissue	Soft Tissue	New Data
Osteological contributions to the natural history of the chimpanzees (*Troglodytes*) and orangs (*Pithecus*). No. IV. Description of the cranium of an adult male gorilla from the river Danger, West Coast of Africa, indicative of a variety of the great chimpanzee (*Troglodytes Gorilla*), with remarks on the capacity of the cranium and other characters shown by sections of the skull, in the orangs (*Pithecus*), chimpanzees (*Troglodytes*), and in different varieties of the human race	Owen, Richard	1853	Yes	Yes	Yes	Cranium		Yes
Mémoire sur les Plis Cérébraux de L'homme et des Primates. Paris: Libraire Arthus Bertrand	Gratiolet, Pierre	1854	Yes	Yes	Yes		Brain	No
Caracteres Anatomiques des Grands Singes Pseudo-Anthropomorphes	Duvernoy, George Louis	1855	Yes	Yes	Yes	Cranium Dentition Axial skeleton Upper limb/girdle Lower limb/girdle		Yes
Prof. Jeffries Wyman gave an account of the dissection of a black chimpanzee, (Troglodytes niger) one of the collection presented by Dr. J. V. C. Smith. The dissection was made particularly with reference to a comparison of the muscular system of Troglodytes with that of man.	Wyman, J.	1855	Yes	No	No	Upper limb/girdle	Muscles	Yes
Osteological contributions to the natural history of the chimpanzees (*Troglodytes*) and Orangs (*Pithecus*). No. V. Comparison of the lower jaw and vertebral column of the *Troglodytes Gorilla*, *Troglodytes niger*, *Pithecus Satyrus* and different varieties of the human race	Owen, Richard	1856	Yes	Yes	Yes	Mandible Axial skeleton		Yes
On the brain of a young chimpanzee	Marshall, John	1861	Yes	No	No	Measurements of upper limb/girdle, lower limb/girdle	Nerves Brain	Yes
On the affinities of the brain of the orang utang	Rolleston, George	1861	Yes	No	Yes	Dentition	Brain	Yes
On the myology of the Orang Utang (Simia Morio)	Church, William Selby	1861	Yes	Yes	Yes		Muscles	Yes
On the myology of the orang utang	Church, William Selby	1862	Yes	Yes	Yes		Muscles	Yes
Contributions to the comparative myology of the chimpanzee	Wilder, Burt G.	1863	Yes	Yes	Yes		Muscles	Yes

(*Continued*)

TABLE 13.1 (*Continued*)

Publications and Reviews of Dissections of *Pan, Gorilla* and *Pongo* Specimens (1699–1871)

Publication	Author	Year	Troglodytes	Gorilla	Pongo	Hard Tissue	Soft Tissue	New Data
Professor Huxley's Hunterian lectures on 'The Structure and classification of the Mammalia'	Huxley, Thomas Henry	1864	Yes	Yes	Yes	Cranium Mandible Dentition Axial skeleton Upper limb/girdle Lower limb/girdle	Muscles Viscera Nerves Brain	Yes
On the os penis of the chimpanzee (Troglodytes niger) and of the orang (Simia satyrus)	Edwards, Crisp	1865	Yes	No	Yes	Splanchnic bone		Yes
Contributions towards a more complete knowledge of the axial skeleton in the primates	Mivart, St. George	1865	Yes	Yes	Yes	Axial skeleton		No
On some points in the anatomy of the chimpanzee	Humphry, G.M.	1866	Yes	No	No	Upper limb/girdle Lower limb/girdle	Viscera Muscles Blood vessels	Yes
On the muscles and nerves of a chimpanzee (*Troglodytes niger*) and a *Cynocephalus Anubis*	Champneys, Frank	1871	Yes	No	No		Muscle Nerves	Yes
On some points in the myology of the chimpanzees and others of the primates	Macalister, Alexander	1871	Yes	No	No		Viscera Muscle Nerves	Yes

three great apes known prior to 1871 were included in each study and in broad terms what type of information about their gross anatomy was presented in each publication.

Not all reports—especially the earlier ones—provide unambiguous evidence about the identity of the cadaver(s) that were dissected, or the skeletal evidence that was examined. One reason is that terminology has changed over time. When the first animals we would now refer to as orangutans appeared in Europe in the 17th century, they were called either ourang-outangs (e.g. Bondt, 1658; Barsanti, 1989), or ourang-utans (e.g. Schouten, 1676; Barsanti, 2009). But later in the 17th century, and in the 18th century, the label orang-outang was used more inclusively to refer to *any* of the apes that would now be assigned to *Gorilla*, *Pan* and *Pongo* (see Barsanti, this volume). It was not until Petrus Camper made the distinction between the red apes of Asia and the black apes of Africa (Camper, 1779) that the differences began to be clarified. Even then, it took a long time for orang-outang and ourang-outang to revert to their original, more exclusive, interpretation and for the spelling to change to the contemporary orangutan in US English, and orang-utan in UK English. Despite Camper making this distinction at the end of the 18th century, those knowledgeable about the literature emphasize that ambiguity and confusion about the names of great ape species persisted well into the 19th century (Barsanti, 2009). For example, when the lowland gorilla was first formally described in 1847, it was referred to as 'a new species of Orang' (Savage & Wyman, 1847: 417).

Given this background, for those publications where there was a possibility of ambiguity in the use of the label 'Orang,' 'orang-outang' or 'ourang-utan,' we made decisions about the likely identity of the dissected cadaver, or the skeletal remains, on the basis of their reported geographical provenance. In the absence of any clues about the external appearance of animals (e.g. 'black' or 'red'), if the animals were obtained from traders in ports that border the Indian Ocean, we assumed they were orangutans *sensu stricto*, and if they were obtained from West African traders, we assumed they were either chimpanzees or gorillas.

The remainder of this review is divided into three parts. The first part is devoted to just one publication, the first comprehensive description of the morphology of a great ape by Edward Tyson. The second and third parts focus on the reports published between 1700 and 1829, and between 1830 and 1871, respectively. We do not review all of the publications in those time intervals included in **Table 13.1**, just those that are, for one reason or another, of particular interest.

13.2 EDWARD TYSON (1650–1708)

Early in 1698, a merchant ship from West Africa arrived in Bristol in the UK with what was almost certainly the first chimpanzee or bonobo to reach the British Isles alive. The creature—an infant—had been injured during the voyage from Angola, and after its death, likely in early April 1698, it was dissected in London by Edward Tyson, with the assistance of William Cowper, a distinguished human anatomist. The illustrations in Tyson's monograph (Tyson, 1699) plus the provenance and the morphology of the preserved skeleton (which is now on display in the Natural History Museum in London) give us confidence that Tyson's meticulous anatomical description was the first scientific report of the anatomy of a chimpanzee or bonobo (see **Fig. 13.1**). The only way to make sure which of these it was would be to analyze the DNA of the skeleton, but morphologically the skeleton looks more like that of a chimpanzee than a bonobo.

Why was the infant chimpanzee dissected and not destroyed, and why was its dissection entrusted to Tyson? We know as much as we do about Edward Tyson's life and career because of Ashley Montagu's meticulously researched biography (Montagu, 1943). According to Ashley Montagu, Edward Tyson was born in Bristol in 1650 to well-to-do parents: his father was a prosperous merchant. He was admitted to Magdalen Hall in Oxford when he was only sixteen years old. Among other subjects, Tyson studied

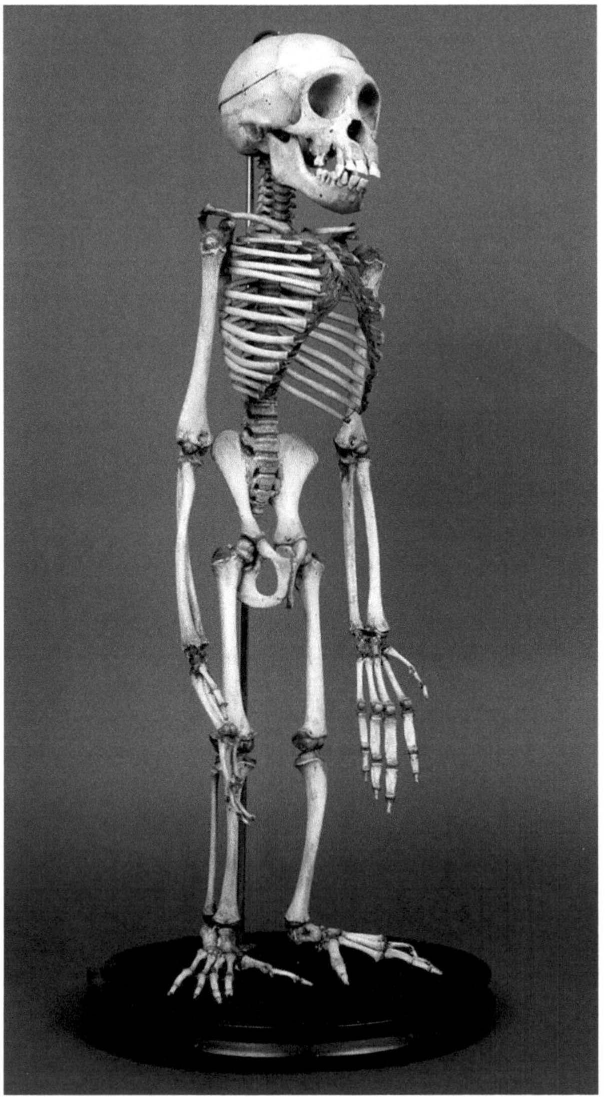

FIGURE 13.1 Edward Tyson's Pygmie. (© The Trustees of The Natural History Museum, London.)

anatomy. His instructor, Robert Plot, held classes in the *Scholar Anatomica*, where the University's modest natural history collection was housed. So, Tyson would have been exposed to what we now refer to as comparative anatomy. After the award of a Master of Arts in 1673, Tyson switched to medicine, and in 1677, after completing his medical studies, he moved to London where he consolidated his reputation as an anatomist; two years later, Tyson was elected to the fellowship of the Royal Society of London.

As well as maintaining a busy practice as a physician, Tyson was the 'go to' person for comparative anatomy, so he was given the opportunity to dissect exotic creatures brought to the attention of fellows of the Royal Society. By 1698, Tyson had accumulated an impressive publication record in human and comparative anatomy as the author not only in the Philosophical Transactions of the Royal Society but also of monographic publications, including careful anatomical descriptions of a porpoise (Tyson, 1680), a rattlesnake (Tyson, 1683a), a tapeworm (1683b), a Mexican musk-hog (Tyson, 1683c), and an opossum (1698). Tyson's connections with Bristol—and his father's status as a prominent merchant—mean that he would likely have been made aware of the arrival of any exotic creatures in Bristol, and his reputation as the pre-eminent comparative anatomist of the time also made it likely he would be entrusted with the dissection of what came to be referred to as the Pygmie—presumably because of the small overall size and short stature of the chimpanzee infant.

Tyson's monograph 'Orang-outang sive Homo sylvestris: *or The Anatomy of a Pygmie compared with that of a* Monkey, *an* Ape, *and a* Man' consists of 105 pages of text and illustrations. Bound together with Tyson's anatomical description are three 'Philological Essays' concerning the ways the Ancients interpreted 'Pygmies,' 'Cynocephali' and 'Satyrs and Sphinges'; for more information about these essays, see Wood (2020).

The first section of the monograph, comprising 58 text pages, consists of Tyson's careful descriptions of the external appearance (see **Fig. 13.2**) and the viscera of the Pygmie. The next section, 26 text pages, is a description of the skeleton, followed by seven text pages devoted to myology (see Wood 2020). Given that Tyson and Cowper were dissecting a whole animal in poor condition, Tyson's descriptions of the muscles are remarkably accurate. Subsequent research has shown more detailed differences in the myology of modern humans and chimpanzees than were recorded by Tyson (see Diogo & Wood, 2012), but these were revealed by dissections limited to the striated muscles and carried out on cadavers in good condition.

The penultimate section of the monograph consists of two lists. The first documents the 48 ways Tyson's Pygmie "more resembled a Man" (p. 92) and the second the 34 ways it "resembled more the Ape and Monkey-kind" (p. 94). In a brief summary on p. 91, Tyson suggests that "our *Pygmie*, more resembled the humankind than *Apes* and *Monkeys* do," yet "our *Pygmie* is no *Man*, nor yet the *Common ape*; but a sort of *Animal* between both." The final

FIGURE 13.2 Chimpanzee-like external appearance of Tyson's Pygmie.

section (pp. 95–108) consists of legends for the exquisite copper-plate illustrations of the dissections prepared by M. Vandergucht, whose name appears on the frontispiece of the second, 1751, edition, but not the first.

Tyson's observations about the Pygmie have proved to be remarkably accurate and perceptive, and many have been confirmed by later research. For example, with respect to the brain, he wrote "when I am describing the Brain of our Pygmie, you may fully suspect I am describing that of Man … by allowing only for the magnitude of its parts" (p. 54), as well as stating that "in all respects the Brain of Pygmie does so exactly resemble a Man's" (p. 55). It is also clear that Tyson had identified the importance of relative brain size, for he goes on to qualify his previous observation by noting that "the brain of Man, in respect of his body, be much larger than what it is to be met with in any other animal" (p. 55). As for the skeleton, Tyson recognized the critical differences in shape between the pelvic bones of modern humans and chimpanzees, with the pelvis as a whole being broader and shorter in modern humans, and taller and narrower in the Pygmie. According to Tyson, "there was no part I think of the whole skeleton where the Pygmie differed more from Man, than in the structure and figure of the Os Ilium" (p. 74). Specifically, he suggested that whereas in modern humans, the ilium is "spread broad" and "hollow on the inside," in the Pygmie, the ilium is "proportionately longer and narrower" and "in all respects conformable to the shape of the same bones in the Ape and Monkey-kind" (p. 74). Tyson's pioneering contribution is still one of the most comprehensive reports of the dissection of an ape (Wood, 2020), and we know Huxley was aware of Tyson's monograph because he refers to Tyson's work (see below).

The reports of subsequent dissections and examinations of dental and skeletal morphology that contributed new knowledge of the anatomy of the orangutan, common chimpanzee and gorilla, plus a few reviews, are listed and summarized in **Table 13.1**. The pre-1830 publications report the results of the dissection of creatures or cadavers that were acquired from merchants trading in southeast Asia, for orangutans, or West or Central Africa, for chimpanzees. Later publications report the dissection of cadavers obtained from zoological gardens or menageries. Almost all of the dissections were carried out by physicians or surgeons, and many were carried out by medically qualified individuals who were also distinguished comparative anatomists (e.g. Camper, Cuvier, Duvernoy, Huxley and Owen).

13.3 1700–1829

There is no simple explanation for the just over one-hundred-year gap between the publication of Tyson's monograph and the next report of a dissection of a great ape, in this case, the dissection of an orangutan (Camper, 1803). In fact, the actual gap was not that long, because while Camper died in 1789, the 1803-posthumous publication reported the results of dissections he carried out in 1777, more than two decades before. The next publication in the historical series reported the dissection in Liverpool of what was almost certainly a chimpanzee (Traill, 1821). Thomas Traill (1781–1862) was a distinguished physician and academic with wide interests in natural history and medical jurisprudence, but, unlike Tyson and Camper, he had no particular reputation as a comparative anatomist.

Around this time, we see the publication of the first report of a dissection of an ape by a researcher working outside of Europe. It communicated the results of the dissection of an immature orangutan that died the day after it reached the port of Boston on the eastern seaboard of the United States (Jeffries, 1824–5). John Jefferies II (1796–1876), who became a distinguished ophthalmic surgeon, carried out his orangutan dissection at the beginning of his surgical career. Like Traill, there is no evidence that Jefferies had any long-term interest in comparative anatomy.

13.4 1830–1871

The publications during this interval mark a shift from mostly opportunistic investigations to more systematic ones, in the sense that they either focussed on multiple dissections of one of the apes (e.g. Vrolik, 1841; Savage & Wyman, 1847), or they compared the apes (e.g. Owen, 1835, 1856; Huxley 1864). Others were even more specialized in that they focussed on just one region or system. For example, Church (1861, 1862), Wilder (1863) and Macalister (1871) only dealt with the musculature.

The expansion of international trade at the beginning of the 19th century resulted in more opportunities to import exotic animals, initially to private menageries and subsequently for public display. The founding of the Zoological Society of London (ZSL) in 1826—and the opening of the London Zoological Gardens two years later—marked a significant shift in where and how great apes were displayed (Keeling, 2000: 69; Plumb 2010: 16). Great apes imported into Europe during this period were in poor condition, usually arriving as juveniles and dying young (Corbey, 2005; van Wyhe & Kjærgaard, 2015). For example, the first live ape to be acquired by ZSL (van Wyhe & Kjærgaard, 2015) was an oran-gutan that died after only three days at the Zoo (N.B., this was the individual animal dissected by Owen in 1830). In 1835, the ZSL acquired its first chimpanzee, a juvenile that only survived for several months before dying in March 1836 (Flower 1929).

The first oran-gutan to live long enough to be exhibited at the ZSL was a juvenile purchased in November 1837. This juvenile, a young female (c. three years) named Jenny, was housed at the ZSL until her death in May 1389. She was one of the several oran-gutans visited regularly by Darwin in 1838. After returning from his voyage on the *Beagle*, Darwin frequented the ZSL's Gardens while developing his theory of transmutation or evolution. During these visits, Darwin made a series of observations of the oran-gutans, recording any purportedly human-like behaviours or emotions in these apes (van Wyhe & Kjærgaard, 2015). His notes made at the time from these 'up-close' encounters suggest that he saw the oran-gutans as close relatives and not direct ancestors. The anatomical and behavioural resemblances between these apes and modern humans contributed to Darwin later arguing that the similarities—and differences—could be explained by his theory of evolutionary descent (Barrett et al., 1987; van Wyhe & Kjærgaard, 2015).

It was during this period that the publication formally recognizing the gorilla (aka *Troglodytes gorilla*) was published (Savage & Wyman, 1847). That publication also included a systematic comparison of the gorilla partial skeletons brought back from Gabon with what was known of the chimpanzee skeleton. The first gorilla to be brought back live to Europe was imported from the Congo to Liverpool in 1855. Misidentified as a chimpanzee, until its death early in 1856, it was briefly exhibited in the travelling menagerie of George Wombwell. The body was preserved and put on display at Walton Hall, the estate of the naturalist Charles Waterton. In 1861, it was seen at Walton Hall by Abraham Dee Bartlett, the superintendent of the London Zoological Gardens, who identified it as a juvenile gorilla (Cooper & Hull, 2017: 547). The ZSL acquired its first gorilla in 1887 (Flower, 1929). Towards the end of this time interval, the surgeon and biologist Thomas Henry Huxley published his review of the comparative anatomy of the great apes (Huxley, 1863). This was the evidence Darwin refers to in the quote from *The Descent of Man* that begins this contribution.

Huxley's *Evidence as to Man's Place in Nature* consists of three related essays. The first reviews the history of discovery of the great apes and what was known at the time about their distribution and behaviour. The second considers how 'Man' (i.e. modern humans) is related to the rest of the animal kingdom. The third reviews what little was known at the time about the fossil evidence for human evolution. Huxley begins the second essay, a comparative review of the morphological evidence entitled 'On the relations of Man to the lower animals,' with the following:

> The question of questions for mankind—the problem which underlies all others, and is more deeply interesting than any other—is the ascertainment of the place which Man occupies in nature and of his relations to the universe of things. (Huxley 1863: 57)

In the preamble to his review, Huxley makes it clear that "the Ape which most nearly approaches man, in the totality of its organization, is either the Chimpanzee or the Gorilla; and it makes not practical difference, for the purposes of my present argument, which is selected for comparison" (*ibid*: 70). Yet, despite the relative dearth of evidence about the gross morphology of the gorilla compared to what was known at the time about the chimpanzee and the orang-utan, Huxley opted for the gorilla and not the chimpanzee as the comparator. He explains that this is not for any scientific reason, but because he thinks gorillas "are now so celebrated in prose and verse, that all must have heard of him, and have formed some conception of his appearance" (*ibid*: 71).

Huxley summarizes his comprehensive review of the comparative evidence (*ibid*: 71–102) as follows:

> whatever system of organs be studied, the comparison of their modifications in the ape series leads to one and the same result-that the structural differences which separate Man from the Gorilla and Chimpanzee are not so great as those which separate the Gorilla from the lower apes. (*ibid*: 103)

As for how those structural differences should be reflected taxonomically, Huxley suggests the "structural differences between Man and the Man-like apes certainly justify our regarding him as constituting a family apart from them," but "there can be no justification for placing him in a distinct order" (*ibid*: 104).

It is not difficult to understand why this forceful and unambiguous conclusion influenced Darwin to subscribe to Huxley's view that the gap between modern humans and the great apes, especially the chimpanzee and gorilla (N.B., bonobos were not recognized as being distinct from common chimpanzees until the 20th century), was not as wide as was commonly believed in the second half of the 19th century.

13.5 CONCLUSION

Huxley's 'downgrading' of the differences between modern humans and the great apes to the level of the family was not novel. Gray (1825) had gone further, recommending that the great apes and modern humans should be grouped together in the same family, the Hominidae, with *Homo* (i.e. modern humans) in the sub-family Hominina, and *Troglodytes* and *Simia* (plus *Hylobates*) in the sub-family Simiina (*ibid*: 338). But until relatively recently, there was little in the way of consensus about the taxonomic level that should be used to distinguish modern humans from the great apes. For example, in his monumental three-part review of the taxonomy of the primates, Elliot (1913) placed *Pongo*, *Gorilla*, *Pseudogorilla* (likely a juvenile mistaken as a separate taxon) and *Pan* within the family Pongiidae, separated at the family level from modern humans, whereas Wilder (1926) was of the opinion that "the establishment of all (i.e. 'anthropoid apes' and 'men') under the bounds of a single Family (Hominidae) is now an obvious step" (*ibid*: 47).

The next chapter explores the ways additional gross morphological evidence, and various categories of non-morphological evidence, have influenced hypotheses about the relationships among modern humans and the great apes.

REFERENCES

Barrett, P. H., Gautrey, P. J., Herbert, S., Kohn, D. & Smith, S. (Eds.). 1987. *Charles Darwin's Notebooks, 1836–1844: Geology, Transmutation of Species, Metaphysical Enquiries*. London: British Museum.

Barsanti, G. 1989. L'orang-outang déclassé. Histoire du premier singe à hauteur d'homme (1780-1801), in Histoire de l'anthropologie: hommes, idées, moments, a cura di C. Blanckaert et al. *Bulletins et Mémoires de la Société d'Anthropologie de Paris*, n.s., I 3–4: 67–104.

Barsanti, G. 2009. *L'Uomo dei boschi: Piccola storia delle grande scimmie de Aristotle a Darwin*. Rome: La Sapienza University Publishing House.

Bondt, J. D. 1658. *Historiae naturalis et medicae Indiae orientalis libri sex*. Amsterdam: Willem Piso.

Browne, J. 2021. Introduction. In *A Most Interesting Problem: What Darwin's Descent of Man Got Right and Wrong about Human Evolution*, ed. J. DeSilva, 1–23. Princeton: Princeton Univ. Press.

Camper, P. 1779. Kort berigt wegens de ontleding van verscheidene orang outangs, en inzonderheid van die in de diergaarde van Zyne Doorluchtigste Hoogheid, den Heere Prinse van Orange, Erfstadhouder, enz. Enz. Enz. Gestorven is, in den jaare 1777. *Algemeene Vaderlandsche Letter-Oefeningen* 1: 18–36.

Camper, P. 1803. *Oeuvres de Pierre Camper, qui ont pour objet l'histoire naturelle, la physiologie et l'anatomie comparee*. Paris: Chez H.J. Jansen.

Champneys, F. 1871. On the muscles and nerves of a chimpanzee (*Troglodytes niger*) and a *Cynocephalus anubis*. *Journal of Anatomy and Physiology* 6, no. 1: 176–211.

Church, W. S. 1861. On the myology of the orang utang (Simia morio). *Natural History Review: A Quarterly Journal of Biological Science* 1, no. 4: 510–516.

Church, W. S. 1862. On the myology of the orang utang. *Natural History Review: A Quarterly Journal of Biological Science* 2, no. 5: 82–94.

Cooper, J. E. & Hull, G. 2017. *Gorilla Pathology and Health with a Catalogue of Preserved Materials*. Amsterdam: Elsevier Science.

Corbey, R. 2005. *The Metaphysics of Apes: Negotiating the Animal-Human Boundary*. New York: Cambridge Univ. Press.

Crisp, E. 1865. On the os penis of the chimpanzee (*Troglodytes niger*) and of the orang (*Simia satyrus*). *Proceedings of the Zoological Society of London* 33, no. 1: 48–49.

Cuvier, M. F. 1825. *Des dents des mammifères, considérées comme caractères zoologiques*. Paris: F. G. Levrault.

Cuvier, G. 1837. *Du crane en général, et de ses proportions avec la face. Leçons d'anatomie comparée (2nd edition)*. Paris: Crochard.

Darwin, C. 1859. *On the Origin of Species by Means of Natural Selection, or the Preservation of Favoured Races in the Struggle for Life*. London: John Murray.

Darwin, C. 1871. *The Descent of Man, and Selection in Relation to Sex*. London: John Murray.

Diogo, R. & Wood, B. 2012. Soft-tissue anatomy of the primates: phylogenetic analyses based on the muscles of the head, neck, pectoral region and upper limb, with notes on the evolution of these muscles. *Journal of Anatomy* 219: 273–359.

Duvernoy, G. L. 1855. Caracteres anatomiques des Grands Singes Pseudo-anthropomorphes. *Archives du Muséum National D'histoire Naturelle* (*Paris*) 8: 1–248.

Elliot, D. G. 1913. *A Review of the Primates. Monographs of the American Museum of Natural History monographs Series, Volumes I–III*. New York: American Museum of Natural History.

Flower, S. S. 1929. *List of the Vertebrate Animals Exhibited in the Gardens of the Zoological Society of London, 1828–1927. Centenary Edition in Three Volumes. Volume I: Mammals*. London: Zoological Society of London.

Gratiolet, P. 1854. *Mémoire sur les plis cérébraux de l'homme et des primates*. Paris: Libraire Arthus Bertrand.

Gray, J. E. 1825. An outline of an attempt at the disposition of Mammalia into tribes and families, with a list of the genera apparently appertaining to each tribe. *The Annals of Philosophy* 10: 337–344.

Humphry, G. M. 1866. On some points in the anatomy of the chimpanzee. *Journal of Anatomy and Physiology* 1, no. 2: 254–268.

Huxley, T. H. 1863. *Evidence as to Man's Place in Nature*. London: Williams & Norgate.

Huxley, T. 1864. Professor Huxley's Hunterian lectures on "the structure and classification of the Mammalia," delivered at the Royal College of Surgeons. Lectures X-XVIII. Delivered between February 23rd and March 12th. Published in *Medical Times and Gazette* between April 9th and June 4th, pp. 398–646.

Jeffries, J. 1824–1825. Some account of the dissection of a Simia satyrus, ourang-utang, or wild man of the woods. *Webster & Treadwell's Boston Journal of Philosophy and the Arts* 2: 570–580.

Keeling, C. H. 2000. Zoological gardens of Great Britain. In *Zoo and Aquarium History*, ed. V. N. Kisling. Boca Raton: CRC Press.

Knox, R. 1840. Inquiry into the present state of our knowledge respecting the orang-outang & chimpanzee. *The Lancet* 34, no. 873: 289–296.

Macalister, A. 1871. On some points in the myology of the chimpanzees and others of the primates. *Annals and Magazine of Natural History Series* 4 7, no. 41: 341–351.

Marshall, J. 1861. On the brain of a young chimpanzee. *The Natural History Review* July: 296–316.

Mivart, S. G. 1865. Contributions towards a more complete knowledge of the axial skeleton in the primates. *Proceedings of the Zoological Society of London* 33, no. 1: 545–592.

Montagu, A. 1943. *Edward Tyson, M.D., F.R.S., 1650–1708 and the Rise of Human and Comparative Anatomy in England; a Study in the History of Science*. Philadelphia: American Philosophical Society.

Owen, R. 1830. On the anatomy of the *orang utan (Simia Satyrus, L). Proceedings of the Committee of Science Correspondence of the Zoological Society of London* 1: 4–5, 9–10, 28–29, 67–72.

Owen, R. 1835. On the osteology of the chimpanzee and orang utan. *The Transactions of the Zoological Society of London* 1, no. 4: 343–379.

Owen, R. 1853. Osteological contributions to the natural history of the chimpanzees (*Troglodytes*) and orangs (*Pithecus*). No. IV. Description of the cranium of an adult male gorilla from the River Danger, West Coast of Africa, indicative of a variety of the great chimpanzee (*Troglodytes Gorilla*), with remarks on the capacity of the cranium and other characters shown by sections of the skull, in the orangs (*Pithecus*), chimpanzees (*Troglodytes*), and in different varieties of the human race. *Transactions of the Zoological Society of London* 4, no. 3: 75–88.

Owen, R. 1856. Osteological contributions to the natural history of the chimpanzees (*Troglodytes*) and orangs (*Pithecus*). No. V. Comparison of the lower jaw and vertebral column of the *Troglodytes Gorilla, Troglodytes niger, Pithecus Satyrus*, and different varieties of the human race. *Transaction of the Zoological Society of London* 4, no. 4: 89–116.

Plumb, C. 2010. Exotic animals in eighteenth-century Britain. PhD diss., Univ. of Manchester.

Rolleston, G. 1861. On the affinities of the brain of the orang utang. *Natural History Review: A Quarterly Journal of Biological Science* 1, no. 2: 201–217.

Savage T. S. & Wyman, J. 1847. Notice of the external characters and habits of *Troglodytes gorilla*, a new species of orang from the Gaboo River, by Thomas S. Savage; osteology of the same, by Jeffries Wyman. *Boston Journal of Natural History* 5: 417–442.

Schouten, W. 1676. *Wouter Schoutens Oost-Indische Voyage*. Amsterdam: Jacob Meurs & Johannes van Someren.

Tiedemann, F. 1827. Das hirn des orang-outangs mit dem des menschen verglichen. *Zeitschrift fur Physiologie* 2: 17–28.

Tiedemann, F. 1836. XIII. On the brain of the negro, compared with that of the European and the orang-utang. *Philosophical Transactions of the Royal Society of London* 126: 497–527.

Traill, T. S. 1821. Observations on the anatomy of the orang outang. *Memoirs of the Wernerian Natural History Society* 2: 1–49.

Tyson, E. 1680. *Phocæena, or the Anatomy of a Porpess Dissected at Gresham College: With a Præliminary Discourse Concerning Anatomy, and a Natural History of Animals*. London: Benjamin Tooke.

Tyson, E. 1683a. Vipera caudi-sona Americana, or the anatomy of a rattles-snake, dissected at the repository of the Royal Society in January 1682. *Philosophical Transactions of the Royal Society of London* 13, no. 144: 25–58.

Tyson, E. 1683b. Lumbricus latus, or a discourse read before the Royal Society, of the jointed worm, wherein a great many mistakes of former writers concerning it, are remarked; its natural history for more exact observations is attempted; and the whole urged, as a difficulty against the doctrine of univocal generation. *Philosophical Transactions of the Royal Society of London* 13, No. 146: 113–144.

Tyson, E. 1683c. Tajacu seu aper Mexicanus moschiferus, or the anatomy of the Mexico musk-hog. *Philosophical Transactions of the Royal Society of London* 13, no. 153: 359–385.

Tyson, E. 1698. Carigueya, seu Marsupiale Americanum. Or, the anatomy of the opossum, dissected at Gresham-College by Edw. Tyson, M.D. Fellow of the College of Physicians, and of the Royal Society, and reader of anatomy at the Chyrurgeons-Hall, in London. *Philosophical Transactions of the Royal Society of London* 20, no. 239: 105–164.

Tyson, E. 1699. *Orang-outang sive Homo sylvestris: Or the Anatomy of a Pygmie Compared with That of a Monkey, an Ape, and a Man*. London: Thomas Bennet and Daniel Brown.

van Wyhe, J. & Kjærgaard, P. C. 2015. Going the whole orang: Darwin, Wallace and the natural history of orangutans. *Studies in History and Philosophy of Science Part C: Studies in History and Philosophy of Biological and Biomedical Sciences* 51: 53–63.

Vrolik, W. 1841. *Recherches d'anatomie comp arée sur le chimpansé*. Amsterdam: Johannes Müller.

Wilder, B. G. 1863. Contributions to the comparative myology of the chimpanzee. *Boston Journal of Natural History* 7: 325–384.

Wilder, H. H. 1926. *The Pedigree of the Human Race*. London: George G. Harrap.

Wood, B. 2020. Birth of primate anatomy. *Evolutionary Anthropology* 29, no. 1: 9–13.

Wyman, J. 1855. Prof. Jeffries Wyman gave an account of the dissection of a black chimpanzee, (Troglodytes niger) one of the collection presented by Dr. J. V. C. Smith. The dissection was made particularly with reference to a comparison of the muscular system of Troglodytes with that of man. *Proceedings of the Boston Society of Natural History* 5: 274–276.

14 How Old and New Lines of Evidence Have Contributed to Our Understanding of the Relationships among Modern Humans and the Great Apes

1900–2023

Bernard Wood
George Washington University

Rowan M. Sherwood
University of Michigan

14.1 INTRODUCTION

Before the beginning of the 20th century, the only evidence available for determining the closeness, or otherwise, of the relationship between any two species of living animals was their gross morphology (i.e., how similar they were in terms of what the naked eye could tell from their appearance, behavior, organs, bones, teeth, muscles, etc.). The assumption was that the more closely two living taxa resembled each other phenotypically, the closer they are on the "Tree of Life". The previous chapter explored how this type of morphological evidence had caused researchers to embrace the notion of a relatively close relationship between modern humans and the great apes.

During the first half of the 20th century, researchers began to exploit advances in biology, and particularly in what we now refer to as immunology, to develop new lines of evidence for investigating the relationships among living animals (Strasser, 2010). Initially, methods—such as the precipitin reaction developed by Nuttall (1901)—which we might now describe as crude but at the time were novel and insightful, the mechanisms then being poorly understood. This changed in the middle of the 20th century, when researchers made critical advances in protein chemistry and immunology. At about the same time, the discovery of the role played by DNA in inheritance and development, and the subsequent elucidation of the genetic code, held out the possibility that the detailed structure of DNA could be used to investigate relationships among living animals. Potentially, the ability to compare DNA would eliminate the need to rely on proxies, be they traditional phenotypic morphology (i.e., gross anatomy), immunology, or protein structure, for information about relatedness.

This chapter reviews how serological, biochemical, immunological, and DNA evidence, plus additional comparative morphological evidence, have contributed to refining our understanding of the relationships between modern humans and the great apes.

14.2 SEROLOGY

Serology refers to the use of blood serum as a taxonomic tool. In simple terms, fresh blood consists of plasma and red blood cells. When fresh blood coagulates—or clots—the clear fluid that is left is referred to as serum; serum is plasma minus the various factors involved in coagulation. Before the clotting mechanism was well enough understood so that whole blood could be stored without coagulating, researchers interested in experimental systematics focused on serum. This was because serum is relatively easy to manipulate and can be stored for long periods in a cold room.

The precipitin reaction is the final step in an elaborate procedure. The serum of each of the animals being studied was injected multiple times into the peritoneal cavity of a rabbit. This resulted in the rabbit producing what was referred to as "anti-serum". This "anti-serum" was then mixed in a test-tube with the diluted serum of the animal with which it was being compared. If the animals were closely related, a white granular precipitate would form and collect in the bottom of the tube. The closer the relationship between the animal from whom the anti-serum was

DOI: 10.1201/b21963-18

generated and the animal whose diluted serum was mixed with it the greater the volume and granularity of precipitate. The procedure, which was known as the precipitin test or precipitin reaction, was devised by Kraus (1897), but the first person to exploit it systematically to investigate the relationships among animals was Nuttall (1901). In Nuttall's own words, if the rabbit had been injected with modern human blood:

> the serum of this rabbit will be found to have acquired the remarkable property of producing a precipitation immediately on its being added in small quantity to a dilution of human blood-serum. If allowed to rest, the precipitated substance gravitates to the bottom of the tube. I have now tested upwards of 230 bloods obtained from animals of all classes of vertebrates with such anti-serum for human blood, and have, with the single exception of monkey bloods, obtained negative results throughout. (Nuttall 1901: 150–151)

Only when human anti-serum was mixed with diluted human serum was there a granular precipitate.

The next few years saw considerable activity by Nuttall and his collaborators, and the results of their work—they carried out 16,000 tests on a total of 586 species—were summarized in a volume (Nuttall 1904) entitled *Blood immunity and blood relationship. A demonstration of certain blood-relationships amongst animals by means of the precipitin test for blood*. In the book, Nuttall suggests that the first researcher to report the use of the precipitin reaction to assess the relationships among the great apes and modern humans was Albert Grünbaum, a physiologist and a clinician at The Liverpool Hospital for Consumption and Diseases of the Chest in the UK. In a short article that took little more than a single column of the medical journal *The Lancet* (see **Fig. 14.1**), Grünbaum laid a strong claim to be the pioneer of comparative primate serology. In his letter to *The Lancet*, Grünbaum explained that he had "the good fortune to obtain considerable quantities of the blood of three anthropoid apes", and that he injected human, chimpanzee, gorilla, and orangutan serum into rabbits to generate, respectively, human, chimpanzee, gorilla, and orangutan anti-serum. He reported that when he mixed human anti-serum with diluted serum from the three apes, the mixture:

> gives a precipitate with all three species of anthropoid apes already named, and to me this precipitate is practically indistinguishable from that obtained with human blood either in quality or quantity. Occasionally it has seemed that the blood of the orang gave a more gelatinous precipitate as compared with the granular precipitate of the other bloods, but this may have been due to accidental circumstances. (Grünbaum 1902: 143)

He continued that "in the absence of definite standards, I am unable to assert that there is any difference of reaction amongst the many combinations of anti-serum and blood which can be made with the four above-mentioned bloods and sera" (*ibid*: 143). This is almost certainly the first use of

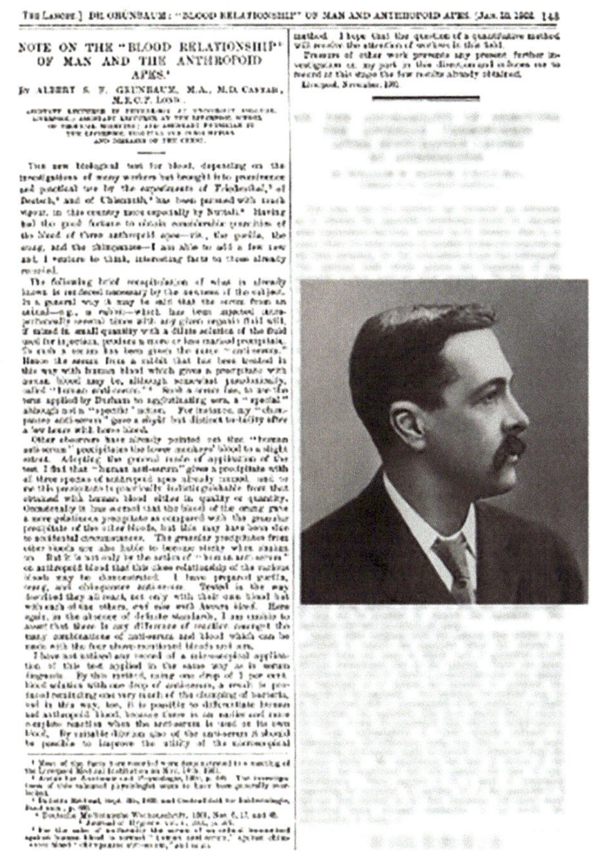

FIGURE 14.1 Grünbaum's 1902 *The Lancet* article, "Notes on the 'blood relationship' of man and the anthropoid apes." (Photo of Albert S. Leyton (born Albert S.F. Grünbaum).)

Photo of Grünbaum from: Stewart, M. J. (1922). Albert Sidney Frankau Leyton, M.A., M.D., Sc.D., F.R.C.P. Born 5th January 1869-Died 21st September 1921. The *Journal of Pathology and Bacteriology, 25*(1), 109–112 (N.B., Albert Grünbaum changed his name to Albert Leyton during WWI).

serology to compare modern humans with *Pan*, *Gorilla* and *Pongo*, and it is remarkable that Grünbaum detected the difference between the stronger (i.e., more granular) reactions of *Pan* and *Gorilla*, and the weaker (i.e., more gelatinous) reaction with *Pongo*. Grünbaum closed his brief article by hoping that others would pursue "a quantitative method" because "pressure of other work prevents any present further investigation on my part" (*ibid*: 143).

For many years, serology was the only alternative to morphology for assessing the relationships among living animals. Following Nuttall's experiments at the beginning of the 20th century, serology was used more systematically and became known as serological systematics (Strasser 2010). This was largely due to the efforts of Alan Boyden, a zoologist at Rutgers University in New Jersey. Boyden, like Nuttall, believed that the degree of the reaction between an anti-serum and what by then had been called antigens in the target serum, was "in proportion to the degree of the relationship of the proteins to each other" (Boyden, 1926: 75). Boyden attempted to improve the reliability of the precipitin reaction by using his own "ring test," which involved finding the minimum concentration of anti-serum that would elicit a reaction and produce a visible ring in the test tube. This way, Boyden could more accurately measure the sensitivity of an anti-serum to a given antigen (Boyden, 1932, 1934; Strasser, 2010). Boyden continued to test and refine his methodology for years in an attempt to identify sources of experimental error, reduce subjectivity, and improve reproducibility.

In 1948, Boyden established a Serological Museum within the Zoological Laboratory at Rutgers University in New Jersey. The Museum was created to collect, classify, preserve, and study serum samples from all types of animals. Instead of the skins or skeletons of animals, the Serological Museum consisted of a cold room where glass containers containing serum were stored on shelves. According to Strasser (2010), by 1950 the Serological Museum held samples from more than four hundred species, and by 1960 Boyden estimated that the specimens in the museum numbered in the thousands. But, by that time, other more specific methods for assessing relationships were being developed.

14.3 HEMOGLOBIN AND ALBUMIN

By the early 1960s, a new generation of analytical methods had been developed, and the initial results of applying these methods to primate relationships—and specifically the relationships among the great apes and modern humans—were reported by Emile Zuckerkandl, using hemoglobin, and Morris Goodman, using albumin.

Zuckerkandl and colleagues (1960) and Zuckerkandl (1963) exploited knowledge about the structure of proteins to study the differences between hemoglobin molecules from modern humans and non-human primates. They used enzymes to break down the protein component of red blood cells into its component peptides. Then, they used a method known as two-way electrophoresis—which exploits differences in the mass and the weak electric charge of each peptide—to show that the patterns in starch gels made by the globin-derived peptides from a modern human, a gorilla, and a chimpanzee were very similar, and all three differed from the orangutan pattern (see **Fig. 14.2**).

Goodman (1962, 1963) used a different method, immunodiffusion—which exploits the same system that was much more crudely investigated by Nuttall, Grünbaum, and Boyden—to study the affinities of a serum protein called albumin. He injected small amounts of chimpanzee albumin into monkeys, bled them, and extracted serum that now included the antibodies developed in response to the injected chimpanzee albumin. Goodman used the "trefoil Ouchterlony procedure" to compare the various combinations of albumins and the anti-chimpanzee albumin antibody. This consists of a series of three wells excavated in plates of agar. The anti-chimpanzee albumin serum was placed in the bottom well, and albumin from pairs of species (e.g., modern human/chimpanzee, chimpanzee/gorilla, chimpanzee/orangutan, etc.) was placed in the wells on the top right and top left. The molecular weights of the anti-chimpanzee albumin antibodies, and the albumin molecules, were small enough so that they could both diffuse through the agar and where they met, they interacted to form an opaque coagulate. If the coagulate was a smooth curve, the assay had failed to detect any difference in the responses between the anti-serum and the two albumins. If a spur projected from the curve in the direction of one of the top two wells, this indicated that the assay had detected a difference in the responses. The bigger the difference, the bigger the spur. Goodman's method was unable to detect any difference in the albumins of modern humans and the chimpanzee; in both, the coagulate was a smooth curve with no detectable spur. But, when gorilla albumin was substituted for the human albumin, there was a detectable spur in the direction of gorilla (see **Fig. 14.3**). Goodman concluded that this was because the albumin molecules of modern humans and chimpanzees were—to all intents and purposes—identical, whereas there was a difference between the albumins of chimpanzees and gorillas. Zuckerkandl and Pauling proposed the term "molecular anthropology" for research using molecules such as hemoglobin and albumin to investigate the relationships among modern humans and closely related living primates (Zuckerkandl & Pauling, 1962; Zuckerkandl, 1963).

Proteins are made up of a string of amino acids, and in many instances, one amino acid can be substituted for another without affecting the function of the protein. Subsequent analyses focused on amino acid sequences instead of the physical and chemical properties of whole proteins. Sarich and Wilson (1967) used cross-reactions between proteins of the target species and antisera of

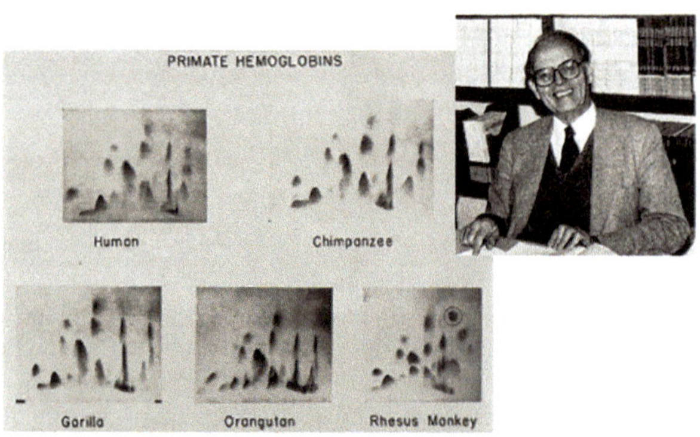

FIGURE 14.2 Tryptic peptide patterns of primate hemoglobins (Zuckerkandl et al., 1960).

Photo of Zuckerkandl from: The Pauling Blog, Oregon State University Libraries' Special Collections & Archives Research Center.https://paulingblog.wordpress.com/2013/12/20/emile-zuckerkandl-1922-2013/.

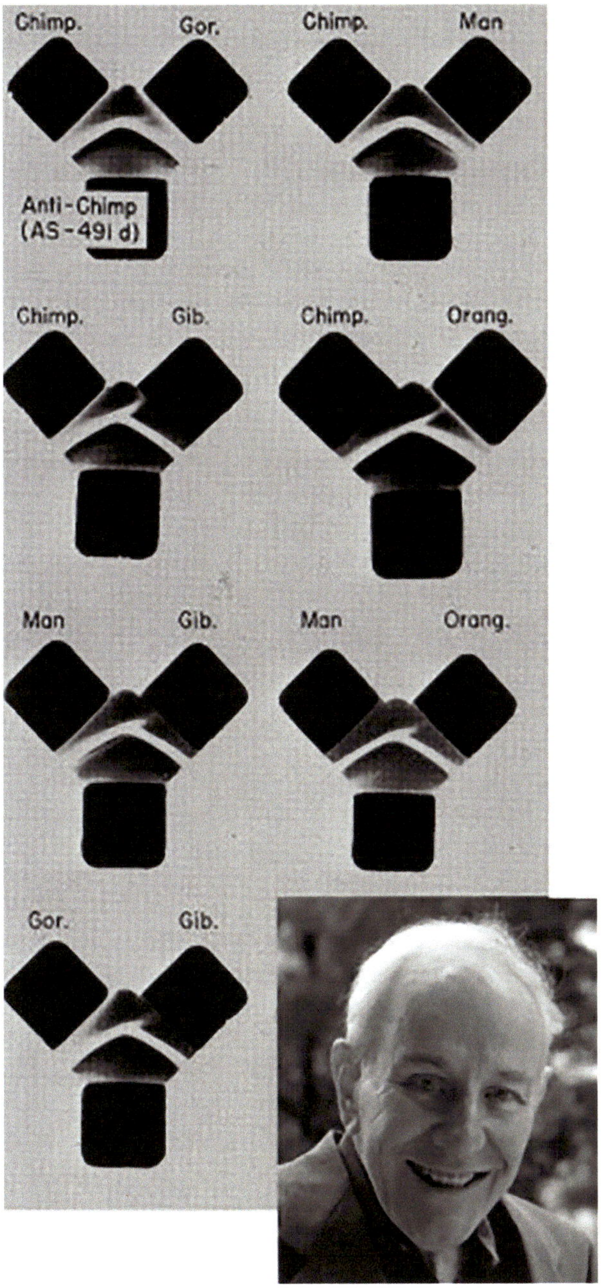

FIGURE 14.3 Patterns of coagulation produced when anti-chimp albumin, in the bottom well, is allowed to diffuse and interact with albumin diffusing from the top two wells (Goodman, 1962).

Photo of Goodman from: Interview with Morris Goodman, The Dibner Institute for the History of Science and Technology. https://authors.library.caltech.edu/5456/1/hrst.mit.edu/hrs/evolution/public/goodman.html.

comparands as a proxy for the differences in amino acid sequences to provide the first molecular clock estimate of modern human-ape divergence, which dated the split to approximately 5 million years ago, a date that was less than half of contemporary estimates of modern human-ape divergence times based on fossil evidence. Subsequent analyses focused on amino acid sequences instead of the physical and chemical properties of whole proteins. Later, Mary-Claire King and Allan Wilson suggested that 99% of the amino acid sequences of chimpanzee and modern human blood proteins were identical. King and Wilson's (1975) paper reporting this similarity was a landmark in evolutionary developmental biology (evo-devo) because it proposed the hypothesis that turning genes on or off during development (i.e., gene expression) could be a powerful driver of phenotypic evolution, and they even suggested it may be a more important influence on evolution than changes to the genes themselves. This hypothesis has gained substantial support from subsequent genomic and transcriptomic studies.

14.4 DNA

The discovery by James Watson and Francis Crick—with unwitting help from Rosalind Franklin—of the basic structure of deoxyribonucleic acid (DNA), and the subsequent discovery by Crick and others of the nature of the genetic code, meant that the structural details of DNA could be used to help determine the relationships among the great apes and modern humans. Researchers had shown that the sequence of bases in the DNA molecule determined the nature of the proteins manufactured within the cells of the body. This meant that by comparing their DNA, the relationships among organisms could be pursued at the level of the genome, thus potentially eliminating the need to rely on proxies, be they traditional phenotypic morphology (i.e., gross anatomy) or indirect measures of the structure and shape of proteins (i.e., Zuckerkandl & Goodman's research) for information about relatedness.

The DNA in cells is located either within the nucleus (aka nuclear DNA or nDNA) or within organelles in the cytoplasm called mitochondria (aka mitochondrial DNA or mtDNA). The latter is faster evolving, but the changes are less stable. Initial attempts to compare the DNA of organisms were made using a method called DNA hybridization, which compared all the mitochondrial or nuclear DNA, but at a relatively crude level (Caccone and Powell 1989). Initial attempts to use DNA sequencing to generate hypotheses about the relationships among taxa were made using the base sequences of relatively small sections of a DNA molecule. In effect, DNA hybridization provided researchers with a modest amount of information about all of the DNA molecules, whereas the early attempts at DNA sequencing provided researchers with more information about a very small part of the DNA molecule.

Today, technological advances mean that whole genomes can be sequenced with degrees of accuracy and at costs that were unattainable just a few years ago. Sequencing takes place at various levels of detail—the technical term is coverage—and thanks to ingenious new technology, genomes are being sequenced in more detail. Sequence data allow researchers to generate hypotheses about the nature of the differences among the base sequences. Base changes from A to G and from T to C (aka transitions) switch relatively readily back and forth, so they provide less reliable estimates of genetic distance than the less common type of base change from A to C and from T to G (aka transversions).

Both DNA hybridization (e.g., Caccone & Powell, 1989) and sequencing (e.g., Bailey et al., 1992; Arnason et al., 1996) were used to generate hypotheses about the relationships among living hominoids (Ruvolo 1997), and new studies appear regularly (see Prado-Martinez et al., [2013] for a review). Over the last two decades, researchers have published good-quality draft sequences of the nuclear genomes of the chimpanzee (Chimpanzee Sequencing and Analysis Consortium 2005), orangutan (Lock et al. 2011), gorilla (Scally et al., 2012), and bonobo (Prüfer et al., 2012), with better quality data from more animals being accumulated steadily (e.g., Gordon et al., 2016). Das et al. (2014) used the differences between the mitochondrial genomes of one eastern (*Gorilla beringei*) and one western gorilla (*Gorilla gorilla*) to date their divergence, and even more recently, de Manuel et al. (2016) published an additional 40 complete common chimpanzee/bonobo genomes with a 25-fold sequence coverage. These additional data have clarified both the timing of the split between common chimpanzees and bonobos, and the patterns of subsequent gene exchange. These methods have not only allowed researchers to investigate the relationships among living hominoids, but they have also allowed researchers to investigate the recent evolutionary history of each of the great apes. For example, Takemoto et al. (2015) reviewed the various hypotheses about the role that the Congo River played in the origin of bonobos as a separate species; Nater et al. (2017) used genomic evidence as a part of their argument for recognizing a new species of orangutan on Sumatra, and Kuhlwilm et al. (2016) provide an accessible summary of how DNA evidence has informed our knowledge of great ape evolution and demography.

Scally et al. (2012) sampled three gorillas—two western lowland (*G. gorilla gorilla*) and one eastern lowland (*G. beringei graueri*)—and their analysis supported the results of earlier analyses of both nuclear and mtDNA (e.g., Ruvolo, 1997; Li & Saunders, 2005), which suggested that modern humans and chimpanzees are more closely related to each other than either is to the gorilla. But, Scally et al. (2012) also found that "in 30% of the genome, gorilla is closer to human or chimpanzee than the latter are to each other" (*ibid*: 169). This is an example of a phenomenon termed "incomplete lineage sorting" (ILS), and Kronenberg et al. (2018) subsequently suggested that the percentage of the gorilla genome that is closer to the human or chimpanzee genomes may be as high as 36%. Prüfer et al. (2012) showed that bonobos and common chimpanzees are 99.7% alike, whereas bonobos and modern humans are 98.7% alike. They also provided evidence of ILS, finding that 1.6% of the modern human genome is more similar to bonobos than to common chimpanzees, and 1.7% is more similar to common chimpanzees than to bonobos. Thus, approximately 3% of the modern human genome is more similar to either bonobos and common chimpanzees, than bonobos and common chimpanzees are to each other. Prüfer et al. (2012) also showed that 25% of all genes contain regions that are affected by ILS. Mailund et al. (2014) provide an accessible review of lineage sorting in apes.

In summary, the evidence from mitochondrial, but more importantly, nuclear DNA strongly supports the hypothesis that modern humans are more closely related to common chimpanzees and bonobos than common chimpanzees and bonobos are to gorillas. This is additional support for the view that it is no longer possible to support a family-level split between modern humans and the African apes. Researchers who take the view that branching sequence should be reflected in category nomenclature now downgrade the division between modern humans and common chimpanzees and bonobos to the level of a tribe, or even sub-tribe.

The DNA differences among modern humans and the great apes can be calibrated using paleontological evidence. The traditional calibration reference has been the split between the apes and the Old World monkeys, so if you have a date for that divergence based on fossil evidence, and you know how different their DNA is, and if you assume that mutation rates are constant across time, you can estimate how long it took to generate the differences between the DNAs of modern humans and chimpanzees/bonobos. If we assume that most of the DNA differences are neutral, the differences in their DNAs predict that the hypothetical ancestor of modern humans and chimpanzees/bonobos lived between about 8 and 6 Ma (Bradley, 2008; Stone et al., 2010). But when different, older, divergences are used to calibrate the molecular clock, the predicted date for the split is somewhat older (e.g., >10 Ma Arnason and Janke 2002). Generation time also influences these estimates, and when slightly longer generation times are used (e.g., Langergraber et al., 2012), they suggest the date of divergence may be closer to 8 than to 5 Ma (Arnason et al., 2008). Prado-Martinez et al., (2013) concluded that the divergence date of modern humans and chimpanzees/bonobos is closer to the original estimate of approximately 5 million years ago, but Moorjani et al. (2016) suggested that some mutations behave in a more clock-like manner than others, and when they use these to estimate the modern human and chimpanzees/bonobo splitting time, they come up with an estimate that is at the older end of the 8–6 Ma spectrum.

14.5 MORPHOLOGY AGAIN

These DNA-based hypotheses about the relationships among modern humans and the living apes have received support from recent studies that once again focused on traditional gross morphological evidence (e.g., Pilbeam, 1996; Shoshani et al., 1996; Gibbs et al., 2000, 2002; Pilbeam, 2000; Diogo & Wood, 2011). Research that used the gross morphology of striated muscles showed that the relationships based on molecular evidence can be reliably recovered (Gibbs et al., 2000, 2002; Diogo & Wood 2011). Subsequently, Diogo et al. (2017) published the first analysis of soft-tissue data for bonobos, and showed that bonobos have slightly fewer departures than chimpanzees from what is inferred to be the phenotype of the most recent common ancestor of common chimpanzees and bonobos.

14.6 CONCLUSIONS

There is now overwhelming genetic (i.e., DNA) and phenotypic evidence that the African apes are more closely related to modern humans than the only Asian great ape, the oran-gutan. There is also compelling genetic and soft-tissue morphological evidence that chimpanzees and bonobos are the closest living relatives of modern humans, and it is very likely that the modern human twig, or clade, separated from the rest of the organisms in the Tree of Life *c.*8 million years ago. Because chimpanzees and bonobos are only found in Africa and the earliest evidence for creatures that might belong to the modern human clade also comes from Africa (e.g., Brunet et al., 2002; Senut et al., 2001; White et al., 2009), and as Darwin predicted—Africa is likely to have been the continent where the modern human clade emerged.

The lines of evidence supporting a closer relationship between *Pan* and *Homo* than between *Pan* and *Gorilla* have implications for taxonomy. Most researchers have concluded that the modern human clade should be distinguished in the Linnaean hierarchy beneath the level of the family, which was the level recommended by Huxley (1863). Researchers who have come to that conclusion mostly now interpret the family Hominidae more inclusively than in the pre-molecular era, so that it includes all of the great apes (including modern humans). They use the subfamily Homininae for modern humans and chimpanzees and bonobos, and they distinguish modern humans and chimpanzees and bonobos at the level of the tribe. In this latter scheme, chimpanzees and bonobos, and any extinct taxa more closely related to chimpanzees/bonobos than to any other living taxon, have their own tribe, Panini. Its members are referred to informally as panins. Modern humans and any extinct taxa more closely related to modern humans than to chimpanzees/bonobos have their own tribe, Hominini. Its members are referred to informally as hominins.

The effective resolution of the relationships among the extant great apes and modern humans also has implications for the interpretation of the hominin fossil record. First, it helps generate more reliable alpha taxonomic hypotheses by knowing which taxa are the most relevant comparators for assessing the taxonomic significance of variation in the hominin fossil record. Second, it helps resolve the polarity of the morphological characters used to generate hypotheses about relationships among extinct hominin taxa. Resolving relationships among extant taxa helps in the task of resolving relationships among extinct taxa.

REFERENCES

Arnason, U. and A. Janke. 2002. Mitogenomic analyses of eutherian relationships. *Cytogenetic and Genome Research* 96, nos. 1–4: 20–32.

Arnason, U., J. A. Adegoke, A. Gullberg, E. H. Harley, A. Janke and M. Kullberg. 2008. Mitogenomic relationships of placental mammals and molecular estimates of their divergences. *Gene* 421, nos. 1–2: 37–51.

Arnason, U., X. Xu and A. Gullberg. 1996. Comparison between the complete mitochondrial DNA sequences of *Homo* and the chimpanzee based on nonchimeric sequences. *Journal of Molecular Evolution* 42: 145–152.

Bailey, W. J., K. Hayasaka, C. G. Skinner, et al. 1992. Reexamination of the African hominoid trichotomy with additional sequences from the primate B-globin gene cluster. *Molecular Phylogenetics and Evolution* 1, no. 2: 97–135.

Boyden, A. A. 1926. The precipitin reaction in the study of animal relationships. *Biological Bulletin* 50, no. 2: 73–107.

Boyden, A. A. 1932. Precipitin tests as a basis for quantitative phylogeny. *Experimental Biology and Medicine* 29, no. 8: 955–957.

Boyden, A. A. 1934. Precipitins and phylogeny in animals. *The American Naturalist* 68, no. 719: 516–536.

Bradley, B. 2008. Reconstructing phylogenies and phenotypes: A molecular view of human evolution. *Journal of Anatomy* 212, no. 4: 337–353.

Brunet, B., F. Guy, D. Pilbeam, et al. 2002. A new hominid from the Upper Miocene of Chad, Central Africa. *Nature* 418: 145–151.

Caccone, A. and J. R. Powell. 1989. DNA divergence among hominoids. *Evolution* 43, no. 5: 925–942.

Chimpanzee Sequencing and Analysis Consortium. 2005. Initial sequence of the chimpanzee genome and comparison with the human genome. *Nature* 437: 69–87.

Das, R., S. D. Hergenrother, I. V. Soto-Calderon, J. L. Dew, N. M. Anthony and M. I. Jensen-Seaman. 2014. Complete mitochondrial genome sequence of the eastern gorilla (*Gorilla beringei*) and implications for African ape biogeography. *Journal of Heredity* 105, no. 6: 752–761.

de Manuel, M., M. Kuhlwilm, P. Frandsen, et al. 2016. Chimpanzee genomic diversity reveals ancient admixture with bonobos. *Science* 354, no. 6311: 477–481.

Diogo, R. and B. Wood. 2011. Soft-tissue anatomy of the primates: Phylogenetic analyses based on the muscles of the head, neck, pectoral region and upper limb, with notes on the evolution of these muscles. *Journal of Anatomy* 219: 273–359.

Diogo, R., J. L. Molnar and B. Wood. 2017. Bonobo anatomy reveals stasis and mosaicism in chimpanzee evolution, and supports bonobos as the most appropriate extant model for the common ancestor of chimpanzees and humans. *Scientific Reports* 7: 608.

Gibbs, S., M. Collard and B. Wood. 2000. Soft-tissue characters in higher primate phylogenetics. *Proceedings of the National Academy of Sciences* 97, no. 20: 11130–11132.

Gibbs, S., M. Collard and B. Wood. 2002. Soft-tissue anatomy of the extant hominoids: A review and phylogenetic analysis. *Journal of Anatomy* 200, no. 1: 3–49.

Goodman, M. 1962. Immunochemistry of the primates and primate evolution. *Annals of the New York Academy of Sciences* 102, no. 2: 219–234.

Goodman, M. 1963. Man's place in the phylogeny of the primates as reflected in serum proteins. In *Classification and Human Evolution*, ed. S. Washburn, pp. 204–34. Chicago: Aldine Publishing Company.

Gordon, G., J. Huddleston, M. J. P. Chaisson, et al. 2016. Long-read sequence assembly of the gorilla genome. *Science* 352, no. 6281: aae0344.

Grünbaum, A. S. F. 1902. Notes on the "blood relationship" of man and the anthropoid apes. *The Lancet* Jan. 18: 143.

Huxley, T. H. 1863. *Evidence as to Man's Place in Nature*. London: Williams and Norgate.

King, M. -C. and A. C. Wilson. 1975. Evolution at two levels in humans and chimpanzees. *Science* 188, no. 4184: 107–116.

Kraus, R. 1897. Ueber specifische reactionen in keimfreien filtraten aus cholera, typhus, und pestbouillonculturen, erzeugt durch homologous serum. *Wiener Klinische Wochenschrift* 10: 736–738.

Kronenberg, Z. V., I. T. Fiddes, D. Gordon, et al. 2018. High-resolution comparative analysis of great ape genomes. *Science* 360, no. 6393: 1085–1096.

Kuhlwilm, M., M. de Manuel, A. Nater, M. P. Greminger, M. Krützen and T. Marques-Bonet. 2016. Evolution and demography of the great apes. *Current Opinion in Genetics and Development* 41: 124–129.

Langergraber, K. E., K. Prufer, C. Rowney, et al. 2012. Generation times in wild chimpanzees and gorillas suggest earlier divergence times in great ape and human evolution. *Proceedings of the National Academy of Sciences* 109, no. 39: 15716–15721.

Li, W.-H. and M. A. Saunders. 2005. The chimpanzee and us. *Nature* 437, no. 7055: 50–51.

Lock, D. P., L. W. Hillier, W. C. Warren, et al. 2011. Comparative and demographic analysis of orang-utan genomes. *Nature* 469: 529–533.

Mailund, T., K., Munch and M. H. Schierup. 2014. Lineage sorting in apes. *Annual Review of Genetics* 48: 519–535.

Moorjani, P., Amorim, C.E.G., Arndt, P.F., & Przeworski, M. (2016). Variation in the molecular clock of primates. *Proceedings of the National Academy of Sciences, 113*(38), 10607–10612.

Nater, A., M. P. Mattle-Greminer and A. Nurcahyo. 2017. Morphometric, behavioral, and genomic evidence for a new orangutan species. *Current Biology* 27, no. 22: 3487–3498.

Nuttall, G. H. F. 1901. The new biological test for blood in relation to zoological classification. *Proceedings of the Royal Society of London* 69: 150–153.

Nuttall, G. H. F. 1904. *Blood Immunity and Blood Relationship. A Demonstration of Certain Blood-Relationships amongst Animals by Means of the Precipitin Test for Blood.* Cambridge: Cambridge Univ. Press.

Pilbeam, D. (1996). Genetic and morphological records of the hominoidea and hominid origins: a synthesis. *Molecular Phylogenetics and Evolution, 5*(1), 155–168.

Pilbeam, D. 2000. Hominoid systematics: The soft evidence. *Proceedings of the National Academy of Sciences* 97, no. 20: 10684–10686.

Prado-Martinez, J., P. H. Sudmant, J. M., et al. 2013. Great ape genetic diversity and population history. *Nature* 499: 471–475.

Prüfer, K., K. Munch, I. Hellmann, et al. 2012. The bonobo genome compared with the chimpanzee and human genomes. *Nature* 486: 527–531.

Ruvolo, M. 1997. Molecular phylogeny of the hominoids: Inferences from multiple independent DNA sequence data sets. *Molecular Biological Evolution* 14, no. 3: 248–265.

Sarich, V. M. and A. C. Wilson. 1967. Immunological time scale for human evolution. *Science* 158, no. 3805: 1200–1203.

Scally, A., J. Y. Dutheil, L. W. Hillier, et al. 2012. Insights into hominid evolution from the gorilla genome sequence. *Nature* 483: 169–175.

Senut, B., M. Pickford, D. Gommery, P. Mein, K. Cheboi and Y. Coppens. 2001. First hominid from the Miocene (Lukeino Formation, Kenya). *Comptes Rendus de l'Académie des Sciences* 332: 137–144.

Shoshani, J., C. P. Groves, E. L. Simons and G. F. Gunnell. 1996. Primate phylogeny: Morphological vs. molecular results. *Molecular Phylogenetics and Evolution* 5, no. 1: 102–154.

Stone, A. C., F. U. Battistuzzi, L. S. Kubatko, et al. 2010. More reliable estimates of divergence times in *Pan* using complete mtDNA sequences and accounting for population structure. *Philosophical Transactions of the Royal Society of London B: Biological Sciences* 365, no. 1556: 3277–3288.

Strasser, B. J. 2010. Laboratories, museums, and the comparative perspective: Alan A. Boyden's quest for objectivity in serological taxonomy, 1924–1962. *Historical Studies in the Natural Sciences* 40, no. 2: 149–182.

Takemoto, H., Y. Kawamoto and T. Furuichi. 2015. How did bonobos come to range south of the Congo River? Reconsideration of the divergence of *Pan paniscus* from other *Pan* populations. *Evolutionary Anthropology* 24, no. 5: 170–184.

White, T. D., B. Asfaw, Y. Beyene, et al. 2009. *Ardipithecus ramidus* and the paleobiology of early hominids. *Science* 326, no. 5949: 64–86.

Zuckerkandl, E. 1963. Perspectives in molecular anthropology. In *Classification and Human Evolution*, ed. S. L. Washburn, pp. 243–272. Chicago: Aldine.

Zuckerkandl, E. and L. B. Pauling. 1962. Molecular disease, evolution, and genetic heterogeneity. In *Horizons in Biochemistry*, eds. M. Kasha and B. Pullman, pp. 189–225. New York: Academic Press.

Zuckerkandl, E., R. T. Jones and L. Pauling. 1960. A comparison of animal hemoglobins by tryptic peptide patterns analysis. *Proceedings of the National Academy of* Sciences 46, no. 10: 1349–1360.

Glossary

abranas: Celtic name for monkey.
aff, pl. affen: monkey in German.
akarima: the squirrel monkey, *Saimiri sciureus*, in the Arawak language.
Amemangamanga: term used by the Mbuti (African people) for a "mythical big ape living in the deep forest".
Anthropomorpha: name of the order containing modern humans, simians, and sloths used by Linnaeus in the first edition of the *Systema Naturae* (1735).
amienu: term used by the people of Sierra Leone for western chimpanzees.
ape: term now used for the anthropoid apes (Family Hominidae, Subfamilies Homininae, and Ponginae) and for gibbons (Family Hylobatidae). Also used as a symbol of vanity and lust and for the ability to imitate, 'aping' is synonymous with imitation. See also pithekos.
apan (plural of apa): term used in the *Physiologus* for apes. See also ape.
aquiqui: indigenous name for the howler monkey in Brazil.
aranata: indigenous name still used in Venezuela for the howler monkey. See also araguato.
arim or arimi: term for monkey in the ancient Etruscan language. See also harim.
arimo: See arim.
baboino: term for an anthropomorphic creature in the 16th century. Equivalent to "cercopithecus", an Asian creature half-human and half-monkey, later named *Homo agrestis*, literally "rural man".
babuini/babbuini: term for baboons in Italian, probably a transliteration from the Arabic maymun.
baris: name given to the chimpanzee in Western Africa. See also barris and daris.
b'atz: term for howler monkey. Together with *Chouen* (the spider monkey), the howler represents the fallen monkeys from the original 'humans – monkeys' (in the Popol Vuh, a sacred book of Maya).
bertuccia, pl. bertucce: Italian word for the Barbary Macaque, in the past also used for other kinds of monkeys.
bimanes: a category introduced in the 18th century that separates modern humans from other animals.
botekpa-ebobo: term for the "palm" of a gorilla.
bogio: term of North-African origin for monkey. Related to Spanish "buhar" (to tell or snitch), Italian "bugia" (to lie), and Portuguse "bugiar" (to tell lies). See also bugio, bugia, bogia, bugiot, or bogiot.
brutus pl. bruti: term for "beasts" (non-human animals) in the Medieval and Modern periods.
Buja: part of northern Africa. See also bogio.
bugia: see bogio.
bnti: a baboon, also the name of the son of Re.
bnti.wi: two deified baboons, also sons of Re.
buriqui: indigenous word for *Brachyteles arachnoides*.
cai or **cay:** indigenous word for capuchin monkeys.
calatrice: see callithrix.
callithrix: a term in Greek used in antiquity for black and white colobus monkeys. Literally, "beautiful hair".
callitriche: name used by Buffon to refer to monkeys of the genus *Chlorocebus*.
callithriches: see callithrix.
çagui: See sagoin.
caierê: a species of monkey eaten by the Kaingáng people in the Atlantic Forest of Brazil.
caitaia: indigenous word for "cercopithecus with a yellow and lighter-colored coat". Likely *Sapajus flavius*.
cay: indigenous word for capuchin monkey.
cay-ouässou: indigenous word for capuchin monkey.
cay-ou: indigenous word for *Chiropotes* spp.
cay-mirim: indigenous word for *Saimiri sciureus*.
cebos/cebus: See kepos.
cefusa: term used by Albertus Magnus for a monkey genus. See *Erythrocebus* sp.
cenopha: term for cynocephalus (dog-head). See kynokephalos.
cercopithecus pl. cercopitheci: an informal term used for tailed monkeys since antiquity. *Cercopithecus* is now a genus of African primates (Family Cercopithecidae).
cercopitheques: non-specific monkey, probably refers to guenons.
cercops: term in Greek mythology for the "tailed one", but in later classical times used for any kind of monkey. See simia.
chadi pl. chuada: term used by the Berbers of Algeria and Morocco for monkeys. See also querd pl. querud; sàdan, pl. sàadin; and shadi.
chimpanzee: derived from "kivili-chimpenze" in the Tshiluba language. Now used as the informal term for animals in the species *Pan troglodytes sensu lato.*
chouen: term used in the Popol Vuh, a Mayan sacred text for spider monkey. Together with the **b'atz** (howler monkey), they are fallen humans.
choiropithecos: term used by Aristotle, whose meaning is controversial.
cicholiticas: term for a tailed monkey "with a handsome face and with a beard". Probably the white and black colobus.
cynocephalus: see kynokephalos.
cynocéphales: see kynokephalos.
cythosica: see spinga.

classical period: between the 5th and 4th centuries BCE in Ancient Greece.

cluna: a monkey with a tail. See also clura.

daris: term used for the chimpanzee in Guinea-Bissau. See also baris.

dawa: inclusive term used by the Banganda people of South Eastern Cameroon for monkeys, chimpanzees, and gorillas.

dege: term used by the Dogon people for the black monkey.

devas: celestial gods who come to earth as monkeys to assist Vishnu, incarnate as the human Rama, to slay Ravana.

djinn: natural spiritual entities common in the Arabic world that influence human life.

dongo: term used by the Teda people of Tibesti for baboons. See also dounbou.

dog-headed: monstrous anthropomorphic creature with a dog-like head. See kynokephalos.

drill: a great ape in 17th century English. The derivative mandrill indicated an anthropomorphic creature.

engai: a word used in contemporary Japanese media to indicate "monkey damage".

engeco: the word used by indigenous people for what we now know as the chimpanzee. Andrew Battell wrote about "two kinds of monsters, which are common in these woods"…. "the greatest of these two monsters is called *Pongo* [*Mpungu*] in their language, and the lesser is called *Engeco*".

exquima: the bearded *Cercopithecus* of Guiné, now known as *Cercopithecus* diana. Name used by Buffon for *Ateles* spp.

fanafody: a type of medicine practiced in north and east Madagascar that uses lemurs.

fera upali: a dog-headed monkey.

finga: see spinga.

galeopithecus: name for the common marmoset that suggests it is 'half-monkey, half-squirrel'.

galetrices: term used in the 13th century for an Ethiopian monkey; probably a black and white colobus.

gato maimone: term borrowed from Arabic used by the first Italian navigators reaching the Americas for tailed monkeys. Also, gattomammone, gatto mammone, and gato maimone. See also maimun.

gorilla: term used for the first time in the Periplus, which tells about Hanno II, king of Carthage from 480 until 440 BCE. Now refers to animals belonging to the genus *Gorilla*.

gato monillo: term used by Spanish naturalists and writers in the 16th century for New World monkeys. Literally "monkey with tail".

gatos paules: term used by Marco Polo for monkeys and by Columbus when he saw New World monkeys for the first time.

gbèkrè: term used by the Dogon people for monkey figures used for the divination cult Mbra.

gemara: component of the Talmūd (one of the sacred texts of Judaism) comprising a collection of rabbinical analyses and commentaries on the Mishnah (the first major written collection of the Jewish oral traditions).

geometric style: term used for human and semi-human figures in ancient (from 900 to 700 BCE) Greek art.

ghorilla or **ghol:** term used in 20th century Arabic texts for a gigantic, strong, and aggressive monkey with no tail.

gird: see quird.

gonguê: term used by the Kaingáng people living in the Atlantic Forest of Brazil for a monkey species they consume.

gothic style: architecture style that first appeared in northern France in the early 12th century. Rapidly spread to sculpture, textiles, and painting, including frescoes, stained glass, and illuminated manuscripts.

guariba or goariba, or uvariua: indigenous term for red-handed howler monkeys (*Alouatta belzebul*) in Brazil.

guenon: French word for monkeys with a tail since the 16th century.

habbar or hobr: term used in 20th century Arabic texts for a beautiful small monkey with thick hair and a long tail.

halāl: Arabic term, meaning 'permissible' in English. In the Quran, the term halāl is contrasted with the term harām ('forbidden').

hanu-man: a monkey with a prominent or disfigured jaw. An allusion to simian prognathism.

hapi: the human head of Hapi, one of the four sons of Horus, represented on canopic jars. From the 19th Dynasty onward it was replaced with the image of a hamadryas.

harām: Arabic term, meaning 'forbidden' in English. Referring to any behavior or situation prohibited by the Islamic faith. See also halāl.

hari: term in Sanskrit for "monkey". Literally a "tawny-colored" creature.

hazar: the leader of baboons in ancient Yemen.

Hedj-Wer: the earliest identifiable baboon deity in ancient Egypt (c.2900 BCE). Literally, "the Great White One".

***Homo africanus niger*:** a taxon introduced by Linnaeus.

***Homo agrestis*:** see baboino.

***Homo diurnus*:** binomial for modern humans, introduced by Linnaeus.

***Homo caudatus*:** binomial introduced by Linnaeus for an ape-man creature. Linnaeus was following Gessner's *Cercopithecus formae rarae*, also called *Homo agrestis* by Aldrovandi in the 16th century. Linnaeus did not classify it because he could not decide whether it was a human being or a monkey; he referred to it as Lucifer.

***Homo nocturnus*:** binomial introduced by Linnaeus to contrast with *Homo diurnus*.

***Homo sylvestris*:** literally "wild man of the woods". It was Linnaeus' translation of 'orang-utan', which indicated populations intermediate between human beings and apes.
***Homo troglodytes*:** the orang-utan of Linnaeus, synonym of *Homo sylvestris.*
hou: term for "monkey," usually referring to macaques.
hylophagi: term used by Agatharchides of Cnidos, probably referring to great apes. Literally "wood-eaters".
ibiddauen: term used by the Tuareg people for the Barbary macaque.
iconography: branch of art history concerned with the identification, description, and interpretation of images.
irans: term for parts of chimpanzees traded to appease animistic entities.
ikeiñu: term used by the Wampi people who inhabit the northern part of the Peruvian Amazon for monkeys, squirrels (*Sciurus* spp.), porcupines (*Coendou* spp.), kinkajous, and olingos. Literally "arboreal mammals".
jocko: the name used by Buffon for chimpanzees.
kaiju: term used by the Kalapalo people of the Upper Xingu River in Brazil for monkeys, squirrels (*Sciurus* spp.), porcupines (*Coendou* spp.), kinkajous, and olingos.
kapi: one of the terms in Sanskrit for "monkey". Literally, a "tawny-colored" creature.
kayouroure: term in the Galibi-Karibe language of Amazonia for "white monkey".
kaya, pl. kayatene: term used by Tuareg people for chimpanzees.
kebos: see kēpos.
kephos: see kēpos.
kēpos: Greek term, Latinized as *cebus,* for long-tailed (or tailed) monkeys. May refer to *Erythrocebus patas.* The name may derive from the Hebrew Kos, Gôf, or Koph. See also kapi.
kerkopithecus: see pithekos.
khof pl. kophin: see kôf.
kiber: term for baboon.
kiboz: term for a creature that "has a tail, but its front part is entirely like that of a woman".
kird: term for a monkey
ko: the red monkey mask of the Dogon people of Mali.
kôf (or gôf), pl. kofin: a Jewish term found in the Bible that may derive from Egyptian Kafu (or kif) or the Indian kapi. In Greek kebos and Latin cephus.
kynokephalos (pl. kynocephaloi): Greek term, Latinized as cynocephalus (pl. cynocephali). Literally, a "dog-like face".
kynoprosopos: see kynokephalos.
kyboz: term used for monkeys with tails. Derived from cepos or kebos.
lemur: component of Linnaeus's order primates. Refers to an adaptive radiation of primates in Madagascar.
lowando: name used by Buffon for *Semnopithecus* spp.
lucifer: see *Homo caudatus.*
lynx: an unidentified monkey dissected by Galeno.
mbra: see *gbèkrè.*
macaco: see monkey.
macaquo: term used by Marcgrave 17th century for *Papio* spp.
mahakapi: term used in the Jataka literature for a "great monkey" who used his body to form a bridge so his companions could escape danger by reaching a mountain.
mahavira: "great hero", one of Hanuman's epithets.
makaque: term used by Marcgrave 17th century for African guenons.
maki: name used by Buffon for lemurs.
malbrouk: name used by Buffon for *Chlorocebus cynosuros.*
mamonetus: term, probably derived from the Arabic maimon, used by Albert the Great for an unidentified monkey.
mandrill: see drill.
mangabey: name used by Buffon for *Cercocebus torquatus.*
marytonmorios: term used by Albert the Great for a monkey that could talk.
marikina: name used by Buffon for *Leontopithecus rosalia.* See also mirikina.
marmusettes: term first used in the 13th century for monkeys and later used for grotesque figures.
maucauco: term used by Pennant for prosimians [plus cologus].
maymūn pl. mīmūn: term for a monkey with a tail, likely a baboon, used by Spanish Arabs in the Middle Ages and still used today.
Meerkatzelein – meercatzen: see sea-cats.
Middle Ages: a period of Western history between the late 5th century and the late 15th century.
mico: term in the language of the Arawak people of Venezuela (first description, 16th century.) meaning "small monkeys", such as tamarins and marmosets.
michi: see mico.
mimō: term for monkey, taken from the ancient Greek word μιμώ meaning "to mimic".
mimoon: term in Arabic for a monkey with a giant body, a short tail, and an ugly face, most likely referring to baboons.
mirikina: term in the Tupi language most likely referring to the night monkey, *Aotus* spp. In Tupi myraqui means dirty people, and muriki, buriqui, and other variants refer to different species of monkey. See also marikina.
moccoco: name used by Buffon for *Lemur catta.*
mockman: term used for chimpanzees.
mona: see monkey.
monkey: English term whose origin is unclear. Equivalents in other languages are mona or mono (Spanish), scimmia (Italian), singe (French), and macaco (Portuguese).

monnes: term for monkey in ancient French.
moustac: name used by Buffon for *Cercopithecus cephus.*
mpungu: see *Pongo.*
mumo: term used by the Nthangoni people of eastern Kenya for a lone elderly male baboon with supernatural powers.
muriki: indigenous term for *Brachyteles aracnoides.*
nam-nam: term used by the Ma'āza Bedouin, living in the northern half of the Eastern Desert of Egypt, for chimpanzee.
nas-nās, nasnās or **nisnas** pl. **nasanis:** mysterious creature of the medieval Islamic tradition associated with strange anthropomorphic beings supposed to live in mysterious foreign countries. In the 20th century Arabic literature considered a small monkey with a long tail (grivet or vervet monkey).
nshiégo-mbouwé: likely an indigenous term for the bonobo (Du Chaillu). See *Troglodytes calvus.*
ntchilo: term used by the Mpiemu people of the middle and upper Sangha basins in the Central African Republic for gorillas.
omono: term used by the Dogon people (literally 'white monkey), but it may refer to the green monkey, *Chlorocebus sabaeus.*
onokentauros: mythical creature with the characteristics of a chimpanzee. Latinized as onocentaurus**.**
oosaa: term used by indigenous people for the Western chimpanzee.
orang-outang: term used in Malay by the coastal populations of Borneo and Sumatra for the "non-civilized beings of the interior" (i.e., the Asiatic great apes in the modern genus *Pongo*). Literally, the "man of the woods".
orango: term used in the 19th century for a generic anthropoid ape.
ouanderou: name used by Buffon for *Macaca silenus.*
ouistitì: onomatopoeic name used by Buffon for *Callithrix jacchus.*
ourang outang: See orang-outang. See also satyrs and *Homo sylvestris.*
ouäriue: term for the howler monkey in Tupi. See guariba.
papio: common name for the baboon, presumably derived from the name of the Egyptian god Babi or Baba. Provides the genus name *Papio* for baboons.
papione: term used for the dog-headed primate (*Mandrillus* sp.) described by Aldrovrandi.
patristics: the period of dominance of the early Christian scholars known as "Church Fathers", from the 2nd century until the 7th century or 8th century.
pilosus: a composite monkey, human-like above and goat-like below, that walked erect and lived in the desert regions of Ethiopia.
pigmaeus: a person one cubit (about 50 cm or 20 inches) tall. Throughout history, small-statured, tailless persons have been identified with hominoid primates or with people. Homer (*Iliad* III, 3–7) spoke of "short men" living near the mythical river Ocean, and Aristotle thought they were a tribe in Upper Egypt. Later interpretations continued to confuse pygmies with apes or considered them as creatures intermediate between modern humans and other animals. The first chimpanzee to arrive in Europe in the 17th century was called a 'pygmie'.
pinche: name used by Buffon for *Saguinus oedipus.*
pingalaksh*a*: term in Sanskrit for "monkey". Literally "yellow-eyed".
pithecu*s*: See pithekos.
pithekos: initially a term in Greek for a monkey without a tail, but later used for monkeys with or without tails. Latinized as pithecus. Its origin is obscure.
plavaga: term in Sanskrit for "monkey". Literally "leaper".
pongo: term used by indigenous people for what may be the animal we now know as the gorilla. Andrew Battell wrote about "two kinds of monsters, which are common in these woods".... "the greatest of these two monsters is called *Pongo* [*Mpungu*] in their language, and the lesser is called *Engeco*". Although the term was used for different anthropoid apes in the course of the modern period, today it is the name of the genus, *Pongo*, which contains orang-utans from Borneo and Sumatra.
Ptolemaic period: historical period extending from when Alexander the Great defeated the Persians in Egypt in 332 BCE–30 BCE.
pricky: term used in the 16th century for a group of neotropical monkeys, probably the muriquis.
pygmaios: see pigmaeus.
pygmie: term used by Edward Tyson in the 17th century for what we now know, from its preserved skeleton, was a chimpanzee.
pygmy: see pigmaeus.
qouata: term in Arawak language for monkeys of the genus *Ateles.*
qofim: see gof.
quadrupedia: an outdated taxon that included all the mammals (later renamed Mammalia).
quadrumanes: four-handed animals. Term used by Geoffroy Saint Hilaire in the 19th century to distinguish primates from other mammals.
querd pl. querud: term in Arabic for monkeys that is also used by modern Berber people. See also qurduh and qarduh.
quird: Arabic modern name for monkeys. See also kird.
quoias-morrou: term used by indigenous people for the orang-utan.
rubah: term used in Yemen for baboon and is still used in Arabic for the hamadryas baboon.
roubbah: See rubah.
sàdan, pl. sàadin: term in Arabic for monkey. See shadi.
sagoin (also saguin, sagouin, sagui, saguis, çagui): term used by indigenous people in the Atlantic Forest of Brazil for marmosets and tamarins. Still used today in Portuguese for marmosets.

saki: name used by Buffon for *Pithecia pithecia.* See also yarqué.

san: suffix in Japanese normally reserved for humans but also applied to the Japanese macaque or "snow monkey".

sapajous: term used in the 17th century for capuchin monkeys, probably originating from the Tupi language family.

saru: term used in Japanese for "monkey".

saru gami: term in the Japanese Shinto religion for the "monkey deity".

satyrs: mythical creatures, half-human and half-animal. In Greek mythology, a male nature spirit with an exaggerated erection. Satyrus was the name of a monkey dissected by Galeno. In Greek literature, satyr is also used for gibbons, orang-utans, or langurs. From the Gk. sátyros and L. satyrus. Linnaeus used *Satyrus* for chimpanzees.

say: name used by Buffon for *Cebus capuchinus.* See also cay.

sea-cat: term used in the early Modern period for long-tailed monkeys coming to Europe from overseas. It corresponds to the French guenon, the Latin cercopithecus, the Italian (gato maimone, gato maimon, gatto mammone, or gato mamone), and the German meerkatze. The latter is literally the "cat (katze) of the sea (meer)".

shakhamriga: term used in Sanskrit in the *Ramayana* for monkey. Literally "branch-animal".

shadi: term in Arabic for a small monkey with a long tail; likely a grivet or vervet monkey. See also sàdan.

silʿawi: term used by the Maʿāza Bedouin living in the northern half of the Eastern Desert of Egypt for gorillas.

simia pl. simiae: term in Latin used for monkeys (with or without a tail), but in historical literature it usually refers to the Barbary macaque (*Macaca sylvanus*), which has been imported into Europe since ancient times. In the Medieval and Renaissance periods, simia was synonymous with the modern term simian (i.e., all primates except prosimians). Since the late Middle Ages, simia has been translated in Spanish as mono/mona; in French as mon(n)e; in Portuguese as bogio/bugio; in Italian as scimia/scimmia, bertuccia, and gato mammone/maimone; and in English as monkey, **ape**. Literally, "flat nose".

***Simia troglodytes*:** binomial for chimpanzee in the last edition of Linnaeus' *Systema Naturae*. Earlier editions used *Homo troglodytes.*

singe pl. singes: term for monkey in French.

smit pl. smitten: one of the terms attributed to chimpanzees in the 17th century.

spermatophagi: term used by Agatharchides of Cnidos, probably referring to great apes. Literally, "seed-eaters". See hylophagi, literally "wood-eaters".

sphinga: see spinga.

sphinges pl. sphinga: see spinga.

spingia: see spinga.

spinga: term used in the Greco-Roman world for an African-tailed monkey, probably the gelada baboon (*Theropithecus gelada*). Also known as cythosica or mamonetus.

sphinx: see spinga.

sub-fossil lemurs: refers to remains found in Madagascar dating from *c.*30 thousand to approximately 560 years ago.

taboo: term in western anthropology for a social or religious custom prohibiting or restricting a particular practice or forbidding association with a particular person, place, or object.

talapoin: name used by Buffun for *Miopithecus talapoin.*

tamary: term for small monkeys, likely tamarins (e.g., *Saguinus* spp.).

tretretretre or tratratratra: a mysterious large-sized animal described in Madagascar.

troglodytes calvus: the smaller form of chimpanzee described by Du Chaillu.

troglodytes niger: the common chimpanzee.

vanara: term used in Sanskrit for "monkey". Literally "forest being".

vari: name used by Buffon for *Varecia varecia.*

vrishakapi: term used in Sanskrit. Literally "bull monkey" or "virile monkey".

yuan: term used in Chinese for "ape". Likely refers to gibbons, but could also refer to other primates.

wago: term used by the Mpiemu people of the middle and upper Sangha basins of the Central African Republic for chimpanzees. Local lore suggests they share similar origins with people, but they live in the forest, not in villages.

wolly jerboa: name used by Buffon for *Tarsius.*

xapiri: term used by the Yanomami for the double of each human being (image/spiritual entity).

ximia: see simia.

ximio: see simia.

zoophilia/zooerastia: the use of animals as sexual objects.

Index

Note: **Bold** page numbers refer to tables; *italic* page numbers refer to figures and page numbers followed by "n" denote endnotes.